Unsaturated Zone Hydrology for Scientists and Engineers

James A. Tindall, Ph.D.

United States Geological Survey, National Research Program,
Unsaturated Zone Field Studies;
Department of Environmental Science, University of Colorado at Denver

James R. Kunkel, Ph.D., P.E.

Knight Piésold, LLC, Denver, Colorado;
Department of Geology and Geological Engineering, Colorado School of Mines

with

Dean E. Anderson, Ph.D.

United States Geological Survey, National Research Program,
Unsaturated Zone Field Studies

PRENTICE HALL
Upper Saddle River, New Jersey 07458

Library of Congress Cataloging-in-Publication Data

Tindall, James A.
    Unsaturated zone hydrology for scientists and engineers / James A.
Tindall, James R. Kunkel with Dean E. Anderson.
        p.   cm.
    Includes bibliographical references and index.
    ISBN 0-13-660713-6 (casebound)
    1. Groundwater flow.   2. Zone of aeration.   3. Soil chemistry.
4. Soil mechanics.   I. Kunkel, James R.   II. Anderson, Dean E.
III. Title.
GB1197.7.T56   1999
551.49—dc21   98-36758
                                                                        CIP

Executive Editor: Robert A. McConnin
Manufacturing Manager: Trudy Pisciotti
Art Director: Jayne Conte
Cover Illustrator: John M. Evans
Cover Designer: Bruce Kenselaar
Editorial Assistant: Grace Anspake
Production Supervision/Composition: Interactive Composition Corporation

 © 1999 by Prentice-Hall, Inc.
Simon & Schuster/A Viacom Company
Upper Saddle River, New Jersey 07458

Printed in the United States of America

10 9 8 7 6 5 4 3 2 1

ISBN 0-13-660713-6

Prentice-Hall International (UK) Limited, *London*
Prentice-Hall of Australia Pty. Limited, *Sydney*
Prentice-Hall Canada Inc., *Toronto*
Prentice-Hall Hispanoamericana, S.A., *Mexico*
Prentice-Hall of India Private Limited, *New Delhi*
Prentice-Hall of Japan, Inc., *Tokyo*
Simon & Schuster Asia Pte. Ltd., *Singapore*
Editora Prentice-Hall do Brasil, Ltda., *Rio de Janeiro*

Contents

Preface   xiii

1 INTRODUCTION AND BRIEF HISTORY   1

   Introduction   1
1.1   A Brief History of Unsaturated Zone Hydrology   2

2 PHYSICAL PROPERTIES AND CHARACTERISTICS OF SOILS   7

   Introduction   7
2.1   Mineralogical Composition   7
      Clays   8
      Silicate Clays   9
      Amorphous Clays   10
2.2   Soil Profiles   13
2.3   Soil Texture   15
2.4   Soil Classes   18
2.5   Particle Size Analysis   19
2.6   Soil as a Phase System   24
2.7   Density and Volume–Mass–Weight Relations   24
      Particle Density   24
      Void Ratio   26
      Porosity   26
      Bulk Density   30
      Water Content   30
      Air-Filled Porosity   32
2.8   Specific Surface Area   33
   Summary   36

3 BEHAVIOR OF CLAY–WATER SYSTEMS   39

Introduction   39

3.1   Electrochemical Properties of Clay–Water Systems   40

Units   40
Coulomb's Law   41
Heat of Wetting   42
Interaction between Uncharged Soil Particles and Water   43

3.2   Electrochemical Phenomena of Clays   46

Surface Charge of Clay Minerals   46
Diffuse Electrical Double Layer Theory   47

3.3   Electrokinetic Phenomena   50

Electrophoresis   50
Electroosmosis   51
Streaming Potential   54

3.4   The DLVO Theory of Colloid Stability   54

Interactive Repulsive Forces between Platelets   55
Potential Energy Due to Constant Potential versus Constant Charge   56
Van der Waals–London Forces   58
Potential Energy Due to van der Waals–London Forces   59
Total Potential Energy   60
Non-DLVO Forces   62
Limitations of the DLVO Theory   62

3.5   Ion Exchange   63

Cation Exchange Capacity   63
Sodium Adsorption Ratio   64

3.6   Hydration and Shrinking of Clays   65

Hydration   65
Shrinking   70

3.7   Flocculation and Dispersion of Clays   70

Flocculation   70
Dispersion   72

3.8   Humus in Soil   73

Humic and Fulvic Acids   73
Sorption of Contaminants   76

3.9   Aggregation   79

3.10   Aggregate Formation and Characterization   81

Biological   81
Chemical   82
Physical   83

3.11   Soil Crusting   84

3.12   Soil Cracking   88

Summary   90

4 POTENTIAL AND THERMODYNAMICS OF SOIL WATER   95

Introduction   95

4.1   The Structure of Water   95

4.2   Air–Water Interface: Contact Angle   97

4.3   Capillarity   98

4.4   Capillary Potential   102

4.5   Components of Soil Water Potential   103

Gravitational Potential (ψ_g)   104
Osmotic Potential (ψ_o)   105
Vapor Potential (ψ_a)   107
Matric Potential (ψ_m)   109
Hydrostatic Pressure Potential (ψ_p)   110
Overburden Pressure Potential (ψ_b)   110
Intergranular Pressure (σ_g)   111

4.6   Chemical Potential of Soil Water   113

4.7   Hysteresis   117

Ink Bottle Effect   118
Contact Angle   120
Entrapped Air   120
Shrinking and Swelling   120

Summary   121

5 CHEMICAL PROPERTIES AND PRINCIPLES OF SOIL WATER   126

Introduction   126

5.1   Organic Compounds and Constituents   126

Transport of Organic Constituents in the Unsaturated Zone   127
Coupled Processes   128

5.2   Mass Action and Governing Equations   129

5.3   Activity Coefficients   131

Standard Methods   131
Infinite-Dilution Activity Coefficients   133

5.4   Equilibrium and Free Energy   134

5.5   Electroneutrality   136

5.6   Acid Dissociation   137

5.7   Hydrolysis   139

5.8   Ion Complexes and Dissolved Species   140

5.9   Diffusion   142

5.10   Solubility   143

5.11   Chemical Saturation   145

5.12   Oxidation and Redox Reactions   145

5.13   Microbial Mediation and pH   147

Summary   148

6 PRINCIPLES OF WATER FLOW IN SOIL   152

   Introduction   152
6.1   Bernoulli's Equation   153
6.2   Torricelli's Law   155
6.3   Poiseuille's Law   158
6.4   Flow Characteristics: Laminar and Turbulent Flow   159
   Summary   161

7 SATURATED WATER FLOW IN SOIL   165

   Introduction   165
7.1   Darcy's Law   165
7.2   Hydraulic Conductivity and Permeability   166
7.3   Hydraulic Conductivity Values of Representative Soils   168
7.4   Factors Affecting Permeability and Hydraulic Conductivity   169
7.5   Limits of Darcy's Law   170
7.6   Darcy's Law and Water Flow through Soil Columns   171
7.7   Homogeneity, Heterogeneity, Isotropy, and Anisotropy of Soils   172
7.8   Saturated Flow in Layered Media   173
7.9   Laplace's Equation   175
7.10 Diffusion Equation   178
   Summary   179

8 UNSATURATED WATER FLOW IN SOIL   183

   Introduction   183
8.1   Validity of Darcy's Law for Unsaturated Conditions   185
8.2   Factors Affecting Unsaturated Hydraulic Conductivity   185
8.3   Development of Unsaturated Flow Equations   189
8.4   Hydraulic Diffusivity   192
8.5   Boltzmann's Transformation   195
   Summary   197

9 TRANSPORT OF HEAT AND GAS IN SOIL AND AT THE SURFACE   200

   Introduction   200
9.1   Basic Concepts and Definitions   201
      Energy Transfer and Heat Content   201
      Evaporation and Condensation   202
      Quantifying Water Content in Air   204
9.2   Energy Exchanges at the Surface   210

Radiation Balance   212
Energy Balance   215

9.3   Soil-Heat Transfer   221

Heat Capacity, Conductivity, and Diffusivity   221
Soil-Heat Flux   223
Temperature Distribution in Soil   226

9.4   Soil Moisture Evaporation and the Stages of Soil Drying   230

9.5   Evapotranspiration and the Influence of Vegetation on Soil Moisture   234

Plant Physiology   234
Model Estimates of Evapotranspiration   238

9.6   Measuring Energy Budget Terms   240

9.7   Soil-gas Transport   245

General Properties of Gases   245
Porosity and Permeability   246
Physical Mechanisms Responsible for Soil-gas Transport   247
Mathematical Description of Gas Transport   250

9.8   Coupled Transport of Heat, Water, and Water Vapor   258

9.9   Multiphase Transport of Volatile Compounds in Soil   260

9.10   Composition of Soil-air   262

9.11   Measuring Soil-Gas Flux   267

Soil-atmosphere Exchanges   267
Gas Flux within Soil   269

Summary   269

10 CONTAMINANT TRANSPORT   273

Introduction   273

10.1   Physical Processes and Movement of Solutes   273

10.2   Types of Fluid Flow   275

10.3   Breakthrough Curves, Piston Flow, and Hydrodynamic Dispersion   276

Piston Flow   277
Hydrodynamic Dispersion   277
Mechanical Dispersion   278
Molecular Diffusion   279
Relation of a Breakthrough Curve to the Solution of the ADE   280
A General Solution for the Dispersion of a Displacing Solute Front   280
Determining the Error Function   282
Calculating the Displacing Front of a Breakthrough Curve   283
Calculating the Concentration in a Moving Slug of Fluid for a
    Breakthrough Curve   284
Calculation of the Dispersion Coefficient (D)   285

10.4   Nonreactive Solutes   288

10.5   Sorption Reactions   289

10.6   Equilibrium Chromatography   292

10.7   Mathematical Modeling of Transport Phenomena in Soils   294

10.8   Further Solutions to the ADE: Initial and Boundary Conditions   295

10.9   Numerical Solutions of Equilibrium Exchange   299

10.10   Nonequilibrium Conditions   301

10.11   Combined Effects of Ion Exchange and Dispersion   305

10.12   Mobile and Immobile Regions in Soils   305

Aggregated Soils   306
Fractured Media   309
Layered Media   310

10.13   Preferential Flow Paths, Macropores, and Fingering   312

Modeling Chemical Transport Under Conditions of Preferential Flow   315
Problems Encountered in Modeling Preferential Flow   316

10.14   Colloidal-Facilitated Transport   318

Colloid Migration   318
A Conceptual Model for Colloid Transport   322

10.15   Sources of Contamination   325

Municipal Sewage-Sludge Disposal and Wastewater Irrigation   325
Radioactive Waste Disposal   328
Landfills   331
Other Sources of Contamination   331

10.16   Case Study: Radionuclide Distribution and Migration
        Mechanisms at Maxey Flats   332

Summary   338

11 EFFECTS OF INFILTRATION AND DRAINAGE ON SOIL-WATER REDISTRIBUTION   346

Introduction   346

11.1   Profile-Moisture Distribution   347

Soil Characteristics   349
Liquid Properties   350
Rainfall or Other Liquid-Arrival Factors   350
Other Soil-Surface Factors   352

11.2   Infiltration Theories   352

Green-Ampt Approach   353
Horton and Kostiakov Equations   356
Holtan Model   358
Philip Model   360
Morel-Seytoux and Khanji Model   361
Smith–Parlange Model   361
Comparison of the Various Physically Based Infiltration Models   362

11.3   Effects of Macroporosity   362

11.4   Layered Soils   363

11.5   Crusted Soils   365

11.6   Runoff   367

11.7   Redistribution and Internal Drainage   369

> Water Redistribution   369
> Internal Drainage   370

11.8   Field Measurements   371

> Areal Measurements or Data Analysis   372
> Point Measurements   373

Summary   375

12   FIELD WATER IN SOILS   379

Introduction   379

12.1   Field Water Balance   379

> Precipitation, Air Temperature, and Solar Radiation   380
> Runoff   383
> Evaporation/Transpiration   389
> Change in Surface Storage   399
> Change in Soil-Moisture Storage   399

12.2   Field Radiation and Energy Balance   401

> Radiation Balance   401
> Energy Balance   402

12.3   Water- and Energy Balance Methodology   403

Summary   404

13   APPLIED SOIL PHYSICS: MODELING WATER, SOLUTE, AND VAPOR MOVEMENT   407

Introduction: Modeling Approaches   407

> Analytical Models   407
> Numerical Models   408

13.1   One-Dimensional Deterministic Liquid-Flow Models   409

> Analytical Models   409
> Numerical Models   427

13.2   Three-Dimensional Deterministic Liquid- and Vapor-Flow Models   429

> Analytical Models   429

13.3   Three-Dimensional Deterministic Immiscible Liquid-Flow Model   448

> Analytical Model   448
> Numerical Models   459

13.4   Use of Tracers in Unsaturated Soil Studies   460

> Theory of Unsaturated Liquid Movement Using Tracers   460
> Environmental Tracers   462
> Applied Tracers   464
> Tracer Flux Through the Root Zone   465

13.5   Stochastic and Transfer Function Models   470

Analytical Model   470
Numerical Models   474

Summary   474

14   DRAINAGE IN SOIL WATER AND GROUND WATER   476

Introduction   476

14.1   Problems Associated with Drainage   476

14.2   Typical Drainage Situations Under Field Conditions   478

Flow in an Unconfined Aquifer   478
Flow toward a Well   479
Flow between Parallel Ditches   480
Flow with Uniform Recharge   481
Evaluation of Falling Water Tables   482

14.3   Ground Water Drainage   483

Water Tables and the Capillary Fringe   483
Equipotential Lines, Streamlines, and Potential Flow   484
Construction of Flow Nets   485

14.4   Drainage Design   488

Elements of Drainage Design   488
Development of Drainage-Design Criteria   488
Drainage Coefficients   489
Theoretical Analysis   489

14.5   Case Study: The Florida Everglades   491

Summary   493

15   SOIL REMEDIATION TECHNIQUES   494

Introduction   494

15.1   Soil Corrective-Action Criteria   494

Background   494
General Principles of Application   495
Fate and Transport Models for Evaluating Migration to Ground Water   496

15.2   Alternative Technologies for Soil Remediation   510

No Action   510
Capping   511
Soil Venting   513
Air Sparging   515
Soil Flushing   517
Biological Treatment   518
Soil Excavation   518
Soil Washing   519
On-Site Thermal Treatment   520
In-Situ Vitrification   521

Summary   521

16 SPATIAL VARIABILITY, SCALING, AND FRACTALS   524

Introduction   524

16.1   Frequency Distributions of Soils   524

Geostatistics   529
Semivariogram   530
Kriging   530
Power-law Distributions   531

16.2   Scaling as a Tool for Data Analysis of Physical Properties   534

16.3   The Fractal Dimension   539

Triadic Von Koch Curves   540
Self-Similarity and Scaling   540
Box Dimension   541

16.4   Fractal Construction   543

Fractal Dimension $0 < D < 1$ 543
Fractal Dimension $1 < D < 2$ 543
Fractal Dimension $2 < D < 3$ 545

16.5   Self Similar Fractals: Estimating the Fractal Dimension   545

Grid Dimension   546
Point Dimension   546
Ruler (Divider) Dimension   547
Perimeter–Area Dimension   547

16.6   Self-Affine Fractals   548

Hurst's Empirical Law (Exponent H)   548
Brownian Motion and Random Walks in One Dimension   549
Scaling Properties of Random Walks in One Dimension   550
Fractional Brownian Motion: Mathematical Models   551
Global Dimension of Self-Affine Fractals   551
Measurement of Self-Affine Fractals   552

16.7   Viscous Fingering and Diffusion-Limited Aggregation   554

Diffusion-Limited Aggregation (DLA)   555
Fractal Diffusion Fronts   556
Viscous Fingering in Soil   556

Summary   559

Appendix 1   SITE CHARACTERIZATION AND MONITORING DEVICES   561

Introduction   561

Site Characterization   561

Monitoring Devices   563

Factors Influencing Choice of Devices   563
Water Content Measuring Devices   565
Matric Potential Measuring Devices   567
Soil Solution Sampling Devices   569
Pressure Measurement Devices: Differential Pressure Transducers   570

Appendix 2 MATHEMATICS REVIEW   571

Introduction   571

Algebra   571

Some Basic Definitions   572

Trigonometric Relations   574

Trigonometric Functions   574
Law of Cosines   575
Law of Sines   575

Geometric Relations   575

Powers and Roots   576

Logarithms   577

Calculus   577

Differential and Integral Calculus   577
Vector Calculus   580
Differential Equations   581

Vectors   582

The Taste Function   583

Appendix 3 TABLES   586

Complementary Error Function (erfc)   586

Series Expansions   586

Exponential Integral   586
Error Function   587

Conversion Factors   587

Physical Constants   590
Useful Conversion Factors   590
Statistics on Water   591
Converting X number of ppm_v to density ($g\,m^{-3}$) units   591
Converting water in chemical potential units [$J\,kg^{-1}$] to pressure
  potential units [Pa]   591
Converting X number of pCi (pico Curies) of a radioactive gas to
  its concentration in parts per million (ppm_v) units   591

The International System of Units (SI)   592

SI Units for Use in Unsaturated Zone Hydrology   595
Units for Water Flow Applied to Darcy's Law   595

List of Symbols   596

References   606

Index   621

Preface

Unsaturated zone hydrology has become a vital part of the environmental sciences and of engineering, especially in relation to the characterization, modeling, cleanup, and monitoring of residual solid and liquid storage and disposal. Current and future trends in the environmental sciences and engineering (including multidisciplinary investigations, increasing technological capabilities, more restrictive regulations, and increased hazards to the environment imposed by a growing population), and the viability of employment as scientists, technicians, consultants, environmental engineers, regulators, and so on, in the area of unsaturated zone hydrology, requires a more thorough working knowledge for this field of science. Some universities teach related material in courses on soil physics and agricultural engineering. These courses, however, often do not encompass many environmental fields, and do not always include chemical properties of soils, modeling, monitoring, and contaminant transport. All of these topics, as well as the traditional soil physics topics, are extremely important in the environmental sciences, engineering consulting, and regulatory oversight.

During the past seven years, at both the University of Colorado and Colorado School of Mines, we have taught many students from a wide variety of backgrounds and skill levels in our unsaturated zone hydrology classes. Typical prerequisite classes have not prepared them well for investigating phenomena in the unsaturated zone. Because a knowledge of the unsaturated zone is multidisciplinary, we have been urged by our students to write a book of this nature—beginning with basic physical properties and the behavior of clays, and continuing to include contaminant transport and other parameters such as spatial variability, scaling, and fractals in the earth sciences.

This text was written in response to our students' concerns. It is designed for upper-level undergraduate and beginning graduate students in the fields of environmental science, geology, hydrology, engineering, soil science, soil mechanics, soil physics, and agricultural engineering. While it is possible to cover most of this text in a single semester, some sections are quite lengthy—and perhaps more detailed than necessary. Instructor discretion is required in shaping a meaningful unsaturated zone course, to meet specific objectives. However, the more lengthy and advanced sections can be used as a reference. We have used this book in both basic unsaturated zone hydrology courses at a more applied level, and advanced courses. The primary objective of the basic course is to teach the major processes that occur in the unsaturated zone, and to prepare students to design and install instruments (such as thermocouples, lysimeters, dataloggers, and so on) in order to collect and analyze data under unsaturated zone conditions. In addition to developing greater expertise in devising an unsaturated zone study, the advanced course focuses on modeling the collective data on a computer and making predictions, through monitoring and other risk-assessment techniques, concerning times of transport of a contaminant to ground water and potential hazards to the environment. For either course, the student should have a good mathematics background (including at least two semesters of calculus and, preferably, partial differential equations). In addition to mathematics, a study of the unsaturated zone requires an understanding of many disciplines; thus, it is helpful if the student has studied geology/hydrology/hydrogeology, ground water, soil science, physics, and chemistry at the college freshman or sophomore level.

Chapter 1 defines the unsaturated (vadose) zone, describes current environmental problems in this area, and gives a brief history of unsaturated zone hydrology. Chapter 2 introduces physical properties of soils, and chapter 3 concentrates on the behavior of clay–water systems.

Chapters 4 and 5 define the energy, thermodynamic, and chemical states and properties of soil-water systems. These chapters define the structure of water, the theory of potential as applied to the soil-water system, hysteresis, and organic and inorganic reactions in the soil-water system.

Chapters 6 through 8 present the principles of water flow in soil, water flow in saturated soils, and water flow in unsaturated soils, respectively, from a theoretical perspective. Chapter 9 (written by Dean Anderson) presents the theory of gaseous transport in soil systems.

Chapters 10 through 15 use the definitions, concepts, and theories from the previous nine chapters to investigate applied modeling of fluid movement and contaminant transport. Modeling concentrates on analytical solutions that are seldom presented in advanced texts, because we believe that analytical solutions are generally more available to the prospective student, and also more easily understood. However, a list of numerical unsaturated zone models is presented, along with references related to each. Chapter 16 is a unique exploration of fractals in soil physical properties, and presents a detailed discussion of soil spatial variabilitiy and water movement using fractals.

The text has been written primarily with the student in mind. Throughout the chapters, questions are given at the end of sections when we felt they were needed. These questions are answered completely at the end of each chapter. We did this for two reasons: completing work at the end of each section reinforces the material learned; and seeing the answer to the question increases the students' confidence in their ability to perform the work. Some instructors may feel that we are pampering the students with the complete answer, however; to offset this, and for the instructors' benefit, additional questions are asked at the end of each chapter. Answers to these questions will be found only in an instructor's solutions manual available from the publisher. In certain chapters, sufficient worked examples are given so that the student should not need direct questions with their respective answers.

There are many references that could have been used for the various principles presented in the text. We have made no attempt to indicate the first or primary author of a particular subject, nor to give the subject a complete citation. In some instances, where there has been little significant change in a specific area, we have chosen original references that date back many years; but we have also chosen those references that we believe will be of most value to the student, will represent the most important point of view, or will be most accessible.

ACKNOWLEDGMENTS

We would like to acknowledge Dean E. Anderson of the U.S. Geological Survey, National Research Program, Unsaturated Zone Field Studies, for contributing chapter 9 on gaseous diffusion in soils, and William Beeman for his assistance with the math review in appendix 2.

We also would like to acknowledge the many reviewers who provided helpful comments, which have significantly contributed to the improvement of individual chapters or the entire text: Stephen Anderson, Dennis Baldocchi, Myron Brooks, Brett Bruce, Hsiu-Hsiung Chen, Shih-Chao Chiang, Sally M. Cuffin, Katie Walton-Day, John Dowd, David Eckhardt, Richard Healy, Jerry F. Kenny, Ed Kwicklis, Kenneth Lull, Peter B. McMahon, R. D. Miller, Sheldon Nelson, David Nielsen, John Nimmo, Edmund Perfect, David Radcliffe, P. Rengasamy, Gary Severson, Dave Stannard, Timothy D. Steele, Ken Stollenwerk, David Stonestrom, Scott Tyler, and Jack Weeks.

We would also like to thank the following graduate students for their helpful comments: Tim Axley, Andy Beck, Shannon Boots, Rod Carroll, Peter Cutrone, Rob Fishburn, Holly Hodson, Karen Maestas, Dave Mau, Dave Mennick, Joette Miller, Richard Smajter, and Judy Williams.

Special thanks goes to Edwin P. Weeks, Project Chief, Unsaturated Zone Theory and Field Studies, U.S. Geological Survey, National Research Program. Ed's forty years of experience in this field of science include theory, teaching, and research. He reviewed each chapter and provided invaluable advice, numerous comments, and suggestions that often injected reality into some otherwise rather obtuse concepts. The authors owe a great debt to Ed for the exceptional thoroughness of his reviews, and for the time spent on behalf of this text, working to enhance both presentation and clarity for the student of unsaturated zone science. We would also like to acknowledge and thank John Flager, Regional Reports Specialist, U.S. Geological Survey, Water Resources Division, for both his editorial and technical reviews. Also, a special thanks to John M. Evans, Presentation Graphics Chief, Technical Publications Unit, U.S. Geological Survey, Water Resources Division, for designing the cover of the text and assisting, along with Shannon Boots, with the graphics contained within.

We sincerely hope that this text makes a practical contribution to the teaching of unsaturated zone hydrology, and fills an important gap in the hydrologic curriculum at the university level. We especially hope that, as a reference, it can be used by practicing environmental scientists and engineers in solving real-world problems. A companion text that covers the saturated zone, *Groundwater*, by R. Alan Freeze and John A. Cherry, is also available from Prentice Hall.

James A. Tindall
James R. Kunkel
Denver, Colorado

1

Introduction and Brief History

INTRODUCTION

Unsaturated zone hydrology is the study and investigation of the physical state of the soil; specifically, the transport of all forms of matter and energy within the unsaturated zone of soil. Geologically, the topmost layer of the earth's crust comprises a three-phase system, which includes solid, liquid and gas. The solid phase contains mineral grains of varying types and organic matter, which represents the remains of animals and plants in varying stages of decay. The liquid phase is composed of water that may contain a wide variety of solutes. The gas phase includes air, water vapor, and other gases, all of which may be in different stages of equilibrium. The earth layer that contains all three phases of matter has been termed the unsaturated zone. Other common names given to this zone are the *zone of aeration* and the *vadose zone*.

The unsaturated zone is typically defined to extend from land surface to the underlying water table or saturated zone within porous media. In some areas, such as parts of the eastern United States, the unsaturated zone can be very shallow, ranging from a few centimeters to several meters. More typically, in the western United States, the unsaturated zone can be more than a hundred meters thick. In wetlands, the unsaturated zone may fluctuate seasonally or not exist at all. By definition, the unsaturated zone also includes the root zone of the overlying vegetation, usually about one meter thick. Roots can play a major role in chemical transport within agricultural and undeveloped settings.

The unsaturated zone has a significant influence on the movement and transport of both water and chemicals. Variables of state that affect the unsaturated zone include pressure, volume, and temperature. Soil phenomena such as capillary flow, adsorption, chemical interactions, and matric potential could be said to be dependent on variables of state. A soil system is governed by thermodynamic principles and, in most instances, follows the laws of physical chemistry. A thermodynamic system is a specific part of the physical universe under investigation. A system is defined to be separated from the rest of the universe by an imposed boundary; that part of the universe outside the boundary is referred to as the surroundings. With regard to soil, the surroundings would include the atmosphere and aquifers. Though a system is enclosed by a boundary, in unsaturated zone hydrology matter is usually transferred between the system and its surroundings, and the surroundings may do work on the system, or vice versa. If the boundary around a system prevents matter interchange, or mechanical or thermal contact, the system is called an isolated or closed system.

Most investigations in unsaturated zone hydrology verify the fact that the soil system is not an isolated system. Chemicals in the form of liquid, solids, and gases escape into the atmosphere and leach into ground-water boundaries surrounding the soil system, causing contamination from agricultural, industrial, residential, and recreational uses of land and water. Through the application of the principles of kinetics, in conjunction with thermodynamics, we are able to interpret certain thermodynamic soil properties in molecular terms that make it possible to explain physical processes, and to determine the rates of these various processes.

Soil is highly heterogeneous in both space and time, and is the basis for all terrestrial life. It exchanges water with the atmosphere in the hydrological cycle and influences the atmospheric content of greenhouse gases. It also serves as a medium for microbial activity, plant growth and, unfortunately, storage of various contaminants including radioactive wastes. The soil consists of fragmented components of varying shapes and sizes, which are arranged in complex patterns. Depending upon its mineralogical, chemical, and physical properties, a soil can freeze and thaw, shrink and swell, disperse and flocculate, precipitate and dissolve salts, crack, compact, and exchange ions during alternate wetting and drying processes. Consequently, it can be said that a soil system is never in a state of equilibrium. Due to the nature and complexity of porous media, unsaturated zone hydrology is an inherently complex science that requires a knowledge of chemistry, physics, mathematics, mineralogy, engineering, computer science, soil chemistry, hydrology, and soil–plant–water relations to ascertain answers to contemporary environmental issues. This is especially true in the areas of water quality and treatment, agricultural and waste management, and the recreational and industrial use of water.

1.1 A BRIEF HISTORY OF UNSATURATED ZONE HYDROLOGY

The origins of soil science are inextricably linked to the development of agriculture, irrigation, chemistry, and physics. The earliest writings that mention agriculture include many accounts concerning the physical aspect of soils. Examples of these include Sumerian cuneiform writings dating from circa 1700 B.C. that give instructions on the preparation of land and planting of grain crops in the Euphrates River valley (Kramer 1958, 1963), and other writings from circa 1000 B.C. that mention erosion by wind and water and its effect on crop culture (Bennett 1939). Some of the greatest philosophers, scientists, and educators have written about soil; Buol, Hole, and McCracken (1973) mention writings that involve the soil by Aristotle (384–322 B.C.), Theophrastus (ca. 372–ca. 286 B.C.), Cato the Elder (234–149 B.C.), and Varro (116–27 B.C.).

Democritus (ca. 460–ca. 370 B.C.) wrote that plant growth involved the cycling of indestructible elements. Aristotle taught that plants absorbed their nutrients from humus through the root system (Salmon and Hanson 1964). Sir E. John Russell (1957), in his discussion of the physical properties of the soil and how they are affected by biological activity, quotes from Pliny the Younger (61–ca. 113 A.D.) on the use of marl for land application. The *USDA Yearbook of Agriculture: Soils and Men* (1960, quoting Kellog 1938) refers to the Bible (1 Sam. 13:20) directing the Israelites to sharpen what would be considered their agricultural implements; also, to Homer (ca. 800 B.C.) describing Odysseus the wanderer returning to find his dog lying on a pile of refuse that was to be used to manure the land. However, when Moses says to the Israelites (Deut. 11:10–11), "For the land . . . wateredst it with thy foot . . . and drinketh water of the rain of heaven," he is referring to the irrigation practices of Egypt. Was Moses the first vadose zone hydrologist? In fact, the first Biblical reference to tilling is Gen. 4:2, which refers to Cain as a "tiller of the ground."

Unsaturated zone hydrology also went through a period of stagnation, with little literature to be found between the end of the first century A.D. and the early 1200s. It was in the mid-thirteenth century that Petrus de Crescentiis published a book on agriculture, *Opus ruralium commodum,* in Rome. Keen (1931) states that Fitzherbert's "Book of Husbandry" (1523) was the earliest work in English dealing with practical agriculture. Keen also provides an extensive bibliography of contributions to unsaturated zone hydrology.

LaRocque (1957) notes that the Frenchman Bernard Palissy (1510–1589), in his *Discours Admirables* (1580), described what was probably the first "soil auger," including a description of its construction and the mention of detachable handles. He constructed this device to aid in his exploration of soil. In his *Recepte Veritable* (1563), Palissy challenged the view of Plato (428–348 B.C.), Aristotle, and other important natural philosophers that water moved from the oceans, beneath the land, and into rivers and streams. Palissy said that "for this to happen, it would require an enclosed pipe flowing from the oceans to the mountains," since the mountains are higher than the oceans. Palissy's theory was that ground water and natural springs originated from rainfall that had infiltrated the soil; he was correct and ahead of his time in his theories. Palissy believed strongly in what he could "dig out of the bowels of the earth," and amassed entire experimental collections of various items that he used to challenge famous philosophers of his day. Palissy was a self-made scientist, and quite possibly the earliest to observe and record data from his experiments and explorations.

Sir Francis Bacon (1561–1626) believed that water was the source of nourishment for plants (Daumas 1958). This was later confirmed by Jan Baptiste van Helmont (1577–1644). According to Tisdale and Nelson (1975), in 1629, about seventy years after Palissy's book, Jan Baptiste van Helmont performed a famous experiment in which he grew a willow tree in an earthen container of soil that originally weighed 91 kg. After five years he removed the soil from the pot and weighed the soil and tree separately. The tree that had originally weighed 3 kg was found to weigh about 77 kg, and all but 57 g of the original soil was accounted for. Since the pot had been shielded from the atmosphere and only rain or distilled water had been added, van Helmont concluded that water was the sole nutrient of the plant, for he attributed the 57 g soil loss to experimental error. This idea was followed in 1673 by John Evelyn (1620–1706), a secretary of the Royal Society of London, and appeared to be supported with experimental evidence a few years later by Robert Boyle (1627–1691), also of England. Boyle confirmed the findings of van Helmont, but went one step further, stating that plants contained salts and other materials that were formed from water.

Johann Rudolf Glauber (1604–1668), a German chemist, suggested that potassium nitrate and not water was the "principle of vegetation"; John Mayow (1641–1679), an English chemist, supported the views of Glauber. About 1700, however, the Englishman John Woodward grew spearmint in samples of water he had obtained from various sources, and concluded that earth rather than water was the principle of vegetation. Jethro Tull (1674–1741), an Oxford-educated lawyer and author of *The New Horse Houghing Husbandry* (1731), concluded that plants derived their nourishment from "highly pulverized soil," or humus (Keen 1931). Arthur Young (1741–1820), one of the better known agriculturists of his time, and a prolific writer, edited a fourty-six-volume work entitled *Annals of Agriculture.* That work was supported by many experimental trials utilizing potted containers exposed to various influencing parameters, such as temperature, air, and water. Previous work by Young and others led Francis Home, in about 1775, to conclude that not just one parameter, but many, are required to describe the principles of plant growth.

At roughly the same time, Joseph Priestley (1733–1804) discovered that sprigs of mint "purified" the air, which led him to suggest that plants reversed the effect of breathing (i.e., taking in carbon dioxide and giving off oxygen). At this time he had not yet discovered what we now call oxygen, and when he did a few years later, he failed to recognize its relation to plants. However, the discovery of oxygen was a milestone, and helped unlock a great deal of mystery concerning plant life and the physical properties governing plant growth and soil interactions. This was a great help to Jean Senebier (1742–1809), a Swiss natural philosopher and historian who discovered the basics of photosynthesis in 1782 (Russell 1912). These early discoveries stimulated Nicolas-Theodore de Saussure (1767–1845) to experiment with the effect of air on plants and on the origin of salts in plants. De Saussure demonstrated that plants absorb oxygen and liberate carbon dioxide, but wrongly concluded that the soil furnishes only a small fraction of the nutrients needed by plants.

Sir Humphrey Davy (1778–1829), an English chemist, published *The Elements of Agricultural Chemistry* in 1813. Highly respected, Davy's book represented the best accepted knowledge of the time. He insisted on the importance of the physical properties of soils and their relations with heat, water, and other parameters, and was the first to apply the sciences of chemistry and physics to soil research—which marks the beginning of unsaturated zone hydrology (Wild 1988). As a result, Davy could properly be called its father. Building on the work of Davy and others, Gustav Schubler (1787–1834) greatly furthered the development of the field. Keen (1931) credits Schubler with the first technical investigations in unsaturated zone hydrology that involved systematic studies of the influence of soil physical properties on productivity of soils. Schubler also ascribed the crumbling of calcareous clay soils to the difference in the contraction of the calcareous sand and the clay substance, and has since been referred to as the father of agricultural hydrology (Hilgard 1906).

Contributions to unsaturated zone hydrology have come—and continue to come—from other fields, especially chemistry, mathematics, and physics. Many renowned scientists have made contributions to this area that are still in use today, including Ohm (1789–1854), Faraday (1791–1867), Fourier (1768–1830), Laplace (1749–1827), Helmholtz (1821–1894), Lord Kelvin (1824–1907), Boltzmann (1844–1906), Planck (1858–1947), Joule (1818–1889), and many others.

Wilhelm Schumacher used Schubler's original data to develop theories on the movement of water and air in soil, and introduced the principle of porosity. A contemporary of Schumacher and perhaps the most famous of the European scientists was Henry Darcy (1803–1858). His research in Dijon, France, in which water flux through sand filter beds was indicated to be proportional to the gradient of the hydraulic head, has become a cornerstone in water-flow principles under saturated and unsaturated conditions. This principle of water movement is known as Darcy's law, which will be discussed in detail in chapter 7.

Two other great scientists developed theories of water flow. These were Gotthilf Heinrich Ludwig Hagen (1797–1884) and Jean-Léonard-Marie Poiseuille (1797–1869), who in 1839 and 1840 started with Newton's law of viscosity and independently derived an equation for water flux in capillary tubes in terms of tube radius, pressure gradient, and fluid viscosity. That equation is currently known as the Hagen–Poiseuille equation. It expresses the same relation as Darcy's law, but with a conductivity term used for the measurable quantities of pore radius and viscosity, which adds further meaning to the equation. This relation will be discussed in detail in chapter 6.

Sir George Gabriel Stokes (1819–1903) developed an equation, known as Stoke's law, that is the relation of resistance to flow around a spherical particle. It is the basic equation used in both the hydrometer and pipette methods for particle size analysis.

Martin Edward Wollney (1846–1901) was probably the first soil scientist to be called a soil physicist. He published many articles and abstracts on the principles of unsaturated zone hydrology and agricultural meteorology (see, for example, Wollney 1878). During the late 1800s, Alphonse Theophile Schloesing and Jakob Maarten van Bemmelen (1830–1911) became interested in clays and their colloidal properties. During work with these properties it was discovered that there was a definite microbial role in soils (Russell 1912).

Due to progress in unsaturated zone hydrology as a result of advances in chemistry and physics during this time, scientists became interested in research in other areas within the field. In 1901, Eilhard Alfred Mitscherlich (1874–1956) concluded that the amount of water vapor absorbed by the soil would be proportional to the total surface area, and attempted to calculate this area based on the fact that any water present would be in the form of a monomolecular layer (Atanasiu 1956). Subsequently, the role of hygroscopic moisture was investigated by Lyman J. Briggs and Homer L. Shantz in Akron, Colorado (1912), among others.

Eugene Woldemar Hilgard (1833–1916) in California, Franklin Hyrum King (1848–1911) in Wisconsin, Thomas Burr Osborne (1859–1929) in Connecticut, and Milton Whitney (1860–1927) with the U.S. Department of Agriculture in Washington, D.C., were the best known soil physicists (i.e., unsaturated zone hydrologists) in the United States during this era. King's interest in quantitative aspects of water flow led to his collaboration with a colleague, Charles S. Slichter (1864–1946); Slichter published the results in 1898 as "Theoretical investigation of the motion of ground waters," a valuable paper describing soil water potential, in which he was able to derive the velocity, flow direction, and pressure at various points within porous media. Later, Slichter applied these principles to water flow in horizontal planes and water wells.

Edgar Buckingham (1867–1940) was a prominent scientist who worked closely with principles regarding soil moisture. In 1907, he published a significant paper dealing with capillary potential that was far ahead of its time, but which received little recognition for more than a decade. During this time two well-known Australian scientists, W. Heber Green (1868–1932) and G. A. Ampt (1887–1953), made significant contributions to unsaturated zone hydrology on the topic of infiltration. With the help of R. J. A. Barnard (1865–1945) and T. R. Lyle, they derived equations for vertical and horizontal flow of water in soils based on the Hagen–Poiseuille equation (Green and Ampt 1911).

John A. Widtsoe (1872–1952), a prominent educator in Utah and the intermountain west, made contributions to the field in the form of texts on irrigation (1914). He undoubtedly had a great influence on the progress of irrigation throughout the intermountain west. Widtsoe excelled at the application of science to logical, well-considered problems (Widtsoe and McLaughlin 1920; Gardner and Widtsoe 1921). At about this time, Willard Gardner and his colleagues expanded on Buckingham's ideas regarding capillary potential and moisture flow, and developed the tensiometer (Gardner et al. 1922). This was one of the great advances in the measurement of soil water potential. Thomas L. Martin, a contemporary of Gardner and a chemist, also contributed to soil science and its professional organizations (Martin 1921, 1925). These two men were influenced to some degree by Widtsoe.

George J. Bouyoucos (1890–1981) made major contributions to the field through the development of the hydrometer method for particle size analysis (1927a,b) and gypsum moisture blocks (1947). Bouyoucos immigrated with his family from Greece, and earned his Ph.D. in 1911, at the age of 21. It was in this same period of time that Lorenzo Adolph Richards (b. 1904), also from Utah and a student for B.S. and M.S. degrees under Willard Gardner, began making significant contributions to unsaturated zone hydrology. Richards is best known for the development of the Richards equation (1931) and its use in unsaturated flow. He is also well known for extending the idea of the tensiometer to the porous plate and pressure membrane apparatus (1941), commonly referred to as the "pressure plate apparatus." The development of the tensiometer by Gardner and the extension of this idea to pressure plates by Richards constitute two of the most significant developments in unsaturated zone hydrology.

Another prominent scientist who came under Gardner's influence is Don Kirkham (b. 1908). Kirkham taught physics until World War II, when he began naval research. Kirkham is perhaps best known for his text on soil physics (Kirkham and Powers 1972), extensive work in saturated flow and drainage, and the number of prominent students he has graduated in unsaturated zone hydrology.

Contributions to unsaturated zone hydrology by prominent European soil physicists include *Physical Properties of Soils* (1931) by Sir Bernard Keen (1890–1981). Ernest C. Childs (1907–1973) was known for his work in electric analogs applied to drainage studies, and H. L. Penman (1909–1984) worked extensively in the study of evaporation and evapotranspiration at the Rothamsted Laboratory. Another Englishman who worked at Rothamsted as a contemporary of Penman was R. K. Schofield (1901–1960), known for his introduction of pF of soil moisture. Since 1950, the field of unsaturated zone hydrology has steadily progressed, with significant work in the area of soil moisture diffusivity by Arnold Klute and one of his students, R. R. Bruce.

Eshel Bresler (1930–1991), director of the Soils and Water Institute of the Agricultural Research Organization at Bet Dagan, Israel, made significant contributions in the area of salt movement and irrigation in soils. He worked extensively in the simultaneous transport of energy and water in soils, anion exclusion, and infiltration. Bresler also published numerous texts and scientific articles, including *Saline and Sodic Soils: Principles—Dynamics—Modeling* (1982), coauthored with B. L. McNeal and D. L. Carter. Bresler's work was characterized by his direct approach to research (a result of practical experience gained in cooperative extension work in agriculture), and the interconnecting of water and soil-science disciplines to solve problems.

Limitations of space preclude a more thorough and complete listing of the many other scientists who have made contributions to unsaturated zone hydrology, and the other topics currently under study in the field. It is safe to say, however, that exciting developments to come will further enrich the fascinating history of this most terrestrial of sciences.

2

Physical Properties and Characteristics of Soils

INTRODUCTION

The role of physical properties in unsaturated zone processes—in relation to water content, contaminant transport, porosity, bulk density, saturation, and so on—cannot be overstated. A sampling of the areas within this subject that have been extensively researched might include the rate of water intake in soils (Bertrand 1965), the relation of exchangeable cations to physical properties of soils (Baver 1928), the infiltration of water in layered soils (Bruce and Whisler 1973), soil aeration (Lemon 1962; Lemon and Wiegand 1962), mechanical stress on root growth (Barley 1962; Russell 1952; Wild 1988), X-ray diffraction techniques for identifying soil minerals (Whittig 1986), and many others. This chapter will present a succinct discussion of the mineralogical composition of soils and various clays, soil profile development, and texture, and various physical characteristics and relations of soil constituents. A short list of suggested readings is given at the end of the text for the reader who desires greater detail or additional background on soils and their properties, followed by a complete list of references.

2.1 MINERALOGICAL COMPOSITION

Porous media, of which soil is an example, consist of three separate phases: solid, liquid, and gas. Due to the simultaneous presence of more than one phase, a porous medium is a heterogeneous system (also called a disperse, polyphase system), with each phase varying both mechanically and chemically from the others. Other examples of porous media include sedimentary rocks, stratigraphic columns, glacial deposits, gravel fills, mine tailings, and manmade media such as greenhouse mixes used in the propagation of various flora.

Soils are formed by weathering of rocks on the earth's surface, which causes their decomposition through physical and chemical deterioration. The physical processes involved, determined by climate, include freezing and thawing, wetting and drying, flowing water, moving ice, and abrasion by sand particles moved by wind or water. The small, loose rock particles that result from physical weathering become the parent material of soil. Solubilization, hydration, oxidation, reduction, precipitation, leaching, and other physicochemical processes further decompose the minerals that make up the rock particle. Because of the presence of populations of microbes and other fauna, a continual microbial and biochemical decomposition of some parts of the parent material continues as organic matter is added through the decay of animals and plants. The original rocks become the soil as it is seen today.

The mineralogical composition of soil is determined by the rocks or parent material from which it is formed. Other soil-forming factors include climate, biospheric parameters, effects of topography, and time. The solid phase of a soil includes sand, silt, and clay fractions commonly expressed as a percentage of the whole: 30 percent sand, 35 percent clay, and 35 percent silt, for example, for a total of 100 percent. The sand and silt fractions consist mainly of primary minerals that do not normally occur in the clay fraction. The most important and well-known of these primary minerals are quartz and feldspars.

Soils are usually deposited or formed in layers conditioned by geographic location, climate, and the formation processes that these layers have undergone. The accumulated layers in any given soil are called a profile. Each layer (called a *horizon*) within this profile has unique morphological characteristics.

Quartz is present in nearly all soils and makes up 50 to 90 percent of the sand and silt fractions. The quartz crystal is composed of linked silica tetrahedra in which the four oxygen atoms present are shared with adjacent silica tetrahedra, leaving a lack of cleavage planes. Feldspars are aluminosilicates of calcium, potassium, and sodium. Similar to quartz, the structure of feldspar is made up of silica- and aluminum-containing tetrahedra linked together by the sharing of each oxygen atom between adjacent tetrahedra. The result is a three-dimensional tectosilicate structure. The common minerals of granite and gneiss rocks are composed of potassium feldspars that include adularia, microcline, orthoclase, and others.

Accessory minerals developed during soil formation occur in the sand and silt fractions. Depending on the soil type, these minerals can be present in appreciable amounts and include amphiboles, pyroxenes, apatite, olivines, magnetite, ilmanite, and tourmaline. Occasionally, carbonate and sulphur-bearing minerals occur in soils in the form of dolomite and gypsum, respectively.

QUESTION 2.1

Are there situations in which only two of the three phases might occur in a soil?

Clays

Clays are colloidal and crystalline in nature and are the active mineral portion of soils. Clay minerals are commonly formed from the soluble products of primary minerals, and are usually referred to as secondary minerals. Most clays are crystalline and have a definite, repeating arrangement of atoms, dominated by layered planes of oxygen atoms. Silicon and aluminum atoms bond the oxygen atoms together through two kinds of bonds: ionic (attraction of positively and negatively charged atoms) and covalent (sharing of paired electrons). Three or four planes of oxygen atoms, with intervening silicon, aluminum, or other atoms (depending on clay type), make up one layer (see figures 2.1 and 2.2).

The crystal lattices of most clay minerals are composed of two structural units, namely tetrahedra and octahedra. The tetrahedral units consist of a silica tetrahedron in which the silicon atom is equidistant from four ionically bonded oxygen atoms at the corner of the structural unit. The planes of oxygen atoms held together by the silicon atoms are tetrahedrally oriented, and are referred to as silica tetrahedral sheets. These tetrahedral groups are arranged in a hexagonal network. The octahedral units consist of two sheets of closely packed oxygen atoms or hydroxyl units bound at the edges to an aluminum, iron, or magnesium atom that shares the oxygen atoms (see figure 2.1). Clays in which the oxygen and other atoms are less regularly oriented are called *amorphous* materials.

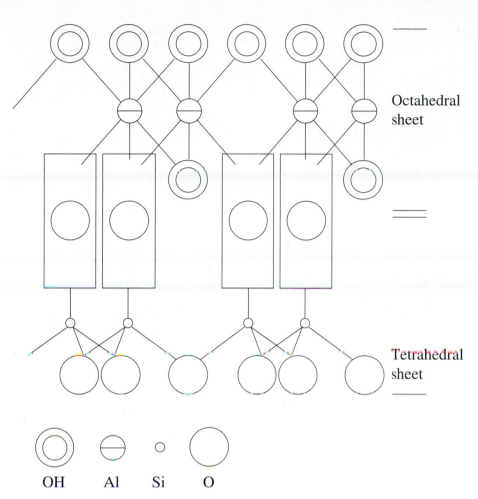

Figure 2.1   Diagram of condensed octahedral and tetrahedral sheets of clays (1 : 1 type); data from Mott (1988)

Octahedral sheet

Tetrahedral sheet

OH   Al   Si   O

Silicate Clays

The most common clays are discussed here, and in the next sub-section. The silicate clays represent well-structured materials. Generally, clays are referred to by the designations 1:1, 2:1, 2:2, and so on. This refers to the ratio of tetrahedral silica to octahedral alumina sheets in a layer of clay. In some clays, a silicon ion can be substituted by aluminum, and vice versa. Likewise, any ion of similar size to aluminum may also substitute, giving the clay very different properties, due to the ionic charge differences between ions. These ions may include iron, magnesium, and zinc.

When clay is examined closely under a microscope, the structure appears very much like a deck of cards. Each "card" represents a layer of clay that is an exact replica of the next layer, with each layer being held together by van der Waal's–London forces (*as explained in chapter 3*). This "deck" can both expand and/or shrink with the addition or loss of water. The amount of expansion between each card or layer and, thus, the thickness of the total number of layers, depends on the ion of substitution; this is *discussed* in section 3.6. Formation of layer after layer forms a *tactoid;* that is, an aggregated platelet. These tactoids are formed primarily by ion exchange in which calcium typically begins to segregate in the interlayer

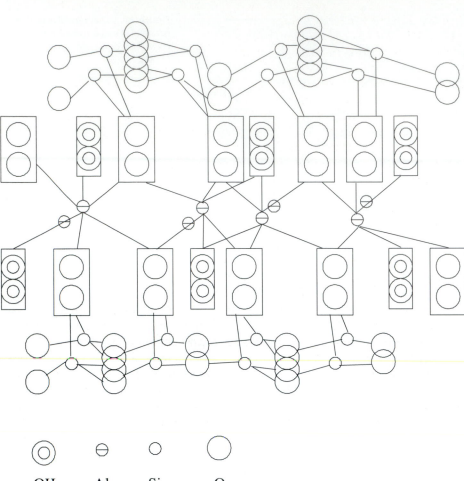

Figure 2.2   Diagram of condensed octahedral and tetrahedral sheets of 2 : 1 layered montmorillonite clays in which no substitution has occurred; data from Borchardt (1989)

OH   Al   Si   O

regions and sodium ions on the external surfaces of the tactoid. Depending on the characteristics of the exchanging ions, tactoids can range in thickness from 1 to over 1000 Å. Additionally, selectivity coefficients play a major role in tactoid formation. A detailed discussion of tactoids is beyond the scope of this text. The reader is referred to McBride (1989) for in-depth discussions of tactoids and the surface chemistry of clay minerals.

Amorphous Clays

Amorphous clays are mixtures of alumina and silica that have not formed a well-oriented crystalline structure. Iron oxides and mixtures of other weathered oxides may also be part of this mixture. Such clays are usually found in areas where copious amounts of weathered products are present but have not had sufficient time or the proper conditions for crystal formation.

Amorphous clays are most common in soils that have formed from volcanic ash, which is also a porous medium. These clays are unique in that their charge originates from hydroxyl ions (OH^-) on the clay surface, which can either gain or lose a hydrogen ion (H^+); thus, the clays can be either positively or negatively charged. They are referred to as variable charged clays, and are more difficult to manage agronomically than other clays. Descriptions of the most important varieties of amorphous clays follow.

Kaolinite is a 1:1 silicate mineral commonly occurring in highly weathered, well-drained soils where leaching has removed Ca^{2+}, Mg^{2+}, K^+, and Na^+. It is more common than montmorillonite in acidic soils. It occurs as a crystal of hexagonal shape having a triclinic symmetry. Kaolinite clays have such strong hydrogen bonding that almost no substitution of silicon or magnesium for aluminum can occur. This is actually due to the bonding between the O layer of the tetrahedral surface of the clay platelet to the OH ions on the octahedral surface of the adjoining platelet, which prevents swelling in the presence of water. The net negative charge is low, resulting in a low cation exchange capacity (CEC) of 3–15 eq kg^{-1}, which varies with pH. The thickness of layers is 7.2 Å. In addition to hydrogen bonding, the smallest hydrated diameter for cations is 7.6 Å for potassium; thus, cations can only attach to the edge or face and not to the interlayer. This explains the lack of swelling and low CEC. Tactoids for kaolinite are 500 Å thick, which results in a low specific area (80,000 m^2 kg^{-1}).

Illite is a 2:1 young (limited weathering) clay that has a structure similar to micas, consisting of two tetrahedral layers separated by an octahedral layer, and is sometimes called a hydrous mica. Illite is characterized by high substitution of Si^{-4} by Al^{-3}, or Fe^{-3}, in the tetrahedral layers. The resulting large negative charge on the tetrahedral surfaces is balanced by sorption of a potassium ion, which fits into the "oxygen hole" in the tetrahedral lattice. This same potassium ion also attaches to an oxygen hole in an adjoining platelet or interunit layer, resulting in adjacent platelets being so tightly bound that water cannot penetrate between them. Consequently, illite has low swelling ability, and also is an important source of potassium in agriculture. The structure is weakened to an extent by the incorporation of water between some lamellae; that is, the exchange of potassium by other cations during the weathering process results in some interlayer hydration, which weakens the structure of the clay aggregate. Thus, the potassium ions locked within the lamellae are slowly released through the course of the growing season. Illite has an intermediate CEC (15–40 eq kg^{-1}). Illite is found in soils high in primary minerals that are not extensively weathered, and may occur in the same environment as montmorillonite. The layer thickness of illite (from top to top) is 10 Å, with a tactoid thickness of 50–300 Å and a medium specific surface area of 80,000–120,000 m^2 kg^{-1}.

Vermiculite is a 2:1 intermediate-age clay that is similar in structure to illite, and is also characterized by substitution of silicon by aluminum in the tetrahedral layers, but the resulting charge imbalance is compensated by interlayer sorption of hydrated magnesium ions, rather than potassium ions. The large hydrated radius of the magnesium ions results in larger interlayer separation. As a result, this is a moderately swelling clay, with a high CEC (120–150 eq kg^{-1}) and a high specific surface area (600,000 m^2 kg^{-1}). The structure of vermiculite is composed of six water molecules in octahedral coordination with magnesium. Layer thickness is 14 Å, which collapses to 10 Å with potassium saturation; tactoids are 50 Å thick.

Montmorillonite (also termed a smectite) is a 2:1 intermediate-age clay, in which water easily penetrates between the layers, causing free swelling of the clay particle. Thus, minerals of montmorillonite or smectite isomorphous series are termed freely expandable layer silicates (other examples are beidellite, nontronite, and saponite). These minerals occur as very small particles ranging in diameter from about 0.01 to 1 μm. Interlayer spacing ranges from about 12 to 18 Å, depending on the exchangeable cation species present and the degree of hydration. Complete drying causes a spacing of less than 10 Å, whereas total hydration can literally "float" the layers apart so that they are independent of each other. The general structural formula is $X_{0.8}(Al_{0.3}Si_{7.7})Al_{2.6}Fe_{0.9}^{+3}Mg_{0.5}) \cdot O_{20}(OH)_4 \cdot nH_2O$; however, without consideration to lattice substitutions, the theoretical formula, according to Grim (1953), is $Si_8Al_4 \cdot O_{20}(OH)_4 \cdot nH_2O$ (interlayer).

Montmorillonite is comprised of clay platelets made up of one aluminum octahedral layer sandwiched between two silicon tetrahedral layers. Charge imbalances in montmorillonite arise due to substitution of aluminum by divalent ions in the octahedral layer and silicon–aluminum substitution in the tetrahedral layer. This results in a net negative charge that is offset by sorbed cations to the surface of the tetrahedral layers. Both tetrahedral and octahedral layers are combined, so that the tips of the tetrahedrons in each silicon layer and one of the hydroxyl (OH) layers of the octahedral layer form a common layer. The atoms which are common to both the tetrahedral and octahedral layers are then O instead of OH. In continous layers—that is, where O layers of each unit tactoid are adjacent to O layers of neighboring units—weak bonds are formed that cause excellent cleavage planes (Grim 1953). Figure 2.2 is an example of the structure of montmorillonite in which no substitution has occurred.

Formation can occur directly from parent material in poorly drained areas, or indirectly from weathering of illite $\rightarrow$ vermiculite $\rightarrow$ montmorillonite. Montmorillonite is characteristically present in soils known as vertisols. The bonding between units varies depending on the adsorbed cation. Montmorillonite has a high CEC (80–100 eq kg^{-1}), and usually falls into one of three types: sodium, calcium, or aluminum montmorillonites. There is no bonding between layers in sodium montmorillonite. Bonding occurs in calcium montmorillonite due to a thinner electrical double layer (discussed in chapter 3), and there is more bonding in aluminum montmorillonite due to a higher valence of the charge species. The layer thickness is 10–20 Å, with a tactoid thickness of 10–100 Å and a high specific surface area (700,000–800,000 m^2 kg^{-1}), resulting in a highly reactive soil. Bentonite is an impure deposit of montmorillonite, and is commonly used as a low-permeability layer beneath landfills and for sealing around scientific instruments in the field, as well-drillers mud, and is also used in cosmetics.

Chlorite is a 2:2 type clay. In chlorite, two silica tetrahedra, an alumina octahedron, and a magnesium octahedral sheet make up the structure. Chlorite is a highly weathered clay that may form from vermiculite or montmorillonite. Some substitution in the tetrahedral and octahedral layers may occur. Consequently, chlorites do not swell upon wetting and have a low CEC (20–40 eq kg^{-1}). Layer spacing is 14 Å, with tactoids 100–1000 Å thick and a medium specific surface area (80,000 m^2 kg^{-1}). Primary chlorites are unstable in acidic weathering environments; thus, they are almost always found in more recent deposits, such as aridisols and mollisols. While we have classified chlorite as a 2:2 clay, there is some disagreement on this matter. A more detailed discussion can be found in *Barnhisel and Bertsch (1989)*.

Sesquioxides are mixtures of aluminum ($Al(OH)_3$), iron hydroxides ($Fe(OH)_3$), and iron oxides (Fe_2O_3) that form under conditions of intense weathering and heavy rainfall. They occur most extensively in warm, humid, well-drained soils in tropical regions and are generally termed unstructured clays. Most of the silica and some alumina have been washed from the soil; as a consequence, these clays are not sticky compared to montmorillonitic types. These clays can be either amorphous or crystalline. Sesquioxide soil types are easily identified, due to the characteristic red and yellow colors developed through the weathering process. Sesquioxides do not shrink or swell; however, they do coat large soil particles and form stable aggregates. Due to the high content of iron and aluminum oxides within the clay, it has a very high specific surface area and can readily adsorb, thus fixing nutrients such as applied phosphorous and forming insoluble phosphates. Soils containing sesquioxide clays are sometimes referred to as oxide minerals or hydrous oxides and contain such minerals as gibbsite and laterites, including hematite and goethite.

Allophanes are another hydrous oxide soil type worth mentioning. They are amorphous aluminosilicate gels having silicon in a tetrahedral coordination, and the metal ion in

TABLE 2.1   Summary of Physical Characteristics for Various Silicate Clays

Silicate clays	Type	Degree of weathering	Substitution	Swelling	CEC (eq kg^{-1})	Layer thickness (Å)	Tactoid thickness (Å)	s* (m^2 g^{-1})
Chlorite	2 : 2	high	some	none	20–40	14	100–1000	80
Illite	2 : 1	limited	high	low	15–40	10	50–300	80–120
Kaolinite	1 : 1	high	none	low	3–15	7.2	500	80
Montmorillonite	2 : 1	intermediate	high	high	80–100	10–20	10–100	700–800
Vermiculite	2 : 1	intermediate	high	moderate	120–150	14	50	600

*Multiply by 1000 to obtain m^2 kg^{-1}

octahedral coordination. They are stable and highly porous and as such are very permeable. Extensive leaching has removed most of the primary minerals and nutrients, leaving the soil infertile. Allophanes also have a high specific surface area, due to the presence of large amounts of aluminum and iron, which results in a high phosphorous-fixing capacity. Allophanes are essentially noncrystalline aluminosilicate compounds that are associated with weathered volcanic ash. They are members of a series of naturally occurring minerals that have been classed as hydrous aluminum silicates. These silicates have a wide range of chemical composition that is characterized by the presence of Si-O-Al bonds, short range order, and a differential thermal analysis curve that displays a low temperature endotherm and high temperature exotherm. Since these criteria are relatively strict, allophanes are limited to a small sector of the total spectrum of noncrystalline (and also paracrystalline) aluminosilicates which have been developed by the weathering of volcanic ash, pumice, and other soil materials. See table 2.1 for a summary of the physical characteristics of various clays.

QUESTION 2.2

What is the primary binding (attractive) force between adsorbed cations and a clay platelet?

QUESTION 2.3

Why do clay particles have different specific surface areas (assuming that they have equal planar extensions)?

2.2   SOIL PROFILES

The clays and minerals discussed in the previous section, organic matter and other solids, along with air and water are the essential elements that compose the soil. Soil formation results in the natural development of layers within a profile. Each of these layers is composed of structural units consisting of various fractions of sand, silt, clay, and organic matter. When formed in place by pedogenic processes, the layer is called a horizon. The soil formation process may take many thousand years or as little as a few decades. The development of each distinguishable horizon is due to physical, chemical, and biological disintegration in the weathering processes that have caused organic matter accumulation, colloid transport, and the deposition of clays, humus, and soluble minerals such as carbonates and gypsum. Most soil horizons noticeably differ from each other in color, content of organic matter, texture, and structure or chemical properties. Horizons are parallel to the soil surface; if a vertical cross-section were exposed, say, along the side of a pit, each horizon could be clearly seen and the differences between horizons would be immediately noticeable.

Horizon development within the soil profile has occurred under four broad categories: addition, removal, transport, and transformation of soil material. The horizons of a soil are usually characterized by soil morphologists into four or five major designations, each composed of subdesignations labeled by Arabic numerals or letters (*U.S. Soil Conservation Service, 1985*). The major horizon designations (from the soil surface downward) consist of: (1) the O horizon, which is characterized by organic matter lying on the surface; (2) the A horizon, which is considered the mineral horizon and may have lost appreciable amounts of clay, iron, and aluminum oxides by eluvial (washed-out) processes; and (3) the B horizon, which contains illuvial (washed-in) concentrations of clay, sesquioxides, and other colloids and materials transported within the profile. The remaining two major designations are: (4) the C horizon, which is considered the parent material and has been affected very little by ongoing biological activity; and (5) the R horizon, which is consolidated bedrock underlying the C horizon. The C horizon does not necessarily have to be the parent material, because the soil profile could have been formed by wind deposition (aeolian) or water sedimentation (alluvial) processes. Likewise, the R horizon may not be like the parent material from which the overlying horizon was formed. If this is the case the R is preceded by a Roman numeral denoting the horizons' lithologic discontinuity. The O horizon is not always present, in which case the A becomes the upper horizon.

Three other horizon designations that are not always present in soil are the AB, A&B, and AC. Within the A and B horizons are subdesignations: A1, A2, A3, B1, B&A, B2, and B3. The subdesignations may not all be present in a particular soil, depending on the developmental processes and geographic location of the soil. The AB horizon is a transitional horizon between A and B, which has the upper part dominated by properties of A and the lower part dominated by properties of B. The A&B could qualify as an A2 subdesignation, except that part of this horizon, constituting less than 50 percent by volume, would be classified as a B horizon soil. The AC horizon is transitional between A and C; it has properties of both A and C but is not dominated by either. This designation is used only where there is no clearly discernible B horizon. An example of a soil profile is illustrated in figure 2.3, along with a description of the major horizon designations.

A "typical" soil profile cannot be illustrated, because there is so much variety from one geographic region to the next and, for that matter, even within a single region. A soil profile consists of a succession of horizons, due primarily to the combination of soil-forming factors under which the profile develops. Climate is usually the dominant factor in soil formation, although many soil-forming processes are interrelated. The detectable layers of clays, organic matter, and salts that have moved downward into the profile are a result of both the amount and pattern of precipitation. Soil profiles tend to develop faster under warm, humid, forested conditions, where there is adequate water to transport soil colloids, and where there are considerable amounts of organic matter to decompose. Other climate-related factors that have a direct effect on soil formation are temperature and the types and amounts of vegetation present. For example, soil profiles that develop under forests have more horizons than profiles that develop under grasslands. Likewise, profiles in arid climates exhibit little horizon development due to lack of organic matter, absence of leaching, and inadequate soil protection by sparse vegetation (against erosion by water and wind). For an illustration of soil profiles developed in different geographic regions and under various climatic regimes, the reader is referred to Hausenbuiller (1978).

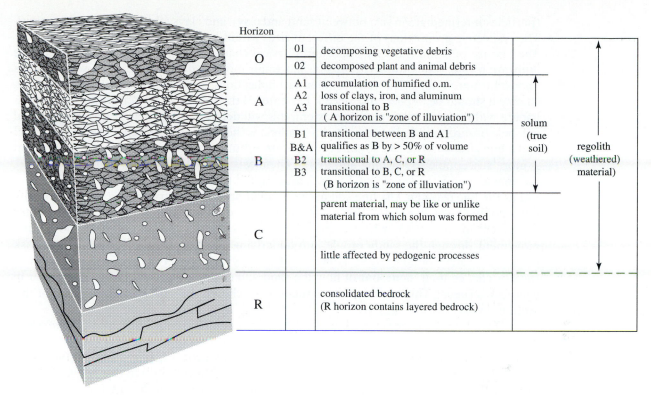

Horizon			solum (true soil)	regolith (weathered) material)
O	01	decomposing vegetative debris		
	02	decomposed plant and animal debris		
A	A1 A2 A3	accumulation of humified o.m. loss of clays, iron, and aluminum transitional to B (A horizon is "zone of illuviation")		
B	B1 B&A B2 B3	transitional between B and A1 qualifies as B by > 50% of volume transitional to A, C, or R transitional to B, C, or R (B horizon is "zone of illuviation")		
C		parent material, may be like or unlike material from which solum was formed little affected by pedogenic processes		
R		consolidated bedrock (R horizon contains layered bedrock)		

Figure 2.3   Representation of a complete soil profile (left) with horizon descriptions (right); data from Hillel (1982)

QUESTION 2.4

What are the five major processes involved in physical weathering in soil profile formation?

QUESTION 2.5

Decomposition of soil is usually referred to as chemical weathering. What are the two major processes associated with this phenomenon?

QUESTION 2.6

Associated with chemical weathering is structural change caused by chemical composition. What are the three major chemical processes affecting structural change?

2.3 SOIL TEXTURE

Soils are composed of particles of varying size, from large rocks to microscopic clays. However, the texture of soils is limited (by definition) to the fine earth fraction of the sand, silt, and clay sized particles. These different size groups, called *separates,* are sands (which usually consist of quartz, but may also be fragments of feldspar, mica, and heavy minerals), silts (particles intermediate in size between sand and clay), and clays (the smallest size fraction). The respective proportions of the different sizes within a specific soil determine the *texture*. The texture of a soil is a very important physical characteristic, because it determines the infiltration rate, water storage, porosity, and water movement rate within the soil. This has important agronomical and environmental implications, such as considerations of how a chemical should be applied and, once it has entered the soil, how it will transport and redistribute within the profile. To a large extent, this will determine whether or not the chemical will be a hazard to the environment. An example is the contrast between a coarse sandy soil and a fine clay soil. The coarse soil is easily wetted, but it also drains much more rapidly and is more susceptible to leaching and loss of chemicals and nutrients than the fine soil. Also, the coarse sand has a very low CEC and, thus, cannot usually adsorb as much organic chemical as the fine clay can. The fine soil has very small particles that provide a greater porosity but with much smaller pores. This results in decreased infiltration and retardation of water movement in comparison with the coarse sand. Thus, there is the potential for more rapid movement through the sandy profile into underlying aquifers, and therefore a greater risk of ground-water contamination. However, these physical properties are not a function of sand or clay content alone, but of how the sand and clay fractions are grouped together in aggregates (peds). This is termed soil structure, which is briefly discussed at the end of this section.

The method used to sort soil separates (including soils) is called *particle size analysis* or *mechanical analysis*. Various methodologies are used to determine size, such as *pipetting, hydrometer, centrifugation, elutriation,* and *sieving* (Bouyoucos 1927a&b; Chepil 1962; Day 1953). These methods are used to separate soil particles into the three distinct size ranges previously mentioned: sands, silts, and clays. Standardized classifications for particle size analysis (shown in figure 2.4) have been established by the United States Department of Agriculture (USDA), United States Public Roads Administration (USPRA), British Standards Institute (BSI), Massachusetts Institute of Technology (MIT), U.S. Bureau of Soils (USBS), and the International Soil Science Society (ISSS). However, while each of these

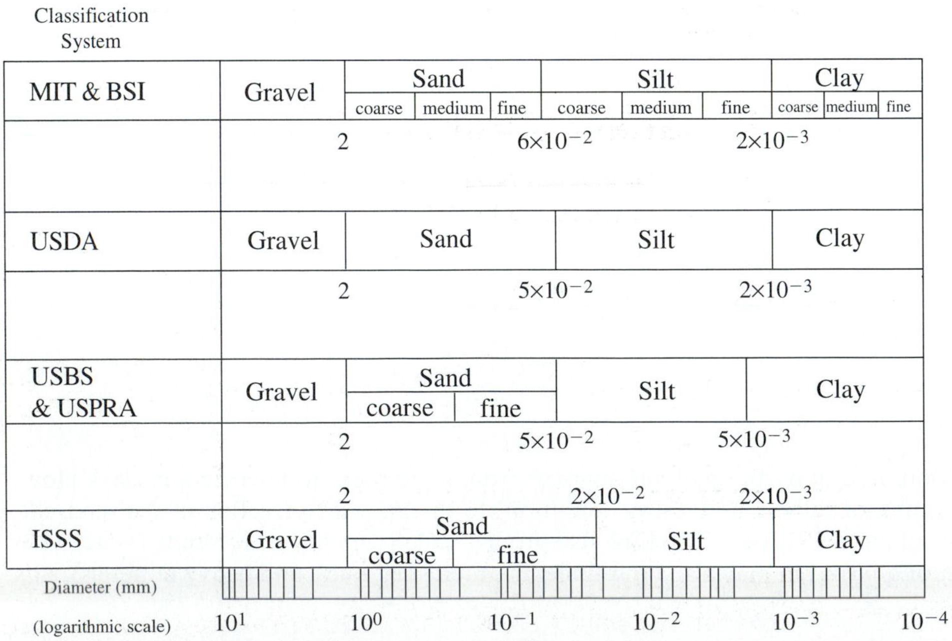

Figure 2.4   Soil classification system based on standards of the Massachusetts Institute of Technology (MIT), British Standards Institute (BSI), U.S. Department of Agriculture (USDA), U.S. Bureau of Soils (USBS), U.S. Public Roads Administration (USPRA), and International Soil Science Society (ISSS)

Classification System

System	Gravel	Sand			Silt			Clay		
MIT & BSI	Gravel	coarse	medium	fine	coarse	medium	fine	coarse	medium	fine
		2		6×10⁻²			2×10⁻³			

USDA	Gravel	Sand	Silt	Clay
		2	5×10⁻²	2×10⁻³

USBS & USPRA	Gravel	Sand (coarse / fine)	Silt	Clay
		2	5×10⁻²	5×10⁻³
		2	2×10⁻²	2×10⁻³

ISSS	Gravel	Sand (coarse / fine)	Silt	Clay

Diameter (mm) (logarithmic scale) 10^1 10^0 10^{-1} 10^{-2} 10^{-3} 10^{-4}

agencies has a standardized classification for soil separates, they vary slightly. This is especially true in the silt and sand separates.

While there is some confusion concerning which system to use, and what the upper and lower limits of particle size are, most members within a particular society or organization tend to use the same classification system; for example, agronomists and soil scientists who belong to the American Society of Agronomy (ASA) and Soil Science Society of America (SSSA) tend to use the USDA classification system. Members of other organizations, such as the American Society of Civil Engineers (ASCE), might use a different classification system.

An additional separate or fraction is important in some soils. This is gravel, or stones of size > 2mm (particle size < 2mm is classified as soil material). If this gravel or stony fraction makes up a significant weight of a soil being studied, and influences the physical and chemical processes such as transport and cation exchange, this size fraction should be reported as well.

The combination of all three fractions constitutes the matrix of the soil, sometimes referred to as its *structure*. Soil structure describes how the various size fractions are grouped together into stable aggregates. Aggregates are secondary units composed of soil particles bound together by organic substances, iron oxides, clays, carbonates, or silica. Natural aggregates that vary in their water stability are called *peds,* while the term *clod* refers to a soil mass broken into virtually any shape by artificial means, such as plowing, earth moving, or tillage. If a soil ped is broken apart, a broken piece of this ped is called a *fragment*. Another term

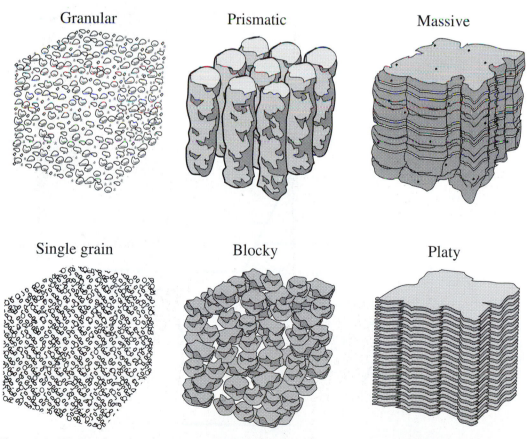

Granular	Prismatic	Massive
Single grain	Blocky	Platy

Figure 2.5   Depictions of soil structures; data from USDA, Agriculture Information Bulletin, No. 199, 1959

often confused with ped is *concretion,* which is a coherent mass formed within soil by chemical precipitation from percolating ground water. An example of this phenomenon is "iron stone" formation in the southern Coastal Plains near Tifton, Georgia. Sometimes referred to as "shot," these concretions become very hard. Upon crushing these pellet shaped concretions, it is observed that they are very hard on the outside with a small content of normally appearing soil on the inside.

As mentioned, structure determines the rate of infiltration of precipitation into the soil profile. Generally, as structure changes from granular or blocky to massive or platy, infiltration decreases. Examples of media structure (shape and arrangement of peds) can be seen in figure 2.5.

2.4 SOIL CLASSES

The texture class of a soil is determined by the proportion of the three size fractions. The soil texture triangle in figure 2.6 depicts the various texture classes a soil can be assigned, based upon the size fraction ratios. The texture triangle is based mainly on the USDA classification, with clay size < 0.002 mm, silt 0.002–0.05 mm, and sand 0.05–2.0 mm. Use of the triangle is

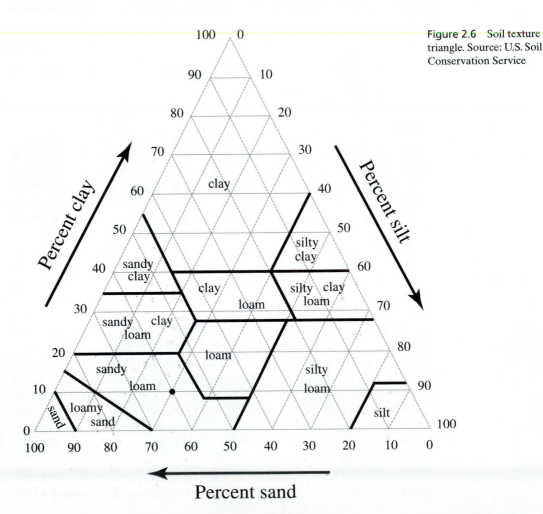

Figure 2.6 Soil texture triangle. Source: U.S. Soil Conservation Service

fairly straightforward. Suppose one had a soil that was 30% silt, 60% sand, and 10% clay; which texture class would it belong to? Proceeding clockwise from the apex of the triangle, we see that the right side of the triangle is 0–100% silt, the bottom is 0–100% sand (from right to left), and the left side is 0–100% clay (from bottom left corner back to the apex). Begin by finding the line on the right side that coincides with 30% silt (parallel to 70% clay), and intersect this line with the 60% sand line (parallel to 0% clay at the bottom of the triangle). By intersecting these two lines with the line that coincides with 10% clay from the left, we see that this soil is a sandy loam (see "black dot" in lower left corner of figure 2.6).

Examination of the texture triangle shows that finer soils (clays) are located at the top of the triangle, coarser soils (sands) in the lower left corner, and intermediate soils (silts) in the lower right corner. The center of the triangle is composed of clay loam and loam soil types. These two soil types, especially the clay loam, have an even ratio of fine, intermediate, and coarse size fractions. Their infiltration rates, CEC, specific surface area, and other physical characteristics will be between those soils that border them on either side; that is, a clay loam will conduct more water than a silty clay loam, but less than a sandy clay loam. Agriculturally, this clay loam would be a preferable medium under most circumstances, because it would retain more moisture and have better aeration and drainage than other soils, and would also have an ample nutrient supply.

QUESTION 2.7

If a soil contains approximately 20% clay, and the remaining fractions are equal amounts of sand and silt, what is its texture?

2.5 PARTICLE SIZE ANALYSIS

Particle size distribution is related to the physical and chemical activity of soils, and represents a stable soil characteristic. As discussed in section 2.8, particle size distribution indirectly relates to specific surface area. Because many chemical and physical properties of porous media are associated with surface activity, particle size analysis is the standard means for characterizing and classifying fractions. The proportioning of solids into different size ranges is termed *particle size analysis* or *mechanical analysis.*

Soil solids are composed of discrete units. To quantitatively determine separate particle size, the solids must be separated into these discrete units by both chemical and physical means. The complete particle size analysis of a medium involves two main processes: *dispersion,* the chemical or physical separation of a medium mass into its discrete units by the removal of cementing agents (such as organic matter, carbonates, and oxides) and subsequent disaggregation of clays; and *fractionation,* the grading or separating of particles into different particle size groups. Prior to dispersion, removal of organic matter and other cementing agents is accomplished by the addition of hydrogen peroxide in the case of organic matter, hydrochloric acid for carbonates, and oxalic acid–sodium sulphide for the ferric state of iron and iron and aluminum oxides. After removal of the cementing agents, *deflocculation* (separation of clay particles) is performed by the addition of a chemical dispersing agent and some form of mechanical agitation.

Mechanical agitation is usually accomplished by shaking or stirring and, occasionally, ultrasonic vibration. Sodium hexametaphosphate $[Na_3(PO_4)_6]$ is commonly used as the dispersing agent. The sodium monovalent cation replaces the polyvalent cations adsorbed on porous media, thereby breaking interparticle linkages. Activity of the polyvalent cations is also reduced by precipitation by phosphorus. Adsorbed sodium cations raise the electronegativity of colloids until these particles repel each other, and complete dispersion or deflocculation occurs (this causes a thickening of the electrical double layer). Care must be taken to insure complete dispersion. Incomplete dispersion results in small particles, flocs, or aggregates settling as larger (sand) or intermediate (silt) sized particles that may bias results of analysis by causing lower values for both silt and clay.

After complete dispersion of a sample, fractionation is used to separate the primary particles into their respective size groups. Separation of coarse particles can be achieved by passing them through a nest of graded sieves to a particle diameter > 0.05 mm, which corresponds to a 270-mesh sieve. This sieving process can be performed by hand or with a mechanical shaker. The sieve nest is stacked in descending order, so that the coarsest sieve is on top, progressing to the finest sieve on the bottom. Sieving can be done under wet or dry conditions. Standard sieve and mesh sizes used for fractionation are listed in table 2.2.

Sedimentation is required for classification of particles finer than 0.05 mm. In 1851, Sir George Gabriel Stokes formulated what has become known as *Stokes' law,* which is used in the calculation of fine particle fractions by sedimentation, in both the micropipette and hydrometer methods of particle size analysis. The theory of Stokes' law is that the resistance offered by a liquid of given density to the fall (terminal velocity) of a spherical particle is proportional to the square of the particle's radius and not to its surface area. The velocity of a particle falling without resistance will increase due to acceleration by gravitational forces. However, a particle falling through a fluid of given viscosity will encounter resistance, and the terminal velocity of the particle will be slowed as stated by Stokes' law (due to the product of the particle's radius and initial velocity and the viscosity of the fluid).

The resistance, or drag force (F_d), acting upon an individual particle as denoted by Stokes' Law, is given by

$$F_d = 6\pi\eta r v \tag{2.1}$$

where η is fluid viscosity ($kg\ m^{-1}\ s^{-1}$), and r (m) and v ($m\ s^{-1}$) are radius and velocity of the soil particle, respectively. In the cgs unit system, F_d is expressed in dynes ($g\ cm^2/sec^{-2}$). The force of gravity (F_g) acting upon the soil particle is expressed as

$$F_g = \frac{4}{3}\pi r^3 (\rho_s - \rho_f)g \tag{2.2}$$

TABLE 2.2   Sieve Series Specifications

Size of sieve (μ)	Sieve number (mesh per in.)	Sieve opening (mm)	Nominal wire (diameter, mm)
4000	5	4.000	1.370
2000	10	2.000	0.900
1190	16	1.190	0.650
1000	18	1.000	0.525
840	20	0.840	0.510
500	35	0.500	0.315
250	60	0.250	0.180
210	70	0.210	0.152
177	80	0.177	0.131
149	100	0.149	0.110
74	200	0.074	0.053
53	270	0.053	0.037
37	400	0.037	0.025

where $(4/3)\pi r^3$ is the particle volume (cm^3), ρ_s is the particle density ($g\ cm^{-3}$), ρ_f is fluid density ($g\ cm^{-3}$), and g is gravitational acceleration ($cm\ s^{-2}$). Setting $F_d = F_g$ and solving for v, we obtain Stokes law:

$$v = \frac{2}{9} \frac{r^2(\rho_s - \rho_f)}{\eta} g \tag{2.3}$$

However, we are usually interested in both the time of settling and the particle diameter. Rewriting equation 2.3 in terms of diameter rather than radius, we have

$$v = \frac{d^2(\rho_s - \rho_f)g}{18\eta} \tag{2.4}$$

Because $v = h/t$ (distance over time), we can substitute this relation into equation 2.4, and solve for the time (t) of settling to obtain

$$t = \frac{18h\eta}{d^2(\rho_s - \rho_f)g} \tag{2.5}$$

A rearrangement of equation 2.5 will yield the diameter of the particle, as follows.

$$d = \left[\frac{18h\eta}{tg(\rho_s - \rho_f)} \right]^{1/2} \tag{2.6}$$

Assuming a constant temperature (for constant liquid and particle densities), the right hand side of the equation is a constant (k) except for time, and can be rewritten as

$$d = \frac{k}{t^{1/2}} \tag{2.7}$$

Several limitations must be assumed for Stokes' law to reflect reality: (1) The particles should be large enough (> 0.0002 mm) to be unaffected by Brownian (thermal) motion of the fluid molecules by which they are surrounded. (2) The particles should be spherical, smooth, and rigid. (3) Free fall of the particles should be free from hindrance by other particles, that is, the solution must be sufficiently dilute to prevent one particle from interfering with another. (4) Laminar flow must exist around each settling particle. Practical limitations arise from the fact that the viscous drag on individual particles begins to increase nonlinearly with velocity, once the Reynolds number becomes about 1.0 (Schlicting 1968). This results in a particle of larger diameter falling more slowly than predicted by Stokes' law. Gibbs, Matthews, and Link (1971) show the specific nature of these effects for soil-type porous media; their results also indicate that soil particles larger than 100 μm do not follow Stokes' law and, therefore, 100 μm should be the particle size limitation when using Stokes' law. Particles larger than 2 mm should be removed by sieving, since they fall rapidly and can create turbulent flow. (5) Particles should be of the same density. The average density of most soil particles is 2.65 $g\ cm^{-3}$, while that of iron oxides and other minerals may exceed 5 $g\ cm^{-3}$. Additionally, hydrated particles will settle faster than dehydrated ones. (6) The temperature of the liquid must remain constant, since fluid viscosity is temperature dependent. Increases or decreases in temperature can increase or decrease terminal velocity, causing error in results.

Essentially, all of these parameters change with each soil sample. Particles are not smooth but have jagged edges, the shapes are not spherical but may be rod shaped or disk shaped or flat. Hence, the data gathered from particle size analysis are approximations, although they serve as a useful index for comparing one media to another.

Stokes' law applies to both the pipette and hydrometer methods. For these methods, the portion of sample collected in the sieve pan ($< 50\ \mu$m) is suspended in a cylinder of water in a constant-temperature bath. After the suspension thermally equilibrates, the cylinder is stirred thoroughly, and time is measured beginning at the end of stirring. Regardless of the method used, a reference depth below the suspension surface, h_1 (generally 10 cm), is selected. Equation 2.5 is used to solve for time of settling to this point. Particles larger than a given size will all have settled by a selected sampling time, so the fluid above the reference point contains only materials of cumulative grain size smaller than this given size. By selecting sampling times for incrementally smaller grain sizes, concentrations of particles of the size increment may be determined by difference. For the hydrometer method, the percent sand is obtained directly from the first hydrometer reading (nominally 40 seconds), the percent silt is obtained from a two-hour corrected hydrometer reading, and the percent clay is obtained by difference (see equation 2.9 and the surrounding discussion). The pipette method utilizes direct sampling. A small subsample of the fluid in question is taken by inserting a pipette into the sol at a depth h_1 and time t. As with the hydrometer method, all coarser particles than those sampled for at the measured depth and time will have settled beyond that point. The soil fraction of interest in the pipette method is the clay fraction (< 2 μm). Settling time is dependent on temperature as well. When using the pipette method, care must be used so that the suction applied to the pipette is not sufficient to disturb the solution surrounding its tip.

Hydrometers come in two types: specific gravity, and gram per liter. The specific-gravity hydrometer measures the percentage of sample remaining in suspension at the reference point discussed previously. This percentage may be calculated by the following equation:

$$P = \left[\frac{V}{W} \frac{\rho_s}{\rho_s - \rho_f} \right] (R_a - \rho_f) 100 \tag{2.8}$$

where P = percentage media remaining in suspension at the reference point; V = volume of suspension (cm^3); W = weight of oven dry soil (g); ρ_s = particle density (g cm^{-3}); ρ_f = fluid density (g cm^{-3}); and R_a = corrected hydrometer reading. Corrections to the reading obtained must be applied due to presence of dispersing agent, temperature, and position of meniscus. The dispersing agent changes the specific gravity of the deionized or distilled water used; hydrometers are usually calibrated in the factory at 20 °C, thus changes in temperature will affect the reading, due to change in the viscosity of the solution; and the hydrometer is designed to be read at the bottom of the meniscus (if the reading cannot be taken at the bottom of the meniscus a correction factor must be applied). These correction factors require the recalibration of the hydrometer for present conditions.

The gram-per-liter hydrometer directly reads the concentration of suspension at the reference point in gL^{-1}. The percentage of soil remaining in suspension can be obtained by

$$P = \frac{Ra}{W} 100 \tag{2.9}$$

where a = correction factor due to particle density (see table 2.3). For vessels less than ten times the particle diameter the following equation should be used:

$$v = \frac{2}{9\eta} \frac{r^2(\rho_s - \rho_f)g}{\left(1 + 8\dfrac{r}{R_a}\right)\left(1 + 3\dfrac{r}{L}\right)} \tag{2.10}$$

where L = point of measurement within hydrometer.

Two additional methods for measuring soil texture are centrifugation and elutriation. Centrifugation is similar to the two previously discussed sedimentation methods, except that

TABLE 2.3   Correction Factor for Varying
Particle Density

Particle density (kg m^{-3}) ρ_s	Correction factor a
2950	0.94
2900	0.95
2850	0.96
2800	0.97
2750	0.98
2700	0.99
2650	1.01
2600	1.02
2550	1.03
2500	1.04
2450	1.05

the particles are forced to settle much more rapidly through the suspension. Modification of Stokes' Law allows calculation as follows.

$$\frac{dR_r}{dt} = \frac{R_r \omega^2 (\rho_s - \rho_f) d^2}{18\eta} \tag{2.11}$$

where R_r = radius of rotation; t = time; and ω = angular velocity.

The radius of rotation (R_r) is taken as the distance from the center of the centrifuge to the surface of the suspension within the tube. Integrating equation 2.11 and solving for time gives

$$t = \frac{18\eta \ln \dfrac{R_{r_2}}{R_r}}{\omega^2 d^2 (\rho_s - \rho_f)} \tag{2.12}$$

where R_{r_2} = depth to point of measurement. As with equation 2.10, this can also be solved for particle diameter (d). The centrifugation method is used where particle size is < 50 nm. This size particle will be affected by Brownian motion, and might not settle if not for the added centrifugal force.

The final method to be discussed is that of elutriation. Soil fractions are determined by letting particles wash out in a flowing current of water at a constant velocity (v_1). Water is passed through a vertical cylinder from bottom to top. A velocity is chosen, which will carry off any particle whose velocity of fall in still water is less than the set velocity. When the flowing water is clear and free of particles, the velocity is increased to v_2 and the process repeated. An easier way to perform this analysis is to pass the porous media through a series of flasks which decrease in size with each flask. In this manner each flask contains different particle sizes.

As a final note on the determination of texture, there is a quick, albeit "dirty" method for rapid determination in the field. This is by feel, a method in which many soil scientists have been trained. It involves molding soil in the hand and rubbing it between the thumb and index fingers, under both moist and dry conditions. Based on the length and size of the ribbon which can be formed, the soil texture is determined. With some soil (such as sands) no ribbon will be formed, and individual particles will feel gritty; the soil will also lack any plasticity. As clay and silt contents increase, there will be a smoother feeling and longer ribbons can be formed. This method is mostly used in soil mapping and surveying. A trained individual can be very adept at judging soil texture by feel. In fact, the American Society of Agronomy has both regional and national student competitions each year that utilize this method.

QUESTION 2.8

The pipette method of texture analysis is being used to determine the time required for a particle ($d = 2$ μm) to settle to a depth of 10 cm. Calculate t and v using Stokes' law. Assume $\eta = 1.002 \times 10^{-2}$ g cm^{-1} s^{-1} and g = 981 cm s^{-2}.

QUESTION 2.9

Calculate F_d and F_g acting upon the above particle. Assume the same properties as in the previous question, and that $\rho_f = 1.39$ g cm^{-3} and $\rho_s = 2.65$ g cm^{-3}.

2.6   SOIL AS A PHASE SYSTEM

A phase is one part of a system, uniform throughout in physical properties and chemical composition. It is separated from other homogeneous parts of the system by boundary surfaces. Systems that exist in nature can be mono- or polyphasic. As described in the introduction to chapter 1, soil is a three-phase system composed of solid, liquid, and gas. The solid phase is composed of mineral and organic matter that make up the soil matrix; the liquid phase (or "soil water") is composed of water and dissolved substances; and the gas phase (or "soil air") contains several gases in dynamic equilibrium. A phase within a system can be homogeneous or heterogeneous (whether gaseous, liquid, or solid). If the chemical composition within the phase does not vary from one location to the next the phase is homogeneous, else it is heterogeneous; it can be homogeneous physically but heterogeneous chemically, or vice versa. An example of this would be ice and water, which is chemically homogeneous but physically heterogeneous.

A system can be completely heterogeneous. Within the soil system, the interfaces between solid, liquid, and air are continually changing, resulting in extensive interactions of adsorption, ion exchange, displacement, surface tension, and friction between each phase or combination of phases. It is important to note here that the solid phase of soils is relatively stable in comparison to the very dynamic liquid and gaseous phases. Generally, the greater the interaction, the larger the interfacial area existing between the phases. This will become more apparent with the discussion of specific surface area in section 2.8. A volume composition denoting the three phases in a typical loam soil is shown in figure 2.7.

2.7   DENSITY AND VOLUME–MASS–WEIGHT RELATIONS

Particle Density

The density of grains composing porous media is known as particle density (or mean particle density). Particle density is defined as the mass of solids per unit volume, and can be obtained from the following equation.

$$\rho_s = \frac{M_s}{V_s}$$

(2.13)

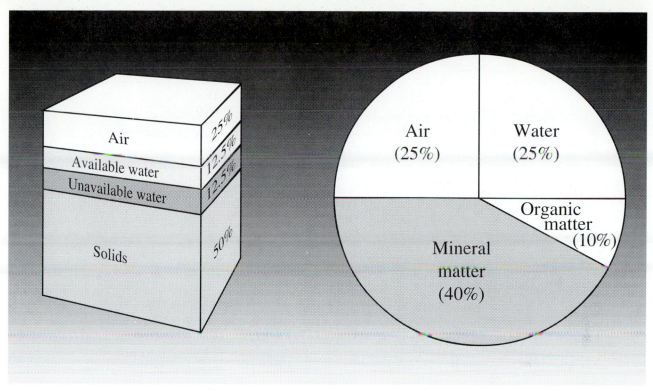

Figure 2.7   Cubical and pie-graph representations, illustrating a typical composition for a loam soil

where ρ_s = particle density (kg m^{-3}); M_s = mass of the solids (kg); and V_s = volume of the solids (m^3). Particle densities for various mineral constituents are given in table 2.4. The particle density of most quartz-type minerals such as sands, which are very common, is about 2650 kg m^{-3}; nonetheless, particle density for soils is quite variable. Tropical soils and other media high in iron and additional heavy minerals have a higher particle density, whereas soils containing substantial amounts of organic matter have lower particle densities.

The most common method of measuring particle density is the use of a pycnometer. The mass is obtained by weighing the sample; the volume is obtained from the mass and density of water displaced within the apparatus. Particle density from the pycnometer is calculated as

$$\rho_s = \frac{\rho_f(m_s - m_a)}{(m_s - m_a) - (m_{sw} - m_w)} \tag{2.14}$$

where ρ_f = fluid density (g cm^{-3}); m_s = mass of pycnometer filled with oven-dry soil sample (g); m_a = mass of pycnometer filled with air (g); m_{sw} = mass of pycnometer filled with soil and water (g); and m_w = mass of pycnometer filled with water (g). Because soil particles vary in density, the data obtained from the pycnometer is a weighted average that will yield a mean particle density. Occasionally, density may be expressed as specific gravity, which is the ratio of the density of the material compared to water at atmospheric pressure and a temperature of 4 °C, rendering the value dimensionless. Other methods for determining density have also been developed, such as using radioactive isotopes (Bernhard and Chasek 1953).

TABLE 2.4   Particle Density of
Various Soil Constituents

Mineral constituents	Particle density $(kg\ m^{-3})$
Humus	1300–1500
Clay	2200–2600
Orthoclase	2500–2600
Quartz	2500–2800
Calcite	2600–2800
Dolomite	2800–2900
Muscovite	2700–3000
Biotite	2800–3100
Apatite	3200–3300
Pyrite	4900–5200
Hematite	4900–5300

QUESTION 2.10

A hydrologist obtains a soil sample of 76 g (after oven drying) from a field site, which is placed in a pyc-nometer (100 cm^3 volume). Deaerated water is added to the sample to drive off all air at 20 °C. The mass of water and soil is 144 g. **(a)** Why was it necessary to remove all air from the sample? **(b)** What is the particle density $^b\rho_s$?

Void Ratio

The void ratio, e, is defined as the ratio between the volume of voids and the volume of solids. It is expressed as

$$e = \frac{(V_a + V_w)}{V_s} \tag{2.15}$$

This relates the fractional volume of soil pores to the volume of solids, rather than to the total volume of soil as in the definition of porosity, discussed next. The volume is composed of $V_a + V_w$, which are the volumes of air and liquid phases, respectively, and V_s, which is the volume of solids. Since porosity is usually expressed as ϕ, one will often see $V_a + V_w$ used when discussing void ratio.

QUESTION 2.11

An intact soil core (30 cm diameter × 40 cm height) is obtained from a research facility in Nevada. The core is weighed and placed in an oven to dry, after which it is reweighed. Water loss is 9766 cm^3 and ρ_b (dry bulk density) = 1.30 g cm^{-3}. Assume a particle density of 2.65 g/cm^3. **(a)** Determine V_a and V_w. **(b)** What is the void ratio?

Porosity

Soils are composed of mixtures of discrete large and small particles that may be loose single grains or bound in the form of aggregates, but the quantity of smaller particles and the aggregate size within a given soil has a marked effect on porosity (Diamond 1970; Klute 1986; Marshall 1958; Page 1948). A soil physicist is mainly concerned with unconsolidated soil, in which porosity depends on the packing of the grains and aggregates, their size distribution, shape, and arrangement. An engineer, on the other hand, may be concerned with consolidated media, where the porosity depends upon the degree of particle cementation within the materials. Porosity is expressed as a volume percent, ranging in most soils from 30–60%. Like many other media properties, soil porosity is governed more by particle size than any other parameter. Generally, the smaller the particle size the smaller the pores, but the greater the porosity; hence, coarse soils have a lower porosity than fine soils. However, particle sorting also has an affect, so this tendency is not absolute.

Porosity is expressed by

$$\phi = \frac{V_f}{V_t} = \frac{(V_a + V_w)}{(V_s + V_a + V_w)} \tag{2.16}$$

To understand the manner in which porosity depends upon structure and sorting arrangement (mode of packing), we should first consider the normal packings of uniform rods or spheres. Figure 2.8 illustrates simple models of porous media and their equivalent porosity. It can be seen that the least compact arrangement of uniform spheres is achieved through cubical packing ($\phi = 48\%$), and the most compact arrangement is represented by a rhombohedredal packing in which each sphere is tangential to twelve neighboring spheres ($\phi = 26\%$). Other packings of intermediate arrays fall somewhere between these two. Because the packings are of uniformly sized spheres, porosity is independent of particle size. If particles vary in size, porosity is dependent on particle size and distribution.

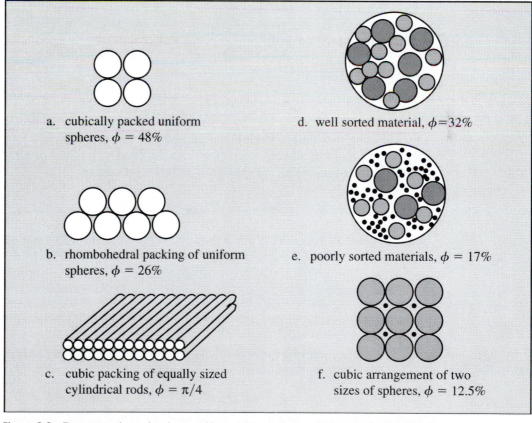

a. cubically packed uniform spheres, $\phi = 48\%$

b. rhombohedral packing of uniform spheres, $\phi = 26\%$

c. cubic packing of equally sized cylindrical rods, $\phi = \pi/4$

d. well sorted material, $\phi = 32\%$

e. poorly sorted materials, $\phi = 17\%$

f. cubic arrangement of two sizes of spheres, $\phi = 12.5\%$

Figure 2.8   Representations of various packings and arrangements for porous media, showing porosity (ϕ) for each arrangement

TABLE 2.5 Bulk Density and Porosity Values of Selected Soils

Porous media	Bulk density* $(kg\ m^{-3})$	Porosity value (%)
Limestone and shale[†]	2780	1–20
Sandstone	2130	10–20
Gravel and sand	1920	30–35
Gravel	1870	30–40
Fine to medium mixed sand	1850	30–35
Uniform sand	1650	30–40
Medium to coarse mixed sand	1530	35–40
Silt	1280	40–50
Clay	1220	45–55
Ideal soil (even particle size distribution)	1310	50–60
Peat ("muck")	0.65–1.1	60–80

Sources: Data from Bear 1972; Day 1965; and Gill 1979
*Values for bulk density reflect average values, and are not specific for a given media type
[†]Includes dolomite

Other factors that affect the porosity of soils are compaction, consolidation, and cementation. Since compacting (overburden) forces vary with depth, porosity will also vary with depth (this is especially true of clays and shales). Typical porosity values for common porous media are given in table 2.5. Consolidated material such as sedimentary rock is initially formed from loosely packed grains of sand, which become cemented by chemical precipitates at the contact point. As cementation progresses, pore space is filled and the porosity is significantly reduced.

Porosity, ϕ, and the void ratio, e, are related in the following manner: Given equation 2.15 the void ratio, e, may be expressed as

$$e = \left(\frac{V_t - V_s}{V_s}\right) = \left(\frac{V_t}{V_s} - 1\right) \tag{2.17}$$

by assuming that both air and liquid fill voids in a medium; then, $V_a + V_w = V_f$, where the subscript f denotes all void space (for example, filled with liquid or air). Thus, in the case of unit volume of soil solid, the void ratio can be expressed as

$$e = \frac{V_f}{1} = V_f \tag{2.18}$$

and, because $V_t = 1 + V_f = 1 + e$ (where 1 represents unit volume of solid), then

$$\phi = \frac{e}{(1 + e)} \tag{2.19}$$

and

$$V_f + V_s = 1; \quad V_s = 1 - V_f \tag{2.20}$$

As a result, considering the soil volume as unity, $\phi = \phi_v/1 = \phi_v$; and, due to equation 2.15, $V_f = \phi = eV_s$; thus, (remembering that $\phi = (V_t - V_s)/V_t$)

$$e = \frac{\phi}{V_s} = \frac{\phi}{(1 - V_f)} = \frac{\phi}{1 - \phi} \tag{2.21}$$

Typically, porosity is the measure preferred by hydrologists and soil scientists, whereas void ratio is usually used in geochemical engineering. Porosity also has a relation with bulk density (discussed in the following subsection), $\phi = [1 - (\rho_b/\rho_s)]$.

Several methods have been developed for the determination of porosity (Bear 1972; Klute 1986). This involves finding V_t, and either V_s or V_f. Here, V is volume and the subscripts t, s, and f refer to total, solid, and fluid respectively. Most of these methods are termed direct methods. The total volume of a sample can be determined by simple measurement if the sample was obtained with a cylindrical or cubical sampling device. However, any device or sample of irregular shape may also be used for determination of V_t (total volume). The most common method for determining V_t is to measure the volume of a fluid displaced by an immersed sample, taking care that the fluid does not penetrate the pores within the sample. The most common direct methods for measuring porosity are: **(1)** *Pycnometer*—V_t is determined first, then the sample is dried and crushed to remove all pores and placed in a pycnometer, where V_s is determined by fluid displacement. A pycnometer is filled with the sample and mercury, and then with mercury alone, to obtain V_t by weight difference (Klute 1986). **(2)** *Mercury Injection*—The sample is placed in a chamber filled with mercury at atmospheric pressure. The volume of mercury displaced gives V_t. Next, the pressure within the chamber is gradually increased, to force the mercury into the pores. This will yield not only V_f (volume of fluid forced into pores), but also a capillary pressure curve. If very high pressures are required, however, this method will not be suitable (Klute 1986). **(3)** *Compression Chamber*—Two chambers are connected by a valve. The sample is placed in the first chamber, and the gas is evacuated from the second chamber. A valve between chambers is then opened to allow isothermal expansion of the gas remaining in the first chamber. Considering an initial pressure of p_1 and a final pressure of p_2, we have

$$(V_1 - V_s)p_1 = (V_1 - V_s + V_2)p_2 \tag{2.22}$$

or

$$V_s = V_1 - \left(\frac{p_2}{p_1 - p_2}\right)V_2 \tag{2.23}$$

where V_1 = volume of the first chamber; V_2 = volume of the second chamber; V_s = volume of solid. As a result, the porosity $\phi = 1 - V_s$. **(4)** *Gas Expansion*—This method is based on Boyle–Mariotte's gas law, and is carried out at constant temperature. A sample is placed in one of two chambers. The first chamber is at V_1, with initial pressure of P_1. Gas is allowed to expand isothermally into the second chamber (by opening the interconnecting valve), which had previously been evacuated and now has a pressure of p_0. The first chamber equilibrates to a pressure of p_2 and, from Boyle-Mariotte's law, we have

$$\frac{p_1(V_1 - V_s)}{Z(p_1)} + \frac{p_0 V_2}{Z(p_0)} = \frac{p_2(V_1 + V_2 - V_s)}{Z(p_2)} \tag{2.24}$$

where $Z(p)$ is the compressibility factor. **(5)** *Washburn–Bunting Porosimeter*—V_f is determined by measuring the volume of air extracted at atmospheric pressure from a sample, by creating a partial vacuum in the porosimeter. A partial volume is obtained by manipulating mercury in the porosimeter reservoir.

An example of an indirect method is that developed by Norel (1967), based on the absorption of radioactive particles by a fluid allowed to saturate a sample. Because of the radiation involved, this has not gained much popularity.

QUESTION 2.12

Using equation 2.21 and the void ratio obtained from question 2.11, what is ϕ for the core described in question 2.11?

Bulk Density

The bulk density of a medium is usually expressed in terms of dry soil, and is the ratio of media mass to total volume, which includes all three phases of the soil system: $\rho_b = V_s\rho_s + V_w\rho_w + V_s\rho_s$; V_α is the volume fraction of phase α, where the phase is either liquid, solid, or gas (Davidson, Biggar, and Nielsen 1963; Gardner and Calissendorff 1967). Average bulk density values for selected media are listed in table 2.5. Since the volume fraction is defined as a ratio of the volume of a phase to the total volume of media, the sum of the ratios of all phases represents that total volume and, thus, equals 1. Dry bulk density, expressed in kg m^{-3}, is calculated as follows:

$$\rho_b = \frac{V_s\rho_s}{V_t} = \frac{M_s}{V_t} \tag{2.25}$$

where M_s = dry mass of porous media in kg (usually considered "dry" after oven drying at 105 °C for twenty-four hours). Bulk density can range from 1220 kg m^{-3} in clay soil to 1850 kg m^{-3} in sands. Bulk density, like porosity, is affected by the uniformity of packing, amount of compaction, size distribution, and clay type. Bulk density is always lower than particle density, because bulk density considers both air and liquid fractions that have large volumes compared to the density of a single particle; it is also dependent on clay and water content in shrink–swell porous media. Particle density is independent of water content. A wet soil cannot be completely compacted and remain porous. The bulk density will still be moderately lower than particle density, even under extreme compaction.

QUESTION 2.13

Calculate ρ_b of 1 m^3 of soil assuming volume fractions of solid, liquid, and gas of 0.5, 0.34, and 0.16, respectively. Assume ρ_l = 1.0 g cm^{-3} and ρ_g = 1.3 kg m^{-3}.

QUESTION 2.14

Assume the soil in question 2.13 is dry. **(a)** What is the contribution of each of the phases to bulk density? **(b)** What is the bulk density?

QUESTION 2.15

Consider the relation of the bulk density of a soil to all three phases, as previously discussed. **(a)** Derive an equation for a dry soil analogous to this. **(b)** With this answer, express the difference between a dry bulk density (ρ_b) and a wet bulk density (ρ_w).

Water Content

Water content, or wetness of a soil, is the volume or mass of water V_w occupying space within the pores (Bruce and Whisler 1973; Davidson, et al. 1969; Gardner 1986). Water content is typically expressed in one of three ways: mass basis, volume or depth basis, and percent or degree saturation.

Mass wetness (sometimes referred to as gravimetric water content) is determined by extracting a soil sample, oven-drying it (generally 105 °C for 24 hours), and determining the amount of water lost through the drying process. Oven drying is necessary to remove hygroscopic water adhering to particles that cannot be removed by air drying. This is sometimes a significant amount, depending on clay content and specific surface area; however, not all hygroscopic water can be removed. Mass wetness is described by

$$w = \frac{V_w \rho_w}{V_s \rho_s} \tag{2.26}$$

where V_s and V_w are the volume fractions of the solid and liquid phases, ρ_s is the density of the solid phase, and ρ_w and ρ_b are the wet and dry bulk densities.

Volumetric water content is generally more useful for field and laboratory studies, because it is the form in which soil water content (as a fractional basis, e.g., 0.34, 0.48, etc.) is reported—in the results from gamma attenuation, neutron probe, and time domain reflectometry (TDR) water content measuring devices. Like mass wetness, it is generally reported as a percentage, but is compared to total volume rather than the volume of solids present. Because both the V_w and V_t units of measure are cm^3, they cancel in the equation, leaving θ unitless. Volumetric water content is given by

$$\theta = \frac{V_w}{V_t} \tag{2.27}$$

Volumetric water content may also be obtained by its relation to mass wetness and bulk density, as follows.

$$\theta = w\left(\frac{\rho_b}{\rho_w}\right) \tag{2.28}$$

where ρ_w = density of water at measured temperature. A diagrammatic representation of the volume–mass relation within soil is shown in figure 2.9.

Equivalent depth of water is a measure of the ratio of depth of water per unit depth of porous media, described by volumetric water content:

$$d_w = \theta d_t \tag{2.29}$$

where d_w = equivalent depth of soil water if it were extracted and ponded over the surface (cm), and d_t = total depth of soil under consideration (cm). For example, water ponded on one hectare to a depth of 2.54 cm has a volume of 254 m^3. Depth of water is a very important concept in agricultural and irrigation practices.

Degree of saturation, s, expresses the volume of water in relation to the volume of pores. In arid-zone soils that remain perpetually dry, saturation can be as low as 0%; but it may be as high as 100% in lowland or wetland soils that remain constantly moist. However, most soils above the water table will never reach 100% saturation because of air entrapped in "dead end" pores. Degree of saturation is expressed by

$$s = \frac{V_w}{V_a + V_w} \tag{2.30}$$

(For groundwater reservoirs, entrapped air will normally dissolve within several months, resulting in water that is supersaturated with N$_2$ and air.)

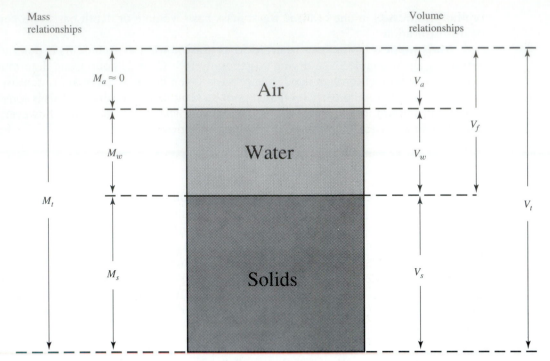

Figure 2.9   Diagrammatic representation of volume–mass relation within porous media

---

QUESTION 2.16

A soil physicist decides to apply water to a research area to investigate the movement of a conservative tracer. If $\rho_b = 1.37$ g cm^{-3} and gravimetric water content (w) is 0.19, what is the volumetric or volume fraction of water (θ) of the research area?

QUESTION 2.17

The scientist in question 2.16 wants to move the chemical tracer to a depth of 40 cm. **(a)** What volume of water must be applied per m^2 of soil surface? **(b)** If the scientist uses 23% less water than needed, to what depth would the tracer move? Assume that the tracer will only move to the maximum depth of the wetting front, although in field situations there would be great variability.

QUESTION 2.18

A soil sample is collected with an initial weight of 136 g. After oven-drying at 105 °C for 24 h, it weighs 118 g. Assuming $\rho_b = 1.32$ g cm^{-3} calculate: **(a)** w. **(b)** θ.

Air-Filled Porosity

We have discussed void ratio and porosity, which consider both air volume and water volume that occupy soil pores. The term air-filled porosity is used to specify the volume ratio of the air-filled pores. Air-filled porosity is expressed as

$$V_f = \frac{V_t - V_s - V_w}{V_t} \tag{2.31}$$

where V_f = volume fraction of air occupying soil pores. Equation 2.31 is a measure of the relative air content of soils. As such, it is an index of soil aeration, which is important in an agricultural setting. For example, an unsaturated zone hydrologists seeking to determine phase transport of herbicides would consider that air-filled porosities < 10% are not conducive to gas transport (since the soil is near saturation at that point). Porosity is also related negatively to the degree of saturation s by $V_f = \phi - s$ (remember that, in this case, $\phi = (V_a + V_w)/V_t$).

QUESTION 2.19

A 1 m^3 volume of porous media has $\theta = 0.39$ (see equation 2.27). **(a)** What is the air-filled porosity V_f? **(b)** If ρ_b (the oven-dry bulk density) = 1.32 g cm^{-3}, what is V_a? **(c)** What causes the difference between **(a)** and **(b)**? **(d)** What is the difference between the void ratio and the air-filled porosity of a porous media?

2.8   SPECIFIC SURFACE AREA

The specific surface area (s) of a soil is generally defined as the total interstitial surface area of pores (A_s) per unit bulk volume (V_t), per unit mass (M_s), or per unit volume of platelets (V_s) (Bower and Goertzen 1959; Carter, Mortland, and Kemper 1986). In equation form, they appear as follows.

$$s_b = \frac{A_s}{V_t}; \qquad s_m = \frac{A_s}{M_s}; \qquad s_v = \frac{A_s}{V_s} \tag{2.32}$$

where s_b, s_m, and s_v = specific surface area per unit bulk volume of soil, per unit mass, and per unit volume of particles, respectively; A_s = total surface area per specified unit; V_t = bulk volume of soil; M_s = mass of solid particles; and V_s = volume of solid portion.

Specific surface area is normally expressed in square meters per gram of soil (m^2 g^{-1}). The physical behavior of soil—water retention, hysteresis, or cation exchange, for example—is highly dependent upon the surface area. Particle size as well as shape are the parameters that have the greatest influence on the surface area of a porous media. Clays and organic colloids have very small particles; because of this, they have a high specific surface area compared to sands. This is especially true for clays with platy, disk, or rodlike shapes that have a far greater surface area per unit volume than spherical sand particles of the same volume. Not only do clay particles have a greater outer surface area but, as discussed previously, clays that shrink and swell (such as montmorillonite) expose a large internal surface area. The surface area of a typical sand is about 1 m^2 g^{-1}, whereas that of montmorillonite is about 800 m^2 g^{-1}. Mathematically, we can compare the surfaces of a sphere and of a thin rectangular particle to their volumes. For a sphere, we obtain

$$s_v = \frac{A_s}{V_s} = \frac{\pi d^2}{\frac{\pi}{6} d^3} = \frac{6}{d} \tag{2.33}$$

Relating this to a mass basis to obtain proper units, we get

$$s_m = \frac{s_v}{\rho_s} \tag{2.34}$$

Assuming $\rho_s = 2.65$ g cm^{-3}, and the diameter of a spherical particle is 0.05 mm, the surface area is 0.045 m^2 g^{-1}. For a thin rectangular particle such as a clay, the relation is

$$s_v = \frac{A_s}{V_s} = \frac{2L^2 + 4L^1 l}{L^2 l} = \frac{2(L + 2l)}{L^1 l} \tag{2.35}$$

Assuming that l is insignificant in the numerator (in this case, l is the thickness of the clay particle and $l \ll L$), due to properties of addition, we have $s_v = 2/l$ with units cm^2 cm^{-3}.

$$s_m = \frac{s_v}{\rho_s l} = \frac{2}{(2.65)l} \tag{2.36}$$

Further, assuming that $L = 1$ cm and the thickness (l) of the rectangular particle is 10 Å (10×10^{-10} m), the surface area will equal 750 m^2 g^{-1}. The specific surface area for selected soils are given in table 2.6.

The specific surface area of a soil is probably one of its most important properties. For unsaturated zone hydrologists investigating contaminant transport, the surface area plays a direct role in the movement of chemicals within the profile: how chemicals are retained and exchanged, released and retarded, by the soil surface. The greater the surface area of the soil, the more reactive it will be—especially in the presence of nonconservative chemicals and tracers. In such studies, surface area plays a very important role, and is a more useful index than texture for describing soil.

Specific surface area can be measured by various methods, the most common of which is gas adsorption as developed by Brunauer, Emmett, and Teller (1938). Nitrogen is adsorbed onto the soil surface in a monomolecular layer at low temperature (near the boiling point of liquid nitrogen). The adsorption of the gas takes place when a concentration gradient of the gas exists at the surface of the solid. The gas molecules are attracted by their degree of polarity to the force fields of the atoms at the solid surface.

TABLE 2.6 Specific Surface Area of Common Soil Types

Porous media	Surface area (m^2 kg^{-1})
Montmorillonite	700,000–800,000
Clay soil	150,000–250,000
Silty clay loam	120,000–200,000
Silt loam	50,000–150,000
Illite	80,000–120,000
Loam	50,000–100,000
Kaolinite	80,000
Sandy loam	10,000–40,000
Silt soil	5,000–20,000

Sources: Data from Gill 1979; Hillel 1980; and Carter, Mortland, and Kemper 1986

As Brunauer, Emmett, and Teller described the method, at sufficiently low gas pressure, the amount of nitrogen adsorbed per unit area is related to pressure, temperature, and heat of adsorption. This relation is given as follows.

$$\sigma = K_1 p \exp\left(\frac{Q_h}{RT}\right)$$

(2.37)

where σ = adsorption per unit area (mol m^{-2}); K_1 = constant (mol/N); P = gas pressure (Pa); Q_h = heat of adsorption (cal mol^{-1}); R is the gas constant (1.987 cal mol^{-1} K^{-1}); and T = temperature (K).

Equation 2.37 indicates that as pressure increases, the adsorption increases, and as temperature increases, adsorption decreases. The Langmuir equation describes this phenomena more precisely, because the soil surface has a finite area. It is given as follows.

$$\frac{P}{V} = \frac{1}{K_2 V_m} + \frac{P}{V_m}$$

(2.38)

where P = pressure; V = volume of adsorbed gas per gram of soil; K_2 = constant (1/Pa); and V_m = volume of adsorbed gas forming a monomolecular layer on adsorbent. Plotting P/V versus P will yield a straight line with a slope of $1/V_m$. Assuming a monomolecular layer, the specific surface area of the soil can be obtained by multiplying the cross-sectional area by the number of molecules in V_m.

As described by Brunauer, Emmett, and Teller (1938), and now known as the BET equation, the heat of nitrogen adsorption decreases with each successive layer of molecules adsorbed; however, it is assumed that after the first layer, subsequent layers are equal to the heat of condensation of the gas. The BET equation calculates the number of molecules in a monomolecular layer of adsorbent as follows.

$$\frac{P}{V(P_0 - P)} = \frac{1}{V_m C} + \frac{(C - 1)P}{V_m C P_0}$$

(2.39)

where P_0 = the pressure of gas required for a monolayer saturation at experimental temperature, $C = \exp[(Q_{h1} - Q_{h2})/RT]$; Q_{h1} = heat of adsorption of the first layer of adsorbed gas, and Q_{h2} = heat of liquification of gas. For this equation, V_m can be obtained by plotting $P/V(P_0 - P)$ versus P/P_0. Because of the heterogeneity of the active surfaces of most adsorbents, the Langmuir equation is not strictly observed; the BET equation is most useful for pressures (P/P_0) of 0.05 to 0.45.

Other methods for determination of specific surface area, though not commonly used, include a statistical method developed by Chalky, Cornfield, and Park (1949); a heat of wetting method; a fluid flow method developed by Kozeny (1927); and an adsorption method using cetyl pyridinium bromide as the adsorbate. Several of these methods were compared by Brooks and Purcell (1952). A method that has gained attention is one in which ethylene glycol is added to a soil and allowed to evaporate until the rate of loss decreases and equilibrium is presumed to be reached, at which time it is assumed that only a monomolecular layer of gas remains on the surface. A more extensive review of this method, in which the adsorbate has been replaced by ethylene glycol monoethyl ether, may be found in Carter, Mortland, and Kemper (1986).

QUESTION 2.20

Calculate the specific surface area (s) of 35 g of soil that has an average particle diameter (d) of 20 nm.

SUMMARY

In this chapter we have discussed the mineralogical composition of a porous medium consisting of solid, liquid, and gaseous phases, and the various processes that affect that composition—such as solubility, hydration, oxidation, reduction, leaching, and so on. Also discussed were how soil profiles and horizons develop, and the colloidal and crystalline nature and properties of soils. Various types of silicate and amorphous clays were described in some detail. We illustrated the various fractions of soil (specifically, sand, silt, and clay) and how they are separated. The analysis of soil fractions is critical to many environmental reports and research investigations, because it is these fractions which primarily determine how a chemical will adsorb, degrade, transport, and contaminate a given site (contaminant transport is covered in detail in chapter 10). We presented precise density, volume, mass, and weight relations between soil phases that describe characteristics of various media. In particular, the specific surface area is very important in understanding physical behavior, because water retention, hysteresis, cation exchange, adsorption, and so on, are highly dependent on surface area. Chapter 3 will focus on the microscopic properties associated with soil particles—such as the electrical double layer, ion exchange, hydration, flocculation, and cracking.

ANSWERS TO QUESTIONS

2.1. When all media pores are filled with liquid, only two phases would exist (liquid and solid). Also, when a soil is very dry, all measurable liquid has been removed, leaving only the solid and gaseous phases.

2.2. The primary force is the electrostatic force, in which the cations are electrically attracted to soil particles. This is discussed in further detail in chapter 3.

2.3. Clays are not spherical; in general, they have a platelike shape. As plate thickness (δ) increases, s decreases. Thus, a stacking of clay layers occurs within crystals, which forms thicker particles. This is the most logical reason for differences in s among clay minerals.

2.4. The major processes involved in physical weathering are freezing and thawing, heating and cooling, wetting and drying, grinding actions, and microbial activities leading to decomposition.

2.5. The two major processes are *hydrolysis* and *carbonation.* In hydrolysis, minerals react with water to form hydroxides, such as orthoclase feldspar with water:

$$KAlSi_3O_8 + H_2O \rightarrow HAlSi_3O_8 + KOH$$

Hydrolysis is one of the most important processes in soil profile development and change. Carbonation is also important, because carbonic acid (H_2CO_3) dissolves minerals more readily than water alone, and also forms more soluble bicarbonates.

2.6. The three major chemical processes affecting structural change are: (1) *hydration,* in which water changes mineral structure and makes the mineral softer, more stressed, and more easily decomposed; (2) *oxidation,* where a chemical combining with oxygen changes in oxidation number of the element (usually increasing it), also increasing element volume and making it softer; and (3) *reduction,* which usually takes place in zones of decreased aeration that results in electrically unstable compounds, more soluble compounds, or compounds which are stressed more internally, all of which cause more rapid decomposition.

2.7. By examining the texture triangle (figure 2.6), one may determine that a mixture of 20% clay, 40% sand, and 40% silt is classified as a loam.

2.8. Using equation 2.5, we obtain $t = [18(10 \text{ cm})(1.002 \times 10^{-2} \text{ g cm}^{-1} \text{ s}^{-1})/(2 \times 10^{-4} \text{ cm})^2$ $(2.65 \text{ g cm}^{-3} - 1.00 \text{ g cm}^{-3})(981 \text{ cm s}^{-2})] = 2.79 \times 10^4 \text{ s} = 7.7 \text{ h}.$

2.9. First, solve for v using equation 2.3. Assume $\rho_s = 2.65 \text{ g cm}^{-3}$ and $\rho_f = 1.39 \text{ g cm}^{-3}$; then, $v = 2/9[(2 \times 10^{-4} \text{ cm}/2)^2(2.65 \text{ g cm}^{-3} - 1.39 \text{ g cm}^{-3})(981 \text{ cm s}^{-2})/1.002 \times 10^{-2} \text{ g cm}^{-1} \text{ s}^{-1})] =$

2.74 cm s^{-1}. Using this answer and equations 2.1 and 2.2, $F_d = 6\pi(1.002 \times 10^{-2}$ g cm^{-1} s$^{-1}) \times (2 \times 10^{-4}/2)(2.74$ cm s$^{-1}) = 5.18 \times 10^{-5}$ g cm s^{-2}; and $F_g = 4/3\pi(2 \times 10^{-4}$ cm/2)^3(2.65 g cm^{-3} − 1.39 g cm^{-3})(981 cm s^{-2}) = 5.18 \times 10^{-9}$ g cm s^{-2}.

2.10. **(a)** Without driving off the air, the mass of soil and water in the pycnometer will be too small, resulting in a larger volume of solid than is actually present. This will yield a $^b\rho_s$ which is smaller than it should be. **(b)** Mass of solids = 76 g, mass of water + solids = 144 g, mass of water = (144 g − 76 g) = 68 g (this would also be the volume of water, assuming $\rho_f = 1.0$ g cm^{-3}). The volume of solids = (100 − 68) cm^3 = 32 cm^3, thus, $^b\rho_s = M_s/V_t = $ 76 g/32 cm^3 = 2.38 g cm^{-3}.

2.11. **(a)** Because we know the water loss, $V_w = 9766/V_t = 0.34$. **(b)** The total mass of soil is 28.724 × 10^3 cm^3 × 1.30 g cm^{-3} = 37.34 kg. To determine V_s, we must determine the total volume of soil and then divide it by the total volume of the intact core (28,724 cm^3). Thus, (37.34 kg/2650 kg m^{-3})(1 × 10^6 cm^3 m^{-3}) = 14,091 cm^3 soil. Divide this by V_t to obtain V_s = 0.49. Since all phases sum to unity, $V_g = (1 − V_s − V_w) = (1 − 0.49 − 0.34) = 0.17$. Hence, from equation 2.15, the void ratio = (0.17 + 0.34)/(0.49) = 1.04. This is a midrange value, since void ratios vary between 0.30 to 2.0.

2.12. Using the first part of the equation, $\phi = (1.04)(0.49) = 0.51$. This is slightly higher than one would expect, but has been measured in highly expanding soils.

2.13. Assuming standard values for ρ_a, the $\rho_b = (V_s\rho_s + V_w\rho_w + V_g\rho_g) = (0.50 \times 2650 + 0.34 \times 1000 + 0.16 \times 1.3)$kg m^{-3} = 1665.21 kg m^{-3}.

2.14. **(a)** The contribution would be 1325 kg, 340 kg, and 0.21 kg for the solid, liquid, and gas phases, respectively. There is very little contribution by the liquid and gas phases. **(b)** $\rho_b = 1665$ kg m^{-3}.

2.15. **(a)** Because the soil is dry, the liquid phase is removed, leaving $\rho_b = (V_s\rho_s + V_g\rho_g)$. **(b)** Considering a wet soil $\rho_w = (V_s\rho_s + V_w\rho_w + V_g\rho_g)$, by setting this equal to the value for $^b\rho_s$ in part (a), we obtain $\rho_w − \rho_s = (V_w\rho_w − \rho_g(\phi − V_g))$.

2.16. By assuming $\rho_w = 1.0$ g cm^{-3}, we can use equation 2.30 to obtain $\theta = 0.19(1.37$ g cm$^{-3}/1.00$ g cm$^{-3}) = 0.26$.

2.17. **(a)** The scientist wants water to penetrate to a 40 cm depth, so a volume of 40 cm^3 of water will be required for each cm^2 of soil. Using the relation $\phi = 1 − (\rho_b/\rho_s)$, we have $\phi = 1 − (1.37/2.65) = 0.48$. Because the liquid portion $\theta = 0.26$ (from question 2.16), $V_g = 0.22$. Thus, if we want only the gas-filled portion of soil to be filled with water, $0.22 \times 40 = 8.8$ cm^3 of water would be required to move the wetting front to the desired depth. However, because infiltrating water will tend to displace resident water in the soil pores through piston-type displacement, we would need to replace the entire porosity of the soil with new, infiltrating water. The required volume would be $0.48 \times 40 = 19.2$ cm^3 of water. **(b)** If 23% less water is applied 0.77 × 40 cm (water needed) = 30.8 cm, which would be the depth of maximum water movement.

2.18. **(a)** $w = M_w/M_s = (136 − 118)/(118) = 0.15$. **(b)** From equation 2.29, $\theta = 0.15(1.32$ g cm$^{-3}/1.0$ g cm$^{-3}) = 0.20$.

2.19. **(a)** θ is the liquid-filled porosity (V_w). Assuming an ideal soil (which is not necessarily the case under field conditions, but which will serve for illustration), $V_s = 0.50$, and $V_s + V_w + V_a = V_t$; thus, $V_a = (1 − 0.50 + 0.39) = 0.11$. **(b)** Using the relation $V_a = (1 − \rho_b/\rho_s)$, then $\phi_a = (1 − 1.32$ g cm$^{-3}/2.65$ g cm$^{-3}) = 0.50$. Accounting for the ϕ_l portion in the pores, $V_a = (0.50 − 0.39) = 0.11$. **(c)** In this instance there is no difference. Differences that might occur could be due to several factors: ρ_s as given in the general relation as 2.65 g cm^{-3} is usually true only of mineral soils, and actual particle densities in other soil types are somewhat lower; spatial variability (because any value obtained would be an average for the area selected); and accuracy and dependability of laboratory techniques. **(d)** The void ratio is defined as the ratio between the volume of voids and the volume of solids, whereas porosity is defined as the volume of void space per bulk volume of soil.

2.20. Assuming a spherical particle, $s = s_v/\rho_s = ((6/d)/2.65$ g cm$^{-3}) = (2.26$ cm$^2/2 \times 10^{-6}$ g) = 1.13 × 10^6 cm^2 g^{-1} = 113 m^2 g^{-1}. Thus, total surface area for the sample = (35 g)(113 m^2 g^{-1}) = 3.96 × 10^3 m^2. If we assume a platelet shape, $s = s_v/\rho_s l$ (where l is the thickness of the platelet) = [2/(10^{-10} m)(2.65 g cm^{-3})(10^6 cm^3/m^3)] = 754.7 m^2 g^{-1}. This gives a total surface area for the sample of 26.415 × 10^3 m^2.

ADDITIONAL QUESTIONS

2.21. Consider a 1 m^3 sample of a soil that contains 38% pore space.
   (a) What is the mass of water required to saturate it?
   (b) What is the mass of air when the soil is completely dry?
   (c) What is the mass of the solid phase?
   (d) What is the mass of the soil when saturated?
   (e) What would be the mass of soil if the volume fraction of air were 16%?

2.22. An intact soil core (30 cm diameter by 40 cm high) requires 11,500 cm^3 of water to saturate it. What is the porosity of the core?

2.23. To determine particle density (ρ_s), a scientist takes 88 g of granular (dry) soil at random and puts it into a flask > 100 cm^3 in volume. Next, the scientist adds deaerated water at 25 °C to drive out all air in the soil. The scientist then fills the bottle to exactly 100 cm^3. The mass of water and soil is 163 g.
   (a) What is ρ_s?
   (b) Why was it necessary to drive out all the air?

2.24. A 100 cm^3 soil sample has a wet mass of 183 g. After drying at 105 °C, its mass is 141 g.
   (a) Find ρ_w and ρ_b.
   (b) Find V_s, V_w, and V_g.
   (c) Is the soil well aerated?
   (d) Are the values obtained for ϕ true for every part of the soil sample?

2.25. Express w in terms of ρ_w and ρ_b.

2.26. Express θ in terms of ρ_w and ρ_b.

2.27. A soil scientist obtains a soil sample that has a $\rho_w = 1820$ kg/m^3 and $\rho_b = 1528$ kg/m^3. Calculate the following.
   (a) The volume fraction of water (θ).
   (b) The wetness (w).
   (c) Air filled porosity (V_g).

2.28. Soil samples are collected with an extraction device that has a volume = 90 cm^3, mass = 120 g, mass + moist soil = 243 g, mass + oven-dry weight = 207 g, mass + soil after heating to 900 °C = 167 g. Calculate ρ_w, ρ_b, w, θ, V_{min}, V_{org}, V_s, V_g, V_T, and ρ_s. What is the significance of ρ_s after heating to 900 °C?

2.29. Calculate the specific surface area of a soil that has 80% fine sand (average $d = 0.1$ mm); 10% silt (average $d = 10$ μm); 10% illite clay (average $a = b = 200$ nm and $\delta = 5$ nm). (The density of the solid phase for sand and silt is 2660 kg/m^3, and for the illite clay is 2750 kg/m^3.)

2.30. Consider the situation in question 9.
   (a) What is the contribution of each of the texture separates to the specific surface area of the soil?
   (b) What is the texture class of this soil?
   (c) To which component can most of the physical and chemical properties of the soil be attributed?

2.31. Assume that the average volumetric water content in the top meter of a soil is 0.17, a rainfall of 2.5 cm occurs, and none of the rain goes below 1 m in the soil.
   (a) How much water (L) falls on an area 1 ha in size?
   (b) How much water (L) is in the soil before the rain?
   (c) How much water (L) is in the soil after the rain?
   (c) What is θ after the rain?

2.32. Assume the same situation as in the previous problem, but with a 3 cm rain.
   (a) What is d_w (cm) in the top meter of soil before the rain?
   (b) What is d_w (cm) after the rain (in the top meter)?
   (c) What is θ in the soil after the rain?

2.33. If a mass of soil is immersed in water, will it become completely saturated?

3

Behavior of Clay–Water Systems

INTRODUCTION

Dispersion and flocculation play an important role in determining the physical and chemical behavior of soil colloidal fractions. This is particularly true in the areas of colloidal facilitated transport, crusting, infiltration, cracking, and swelling (Aly and Letey 1988; Babcock 1963; Goldberg and Forster 1990).

The process of dispersion and flocculation can be understood as occurring in four stages. Stage I is the dry state of the soil. We assume that the initial dry aggregate contains clay particles, which have their own forces of attraction. When this aggregate is "wetted up," the reactions due to hydration separate the individual particles; the distance of separation depends on the number of water molecules bound between cations and the clay surface to which they are attracted.

The process by which the particles are hydrated and pushed apart is known as stage II, or the swelling/slaking stage. Monovalent cations are normally bound ionically, whereas divalent ions such as calcium and magnesium are typically bound to the clay particle by polar covalent bonds. Increases in ionic bonding cause hydration and, thus, soil swelling. When polar covalent bonding is dominant, hydration is limited; generally, only crystalline swelling occurs. Such is the case with calcium clays, which can be dispersed when the soil water content is high and an external mechanical stress is applied.

When sufficient water has been added that the clay particles are separated by a distance of approximately 7 nm, the cations are no longer linked to the clay particle surface, because the cationic charge is shielded by water molecules. The clay is then completely dispersed, and stage III has been reached. Repulsive forces dominate; the repulsive forces are proportional to the charge of the individual clay particle. It is only at this stage that the DLVO theory (explained in section 3.3) is applicable.

Once a system is dispersed in stage III, it can become dehydrated and flocculated, reaching stage IV. Although both divalent and monovalent clays can exist in dispersive conditions, the reformation of polar covalent bonding can occur in divalent cationic clay systems. Such a process will normally be assisted by dehydration due to osmotic effects (salts present in solution). As a result, flocculation occurs due to the combined effects of the nature of bonding in the system (ion type, charge, and concentration) and of dehydration. The type of bonding that occurs depends primarily on the charge structure of the cations present.

In summary, hydration occurs between stage I and stage II, and swelling or slaking of the system has occurred at the attainment of stage II. For typical clay systems, mechanical separation occurs from stage II to stage III, and when stage III is reached, dispersion has

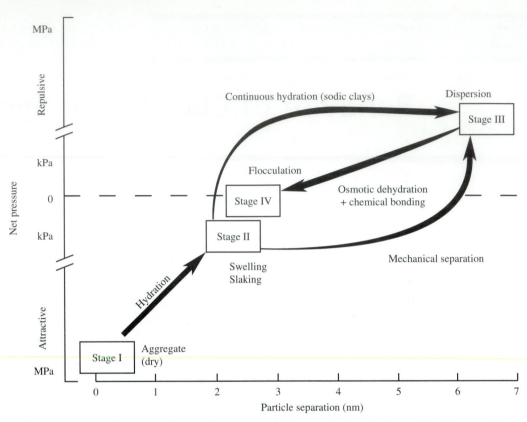

Figure 3.1 Stages of flocculation. (*Source:* Rengasamy and Sumner (forthcoming), reproduced by permission)

occurred. (For sodic clays, however, a continuous hydration occurs between stage II and stage III.) Once the system has been dispersed, flocculation (stage IV) can occur due to osmotic dehydration and chemical bonding. Progression from one stage to the next depends on the amount of water in the system (see figure 3.1).

The formation of aggregates is a result of physical, chemical, and biological processes occurring in soils (Lyklema 1978; Marshall 1964). Crusting and cracking of soils result primarily from flocculation and dispersion processes. A full understanding of aggregate formation, crusting, and cracking requires an understanding of the primary causes of separation, chemical bonding, hydration, swelling, slaking, flocculation, and dispersion of clay–water interactions (McBride 1989; Sposito 1984, 1989). These phenomena are discussed in detail in this chapter.

3.1 ELECTROCHEMICAL PROPERTIES OF CLAY–WATER SYSTEMS

Units

To avoid confusion, we will briefly digress, with a description of the units and unit systems that will be used in the following equations. Coulomb's law, as given in equation 3.1, is a fundamental law of nature and is not readily expressed in the SI unit system without adjusting the permittivity constant, ε_0. Permittivity is the ability of a dielectric to store electrical potential energy under the influence of an electric field, measured by the ratio of the capacitance of a condenser with the material as dielectric to its capacitance (i.e., with vacuum as dielectric). Many physical chemists prefer the cgs unit system to the SI unit system for the description of electrochemical phenomena. Since the aspects of clay–water interactions greatly depend on electrochemistry, the authors share this preference. The permittivity (of a vacuum), $\varepsilon_0 = 8.854 \times 10^{-12}$ C^2 N^{-1} m^{-2}, is of great use in converting dynes to Newtons, centimeters to meters, and electrostatic units (esu) to coulombs (C).

Many of the following equations are expressed in units of one or the other system (i.e., cgs/esu or SI); but not all are written in SI units. We hope by this to encourage equal familiarity with both unit systems. As an additional help, the term $e\psi/kT$, which represents the ratio of electrical to thermal energy, is the same in both unit systems, a dimensionless number; kT is the Boltzmann constant multiplied by the absolute temperature. For example, in the cgs/esu system, e has units of esu/ion. If ψ is in esu, the product of $e\psi$ yields energy in ergs/ion with a corresponding Boltzmann constant, $k = 1.3805 \times 10^{-16}$ ergs/ion-K. One erg/esu of potential is 300 V. A useful number to remember is that $kT/e = 25.69$ mV at 25 °C. For the SI system, e is in coulombs, 1.6021×10^{-19} C/ion. In the SI system, the product of $e\psi$ yields energy in joules (J), assuming that ψ is expressed in volts (J/C). The Boltzmann constant in SI units is 1.3805×10^{-23} J/ion-K and the value for kT/e is the same, 25.69 mV at 25 °C. For conversion purposes, two useful numbers to remember are Avogadro's number (6.023×10^{23} ions/mole, etc.) and electronic charge, e (4.803×10^{-10} esu/ion).

The dielectric constant, D, is the ratio of the permittivity of the medium, ε, to that of a vacuum, ε_0, and is dimensionless: $D = \varepsilon/\varepsilon_0$. However, ε has units of $esu^2/dyne \cdot cm^2$ or $esu^2/erg \cdot cm$ in the cgs system, and of $C^2/N \cdot m^2$ or $C^2/J \cdot m$ in the mks system. As a result, the cgs system has the advantage that permittivity is a simple number, but the disadvantage that the electric potential unit of erg/esu is not in common use.

Coulomb's Law

Coulomb's law expresses the attractive or repulsive force exerted on each other by two charged particles (say, q_1 and q_2) in a dielectric medium. If that medium is a vacuum, the law may be expressed as

$$F = \frac{kq_1q_2}{x^2} \tag{3.1}$$

where k is the dimensional proportionality constant. Physical chemists have defined the unit of electrostatic charge (esu), such that $k = 1$ dyne cm^2 esu^{-2}. The k term thus becomes numerically unimportant in equation 3.1 if cgs units are used, although it remains dimensionally necessary. In SI units, the unit of charge in common usage is C, the coulomb, which was developed from the theory of electromagnetics, rather than from electrostatics. For practical purposes, the coulomb has been determined experimentally to equal 3×10^9 esu, so that in SI units, $k = 9 \times 10^9$ Nm^2 C^{-2}.

From the theory of capacitors, the dielectric property of a vacuum has been defined as its permittivity, $\varepsilon_0 = 1/4\pi k$. Thus, in SI units, $\varepsilon_0 = 8.85 \times 10^{-12}$ C^2 Nm^{-2}. All dielectric materials have a larger permittivity than a vacuum. This property is generally characterized for a given medium by the dimensionless dielectric constant, $D = \varepsilon/\varepsilon_0$. Thus, Coulomb's law for the force exerted by charged particles in any dielectric medium can be expressed as

$$F = \frac{kq_1q_2}{Dx^2} \tag{3.2}$$

or

$$F = \frac{kq_1q_2}{4\pi\varepsilon x^2} \tag{3.3}$$

or

$$F = \frac{q_1q_2}{4\pi D\varepsilon_0 x^2} \tag{3.4}$$

Equation 3.2 is commonly written for cgs units as

$$F = \frac{kq_1q_2}{Dx^2} \tag{3.5}$$

where k, with a value of 1 dyne cm^2 esu^{-2}, is implied. To convert equations in cgs/esu to equations in SI, replace ψ by $(4\pi\varepsilon_0)^{1/2}\psi$ and q by $q/(4\pi\varepsilon_0)^{1/2}$.

In terms of the Poisson equation $\nabla^2\psi = -4\pi\rho k/D$, which relates the divergence of the gradient of the potential of a specific point to the charge density of that point, we can write

$$\nabla^2\psi = -\frac{\rho}{\varepsilon_0 D} \tag{3.6}$$

where ∇^2 is the Laplace operator $(\partial^2/\partial^2 x^2 + \partial^2/\partial^2 y^2 + \partial^2/\partial^2 z^2)$, D is the dielectric constant, ψ is the potential (volts), and ρ is esu/erg or charge per cubic meter.

Heat of Wetting

Heat is evolved when a dry soil is submerged in water or water is added to it, because water molecules lose kinetic energy in changing from a bulk form to a water film (hydration shell) on soil particle surfaces, and around cations or uncharged surfaces (Stumm 1992). This loss of energy is induced by the electric field surrounding the solid surface, which reduces the internal energy of the water molecules within this field. The water molecules are adsorbed until an equilibrium is established between the water–cation or water–clay particle. Heat of wetting increases as particle size decreases, and represents a measurement of the surface activity of the clay. Measurement of heat of wetting is used as a method to determine specific surface area (Aomine and Egashira 1970). A dry clay also exhibits a heat of wetting with solvents other than water. Likewise, heat applied to a hygroscopic body drives moisture from the body. Heat released at constant temperature is equivalent to the work done to separate the water molecules from the surface. The moisture content at which soils no longer exhibit heat of wetting is defined as hygroscopicity.

There are two types of heat of wetting: integral, and differential. Heat continues to be released as several layers of water molecules are sorbed to soil particles and their associated cations, but the outer layers are less tightly bound, and release progressively less heat. Hence, as water is added incrementally to a soil, the amount of heat produced per unit mass of added water (defined as the differential heat of wetting, i.e., the ratio of the increment of heat evolved, ∂q, to increment of water added, ∂w) decreases. The total heat produced as the medium is wetted from dryness to hygroscopy is termed the integral heat of wetting.

Several concepts can be of assistance in anticipating a range of heat of immersion values: **(1)** particles that readily wet up upon contact with water tend to have high heats of immersion and are characterized by negative values of ΔG (change in Gibbs free energy), whereas particles that do not (organic soil particles or mineral particles with organic coatings) tend to have low heats of wetting; **(2)** surfaces that possess the highest free energy have the most to gain in terms of decreasing the free energy of their respective surfaces by adsorption; **(3)** a surface energy of about $100 \, \text{mJ m}^{-2}$ is normally the cutoff value between high- and low-energy surfaces (sand, glass, metal oxides, metal sulfides, metals and inorganic salts are good examples of high-energy surfaces); and **(4)** generally, the greater the surface area of a soil, the greater the heat of wetting can be. Solid organic compounds tend to have low-energy surfaces. In general, the harder the compound (solids), the higher the surface energy.

The heat of wetting is due to the total energy released by the adsorption of water to the surface, and is proportional to the force with which water molecules are attracted to the surface. This heat of wetting of particles may be mathematically described by

$$-\Delta H_{im} = \gamma_{LV} \cos \theta - T \cos \theta \frac{d\gamma_{LV}}{dT} - T\gamma_{LV} \frac{d \cos \theta}{dT} \qquad (3.7)$$

(Adamson 1990) where γ_{LV} is the surface tension of the liquid (ergs cm^{-2} in cgs and J m^{-2} in SI), normally discussed in terms of equilibrium of the liquid with vapor; θ is the contact angle (discussed in section 4.2); and T is temperature (K). Adamson (1990) gives a value of $d\gamma/dT$ for water at 20 °C of $-0.16 \, \text{J m}^{-2}$. The contact angle, θ, is about zero for many earth materials, indicating that the surface energy of water in contact with such materials is about 120 ergs cm^{-2} or $0.12 \, \text{J m}^{-2}$. The surface area per gram of a relatively neutrally charged material, such as silica flour, could be multiplied by this factor to obtain a very rough estimate of its heat of immersion. Consequently, the predictive capability of equation 3.7 is approximately three to four times less than the experimental evidence of Van Olphen (1969) suggests (see table 3.1).

Interaction between Uncharged Soil Particles and Water

The potential energy of the interaction, or intermolecular potential, between an uncharged clay surface and a water molecule is difficult to calculate with current technology. However, a rough approximation can be obtained by

$$Q = \sum_{n=1}^{\infty} \left(-\frac{RT}{M} \right) \left(\frac{2.3x \, 0.88}{n^{1.80}} \times \frac{s}{(3.286 \, \text{Å})^2} \times \frac{18}{N_A} \right) \qquad (3.8)$$

where Q represents the heat of wetting per unit mass of clay in the same units as s, surface area per unit mass, R is the gas constant ($8.31 \, \text{J K}^{-1} \, \text{mol}^{-1}$), T is temperature (K), M is molecular weight of water, and n is the number of layers of water molecules sorbed to the clay surface (i.e., mass of water divided by water adsorbed onto clay to provide one molecule

TABLE 3.1   Average Heats of Wetting of Mg-Vermiculite

P/P_0^*	Adsorbed water (mg g^{-1})	Heat of wetting J g^{-1}	
0	0	232.51	
0.004	0.96	225.30	
0.018	62.58	110.91	one-layer hydrate ($64.5 \, \text{mg g}^{-1}$)
0.034	83.8	89.97	
0.356	188.6	4.635	two-layer hydrate ($179 \, \text{mg g}^{-1}$)
0.808	207.25	1.647	two-layer hydrate ($198 \, \text{mg g}^{-1}$)

Source: Van Olphen (1969).
*Refers to relative vapor pressure (P_0 generally refers to saturated vapor pressure).

thickness). Equation 3.8 assumes that the adsorbed area of one water molecule is 10.8 Å^2, and can be written more simply as

$$Q = \left(-kT(2.3 \times 0.88) \times \frac{s}{(3.286 \text{ Å})^2}\right)\left(\sum_{n=1}^{\infty} \frac{1}{n^{1.80}}\right) \tag{3.9}$$

where k is the Boltzmann constant. The summation represents a hyperharmonic series which, for an exponent $k > 1$, converges to

$$S = \frac{2^{k-1}}{2^{k-1} - 1} \tag{3.10}$$

where S is the summation value, and for $k = 1.80$ is 2.35. Consequently, the equation can be simplified to

$$Q = (4.42 \times 10^{-19})kTs \tag{3.11}$$

with the effective area for a molecule of water incorporated into the constant. At 25 °C, this equation results in Q = 0.18 J m^{-2}. Jurinak (1963) states that, in his experiment, w_m, the mass of water sorbed to provide one molecular thickness of coverage, is 2.56 mg/g clay. This represents 8.6×10^{19} molecules of water, which would cover 9.25 m^2 of surface, which is presumably the specific surface area of his clay. Thus, the integral heat of wetting at an ambient temperature of 25 °C would be 1.7 J/g clay. This value is substantially smaller than measured values of the heat of wetting for kaolinite listed by Grim (1953), which range from 5 to 10 J/g. Thus, as with equation 3.7, the value will normally be about three to six times less than experimentally measured values, and is a rough approximation. While the above equations can give an approximation for heat of wetting, the most efficient way to obtain accurate values is to measure the phenomena in the laboratory. For details on laboratory methodology, the reader is referred to Anderson (1986).

Heat is also produced by wetting of clays because of hydration of the adsorbed cations. The heat of hydration of ions in free solution can be determined by the Voet (1936) equation as

$$W = -\frac{(ze)^2}{2(r + 0.7)}N_A\left(\frac{D - 1}{\varepsilon}\right) \tag{3.12}$$

where W is the hydration energy (sometimes written as $\Delta G^{\circ}_{H_2O}$), z is the valence of the ion, e is 4.8×10^{-10} esu, $N_A = 6.023 \times 10^{23}$ (Avogadro's number), D is the dielectric constant of water, which varies with temperature (78.8 at 25 °C), r is the hydrated radius of the ion (Å), and ε = permittivity. This equation is valid when the only electric field present is that induced by the cation, and when sorbed water completely surrounds the ions in solution. Generally, heats of hydration of ions in free solution increase with ion valence, and are 86, 106, 399, 477, and 1141 kcal mol^{-1} for K^+, Na^+, Ca^{2+}, Mg^{2+}, and Al^{3+}, respectively, to name a few (Friedman and Krishman 1973).

Cations associated with clay platelets, particularly those in the Stern layer (see section 3.2), are partly bound to the clay particle, and are only partly hydrated. Hence, that portion of the heat of wetting due to ion hydration is less than that for ions in free solution. Janert, according to Grim (1953), determined efficiency ratios for various sorbed cations to be about 1/5 to 1/12, depending on ion species. Thus, the Voet equation will provide greatly erroneous (high) values for heat of immersion of clays. For example, Grim (1953) shows that kaolinite with a CEC of about 4 meq/100 g has a measured heat of immersion of 6 J/g. Assuming that the exchangeable ion is sodium, the Voet equation indicates that the heat of hydration in free solution is about 18 J/g. Using Janert's estimate that the heat should be divided by 4.9 yields a heat of about 4 J/g. The heat of wetting varies with the nature of adsorbed cations, being higher for divalent than for monovalent, as $Ca^{2+} > H^+ > Na^+ > K^+$.

Several researchers have measured heats of wetting for various clays (Aomine and Egashira 1970). The heat of adsorption of water over varied ranges of coverage with adsorbed water for a Mg-vermiculite clay is given in table 3.2. Generally, as the initial water content increases, both the integral heat of immersion and average heat of adsorption rapidly decrease. Figure 3.2 shows the relation between heat of immersion and water content for different clays; figure 3.3 shows the heat of immersion versus surface area for various calcium-saturated clays.

TABLE 3.2   Heats of Adsorption (Average) for Mg-Vermiculite per Mole of Water in Various Ranges of Water Coverage[†]

Coverage, n (mg g^{-1})	Heat of adsorption		
	J mol^{-1}	kcal mol^{-1}	J g^{-1} day^{-1}
0 and 0.96	13.53×10^4	32.31	7.5
0 and 62.58	3.50×10^4	8.36	12.2
0.96 and 62.58	3.34×10^4	7.99	
62.58 and 188.6	1.52×10^4	3.63	
62.58 and 83.8	1.78×10^4	4.25	
83.8 and 188.6	1.47×10^4	3.50	
188.6 and 207.25	0.29×10^4	0.69	

Source: Data from Van Olphen (1969).
[†]Coverage refers to the amount of water (mg) per gram of soil. As the initial water content increases, the heat of adsorption decreases rapidly.

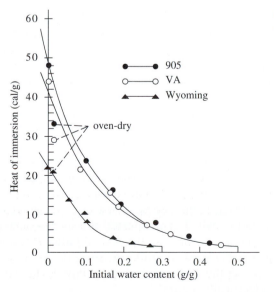

Figure 3.2   Relation between heat of immersion and water content; data from Aomine and Egashira (1970)

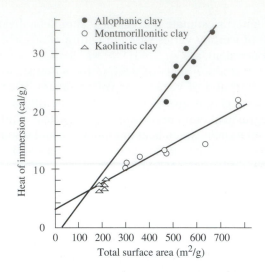

Figure 3.3   Relation between heat of immersion and total surface area for calcium saturated clays; data from Aomine and Egashira (1970)

3.2   ELECTROCHEMICAL PHENOMENA OF CLAYS

In chapter 2, we discussed how soil aggregates can be broken down into individual particles by various processes. As the aggregate is wetted, internal forces are produced. Some aggregates remain stable upon wetting, while others exhibit varying degrees of swelling, slaking, or complete dispersion. Once an aggregate has been broken down into individual particles, ionic charges associated with each particle begin to interact with neighboring particles. The charge on the particle or colloid surface tends to attract ions of opposite charge, or counterions (Grove, Fowler, and Sumner 1982; Verwey and Overbeek, 1948). Such ions tend to diffuse through the soil water in an attempt to equilibrate concentration. As a result, the colloid becomes surrounded by a diffuse cloud of ions, which has come to be known as the diffuse electrical double layer. Various theories that have been proposed to predict the behavior of colloids as affected by the double layer, including van der Waals forces, will be presented in this and subsequent sections.

QUESTION 3.1

An electrical field is induced by potassium. What is the heat of hydration of this ion? (Assume a dielectric constant of 80.)

Surface Charge of Clay Minerals

As clay minerals form, the octahedral and tetrahedral layers (as discussed in chapter 2) bind together one layer at a time, incorporating various ratios of Al, Mg, and Si. Because of the variable ratio of these cations, pure clay minerals rarely exist in nature. During mineral formation, Al^{3+} ions may occupy sites in the crystal lattice usually taken by Si^{4+} ions, and Mg^{2+} or Ca^{2+} may displace Al^{3+} ions. Overall, this process causes an excess negative charge of the clay particle within the crystal lattice, which is manifested as a specific surface charge density. These particular ions can replace each other on the crystal lattice because of their similar size; the process is known as isomorphous substitution. Since the ions are of like size, their substitution does not usually affect the shape of the crystal lattice. If no substitution occurs, the clay particle will remain electrically neutral.

Diffuse Electrical Double Layer Theory

Clay particles are electronegatively charged for three reasons: **(1)** There is an isomorphous substitution of Al (trivalent) for Si (tetravalent) that leaves an excess negative charge

$$O^{--}Si^{++++}O^{--} \rightarrow O^{--}Al^{+++}O^{--}$$

(2) There are broken bonds on crystal edges or ionization of hydroxyl groups attached to silicon of broken tetrahedral planes (i.e., silicic acid),

$$Si\text{-}OH + H_2O \rightarrow SiO^- + H_3O^+$$

and, **(3)** The presence of silicic or phosphoric acid can form an integral, clay particle surface giving rise to a negative charge as well.

The electronegative charge of clay particles implies that a clay–water system is capable of doing work on the anion; thus, potential energy will vary between points within the bulk solution (Sposito 1984). The potential at a point near a clay particle is defined as the work that must be done to induce a unit negative charge from the bulk solution up to that point. Because clay is negatively charged, the unit negative charge is used rather than the conventional positive charge of electrostatic potential.

This overall negative charge is compensated by exchangeable cations held to the clay surface by coulomb forces. The combination of a negative charge on the clay surface and a counteracting charge from the cations creates an electrical double layer around individual clay particles. The effect of this double layer is to make the clay–water system electrically neutral. An anion placed near a clay surface is pushed away by electrostatic repulsion forces between the clay particle and the anion.

Several scientists have derived mathematical expressions for potential as a function of distance from a clay particle, among them Helmholtz (1879), Gouy (1910), Chapman (1913), Debye and Huckel (1923), and Stern (1924). The most commonly discussed expressions or models include the Helmholtz parallel-plate capacitor, the Gouy–Chapman diffuse double layer, and Stern's double layer theory.

Helmholtz suggested that the electrical double layer had a fixed thickness of one molecule that was generally formed at the particle/solution interface. He believed that the inner, negatively charged layer was rigid, adhering firmly to the solid surface, whereas the outer layer of oppositely charged (+) ions in the solution was mobile. Helmholtz's theory represents the clay platelet as a simple parallel-plate capacitor. Because of thermal motion, counterions are distributed within a certain space forming a diffuse layer; consequently, this does not permit a rigid formation at the interface as Helmholtz suggested.

Guoy–Chapman proposed a diffuse double layer in which the negative charges are adsorbed primarily near the solid surfaces and the positively charged ions are distributed away from the surface. The cation concentration near the surface is dense and decreases exponentially with distance, until the net charge density is zero. This exponential distribution occurs because the high cation concentration near the particle is partially counteracted by a tendency for diffusion away from the colloid surface. Local cation concentration is calculated by the Boltzmann equation:

$$n^+ = n_0 \exp\frac{-ze\psi}{kT} \tag{3.13}$$

where n^+ is ion concentration at a specified distance from the charged surface, in ions cm^{-3} (figure 3.4); n_0 is ion concentration in bulk solution (ions cm^{-3}); z is ion valence (dimensionless); e is the unit of electronic charge (coulomb/ion in the SI system, or esu/ion in the cgs/esu system); ψ is the electrical potential of the colloid at the specified distance (ergs/esu or volts);

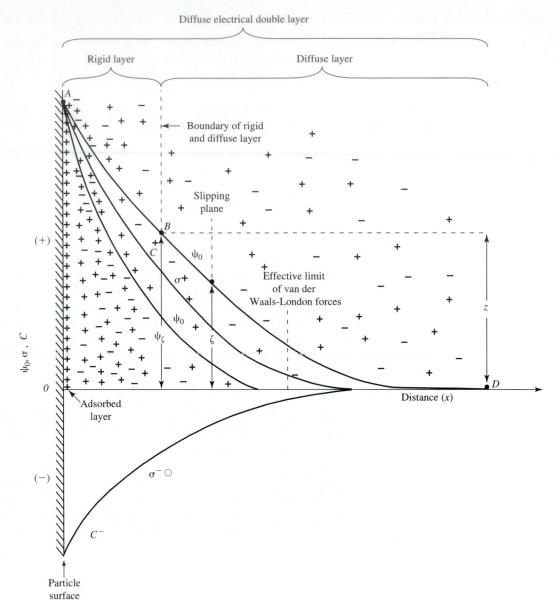

Figure 3.4 Representation of the diffuse electrical double layer

k is the Boltzmann constant (gas constant per molecule, J; or ergs per molecule, K or J K^{-1}); and T is absolute temperature, K.

Anions are repelled by the negative charge of the particle, resulting in a sparse distribution near the particle that increases exponentially with distance, also given by Boltzman's equation:

$$n^- = n_0^- \exp \frac{+ze\psi}{kT} \qquad (3.14)$$

These equations for cation and anion distribution may be combined to develop a relation between the character and concentration of an equilibrium solution and the electrical charge induced by the surface of a colloid. This relation is expressed as

$$\sigma = \left(\frac{2\varepsilon nkT}{\pi}\right)^{1/2} \sinh\left(\frac{ze\psi_0}{2kT}\right) \qquad (3.15)$$

where σ is the charge on the colloid surface (esu cm^{-2} or, if k is replaced by R, meq cm^{-2}); ε is the permittivity of water (esu^2 dyne^{-1} cm^{-2}); n is the electrolyte concentration in solution (ions cm^{-3}—this unit is achieved by converting with Avogadro's number); k is the Boltzmann constant; T is absolute temperature; z is the valence of counterions; e is the electric charge (esu/ion); and ψ_0 is the electric potential (ergs/esu) at the colloid surface.

The diffuse double layer into the surrounding equilibrium solution can be thought of as having an effective thickness (κ^{-1}). Theoretically, due to the exponential nature of the relation between ion concentration and distance, the double layer is infinite in thickness. However, if we assume that a cloud of ions, concentrated as a point, planar charge, is at a distance from the colloid surface similar to that of an electrical condenser (the Helmholtz theory), double layer thickness then can be expressed as

$$\kappa^{-1} = \left(\frac{\varepsilon kT}{8\pi z^2 e^2 n}\right)^{1/2} \qquad (3.16)$$

where all units are as defined for equation 3.15. This effective thickness is analogous to the atmospheric scale height; and for small values of surface potential, the Guoy distribution of potential with distance x from the surface is given by $\psi x = \psi_0 e^{-\kappa x}$. This relation is shown in figure 3.4. Van Olphen (1963) used the preceding equation to compute κ values for solutions of 10^{-1}, 10^{-3}, and 10^{-5}, M concentration with $1/\kappa$ ranging from 10^{-5} to 10^{-7} cm for single valent ions.

Stern showed that neither the sharp, rigid Helmholtz theory, nor the Guoy-Chapman diffuse double layer theory were adequate, and proposed a theory that combined features of the two. Stern proposed a negative charge rigidly attached to the surface of the clay particle, with an adjacent layer of positively charged cations (Stern believed this layer to be about the diameter of one cation in thickness with a sharp drop in potential, as in the Helmholtz theory). As distance increases towards the bulk solution, the remaining part of the double layer becomes diffuse in character (as in the Gouy-Chapman theory). In the rigid portion of the double layer, the ions are mostly immobile, whereas in the diffuse layer, thermal agitation permits the ions to move freely. Within the rigid layer, cations are preferentially adsorbed, which results in a gradual decrease in potential in the bulk solution that contains a uniform charge distribution (figure 3.4). The actual thickness of the diffuse layer is undefined since it depends on both the type and concentration of the ion present. However, common thickness is measured in angstroms, and can vary from < 10 Å to > 400 Å.

The interface between the fixed and diffuse layers is not sharp, and should be thought of as a boundary where some ions are migrating to the clay particle and others are shearing away. This becomes clear if it is assumed that a clay particle has its own ionic atmosphere, and is within an electrical field. Readers should note that, in the following discussion, we use the Gouy-Chapman theory, due to the difficulty of applying the Stern theory (which can lead to ludicrously high local counterion concentrations, because the ions are considered as point charges and are presumed present at the colloid wall where electric potential is highest).

QUESTION 3.2

What is the cation concentration in a solution 12 Å from a clay surface? Assume a monovalent ion, $C_0 = 2.5 \times 10^{-2}$ M, $T = 25\,°C$, $\psi = 100$ mV, and $e = 4.803 \times 10^{-10}$ esu ion^{-1}.

QUESTION 3.3

What is the electrical charge induced by the surface of a clay platelet in a solution of 3.1×10^{-2} M NaCl solution at 25 °C? Assume $\varepsilon = 78.8$ esu²/erg cm and $\psi_0 = 75$ mV.

3.3 ELECTROKINETIC PHENOMENA

Electrophoresis

The term electrophoresis refers to the movement of charged particles in relation to a stationary solution, that is, in an electric field. Envision a clay particle and associated charge placed into an electric field; since the clay particle is negatively charged, it will tend to migrate in the positive field direction. Ions near the clay particle surface migrate with the particle, whereas those further away in the solution will slip away from the particle surface and migrate in the opposite direction of the electric field.

The slipping plane indicated in figure 3.4 divides ions in the double layer into those migrating with the clay particle and those shearing away from it. The potential drop between the slipping plane and bulk solution is the zeta potential, which is the work per unit charge required to move an anion from the bulk solution to the slipping plane (Sennett and Olivier 1965). The zeta potential is represented by the voltage difference between point D and point B in figure 3.4. The zeta potential is affected by cation density in solution: if cation density is decreased, charge density is also decreased, which results in a larger thickness of the cation layer, resulting in a thicker double layer. Likewise, if the cation concentration present in solution increases, a thinner double layer will result. The thicker the double layer becomes (assuming charge density remains constant), the higher the zeta potential will be. This is because the further the cations extend from the particle surface into the diffuse layer, the greater will be the number of cations to the right of the slipping plane (figure 3.4). This hypothesis assumes that the number of cations increases faster in the bulk solution with distance from the particle than the electric field of the capacitor, and that the zeta potential is not too large. The zeta potential is given by

$$\zeta = \frac{4\pi\sigma d}{\varepsilon} \tag{3.17}$$

where ζ = zeta potential (volts), σ = surface charge density (esu cm^{-2}), d = distance (cm), and ε = permittivity of the solution (esu² erg^{-1} cm^{-1}) Equation 3.17 is greatly simplified, and indicates a linear increase in potential from the charged surface when it should reflect an exponential decrease. Mitchell (1993) expresses the zeta potential as $\zeta = \sigma\delta/D$, where δ is the distance between the wall and the center of the plane of mobile charge, and D is the relative permittivity or dielectric constant of the pore fluid. To rigorously derive this equation one must begin with Poisson's equation and the Boltzmann distribution. Additional theory and applications of zeta potential may be found in Sennett and Olivier (1965), and Babcock (1963).

The velocity of particle movement during electrophoresis is that at which coulombic force, $\zeta\varepsilon E_s$, is balanced by viscous force, $4\pi\eta v$, as described by Stoke's law (see section 2.5). Thus,

$$V = \frac{\zeta\varepsilon E_s}{4\pi\eta} \tag{3.18}$$

where V is electroosmotic velocity, E_s is potential gradient, ε is the permittivity of the solution, the numeral 4 is for cylindrical shapes (this would be replaced by 6 for spherical shapes), and η is the fluid viscosity, Pa s. The Dorn effect is the reverse of electrophoresis, and is the potential difference resulting when particles fall through a solution due to gravity.

QUESTION 3.4

Assuming a parallel-plate capacitor, calculate the zeta potential 23 Å from particle surface. Assume $\varepsilon = 78.8$, $e = 4.8 \times 10^{-10}$ esu ion^{-1}, and that the media is montmorillonite with a CEC of 1.5×10^{-7} meq cm^{-2} and a surface area of 650 m^2/g.

Electroosmosis

During the process of electroosmosis, water movement occurs when a moist soil is placed between two electrodes and subjected to an external electrical potential difference; cations in the double layer migrate (by electrostatic attraction) towards the cathode, or negative pole, while the anions migrate toward the anode, or positive pole (Adamson 1990; Ghildyal and Tripathi 1987; Stumm 1992). (See figure 3.5.) The cations translocate in the electric field and, while doing so, pull with them the oriented water molecules. This water layer is now positively charged and becomes mobile, dragging the remaining water along the immobile part of the liquid film. The rate of flow depends on the magnitude of the electrical potential difference and the viscosity of the liquid; as with electrophoresis, the velocity of water movement is determined as that at which viscous drag equals the coulombic force exerted by the electric field. Microscopically, the electric force in the capillary is applied to the ions within the fluid and not directly to the water.

During electroosmosis, the electrolyte in the capillary is acted on by electric and viscous forces; there are other forces, however, including mechanical, pressure, inertial, and surface tension, that produce effects that act on the electrolyte, and that are often inseparable from those of electroosmosis. The movement of water in the capillary is determined by the effects of all forces, with the primary forces being electrical, thermal, and viscous. Esrig and

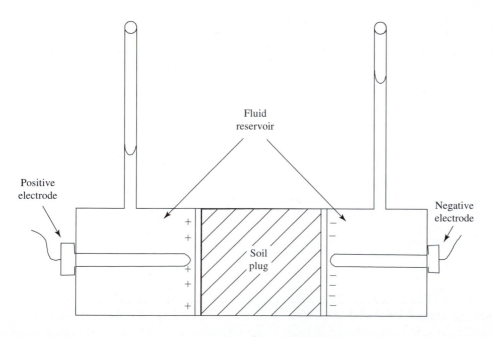

Positive electrode

Negative electrode

Fluid reservoir

Soil plug

Figure 3.5   Apparatus for measuring electroosmotic pressure; data from Adamson (1990)

Majtenyi (1965) describe the total electric field as the negative gradient of the sum of existing potentials. For electroosmosis, electrical potentials are of three types: potentials applied by electrodes, potentials created by ions other than the one being considered, and potentials created by dipoles. For example, the electrical force on a dipole (water molecule) depends on the dipole moment and gradient of the electric field.

Thermal phenomena can influence electroosmotic water flow by causing either geometrical or physical changes in the capillary system. Such changes generally correspond to variations in temperature. If a concentration gradient exists, thermal agitation of molecules and particles within the capillary system will produce a tendency towards equilibrium. Changes in physical conditions may not be accompanied by changes in geometry if the temperature of the capillary system remains constant. Geometric changes occur due to volume expansions or contractions of the capillary fluid, or of the solid portion of the medium when subjected to a temperature fluctuation.

Viscous forces depend on the velocity gradient, dv/dn, and the coefficient of viscosity, μ. The viscous force, V_f, for a volume element, dv, on a surface element, dS, is given as

$$dV_f = \mu \frac{\partial^2 V}{\partial n^2} \, dv \qquad (3.19)$$

where n is normal to the surface and V is velocity.

Electroosmotic hydraulic conductivity, k_e, indicates flow velocity under a unit electrical gradient. Three common theories are used to predict k_e: the Helmholtz and Smoluchowski theory (pore water flow occurs in a large soil pore); the Schmid theory (pore water flow occurs in a small soil pore); and the Spiegler friction model (flow is considered as a result of interactions of mobile water molecules and ions). These theories are discussed in detail in Mitchell (1993). In contrast to soil water hydraulic conductivity, which varies with the square of effective pore size, k_e is relatively independent of pore size. Mitchell (1993) shows a general range of 1×10^{-9} to 1×10^{-8} m²/s/volt (m/s per volt/m); and that regardless of soil type, k_e is on the same order of magnitude. However, hydraulic conductivity changes by several orders of magnitude for the same material (see table 3.3). A knowledge of k_e can sometimes provide an efficient means of dewatering soils, which has proven useful for temporary stabilization during excavation. This technique was first used in the 1930s by Casagrande, who compiled information on various applications of the process, which is little used due to its expensive nature (Bjerrum, Moum, and Eide 1967; Casagrande 1959).

The common principle for electroosmotic dewatering is to apply an electrical gradient to produce a flow of water through the soil. Increases in effective stress resulting from the flow of water produced will often increase the shear strength (Bjerrum, Moum, and Eide

TABLE 3.3   Coefficients of Electroosmotic Permeability

Material	Water content (%)	k_e in 10^{-5} (cm²/sec-V)	K (cm/sec)
London clay	52.3	5.8	10^{-8}
Kaolin	67.7	5.7	10^{-7}
Clayey silt	31.7	5.0	10^{-6}
Rock flour	27.2	4.5	10^{-7}
Na-Montmorillonite	170.0	2.0	10^{-9}
Na-Montmorillonite	2000.0	12.0	10^{-8}
Mica powder	49.7	6.9	10^{-5}
Fine sand	26.0	4.1	10^{-4}

Source: Data from Mitchell (1993).

1967). This process typically has been used for silt or silty soils of medium permeability, but Bjerrum, Moum, and Eide (1967) showed that it could be used for soft clays as well. In their experiment, they increased soil strength, due to consolidation, from an initial value of less than 1 t/m^2 to an average value of 4 t/m^2. In electroosmosis, dewatering occurs when water is drawn to a cathode, where it is drained away, and no water is allowed to enter at or near the anode. This results in a consolidation and, therefore, stabilization of the soil between the electrodes, which is equal to the volume of water removed.

In chapter 2, we discussed soil profiles, composition, and geometry. In a typical soil, some particles are in contact due to consolidation pressure. Such particles can be considered to be the "skeleton" of a soil. Other particles may be surrounded by water. If we apply an electric field to this system, a migration of charged particles might result; that is, electrophoresis. The occurrence of electrophoresis indicates that soil particles are migrating against the flow of water. This is especially true in clays. Such particles can be obstructed by the soil skeleton, resulting in clogged capillaries and a restriction of water flow. This is an undesirable condition when electroosmotic dewatering is attempted in soils. In this regard, particle size distribution is very important and should be examined carefully, particularly the colloidal fraction as it exhibits the greatest electrochemical activity. These conditions can geometrically alter the capillary system.

Changes in matric potential or soil pressure can alter the geometry of the capillary system (Sposito 1989). For example, when a dry clay comes in contact with water, an increase in pressure results, which may cause the displacement of soil particles and alteration of pore geometry. This will create a new condition for electroosmotic flow. All soil characteristics must be known before attempting to calculate electroosmotic flow. Various laboratory tests will provide information on soil porosity, mineralogy, surface area, water content, degree of saturation, pore size distribution, charge density, hydraulic conductivity, particle size analysis, and other chemical and physical parameters, including capillary orientation with respect to the direction of the applied electric field.

Esrig and Majtenyi (1965) have given the total outflow per unit time (assuming cylindrical capillary), Q, from a soil mass as

$$Q = C_r \frac{\pi}{2\mu} \left(\sum_n m_n \overline{\rho_n} \left(R_n^3 \cdot d_n - 0.75 \cdot R_n^2 \cdot d_n^2 \right) \right) \cdot A \cdot E \qquad (3.20)$$

where C_r is a coefficient to account for assumptions in double layer thickness and effective capillary radius, which will vary between 0 and 1 and will be less than unity for all soils; m_n is the number of capillaries of radius R_n per unit cross-sectional area A; d is double layer thickness; ρ_n is the average mobile charge density ($q\ L^{-3}$); and E is electric field strength ($M\ L\ T^{-2}\ q^{-1}$). By analogy to Darcy's law, $Q = k_e A E$, where k_e, the electroosmotic permeability of a soil, is expressed as

$$k_e = C_r \frac{\pi}{2\mu} \left(\sum_n m_n \overline{\rho_n} \left(R_n^3 \cdot d_n - 0.75 \cdot R_n^2 \cdot d_n^2 \right) \right) \qquad (3.21)$$

This equation can take several different forms, depending on soil physical characteristics. The velocity of electroosmotic flow is independent of capillary radius and cross-sectional area, if the capillary is large enough to ensure that the double layer along each wall does not interact. For large capillaries, the volume outflow per unit time can be related to porosity (Esrig and Majtenyi 1965). If capillary radius is small and double layers interact, electroosmotic flow velocity will depend on the curvature of the capillary wall and upon capillary radius; the volume of outflow per unit time for such capillaries cannot be related to soil porosity. For a more detailed discussion of electroosmosis, the reader is referred to Adamson (1990), Mitchell (1993), and Esrig and Majtenyi (1965).

Streaming Potential

Unlike electroosmosis, in which a liquid flows along a charged surface when an electric field is applied parallel to the surface (in the liquid), the streaming potential is a measure of the electric potential created when a liquid is forced to flow along a charged surface. As water in a soil moves near a fixed clay surface, some of the cations in the diffuse layer will be carried with it. However, these cations resist the movement due to their attraction to the negatively charged surface of the clay particles. This resistance creates a drag force (F_d) on the water, causing it to move less rapidly. As explained in the description of electrophoresis, F_d is affected by cation density in solution.

Consider a column of soil: as water percolates through the column, the cation concentration at the outflow end of the column will be greater than at the inflow end. Consequently, as water continues to move through the column, increased pressure will be needed at the inflow end to maintain the same rate of flow, because the water must move cations in the solution against the repulsive force of the increased number of cations (increased charge density) at the outflow end. Generally, this will create a measurable potential difference across ends of the column, known as the streaming potential. The streaming potential is the reverse of electroosmosis. Because streaming potential can be measured directly during a measurement of hydraulic conductivity (using a high impedance voltmeter and reversible electrodes), one can obtain an estimate of electroosmosis using Saxen's law (Mitchell 1993).

The streaming potential may be expressed as

$$E_s = \frac{\zeta P \varepsilon}{4 \eta \pi \lambda'} \tag{3.22}$$

where E_s is the induced streaming potential (volts), P is the net pressure necessary to stop the water flow (Pa), η is fluid viscosity (Pa s), and λ' is the specific or electrical conductivity of water (ohms or μohms cm^{-1} in the cgs system and dS m^{-1} in the SI system).

According to Saxen's law, the prediction of electroosmosis from streaming potential is given by

$$\left(\frac{q_h}{I}\right)_{\Delta P = 0} = -\left(\frac{\Delta E}{\Delta P}\right)_{I = 0} \tag{3.23}$$

where q_h is the hydraulic flow rate, I is the electric current, ΔP is the pressure drop, and ΔE is the electrical potential drop. Equation 3.23 was first shown experimentally by Saxen (1892), and has been verified for clay–water systems.

QUESTION 3.5

What is the double layer thickness at the surface of a clay platelet suspended in a solution of 10^{-2} M NaCl? Assume $T = 25\,°C$ and $\varepsilon = 78.8$. Use Avogadro's number for conversion units.

QUESTION 3.6

What is the linear electroosmotic velocity of flow for a liquid plug in a column 1 m long? Assume $\zeta = 60$ mV and an applied voltage drop of 100 V across the column.

3.4   THE DLVO THEORY OF COLLOID STABILITY

The DLVO theory was proposed independently by Derjaguin and Landau (1941) and Verwey and Overbeek (1948); hence, the theory goes by their combined initials. Derjaguin and Landau (1941) describe the repulsive force between particles, whereas Verwey and

Overbeek (1948) consider the repulsive energy. The end result is basically the same, however. Both theories assume that clay particles (colloids) are planar in shape. The DLVO theory describes the repulsion of clay particles in close proximity that have an interacting double layer and in which there is an associated interaction (both attraction and repulsion) between van der Waals–London forces associated with each particle. Because the diffuse double layer extends some distance from a particle, as particles are brought together, their respective double layers interact. Consequently, a repulsive force exists, and a potential energy has to be overcome to bring particles together. The DLVO theory predicts the energy associated with interaction as a function of interparticle distance. Given a repulsive force, as a function of the distance $2d$ between two clay platelets, the potential energy, V_R, is reversible and isothermal (Gibbs free energy), and is described by

$$V_r = 2 \int_{-\infty}^{d} P \, dz \qquad (3.24)$$

where V_r is the potential energy (ergs cm^{-2}), d is the particle diameter (cm), and P is the pressure exerted on the plane midway between the two particles (in ergs for the cgs system, and Pa for SI). This pressure can also be considered the external pressure (P_e) applied to the outside surface of each particle (figure 3.6).

Interactive Repulsive Forces between Platelets

From the Kelvin–LaPlace equation, the chemical potential of soil water at point X (figure 3.6) in the solution away from the particles or plates is determined by

$$\mu_X = \mu_0 + Z_X g \rho = -\frac{RT}{M} \sum x_i + Z_X g \rho = -\frac{RT}{M}\left(\frac{2n_0}{1M^{-1}} \frac{1}{N_A}\right) + Z_X g \rho$$
$$= -2kTn_0 + Z_X g \rho \qquad (3.25)$$

where μ_0 (n_0 in figure 3.6) is the decrement in chemical potential due to ions in solution; M is the molecular weight of water (g); x_i is the mole fraction of ith ions; $2n_0$ is the number of ions per cm^3 in the solution; k is the Boltzman constant; Z_x is the increment in distance from standard height to height of point X; g is the gravitational constant; and ρ is the density of

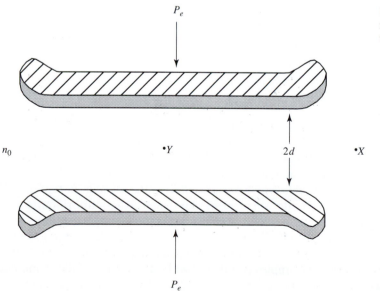

Figure 3.6 Representation of repulsive force acting between two charged clay platelets

solution. The term $2kTn_0$ will have units of dynes/cm^2 (cgs system) or Pa (SI system). The chemical potential at point Y is given as

$$\mu_Y = \mu_f + \mu_0 + \int_0^{P_i} \overline{v_w}\, dP + Z_Y g\rho \tag{3.26}$$

where P_i is the internal pressure between the particles and v_w is the partial volume of water. This equation can be simplified by assuming that $\mu_f \ll \mu_0$ (negligible), and $v_w = 1.0$ and is independent of pressure. Thus,

$$\mu_Y = \mu_0 + P_i + Z_Y g\rho \tag{3.27}$$

Because $P_i = P_e$, cations and anions at point Y are expressed as

$$n^+ = n_0 e^{-ze\psi_d/kT} \qquad n^- = n_0 e^{ze\psi_d/kT} \tag{3.28}$$

where n^+ represents cations, n^- represents anions, and z is ion valence. To simplify the following equations, we will let $y_d = ze\psi_d/kT$. Consequently, the potential at point Y can be expressed as

$$\mu_Y = kTn_0(e^{-y_d} + e^{y_d}) + P_e + Z_Y g\rho \tag{3.29}$$

assuming that the system is at equilibrium $\mu_X = \mu_Y$ and $Z_X = Z_Y$; so

$$-2kTn_0 = -kTn_0(e^{-y_d} + e^{y_d}) + P_e \tag{3.30}$$

To calculate the external pressure (P_e) on the two plates (assuming $P_e = P_i$), equation 3.30 can be rearranged to obtain

$$P_e = n_0 kT(e^{-y_d} + e^{y_d} - 2) = 2n_0 kT[\cosh(y_d) - 1] \tag{3.31}$$

which also assumes that the distance between the two plates is $2d$ (two times the particle thickness).

Note: The equations just developed also apply to clay swelling. Thus, the equations developed to describe the pressure field generated by coulombic forces are exactly the same whether we consider the particles as separate entities or as layers of a montmorillonite grain in which internal layers are subject to formation of the double layer.

QUESTION 3.7

What is the external pressure (P_e) on two platelets at a distance $2d$ apart? Assume $n = 0.01$ N, $z = 1$, $k = 1.38 \times 10^{-23}$ J K^{-1} or 1.363×10^{-22} atm cm^3 K^{-1} ion^{-1}, $T = 25$ °C, $e = 1.6 \times 10^{-19}$ C or 4.803×10^{-10} esu, and $\psi_d = 100$ mV or 3.366×10^{-4} erg esu^{-1}.

Potential Energy due to Constant Potential versus Constant Charge

In the following discussion, one must determine whether the particle being considered satisfies a condition of constant surface potential or of constant charge. Constant surface potential implies that the electric potential at the surface of a particle is constant despite the distance between plates, although the amount of charge at the surface will depend on this distance. For particles with constant charge, the surface potential will vary with distance between plates. The flat surfaces of clays such as montmorillonite satisfy the condition of constant surface charge, due to their permanently charged nature. Also, the assumption of a constant surface potential is considered valid for particle interaction where the surface charge density is due to the concentration of a potential-determining ion (such as H$^+$) in the equilibrium solution. Examples of this would be the edges of allophane and kaolinite particles.

Mathematically, the interactions between constant potential and constant charge surfaces are very different. However, for weak interactions and large distances there is little difference.

Equation 3.24 is difficult to integrate. To obtain the potential energy due to repulsive forces, V_R (erg cm^{-2}), we may substitute equation 3.31 into equation 3.24:

$$V_R = -2 \int_{\infty}^{d} [2n_0 kT (\cosh(y_d) - 1)] \, dz \tag{3.32}$$

Care must be taken when integrating equation 3.32, as its calculation will vary depending on constant charge or constant potential; units will be in ergs/cm^2. For the case of constant potential, the potential at the surface of the colloid remains constant irrespective of distance between particles; the repulsive energy between platelets separated by a distance, d, is

$$V_R^{\psi} = 2 \frac{n_0 e^2 (\psi_0^{\infty})^2}{kT\kappa} e^{-2\kappa d} \tag{3.33}$$

where κ (m) is as expressed in equation 3.16, and ψ_0^{∞} represents constant surface potential at infinite particle separation. For spherical colloids at constant potential, assuming the electric potential at the particle surface is < 3 mN/m^2, the repulsive energy is given by

$$V_R^{\psi} = \frac{\varepsilon r (\psi_0)^2}{2} \ln[1 + \exp(-\kappa H)] \tag{3.34}$$

where ε is the permittivity of the solution, r is the radius of the particle, and H is the minimum separation of the two particle surfaces.

Under constant charge, potential varies with distance between particles. This condition is satisfied in soil systems of typical clays (such as kaolinite or montmorillonite) that are platelike in shape. The potential for the surface of constant charged particles depends on whether $e\psi_d/kT$ (dimensionless) is small or large. For cases where the potential is < 0.1 J/C or V, the repulsive force is expressed as

$$V_R^{\psi} = \frac{2n_0 kT}{\kappa} (y_0^{\infty})^2 \left(\coth \frac{\kappa d}{2} - 1 \right) \tag{3.35}$$

where d is distance between plates. When the potential > 0.1 J/C or V, the repulsive force is expressed as

$$V_R^{\psi} = \frac{2n_0 kT}{\kappa} \left\{ 2y_0^{\infty} \ln \left[\frac{B + Y_0^{\infty} \coth\left(\frac{\kappa d}{2}\right)}{1 + y_0^{\infty}} \right] - \ln[(y_0^{\infty})^2 + \cosh(\kappa d) + B \sinh(\kappa d) + \kappa d] \right\} \tag{3.36}$$

where $y_0^{\infty} = e\psi_0^{\infty}/kT$ and $B = \sqrt{1 + (y_0^{\infty})^2 \operatorname{csch}^2(\kappa d/2)}$ (Verwey and Overbeek 1948; Wilemski 1982).

The repulsive force for energy between two spherical particles of equal dimension is given by

$$V_R = V_R^{\psi_0^{\infty}} - \frac{\varepsilon r (\psi_0^{\infty})^2}{2} \{\ln[-\exp(-2\kappa H)]\} \tag{3.37}$$

where $V_R^{\psi_0^\infty}$ is the repulsive energy between particles of constant surface potential ψ_0^∞ (which represents the electric potential at the particle surface when the particle separation is infinite) and r is the particle radius. Thus,

$$V_R^{\psi} = \frac{\varepsilon r(\psi_0)^2}{2} \ln[1 + \exp(-\kappa H)] \tag{3.38}$$

When dealing with constant charge parameters, the left side of equations 3.32–3.37 might be more appropriately labeled with the superscript σ to denote constant charge density (i.e., V_R^{σ}). A general relationship of V_R versus distance is given in figure 3.7.

QUESTION 3.8

Determine the repulsive force (V_R) for a spherical colloid, assuming a small electrical potential. Also, determine this force for a platelet-shaped particle.

Van der Waals–London Forces

Van der Waals–London forces exist between neutral nonpolar molecules; they do not depend on a net electrical charge, and are independent of ionic strength of solution in most aqueous systems. Thus, they can be thought of as short-range, electrostatic attractive forces. In 1930, Fritz Wolfgang London (1900–1954) used quantum mechanics to quantitatively express this force. Neutral atoms constitute systems of oscillating dipole charges ($> 10^{15}$ Hz) resulting from the variable orbital location of negatively charged electrons about the positively charged nucleus. At distances > 100 Å, quantum mechanics predicts that an atom cannot polarize an adjacent atom. Van Olphen (1963) and others report that van der Waals–London forces are additive. Because these forces are additive, the attraction between particles which contain a large number of atoms is equal to the sum of the attractive forces for each atom pair; the total attractive force can become quite large. Force decays less rapidly with distance for large particles containing large numbers of atom pairs. Van der Waals–London forces between colloidal particles vary between the inverse of the 3rd and 7th power of the distance between the particles. These forces strongly affect the flocculation of clay particles in soil–water systems, but contribute little to the attraction between water–clay interfaces.

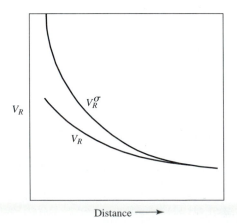

Figure 3.7   General relation of potentials; constant potential, V_R, and constant charge, V_R^{σ}.

Potential Energy Due to van der Waals–London Forces

As for repulsive forces, the concepts of energy due to van der Waals–London forces apply to both platelet and spherical shapes. For particles of platelet shape, the attraction energy in ergs cm^{-2} (V_A) is given as

$$V_A = -\frac{A}{48\pi}\left[\frac{1}{d^2} + \frac{1}{(d+\delta)^2} - \frac{2}{(d+\delta/2)^2}\right] \tag{3.39}$$

(van Olphen 1963), where A is the Hamaker constant $[\pi^2 n^2 (3/4) h v_o \alpha^2]$ (Adamson 1990); n is molecules per cm^3; h is the Planck constant (ergs · s); v_o is the activation or London frequency (s^{-1}); α is the polarizability (cm^3); and δ is the plate thickness at distance $2d$ (two times the plate thickness or diameter). The potential of the van der Waals forces, ψ, between two atoms is

$$\psi = -\frac{c}{r^6} \tag{3.40}$$

where r is the distance between atoms and c, as given by Slater and Kirkwood (1931), is

$$c = \frac{3ekT}{4\pi\sqrt{m}}\frac{\alpha_1\alpha_2}{\sqrt{\alpha_1/N_2} + \sqrt{\alpha_2/N_2}} \tag{3.41}$$

where e is the elementary electric charge, k is the Boltzmann constant, m is mass of the electron, and N is the number of electrons in the outermost shell. Differentiating equation 3.40 with respect to r yields the force f as

$$f = \frac{6c}{r^7} \tag{3.42}$$

where r is the distance between atoms and n is the proportionality of force. For example, Casimir and Polder (1948) show that the force is proportional to r^{-8} for retarded van der Waals forces. The typical equation given in the literature for the attractive energy between two platelets is

$$V_A = -\frac{A}{48\pi}\frac{1}{d^2} = -\frac{A}{12\pi}\frac{1}{D^2} \tag{3.43}$$

In this instance $D = 2d$, and it is assumed that $\delta \gg d$. To obtain the average attractive force per unit of the material (V_A), equations 3.39 and 3.43 must be differentiated for d and D, respectively, to give

$$V_A = \frac{A}{24\pi}\left[\frac{1}{d^3} + \frac{1}{(d+\delta)^3} - \frac{2}{(d+\delta/2)^3}\right] \tag{3.44}$$

and

$$V_A = \frac{A}{6\pi}\frac{1}{D^3} \tag{3.45}$$

These two equations are generally used for distances < 200 Å. For distances > 200 Å (Tabor and Winterton 1968), the attractive energy (V_A) would be R/D^4, where R is the retardation constant of the soil being used (see chapter 10). The value of parameter R (as used for V_A) for retarded attractive forces of platelet shapes (clay–air–clay) has been shown by various researchers to vary from 0.77×10^{-26} to 2.0×10^{-26} J m^{-1} (Tabor and Winterton 1968, and others). The lower value represents micas, and the higher value quartz. Values of A calculated from theoretical analysis of coagulation measurements are 3.1×10^{-20} J for kaolinite, 2.5×10^{-20} J for illite, 2.2×10^{-20} J for montmorillonite, and 1.63×10^{-19} J for palygorskite (Novich and Ring 1984).

For spherical particles, the attractive energy due to van der Waals–London forces is

$$V_A = -\frac{A}{6}\left(\frac{2}{S^2 - 4} + \frac{2}{S^2} + \frac{S^2 - 4}{S^2}\right) \tag{3.46}$$

where A is the Hamaker constant, S is R/a which reduces to $2 + H/r$, R is the distance from sphere to sphere (center), r is sphere radius, and H is the minimum distance between spheres.

The Hamaker constant as normally written applies to the interaction between particles of a medium in a vacuum. For soils not in a vacuum, $A = A_{11} + A_{22} - A_{12}$, where A_{11}, A_{22}, and A_{12} are the Hamaker constants for particle–particle, water–water, and particle–water interactions. The Hamaker constant for mica–air–mica has been measured to range from 1.02×10^{-19} to 1.50×10^{-19} J (Tabor and Winterton 1968; Israelachvili and Tabor 1972). For a clay, A_{11} is approximately equal to that for mica–air–mica. Tabor and Winterton (1968) obtained 2.4×10^{-19} J for A_{22} while Verwey and Overbeek (1948) obtained 6×10^{-19} J. Assuming this latter value, A for a clay–water–clay system is approximately 10^{-19} J; assuming the value reported by Tabor and Winterton, A is about 2.0×10^{-19} J. Israelachvili and Adams (1978) obtained 2.2×10^{-20} J for the Hamaker constant using a mica–aqueous electrolyte solution of 10^{-1} to 10^{-3} mol KNO_3.

Schenkel and Kitchener (1960) derived an equation corrected for the effects of retardation when H < 150 Å,

$$V_A \approx -\frac{Ar}{12H}\frac{\lambda}{\lambda + 3.54\pi H} \tag{3.47}$$

and another when $H > 150$ Å,

$$V_A = \frac{Ar}{\pi}\left(\frac{2.45\lambda}{120H^2} - \frac{\lambda^2}{1045H^3} + \frac{\lambda^3}{5.62 \times 10^4 H^4}\right) \tag{3.48}$$

where λ is the wavelength of the London frequency. Both equations 3.47 and 3.48 are for equal spheres; equation 3.48 is accurate to about 5%.

QUESTION 3.9

What is the difference (if any) between van der Waals–London attractive force using equations 3.44 and 3.45? How does this compare to equation 3.46, assuming a spherical shape?

Total Potential Energy

The addition of attractive and repulsive forces yields the total potential energy (V_T) between charged particles (i.e., $V_T = V_R + V_A$). Since V_R depends on d, n_0, T, z, and δ, V_T is also a function of these. Forces of attraction (V_A) are negative; forces of repulsion (V_R) are positive and are usually expressed in J m^{-2}. Honig and Mul (1971) calculated the repulsive energy of a montmorillonitic soil under constant charge (σ) and constant potential (ψ) conditions (table 3.4). The repulsive energies in both instances are a function of κd and surface potential, assuming infinite distance of separation. The reduced repulsive energy between two parallel plane double-layer systems as they approach is determined by

$$V_R^{\psi} = \frac{64n_0 kT}{\kappa}\gamma^2 e^{-2\kappa d} \tag{3.49}$$

TABLE 3.4   Repulsion Energy at Constant Potential (ψ)
and Constant Charge Density (σ); $y_0 = 2 = e\psi_0/kT$

$d*$	$W^{\psi\dagger}$	$1W^{\sigma\dagger}$
0.0	0.1358	∞
0.3	0.0873	0.1339
0.6	0.0522	0.0641
0.8	0.0365	0.0417
1.0	0.0253	0.0277
1.2	0.0174	0.0185

Source: Honig and Mul (1971).
*$d = (\kappa/2)H_0$, where H_0 is the shortest distance between
2 plates.
$^\dagger W = (1/b)V_R; b = 64n_e kT/\kappa$ (J m^{-2}).

(Adamson 1990), where

$$\gamma = \frac{e^{y_d/2} - 1}{e^{y_d/2} + 1} \tag{3.50}$$

and κ is the Debye kappa. Equation 3.50 applies when surfaces are far apart, the potential midway between the plates is small, and there is weak interaction.

This repulsion plays a major role in determining colloidal stability against flocculation. Irving Langmuir (1881–1957) stated in 1938 that the total energy acting on the parallel planes may be regarded as the sum of an osmotic pressure force and an electrical field (Derjaguin and Churaer 1978). Total energy must be constant in the space between the parallel planes, and because the field $d\psi/dx$ is zero at the midpoint, total energy is given by the net osmotic pressure at that point. When the potential is large, the osmotic pressure, P, may be determined by

$$P = \frac{\pi}{2}D\left(\frac{kT}{ed}\right)^2 \tag{3.51}$$

Quite often, van der Waals–London forces of attraction are balanced by the electrical double-layer repulsion, such as in the flocculation of lyophobic colloids. In a colloidal solution, referred to as a sol, the charged particles will experience both van der Waals–London forces of attraction and double-layer repulsion. The balance of these two forces will determine the rate and ease of flocculation. For solutions of low ionic strength (measured by κ), the double-layer repulsion is very large, except at small separations. However, as κ is increased, a limiting condition of net attraction at all distances is reached (Adamson 1990). At a distance of separation x, almost equal to particle diameter, a critical region of the κ value is reached in which a small potential minimum of about $(1/2)kT$ occurs.

The preceding discussion demonstrates why increased ionic strength in a solution increases flocculation. The approximate net potential for a sol can be expressed, beginning with equation 3.49, as

$$V_R^\psi = \frac{64n_0 kT}{\kappa}\gamma^2 e^{-2\kappa d} - \frac{\left(\frac{1}{12}\pi\right)A}{x^2} \tag{3.52}$$

Adamson (1990) stated that if rapid flocculation is taken to indicate that no barrier exists, $V_R^\psi = 0$ and $dV_R^\psi(x)/dx = 0$ for some value of x. As a consequence, the ionic solution concentration at which rapid flocculation will occur may be found from the following equation.

$$n_0 = \frac{2^7 3^2}{\exp(4)} \frac{\varepsilon^3 k^5 T^5 \gamma^4}{(ze)^6 A^2} \tag{3.53}$$

Thus, there is a straightforward relation between the flocculating concentration and the double-layer potential. Since z is valence, which typically has values of 1, 2, or 3, the concentration of ions of low valence (1–2) increases as double-layer potential increases. For large values of y_d (large values of z and/or ψ_0), concentration is insensitive to values of potential, and it approaches a limiting value because γ approaches unity. This holds for the Schulze-Hardy rule for the effect of valence type on the flocculating ability of ions. Because of the increased tendency for specific adsorption of ions with greater charge, the ability to flocculate depends on the reduction of potential and double-layer thickness. For a more detailed discussion, the reader is referred to Adamson (1990) and Verwey and Overbeek (1948).

QUESTION 3.10

In question 3.3, you were given a surface potential of 75 mV. As a comparison with the Gouy–Chapman theory, calculate the surface potential of the same clay material used in question 3.3 from $\psi_0 = \sigma 4\pi/\varepsilon\kappa$, (you will have to first compute κ).

Non-DLVO Forces

The stability of colloids is affected by forces other than those of DLVO origin, including hydration repulsive forces, hydrophobic attractive forces, and steric forces. During flocculation, particles are closely attracted; ions associated with these particles require a partial loss of their hydration for flocculation to occur. This process requires repulsive forces known as hydration forces, which can be appreciable in aqueous colloidal systems. Hydration forces partially negate van der Waals–London attractive forces, allowing dispersion of soil to occur more easily by reducing the electrolyte concentration. As water surrounds a particle, hydrogen bonding assures a well-structured affect. Normally, as water covers a particle, repulsive forces between particles help force them apart, but in the absence of water, hydrophobic attraction between colloidal surfaces occurs that is stronger than van der Waals–London forces and extends away from the surface. However, Adamson (1990) indicates that van der Waals attraction of water to itself seeks to exclude hydrophobic particles, making them stick together more tightly than they would due only to their own forces. Steric forces, associated with the adsorption of minute amounts of organic compounds on the edges of clay particles and the surfaces of sesquioxide compounds, can lead to charge reversal in the areas of adsorption. This can promote dispersion by preventing interaction between oppositely charged surfaces.

Limitations of the DLVO Theory

The repulsive and attractive forces between double layers in the DLVO theory are calculated considering ions as point charges. Research by Israelachvili and Adams (1978) and Ducker, Senden, and Pachley (1991) has shown good agreement with calculations based upon the DLVO theory. The theory works well when distance between particles is great enough, usually > 30–40 Å, that the magnitudes of ions present can be ignored. The theory fails at shorter distances from the surface. This is partially a result of non-DLVO forces and capillary effects, as well as the presence of bivalent ions that are more likely to approach the colloid surface than monovalent ions. Generally, the electrical intensity or potential at points mid-way between two particles, where the distance from the charged surface is greater than the average distance between adjoining charges on the surface, prevails in the majority of soils. This makes use of the DLVO theory attractive.

3.5 ION EXCHANGE

Ion exchange refers to a reversible process in which both cations and anions are exchanged between solid and liquid surfaces. The active fraction of soils in which ion exchange occurs consists of clay, colloidal organic matter, and silt. Positively charged ions are attracted to the surface of negatively charged particles within soil. Charges in organic matter and humic acid arise from —COOH and —OH groups. The charge associated with ion–surface interaction arises from isomorphous substitution, and from ionization of hydroxyl groups attached to silicon atoms at the lattice defects (i.e., broken edges) of tetrahedral planes. This substitution is balanced by ion exchange between the layers, and charge is usually balanced by the potassium ion. The substitution is generally in the form of a monovalent ion exchanging for a divalent ion in a mass action approach, that is,

$$\text{Ca}X_2 + 2\text{K}^+_{(aq)} \rightleftharpoons 2\text{K}X + \text{Ca}^{2+}_{(aq)} \tag{3.54}$$

Here, X represents the negatively charged surface assuming an adequate number of available exchange sites. Mass action can be described as

$$K_c = \frac{(\text{K}X)^2(\text{Ca}^{2+})}{(\text{Ca}X_2)(\text{K}^+)^2} \tag{3.55}$$

where K_c is the exchange equilibrium constant and the parentheses are activities.

Additional reactions of exchange include the following.

$$\text{CaCO}_3 + \text{Sr}^{2+} \rightleftharpoons \text{SrCO}_{3(s)} + \text{Ca}^{2+}$$

$$\text{Fe(OH)}_{3(s)} + \text{HPO}_4^{2-} + 2\text{H}^+ \rightleftharpoons \text{FePO}_{4(s)} + 3\text{H}_2\text{O} \tag{3.56}$$

There are many such reactions; however, these will suffice for purposes of illustration. If divalent ions exchange for monovalent ions, only half as many will be needed to balance the charge. This is a good example of the Gouy theory, which predicts a greater concentration of divalent ions in the double layer than monovalent ions, and which agrees with experimental data.

Cation Exchange Capacity

The ability of soils to exchange cations is referred to as the cation exchange capacity (CEC), and is expressed as the quantity of exchangeable cations in milliequivalents per 100 g of oven-dry media. Ion exchange occurs entirely within the double layer of the soil solution or suspension liquid. In figure 3.4, the ion exchange capacity corresponds to the area marked as σ^+, which is the charge due to a surplus of cations, whereas σ^- is the charge due to a deficiency of anions. Because ions are preferentially adsorbed electrostatically due to isomorphous substitution and ionization, CEC is pH dependent; that is, charge increases with increasing pH. Since surface mechanisms are not explained by simple CEC models, equations used here are derived from a basic understanding of thermodynamic properties of ions that are independent of process.

Montomorillonitic clays allow water to penetrate into the interlayer spaces and are high-swelling clays (Low 1979). Water penetrates these interlayer spaces due to the difference in osmotic pressure between the interlayer space and the surrounding solution, and interlayer spacing becomes a function of the hydration tendency of the counterions and interlayer forces. As the hydration of the counterion increases, swelling results in increased distance of the interlayers. Half as many divalent ions as monovalent ions are required to balance the charge, so the osmotic pressure between interlayer and solution is reduced, and less swelling will be observed for a solution of divalent ions than if a monovalent ion (such as Na^+) were present.

If coulombic interaction between counterions is less than the ion-induced dipole interactions between counterions and water molecules, the affinity of ions for the exchangeable surface will follow the Hofmeister series (Sposito 1984):

$$Cs^+ > K^+ > Na^+ > Li^+ \tag{3.57}$$

and

$$Ba^{2+} > Sr^{2+} > Ca^{2+} > Mg^{2+} \tag{3.58}$$

(In other words, the ion of larger hydrated radius is replaced with an ion of smaller hydrated radius.)

Sodium Adsorption Ratio

The Gapon equation is used extensively in understanding the exchange of Na^+ for Ca^{2+} in saline soils. The activity of individual ions on the surface is equal to the ions' equivalent fraction there; thus, the Gapon approach to cation equilibria may be written as

$$Ca_{1/2}X + Na^+_{(aq)} \rightleftharpoons NaX + \frac{1}{2}Ca^{2+}_{(aq)} \tag{3.59}$$

(Mott, 1988). Notice the close similarity of the Gapon approach to the commonly used mass action approach, (equation 3.54). Using the Gapon approach, the equilibrium constant is expressed by

$$K_c = \frac{(NaX)\sqrt{(Ca^{2+})}}{(Ca_{1/2}X)(Na^+)} \tag{3.60}$$

Thus, the Gapon constant (K^G) may be expressed using solution concentrations instead of activities as

$$K^G = \frac{\sqrt{[Ca^{2+}]}}{[Na^+]}\frac{[NaX]}{[Ca_{1/2}X]} \tag{3.61}$$

Saline soils usually contain considerable amounts of Mg^{2+}, which is assumed to exchange in a manner similar to Ca^{2+}, and which has the following Gapon relationship expressed in equivalent units.

$$\frac{[NaX]}{[(Ca + Mg)X]} = K^G\frac{[Na^+]}{\sqrt{\dfrac{[Ca^{2+} + Mg^{2+}]}{2}}} \tag{3.62}$$

Generally, the ions of significance in saline soils are calcium, magnesium, and sodium. Considering this, equation 3.62 describes the sodium adsorption ratio (SAR), which is a characteristic of ions dissolved in soil solution; that is, a ratio for soil extracts and irrigation water used to express the relative activity of sodium ions in exchange reactions with soil. The SAR may be expressed by

$$SAR = \frac{[Na^+]}{\sqrt{\dfrac{[Ca^{2+} + Mg^{2+}]}{2}}} \tag{3.63}$$

The exchangeable sodium percentage (ESP) is the degree of saturation of the soil exchange complex with sodium. ESP = [NaX/CEC] $\times$ 100; or, for a soil that is in equilibrium with the irrigation water that has been applied,

$$\text{ESP} = \frac{\dfrac{100\,\text{SAR}}{K^{G}}}{1 + \left(\dfrac{\text{SAR}}{K^{G}}\right)} \tag{3.64}$$

The SAR is an indicator of the suitability of water for irrigation purposes on already irrigated land, or for irrigation during land reclamation. The ESP was originally used as the main criterion for measuring excessive sodium levels in soils, but more emphasis has now been placed on the SAR. This is probably due to the fact that the ESP of surface soil can be predicted reasonably well from the SAR of applied water, because surface soil is being equilibrated with the applied water at each application. When the ESP is initially high, puddling of soils can decrease the percolation rate to the point that an extended period of time is required for land reclamation using water with a low SAR value. The ESP measurement in such cases can indicate how much lime or gypsum is necessary to acquire adequate tilth to begin reclamation. For a more detailed discussion of the SAR and ESP, the reader is referred to Bresler, McNeal, and Carter (1982).

3.6   HYDRATION AND SHRINKING OF CLAYS

Hydration

The imbibition of water by a clay is sometimes accompanied by an increase in volume. This phenomena is termed swelling, which is a macroscopic manifestation of expansion between clay layers caused primarily by an increase in volumetric water content. The crystal layers of some clays, such as kaolinite are bound together by hydrogen bonding; whereas illite layers are bound by potassium within layers (Hunter and Alexander 1963). Both of these elements cause very tight bonding, making it virtually impossible for liquid to enter between the layers. Consequently, hydration of these clays results in little or no swelling. In contrast, layers of montmorillonite clay are bound by cations that exchange easily for other cations in the surrounding aqueous solution. As a result, water molecules can be attracted to the cations and can be allowed to enter between the layers, resulting in a high swelling ability. As water enters an initially dry montmorillonite, the energy of hydration due to ion–dipole electrostatic attractions causes water to bond to cations, increasing the effective size of the cation. If this cation is between platelets, the platelets will normally be pushed apart, allowing other molecules to enter between them. This type of swelling is due to short-range interactive forces between the layers of expanding media, or between the planar surfaces of individual mineral crystals.

The influence of salt on water imbibition by clays varies according to the lyotropic series (Sumner and Stewart 1992). When large amounts of salts are either present in or added to soil, "salting out" may occur (precipitation of salt). Good examples of this occur in soils in the western United States and the arid Middle East. The lyotropic series for cations is

$$\text{Mg}^{++} > \text{Ca}^{++} > \text{Sr}^{++} > \text{Ba}^{++} > \text{Li}^{+} > \text{Na}^{+} > \text{K}^{+} > \text{Rb}^{+} > \text{Cs}^{+} \tag{3.65}$$

and for anions is

$$\text{Citrate}^{---} > \text{Tartrate}^{---} > \text{SO}_4^{-} > \text{C}_2\text{H}_3\text{O}_2^{-} > \text{Cl}^{-} > \text{NO}_3^{-} > \text{ClO}_3^{-} > \text{I}^{-} > \text{CNS}^{-} \tag{3.66}$$

The anions citrate, tartrate, sulfate, and acetate inhibit swelling; whereas chloride, nitrate, and so on favor water imbibition. Swelling depends upon the nature and pH of the electrolyte, as well as on the anionic species present. If we assume that the soil is at equilibrium before the addition of an electrolyte, the swelling pressure is given by

$$P_s = \frac{\Delta\mu}{w_s} \tag{3.67}$$

where P_s is the swelling pressure, $\Delta\mu$ is the difference in chemical potential between the outer and inner solution, and w_s is the specific volume of water in the sample (cm³). Imagine clay platelets separated by distance $2d$ (as in figure 3.6), with a chemical potential of μ_i between the two plates and a chemical potential of μ_o in the outer aqueous solution surrounding the two plates. Initially,

$$\mu_o = \frac{-RT \sum n_i \pi_i w_s}{1000} \tag{3.68}$$

where n_i is the molarity of solute i and π_i is its osmotic coefficient. After adding an electrolyte, equilibrium is disturbed and μ_o changes to a new concentration μ_o', which may be given by

$$\mu_o' = \frac{-RT \sum n_i' \pi_i' w_s}{1000} \tag{3.69}$$

Considering that, initially, $\mu_o = \mu_i$, then $\Delta\mu = \mu_o' - \mu_i$, which may be found by

$$\Delta\mu = \frac{RT \sum (n_i \pi_i - n_i' \pi_i') w_s}{1000} \tag{3.70}$$

This can be calculated if the change in concentration of the outer solution is known.

Thermodynamically, soil absorbs water until the partial molar free energy of the water adsorbed equals that of water in the equilibrium solution. As the particles are hydrated, the ions in the diffuse layer between the two platelets reduce the free energy of the free water. This causes diffusion of water from the outer aqueous solution into the inner solution between the particles. Consequently, as water is absorbed, the platelets push apart, exhibiting a swelling pressure. The platelets will continue to push apart until the increase in the free energy of water (due to hydraulic pressure) equals the decrease in free energy of water (caused by the higher ionic concentration between the particles when compared to the external solution). Mathematically, this is described by

$$V \Delta P = \Delta F = -RT \ln \frac{M_i}{M_o} \tag{3.71}$$

where V is the molar volume of water (18 cm³ mol⁻¹), P is pressure (dynes cm⁻²), F is free energy (ergs mol⁻¹), T is absolute temperature, M_i is the mole fraction of water in the inner solution, and M_o is the mole fraction of water in the outer solution. Then, for dilute solutions ($M < 0.1$)

$$-\Delta M_{\text{solute}} \approx \ln \frac{M_i}{M_o} \tag{3.72}$$

Considering this, equation (3.67) can be rewritten as

$$\Delta P = \frac{RT}{V} \Delta M_{\text{solute}} \tag{3.73}$$

Also, because the ratio of M_{solute} per molar volume of water (cm^3 mol^{-1}) is approximately equal to the difference in concentration of solutes (n_i) in mol cm^{-3}, then

$$\Delta P = RT \sum (n_i - n_{io}) \text{ ergs cm}^{-3} \tag{3.74}$$

where n_i is the concentration of ionic species i (mol cm^{-3}), and n_{io} is the concentration of ionic species in the outer solution (which follows Boltzmann's law of distribution for cations and anions, as given in equation 3.28). Substituting equation 3.28 into equation 3.74, the osmotic pressure difference between any point where the potential is ψ_d and a point in the equilibrium solution where the potential is zero can be written as

$$\Delta P = RTn_o(e^{y_d} + e^{-y_d} - 2) \tag{3.75}$$

which leads back to equation 3.31:

$$P_e = n_0kT(e^{-y_d} + e^{y_d} - 2) = 2n_0kT[\cosh(y_d) - 1]$$

The general relation between potential and anion and cation concentration for overlapping double layers is given in figure 3.8.

At the midpoint d, between the two platelets, $d\psi_d/dx = 0$ ($\psi_d \neq 0$), net dipole attraction of water molecules to either particle surface is essentially zero. Letting $x = y_d$, the difference between osmotic pressure at the midpoint, where $d = x_d$ (d is distance, x_d is dimensionless), and that in the equilibrium solution is a good measure of swelling pressure. From this, the swelling pressure can be calculated as

$$\Delta P = 2RTn_0(\cosh(y_d) - 1) \tag{3.76}$$

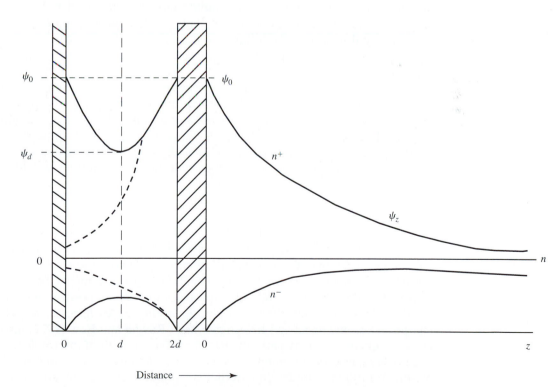

Figure 3.8   Representation of overlapping double layers, showing electrical potential (ψ) and cation and anion concentration ($n^{+/-}$). Data from Kruyt (1952)

TABLE 3.5 Relation of Swelling to Clay Mineral Type and Exchangeable Cations

Clay Mineral	Swelling $cm^{-3}g^{-1}$					
	H^+	Li^+	Na^+	K^+	Ca^{2+}	Ba^{2+}
Beidellite	0.81	4.97	4.02	0.50	0.91	0.85
Halloysite	0.05	—	—	—	—	—
Montmorillonite	2.20	10.77	11.08	8.55	2.50	2.50

Montmorillonite has very prominent swelling properties that are due mainly to inter-particle water interaction. Table 3.5 demonstrates swelling properties in the lyotropic series for both beidellite and montmorillonite clays. The hydration energy of clays is cation dependent. Likewise, the intake of water varies considerably with the nature of the clay. It is initially very rapid but slows with time. Typical clays take about one to three days to become completely hydrated, while bentonite may take up to one week. For some clay minerals, swelling decreases with diminishing hydration, in the following order.

vermiculite > montmorillonite > beidellite > illite > halloysite

Another manner of swelling, caused by hydration, is osmotic swelling. Osmotic swelling is caused by the presence of hydrated ions that surround the hydrated surface of the particle in the diffuse electrical double layer, and that are dissociated from the surface. At this juncture, the osmotic forces are in equilibrium with the van der Waals–London forces of attraction, due to the charge difference between the particle and exchangeable ions. Since the ions are dissociated, the influx of water between the particles is a function of osmotic pressure. Osmotic swelling generates considerable pressure, and is believed to function by the principles of Donnan equilibrium: $X^2 = Y(Y + Z)$, where X is the concentration of each ion in solution, Y is the concentration of each electrolyte between the platelets, and Z is the concentration of dissociated and adsorbed cations (Ghildyal and Tripathi 1987). Generally, osmotic pressure is greater at the particle surface than in the outer solution. Thus, the influx of water from the solution to the space between the platelets counteracts the greater concentration of cations, and swelling occurs. Osmotic swelling is due to long-range particle interaction in which the energy of surface hydration is inoperative, and diffuse electrical double layer repulsion is the major force of particle separation. Osmotic swelling takes place from an interlayer spacing of about 10 to 120 Å at equilibrium; the swelling pressure induced may be > 1 MPa.

Additional factors that determine swelling include those that affect the total potential energy of clay particles in a solution. These include surface charge density, counterion valence, concentration of the aqueous solution, and pH (Winterkorn and Baver 1934). Total potential energy involved in swelling is particularly dependent on pH, because a decrease in pH causes the edge charge of individual particles to become positive, increasing edge-to-face bonding. This has a net effect of reducing swelling. (An increase in pH causes the edge charge to become negative.) Investigations of Ravina and Low (1972) indicate that crystal-lattice configuration may partially determine the strength of intermolecular interaction between clay particles and water molecules. Their findings proved that maximum swelling of Na-montmorillonite under no external constraint decreases linearly with a subsequent increase in the dry-state b dimension (figure 3.9). They also found that the magnitude of the b

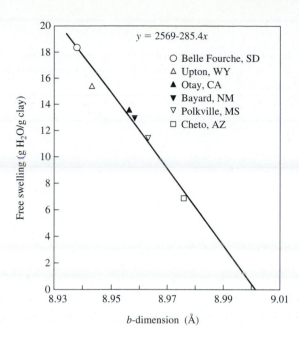

Figure 3.9   Swelling of clays versus the *b*-dimension; data from Ravina and Low (1972)

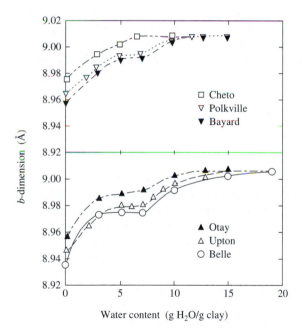

Figure 3.10   Relation between the *b*-dimension and soil water content; data from Ravina and Low (1972)

dimension increases incrementally with water content until a final *b* dimension is reached (figure 3.10). The *b* dimension can be defined as the *b* axis, and is the characteristic width in relation to the depth (*a*-dimension, or axis) and height (*c*-dimension, or axis) of the clay molecule; the *b* dimension has also been defined as the interlayer spacing or basal spacing—the distance between the lowest sheet of one layer and the equivalent sheet in the layer above. Further details about the *b* dimension are given in Grimshaw (1971).

QUESTION 3.11

What is the swelling pressure for a Na-montmorillonitic clay, assuming that $w_s = 1$, $n_i = 10^{-4}$ M, $n_i' = 10^{-2}$ M, $\pi_i = 2$, $\pi_i' = 2$, $v_s = 1$ g cm^{-3}, and $T = 25\,°C$? It may be assumed that NaCl is the electrolyte and that $R = 0.082$ atm L mol^{-1} K^{-1}.

QUESTION 3.12

What is the swelling pressure for the same clay if equation 3.73 is used for the calculation parameters?

QUESTION 3.13

What is the potential energy between a soil solution and an adjoining clay particle where $d = 2.6$ Å, $\varepsilon = 78.8$, and σ is as calculated in question 3.3?

Shrinking

The process of shrinking is basically the reverse of swelling. As a clay loses water, the distance between platelets collapses (due to dehydration) as the medium approaches its original volume before it was wetted up. This indicates a return of the interlayer spacing to about 2.6 Å for 2:1 clays. However, the interacting chemical processes are not necessarily reversible as the medium dehydrates; precipitation and other chemical processes may occur during dehydration. Cracking may also occur during dehydration in some soils where lines of cleavage develop at points of least resistance, usually coinciding with the areas of greatest water content.

3.7   FLOCCULATION AND DISPERSION OF CLAYS

Flocculation

Flocculation of clays results directly from charge relations discussed in previous sections. The term *critical zeta potential,* which has been defined as the time at which flocculation will occur within a colloidal solution (Sumner and Stewart, 1992), is still prevalent in textbooks. While the zeta potential is obtained from the electrophoretic mobility of a soil particle, the magnitude of the mobility is considered a measure of particle repulsion. The zeta potential usually decreases when an electrolyte is added to solution. At flocculation, the electrolyte was considered to have reached a critical value. Below this value, it was assumed that particle repulsion was no longer strong enough to prevent flocculation (van Olphen 1963). At the time of that writing, the nature of forces causing flocculation were not clear. Since that time, it has become clear that the primary force or framework of the zeta potential is the slipping plane.

Because the position of the slipping plane is not known, the zeta potential is the electric potential at some unknown distance from the surface of the double layer. As a result, the zeta potential is not equal to the surface potential, but somewhat comparable to the Stern potential. It is not surprising, therefore, that a relation exists between the stability of colloids and the magnitude of the zeta potential. Due to its ill-defined character, the zeta potential is not useful as a quantitative criterion of colloidal stability.

As colloids lose their stability, they tend to flocculate, which occurs when attractive forces overcome repulsive forces. The flocculation concentration, F_c, at which attractive forces dominate and flocculation will occur, is described mathematically by

$$F_c = \frac{K\left[\tanh\left(\dfrac{ze\psi_0}{4kT}\right)\right]^4}{A^2 z^6} \tag{3.77}$$

(Sumner and Stewart 1992), where F_c is the flocculation concentration value (m mol/L), z is counterion valence, e is the electronic charge, ψ_0 is the electric potential at the surface, A is the Hamaker constant, k is the Boltzman constant (1.38×10^{-23} J K^{-1}), and K is a constant (8×10^{-36} J^2 L^{-1} or 8×10^{-22} ergs^2 L^{-1}—assumed constant for most soil solutions). From this equation, the influence of ionic valence on the precipitating effect of the ion can be seen. This effect is known as the Schulze–Hardy rule, which denotes the dependence of F_c on z^{-6}, indicating that ionic charge has a dominant effect on flocculation. According to this rule, flocculation values range from 25 to 150 m mol L^{-1}, 0.5 to 2 m mol L^{-1}, and 0.01 to 0.1 m mol L^{-1} for monovalent, divalent, and trivalent ions, respectively. Consequently, ions such as Ca^{2+} or Ba^{2+} have a much greater effect on flocculation of clays than K$^+$. As ions of opposite charge than that of clay particles prevail, the numerical value of the potential within the slipping plane and, thus, the zeta potential, will change. Positive ions increase potential in the slipping plane and negative ions decrease it. When potential in the slipping plane decreases, the velocity of individual particles slows and flocculation takes place. Within the electrical double layer, ions effective in precipitation are adsorbed onto the solution side of the fixed part of the double layer (see figure 3.4).

Rengasamy and Sumner (forthcoming) have derived a factor based upon the ratio of the first to the second ionization potential of the cation and its valence. They have suggested that hydration and flocculation effects depend on this factor, termed the flocculative power. The flocculative power $\approx 100(I_z/I_{z+1})^2 z^3$, where I_z is the zth ionization potential, and z is the cation valence.

As an example of the effect of ionic valence on the zeta potential, consider a soil system (colloidal in nature) that is typically negatively charged. The addition of a monovalent electrolyte will increase the zeta potential (BC in figure 3.11). If excess electrolyte is added, the thickness of the double layer will tend to collapse due to a decrease in charge density (Rengasamy 1983); as a result, the zeta potential will decrease as well. In this situation the

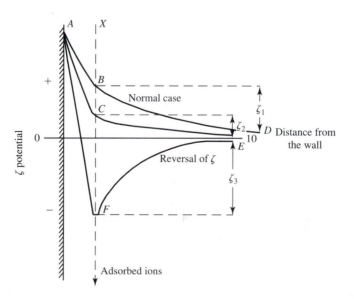

Figure 3.11   Effect of ionic valence on zeta potential

cations will accumulate near the particle surface (*AB* in figure 3.11) or rigid layer (also see figure 3.4), which is the solution side of the electrical double layer. The decrease in zeta potential is shown by line *CD* in figure 3.11. As the valence of the ion in the added electrolyte increases, the smaller the quantity of electrolyte required to effect the same changes in zeta potential. A continual addition of electrolyte will result in a reversal of zeta potential sign, due to surface charge neutralization or decreased electrical double layer thickness. The double layer can actually collapse and reform with a reversed charge (*FE* in figure 3.11).

There are essentially three stages of flocculation (figure 3.12): **(1)** a stable colloidal suspension (zone I); **(2)** slow flocculation (zone II); and **(3)** fast flocculation (zone III). For low concentrations of electrolytes ($n_t/n_0 = 1$), the suspension is stable (i.e., dispersed). Collision of particles in this stage does not form a stable consolidation; however, the zeta potential changes appreciably. During the second stage a slow flocculation occurs because the zeta potential incrementally decreases in a uniform manner while n_t/n_0 falls rapidly. This implies that collisions rise from 0 to 1 in a small range of electrolyte concentrations. During the third stage, n_t/n_0 remains constant while zeta potential decreases. During this stage, collisions between particles are more effective and the suspension becomes unstable (flocculates). Thus, flocculation may occur very rapidly or slowly, dependent on both Brownian motion and particle interaction. For flocculation to occur, the net repulsive forces within the system must decrease. Usually, the smaller particles will flocculate first, since collisions between large and small particles are occurring and the number of small particles is likely to decrease with respect to visible changes in large particles. Arid zone soils contain large amounts of salts that affect both flocculation and dispersion. If an applied salt contains the same cation with which a clay soil is saturated, compression of the double layer will occur, lowering the net negative charge of the system, reducing the repulsive capacity and, thus, inducing flocculation. If the applied salt contains a different cation, compression of the double layer will still occur; however, ion exchange will also occur.

Dispersion

In dispersion (occasionally referred to as deflocculation), the addition of monovalent cations causes increased hydration of the clay micelles, which causes particles to repel each other. In this instance, the zeta potential is increased. The presence of monovalent cations results in a more extensive diffuse layer within the double layer because the extent of the double layer

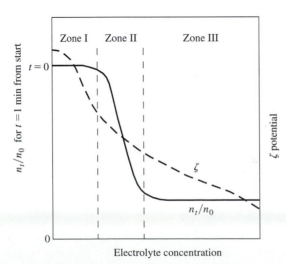

Figure 3.12 The three zones of flocculation. The symbols n_t/n_0 represent the particles per unit volume present after t seconds from the time of addition of the electrolyte, divided by the number of particles present initially. Zone I represents a stable sol, zone II represents slow coagulation, and zone III represents quick coagulation. At low electrolyte concentration, $n_t/n_0 = 1$

varies inversely with cation charge (valence), as described by the Schultze–Hardy rule. As this extensive diffuse layer develops around each particle, it prevents adjacent particles from approaching; thus, they remain as separate particles. In effect, dispersion is the opposite of flocculation as just discussed. In particle size analysis, dispersion is accomplished by mechanical shaking and the addition of a chemical dispersing agent, because most soils are not easily dispersed. However, some soils (such as those found in the southeastern United States) require very little mechanical energy to disperse, and actually have been found to disperse readily by the kinetic energy supplied by rainfall. Miller and Radcliffe (1992) surveyed thirty-five topsoils in the southeastern United States and discovered that a majority contained more than 50% of their total clay content in a water-dispersible form. Soils in the western United States are also easily dispersed, consequently these soils exhibit low infiltration rates. Dispersion greatly affects the sealing and crusting of soils, and hydraulic conductivity (which will be discussed in greater detail in chapter 8).

QUESTION 3.14

Determine the value of F_c that will result in flocculation for a montmorillonite soil that has NaCl as a background electrolyte. Assume $T = 25\,°C$, NaCl $= 0.017$ M, $\varepsilon = 78.8$, $\psi_0 = 25$ mV or 0.025 J/C, $e = 4.803 \times 10^{-10}$ esu ion^{-1}, $z = 1$, $k = 1.381 \times 10^{-16}$ erg ion^{-1} K^{-1}, $A = 2.2 \times 10^{-13}$ erg, and $K = 8 \times 10^{-36}$ J^2/L or 8×10^{-22} erg^2/L.

3.8   HUMUS IN SOILS

Humic and Fulvic Acids

The organic fraction of soil is composed of living organisms and their partially to completely decomposed remains. The term organic matter refers to the heterogeneous mixture of products resulting from chemical and microbial transformation of organic animal and plant debris. This transformation is known as the humification process and gives rise to humus. Humus is a mixture of variously transformed organic debris, which has a high resistance to further decomposition by microbes, and bears little resemblance to structures from which it was derived. Generally, humus is classified into two major groups: amorphous, polymeric humic substances, which are usually differentiated by solubility characteristics as either humic acids, fulvic acids, or humins (these are usually brown in color); and compounds such as polysaccharides, polypeptides, and altered lignins, which can be synthesized by microorganisms or may occur from modifications of comparable compounds from the original organic debris.

Humic and fulvic acids and related constituents are plentiful in soils, as well as ground and surface waters (Stumm and Morgan 1981; Kumada 1987; Oades 1989). Physically, these substances sometimes appear as yellow to black polyelectrolytes. Humic substances occurring in soil and natural waters have similar characteristics. They have large specific surface areas (900 m^2 g^{-1}) and high CEC (1500 to 3000 meq kg^{-1}; Schnitzer 1976). The development of adequate extraction and fractionation processes to isolate separate humic compounds has been difficult. Consequently, humic and fulvic acids, although known to give specific characteristics to soil and natural waters, have been hard to identify, and a satisfactory method for determining the occurrence of humic and fulvic acids in various environments has not been established. Humic substances have molecular weights ranging from approximately 300 to 300,000; in contrast, fulvic acids are relatively low in molecular weight, and contain higher O$_2$ but lower carbon contents. Fulvic acids also contain more acidic (COOH) functional groups. The OH groups associated with humic and fulvic acids are presumed to be phenolic. Both humic and fulvic acids contain a variety of carboxylic and phenolic functional (hydrophilic)

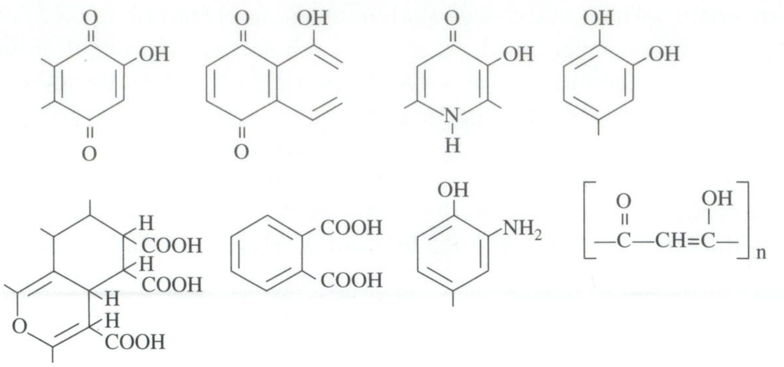

Figure 3.13 Various functional groups associated with humic substances

groups and aliphatic and aromatic moieties, which convey hydrophobic properties to these substances (see figure 3.13). Humic fractions within soil are believed to represent a system of polymers which systematically vary in pH, elemental content, molecular weight, and degree of polymerization (Stevenson 1982a, 1985). Humins, on the other hand, are believed to consist of humic acids that are so closely bound to mineral matter that the two cannot be separated. Soil contents of humic and fulvic acids vary by depth, climate, and geography (Thurman 1985b). Forest soils such as alfisols, spodosols, and ultisols generally are high in fulvic acids, whereas grassland soils such as mollisols are high in humic acids.

Humic substances in soils usually occur in insoluble forms and are bound in three ways: as insoluble macromolecular complexes; combined with clay minerals through polyvalent cations, hydrogen bonding, and van der Waals–London forces; and as macromolecular complexes bound together by cations such as Ca^{2+}, Al^{3+}, and Fe^{3+}. Macromolecular complexes generally occur in organic rich soil, such as peats and organic rich sediments, where clay–metal complexes are present in low amounts compared to humus substances. For forest soils such as spodosols, significant amounts of aluminum, iron, and organic matter have been mobilized and transported deeper into the profile in the B horizon. Consequently, this horizon has an abundant supply of fulvic acids that can be separated from the sesquioxides by mild extractants.

Because clays and organic colloids are normally negatively charged, humus substances associated with bi- and trivalent ions can form clay–humate complexes; the cation satisfies the surface charge, thus linking the colloids together. The trivalent cations form coordination complexes with humic substances in which bonding is very strong. In these instances, extraction and fractionation is difficult. Typically, in soils with organic matter, the CEC is proportional to the amount of organic matter present. About 30–90% of the total CEC of soils is a result of the organic fraction of the media.

The formation of stable complexes with polyvalent cations facilitates the mobilization, transport, segregation, and deposition of trace metals in soils. Because organic complexing agents function as carriers, trace metals found in soils normally occur in organically bound forms. However, not all natural occurrence of trace metal cycling is attributable to humic substances. A schematic denoting organic matter–trace metal interactions is given in figure 3.14. The ability of humic substances, R, to form complexes with metal ions is due to the functional groups present, such as COOH, phenolic OH, enolic OH, and C=O structures. Examples of these include

$$R—\overset{\overset{\displaystyle O}{\|}}{C}—OH\text{----}O{=}C—O.M. \tag{3.78}$$

where R is a polyanionic humic colloid, C is carbon, O is oxygen, OH is the hydroxyl group, and O.M. is organic matter (Hayes and Swift 1978).

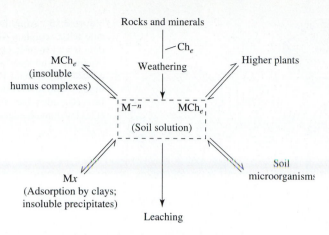

Figure 3.14   Organic–trace metal interactions of humic substances

Many instances occur in which soils exhibit hydrophobic characteristics. These are generally associated with dry spots on golf greens, burned forest areas, and citrus groves, and have been associated with coarse textured sands. This hydrophobic behavior has sometimes been attributed to fats and waxes coating the sand particles, as well as fulvic acids synthesized by fungi (Miller and Wilkinson 1977). This is a result of the effects of the humic substances on the contact angle of water on individual particles (which will be discussed in detail in chapter 5). Both humic and fulvic acids have been classified as hydrophobic substances, due to their adsorption onto resins with pH adjustment (low pH for acids and neutral pH for bases). Due to the characteristics of humic substances to convey hydrophobic properties, the hydrophobic interaction of humic and fulvic acids tends to accumulate at the particle–water interface. Additionally, adsorption is influenced by a coordinative interaction:

$$\equiv\!Al\!-\!OH + RC\overset{O}{\underset{OH}{\Big\langle}} \rightleftharpoons \equiv\!AlO\!-\!\overset{O}{\overset{\|}{C}}\!-\!R + H_2O \tag{3.79}$$

Adsorption of humic substances is multicomponent; thus, the Langmuir equation cannot be used to calculate it, because this equation is based on a single adsorbate. During the adsorption of humic substances to particles, an interface fractionation occurs because larger molecules adsorb preferentially over smaller ones. Thus, the molecular weight of material in solution is lowered. In terms of ion transport (see chapter 10), immobilization and increased residence time within the soil system occurs. The kinetics of these reactions are controlled by diffusion or convection. As with sesquioxides in variable charged soils, humic acids adsorbed to particles alter their chemical properties and, thus, their propensity for metal cations. This adsorption (hydrated surface) can be described by

$$S(H_2O)_m + X(H_2O)_n \rightleftharpoons SX(H_2O)_a + (m + n - a)H_2O; \quad \Delta G^\circ_{ads} \tag{3.80}$$

Assuming ΔG° for hydration of the product is small, the free energy of adsorption is given by

$$\Delta G^\circ_{ads} = \Delta G^\circ_1 - \Delta G^\circ_{solv} \tag{3.81}$$

where ΔG°_1 is given as

$$S(H_2O)_m + X \rightleftharpoons SX(H_2O)_m \tag{3.82}$$

and ΔG°_{solv} is given as

$$X + n(H_2O)\,\Delta X(H_2O)_n \tag{3.83}$$

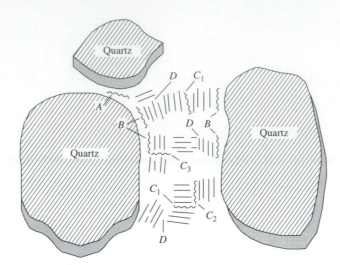

Figure 3.15 Amphipathic organic molecule self-association with particle surfaces; data from Emerson (1959). (A) Quartz–organic colloid–quartz, (B) quartz–organic colloid–clay domain, (C) clay domain–organic colloid–clay domain, (C_1) face–face, (C_2) edge–face, (C_3) edge–edge, and (D) clay domain edge–clay domain face

Equations 3.82 and 3.83 represent the affinity for the surface and solvent, respectively. Cations that are readily hydrated normally remain in solution, whereas large organic ions may be adsorbed at the particle–water interface because they have a low propensity for the liquid phase. This follows Traub's rule, which states that organic substances have a tendency to be adsorbed from solution based upon increasing molecular weight. Amphipathic organic molecules—those which contain a hydrophilic ionic or polar group and a hydrophobic constituent—tend to migrate to the particle surface, and also have a tendency for self-association which results from hydrophobic bonding. This self-association or micelle formation is depicted in figure 3.15.

Sorption of Contaminants

As contaminants (including pesticides, waste products, and organic chemicals) are applied or spilled onto soils, they may be preferentially adsorbed by humic substances present in the media. The longevity and hazard of the contaminant will depend on the chemicals' persistence, half-life (if applicable), leachability, volatility, degradability, and biological activity within the medium. Also, it will depend on soil parameters—such as hydraulic conductivity, infiltrability, pH, and so on. Obviously, the greater the content of humic substance present in a soil, the greater will be its capacity to buffer the deleterious effects of a contaminant. As depicted in figure 3.16, the interaction of humic substances with clay provides an organic surface for adsorption; thus, individual effects of either clay or organic matter alone are not easily ascertained. This is because the clay–humus or clay itself are the two primary surfaces that interact as an adsorbent, and because even clay particles usually have humic substances attached if any are present in the soil. In instances where herbicides are applied, the quantity of humic substances present have a large effect on adsorption (Adams and Thurman 1991; Armstrong and Konrad 1974; Stevenson 1976). Primarily, cationic organic molecules, such as those found in herbicides, are bound to the particle surface by ion exchange mechanisms (examples of these types of herbicides are diquat and paraquat). Other compounds, such as polar organic chemicals (neutral in characteristic), are bound to the surface particle primarily by organic matter, and also by clay.

The phenolic-, aliphatic-OH, COOH, enolic-, amino, and other nucleophilic reactive groups occurring in humic and fulvic acids produce chemical changes in many pesticides

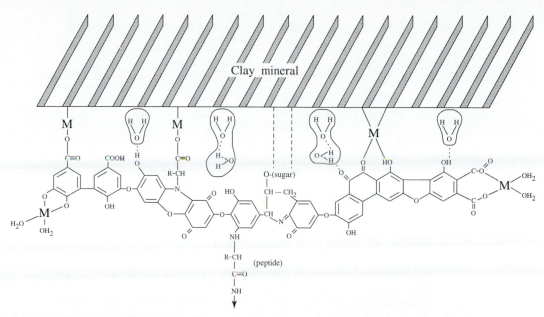

Figure 3.16   Interaction of humic substances with clay minerals; data from Mathur and Farnham (1985)

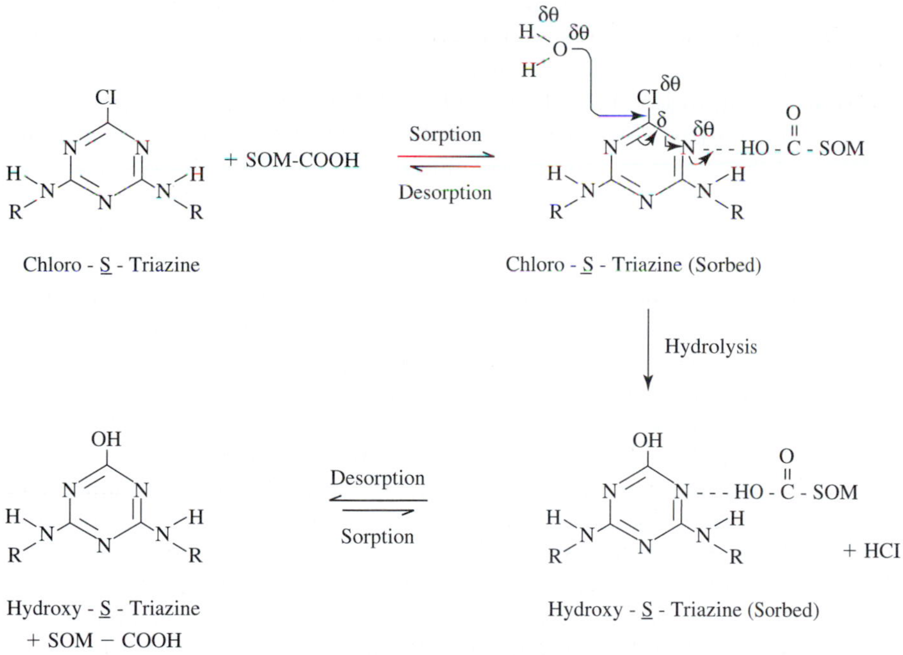

Figure 3.17   Selected reactions that take place in the hydroxylation of chloro-s-triazines

Chloro - S - Triazine

Chloro - S - Triazine (Sorbed)

Sorption

Desorption

Hydrolysis

Hydroxy - S - Triazine
+ SOM − COOH

Desorption

Sorption

Hydroxy - S - Triazine (Sorbed)

+ HCl

applied to soils. These chemical changes promote nonbiological degradation of many organic chemicals applied as pesticides (Armstrong and Konrad 1974). Thus, once the chemical is applied, a chemical transformation can take place. In many instances metabolites of the original chemical will be generated that may be more harmful by persisting in the soil longer than the parent compound. Reduction and hydrolysis, along with other reactions of the pesticide, can take place, all of which alter the parent. Figure 3.17 shows some of the reactions believed to take place in the hydroxylation of chloro-s-triazines.

TABLE 3.6 Sorption of Polar Liquids Compared to Dielectric Constant

Liquid	Total intake $(cm^3 g^{-1})$	ε	S $(cm^3 g^{-1})$	ε/S
H_2O	0.99	76.0	0.58	131
CH_3OH	0.66	31.9	0.25	128
C_2H_5OH	0.60	24.1	0.19	127
$C_3H_7OH_{(n)}$	0.57	20.5	0.16	128
$C_5H_{11}OH_{(n)}$	0.53	14.6	0.12	122
CCl_4	0.41	2.2	—	—

Source: Data from Winterkorn and Baver (1934).

The bonding mechanisms believed to be responsible for retention of organic chemicals by humic substances in soils include hydrogen bonding and van der Waals–London forces. Bonding of chemicals to humic substances through ion exchange is limited to organic chemicals that exhibit ionic or ligand exchange (bonding coordination by attached metal ions). Exchangeable groups may exist as cations that can become positively charged through protonation. Protonation will depend upon the pK_a of the chemical and the proton-supplying capacity of the humic substance.

Other bonding effects include the partitioning into hydrophobic media, where bonding of organics is believed to occur on the active surfaces of fats, waxes, resins, and the aliphatic side chains of humic and fulvic acids. As mentioned earlier, sometimes a metabolite to the parent may be more persistent in soils than the parent itself. This is because stable chemical linkages are formed between the organic chemical and humic substances, which tends to increase the persistence of the chemical. In fact, acylanilides and phenylcarbamates partially decompose to produce chloroamines when added to a soil. The two primary binding mechanisms appear to be a direct attachment of the chemicals to reactive sites on colloidal-size humic substances, and incorporation into newly formed humic and fulvic acids during the humification process. It is believed that organic chemicals are entrapped in the internal voids of the sievelike humus molecules. Organic soils usually require higher rates of pesticide application than mineral soils, both because the chemical is strongly retained onto humic substances and because organic soils contain more water on a volume basis, requiring more solute to obtain a given concentration.

In recent years, there has been increased concern regarding contamination of ground water by pesticides, radioactive wastes, and other organic chemicals. The primary concern is regarding transport of contaminants due to colloidal facilitated conveyance and particulate-matter transport. This transport occurs by attachment of organic and other hazardous chemicals to water soluble organic carbon (WSOC) substances and particulate matter so small that it is sometimes unstrained as it passes through the unsaturated zone, ultimately reaching ground water. Generally, WSOC concentrations are greater in the upper O and A horizons, and decrease with depth to the parent material. Concentrations of these materials are highest during spring and winter, when heavier well-structured soils are nearly water saturated and coarse-grained soils are very moist. Also, humic substances tend to increase in ground-water as the concentration of WSOC increases, primarily due to increased organic matter content within the unsaturated zone through which soil water must pass to reach ground water. Likewise, in ground waters that have a murky or colored appearance, humic substances may account for up to 65% of WSOC present, whereas in clearer water they account for only 10–30%.

Humic substances also play an important role in transport of nitrogen to groundwater (Wallis 1979). Nitrate is one of the largest non–point source pollutants in the United States; humic acids and humin contain about 2–6% N, whereas fulvic acids contain about half this amount. Except in agricultural settings and for leguminous plants, N_2 is primarily supplied to soils by the atmosphere. However, once in the soil, N_2 can be converted to organic-N. In this form, nitrogen can undergo several processes of transformation including amminization, ammonification, nitrification, and denitrification (Bremner 1967; Stevenson 1982b). During the process of amminization, organic-N is converted to various amino groups (the hydrolytic decomposition of proteins and subsequent release of amines and amino acids), which are further converted to NH_3 and NH_4^+. In the ammonium state, the N present may be immobilized within the organic matter or assimilated, or it may be nitrified to NO_3^-. The major steps of nitrogen transformation are as follows.

Microbial N-fixation	$N_2 \rightarrow$ Soil organic-N
Amminization	Soil organic-N $\rightarrow$ R-NH_2 + CO_2 + additional products + energy
Ammonification	R-NH_2 + $H_2O \rightarrow NH_3$ + R-OH + energy
Nitrification	$2NH_4^- + 3O_2 \rightarrow 2HNO_2^- + 2H_2O + 2H^+$
	$2HNO_2^- + O_2 \rightarrow 2NO_3^- + 2H^+$
Denitrification	$NO_3^- \rightarrow NO_2^- \rightarrow NO \rightarrow N_2O \rightarrow N_2$

These reactions are driven by microorganisms in soils which greatly depend on temperature, pH, water content, and the amount of organic matter present. A detailed discussion of these processes can be found in Mengel and Kirkby (1982). For further understanding of nitrogen and its role in soils, the reader is referred to Stevenson (1982b).

3.9   AGGREGATION

Sections 3.9 through 3.12 discuss aggregation, aggregrate formation, crusting, and cracking of soils. To understand each of these sections, it is important to understand the primary causes of separation, chemical bonding, hydration, swelling, slaking, flocculation, and dispersion of clay–water interactions that were discussed previously.

In chapter two, we discussed the three separate phases of a soil (solid, water, and air) and how these phases combine to form the soil matrix. Further discussion revealed that the individual particles combine to make up the soil structure (the arrangement of soil particles). Individual, primary particles of sand, silt, and clay combine by biological, chemical, and physical means to form secondary particles or aggregates. From chapter 2, the reader may recall that aggregates can be defined as a group of two or more primary particles that form a cohesive mass stronger than the surrounding soil mass. Natural aggregates that vary in their water stability are called *peds* (also known as macroaggregates), whereas the term *clod* refers to a soil mass broken into various shapes by artificial means such as earthmoving activities or tillage. Soil peds broken into pieces are termed *fragments*. *Concretion* refers to a coherent mass formed within the soil by chemical precipitation from percolating ground water.

As with soil profile formation, the formation of aggregates is directly affected by the parent material, climate, and presence of vegetation. Generally, the more vegetation, the greater the organic complexing and cementation there will be for aggregate formation, due to biological activity. Many soil properties, including water retention capacity, infiltration, and transport, are directly dependent on the amount of aggregation present. In figure 3.18,

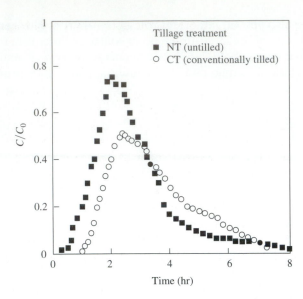

Figure 3.18   Example of a conservative tracer (potassium bromide) pulse (BTC) through aggregated, untilled (NT), and conventionally tilled (CT) Cecil series soils (Georgia, U.S.A.). C is measured and C_0 is initial concentration

chloride movement through a highly aggregated, untilled soil is compared to that through a conventional tilled, little-aggregated soil. Water movement is faster in the untilled soil, due primarily to its greater aggregate content.

The USDA has described and classified soil aggregates based upon their size, shape, and edge–face characteristics. They use three main categories for description: class or size of structure, type or shape of structure, and the grade of the structure and its durability. There are four (as illustrated in figure 2.4): **(1) Blocky**—vertical and horizontal cracking are equally developed, giving peds approximately equal axes. The tops, which may be flat or rounded, fix together upon swelling so as to leave no space or gaps between contiguous peds. **(2) Spheroidal**—these differ from the blocky type in that they are rounded, and when they are wet leave a fairly uniform distribution of pores between contiguous peds. **(3) Laminar**—natural cracking is generally horizontal. **(4) Prismatic**—vertical cracking is often exhibited, with the ped faces appearing relatively smooth and flat. These grades can be further divided into subgroups for more exact classification.

One of the most important criteria for soil classification is the grade of structure, which depends on the durability of aggregates and the proportion of aggregated and nonaggregated soil particles. There are four grades of structure: **(1) Structureless**—no observable aggregation; **(2) Weak**—weakly formed individual peds that are barely discernible in undisturbed soil; **(3) Moderate**—well-formed, distinct peds that are moderately durable but not distinct in undisturbed soil; and **(4) Strong**—durable peds that adhere weakly to each other and are readily evident in undisturbed soil.

As with the classification of types of structure, the various structural units within the soil can be classed into four separate size groups: **(1) Domains**—groups of clay particles ranging in size to about 5 μm, consisting of inner pore concentrations of approximately 20 nm and interdomain pores of about 100 nm; **(2) Granules**—groups of domains, fine sand, and silt particles cemented together, ranging in size to about 0.5 mm; the larger particles do not break apart on wetting (this group is also known as microaggregates or microcrumbs); **(3) Crumbs**—groups of granules ranging in size up to several millimeters; and **(4) Clods**—aggregates > 1 cm that, when broken apart, form crumbs.

The presence of aggregates within the soil is one of the factors affecting the nature of soils. Buildings, roads, bridges, other types of construction, tillage operations, plant growth, and soil mechanical properties are all influenced by soil structure. The growth of plants can be promoted by the presence of aggregates, which form natural cracks for root penetration to greater depths within the soil profile. Water also moves to greater depths, due to the presence of these cracks. Chemicals in this water contact fewer adsorption sites and, thus, have a greater potential for contaminating groundwater. Soil compaction is difficult in the presence of large aggregates, complicating the erection of large structures. Soils with little structure and aggregation confine the movement of water, restrict plant growth, and make drainage difficult.

QUESTION 3.15

Figure 3.18 shows a comparison of breakthrough curves (BTC) for a conventional tilled agricultural soil versus an untilled soil. Much of the literature on solute transport also deals with BTCs. What is the importance of a BTC; that is, why are they useful?

3.10   AGGREGATE FORMATION AND CHARACTERIZATION

Aggregation begins with the binding together of colloidal particles in the form of loose and irregular floccules, due to the presence of the electrical double layer. The collision of particles results in a mutual attraction when the electrokinetic potential is sufficiently low; consequently, a floccule is formed. Through chemical and biological activity, these floccules combine together to form aggregates.

Aggregation is primarily influenced by the amount of clay present. The flocculation of the clay, plus cementation by chemical and especially biological activity within the soil, is what forms aggregates. Clay flocculation does not form a stable aggregate because the clay is usually not strong enough to bind aggregates together into stable masses. However, the addition of cementation by root exudates and biological activity can form very stable aggregates. These cementing agents are usually sesquioxides, organic matter, humus (by adsorption of cations), silicate clays, lime, and polysaccharide gums (Wild 1988).

The expansion and shrinkage of soil upon wetting and drying, freezing and thawing, faunal activity, water movement, and root activity are the primary factors involved in aggregation. These factors vary greatly, dependent on climate and geographical location as discussed in chapter 2. Three primary mechanisms are responsible for aggregate formation: biological, chemical, and physical. Figure 3.15 represents the suggested arrangement of the various particles which compose soil aggregates. Quartz crystals are bound by several types of associations: (A) quartz–organic colloid–quartz, (B) quartz–organic colloid–clay domain, (C) clay domain–organic colloid–clay domain, (C_1) face–face, (C_2) edge–face, (C_3) edge–edge, and (D) clay domain edge–clay domain face (Emerson 1959).

Biological

The biological mechanism of aggregate formation is composed of both plant and animal activity, usually in the form of microbial decomposition and degradation of organic matter, and the decay of dead fauna. As organic matter is added to the soil, the microbial population flourishes, producing intense activity which helps decompose the organic matter. The primary microbial participants are actinomycetes, bacteria, fungi, algae, and protozoa. As the organic matter is decomposed, it is synthesized into various forms of organic material. Both actinomycetes and fungi produce branching and individual filaments termed conidia and mycelia, respectively, which help bind the synthesized organic material together.

While the actinomycetes and fungi are primarily responsible for mechanical binding of soil particles, bacteria, roots, and fauna such as earthworms create polysaccharide gums, casts, and other cementing agents that further bind the organic materials. These functions are dependent upon favorable climate and upon the species of organism present, itself a function of an available energy source and soil nutrients (such as nitrogen) that aid in the decomposition process. A comparison of temperate and tropical region soils to arid zone soils indicates that soils high in organic matter form more water-stable aggregates than soils in arid zones, which are low in organic matter and are sparsely aggregated.

The effects of plant and root growth and distribution affect aggregation. The plant canopy shields the soil surface directly beneath it from mechanical energies originating from raindrop impact, preventing aggregate breakdown and dispersion (physical). Plant root systems, especially those of annuals, continually rejuvenate themselves during early and mid-growth stages. As a result, old roots die, while new roots form to obtain water and nutrients for the plant. This has a dual effect. As the old roots die, they decompose and become food for the microbial population of the soil, which synthesizes them into organic complexes (biological). The new roots excrete polysaccharide and polyuronide gums from epidermal and cortical cells, which serve as additional binding agents for the surrounding soil (chemical). This assists in the formation of microaggregates which can later form macroaggregates. Additionally, roots create channels and large pores as they penetrate the soil, allowing more water and air movement. This increases water and nutrient supply, and biological activity, to greater depths in the soil profile (physical).

Large areas covered in grass such as pastures, sod farms, and golf courses have extensive root system development and usually promote the formation of very stable aggregates, due to the physical action of the roots on the soil and chemical and biological effects. As a result, root exudates and microbial decomposition are responsible for binding together small aggregates into large stable aggregates. In special agronomic circumstances such as rice production, aggregates are destroyed in the process of puddling the soil to prevent or greatly retard water loss due to infiltration. However, the oxidizing ability of roots on iron and other reduced matter, along with root exudates, regenerate aggregates.

Chemical

The formation of aggregates through chemical means is achieved by ionic and hydrogen bonding between carbonyl groups and oxygen atoms within the layered clays. The ionic bonding is accomplished through cation bridging, which bonds the organic complexes to the clay surface. Additional bonding mechanisms are van der Waal's forces and sesquioxide–humus complexes. It is commonly known that calcium tends to flocculate a soil, whereas oversaturation with sodium will disperse the soil and break apart aggregates. Calcium and other divalent cations (such as magnesium) will bind the granules together by acting as a cation bridge between the clay and organic colloids as follows: Clay-Ca^{2+}-OOC-R-COO-Ca^{2+}-Clay. Aggregate formation in this instance depends on the interaction between the exchangeable cations held by the clay particles and soil pore water. Because of the presence of the electrical double layer, the negatively charged clays are surrounded by both fixed and mobile cations, causing water dipole orientation along the lines of force in each ion's electrical force field and the force field of the negatively charged clay. This causes the formation of cation–dipole linkages between separate clay particles. As dehydration occurs around the particles, the ions share water envelopes that form a calcium or magnesium linkage between clay particles, resulting in the formation of stable aggregates. Similar circumstances can occur with polyvalent cations.

Other organic substances, such as humic acids will not attach directly to clay surfaces because of their negatively charged surfaces, they form complexes with soluble sesquioxides that have been deposited on the surfaces of clays (Dawson et al. 1981). Humic acids can also be coordinated with Al^{3+}, Ca^{2+}, Fe^{3+}, and Mg^{2+}, which act as cation bridges between the clay surface and humic acid. Soil aggregates are divided into two groups, based upon their association with divalent or trivalent ions. When associated with divalent ions, the clay–organic complexes are saturated with calcium and magnesium, forming microaggregates. When associated with trivalent ions, soluble sesquioxides are deposited on clays forming complexes with humus, also forming microaggregates. It is the latter group that forms stable soil aggregates.

Additional chemical factors that influence soil aggregation are hydrogen bonding and cation exchange. As mentioned at the beginning of this section, hydrogen bonding can occur between carbonyl groups (on organic molecules or complexes) and the oxygen atoms of the clay crystal. When the charge on organic matter is not satisfied, ionic bonding with hydrogen is very strong, much the same as when polysaccharides with carboxyl groups attached are adsorbed to the clay surface. Due to the nature of charge in soils, the system favors adsorption with complexes and materials not conveying a charge. This could be the effect of van der Waal's forces, which become stronger as dehydration of the soil occurs. The effects of cation exchange can either flocculate or disperse a system, a good example being calcium and sodium. Calcium is known to flocculate soils as (illustrated by the application of gypsum to agricultural fields). In contrast, sodium-saturated soils are more hydrated and dispersed than calcium-saturated soils. This is not true for acid soils where hydrogen is the dominant cation, although other cations may be present in abundant number (such as aluminum and iron); particularly if the soil is of the sesquioxide type. In acid soils, the adsorption of calcium is associated with the organic colloid fraction, because H^+ saturated soils are usually not well flocculated.

Physical

Various aspects of the physical phenomena which cause aggregation have been discussed. These include freezing and thawing, shrinking and swelling, and the effects associated with colloidal clays. Reflecting on the discussion in chapter two about soil formation, we remember that wetting, drying, swelling, and shrinking, which are somewhat interdependent, occur in alternate cycles. As the soil wets and subsequently dries, the soil mass is separated or broken into smaller fragments due to shrinking. Depending on the soil type (whether it be a 1:1 or 2:1 clay), the water content of the soil can have significant effects on aggregate formation. This causes the shape and formation of various types of structural peds. These properties are responsible for slaking (the destruction of aggregates upon wetting), mudslides, and other soil events.

The loss of water from a soil as it dries allows air to enter pores, increasing the total volume of air present while decreasing the amount of water. This decrease of soil volume upon drying is due to both the loss of water and the cementation of soil particles because of a collapse of pore size. The solid soil mass does not change. This means that soil shrinkage is less than the volume of water lost due to evapotranspiration and plant use. However, as the soil rewets, the volume will be greater, because air is occluded in various pores, and because of swelling.

Slaking occurs when a dry soil is rewetted. Rewetting of a dry soil has two effects: occluded air within the soil pores is compressed, which causes minute explosions within the pores, disrupting the clod (this is the major cause of slaking); and absorption of water causes unequal swelling within an aggregate, resulting in fractures along the cleavage planes. The rewetting of a dry soil also involves two forces, namely matric and cohesive. As the soil first begins to wet, matric forces exceed cohesive forces but, as the soil becomes more saturated, destruction of the cementing bonds between particles occurs, causing a loss of cohesion and breakdown of the aggregate.

An example of some of these processes may be found in the instance of mudslides. Mudslides usually occur when a soil has an overlying layer of organic material or loose alluvium from deposition due to erosion, landslides, or winds. The slide will usually occur under a heavy rainfall event. The overlying alluvial soil layer becomes rapidly saturated as the wetting front moves vertically through the soil. This drives soil air ahead of the wetting front, entrapping it between the upper alluvial layer and the second layer of soil. (Air can only be trapped to a pressure equal to the air entry pressure of the soil. At greater pressures, the air will quickly escape, being more than fifty times more mobile than water). Additionally, compression of occluded air causes slaking of aggregates that may be present in the alluvial layer, with an overall loss of slope stability that is also promoted by the loss of cohesive forces as the soil saturates. The upper layer becomes very heavy due to the additional weight of water absorbed and begins to slide: the development of the entrapped air layer maintains dry conditions at the interface of the second soil layer and, of a sudden, the entire slope moves. In mudslides, slaking would be of minor importance; the major factors are entrapped air and loss of cohesive forces. Thus, the breakdown of soil aggregates and instability of soils can be attributed to reduction in cohesive forces, destruction of cementing bonds, entrapped air, and the stress–strain relation in alternate wetting and drying cycles. Erosion is also a major factor, causing slope instability and aggregate breakdown (Davies and Payne 1988).

Freezing and thawing of soils can both form stable aggregates and destroy soil structure. When a soil cools slowly, a smaller number of large ice crystals are formed. These crystals partially melt during the thawing cycle; the remaining crystals serve as a foundation for further freezing. As these crystals form, water is drawn to them during the freezing process, which dehydrates the surrounding soil. The crystals expand in the process, compressing the dry soil in the immediate vicinity. The combination of both these parameters causes aggregation. However, if the soil is frozen rapidly, the formation of a large number of crystals can cause aggregate destruction. When performing laboratory analysis on intact soil columns, care should be taken not to freeze them, as this will cause a breakdown of soil structure not visible to the naked eye and also develop artificial gradients within the soil by dehydrating the soil around larger ice crystals that may form. This consideration can be critical in contaminant transport studies, possibly biasing results.

QUESTION 3.16

We have discussed how aggregates can be broken or dispersed and pores compressed or destroyed by earthmoving and agricultural activities. What environmental forces can cause the same results?

3.11   SOIL CRUSTING

Under conditions of heavy rainfall and irrigation, crust formation on soil surfaces is due to soil aggregates breaking apart because of the kinetic energy of raindrops that may cause compaction in the upper surface layer, and clay and its cementing agents possibly being dispersed from quartz crystals that clog the overlying surface pores due to physiochemical dispersive effects. The larger the raindrop, the more kinetic energy it possesses and the greater the aggregate breakdown and dispersion; thus, larger drops will destroy the original structure to a greater depth (Miller and Radcliffe 1992). Crusts under these conditions may be 5 mm thick. As the aggregates break apart, the finer clay particles slowly wash into the soil pores, reducing infiltration and causing erosion on sloping soils. Generally, the greater the clay content and the heavier the rainfall, the thicker will be the soil crust that forms. As the soil dries, the top layer, which is now composed of very fine clay and silt particles, shrinks. Depending on the major ions present in the soil, the edges of the crust will crack and sometimes curl. If the soil is inhomogeneous parallel to the surface, curling or peeling will occur; inhomogeneity normal to the surface will cause cracking (both occur during the drying stage). In arid regions, the predominant ion is sodium, which causes dispersion and tends to form thin crusts that curl upward at the edges. In more temperate regions (such as the southeastern United States), where aluminum and iron sesquioxides abound, crusts tend to be very hard and thick with little edge curling.

Close examination of a soil crust will reveal what soil scientists call depositional crusting, where an actual structure has developed during crust formation. Normally, a stratification is present in which the larger particles are deposited first, followed by successively finer particles, with the very finest particles deposited at the top of the crust. These form an effective seal or barrier. The crust reduces infiltration while also reducing evaporation and, as mentioned earlier, can be a significant factor in causing erosion. The soil crust is marked by a high bulk density, due to the presence of quartz crystals in the underlying layer of the crust. As the crust dries, the cementing agents present at the initial dispersion of the aggregate enhance cementation of the particles, causing a very hard crust. Most soil crusts are thin (< 2 mm). Typically, a soil crust may be characterized as possessing stratification, high bulk density (compared to surrounding soil), and a low non-capillary pore space. Once dry, the crust is harder than the remaining soil mass, due to particle cementation. Environmentally, soil crusting encourages run-off and erosion which, when contaminants (such as agricultural herbicides, fertilizers, and industrial and municipal wastes) are present, can pose hazardous conditions for both surface and groundwater pollution. Factors commonly associated with soil crusting are high ESP, high silt content, and low organic matter content. Soils that have high organic matter content, and those that are mulched by straw or artificial barriers (such as plastic mesh), have a low susceptibility to crusting.

Soil texture, steepness of slope, and aggregate stability are three of the major physical factors that greatly affect crusting potential. A crust can form on almost all soil textures; textures most likely to form crusts are those with high silt contents. Also, crust formation is more likely to occur on sandy loams than on clay loams. Textures that are unlikely to form crusts are those with low clay, silt, and coarse sand contents. The steepness of slope is usually inversely proportional to surface sealing or crusting, due to several factors. As slope increases: there is usually a decrease in the number of raindrops per unit surface area and, thus, a subsequent decrease in total kinetic energy; interrill erosion also increases, which erodes a large portion of formed crusts; and rill density and depth increase, which increases surface roughness, making formation of crusts less likely. If the slope is resistant to crusting due to textural properties, these factors would not affect sealing or infiltration. Consequently, the general characteristic of slope steepness on crusting is that as slope increases, soil strength decreases, which results in an increase in infiltration. Generally, as aggregate stability decreases, the potential for crust formation increases. Also, if aggregates are large, a longer period of time is required for crust formation than with small aggregates. Additionally, the water content of the aggregate at the time a rainfall event occurs can dramatically affect crusting potential. Normally, initially dry aggregates deteriorate from rainfall effects due to slaking. Such aggregate breakdown causes a rapid filling of interaggregate spaces with smaller aggregates, which break down very rapidly to form a crust after drying. However, if the aggregates have a high water content during rainfall, the slaking process is unlikely to affect their breakdown, and the intensity, duration, and related energy of raindrop impact will be the primary factor affecting aggregate deterioration. Under these conditions, there is a low potential for crust formation. Also, the rougher the soil surface, the less likely a crust will form, despite silt content, unless the aggregates and soil surface are very dry and the initial rainfall event is very intense and of a duration of longer than about forty minutes.

There are two major categories of crusting, structural and sedimentary (Sumner and Stewart 1992). *Structural crusts* are generally formed in response to the direct physical reactions and processes associated with raindrop impact. A soil with a high clay content (20–30%) and the presence of stable aggregates can have several layers within a crust. Each layer is formed by rainfall-induced aggregate breakdown, soil particle or aggregate rearrangement, and the coalescence of aggregates. The entire crust may be relatively thick (1 to > 10 mm). A thin layer or microlayer may be formed at the surface due to aggregate breakdown and deposition from suspension at the end of a rainfall event. This is termed a skin seal, which is more typical in readily dispersible soils and soils high in sodium content. Another microlayer may be what is termed a washed-out layer, which is formed from disjunction of clay particles from quartz crystals. Essentially, the clay is mechanically removed from the soil particle by rainfall, and then the surface of the soil is sealed. This type of action is readily seen on piedmont soils in the southeastern United States. After a heavy rainfall event, individual quartz crystals from which the clay has been removed are readily visible. Occasionally, the micromass removed from the surface particles may form a micromass-enriched layer by translocation of the mass a short distance below the surface. This type of crust is termed a washed-in layer. Also, aggregate coalescence can form a disruptional layer. All of these microlayer crusts can be formed during a rainfall event within one soil. However, it is unlikely that all of these microlayers would be found in a single structural crust.

The skin seal is primarily a dense layer (0.002 to < 0.1 mm) of fine particles at the surface of a structural crust. The most likely mechanism for formation of a skin seal is the deposition of clay particles from suspension at the end of the rainfall event. The basic evidence proposed for this is the occurrence of the skin seal over washed-out layers, and the uniform orientation of the clay particles within the skin seal. Also, it is likely that continual raindrop impact on such a thin layer would destroy it; thus, it is indeed likely that these seals do not form until the end of a rainfall event. Washed-out layers (< 1 mm) occur at the surface or lie under a skin seal and appear as skeletal grains with no associated clay or micromass. If a washed-in layer is present, it will be subjacent to the washed-out layer and generally < 1 mm thick. If runoff is a problem on particular soils, the clay materials stripped from the skeletal grains may move laterally with the runoff and, thus, cause an absence of a washed-in layer. Typically, increases in ESP values from low to high (2–10) can substantially contribute to the formation of washed-out and washed-in layers. Electrolyte concentration and composition in rainwater and clay mineralogy will affect soil dispersibility. The disruptional layer is a function of initial aggregate size. If the media surface is initially dry, the rapid wetting of the aggregates will cause slaking, resulting in swift aggregate breakdown. Also, as aggregates become wetter, aggregate strength is reduced to a point at which the stress imposed by raindrops disrupts the aggregate. At this point, the disruptional layer forms rapidly, releasing particles from individual aggregates into interstitial spaces between aggregates, creating a smooth layer with reduced porosity. As much as a 90% reduction in porosity may result due to structural crust formation.

Sedimentary crusts generally result from the transport of particles from relatively higher to lower topographic locations or positions. This transport is normally induced by runoff, and may cover long distances. Sedimentary crusts are sometimes referred to as depositional crusts. As one would suspect, when the flow velocity slows, larger particles are deposited first, followed by finer particles. This is normally the result of decreasing slope and/or reduced rainfall volume. In fact, for sedimentary crusts to form, the rainfall rate must be greater than the infiltration rate, and the capacity of the runoff to transport must be less than that required to completely remove the individual particles from the site. Sedimentary crusts range in thickness from 0.6–20 mm, and may reduce porosity by as much as 91 percent. A common feature of such crusts are circular voids (vesicles) found directly below the soil surface, which appear to be due to air entrapment because of ponded water during intense rainfall events (West, Chiang, and Norton 1992).

There are a number of cementing agents that are related to aggregation of soil and its structural stability. These include silica, organic materials, iron and aluminum sesquioxides, cations, and other cementing agents. The effectiveness of a cementing agent depends upon its solubility, and whether the solid phase is in equilibrium with the dissolved phase (or dries irreversibly, or rehydrates and redissolves on rewetting). Additional factors include, whether the dissolved phase ionizes, remains as an undissociated molecule, or precipitates as discrete particles or surface coating (such as sesquioxides); and whether it is crystalline or noncrystalline, or can acquire a charge.

The most common cementing agents in arid and semiarid regions are silica, gypsum, and lime. In the humid tropics, noncrystalline hydrated aluminosilicates and sesquioxides are the most likely cementing agents. For cementing action to occur, one soil particle must be brought into close proximity with another. This will depend on the number of particles per unit volume, which depends on soil texture, particle shape, and packing (as discussed in chapter 2). Both the soil water and solutes will determine the spatial arrangement of the soil particles within a crust. Consequently, a nearly saturated or saturated condition is necessary at the surface for a crust to form. Soil water tension increases as the soil dries, bringing soil particles closer together, and soil pore sizes decrease due to clogging. The combined effects of these basic processes stimulate bond formation, thus inducing crust formation. The reverse is also true: as water is added to most crusts, they lose their strength as a result of swelling and softening of cementing agents. Soil inhomogeneities attributed to uneven packing and flocculation influence the stress distribution in a drying soil, which also determines crust characteristics. In lateritic soils, extreme desiccation causes irreversible hardening and crusting. Additionally, although organic matter usually aids in preventing crust formation, it is the overall mixture of organic matter prior to drying that has the greatest influence on soil surface conditions after drying.

One other crust type that deserves mention is prevalent in the western United States and other arid regions throughout the world. This is a chemical crust that is essentially a salt incrustation or deposit on the surface, caused by evaporation. The most common forms are sodium chloride, carbonate, and sodium sulfate chemical crusts. Other minerals, such as bloedite, gypsum, epsomite, and hexahydrite, can also be found in these chemical or evaporative crusts—depending on geographic region and soil type (Timpson et al. 1986).

There are two common methods for measuring crust strength: modulus of rupture, and balloon pressure. The modulus of rupture measures the breaking strength of a crust, and is the maximum stress that a crust will withstand without breaking. Generally, a soil sample is subjected to a bending force. The specific surface area, mechanical composition, and ESP appear to be linearly related to the force required to rupture a specific soil sample. However, there is also an inverse relation between the modulus of rupture and particle size distribution. The modulus of rupture is given by

$$\sigma_r = \frac{3F_bL}{2bd^2}$$

$$(3.84)$$

where σ_r = modulus of rupture (dynes cm^{-2}), F_b = breaking force applied at the center of the soil sample beam span (dynes), L = distance between the sample and support (cm), b = width of sample (cm), and d = sample thickness (cm). It has been established that the modulus of rupture is strongly correlated with hydrous mica content of soil.

Crust strength is also measured by balloon pressure. A balloon is buried in the soil, and the pressure required to inflate the balloon to the point where the crust begins to crack is determined. Cone penetrometers are also used to measure crust resistance or strength. It should be recognized, however, that the various methods of crust strength measurement were originally designed for agricultural purposes. The primary purpose was to imitate the process by which seedlings force themselves upwards through a crust, to determine the minimum strength that would inhibit germination. Environmentally, strength tests are valid for determining possible rates and time of runoff due to rate of crust formation for given soils.

Generally, crust strength is higher in natural soils that have a high silt and fine sand content. These types of soils are structurally unstable when wet. In arid areas where sodic soils predominate, osmotic swelling forces attract water into the diffuse electrical double layer. This causes dispersion, which separates individual clay particles by breaking particle-to-particle bonds. As a result, sodic clays are drawn into a closely oriented, laminated configuration by surface tension during drying. As the surface area increases for certain clay minerals, more stress occurs on the crust during drying; consequently, more fracture planes develop. As a rule of thumb, crust strength will increase by about a factor of three as water potential at the surface decreases from -33 kPa to -1500 kPa. Normally, the process of wetting removes soil air; when this happens, crust strength can significantly increase. Thus, flooding can significantly increase crust strength and cementation properties.

There are environmental hazards associated with crust formation. As crusts form, the hydraulic conductivity of the soil is greatly reduced because of the impermeable crust now overlying it. This will cause increased runoff and possible herbicide contamination of nearby surface waters in industrial and agricultural areas. Erosion also becomes a severe problem in some soils, which can result in the deterioration of surface water quality due to silt and clay content. This can decrease oxygen supplies in the water, causing high death rates of aquatic species. In an environmental setting, the soil acts as a natural filter for chemicals which adsorb, exchange, degrade, and so on, as they pass through the soil matrix. The formation of a crust usually impedes the penetration of these chemicals through the soil and they are subsequently transported by runoff processes.

3.12 SOIL CRACKING

Various soil types are prone to cracking upon loss of water. Vertisols and alfisols both exhibit a high degree of shrinking and swelling; consequently, cracks develop that can expose additional surface area for water loss due to evaporation, accelerating the drying process. These two soil types are characteristically dark, having uniform fine or very fine texture and a low content of organic matter; but perhaps their most important property is the dominance in the clay fraction of expanding lattice clay, usually montmorillonite, that causes large volume changes with small changes in water content. The area where the soil cracks is termed a cleavage plane. Typically, as shrinkage progresses, cleavage planes develop at the points of least resistance, which correspond to the plane of highest water content. Cleavage planes develop both vertically and horizontally, producing irregularly shaped cracks that may be > 1 cm wide and > 50 cm deep (Sumner and Stewart 1992). Such cracks can facilitate the penetration of water and associated contaminants during periods of heavy recharge, inducing a high risk of contamination of groundwater.

The development of cracks can drastically change the hydrology of a given soil and, due to the presence of high clay content ($> 30\%$), the water content in the soil may vary from complete saturation to less than the wilting point. With the onset of the dry season, the soil dries, shrinks, and cracks to form large prisms. Further drying takes place from the crack surface; if there are soluble salts or inorganic contaminants present, they may form an efflorescence on the surface of each prism. With the first significant recharge event, water flows rapidly down the cracks along the surface of the prism, dissolving salts and flushing other chemicals out of the soil, leaching these chemicals deeper into the profile. Frequently, the water table may be within 1.5 m of the surface, which can cause quick contamination; often, the entire soil profile becomes saturated with water until surface ponding occurs.

Normally, the crack pattern developed on the surface of a uniform, drying soil will appear as a group of polygons, essentially similar in both size and shape. This is especially true when the pattern is constricted by the use of devices such as ring infiltrometers or barrels filled with soil for testing. However, under ideal conditions, the polygons will be hexagonal in shape, with the cracks meeting at a 120° angle. Generally, subsequent cracks occur at right angles to existing cracks. As each new crack is formed, the cleavage plane from which it departs allows the release of strained energy only on the side on which the new crack is located. Consequently, new cracks will not normally cross existing cracks. These characteristics are true only for bare soils. On soils that have cultivated crops, cracks will tend to form parallel to the plant rows—both between and around the plants, and within the furrows. In the presence of sporadically placed trees, plants, and grasses, irregular polygonal patterns also tend to form. Major crack formation will tend to form not directly around the plant or its roots but away from the plant in bare soil. This is probably the result of the physical and biological effects of root systems responsible for stable granulation. It is well known that roots secrete organic cementing substances for aggregate formation; this would probably reduce the tendency for crack formation directly around the root.

Cracking takes place when the release of energy per unit area of the crack is greater than the increase of surface energy. There are two separate energy terms, and the balance between them determines the amount of cracking that will occur. These terms are free surface energy due to new surface exposure, and energy released per unit free surface. Free surface energy due to new surface exposure is described mathematically as

$$\frac{d\mu}{dA} = 2\gamma_m \qquad (3.85)$$

where μ = energy of medium under consideration, A = area due to new surface exposure by cracking, and γ_m = surface tension or energy of the material. Energy released per unit free surface is described by

$$\frac{d\mu}{dA} = \frac{\tau\sigma_s^2 D}{E} \qquad (3.86)$$

where τ = tensile stress normal to the plane of the crack, σ_s = limiting stress (dyne cm^{-2}), D = major diameter of the crack (which is assumed to be elliptical), and E = Young's modulus of the material. A combination of equations 3.85 and 3.86 yields the well-known Griffith theory for crack propagation, as

$$\sigma_s = \sqrt{\frac{2E\gamma_m}{\pi D}} \qquad (3.87)$$

where the equation has been modified to make it applicable to plastic materials such as soils, which exhibit a measurable deformation before rupture. Deformation in porous media begins when the limiting stress (σ_s) is applied. When $\sigma = \sigma_s$, the critical strain-energy released rate (G), is given as the relation,

$$G = \frac{\pi \sigma_s^2 D}{E} \tag{3.88}$$

Both σ_s (determined from the modulus of rupture) and E depend on θ_{soil} and $^b\rho_w$. D can be estimated from pore size distribution or particle size analysis. Thus, G/D (dyne cm^{-2}) is the energy released per unit area per unit length of crack, and can be used to characterize soil cracking behavior. The value for Young's modulus (E) has normally been obtained from the measurement of pulse transmission velocity using ultrasonic assembly, in which E is expressed as

$$E = \rho_b v^2 \tag{3.89}$$

where ρ_b = wet bulk density (g cm^{-3}) and v = pulse transmission velocity. Assuming that both σ_s and E are known, the occurrence of cracking as a function of soil water content can be predicted by

$$\sigma(\theta) < \sqrt{\frac{\frac{G}{D}(\theta)E(\theta)}{\pi}} \tag{3.90}$$

where a value of $\sigma(\theta)$ that satisfies the inequality represents a condition in which no cracking can be expected, and where the notation (θ) represents the dependence of the quantity on water content (i.e., E(θ) is read, "E as a function of θ").

QUESTION 3.17

Give several examples of processes in which flocculation and dispersion of clay is important.

SUMMARY

Flocculation and dispersion play an important role in the physical and chemical behavior of soil colloidal fractions, particularly in the areas of colloidal facilitated transport, crusting, infiltration, cracking, and swelling, as related to unsaturated zone investigations.

In this chapter, we discussed van der Waals–London forces and how they can be thought of as short-range, electrostatic attractive forces. The interaction between uncharged soil particles and water, the surface charge of clay minerals, and ion distribution and exchange were also discussed, including how these parameters affect clay–water behavior. This led to a discussion of the double layer theory, including zeta potential and the four types of electrokinetic effects normally observed in association with the double layer (streaming potential, electroosmosis, eletrophoresis, and the Dorn effect). Following a discussion of electrokinetic properties of clays, we presented the reversible process of ion exchange, which is the ability of cations and anions to exchange between the solid and liquid surfaces in a system. This was followed by a discussion of the sodium adsorption ratio of soils, an important characteristic in arid areas such as the western United States, Egypt, Israel, Nigeria, Iran, and Iraq. We also discussed the evolution of heat from a dry particle when immersed in water, termed the heat of wetting, and how it is induced by the electric field surrounding clay particles, because of a decrease of internal energy of the water molecules within the field.

The DLVO theory, hydration and swelling, and flocculation and dispersion of clays were presented in a quantitative and practical-application fashion for a clearer understanding of the forces of attraction and repulsion associated with clay–water systems. Subsequent sections discussed the importance of humus in soils, followed by aggregate formation and the physical, biological, and chemical principles responsible for that formation. The chapter concluded with discussions of crusting and cracking, which relate directly to flocculation and dispersion in soils. Throughout, it has been apparent that both dispersion and flocculation are governed by the attractive and repulsive forces associated with the electrical double layer, and by reactions between adjoining double layers as presented in the DLVO theory. The double layer is essentially composed of a charged clay-particle surface, typically termed a colloid, plus the diffuse cloud of counterions that serve to neutralize that colloidal surface.

ANSWERS TO QUESTIONS

3.1. Assuming a dielectric constant, D, of 80 and a permittivity of 78.8 esu^2/erg · cm, equation 3.12 may be used to determine the heat of hydration of this ion, such that $W = -\,[[(1)(4.803 \times 10^{-10}$ esu/ion]^2/[(2)[(1.33 + 0.7) \times 10^{-8}$ cm/ion]] * [(6.02 \times 10^{23}$ ion/mol)] * [(80 - 1)/(78.8 esu^2/erg · cm)] = -3.43×10^{12} erg/mol. Note: $e = 4.803 \times 10^{-10}$ may be used here for other unit systems, because it readily converts to erg/mol by multiplying by 2.389×10^{-8} cal/erg to obtain kcal/mol. Also, regarding the negative sign, the electric force of the charged sphere on the unit positive charge (i.e., the field) acts in a direction opposite to that in which the charge is being moved. Because both the field and the direction of transport are vectors (quantities with both direction and magnitude), and because the vectors point in opposite directions, their product is negative. For example, the product of two vectors A and B is $AB \cos \theta$, where θ is the angle between the two vectors. If the vectors are in opposite directions, $\theta = \pi$ and $\cos \theta = -1$ and the product is $-AB$. Additionally, Voet's equation (3.12) is considered more appropriate for ions belonging to alkaline metals, alkaline earth metals, and trivalent closed-shell ions. Voet's equation can be converted to Born's equation by replacing $(r + 0.7)$ with r in the denominator of the left-hand side for use with other ions.

3.2. Because of the units of e, $k = 1.381 \times 10^{-16}$ erg ion^{-1} K^{-1} = 4.139×10^{-11} mV esu ion^{-1} K. Thus, $C = C_0 \exp - [(1)(4.803 \times 10^{-10}$ esu ion^{-1})$(-100$ mV)/4.139×10^{-11} mV esu ion^{-1} K^{-1}) × (298 K)] = -3.8925. Hence, $C = 2.5 \times 10^{-2}$ M exp(3.8925) = 1.22 M. If units for e are coulombs, then $e = 1.6021 \times 10^{-19}$ coulombs, and at 25 °C, $kT/e = 25.69$ mV; thus, $e/kT = 38.925$ V^{-1}. The answer is the same; however, the student may wish to check the problem using the other units (remember to convert ψ to potential in these units, rather than mV as in the preceding answer).

3.3. First, n must be converted. This can be accomplished by multiplying by Avogadro's number (N_A). Thus, 3.1×10^{-2} mole L$^{-1}(N_A)$ = 1.867×10^{19} ions cm^{-3}. Hence, $\sigma = [(2)(78.3)(1.867 \times 10^{19}$ ions cm^{-3})(1.381 \times 10^{-16}$ esu^2 ion^{-1} cm^{-1} K^{-1})/(3.1417)] sinh[(1)(4.803 \times 10^{-10}$ esu^2 ion^{-1}) (75 mV)/(2)(4.139 \times 10^{-11}$ esu mV ion^{-1} K^{-1})(298 K)]. Thus $\sigma = (6.19 \times 10^3$ esu cm^{-2}) × sinh(1.4603) = 1.261×10^4 esu cm^{-2} or 2.626×10^{16} meq cm^{-2}. This problem can also be calculated by assuming that $kT/e = 25.69$ mV at 25 °C, so that $e/kT = 38.925$ V^{-1} and $k = 1.381 \times 10^{-16}$ erg K^{-1} ion^{-1}, which is converted to 1.381×10^{-16} esu^2 (cm ion K)$^{-1}$. The answer will be the same. Again, check the problem both ways to become familiar with the units used by different societies and scientists.

3.4. In the text, we stated that equation 3.17 is greatly simplified, and indicates a linear increase in potential from a charged surface when it should reflect an exponential decrease. As a result, the solution to this equation will greatly overestimate the actual potential. We shall prove that now. Assuming a parallel-plate capacitor, calculate the zeta potential 23 Å from particle surface. Using an meq density of 1.5×10^{-7} meq/cm^2, multiplying by $0.001 N_A$ and by e $(4.8 \times 10^{-10}$ esu), the charge density is about 43,200 esu/cm^2. Using this value in equation 3.17, $\zeta_{mont} = [(4\pi)(43,200$ esu · cm^{-2})(23 \times 10^{-8}$ cm)/(80 esu^2 erg^{-1} cm^{-1})] = 0.0016 erg/esu, where the subscript "mont" represent montmorillonite. Since one esu is 300 V practical, the potential is 0.468 V or 468 mV. This is a very high value and makes little sense. Now, we solve the problem in terms of ψ_0, which may be determined using the inverse sinh relation (as applied in equation 3.15) if a bulk fluid ion concentration is assumed. Assuming a 0.001 NaCl solution, ψ_0 is about 250 mV, confirming the results of van Olphen (1963, p. 255); this is substantially less than the streaming potential computed using equation 3.17, and there is no way that ζ can exceed ψ_0. Generally, the distance from the particle surface to the slipping plane is not known, although one gets the impression from van Olphen (1963) that it might be approximately (although clearly not *exactly*) $1/\kappa$. Perhaps, noting that (based on the Guoy theory) the potential at that distance is about $1/e$, or about 0.37 of that at the particle surface, it might make sense to say that, in the absence of data on electrophoresis or electroosmosis, ζ is very approximately $0.4\psi_0$, or for the above value, about 100 mV.

3.5. Using equation 3.10, and rounding for calculation, $\kappa^{-1} = [(80)(1.381 \times 10^{-16} \text{ esu}^2 \text{ ion}^{-1} \text{ cm}^{-1} \text{ K}^{-1})$ $(298 \text{ K})/(8\pi)(1^2)(4.803 \times 10^{-10} \text{ esu ion}^{-1})^2(6.022 \times 10^{18} \text{ ions cm}^{-3})]^{1/2} = 3.0702 \times 10^{-7}$ cm, or 30.702 Å.

3.6. By applying electroosmosis with a voltage drop of 100 volts across the column, we generate a field of strength of 1 V/cm, or 3.33×10^{-3} erg/esu cm. Using equation 3.18, we have

$$V = \frac{\left(\dfrac{6.92 \times 10^{-10} \text{ C}^2}{\text{J} \cdot \text{m}}\right)\left(\dfrac{0.06 \text{ J}}{\text{C}}\right)\left(\dfrac{100 \text{ J}}{\text{C} \cdot \text{m}}\right)}{\left(\dfrac{4\pi 0.001 \text{ J} \cdot \text{s}}{\text{m}^3}\right)} = 3.32 \times 10^{-7} \text{ m/s}$$

However, in practice, the volume rate of flow (electroosmotic) is measured instead of the linear velocity. Assuming the cross sectional area of a capillary is A, the volume rate of flow, $V_E = AV$. Using Ohm's law, $E = i/A\lambda'$ where I is the electric current in amps or C/s, and λ' is the specific conductance. Thus, $V_E/I = \varepsilon\zeta/4\pi\eta\lambda'$ (van Olphen 1963). Using an $\varepsilon = 78.5$ and substituting ζ and λ' we have $[(6.95 \times 10^{-10} \text{ C}^2/\text{J})(0.06 \text{ J/C})]/[(4\pi\ 0.001 \text{ J s/m}^3)(0.1 \text{ C}^2/\text{J m s})]$, which yields 3.33×10^{-8} m^3/C or 0.033 cm^3/C. Multiplying by I to get a water flow rate would require that I be in the range of a few to several units to be realistic.

3.7. Using equation 3.31, we obtain $P_d = \{(2)(6.02 \times 10^{18} \text{ ions cm}^{-3})(1.381 \times 10^{-22} \text{ atm cm}^3 \text{ K}^{-1} \cdot$ ion^{-1})$(298 \text{ K})\} \cosh[\{(4.801 \times 10^{-10} \text{ esu ion}^{-1})(3.336 \times 10^{-4} \text{ erg esu}^{-1})/(1.381 \times 10^{-16} \text{ erg ion}^{-1} \cdot$ K^{-1})$(298 \text{ K})\} - \{1\}] = 11.65$ atm, or 1.177×10^6 Pa. This solution has used the esu system; the other constants given in the question also readily convert to the same answer obtained here in Pa. The student should work the problem both ways for comparison.

3.8. First, κ (thickness of the double layer) must be determined, using equation 3.16. To begin, we will assume $r = 5 \times 10^{-8}$ cm, $z = 1$, $\varepsilon = 80$, $n = 5 \times 10^{-3}$ mol L^{-1}; thus, $n_0 = 3.011 \times 10^{18}$ ions cm^{-3} (this comes from multiplication by N_A), $k = 1.381 \times 10^{-16}$ esu^2 (ion cm K)$^{-1}$, $T = 25$ °C, $e = 4.803 \times 10^{-10}$ esu ion^{-1}, $\psi_0 = 25$ mV (8.339×10^{-5} erg$^{1/2}$ cm$^{-1/2}$), and $H = 240 \times 10^{-8}$ cm. Substituting the required values into the equation, the inverse thickness (κ) of the double layer is 2.303×10^6 cm^{-1}. With this value, we use equation 3.34 to calculate V_R. Hence, $V_R = \{(80)(5 \times 10^{-8} \text{ cm})(8.339 \times 10^{-5} \text{ erg}^{1/2} \text{ cm}^{-1/2})^2/2\} \ln[1 + \exp\{(-2.303 \times 10^6 \text{ cm}^{-1})(240 \times 10^{-8} \text{ cm})\} = 5.521 \times 10^{-17}$ erg. For V_R of the platelet-shaped particle, use equation 3.32 for calculation purposes. Using the same constants as in the preceding, one should obtain 146.2 erg cm^{-2}. However, for learning purposes, calculate the problem using the following constants: $\kappa^{-1} = 30.84$ Å, $d = 150$ Å, $n_0 = 6.02 \times 10^{18}$ ions cm^{-3}, $e = 1.6 \times 10^{-19}$ C, $\psi_0 = 75$ mV. Using these parameters, the answer in SI units is about 7.8×10^{-8} J m^2.

3.9. By letting $D = 2d$, the mathematical difference can be expressed as $[(1/d^3 + 1/(d + \delta)^3 - 2/\{d + \delta/2\}^3] - 1/2d^3$. If $\delta \gg d$, the above expression reduces to $1/d^3 - 1/2d^3 = 1/2d^3$. Therefore, equation 3.44 produces a result that is approximately twice that of equation 3.45. For example, let $A = 1$, $d = 4 \times 10^{-10}$, $\delta = 200$, and $D = 2d$; substitution into equation 3.44 yields 2.1×10^{26} and substitution into equation 3.45 yields 1.04×10^{26}. In the case of a spherical particle, we will let $H = d$, $a = \delta$, and $s = \{(H/a) + 2\}$; substituting these into equation 3.46 yields 4.166×10^{10}. The value for the spherical particle is substantially less than those obtained for platelet shapes (from chapter 2, we can clearly see that this is related to specific surface area). Additionally, forces decay much slower in the spherical case, as the value for "H" decreases. For large distances between spheres, the attractive force decays proportionally to $1/R^6$. This is similar to the case of the platelets, and is a feature of the long-range character of the van der Waals–London forces.

3.10. First, we will assume some parameter values for calculation purposes: $\sigma = 1.261 \times 10^4$ esu cm^{-2} or 3.78×10^6 V cm^{-1}, $T = 25\,°C$, $\eta = 3.1 \times 10^{-2}$ (thus, by multiplication of N_A, $\eta_0 = 1.866 \times 10^{19}$ ion cm^{-3}), $z = 1$, $k = 1.381 \times 10^{-16}$ erg (ion K)$^{-1}$ or 1.382×10^{-16} esu^2 (ion K cm)$^{-1}$, and $e = 4.803 \times 10^{-10}$ esu ion^{-1}. From this information, using equation (3.16), $\kappa = 5.8 \times 10^6$ cm^{-1}. Therefore, $\psi_0 = \{(3.78 \times 10^6 \text{ V cm}^{-1})(4\pi)\}/\{(78.3)(5.8 \times 10^6 \text{ cm}^{-1})\} = 104.6$ mV.

3.11. Using equation 3.68, $\mu_0 = -4.841 \times 10^{-4}$ atm g^{-1} cm^3 (atm per unit of water), or -490.6 erg g^{-1}.

3.12. The most laborious calculation for this problem is that of y_d from equation 3.76 (right-hand side for anions); thus, $y_d = \exp[\{(1)(4.803 \times 10^{-10} \text{ esu ion}^{-1})(4.336 \times 10^{-5} \text{ esu cm}^{-1})\}/\{(1.381 \times 10^{-16} \text{ esu}^2 \text{ ion}^{-1} \text{ K}^{-1} \text{ cm}^{-1} (298 \text{ K})\}] = 1.659$. Thus, using this value in equation 3.73, we obtain $\Delta P = [(2)(0.082 \text{ atm L mol}^{-1} \text{ K}^{-1})(298 \text{ K})(0.01 \text{ mol L}^{-1})][\cosh(1.659) - 1] = 0.841$ atm.

3.13. Use the equation $E_i = (2\Pi\sigma^2 d)/\varepsilon$. First, calculation of Π must be performed. $\Pi = MRT$; thus, given $M = 10^{-2}$ M, $R = 0.082$ atm L mol^{-1} K^{-1}, and $T = 25\,°C$, $\Pi = 0.758$. Thus, $E_i = (2) \times (0.758)(1.26 \times 10^4 \text{ esu cm}^2)(2.6 \times 10^{-8} \text{ cm})/78.3 = 7.99 \times 10^{-2}$ erg cm^{-2}. Consider how this value would compare to that from other equations.

3.14. Converting units and using equation 3.77, $F_c = 18.74$ mmol L^{-1}. Novich and Ring (1984) lists a value of $A = 2.2 \times 10^{-20}$ J for montmorillonite. Recalculate the problem based on their experimental value.

3.15. The primary usefulness of BTCs is that they can reveal a great deal about a given soil: degree of aggregation, presence of macropores, adsorption sites, anion exclusion, and so on. For more details on BTCs, see chapter 10.

3.16. Environmental causes are kinetic energy due to raindrop impact, freezing and thawing, and rapid wetting of dry materials. These effects are greatly exacerbated when the soil is wet.

3.17. Several examples are infiltration, erosivity, particle size distribution, and profile development. Using infiltration as an example, as long as a clay is flocculated, there will be significantly greater infiltration. If the media starts to disperse, the smaller clay particles will clog surface pores and reduce infiltration.

ADDITIONAL QUESTIONS

3.18. What is the primary binding force between adsorbed cations and a clay platelet?

3.19. What is the difference in electrostatic attraction by a clay platelet to Na$^+$ ions and Ca^{2+} ions?

3.20. How does the electrostatic attractive force change, qualitatively, with distance from a clay platelet?

3.21. What is the electrical status of a clay platelet dried out from a soil solution?

3.22. What will be the influence of salt concentration in the bulk solution on the extent of the double layer?

3.23. Suppose the cations in the double layer are Na$^+$ ions, and that they are then replaced by an ionic equivalent of Ca^{2+} ions. Describe, quantitatively, how the equilibrium distribution of ionic equivalents in the diffuse layer will change. Use a sketch.

3.24. **(a)** As a scientist, you would like to know how thick the water film might be around clay platelets that are air dry. Assume air-dry $w = 0.21$ g/g and $a_m = 132.3$ m^2/g. Ignore any water that remains after oven drying (105 °C/24 hr). Express your answer in Å.
  (b) How many water molecules thick is this layer.

3.25. Why is the osmotic pressure in the double layer higher for a clay containing sodium than for a clay containing calcium?

3.26. How is the stability of plate condensation influenced by the amount of salts in soil water?

3.27. Explain how plate condensation and cardhouse-type flocculation take place in clay suspensions. ("Cardhouse flocculation": Some charged surfaces such as the edges of clay and surfaces of sesquioxides can become positively charged under acid conditions, which results in coulombic interaction between particles. This leads to a type of edge-to-face flocculation at low electrolyte concentrations in which the attraction of particles resembles a house of cards. This is also referred to as mutual flocculation.)

3.28. A soil aggregate with a mass of 8.99 g, sealed in wax, is lowered into a container of water on a string attached to a torsion balance. At equilibrium, the balance reads 2.79 g. Assume the mass of wax and string are negligible.

   (a) Calculate the weight of the aggregate.

   (b) What is the dry bulk density of the aggregate?

3.29. (a) Calculate the CEC in meq/100g soil for the following.

montmorillonite	$(Si_8)^{IV}(Al_{3.5}Mg_{0.5})^{VI}O_{20}(OH)_4$
illite	$(Si_7Al_1)^{IV}(Al_4)^{VI}O_{20}(OH)_4K_{0.8}$
vermiculite	$(Si_6Al_2)^{IV}(Al_4)^{VI}O_{20}(OH)_4K_{0.5}$

   (b) What ions have substituted for what other ion, and did this substitution take place in the octahedral or tetrahedral layer?

3.30. By an examination of the capillary rise phenomena, the total energy (E) associated with both interfaces (air and water) is

$$E = \sigma f + \gamma i = \sigma s/\sin\phi + \gamma s/\tan\phi$$

where σ is surface tension (J m^{-2}), which is the interface potential energy divided by interface area; and γ is the potential energy of water molecules near the solid surface, divided by the area of the solid–water interface (J m^{-2}). The angle (ϕ) will adjust itself until E is at a minimum. Describe this mathematically.

4

Potential and Thermodynamics of Soil Water

INTRODUCTION

Water that has entered a soil but has not drained deep into the soil profile will be retained within pores or on the surface of individual soil particles. In the unsaturated zone, soil particles are typically surrounded by thin water films that are bound to solid surfaces within the medium by the molecular forces of adhesion and cohesion. Because of this, a simple measurement of water content is not sufficient to enumerate the complete status of water in a soil. An example of this may be observed in two different field or laboratory soils that have been treated in the same manner: though the same amount of water has been applied to each, they will have different water contents and varying abilities to contain water, due to individual physical and chemical properties. As a result, defining water content alone will not give any indication of chemical, osmotic, or other potentials involving water.

While the quantity of water present in soil is very important (affecting such processes as diffusion, gas exchange with the atmosphere, soil temperature, and so on), the potential or affinity with which water is retained within the soil matrix is perhaps more important. This potential may be defined as the amount of work done or potential energy stored, per unit volume, in moving that mass, m, from the reference state (typically chosen as pure free water). In this manner, one may think of matric potential as potential energy per unit volume, E (J m^{-3}). Energy is work with units N (Newton) per distance (N $\cdot$ m). Consequently, J m^{-3} = N $\cdot$ m m^{-3}, or N m^{-2}, which is also expressed as a Pascal (Pa). A Pascal is a force per unit area or pressure, which explains the use of the term "pressure potential," and is the reason why some unsaturated zone hydrologists refer to matric potential as "soil pressure."

This chapter will discuss the concepts of potential and thermodynamics (chemical potential) of soil water, beginning with the structure of water and the properties of adsorbed water and following with discussions of capillary water and various components of soil-water potential, including the thermodynamics of soil water. Hysteresis will also be addressed.

4.1 THE STRUCTURE OF WATER

Molecules are polarized when a separation exists between the "center of gravity" of the negative and the positive charges in the molecule. With molecules like water, such a condition is always present. The water molecule is arranged so that the oxygen atom is bonded to the hydrogen atoms with an angle of 105° between two bonds (figure 4.1). The center of the negative charge is near the oxygen atom while the center of the positive charge is at a point midway between the hydrogen atoms (represented by x in the figure). Materials composed of molecules

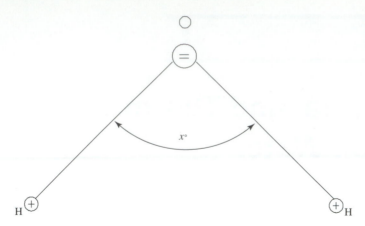

Figure 4.1   The positive charge of the hydrogen atom produces a polarized water molecule (H_2O) that appears bent

that are permanently polarized in this way usually have large dielectric constants. As an example, the dielectric constant (D, unitless) of water is 80.4, compared to 2.24 for carbon tetrachloride (both measured at 20 °C). A symmetrical molecule might have no permanent polarity; however, polarity can be induced by an external electric field. For example, if a linear molecule lies along the x axis, an external electric field in the positive direction of x would cause the center of the positive charge to shift to the right from its initial position; the center of the negative charge would thus shift to the left.

The electric charge structure of an individual water molecule has a basic tetrahedral shape in which the oxygen atom is near the center (figure 4.2). Two corners of this structure are positively charged, due to the partially screened protons of the hydrogen atoms. The remaining two corners of the tetrahedral structure are negatively charged, due to two pairs of lone-pair electrons. Such an arrangement makes the water molecule a dipole; that is, one end of the water molecule tends to be positive while the other end tends to be negatively charged. We can determine the potential energy of an electric dipole as a function of its orientation with respect to an external electric field. To accomplish this, we must recognize that work must be done by an external agent to rotate the dipole through a given angle, θ, in the electric field. The work done is stored as potential energy in the system (i.e., the dipole and the external field), U. The work, dW, required to rotate the dipole through an angle $d\theta$ is given by $dW = \tau\, d\theta$, where $\tau = pU \sin\theta$ (p is a vector).

The electric dipole, using the water molecule as an example, consists of two equal and opposite charges separated by a distance $2a$. The electric dipole moment of the water molecule has a vector p whose magnitude is $2aq$; that is, the separation $2a$ multiplied by the associated charge q. Because the work is transformed into potential energy E, for a rotation from θ_0 to θ, the change in potential energy is

$$E - E_0 = \int_{\theta_0}^{\theta} \tau\, d\theta = \int_{\theta_0}^{\theta} pU \sin\theta\, d\theta = pU \int_{\theta_0}^{\theta} \sin\theta\, d\theta \tag{4.1}$$

and

$$E - E_0 = pU[-\cos\theta]_{\theta_0}^{\theta} = pU(\cos\theta_0 - \cos\theta) \tag{4.2}$$

The term involving $\cos\theta_0$ is a constant that depends on the initial orientation of the dipole. For convenience, we choose $\theta_0 = 90°$, so that $\cos\theta_0 = \cos 90° = 0$. Additionally, we choose $E_0 = 0$ at $\theta_0 = 90°$ as our reference of potential energy. As a result, we may express E as

$$E = -pU \cos\theta \tag{4.3}$$

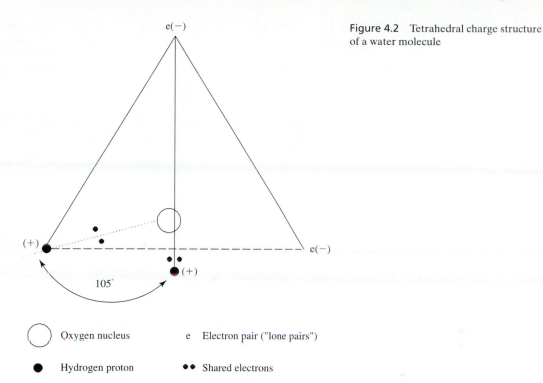

Figure 4.2 Tetrahedral charge structure of a water molecule

◯ Oxygen nucleus	e Electron pair ("lone pairs")
● Hydrogen proton	●● Shared electrons

Equation 4.3 can also be written as the dot product of vectors p and U: $E = -p \cdot U$.

QUESTION 4.1

The water molecule has a dipole moment of 6.3×10^{-30} C · m. A sample contains 10^{21} molecules. The dipole moments of all these molecules are oriented in the direction of an electric field of 2.75×10^{5} N C^{-1}. How much work is required to rotate the dipoles from $\theta = 0°$ to $\theta = 90°$ (i.e., where all the moments are perpendicular to the field)?

4.2 AIR–WATER INTERFACE: CONTACT ANGLE

When a water droplet is formed or placed on a clean surface, like a freshly waxed automobile, the size of the droplet will depend on the attractive forces associated with the air–liquid, solid–liquid, and solid–air interfaces (figure 4.3). Forces, for bulk water, consist primarily of London–van der Waals forces and hydrogen bonding; one-third London–van der Waals and two-thirds hydrogen bonding (Stumm 1992). Molecules at an air–water interface are not subjected to attractive forces from without, but are attracted inward to the bulk phase. This attraction tends to reduce the number of molecules on the surface region of the droplet due to an increase in intermolecular distance. For the area of the interface to be enlarged, energy must be expended. For water at 20 °C, this force, per unit of new area, may be expressed as a force per unit length that has the value 73 mN m^{-1}.

As can be seen in figure 4.3, the curvature of the liquid surface is dependent on the contact angle. The angle that the tangent to the fluid surface makes at the point of contact with the solid surface (inside the fluid) is known as the contact angle, and is generally represented

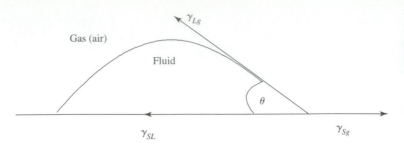

Figure 4.3 Interfacial tension components for the equilibrium of a droplet of fluid on a smooth surface in contact with air. The subscripts S, L, and g refer to solid, liquid, and gas. The contact angle is shown by angle θ.

by the symbol θ (not to be confused with volumetric water content). Normally, this angle is curved, but can range from 0 to 180°. For the majority of fluids on glass it is less than 90°. If the contact angle is greater than 90°, cos θ is negative. For example, if a glass tube is immersed in mercury, which has a contact angle of 140°, the mercury in the tube is pressed below the mercury level directly surrounding and outside the tube. For this same tube immersed in water, the water will rise slightly along the tube: the contact angle, the angle tangent to the water surface and the tube immersed in the water, will be acute. For pure water on clean glass, the contact angle is 0°; any contamination of the glass or water will increase the contact angle. For example, the contact angle for tap water and clean glass is about 15–18°. For practical purposes, when the contact angle is greater than 90°, the liquid does not wet the solid. For water, this condition is termed hydrophobic; drops tend to move about easily, but not enter capillary pores. This is a well known phenomena in organic soils throughout the world and also on golf course greens in the southeastern United States.

The change in surface free energy, ΔG^s, that accompanies a small displacement of the fluid per change of solid area (ΔA) covered, is

$$\Delta G^s = \Delta A(\gamma_{SL} - \gamma_{Sg}) + \Delta A \gamma_{Lg} \cos (\theta - \Delta\theta) \qquad (4.4)$$

and

$$\gamma_{Lg} \cos \theta = \gamma_{Sg} - \gamma_{SL} \qquad (4.5)$$

where γ is the interfacial tension (J m^{-2} or N m^{-1}), the subscripts L, g, and S refer to liquid, gas, and solid, and $\Delta\theta$ is the change in contact angle with change in solid area wetted. Equation 4.5, known as the Young equation, implies that various surface forces can be represented by surface tensions acting in the direction of the surface of the fluid.

4.3 CAPILLARITY

Capillarity, also referred to as surface tension (force per unit length), is related to stretched surfaces in tension. Capillarity concerns interfaces that are mobile and can assume an equilibrium shape. Common examples are water droplets formed in air, drops formed by liquids in liquids, thin films, meniscuses, and so on. Capillarity deals with both the macroscopic and statistical behavior of interfaces, rather than their molecular structure. This phenomenon is extremely important in soil water retention. Surface tension occurs at the molecular level and involves two types of molecular forces: adhesive forces, which are the attractive forces of molecules of dissimilar substances (the force varies depending on the substance), and cohesive forces, which are the attractions between molecules in like substances. Cohesive forces decrease rapidly with distance, and are strongest in the order solids, liquids, gases.

Surface tension is normally expressed by the symbol γ. Because γ is work per unit area, the customary units are ergs cm^{-2} or dynes cm^{-1}. In the SI unit system, γ is J m^{-2} or N m^{-1};

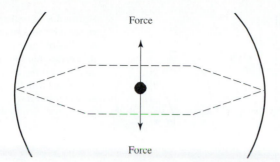

Figure 4.4   Forces inside a soap bubble

surface tensions reported as dynes cm^{-1} and 10^{-3} N m^{-1} have the same numerical value. Surface tension is highly dependent on temperature, decreasing almost linearly as temperature increases. Simply stated, cohesive forces at the surface and inside the fluid phase decrease with thermal expansion; this is generally accompanied by an increase in vapor pressure.

Examination of a simple soap bubble illustrates the significance of surface tension. For a soap bubble to form, a pressure difference must exist between the inside of the bubble and the atmosphere outside the bubble. As one would suspect, the pressure inside the bubble is greater than the pressure outside the bubble. Without this inside pressure, the bubble would collapse as a result of surface tension. Figure 4.4 shows an imaginary central plane inside a soap bubble that separates it into two hemispheres. Letting P (J m^{-3}) represent the excess pressure inside the bubble, there is a tendency for the two hemispheres to be driven apart by a force F. This force is equal to a product of both the pressure and area of the plane where the two hemispheres meet (indicated by the dashed line in the figure), where the area $A = \pi r^2$. As a result, this force can be obtained by calculating $F = PA$. The force is counteracted by the surface tension γ of the bubble, acting along the plane between the two hemispheres, which has a tendency to draw the two hemispheres together and can be represented as $2\pi r\gamma$. Because at equilibrium the two forces must be equal, by equating the two quantities we have

$$P = \frac{2\gamma}{r} \qquad (4.6)$$

The larger the radius of the bubble, the smaller the difference in pressure (i.e., the inside pressure is inversely related to the bubble radius). The radius of curvature of the liquid surface plays an important role in soil water retention for various soil types. For nonspherical bubbles, we may write

$$P = \gamma\left(\frac{1}{R_1} + \frac{1}{R_2}\right) \qquad (4.7)$$

where R_1 and R_2 are the radii of curvature for a specific point on the bubble interface. This curvature of the liquid surface is related to the contact angle.

As water rises in a cylindrical tube (figure 4.5), the meniscus is spherical in shape and concave upward. By letting r equal the tube radius, the excess pressure above the meniscus compared to the pressure directly below it can be described by $2\gamma/r$; that is, the pressure in the liquid below the meniscus will be less than the atmospheric pressure above the meniscus by this amount. In the laboratory, the pressure on the water surface outside the capillary tube will be atmospheric. This will force the fluid up the tube until the hydrostatic pressure of the fluid column within the tube equals the excess pressure of $2\gamma/r$. Because the circumference of the tube is $2\pi r$, the total force (upward) on the fluid is $2\pi r\gamma \cos \theta$. This force supports the

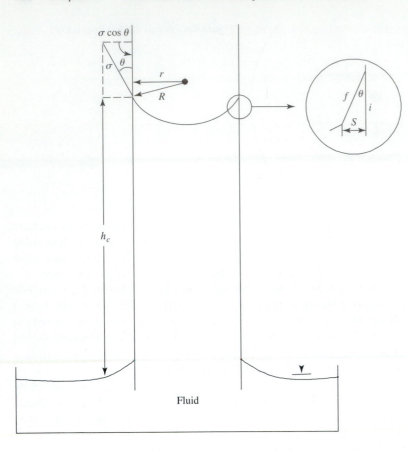

Figure 4.5   Capillary rise in a cylindrical tube, where θ is the contact angle, r is the tube radius, R is radius of curve of the gas-liquid interface, γ is the surface tension, and h_c is the height of capillary rise; i is the area of the solid-water interface associated with the tringular volume, and f is the air-water interface

weight of the fluid column to the height h_c. The total weight of fluid in the tube can be determined by

$$\rho_l g(\pi r^2 h_c + V) \tag{4.8}$$

where ρ_l is liquid density (kg m^{-3}), V is the volume of water in the meniscus (m^3), and all other parameters are as previously defined.

At equilibrium, both the force on the fluid and the weight of the fluid must be balanced, thus

$$2\pi r \gamma \cos\theta = \rho_l g(\pi r^2 h_c + V) \tag{4.9}$$

However, in very small capillaries (such as those found in soils), V is negligible, and the height of capillary rise can be given as

$$h_c = \frac{2\gamma \cos\theta}{\rho_l g r} \tag{4.10}$$

The radius of curvature, R, of the capillary meniscus in figure 4.5 can be determined by $R = r/\cos\theta$. It should also be noted that for a contact angle of zero, $r = R$. An approximation of the height of capillary rise is $h_c = 0.15/r$, which assumes a pure water state at 20 °C and its corresponding parameters for surface tension and density, and a contact angle of zero. Equation 4.10 can be used to calculate the height of water rise in a specific soil for which the largest effective pore size is known, and also to calculate the diameter of the largest effective pore (assuming one knows h_c). These equations are exact for capillary tubes; however, because soils do not behave as a single capillary such as a glass tube, these equations are at best approximations of water behavior in soils.

From the equation of capillary rise (4.10), one can observe that it is both adhesive and cohesive forces that bind water in soil capillaries. Both adhesion and repulsion of water by solid soil surfaces are characterized by the potential energy of water molecules near that surface, divided by the area of the solid-water interface. A positive γ indicates that the energy of water molecules near the soil surface is greater than the energy of molecules in bulk solution. When this situation occurs, water will not adhere to the solid surface and the condition is known as hydrophobic. Solid surfaces with a negative value for γ will readily attract water, because the energy near the soil surface is lower than the energy in the bulk solution. This condition, termed hydrophilic, is typical of field situations for most soils.

Cohesive forces can be described similarly. Within a bulk solution, a molecule is attracted in all directions equally. However, at an air-water interface, molecules are attracted only inwardly because no water molecules exist, in a liquid state, on the air side of the interface. As a result, the energy within molecules at the interface is greater than that in bulk solution, and this energy difference is characterized by the surface tension. Because the potential of water molecules in bulk solution is zero, γ is positive, which means that water in contact with air will attempt to reduce its surface area. This is easily seen in the fact that water droplets (rainfall, spray irrigation, and so on) are spherical in shape.

Considering that all three phases (air, water, and solid) are present in a capillary, the interfacial shape, which is characterized by the contact angle, θ, is dependent on whether or not γ is positive, negative, or zero, and upon the magnitude of γ with respect to σ (surface energy per unit area of fluid-gas interface in J m^{-2}). The inset in figure 4.5 illustrates this more clearly. The triangular volume that is horizontal to the tube has a maximum distance, s, to the solid surface of the tube. This distance is arbitrary, except that it will (or should) be large compared to the distance at which both the cohesive and adhesive forces are active, but should be small in comparison to the radius of curvature of the meniscus, R. The area of the air-water interface associated with the triangular volume is represented by f, while the area for the solid-water interface is represented by i. The total energy associated with both interfaces can be represented by

$$E = \sigma f + \gamma i = \frac{\sigma s}{\sin \theta} + \frac{\gamma s}{\tan \theta} \qquad (4.11)$$

The contact angle, θ, will adjust itself until the interfacial energy is at a minimum, which can be given as

$$\frac{dE}{d\theta} = -\frac{\sigma s \cos \theta}{\sin^2 \theta} - \frac{\gamma s}{\sin^2 \theta} = 0 \qquad \cos \theta = -\frac{\gamma}{\sigma} \qquad (4.12)$$

As a result, we can see that $\theta = 180°$ when $\gamma = \sigma$; $\theta = 0°$ when $\gamma = -\sigma$; and $\theta = 90°$ when $\gamma = 0$, where $\theta = 180°$ and $\theta = 0$ are the physical limits (i.e., $\theta = 0$ for $\gamma \leq -\sigma$ and $\theta = 180°$ for $\gamma \geq \sigma$). For exact solutions to the capillary rise problem and methods of measuring capillary rise, the reader is referred to Adamson (1990).

QUESTION 4.2

Determine the height of capillary rise of a fluid in a clean glass tube of 85 μm inner diameter; assume surface tension is 0.05 N/m.

4.4 CAPILLARY POTENTIAL

Figure 4.6 illustrates capillary potential on the basis of weight, volume, and mass (note bottom scale in figure). As can be seen, the pressure potential of the water in the capillary tube, ψ_p, decreases with height to offset the increasing gravitational potential, ψ_g (both potential components are discussed in section 4.5). As a result, the pressure of the water within the capillary tube is less than atmospheric pressure (which is pressing down on the water in the tray), creating a pressure difference on both sides of the meniscus. Thus, it can be seen that the pressure potential in the air above the concave portion of the meniscus in the tube is 0. Because $R = r/\cos \theta$ (which can be rearranged to $r = R \cos \theta$), a general relation for the pressure difference across the meniscus interface with radius of curvature R is $\Delta p = 2\sigma/R$. It should be noted here that the highest pressure is on the concave side of the meniscus. This results in the pressure difference across the interface being inversely proportional to R, which is why reference points for pressure potential are not chosen at the meniscus, but at the water surface, thus avoiding the influence of a solid surface.

Capillary potential relates directly to the air entry value of porous ceramics, because the porosity of such ceramics mimics that found in soils of similar pore geometry. For example, the capillary tube in figure 4.6 has a meniscus height or capillary rise, h_c, of 0.6 m. If this tube is immersed in a container of water and then slowly raised, R is gradually decreased, but the capillary tube remains filled with water until its top reaches a height above the water level of the container that equals the maximum height of capillary rise (0.6 m, in this case). If

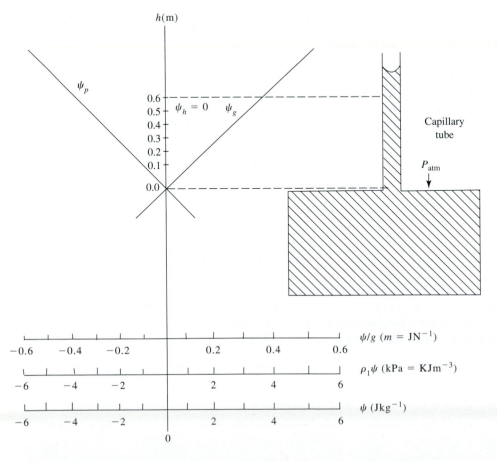

Figure 4.6   Diagram of potential in a capillary tube, where ψ_p is the pressure potential, ψ_h is head potential, ψ_g is gravitational potential, and h is head in meters

the tube is raised further, air enters from the top so that the water column remains at a height of 0.6 m. This is the reason why the pressure equivalent ($-\rho_l g h_c$) of the h_c is termed the air entry value. If water is expelled from a capillary by a positive gas pressure, that pressure is termed the bubbling pressure. Because h_c is the pressure head at the air-water interface of the meniscus, this concept can be extended to soil pores. For example, if the pressure potential in a soil pore is lower than the air entry value, both cohesive and adhesive forces cannot hold the water any longer. This will result in water draining from the pore until the pressure potential at the air-water interface is equal to that of the air entry value. Because soils are of many irregularly shaped "capillaries," and not a clean glass tube, the concepts discussed here can only be applied qualitatively to capillary water in soils.

QUESTION 4.3

Considering equation 4.10, what is a suitable expression for the pressure potential, volume basis (kPa), of water at the meniscus in figure 4.6?

QUESTION 4.4

In a typically irrigated, completely wetted soil, the noncylindrical capillaries have an air entry value of -20 kPa. Calculate the height of capillary rise and the capillary diameter.

QUESTION 4.5

A capillary has a diameter of 18 μm. Assuming complete wetting, what is the air entry value for this capillary?

4.5   COMPONENTS OF SOIL WATER POTENTIAL

The total potential of soil water has been defined as "the amount of work that must be done per unit quantity of pure water in order to transport reversibly and isothermally an infinitesimal quantity of water from a pool of pure water at a specified elevation at atmospheric pressure to the soil water at the point of consideration" (Bolt 1976). This transformation of pure water from the reference state to the soil-water state is actually broken into a series of steps. These steps are generally reversible and isothermal. Because of this, the total potential is actually a sum of each sequential step. These steps have been referred to as force fields (Hillel 1982), but are actually various components constituting the total potential that we shall discuss.

Consider a mass, m, that moves due to an external force acting in the opposite direction. The force applied is termed a static force, F^s. Any force applied must be only infinitesimally larger than the present force for movement to occur. The energy gained or lost in this process is the difference in potential energy of the mass between the starting and ending positions.

If we divide the potential energy difference by mass, we obtain the potential difference, $\Delta\psi$, which can be described by

$$\Delta\psi = -\frac{F^s}{m}\Delta s \tag{4.13}$$

where s is distance. The minus sign arises because potential decreases as the mass moves in the direction of the force. Since potential is potential energy divided by mass, it is expressed in J kg^{-1}.

Equation 4.13 is only valid when F^s is constant over the change in distance, Δs. However, in unsaturated zone studies with soil water, this is not normally true. If Δs is confined to a minute distance over which the force F^s can be assumed constant, equation 4.13 can be expressed as

$$\frac{F^s}{m} = -\lim_{\Delta s \to 0} \frac{\Delta \psi}{\Delta s} = -\frac{d\psi}{ds} \quad \text{or} \quad d\psi = -\frac{F^s}{m} ds \tag{4.14}$$

If we choose a reference point s_0 such that $\psi = 0$, then the potential for any point s_p can be obtained by integrating the right-hand equation in 4.14, yielding

$$\psi = -\int_{s_0}^{s_p} \frac{F^s}{m} ds \tag{4.15}$$

Because we are generally concerned only with differences in potential between two points, absolute values are neglected and we can assign an arbitrary value to the potential at s_0, usually zero.

For a combination of directions of F^s and ds, equation 4.15 can be expressed as

$$\psi = -\int_{s_0}^{s_p} \frac{F^s}{m} \cdot ds = -\int_{s_0}^{s_p} \frac{F^s}{m} \cos\theta \, ds \tag{4.16}$$

where θ is the angle of direction of movement. Because more than one force field or component is generally present in a soil system, each component will contribute its own potential. This can be called a partial potential ψ_i, where the subscript is the ith component. Thus, the total potential, ψ_t, is the sum of all ith potentials. This is written as

$$\psi_t = \sum_i \psi_i = \sum_i -\int \frac{F_i^s}{m} \cdot ds = -\int \frac{\sum_i F_i^s}{m} \cdot ds \tag{4.17}$$

Assuming $\sum F_i^s = 0$ (static equilibrium), then

$$\psi_t = -\int \frac{F_i^s}{m} \cdot ds = -\int 0 \, ds = c \tag{4.18}$$

where c is the constant of integration.

As an example of two of the components, we can examine hydrostatics, the study of pressures or potentials in liquids at hydrostatic equilibrium. In hydrostatics (and in the hydraulic flow of a soil solution), the gravitational potential, ψ_g, and hydrostatic pressure potential, ψ_h, are generally the only two components of total potential that need to be accounted for. The sum of these two potentials is termed the hydraulic potential, and is expressed

$$\psi_h = \psi_g + \psi_p \tag{4.19}$$

However, additional components sometimes need to be measured. The following is a complete list of the potential components: gravitational potential (ψ_g), osmotic potential (sometimes referred to as solute potential) ψ_o, vapor potential (ψ_a), matric potential (ψ_m), hydrostatic pressure (submergence) potential (ψ_p), and overburden pressure potential (ψ_b). A discussion of each component follows.

Gravitational Potential (ψ_g)

All objects on the earth are attracted downward due to gravitational force. This force is equal to the weight of an object, which is a product of mass and gravitational acceleration. Gravitational potential is the energy of water (on a unit volume basis) that is required to move a specific amount of pure, free water from an arbitrary reference point to the soil-water elevation. If the soil water elevation is above the reference point, z is positive ($+$); if the soil water elevation is below the reference, z is negative ($-$), where z is the vertical distance from the reference point to the point of interest. This means that the gravitational potential is independent of soil properties, and is solely dependent on the vertical distance between the arbitrary reference point and the soil-water elevation or elevation in question. Gravitational potential has the value

$$\psi_g = \rho_l g z \tag{4.20}$$

Here, ψ_g has the units of J m^{-3} and is assumed to be positive upward. Suppose one has chosen two soil-water elevation points in a soil column (for example, points 1 and 2 in figure 4.7). What would $\Delta\psi_g$ between the two points be? Because ψ_g is dependent on vertical distance (and not chemical, pressure, or other potentials), the answer is a simple one; it would merely be

$$\psi_g = \rho_l g(z_1 - z_2) = \rho_l g(0.25 - (-0.40)) = \rho_l g(0.65) \tag{4.21}$$

Often, the soil surface is chosen as the reference point. In such cases, the distance, z(m) to each point of interest below the surface is negative by convention.

Osmotic Potential (ψ_o)

The primary effect of salts in soil water is that of lowering the vapor pressure. However, this does not directly affect the mass flow of water in the soil, except in the presence of a membrane barrier of some type and in the case of vapor diffusion. Plants usually make up

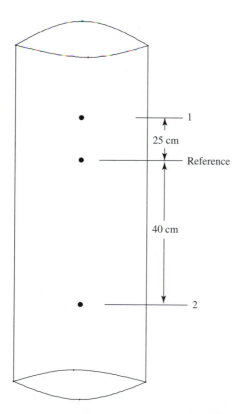

Figure 4.7   Soil column showing two points of soil water elevation in respect to an arbitrary reference

part of a dynamic soil-water-plant system, and their roots are often affected by the presence of salts in the soil solution. The membranes of plant roots transmit water more readily than salts, and are influenced by osmotic forces.

Osmotic potential may be thought of more as a suction than a pressure. This is illustrated in figure 4.8. In the presence of a selective permeable membrane (as found in plant roots), the pressure or energy potential (a suction) on the solution side is less than that of pure, free water, and the water first will pass through the membrane to the solution side. This will cause an initial rise in the solution level in the column (figure 4.8), and a fall in the fluid level on the pure water side. As equilibrium is established, both the concentration and fluid level will be constant on each side of the membrane. One can prevent flow across the membrane and maintain static equilibrium by exerting gas pressure, P, on the soil solution side of the container in excess of atmospheric pressure. This will result in the total potential at the surface of the solution, ψ_t, being

$$\psi_t = \frac{1}{\rho_w} P + \psi_g + \psi_o \tag{4.22}$$

where P is the air pressure (Pa) in excess of atmospheric. The term $1/\rho_w$ is included due to the small difference between the density of the solution and that of pure water, and is expressed in units of J kg^{-1}. However, in practice, it is often neglected, and the simple term P is used (in which case the units are expressed in J m^{-3}). At the surface of the pure water, the total potential is

$$\psi_t = \psi_g + \psi_m + \psi_o = \psi_g + 0 + 0 = \psi_g \tag{4.23}$$

Because the solution is pure water, we can ignore the matric potential, ψ_m, and osmotic

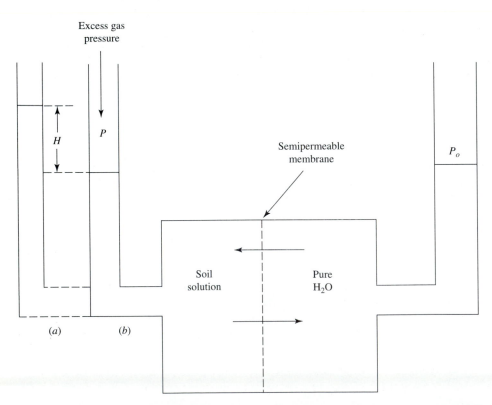

Figure 4.8 Illustration of osmosis in relation to osmotic potential/suction across a permeable membrane. (*a*) illustrates that the flow of water through the membrane into the solution is greater than from solution to water (i.e., osmosis); (*b*) the osmotic pressure of the solution is equal to the hydrostatic pressure that must be applied on the solution side, a (+) pressure on the P_o side, to equalize flow from the solution to the water side, or (−) suction on the P side. Both must be equal to obtain equilibrium.

Excess gas
pressure

P

H

Semipermeable
membrane

P_o

Soil
solution

Pure
H$_2$O

(*a*)   (*b*)

potential, ψ_o. At static equilibrium, $\psi_t = \psi_g$. Consequently, expressed in J m^{-3},

$$P + \psi_o = 0 \tag{4.24}$$

To maintain equilibrium, the excess air pressure needed must be equal to what is commonly referred to as the osmotic pressure, Π, of the solution, resulting in an expression for osmotic potential of

$$\psi_o = -\frac{\Pi}{\rho_w} \tag{4.25}$$

If we assumed the presence of soil and water, equation 4.25 (in J/kg) could be expanded to include the total potential of soil water components, and written as

$$\psi_t = \frac{(\Pi - \psi_m)}{\rho_w} + gz \tag{4.26}$$

Because soil is porous, it acts as a semipermeable barrier for salt solutions. As the reader may recall from chapter 3, each individual particle has an electric field that causes a decrease in concentration of the soil solution near the particle surface. This results in an increased cross sectional area for water transport compared to the area available for transport of dissolved salts, and causes water to move, due to osmotic differences. The amount of flow depends on the particle size of the media and the gradient of osmotic pressure. For example, in a coarse, saturated sand, soil pores are relatively large in comparison to those in clays. Consequently, the thickness of the fluid layer with decreased salt concentration is small when compared to the total area available for transport. As a result, the semipermeability will be small, and the transport of soil solution due to osmotic potential is negligible. In clays, the opposite is true. With many more pores, the solutes in soil solution can be completely repelled from the soil and restricted from flow. In such soils, water movement in response to an osmotic pressure gradient is as significant as when a hydraulic pressure gradient of equal size is applied. Hence, osmotic and hydraulic pressure gradients have equivalent effects on water movement. This can be proven by a modification of Darcy's law, such that $q = -K[dH/dz - (1/\rho g)(d\Pi/dz)]$.

Vapor Potential (ψ_a)

A combination of matric and osmotic forces determines the equilibrium vapor pressure for soil vapor diffusion. Since water vapor consists of pure water, both the osmotic and matric potentials are 0. The total potential in the vapor phase (the soil vapor pressure and the position of the gravitational field) can be determined such that

$$\psi_t(\text{vapor}) = \int \frac{1}{\rho_{\text{vapor}}} \, dP + gz \tag{4.27}$$

where the denominator is the density of the vapor. However, we can make substitutions to obtain the total potential of the water vapor as

$$\psi_t = \int \frac{RT}{PM} \, dP + gz = \frac{RT}{M} \ln P + gz + C \tag{4.28}$$

where R is the gas constant (8.314 J/mol K), T is temperature (K), P is vapor pressure (Pa), M is the molar mass of water (18.013 g), and C is an integration constant.

If we assume the gas-liquid interface, gz, is identical for both the liquid and the water vapor, then it can be expressed as

$$\psi_t(\text{liquid}) = gz = \psi_t(\text{vapor}) = \frac{RT}{M} \ln P_0 + C + gz \tag{4.29}$$

and

$$C = -\frac{RT}{M} \ln P_0 \tag{4.30}$$

By substituting equation 4.30 into equation 4.28, we obtain the total potential of water vapor at a vapor pressure, P:

$$\psi_t(\text{vapor}) = \frac{RT}{M} \ln \frac{P}{P_0} + gz \tag{4.31}$$

where p_0 is the vapor pressure at a free surface of pure water under atmospheric conditions (see figure 4.8). By convention, at static equilibrium, the total potential in the soil solution must be equal to the total potential of water vapor in the soil air; thus,

$$(\Pi - \psi_m)\frac{1}{\rho_l} = \frac{RT}{M} \ln \frac{P}{P_0} \tag{4.32}$$

Assuming standard unit values for pure water density, R, T, and M at 25 °C, equation 4.32 may be rewritten as

$$\ln \frac{P}{P_0} = (\Pi - \psi_m)7.5 \times 10^{-9} \, \text{Pa}^{-1} \tag{4.33}$$

Thus, one can see that the relative vapor pressure, P/P_0, at the interface of the soil water and soil air is determined by both the osmotic potential and the matric potential of the soil. The matric potential is sometimes referred to as tensiometer pressure and denoted as p_t. In field studies, it is generally assumed that soil air is at atmospheric pressure and, as such, can be expressed as

$$\ln \frac{P}{P_0} = (\Pi - p_m)7.5 \times 10^{-9} \, \text{Pa}^{-1} \tag{4.34}$$

where p_m is the pressure (Pa) equal to the matric potential of the soil. The relative vapor pressure may also be expressed in head equivalent form; that is, $\rho_l g$ (approximately $10^4 \, \text{N m}^{-3}$) as

$$\ln \frac{P}{P_0} + (h_m + h_o) \, 7.5 \times 10^{-5} \, \text{m}^{-1}; \qquad h_o = \frac{\Pi}{\rho_l g} \tag{4.35}$$

where h_m is the matric head of soil water (m) and h_o is the osmotic pressure head of the soil water. As a note, the reader should remember that $\Pi = MRT$, where M in this case is the molar concentration of the solution; R and T are as previously defined.

QUESTION 4.6

Why is the total potential of soil water at the water surface $\psi_t = gz$ (see equation 4.26)?

QUESTION 4.7

If the relative vapor pressure in a laboratory is kept at 50%, what is the head value of air measured by a tensiometer in the laboratory?

Matric Potential (ψ_m)

The matric potential of a soil results primarily from both the adsorptive (see van der Waals-London forces in chapter 3) and capillary forces due to soil matrix properties and is, thus, a dynamic soil property. The matric potential is often referred to as capillary or pressure potential, and is usually expressed with a negative sign. By convention, pressure is generally expressed in positive terms. Suppose one had a standing soil column one meter in height with a water table at a depth of 0.80 m (figure 4.9). The matric potential above the water table would be negative, the potential at the water table would be zero, and the potential below the water table would be positive in sign. By convention, ψ_m, the matric potential, becomes ψ_p below the water table, the pressure potential, and is sometimes referred to as the submergence potential. Below the water table, the pressure potential is equivalent to ρgh, where h is distance below the water table. It should be noted that an unsaturated medium has no pressure potential, but only a matric potential with negative units.

The term matric potential works well for rigid, self-supporting soils. In these soil types, the weight of the solid portion of the soil above the specific point of measurement is sustained by the soil matrix. External forces will not influence the matric potential in this case, as it is characterized by the amount of soil water present. However, in swelling soils, solid particles are not in complete contact with each other. Consequently, part of the weight overlying the soil matrix may be exerted on any element of soil water. Also, with a change in water content, a reorientation of solid particles may occur that can influence the interfacial

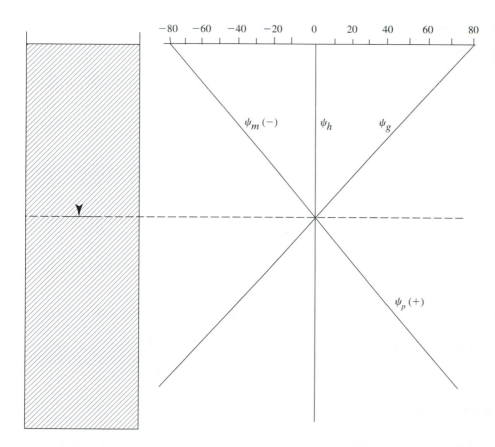

Figure 4.9   Soil column with water table at 80 cm, illustrating relation between soil component potentials

curvature of water in the soil. Variation in the pneumatic potential may also cause reorientation of solid particles. As a result, the matric potential typically is separated into two additional components for swelling soils. These components are: the effects of mechanical forces exerted by the pressure of the solid material (this is generally referred to as the overburden pressure, ψ_b, and is described in a subsequent section); and the effect of soil geometry, such as interfacial curvature and adsorptive forces. The latter component can be adequately described by the wetness potential, ψ_w, which is the value of the matric potential when both the soil stress, P_e, and external air pressure are 0.

Hydrostatic Pressure Potential (ψ_p)

The term pressure potential usually implies that soil water is at a hydrostatic pressure greater than atmospheric pressure and, thus, has a positive sign. Observation of the column in figure 4.9 indicates that the water beneath the free surface (water beneath 80 cm) has a positive pressure potential. Water at the surface—at 80 cm in the column—has a pressure potential of 0. The pressure potential for any point beneath the water table can be written as

$$\psi_p = \rho_l g h \tag{4.36}$$

where h is the distance below the water table. It is sometimes expressed as $z_{wt} - z_{\text{soil}}$ where the subscripts "wt" and "soil" refer to the water table and to a specific point within the soil.

Overburden Pressure Potential (ψ_b)

In many soil physics texts, Marshall and Holmes (1988) for example, the term envelope pressure, P_e, is used in conjunction with the effects of mechanical forces (stress) exerted by the pressure of the solid material. In this section we will discuss the overburden pressure potential, which is often termed intergranular pressure. Soil mechanics describes overburden in terms of compaction, consolidation, or compression. We shall begin by discussing static equilibrium for the soil matrix. We have discussed the total potential of soils, and applied the condition of static equilibrium to the water phase of the geologic medium. These same conditions may also be applied to a soil by properly defining the various potentials.

For static equilibrium to occur, the total soil system potential, ψ_t, must also be uniform. In this case, the soil system includes the soil matrix and all substances within the pore spaces of the matrix. Because ψ_t is composed of both gravitational potential, ψ_g, and pressure potential, ψ_p, the gradient of the gravitational potential must be compensated at an equal but opposite gradient by the pressure potential. This may be expressed as $\psi_t = \psi_g + \psi_p =$ uniform, where the osmotic, matric, and vapor potentials are ignored. As an example, suppose one had a column (figure 4.9) filled with water; using the same reference points for ψ_p and ψ_g, the potential diagram would look the same if the column were filled to the same level with soil (for potentials on a mass basis).

For a volume basis, because of pressure equivalents, the difference between the density of the soil and soil water must be accounted for. Thus, we can express the total potential on a volume basis as $\rho_b \psi_t = \rho_b \psi_g + \rho_b \psi_p =$ uniform. The pressure potential on a volume basis is generally termed the soil stress, σ_s, which can be expressed as $\sigma_s = \rho_b \psi_p$, where ρ_b is the dry bulk density (kg m^{-3}). The soil pressure is simply the vertical force caused by the weight of the overlying material divided by the horizontal surface area over which it lies. It is actually the normal component; that is, that component perpendicular to the plane of interest of the stress at a specific location.

Intergranular Pressure (σ_g)

Intergranular pressure, σ_g, written in terms of stress, is the contact force between individual particles divided by the area. In water-saturated soil, the weight of the overlying medium is supported by both the contact force between particles and also the contact force in the water itself, which may be expressed simply as $\sigma_s = \sigma_g + \psi_p$, where ψ_p is the water pressure (which can also be defined as $\psi_p = p - p_{atm}$). Here, ψ_p is the tensiometer pressure potential of soil water, and p is the absolute pressure potential. In a dry soil, the vertical intergranular pressure will be equal to the soil pressure, σ_s, because the weight of overlying material will be supported solely by the contact forces between individual particles. For a rigid soil, this may be expressed as $\sigma_s = \sigma_g$.

Due to friction between individual particles, soils resist motion of the particles relative to each other that might result from differential loading. Examples of soil movement due to differential loading include compaction of soils due to construction practices, operation of heavy equipment on soils, and agricultural activities such as tillage. The resistance of the soil to each of these activities is determined by friction caused by intergranular pressure.

For a rigid unsaturated medium, ψ_p is negative, simply because this negative pressure is supported by the soil particles pressed against each other with a force that exceeds the weight of the overlying soil. Thus, the ψ_p is active only within water-filled pores. As a result, the soil pressure for an unsaturated rigid soil may be expressed as

$$\sigma_s = \sigma_g + \frac{\theta}{\phi}\,\psi_p \tag{4.37}$$

Remembering that the volumetric water content, θ, cannot exceed the porosity, ϕ, this coefficient for the water pressure will vary from 0 at air dryness to 1 at full saturation. It should be noted that the second term on the right hand side of equation 4.37 has been termed the wet bulk density (Bolt 1976), and often appears as the term ρ_{wb}. Assuming an absence of load at the soil surface (such as machinery or structure), the overburden pressure at a chosen depth is due to the load of soil above that depth and can be expressed by

$$\psi_b = \sigma_g + \int_z^0 \frac{\theta}{\phi}\,\psi_p\,dz \tag{4.38}$$

Intergranular pressure decreases due to upward water transport. This causes a reduction in the bearing capacity of the soil, which is dependent on frictional forces between individual particles. In certain cases of strong upward flux the intergranular pressure may completely disappear; an example of this is quicksand.

In contrast to rigid soils, clay soils have unique properties, especially in regards to the propensity of some clays to swell. The intergranular or overburden pressure in these soils is not supported solely by the grain to grain contact as in rigid soils, but includes the pore water, which forms continuous vertical columns between the grains. In these soils, consolidation plays a major role in the ability of the soil to conduct water, which affects the hydraulic properties because of subsequent reduction in void space with time. Thus, with depth, the pressure or stress of the overlying material increases, reducing the gradient in hydraulic potential (Philip 1971).

The hydraulic potential, $\psi_h = \psi_{mL} + \rho g z$ (where ψ_{mL} is the matric potential of the loaded sample), of a swelling soil can be expressed as

$$\psi_h = \psi_{mu} + \beta\psi_b + \rho g z \tag{4.39}$$

where ψ_{mu} is the matric potential of the unloaded sample, ρ is the density of water, z is depth relative to soil surface (where $z = 0$), and β is the compressibility factor, often termed in soil mechanics the pore compression coefficient. The pressure potential, $\beta\psi_b$, due to overburden pressure expressed in equation 4.39, on a per unit volume of water basis, can be expressed as

$$\beta\psi_b = -\beta g p_{wb} z \tag{4.40}$$

where ρ_{wb} is wet bulk density. To obtain the hydraulic potential, equations 4.39 and 4.40 can be combined such that

$$\psi_h = \psi_{uL} + \rho g \left(\frac{1 - \beta\rho_{wb}}{\rho}\right) z \tag{4.41}$$

where ψ_{uL} is the hydraulic potential of the unloaded sample. Equation 4.41 indicates a reduction in the gravitational component due to overburden pressure.

The β term, or pore compression coefficient, may be expressed as the instantaneous increase in pore-water pressure divided by the increase in total stress of the overburden: $\beta = P_w/\psi_b$. The compressibility of pore water is negligible in fully saturated soils if we assume that the compressibility of the solid particles is 0. In this case, $\beta = 1$. For partially saturated soils, the compressibility of pore fluid is high and $\beta < 1$. As soil water content decreases, β also decreases. Talsma (1977) measured field samples of $\beta < 0.25$ for several swelling clay soils, whereas Croney, Coleman, and Black (1958) reported values of $1, 0.3$, and 0.02, for a clay, silty clay, and sand, respectively, which had unloaded suctions of < 40 cm. While the degree of saturation will vary for each soil type, a typical variation of β with degree of saturation is shown in figure 4.10. The value for β is typically measured in a triaxial test as cell pressure is increased. Note: "loaded samples" refers to samples that have been placed in the triaxial cell, brought to a certain water content, and a pressure then applied. "Unloaded samples" refers to soils extracted from the field, in which the pore pressure is then measured.

For most soils, three cases of soil-water equilibrium can occur. First, if a soil system contains a selective or semipermeable barrier that allows movement of water but not solutes, an

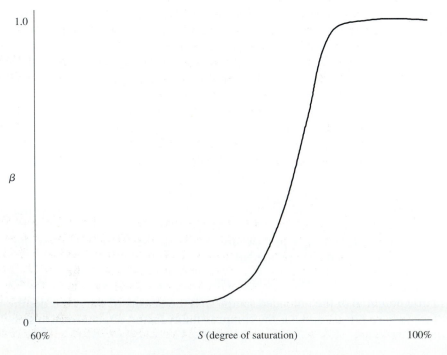

Figure 4.10   The pore pressure coefficient, β, versus degree of saturation

equilibrium will be obtained only if the total potential is uniform; that is, if $\psi_t = \psi_m + \psi_g + \psi_o$. This case would be most likely in dry soils where water is in thin films, small pores, or electrical double layers and movement occurs through the gas phase. Second, solutes and dissolved salts can generally move freely through sandy soils. In this case, if the hydraulic potential is uniform (hydraulic potential is defined as the sum of the pressure potential, ψ_p, and the gravitational potential, ψ_g: $\psi_h = \psi_p + \psi_g$), an equilibrium of the soil solution will occur if $\psi_h = \psi_m + \psi_g$. This assumes that the hydraulic potential does not vary with position. Third, in the case of moist clay soils, salt transport is difficult, especially in fine, clayey soils, because soil water is normally present in thin films on individual particles. Moist clay soils have barriers that obstruct the transport of salts or solutes without completely preventing it. As water diffuses in the direction of the solute, small quantities of solute are transported. Before equilibrium occurs, the difference in hydraulic potential in the water and solute balances the difference in osmotic potential. However, because the solute concentration is still different on both sides (i.e., the water next to the soil particle and that within the bulk solution), total equilibrium will exist only if ψ_h = uniform and ψ_o = uniform. At some time, ψ_t = uniform (i.e., equilibrium is reached). Because of the effects of the electrical double layer, anions are repelled by the electrostatic charge (see chapter 3).

QUESTION 4.8

An experiment is conducted on a dry sandy soil with a bulk density of 1500 kg m^{-3}. Assuming the soil is at static equilibrium, what is the soil pressure 1 m below the surface?

4.6   CHEMICAL POTENTIAL OF SOIL WATER

Prior to reading this section, the reader may wish to review the basic laws and relations of thermodynamics, and the correlation with enthalpy, entropy, and free energy, by consulting a standard physical chemistry text such as Alberty (1987). The increase in internal (or stored) energy of a system must be equivalent to the quantity of heat energy, Q_h, absorbed by the system from its surroundings, less the energy lost by the system to the surroundings as a result of work, W, done by the system due to the absorption of heat. This change of free energy at constant temperature and pressure in an open system is associated with the transfer of mass from one phase to another (i.e., a change in mass from one chemical combination to another, such as condensation, evaporation, and inflow and outflow from a given section). Differentially, the equation for chemical potential, μ_i, of the ith component can be written as

$$d\mu_i = -S_i \, dT + V_i \, dP + \sum_j \left(\frac{d\mu_i}{dM_i}\right)_{T,P,m_j} dM_i \tag{4.42}$$

where S, T, V, and P refer to entropy, temperature, volume, and pressure, respectively; m_i is the mass or number of moles of the various constituents in the system; and M_i is the mole fraction of the ith constituent. However, equation 4.42 describes the thermodynamics of the system involving T, P, and M_i only as variables. The description of water in soil will also require at least one external force field as an additional variable. That force field and variable is gravity. A generalized equation for the internal energy of the system, E, with the new variable, x_n, can be expressed as

$$dE = T \, dS - P \, dV + \sum_i \mu_i \, dm_i + \sum_n \left(\frac{\partial E}{\partial x_n}\right)_{S,V,m_i x_n} dx_n \tag{4.43}$$

The new intensive property X_n can be defined as

$$X_n = \left(\frac{\partial E}{\partial x_n}\right)_{S,V,m_i x_n} \tag{4.44}$$

This allows equation 4.43 to be rewritten as

$$dE = T\, dS - P\, dV + \sum_i \mu_i\, dm_i + \sum_n X_n\, dx_n \tag{4.45}$$

We can specify X_n using only gravity as the external force field. For example, if a mass of soil water, m, is at a height in the soil, z, in a gravitational field above a specific reference point, we can express the potential energy, E_p, associated with this soil water as $E_p = mgz$. This implies the gravitational potential, $\psi_g = E_p/m = gz$. In this case, the gravitational potential is an intensive property which will allow X_n to be specified as $X_n = \psi_g = gz$ (on a mass basis; expressed on a volume basis it is $X_n = \psi_g = \rho gz$), and

$$dX_n = \sum_i M_i\, dm_i \tag{4.46}$$

By substituting these relations into equation 4.45, we obtain

$$dE = T\, dS - P\, dV + \sum_i (\mu_i + M_i \psi_g)dm_i \tag{4.47}$$

and

$$\left(\frac{\partial E}{\partial m_i}\right)_{S,T,m_i} = \mu_i + M_i \psi_g \tag{4.48}$$

Because the intensive properties of substances are not affected by the presence of a gravitational field, and because ψ_g is independent of mass, equations developed for chemical potential in the absence of a gravitational field are also valid if a field is present. Consequently, at constant entropy, volume, and mass and with $\psi_g = 0$, we may write

$$\mu_i = \left(\frac{\partial E}{\partial m_i}\right)_{S,V,M_i,(\psi_g=0)} \tag{4.49}$$

Recalling the relation of the ideal gas law and Raoult's law for an ideal solution, and because the activity of a pure solid or liquid at a chosen reference state at each selected temperature and pressure is unity, the chemical potential for various conditions for the ith component of a soil water system can be expressed as:

$$(\mu_w - \mu_w^0) + RT \ln\left(\frac{P_w}{P_w^0}\right) = RT \ln\left(\frac{e}{e^0}\right) \tag{4.50}$$

$$(\mu_w - \mu_w^0) = RT \ln M_w \tag{4.51}$$

and

$$(\mu_w - \mu_w^0) = RT \ln a_w \tag{4.52}$$

where the subscript w is the soil water component; the superscript 0 indicates the component of pure free water at the same temperature and atmospheric pressure; P indicates pressure, and a is activity; e is the soil water vapor pressure, and e^0 saturated water vapor pressure.

The partial molal free energy, F, of soil water is expressed as

$$\left(\frac{\partial F}{\partial m_w}\right)_{T,P,m_i} = F_w = \mu_w + M_w \psi_g \tag{4.53}$$

However, it should be noted that F_w is under the influence of forces associated with fluid-air and solid-fluid interfaces within the soil matrix. By ignoring gravity, the effect of volumetric water content can be factored into any change in chemical potential of soil water. This can be expressed as

$$d\mu_w = -S_w \, dT + V_w \, dP + \sum_j \left(\frac{\partial \mu_w}{\partial m_i}\right)_{m_j, T, P} dm_i + \left(\frac{\partial \mu_w}{\partial \theta}\right)_{m_w, T, P} d\theta \qquad (4.54)$$

where P is the pressure in the gaseous phase at the fluid-solid interface, V_w is the partial molal volume of soil water, and $\partial \mu_w / \partial \theta$ is the change in chemical potential with respect to the change in volumetric water content. The summation is for all chemical species, but for those for which the mole number does not change, the contribution is 0. Examination of equation 4.54 indicates that under isothermal conditions, the first term on the right-hand side will disappear, and for unsaturated zone conditions, the second term can be ignored as well, because the pressure difference between the soil surface layers up to the active biological zone and the atmosphere generally is small. That leaves the third and fourth terms to obtain.

The pressure in soil water is less than that of pure water due to the presence of solutes; the soil water is brought to equilibrium by applying external gas pressure, P (as described in figure 4.8), to raise the chemical potential of the soil water so that it is in equilibrium with that of pure water. This can be expressed mathematically by

$$\int_{\mu_w(m_i)}^{\mu_w^0} d\mu_w = \int_P^{P_{ext}} V_w \, dP \qquad (4.55)$$

where μ_w^0 is the chemical potential of pure water, $\mu_w(m_i)$ is the chemical potential of soil water, and P_{ext} is the external pressure required to bring the soil-water chemical potential to that of pure water. Because the net volume is constant at equilibrium, one may integrate equation 4.55 to obtain

$$[\mu_w(m_i) - \mu_w^0] = -V_w(P_{ext} - P) = -V_w \Pi = V_w \psi_o \qquad (4.56)$$

To determine the fourth term on the right hand side of equation 4.56, one may use the analogy of a porous cup tensiometer. In unsaturated soils, the porous cup of a tensiometer is permeable to both the soil water and other chemical constituents of the fluid present in the system. When an equilibrium within the porous cup and the soil solution has been attained, the water in the cup will have the same composition as that in the surrounding soil matrix. However, in this case, equilibrium is attained by decreasing the pressure of water in the porous cup from P_o to P. The chemical potential is also correspondingly lowered from μ_w^0 to μ_w. As a result, we may obtain the fourth term from

$$\int_{\mu_w(\theta)^{(B)}}^{\mu_w^{0(A)}} d\mu_w^{0(A)} = \int_P^{P_0} V_w^{(A)} \, dP \qquad (4.57)$$

where A and B refer to solution in tensiometer and soil (see figure 4.11). Because $V_w^{(A)}$ is constant at equilibrium, we obtain

$$[\mu_w(\theta) - \mu_w^0] = V_w^A(P - P_0) = V_w \psi_m \qquad (4.58)$$

This allows us to write equation 4.54, due to isothermal conditions, as

$$d\mu_w = V_w \, dP + V_w \, d\psi_o + V_w \, d\psi_m \qquad (4.59)$$

As a result, the chemical potential of soil water, compared to pure free water, may be expressed as

$$\frac{\mu_w - \mu_w^0}{V_w} = P + \psi_0 + \psi_m \qquad (4.60)$$

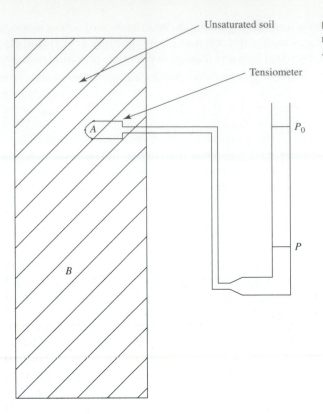

Unsaturated soil

Tensiometer

A

B

P_0

P

Figure 4.11 Chemical potential measurements in tensiometer, A, and unsaturated soil, B, in conjunction with equation 4.57

These equations assume a uniform geometry of particles in the soil matrix. For swelling soils, a geometric term would have to be added to equation 4.54 to obtain the chemical potential. Equation 4.54 would be rewritten as

$$d\mu_w = -S_w\, dT + V_w\, dP + \sum \left(\frac{\partial \mu_w}{\partial m_i}\right)_{m_w, T, P, v} dm_i$$

$$+ \left(\frac{\partial \mu_w}{\partial \theta}\right)_{m_w, T, P, v} d\theta + \left(\frac{\partial \mu_w}{\partial v}\right)_{m_i, T, P, m_v} dv \tag{4.61}$$

where v is the geometric term and can be expressed in volume units as

$$\left(\frac{\partial \mu_w}{\partial v}\right)_{m_i, T, P, m_v} dv = \left(\frac{\partial \mu_w}{\partial V}\right)_{m_i, T, P, m_v} dV \tag{4.62}$$

Pursuant to this discussion we may say that the energy status of the soil can be expressed in both mechanistic and thermodynamic terms. The mechanistic view represents the microscopic viewpoint and isothermal conditions (no change in temperature), where a volume element of water is used to describe the force fields which act upon soil water, but are external to the unit volume of water. (These force fields arise primarily because of solutes, gravity, matrix properties, etc.) The thermodynamic view considers measurable variables such as external pressure, temperature, volume, ions, water, soil particles, and so on. Thermodynamically, the derivation of Gibbs function (equation 4.54) and the chemical potential (i.e., consistent with the first and second laws of thermodynamics), includes temperature and entropy, which allows broader applicability in a description of the soil-water energy status than a mechanistic (microscopic) viewpoint.

The term "free energy" has been widely accepted as the best to use in describing the energy status of soil solutions. This term includes all types of energies: kinetic, chemical, potential, heat, electrical, and so on. Rather than the absolute value of free energy, differences in free energy between two different states are typically determined. By knowing the difference, the direction of a spontaneous energy change between the two states can be obtained. A number of measurements are necessary to characterize the different components of soil water. Once taken, they must be integrated to obtain the difference of free energy between soil water and a reference water. Perhaps the best method of measuring chemical potential is the vapor pressure method (Ghildyal and Tripathi 1987) which, although concise, is also complicated.

4.7 HYSTERESIS

A typical soil-moisture characteristic curve for an initially saturated soil is shown in figure 4.12. As matric potential decreases, that is, becomes more negative (increased suction), the volumetric water content decreases. Generally, these types of curves are obtained to relate soil moisture to matric potential. The moisture curve shown in figure 4.12 is termed a "desorption" curve. However, in typical field situations, not only will there be a drying or drainage cycle as illustrated in the figure, but there will also be a sorption cycle as the soil profile is rewetted through natural or artificial means. The sorption curve is obtained by wetting an initially dry soil while measuring the soil matric potential. Two curves can be obtained, a desorption and a sorption curve, which are illustrated in figure 4.13. The primary curves from saturation to dryness and vice versa are known as main branches. Due to the wetting and drying history of a soil, especially in the field, these two curves normally will not be the same, although they will typically be smooth in shape. One may see that during equilibrium, at any point along the curve, the moisture content for the desorption curve is always greater than for the sorption curve (figure 4.13). The dependence of soil-water

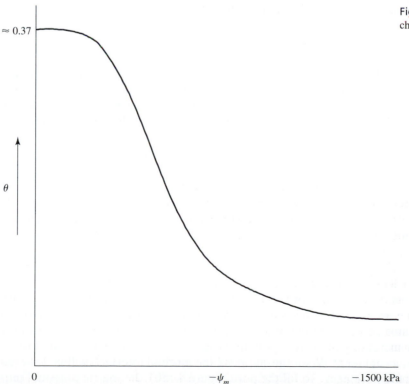

Figure 4.12 Soil-moisture characteristic curve

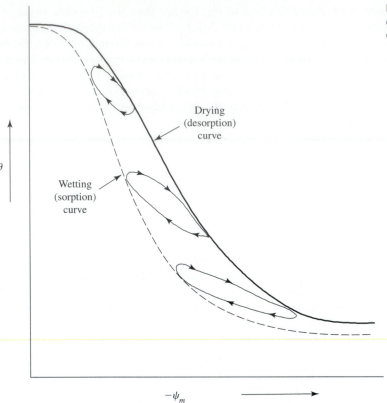

Figure 4.13 Sorption and desorption curves, illustrating effects of hysteresis

equilibrium content and matric potential on the direction of the process leading up to it is termed hysteresis (Topp 1969). The relation between soil-water content and matric potential over a very limited range, when a partially wetted or dried soil is partially drained or rewetted, is known as a scanning curve (illustrated as the small loops inside the main curves in figure 4.13).

The best way to visualize a soil-moisture curve and related scanning curves is to assume you have a sample of dry soil that is as homogeneous as possible; i.e., the soil does not vary from point to point, with little possibility for entrapped air. Place the sample in a container and slowly add water until all pores are filled (i.e., the matric potential is 0). Plotting the results of this process would yield a wetting curve similar to that in figure 4.13. By allowing the sample to dry through gravity drainage and evaporative processes, or bringing the soil into contact with a dry medium, the desorption curve would be obtained. However, if rewetting occurs before the sample is completely dried, the resultant plot would yield a scanning curve. The shapes of the curves would be a result of the hysteresis of the sample. The phenomenon of hysteresis in soils has been attributed to four primary causes: **(1)** the geometric or "ink bottle" effect; **(2)** contact angle; **(3)** entrapped air; and **(4)** shrinking and swelling.

Ink Bottle Effect

Soils contain pores of varying size and shape that are connected by narrow passageways of various sizes. Soil pores can drain and fill with water in a rapid and/or sporadic manner, a result termed the ink bottle effect (figure 4.14). If one allows soil water to drain from a pore by increasing the matric potential (figure 4.15a), once the potential reaches ψ_r, the pore will drain suddenly. When this happens, the air-fluid interface will be lowered to a critical radius of the pore neck. To fill the pore (figure 4.15b), the matric potential must increase to ψ_R, in

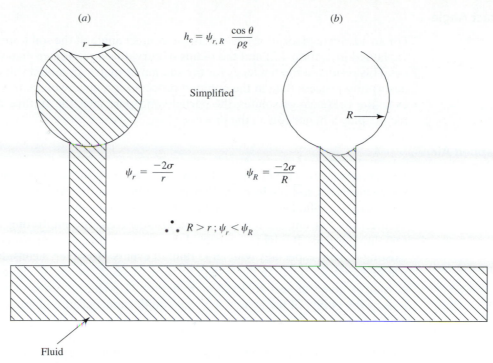

Figure 4.14   The "ink bottle" effect determines the equilibrium height of soil water in a nonuniform width pore. (a) desorption and (b) sorption (data from Hillel 1982)

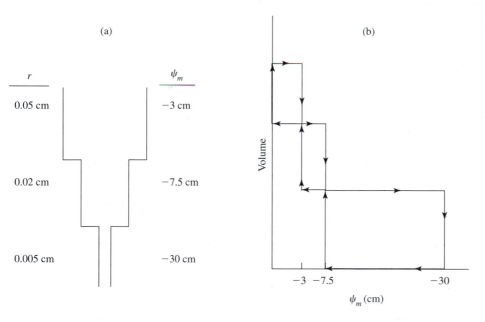

Figure 4.15   Matric potential associated with soil pore radius and hysteresis

which case the pore will fill abruptly. These sudden changes in pore water content are termed Haines jumps. At this point, $\psi_r > \psi_R$. Consequently, we see that the desorption process is dependent upon the smallest radius of the neck or connecting channels, while the sorption process is dependent on the largest diameter of the larger pores. Because of this, as is easily seen in figure 4.13, the equilibrium water content at any matric potential will depend on whether the soil system is draining or wetting.

Contact Angle

For an advancing (sorption) meniscus, the contact angle of the soil water on the pore wall (as explained in section 4.2) and the radius of curvature are greater than for a receding meniscus. This results in the tendency for the suction or matric potential values to be higher in the desorption process than in the sorption process for a given volumetric water content. However, the presence of solutes, the particle and pore size, the surface roughness, and other factors play a major role in the process.

Entrapped Air

Hysteresis can be accentuated when air is entrapped in a soil system. During infiltration, water displaces soil air from various pores. Once the initial water front redistributes, soil pores can be refilled with air from the atmosphere that will be trapped and moved downward by the next wetting front. Also, O_2 will exsolve from the infiltrating water and enrich the remaining soil air. It might be expected that the rewetting process would renew or replace all existing soil air, especially with large rainfall events. However, a considerable amount of entrapped air can remain in the system because of dead-end and occluded pores. (Further discussion is given in section 10.13 and chapter 11.) The presence of entrapped air initially lessens the ability of the system to reach equilibrium, because of greater pressure within the entrapped air pockets versus surrounding pores. This will lessen with time, especially in very wet soils where the air effervesces into the surrounding soil water. Entrapped air will reduce the volumetric water content of the system, making hysteresis more pronounced.

Shrinking and Swelling

Alternate drying and rewetting of soils can cause both shrinking and swelling (see chapter 3). This can cause changes in soil structure, accompanied by changes in pore space, dependent on the wetting and drying history of the soil (Hillel and Mottes 1966). These changes in pore size and distribution may cause significant variation in soil moisture content and equilibrium processes during the drying and rewetting processes, because of subsequent dissolution and release of soil air.

Due to the complexity of the hysteresis process, it is often ignored in field studies. However, it is known to be important in many soils, especially where both drying and wetting occur at the same time at different locations in a profile. The profile may have several layers of the same texture that, because of different wetting and drying histories, will have different water contents. The primary difficulty in simulating hysteresis is obtaining the scanning curves. As a result, the desorption curve is usually reported because it is much easier to obtain.

While hysteresis is not well studied due to its complexity, it is of considerable importance in evaporation, infiltration, and in coupled processes in soils in which both wetting and drying may occur simultaneously. The most pronounced effects of hysteresis occur in coarse soils (usually sands) at low values of matric potential.

The phenomena of hysteresis can be represented for further clarification in a problem-solving manner. Obtaining solutions to questions 4.9, 4.10, and 4.11 will enhance the reader's understanding of hysteresis.

QUESTION 4.9

Figure 4.16 illustrates a "soil pore" with a cylindrical shape, encased in a walled cylinder—that is, like an intact core sample with a smaller radius neck at each end; the respective diameter and height are listed. Since one can calculate both the volume of the inner pore and the total volume of the sample (i.e., the encasement cylinder), what is the total porosity of the soil pore?

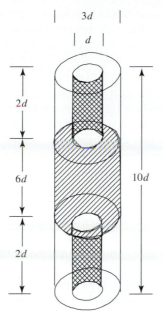

Figure 4.16   Diagram of soil pore in an encasement

QUESTION 4.10

Assume that you completely saturate the sample in question 4.9. **(a)** What is the volumetric water content? **(b)** Sketch a desorption curve of the sample assuming that you are applying a gradual pressure to the bottom end of the sample (assume $\sigma = 7 \times 10^{-2}$ N/m and $d = 15\ \mu$m). *Hint:* ignore gravitational potential due to pore size.

QUESTION 4.11

Draw a sorption curve (on the same graph as you created for question 4.10), assuming that you gradually increase the pressure again—back to 0. Having performed the steps in questions 4.10 and 4.11, you have illustrated the effects of hysteresis on the soil pore!

SUMMARY

In this chapter we have discussed various concepts of soil water potential. Within the unsaturated zone, soil particles are typically surrounded by thin water films that are bound to the solid surfaces of the medium by the molecular forces of adhesion and cohesion. A simple definition of water content is not sufficient to enumerate the complete status of water in a porous medium. We have presented the definition of matric potential, the force with which water is held to individual particles in the soil, which may be defined as the amount of work done or potential energy stored, per unit volume, in moving that volume, m, from the reference state (typically chosen as pure free water). We have also discussed the structure of water: how materials composed of molecules that are permanently polarized, as is water, usually have large dielectric constants. Included was a discussion of the formation of water droplets and the attractive forces associated with the air-liquid, solid-liquid, and solid-air interfaces. Forces for bulk water consist primarily of van der Waals-London forces and hydrogen bonding. Molecules at an air-water interface are not subjected to attractive forces from without, but are attracted inward to the bulk phase.

Capillarity and how it relates to surfaces in tension was explained, and common examples illustrated. Because capillarity deals with macroscopic and statistical behavior of interfaces rather than molecular structure, it is extremely important in soil water retention and in unsaturated zone studies. Adhesive and cohesive properties are the attractive forces of molecules of dissimilar substances, and the attraction between molecules in like substances. A detailed description of capillary potential on a weight, volume, and mass basis was presented, illustrating that the pressure potential (ψ_p) of the water in the capillary tube decreases with height to offset the increasing gravitational potential (ψ_g). This causes the pressure of the water in a capillary tube to be less than atmospheric pressure, and in certain circumstances this is analogous to soil-water systems. We also demonstrated how capillary potential relates directly to the air-entry value of porous ceramics and, thus, soil pores.

The discussion of capillary potential was followed by a discussion of total soil water potential and its definition and various components: gravitational potential (ψ_g), osmotic potential (sometimes referred to as solute potential) ψ_o, pneumatic (air or vapor) potential (ψ_a), matric potential (ψ_m), hydrostatic pressure (submergence) potential (ψ_p), and overburden pressure potential (ψ_b). Additionally, the chemical potential of soil water, which is very complex and often overlooked, was also presented. The increase in internal (stored) energy of a system must be equivalent to the quantity of heat energy, Q_h, absorbed by the system from its surroundings less the energy lost by the system to the surroundings as a result of work, W, done by the system due to the absorption of heat. The change of free energy at constant temperature and pressure in an open system is associated with a transfer of mass from one phase to another, or a change in mass from one chemical combination to another (i.e., chemical potential). Examples of a change in free energy include condensation, evaporation, and inflow and outflow from a section of soil.

Additionally, soil moisture characteristic curves and the relation to hysteresis were discussed, as were the primary curves from saturation to dryness and vice versa (known as the main branches of soil-moisture characteristic curves). The four primary causes of hysteresis were elaborated: **(1)** the geometric or 'ink bottle' effect; **(2)** contact angle; **(3)** entrapped air; and **(4)** shrinking and swelling. While hysteresis is very important, it is often dismissed due to its complexity and an inability to effectually measure its effects on a soil system.

ANSWERS TO QUESTIONS

4.1. The work required to rotate one molecule by 90° is equal to the difference in potential energy between the 90° orientation and the 0° orientation. By using equation 4.3, $W = E_{90} - E_0 = (-pU \cdot \cos 90°) - (-pU \cos 0°) = pU = (6.3 \times 10^{-30} \text{ C} \cdot \text{m})(2.75 \times 10^5 \text{ N/C}) = 1.73 \times 10^{-24} \text{ N} \cdot \text{m}$ or $1.73 \times 10^{-24} \text{ J} \times 10^{20}$ molecules $= 1.7 \times 10^{-3} \text{ J}$.

4.2. Remembering that the contact angle for clean glass is 0 and $r = 42.5 \ \mu\text{m}$, we may use equation 4.10, substituting the necessary conversion factors. Thus,

$$h_c = \frac{2(0.05 \text{ N m}^{-1})1}{(1000 \text{ kg m}^{-3})(9.81 \text{ N kg}^{-1})(42.5 \ \mu\text{m})(10^{-6} \text{ m } \mu\text{m}^{-1})} = 0.24 \text{ m}$$

4.3. Assuming that the reference point is taken at the flat water surface, then for every location where $\psi_h = 0$ and $\psi_p = -\psi_g$, the volume basis pressure potential at the meniscus will be

$$\rho_l \psi_p = -\rho_l \psi_g = -\rho_l g h_c = -\frac{2\sigma \cos \theta}{r}$$

4.4. Noncylindrical capillaries have the same height of rise as cylindrical ones. The air entry value, -20 kPa, is converted to meters, yielding 2.0 m; this value is the air-entry value and also the height of capillary rise. Substitute 2.0 m into equation 4.10, and solve for the radius of the capillary tube. You will find that $r = 7 \ \mu\text{m}$, thus the diameter of the capillary tube is 14 μm.

4.5. Assuming complete wetting, a contact angle of 0°, and a surface tension of 0.07 N/m, we may use equation 4.10 to obtain

$$h_c = \frac{2\sigma \cos \theta}{\rho_l g r} = \frac{(2)(0.07 \text{ N/m})(1)}{(1000 \text{ kg m}^{-3})(9.81 \text{ N/kg})(9 \ \mu\text{m})(10^{-6} \text{ m}/\mu\text{m})} = 1.53 \text{ m}$$

Thus, the air entry value is 1.53 m or $-\rho_l g h_c$, which is -15.3 kPa.

4.6. The reason for this is quite simple: $\Pi = 0$ in pure water, and under atmospheric pressure at the "free" water surface, $\psi_m = 0$. As a result, $\psi_t = gz$.

4.7. First, we assume that $h_o = 0$ (no salt in the atmosphere); then, using equation 4.28 (rearranged), we have

$$h_m + 0 = \frac{\ln 0.50}{7.5 \times 10^{-5} \text{ m}^{-1}} = -9.24 \times 10^3 \text{ m}$$

As one can see, this is a large negative value. A tensiometer can only read about -1.0 m. As a result, the water in the tensiometer will evaporate through the porous cup, the entire tensiometer will fill with air, and no reading will be obtained.

4.8. The soil pressure is easily calculated by $\sigma_s = -\rho_b gz = -1500 \text{ kg m}^{-3} \times 9.81 \text{ N kg}^{-1} \times (-1 \text{ m}) = 14{,}715 \text{ N m}^{-2}$ or 14.715 kPa. Remember that an N m^{-2} is equal to a Pascal unit of pressure.

4.9. The porosity is determined as the volume of the soil pore divided by the volume of sample. Thus, $(2 \times 1/4\pi d^2 \times 2d) + (1.5^2 \pi d^2 \times 6d) = \pi d^3 + 13.5\pi d^3 = 14.5\pi d^3 =$ the pore volume. The volume of the sample $= 2.25\pi d^2 \times 10d = 22.5\pi d^3$. Hence, the porosity, ϕ, is

$$\phi = \frac{14.5\pi d^3}{22.5\pi d^3} = 0.644$$

4.10. (a) The volumetric water content, $\theta = \phi = 0.644$ at $\psi_m = 0$. **(b)** To draw the desorption curve one must calculate the capillary rise potential thus,

$$h_c > \frac{-2\sigma}{\rho_l g r} = \frac{-2(7 \times 10^{-2} \text{ N/m})}{1000 \text{ kg/m}^3 \times 9.81 \text{ N/kg} \times (7.5 \times 10^{-6} \text{ m})} = -1.90 \text{ m}$$

The desorption curve is shown in figure 4.17. Essentially, the soil pore is full as the pressure is

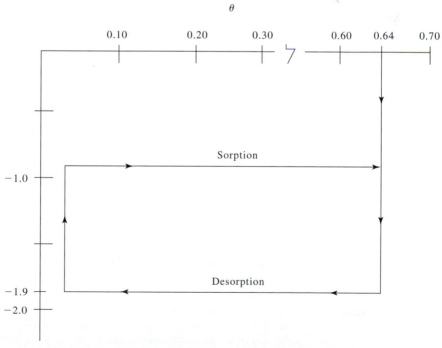

Figure 4.17   Hysteretic effects on soil pore

increased from 0 to -1.90 m. However, there will be a slight decrease in θ due to a change in the shape of the meniscus from flat to concave (hemispherical); but when the pressure reaches -1.90 m, the pore will empty completely.

4.11. To draw the sorption curve, one must consider that the small neck at the bottom of the soil pore will refill with water when the pressure is larger than the pore's air-entry value (-1.87 m), and the main body of the soil pore will fill when

$$h_c > \frac{-2\sigma}{\rho_l g r} = \frac{-2(7 \times 10^{-2}\,\text{N/m})}{1000\,\text{kg/m}^3 \times 9.81\,\text{N/kg} \times (15 \times 10^{-6}\,\text{m})} = -0.95\,\text{m}$$

When the pore is filled, $\theta = 0.444$, but when only the lower neck is filled,

$$\theta = \frac{1/4\pi d^2 \times 2d}{22.5\pi d^3} = 0.02$$

The corresponding sorption curve is shown in figure 4.15b. Thus, one can see the effects of drying and wetting of a soil; in other words, hysteresis.

ADDITIONAL QUESTIONS

4.12. Calculate the total potential (ψ_m, ignoring osmotic effects) of soil water on a volume basis at a depth of 63 cm, where the relative humidity (RH) of air in a soil pore is 98.9% (temperature = 25 °C).

4.13. Explain how a pressure plate apparatus works. Why is the air pressure equal to the negative of the ψ_m of the sample of soil water?

4.14. Calculate the percent pore volume in each of five size categories (pore diameter in microns) for the following.

ψ_m	θ
0	0.495
-60	0.393
-200	0.304
-1000	0.215
-10000	0.159

4.15. (a) Plot the soil-moisture characteristic curve (θ versus P) for the following soil, given that dry weight of soil = 556.0 g and sample volume = 347.5 cm^3.

P (bars)	Sample Weight (g)
0.10	674.4
0.50	658.9
1.00	639.9
2.00	631.1
3.00	617.6
8.00	601.0
15.00	573.2

(b) If one assumes field capacity is 0.10 bar, how much plant-available water is in this soil?

4.16. If you asume a temperature of 20 °C, a contact angle 0, density of $H_2O = 1\,\text{g/cm}^3$, and $g = 978.05$ cm/s^2, and stay in the cgs unit system, the height of rise of water due to capillarity is approximately $0.15/r$. Show this.

4.17. By using the simplified equation for capillarity ($h = 0.15/r$), we can calculate the maximum radius of water-filled pores at various pressures. Do this for each of the pressures in problem 4.15 (a). Express the size of the pore diameter in μm. Remember, the simplified equation requires that h and r be in cm (1 bar = 1022 cm of water).

4.18. In figure 4.18, the vertical distance from the surface of the mercury reservoir to the center of the ceramic cup is 20 cm and the value of z_{Hg} is 14.2 cm. What is ψ_m?

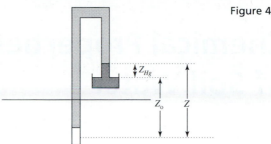

Figure 4.18

4.19. A uniformly-packed (homogeneous), U-shaped soil column has one end suspended in a container of water (see figure 4.19). The water level in the container has been kept constant and sealed against evaporative losses so that the soil column has reached equilibrium. As a result, there is no net water flow in the column. Calculate ψ_h (weight hydraulic potential) and its components (ψ_p, ψ_m, and ψ_z) for points A through F in the soil column.

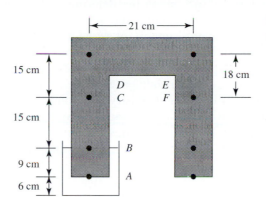

Figure 4.19   Data from Hanks and Ashcroft (1980)

4.20. Given a soil in which the liquid water is in equilibrium with a water table at -70 cm, and the reference level is at soil surface, find the values of ψ_m, ψ_z, ψ_p, and ψ_h throughout the soil profile to -110 cm.

4.21. Water is evaporating at a soil surface; find values of ψ_z, ψ_p, and ψ_h throughout the soil profile to -60 cm. In order to find ψ_h, measured or estimated values of ψ_m must be available. Make estimates of ψ_m for the conditions specified. *Note:* there may be variations in the gradient $\Delta\psi_h/\Delta z$ with depth; but for upward flow, the sign must always be negative.

4.22. This is a revisitation of material discussed in section 4.5. Relative humidity is the ratio of saturated vapor pressure to total vapor pressure. Calculate the relative humidity (in percent) at 25 °C for matric suctions of **(a)** -0.1, **(b)** -2.0, **(c)** -15.0, **(d)** -100.0, and **(e)** -1000.0 bars.

5

Chemical Properties and Principles of Soil Water

INTRODUCTION

Natural waters are never pure, always containing some type of impurity in the form of solids or gases. The quantity of such impurities determines the usefulness of water for various purposes, whether for agricultural, industrial, recreational, or personal use. Thus, the chemical properties of water become as important as the physical properties and quantity. Whereas it is commonly assumed that physical processes within a soil-water system are the dominant influences affecting change and equilibrium, chemical processes in ground and soil water have been found to be the controlling influences (Freeze and Cherry 1979).

A detailed study of water chemistry is beyond the scope of this chapter; the reader is referred to Freeze and Cherry (1979) and Stumm and Morgan (1996) for a comprehensive study of such material. The purpose of this chapter is to present the reader with a basic background in the chemical properties and principles that control the behavior of dissolved constituents in ground and soil water.

5.1 ORGANIC COMPOUNDS AND CONSTITUENTS

As a general rule, most organic compounds are those that contain carbon, with oxygen and hydrogen as the main elemental components for their structure. Humic and fulvic acids make up a large component of organics found in soil systems (see chapter 3); although little is known about them, they are not of great concern as contaminants within the unsaturated zone, even though they may be a source of contamination by complexation of heavy metals. The greatest concern for contamination in the unsaturated zone is caused by man-made organic chemicals. Most of these are in the form of pesticides, volatile organic compounds (VOCs) such as solvents or petroleum products, or semivolatile organic compounds (SVOCs) such as creosote or phenolic compounds. Adsorption, volatilization, chemical degradation, and biological uptake and degradation are mechanisms that help prevent their transport through the unsaturated zone. Most organic compounds have very low solubility in water; however, the constraints offered by low solubility are often insufficient to prevent migration to groundwater. Because many are toxic at low concentrations they often enter the biosphere, causing environmental problems.

Understanding sorption, biodegradation, and transport of organic chemicals through the unsaturated zone is essential in predicting their fate and transport in the environment. Most pesticides and other organics undergo some type of biochemical degradation, it being the dominant mechanism of organic chemical transformation in soils and aquifers. Some organic compounds are subject to microbial degradation but others are resistant; generally, the rate of biodegradation is limited by physical and chemical processes that lower solute concentrations. These include sorption, hydrodynamic dispersion, and factors such as temperature and water content that limit physiological activity of the microbial population in soils, or lower concentrations in the solution phase. Biodegradation is also limited by factors that hinder substrate uptake by microorganisms such as hydrophobicity, and intracellular or biochemical factors that limit utilization of the compound, that is, the presence of enzyme systems. The organics of greatest concern are refractory compounds; compounds not readily degraded by bacteria within the unsaturated zone nor easily removed in water-treatment (sewage) facilities. These compounds are often relatively soluble and nonvolatile. In-situ or on-site bioremediation that uses indigenous microorganisms is a current technology being applied in an attempt to increase biodegradation of such compounds.

Transport of Organic Constituents in the Unsaturated Zone

Organic particles within a soil usually vary in size, are widely dispersed, and are heterogeneously packed (see chapter 2). In addition, the unsaturated zone has both air and water within the pores and may contain cracks, root channels, and faunal tunnels. Consequently, concentration of chemicals in the unsaturated zone may vary locally because of preferential flow, aqueous phase chemical reactions, or phase transfer of the solute (i.e., to or from solid surfaces, or to or from the gas phase).

Examining adsorption–desorption (phase interaction between liquid phase and solid phase), the one-dimensional form of the advection–dispersion equation for homogeneous saturated porous media may be written as

$$-v\frac{\partial C_i}{\partial z} + D\frac{\partial^2 C_i}{\partial z^2} - \frac{\rho}{\phi}\frac{\partial S_i}{\partial t} = \frac{\partial C_i}{\partial t} \tag{5.1}$$

where the first, second, and third terms of the equation refer to advection, dispersion, and adsorption respectively; v is the linear velocity (length/time), D is the dispersion coefficient (length^2/time), C_i is the concentration of species i (mol/L), z is the distance along the direction of flow (in length units), S_i is the concentration of species i sorbed (mol/kg), ρ is the bulk density (kg/m^3), and ϕ is the porosity. For equations 5.1 to 5.5, the porosity, ϕ, is generally expressed in terms of water-filled porosity. As a result, θ is often substituted for ϕ; where θ is water filled porosity. The third term on the left side of equation 5.1 represents concentration change in solution caused by adsorption or desorption. This term can be expressed as

$$\frac{\rho}{\phi}\frac{\partial S_i}{\partial t} = \frac{\rho}{\phi}\frac{\partial S_i}{\partial C_i}\frac{\partial C_i}{\partial t} \tag{5.2}$$

where $(\partial S_i/\partial C_i)$ can be interpreted as the linear adsorption coefficient (K_{oc}), such that

$$\frac{\partial S_i}{\partial C_i} = K_{oc} \text{ (volume/kg)} \tag{5.3}$$

If the dispersion term of equation 5.1 is set to zero so that

$$D\frac{\partial^2 C_i}{\partial z^2} = 0 \tag{5.4}$$

then, the transport–adsorption equation can be rewritten as

$$-v\frac{\partial C_i}{\partial z} = \frac{\partial C_i}{\partial t}\left(1 + \frac{\rho}{\phi}K_{oc}\right) \tag{5.5}$$

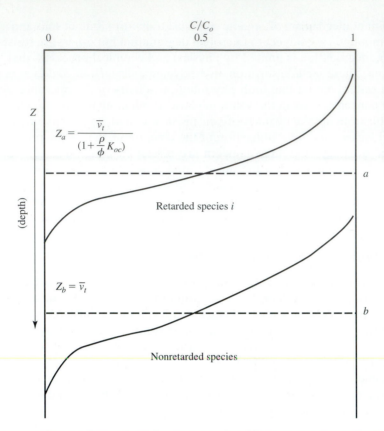

C/C_o

0   0.5   1

Z

$$Z_a = \frac{\overline{v}\,t}{\left(1 + \frac{\rho}{\phi}K_{oc}\right)}$$

(depth)

Retarded species i

-    a

$Z_b = \overline{v}\,t$

-    b

Nonretarded species

Figure 5.1   Advancing fronts of adsorbed and nonadsorbed solutes in a soil column represented by breakthrough curves where K_{oc} is the coefficient of adsorption, C is the measured concentration, C_o is the initial concentration, z is depth (subscripts a and b refer to depth of movement of retarded and nonretarded species), v is average velocity, t is time, ρ is bulk density, and ϕ is porosity

where the term in parenthesis is the retardation factor R. The retardation factor is written simply as the average linear velocity of groundwater v' over the average linear velocity of the retarded constituent v_i' (i.e., v'/v_i' or more simply, $1/R$). Generally, both velocities are measured where C/C_o is 0.5 in the concentration profile as in a breakthrough curve; see figure 5.1. The measured concentration during an experiment is C, while the initial concentration is represented as c_o. This is true only for soils; for water, we really do not have a C/C_o. In general, a conservative tracer is added along with a reactive tracer to determine v'. This is one of many models which have been proposed for investigating the transport of organics in the unsaturated zone since the 1970s. A majority of models have been written to predict biodegradation during transport through soil.

Coupled Processes

Because of the complexity involved, many models are written to include coupled processes. A coupled process is a combination of two or more of the basic flux laws (i.e., Newton's, Fourier's, or Fick's), which state that a nonuniform distribution of a state variable such as temperature produces a flux of the corresponding extensive quantity, such as heat. However, gradients of state variables corresponding to other extensive quantities, such as deformation flow, temperature flow, osmotic potential, and others can also contribute to the flux of an extensive quantity. Such phenomena are referred to as coupled processes or cross-effects. The governing differential equation that serves as the framework for most coupled-transport processes is described by

$$\left[\frac{\partial(\theta C_i)}{\partial t}\right] = [\nabla \cdot D\theta \,\nabla C_i] - [\nabla \cdot q C_i] - \left[\frac{\partial(\rho S_i)}{\partial t}\right] \pm \sum \Phi_i \qquad (5.6)$$

where C_i is the solution phase concentration of species i (M/L^3; L refers to length), S_i is the sorbed phase concentration of species i (M/M), t is time (T), ρ is bulk density (M/L^3), θ is the fractional volumetric water content (unitless), D is the local-scale hydrodynamic dispersion coefficient (L^2/T), q is Darcy's flux for water flow (L/T), ∇ is the Laplacian differential operator, and Φ_i is the rate(s) ($M/L^3 T$) of gain or loss via various sources and sinks. A detailed discussion of coupled phenomena is beyond the scope of this text; the reader is referred to Bear and Bachmat (1991) for a complete and detailed mathematical discussion of the subject.

5.2   MASS ACTION AND GOVERNING EQUATIONS

Most chemical reactions that take place in soil water in the unsaturated zone are spontaneous, and usually proceed until a dynamic equilibrium is attained. At equilibrium, both forward and reverse reactions take place at the same rate, and concentrations then remain relatively constant over time. While this is a simple assumption to make when sampling soil-pore water, it is not necessarily true because the soil-pore water system is dynamic.

All chemical systems tend toward equilibrium. The relation that governs the relative proportions of each of these chemical reactions and their products is simple, usually written as

$$aA + bB \rightleftharpoons eE + fF \tag{5.7}$$

where a, b, e, and f are the number of moles present of each of the reactants A and B and products E and F, respectively. At constant temperature, the condition that is fulfilled at equilibrium is

$$\frac{[E]^e [F]^f}{[A]^a [B]^b} = K_c \tag{5.8}$$

where the quantities in brackets are the equilibrium molar concentrations and K_c is the thermodynamic equilibrium constant. This relation is known as the law of mass action and is very useful in the analysis of soil-water samples. The left side of equation 5.8 is termed the mass action expression. This term is constructed using the molar coefficients in the balanced equation as exponents for the appropriate concentration.

The important point to remember about the law of mass action is that there are no restrictions on the individual concentrations of any reactant or product. The only requirement for equilibrium is that when these concentrations are substituted into the mass action expression, the fraction is equal to K_c. It should be noted that a given reaction does not necessarily reach immediate equilibrium; this may take minutes, hours, even years, depending on the reactant and disturbance to the system. Additionally, any change in temperature, pressure, or both can shift the system toward a new equilibrium. If disturbances are constant, groundwater equilibrium essentially will never be reached, usually the case with soil-pore water.

When chemical reactions involve gases, the concentration of each of the reactants and the products are proportional to the partial pressures. The equilibrium expression or constant for these reactions, K_p, may be expressed similarly to that for reactions in water, except that they are written in terms of partial pressures rather than concentrations, in the form of

$$\frac{P^e E\, P^f F}{P^a A\, P^b B} = K_p \tag{5.9}$$

where P is the partial pressure (atm) and K_p is used to denote equilibrium constants derived from partial pressures, not concentrations. The rate law cannot be predicted for a chemical reaction based on the balanced overall equation, but this does not affect the kinetic interpretation of the law of mass action. Without showing the mathematics of the proof (which can be found in any general inorganic chemistry text), one always arrives at the same requirement for equilibrium, regardless of the mechanism of the reactions.

When working with dynamic systems such as in the unsaturated zone, temperature and pressure are not always constant; they may vary at any time during the course of an experiment, especially during seasonal fluctuations in weather patterns. The basic relations which influence equilibrium due to temperature change are

$$\frac{d \ln K}{dT} = \frac{\Delta H^\circ}{RT^2} \tag{5.10}$$

Integrating from initial temperature (T_1) to final temperature (T_2), we obtain

$$\ln \frac{K_2}{K_1} = \int_{T_1}^{T_2} \frac{\Delta H^\circ}{RT^2} \, dT \tag{5.11}$$

where ΔH° is change in enthalpy (kJ mol^{-1}), K_1 and K_2 are equilibrium constants at initial and final condition (unitless), R is the gas constant (8.314 J K^{-1} mol^{-1}), and T is temperature (Kelvin). If ΔH° is independent of temperature, we may write

$$\ln K = -\frac{\Delta H^\circ}{RT} + C \tag{5.12}$$

or

$$\ln \frac{K_2}{K_1} = \frac{\Delta H^\circ}{R} \left(\frac{1}{T_1} - \frac{1}{T_2} \right) \tag{5.13}$$

where C is a constant. When the heat capacity, C_p°, of a reaction is independent of temperature, then

$$\Delta H_2^\circ = \Delta H_1^\circ + \Delta C_p^\circ (T_2 - T_1) \tag{5.14}$$

By integrating equation 5.10 and combining with equation 5.14, we obtain

$$\ln \frac{K_2}{K_1} = \frac{\Delta H_1^\circ}{R} \left(\frac{1}{T_1} - \frac{1}{T_2} \right) + \frac{\Delta C_p^\circ}{R} \left(\frac{T_1}{T_2} - 1 - \ln \frac{T_1}{T_2} \right) \tag{5.15}$$

or

$$\ln K = B - \frac{\Delta H_0}{RT} + \frac{\Delta C_p^\circ}{R} \ln T \tag{5.16}$$

where B and ΔH_0 are constants. When the heat capacity is a function of temperature and is given for each reactant and product in the form

$$C_p^\circ = a_i + b_i T + C_i T^2 \tag{5.17}$$

then the heat capacity can be mathematically expressed as

$$\frac{d\Delta H^\circ}{dT} = \Delta a + \Delta b T + \Delta c T^2 = \Delta C_p^\circ \tag{5.18}$$

which upon integration will yield

$$\ln K = B - \frac{\Delta H_0}{RT} + \frac{\Delta a}{R} \ln T + \frac{\Delta b}{2R} T + \frac{\Delta b}{6R} T^2 \tag{5.19}$$

where Δa is $\Sigma \, v_i a_i b_i, \ldots,$ B and ΔH_0 are constants as before, and v_i is the stoichiometric coefficient, that is, the relative molar number. For example, one may be working with H_2SO_4; the molar number for H, S, and O would be 2, 1, and 4.

The basic relations which influence equilibrium due to pressure change are

$$\left(\frac{\partial \mu_i}{\partial P}\right)_T = \overline{V}_i \tag{5.20}$$

and

$$\left(\frac{\partial \mu_i^0}{\partial P}\right)_T = \overline{V}_i^0 \tag{5.21}$$

where V_i and V_i^0 are the partial molar volumes of i under actual conditions, as well as under defined standard-state conditions. The rate of change of molar volume with pressure (the standard partial molar compressibility, $\overline{k}_i^0$) is represented mathematically by

$$\overline{k}_i^0 = -\left(\frac{\partial \overline{V}_i^0}{\partial P}\right)_T \tag{5.22}$$

and

$$\Delta V = \sum v_i \overline{V}_i$$

$$\tag{5.23}$$

$$\Delta V^0 = \sum v_i \overline{V}_1^*$$

where ΔV and ΔV^0 are the changes in volume of the reaction under actual and defined standard-state conditions. Also,

$$\left(\frac{\partial \ln K}{\partial P}\right)_T = -\frac{\Delta V^0}{RT} \tag{5.24}$$

In this instance, K is the equilibrium constant. If ΔV^0 is pressure-independent, then

$$\ln \frac{K_p}{K_1} = -\frac{\Delta V^0 (P - 1)}{RT} \tag{5.25}$$

If Δk^0 (where $\Delta k^0 = \sum v_i \overline{k}_i^0$) is pressure-independent, then

$$\ln \frac{K_p}{K_1} = -\frac{1}{RT}\left[\Delta V^0(P - 1) - \frac{1}{2}\Delta K^0 (P - 1)^2\right] \tag{5.26}$$

However, if Δk^0 is a function of pressure, then

$$RT \ln \frac{K_p}{K_1} = -\Delta V^0(P - 1) + \Delta K^0(B + 1)(P - 1) - (B + 1)^2 \ln\left(\frac{B + P}{B + 1}\right) \tag{5.27}$$

Here B is pressure-independent, but depends on temperature.

QUESTION 5.1

The heat capacity of N_2 can be represented by

$$C_p(N_2) = 26.984 + 5.91 \times 10^{-3}T + (-3.377 \times 10^{-7})T^2 \tag{5.28}$$

How much heat is required to heat a mole of N_2 from 300 K to 1000 K at constant pressure?

5.3   ACTIVITY COEFFICIENTS

Standard Methods

For a simple definition as applied to solutions, an activity coefficient, γ, having units of kg/mol, is a correction factor that compensates for non-ideal behavior. Ideal behavior for a solution is defined as a lack of interactions between the molecules of a liquid; that is, the solution would behave as if any molecular interactions present had a negligible effect on the intensive properties of the solution, such as Gibbs' energy, entropy, enthalpy, or volume. However, due to the driving forces for mixing, et cetera, (ionic strength, electrostatic attraction, hydration shells, changes in entropy), very few solutions are ideal. Thus, the activity of an ion is related to its concentration and is adjusted for non-ideal behavior by an activity coefficient. Generally, if molality of a chemical is known, both equilibrium and solubility of a substance can be calculated. The chemical activity of a substance *must* be calculated before the law of mass action can be applied. The general theoretical expressions for calculation of the activity coefficient are the Debye-Hückel:

$$\log \gamma_i = -Az^2 \sqrt{I} \tag{5.29}$$

Extended Debye-Hückel:

$$\log \gamma_i = -Az^2 \frac{\sqrt{I}}{1 + a_i B \sqrt{I}} \tag{5.30}$$

the Güntelberg:

$$\log \gamma_i = -Az^2 \frac{\sqrt{I}}{1 + \sqrt{I}} \tag{5.31}$$

and the Davies:

$$\log \gamma_i = -Az^2 \left(\frac{\sqrt{I}}{1 + \sqrt{I}} - 0.3I \right) \tag{5.32}$$

where I is the ionic strength equal to $\frac{1}{2} \Sigma m_i z_i^2$ (m is the molality of ith ion and z is the ion charge), $A = 1.82 \times 10^6 (DT^{-3/2})$ (D is the dielectric constant), $B = 50.3(\varepsilon T)^{-1/2}$, and a is the effective diameter of the ion in angstroms. By calculation and at a pressure of one atmosphere, the values of A and B are 0.5085 and 0.3281 $\mathring{A}^{-1}$ in water at 25 °C. The range of solution ionic strength (M) for applicability is $< 10^{-2.3}$, $< 10^{-1}$, $< 10^{-1}$, and < 0.5 for the Debye-Hückel, extended Debye-Hückel, Güntelberg, and Davies' equations, respectively. It should be noted that the range given for Güntelberg is considered useful in solutions of several electrolytes, while the other equations are useful only for single-ion activities. The general relation between activity and molality may be written as $\alpha = \gamma m$, where α is the chemical activity (dimensionless), m is the molal concentration, and γ is the activity coefficient. Before $\log \gamma_i$ can be determined, I must be known or calculated. As a note of interest, the Debye-Hückel and its extension presented here are simplifications of the electrochemical equations described in chapter 3, for charged-clay particles.

Quantities that are important to the environmental scientist can be estimated from knowledge of the activity coefficients. Examples of these would include Henry's law constants, solubility limits, solution equilibria, and octanol–water partition coefficients. An often-used example of this shows that the octanol–water partition coefficient, K_{ow}, is proportional to the ratio of activity coefficients of a component α in water and octanol, described mathematically as

$$K_{ow} = 0.151 \frac{(\gamma_w^\alpha)^\infty}{(\gamma_o^\alpha)^\infty} \tag{5.33}$$

where γ_w and γ_o are the infinite-dilution activity coefficients of component α in water and octanol, respectively. Octanol–water activity coefficients range in value from 0.4 to 10^7 and are dimensionless. For inorganic solutes in water at infinite dilution, the activity coefficient would be one and activity will equal molality. For a binary system (a mixture of two fluids in which interaction between ion pairs is not required), the basic equation that describes the activity coefficient is

$$G^M = RT(n_1 \ln x_1 + n_2 \ln x_2) + RT(n_1 \ln \gamma_1 + n_2 \ln \gamma_2) \tag{5.34}$$

where G^M is the energy of mixing, R is the gas constant, T is the temperature, n is the number of moles in compound 1 or 2 and x is the mole fraction, that is, $x_2 = n_2/(n_1 + n_2)$. The first term on the right-hand side of equation 5.34 is normally referred to as G^I (the Gibbs free energy of mixing for an ideal mixture), and the second term as G^E (the excess Gibbs free energy). The expression for G^E can be rewritten to obtain

$$\left(\frac{\partial G^E}{\partial n_1}\right) = RT \ln \gamma_1; \quad \left(\frac{\partial G^E}{\partial n_2}\right) = RT \ln \gamma_2 \qquad (5.35)$$

Consequently, one may relate activity coefficients to composition in mole fraction.

In addition to the standard methods for calculating activity coefficients shown above, two other methods can be of value in calculations. The first is the infinite-dilution activity coefficient, described in the next section; the second is the UNIFAC (UNIQUAC Functional Group Activity Coefficients) method, which estimates the concentration dependence of γ_1 directly (see Nicolaides and Eckert 1978).

Infinite-Dilution Activity Coefficients

The infinite-dilution activity coefficient alone (see equation 5.36 and subsequent definition) can be used to estimate Henry's law constants, octanol-water partition coefficients, and solubility limits. It can also be used to estimate the parameters in any two-parameter equation, such as the van Laar equation, $g^E = [(Ax_1x_2)/\{x_1(A/B) + x_2\}]$ (Reid, Pravsnitz, and Sherwood 1977), which is used to estimate activity coefficients at any composition. The infinite-dilution method requires a knowledge of the molecular structure of the compounds involved, because it relates γ^∞ to the molecular structures of both solvent and solute molecules by using an equation that contains the number of carbon atoms (N) for the solute and solvent; if the solvent were water, with no carbon atoms, $N_2 = 0$. This equation may be written in the form

$$\log \gamma_1^\infty = A_{1,2} + B_2 \frac{N_1}{N_2} + \frac{C_1}{N_1} + D(N_1 - N_2)^2 + \frac{F_2}{N_2} \qquad (5.36)$$

$A_{1,2}$ is a coefficient dependent on the nature of the solute and solvent functional groups; B_2 is a coefficient dependent only on the nature of the solvent functional group; C_1 is a coefficient dependent only on the solute functional group; D is a coefficient independent of both solute and solvent functional groups; and F_2 is a coefficient essentially dependent on the nature of solvent functional groups.

If a secondary or tertiary alcohol is contained within a mixture, the C term is modified so that it becomes $C_1(1/N_1' + 1/N_1'')$ and $C_1(1/N_1' + 1/N_1'' + 1/N_1''')$ for secondary and tertiary alcohols, respectively. The "primed" Ns are the number of carbon atoms of the respective branches counted from polar groupings: such as 6 for benzene, 0 for water. A single value for C_1 will work for most primary alcohols with little loss of accuracy. The F_2 term can be similarly represented if alcohol is the solvent. Additional modifications are needed for acetals, cyclic hydrocarbons, and ketones; see Pierotti, Deal, and Derr 1959, and Reid, Pravsnitz, and Sherwood 1977. For example, a mixture of benzene and water would require a modification of equation 5.36, such that the second term on the right-hand side becomes $B_2(N_1 - 6)$, the third term becomes $C_1[1/(N_1 - 4)]$, and the remaining terms are eliminated (see Reid, Pravsnitz, and Sherwood 1977).

The basic steps to follow when calculating an infinite-dilution activity coefficient are: **(1)** draw the structure of the chemical involved; **(2)** obtain appropriate correlation constants (see table 5.1) and modify equation 5.36 if needed (see Reid, Pravsnitz, and Sherwood 1977); **(3)** substitute constants and modifications into equation 5.36; **(4)** if using an aromatic compound with water as the solvent, calculate $\log \gamma_1^\infty$ for the unsubstituted hydrocarbon, and substitute any corrections to $\log \gamma_1^\infty$; and **(5)** the antilog (corrected if necessary) of the $\log \gamma_1^\infty$ will yield γ_1^∞.

TABLE 5.1   Correlation Constants for Activity Coefficients at Infinite Dilution

Solute	Solvent	Temp. °C	$A_{1,2}$	B_2	C_1	D	F_2
n-primary alcohols	water	25	−0.995	0.622	0.558	0	0
		60	−0.755	0.583	0.460	0	0
		100	−0.420	0.517	0.230	0	0
Secondary alcohols	water	25	−1.220	0.622	0.170	0	0
		60	−1.023	0.583	0.252	0	0
		100	−0.870	0.517	0.400	0	0
Tertiary alcohols	water	25	−1.740	0.622	0.170	0	0
		60	−1.477	0.583	0.252	0	0
		100	−1.291	0.517	0.400	0	0
n-ketones	water	25	−1.475	0.622	0.500	0	0
		60	−1.040	0.583	0.330	0	0
		100	−0.621	0.517	0.200	0	0
n-nitriles	water	25	−0.587	0.622	0.760	0	0
		60	−0.368	0.583	0.413	0	0
		100	−0.095	0.517	0	0	0
n-alkylbenzenes	water	25	3.554	0.622	−0.466	0	0
Water	n-alcohols	25	0.760	0	0	0	−0.630
		60	0.680	0	0	0	−0.440
		100	0.617	0	0	0	−0.280
Water	Benzene	25	3.04	0	0	0	−3.14
Water	n-ketones	25	1.857	0	0	0	−1.019
		60	1.493	0	0	0	−0.730
		100	1.231	0	0	0	−0.557
Ketones	n-alcohols	25	−0.088	0.176	0.500	−0.00049	−0.630
		60	−0.035	0.138	0.330	−0.00057	−0.440
		100	−0.035	0.112	0.200	−0.00061	−0.280

Data from Pierotti, Deal, and Derr (1959) and Reid, Pravsnitz, and Sherwood (1977)

QUESTION 5.2

You have extracted a sample from a 2-m depth within the unsaturated zone using a suction lysimeter, in which the molal concentration of calcium is 0.00633. Calculate the activity coefficient γ_i and chemical activity α. Assume 25 °C, I = 0.0281, and a_i = 6 Å, that is, 6×10^{-8} cm.

QUESTION 5.3

At a chemical spill near an industrial complex, a colleague has extracted an unsaturated zone solution sample and determined it to be a mixture of benzene and water. What are the infinite-dilution activity coefficients for the system?

5.4   EQUILIBRIUM AND FREE ENERGY

When dealing with contaminant transport through the unsaturated zone, it is important to know whether the system in question is at equilibrium, or in the process of immediate change. When a system is at equilibrium, it cannot undergo a spontaneous change; in this state, any small change that may take place is said to be reversible. Any irreversible change at this point would result in an equilibrium shift, that is, the original equilibrium would be displaced. The equilibrium of a system can be described by the entropy of a system. However, for the second law of thermodynamics at constant temperature and pressure, the Gibbs free energy, G, provides a more convenient measure of the thermodynamic property than the entropy. The Gibbs free energy of any chemical reaction represents the driving force of the reaction (the change in internal energy per unit mass), which also yields a measure of the system's ability to perform nonmechanical work.

The relation for the Gibbs free energy of a reaction (ΔG) for mixtures to the composition of the system is obtained by substituting the expression for the chemical potential, μ_i, (equal to $\partial G/\partial n_i$ at constant temperature and pressure) in terms of the species activity $\{i\}$

$$\mu_i = \mu_i^0 + RT \ln \{i\} \tag{5.37}$$

into the expression for ΔG (change in Gibbs energy of reaction)

$$\Delta G = \sum_i v_i \mu_i \tag{5.38}$$

Upon rearrangement, this yields

$$\Delta G = \Delta G^0 + RT \ln \prod_i \{i\}^{v_i} \tag{5.39}$$

ΔG^0 is the standard Gibbs free energy of the reaction. The term $\prod_i \{i\}^{v_i}$ is an algebraic shorthand for a continuation of products, also known as the reaction quotient, Q, and defined as

$$Q = \prod_i \{i\}^{v_i} = \frac{\{C\}^{v_C} \{D\}^{v_{D\ldots}}}{\{A\}^{v_A} \{B\}^{v_{B\ldots}}} \tag{5.40}$$

At equilibrium, $\Delta G = 0$ and the numerical value of Q is K, thus, $\Delta G^0 = -RT \ln K$. For most conditions this may be rewritten in the form

$$\Delta G^0 = RT \ln \frac{Q}{K} \tag{5.41}$$

where K is the equilibrium composition of the mixture. This is the primary relation in chemical thermodynamics of mixtures under most conditions. Thus, comparison of Q (actual composition) with K (equilibrium composition) yields a test for equilibrium conditions ($\Delta G = 0$). When comparing the standard Gibbs free energy of formation ΔG^0 at a standard-state pressure of $P^0 = 1$ bar with the value of the standard Gibbs energy of formation at $P* = 1$ atm, it is important to know that

$$\Delta G^0 = \Delta G* - \left[RT \ln \left(\frac{P*}{P^0} \right) \right] \delta$$
$$\Delta G^0 = \Delta G* - (0.109 \text{ J K}^{-1} \text{mol}^{-1}) T \delta \tag{5.42}$$

where δ is the net increase of moles of gas in the formation reaction of the substance from its elements. The standard Gibbs free energy of formation may therefore be calculated by

$$\Delta G^0 = \Delta H^0 - T \left[S^0 - \sum v_i S_i^0 (\text{element}) \right] \tag{5.43}$$

where $\sum v_i S_i^0$ (element) is the sum of the standard entropies of the elements in the formation reaction. The standard Gibbs free energies of formation, of elements in their reference states, are zero at all temperatures. Thus, for a chemical reaction to proceed spontaneously as written, there must be a net decrease in free energy; ΔG must be less than zero (negative). If $\Delta G > 0$, the reaction will proceed from right to left, when $\Delta G = 0$, the reaction is in equilibrium and will not proceed in either direction.

Occasionally, both liquid and gas samples in the unsaturated zone are difficult to obtain. In the absence of data therefore, the standard Gibbs free energy of formation of a substance may be estimated for the perfect gas state using the Benson group method (Benson 1978; Benson et al. 1969). The estimation of the standard enthalpy of formation ΔH^0 and entropy of mixing ΔS_{mix} is necessary for this procedure. Although not discussed here, the reader is free to peruse a standard physical chemistry text to refresh his or her memory on these principles.

QUESTION 5.4

As a scientist working for an environmental consulting firm, you have been assigned a project working on nitrogen dynamics in the unsaturated zone. To predict the movement of nitrates at your site and estimate the time at which the nitrate will reach ground water, the sophisticated model you are using requires the partial pressure of nitrous oxide (NO) in air. What is the equilibrium partial pressure of NO in air at 298 K using the following equation? Assume the value of $\Delta G = 77.77$ kJ mol^{-1}.

$$\frac{1}{2} N_2(g) + \frac{1}{2} O_2(g) \leftrightharpoons NO(g) \tag{5.44}$$

QUESTION 5.5

Estimate the standard Gibbs energy of formation of n-pentane at 298.15 K. (In lieu of insubstantial data, one may substitute the Gibbs free energy of a closely related alcohol for this value.)

5.5 ELECTRONEUTRALITY

Equilibrium conditions between acids and bases can be described by various fundamental equations involving the law of mass action. The equations define both acidity constants and the ion product of water for a given reaction; additional equations are required to describe the concentration and the electroneutrality condition. The underlying principle of electroneutrality is that, on a large or macroscopic scale, an equal number of positive and negative charges usually exist, so that a neutral electric-charge condition exists in the solution. Perhaps the simplest example of this would be a mixture of salt and water

$$Na^+ + 2H^+ = Cl^- + O^{-2} \tag{5.45}$$

Thus, the positive charge for one species balances the negative charge of the other, to yield a condition of electroneutrality. This is more aptly expressed by

$$\sum zm_{cat} = \sum zm_{an} \tag{5.46}$$

where z is the valence of the species and m_{cat} and m_{an} represent the molality of the cation and anion species, respectively; this is the electroneutrality equation.

As a general rule, although the negative and positive charges balance out on a macroscopic scale, they will not necessarily balance on a microscopic scale that is more indicative of conditions in the unsaturated zone. For example, at low-water contents in the unsaturated zone, all charges on a particle's surface may not be satisfied through exchange with opposite charges in soil solution; complete wetting does not allow exchange to occur, or there may be occluded pockets where no exchange occurs due to lack of solution. Such conditions can cause a charge imbalance. The charge-balance error E_c, is a measure of the deviation from electroneutrality. This is normally expressed as

$$E_c = \frac{\sum zm_c - \sum zm_a}{\sum zm_c + \sum zm_a} \times 100 \tag{5.47}$$

where all components have been described previously. Chemicals with no net charge, such as silicon, do not affect this relation. In certain cases a significant deviation from equality of cations and anions may occur. This is normally caused by laboratory errors in analysis, or because of high concentrations of species that may not have been included in the charge-balance calculation. Consequently, the charge balance is a good manifestation of the accuracy of data extracted from water-sample analyses. Normally, an $E_c < 5\%$ is an acceptable limit of error for many laboratory analyses, however, this is dependent on the protocol being followed and tolerance limits allowed. The charge balance applies just as well to surface-water or precipitation samples.

5.6 ACID DISSOCIATION

Chemical properties that determine acid–base interactions between a chemical and the soil matrix exert a major influence on partitioning between gaseous, solid, and solution components. The ionization of an organic acid or base can be markedly different from the corresponding neutral molecule in adsorption, bioconcentration, solubility, and toxicity characteristics. For organic chemicals that are weakly acidic, HA, the acid dissociation constant, K_a, is defined as the equilibrium constant for the reaction. For example,

$$HA + H_2O \rightarrow H_3O^+ + A^- \tag{5.48}$$

Polyprotic chemicals (chemicals that have more than one acidic proton) undergo a successive dissociation. Each dissociation has its own equilibrium constant, K. The dissociation is

$$H_XA + H_2O \rightleftharpoons H_3O^+ + H_{X-1}A^-$$
$$H_{X-1}A^- + H_2O \rightleftharpoons H_3O^+ + H_{X-2}A^{-2}$$
$$\vdots \tag{5.49}$$
$$HA^{-(X-1)} + H_2O \rightleftharpoons H_3O^+ + A^{-X}$$

As a result,

$$K_a = \frac{a_{H_3O^+} \, a_A}{a_{HA} a_{H_2O}} \tag{5.50}$$

where a_i is the activity of species, i, in an aqueous solution. It is generally assumed that the solution is dilute enough that the activity of water is unaffected by the presence of solute(s). This is equivalent to choosing pure water at standard state, and is useful for concentrations < 0.1 M. Thus, the activity of water would be unity and equation 5.50 may be rewritten

$$K_a = \frac{(\gamma_{H_3O}[H_3O^+]) \, (\gamma_{A^-}[A^-])}{\gamma_{HA}[HA]} \tag{5.51}$$

where γ_i is the molar activity coefficient and the brackets are the molar concentration (mol L^{-1}) of species i. Further simplification of equation 5.51 can be obtained by assuming that all of the activity coefficients are unity and activities equal concentrations. Equation 5.51 then becomes

$$K_a = \frac{[H^+][A^-]}{[HA]} \tag{5.52}$$

If the ionic strength is maintained constant, it is convenient to use the apparent acid dissociation constant, K_{app}, that is expressed in terms of concentrations, such that

$$K_{app} = \frac{K_a \gamma_{HA}}{\gamma_A^-} = \frac{10^{-pH}[A^-]}{[HA]} \tag{5.53}$$

By taking the negative logarithm of both sides of equation 5.52, we get

$$-\log K_a = -\log [H^+] - \log \frac{[A^-]}{[HA]} \quad \text{also} \quad pK_a = pH - \log \frac{[A^-]}{[HA]} \quad (5.54)$$

where $pK_a = -\log K_a$. One may observe that the concentration of an organic acid in the dissociated $[A^-]$ and free $[HA]$ forms are equal when $pH = pK_a$. The ratio of A^- to HA will increase by an order of magnitude for each unit of pH above pK_a. The acid–base behavior of a weakly basic organic compound can be treated in much the same way when adding a base to water $[B + H_2O \rightarrow BH^+ + OH^-]$, so that the K_b of the base may be written as

$$K_b = \frac{\gamma_{HB} + \gamma_{OH^-}}{\gamma_B \gamma_{H_2O}} \quad (5.55)$$

The relation between the dissociation of an acid and a base can be mathematically described as

$$K_a = \frac{K_w}{K_b} \quad (5.56)$$

where K_w is the autodissociation constant of water. Consequently, a decrease in K_a for BH^+ is reflected in an increase in K_b for B. A stronger base automatically corresponds to a weaker conjugate acid. A uniform scale can be applied over the complete range of organic acid–base behavior in aqueous media by using the pK_a of the conjugate acid to measure the strength of the base.

The Hammett correlation can be used for aromatic acids to estimate K_a, and is expressed as

$$\log \frac{K_a^x}{K_a^o} = \sigma\rho \quad (5.57)$$

where the numerator is the acid dissociation constant of the substituted compound (x), the denominator is the dissociation constant of the parent compound (O), σ is the substituent constant and ρ is the reaction constant. The substituent constant is derived from the groups attached to the parent compound. For example, p-tert butyl benzoic acid has the tert butyl group $C(CH_3)_3$ attached to it, which is a hydrocarbon group: $\sigma = -0.197$. Other basic groups include carbonyl, nitrogen, halogens and alkyl halide, hydroxy and alkoxy, sulfur, phosphorus, and other miscellaneous groups. The correlation can be rewritten for problem-solving convenience such that

$$K_a^x = K_a^o 10^{\sigma\rho} \quad \text{also} \quad pK_a^x = pK_a^o - \sigma\rho \quad (5.58)$$

To estimate the dissociation of a substituted acid, one must select an appropriate parent compound with known K_a^o and ρ values; select substituent constant values, and calculate K_a^x or pK_a^x. The substituted acid will provide a suitable comparison to the compound in question, especially if you do not know the reaction constant for the compound being tested. Selecting such parameters from a similar, known compound may be prudent. For aliphatic systems, the Taft relation (Shorter 1972) can be used for estimating acid dissociation constants. Both of these methods are accomplished by applying a linear free energy relation (ΔF), which is an empirical correlation between the standard free energies of reaction (ΔF°) or activation $(\Delta F^\ddagger)$. One may remember that for an equilibrium process $\Delta G^\circ = -RT \ln K$, and for a kinetic process $\Delta G^\ddagger$ is proportional to $-RT \ln K$. For greater insight on the process, the reader may wish to consult Wells (1968).

QUESTION 5.6

What is the dissociation constant (estimated) for 3-chloro-4-methoxyphenyl-phosphonic acid? Choose an appropriate parent compound, such as phenylphosphonic acid. Thus, $\rho = 0.755$, $\mathrm{p}K_a^o = 1.83$, $K_a^o = 1.46 \times 10^{-2}$, and the substituent constant, σ, for 3-chloro-4-methoxy is 0.268.

5.7   HYDROLYSIS

The chemical transformation process in which an organic molecule, R—X, reacts with water to form a new carbon–oxygen bond is termed hydrolysis. When the new bond is formed, a C—X bond from the original organic molecule is detached. The reaction is typically written as

$$R\!-\!X \xrightarrow{\ H_2O\ } R\!-\!OH + X^- + H^+ \tag{5.59}$$

Hydrolysis is probably one of the most important reactions of organic compounds in aqueous solutions. It is a very significant environmental-fate process for organic chemicals. The process is similar to the transformation of nitrogen in that many reaction steps take place, thus changing the organic chemical from its original state. Some of the functional groups which can hydrolyze under environmental conditions are: amides, alkyl halides, amines, carbamates, carboxylic acid esters, nitriles, phosphonic acid esters, and sulfonic acid esters. During hydrolysis, a water or hydroxide ion (nucleophile) attacks a carbon atom, phosphorus atom, or other electrophile, which then displaces some groups such as chloride, phenoxide, and others. The displaced groups are called leaving groups.

Hydrolysis reactions usually fit within two unique patterns; these are termed S_N1 (substitution, nucleophilic, unimolecular) and S_N2 (substitution, nucleophilic, bimolecular). The S_N1 process (kinetic) is characterized by a rate independent of concentration and nucleophile nature; a rate enhancement by electron-donor substituents on the central atom; and the formation of racemic products from optically active material. The term optically active refers to the rotation of the plane of plane-polarized light when it passes through a solution. It is believed that the S_N1 pattern is a two-step process. The first step is the ionization of R—X which yields a planar carbonium ion and is the rate-determining step; the second step is a rapid nucleophilic attack on the product of the first step to form an R—OH group. This can be described by equations such that

$$R\!-\!X \xrightarrow{\ slow\ } R^+ + X^-$$
$$R^+ + H_2O \xrightarrow{\ fast\ } R\!-\!OH + H^+ \tag{5.60}$$

As a result, the limiting S_N1 process is favored by R— systems that form a stable carbonium ion (such as tributyl systems), by X— systems that are good leaving groups such as halide ions and sulfonate ions, and by high-dielectric-constant solvents, such as water.

For the S_N2 process, the rate is dependent on both the concentration and identity of the nucleophile and the optically active starting material. The process is essentially one step that can be described by

$$H_2O + R\!-\!X \rightarrow [H_2O \cdots R \cdots X] \rightarrow H^+ + R\!-\!OH + X^- \tag{5.61}$$

As may be seen, the S_N2 process involves a nucleophilic attack on the central atom at the side opposite the leaving groups. This limiting process is favored by R— systems with low steric hindrance and a low carbonium-ion stability (such as methyl and other alkyl (primary) systems), by X— systems that are poor leaving groups such as CH_3CH_2O—, and by organic solvents like acetone.

The rate of disappearance of R—X is proportional to the compound concentration, such that

$$\frac{-d[\text{R—X}]}{dt} = k_T[\text{R—X}] \tag{5.62}$$

where k_T is the hydrolysis rate constant. Thus, the hydrolysis of the organic chemical in water is first-order and implies that the hydrolysis half-life of R—X is independent of the R—X concentration. As a result, data obtained at one concentration (whether high or low) may be extrapolated to another concentration assuming pH and temperature are constant. The hydrolysis half-life may be determined from

$$t_{1/2} = \frac{0.693}{k_T} \tag{5.63}$$

The typical procedure for estimating the rate of hydrolysis for an organic chemical is to: **(1)** categorize the chemical in terms of functional groups; **(2)** determine if k_H (rate constant for specific acid–catalyzed hydrolysis) and/or k_{OH} (rate constant for specific base–catalyzed hydrolysis) in the groups which can be hydrolized are significant in a pH range of 5–8; **(3)** estimate k_H, k_{OH}, and/or k_o from correlations; **(4)** calculate k_T for the pH(s) of interest from

$$k_T = k_H[\text{H}^+] + k_o + k_{OH}[\text{OH}^-] \tag{5.64}$$

and **(5)** adjust for temperature, if not 25 °C. This is done by

$$\log k_{25\,°\text{C}} = \log k_{T_2} - 3830\left(\frac{T_2 - 298}{298T_2}\right) \tag{5.65}$$

where T_2 is a temperature other than 25 °C for k_H, k_{OH}, and/or k_o. Also, T_2 is in K and an average of $\Delta H^{\ddagger}$ or E_A value of 17.5 kcal mol^{-1} is also inherently assumed.

QUESTION 5.7

What is k_H and the estimated hydrolosis half-life for ethyl p-nitrobenzoate? (Assume $k_H^o = 10^{-7}$ M^{-1} s^{-7} for ethyl benzoate in water at 25 °C, $\rho = 0.11$ for ethyl benzoate hydrolysis, $\sigma = 0.778$ for a p-nitro substituent.)

QUESTION 5.8

What is the rate constant for the hydrolysis of methyl p-nitrobenzoate? (Assume $k_{OH}^0 = 7.2 \times 10^{-3}$ M^{-1} s^{-1}, $\rho = 2.38$, and $\sigma = 0.778$.)

QUESTION 5.9

Estimate the rate constant for neutral hydrolysis, k_o, for p-methylbenzyl chloride. (Assume $k_o^0 = 6.2 \times 10^{-6}$ s^{-1}, 25 °C, $\rho = -1.31$, and $\sigma = -0.17$.)

5.8   ION COMPLEXES AND DISSOLVED SPECIES

In soil-pore water and ground water, many ion complexes may form. These may be familiar from the study of ground water and saturated water flow. A typical example would be that of CaSO_4^0, which forms as a result of the presence of Ca^{2+} and SO_4^{2-} within soil-pore water. Other examples would be CO_3^{2-} complexes which may include CaCO_3^0, MgCO_3^0, or NaCO_3^-, depending on the constituents present.

Complexations of metal ions released to the unsaturated zone through agricultural and industrial activities is of primary concern to the unsaturated-zone hydrologist. Such complexation occurs between dissolved metal ions and mineral or organic hydroxyl groups exposed at soil surfaces. As an example, dissolved Cu^{2+} might react with such a group, designated S—OH according to the reaction

$$S—OH + Cu^{2+} \rightleftharpoons S—OCu^+ + H^+ \tag{5.66}$$

with a resulting equilibrium constant of (based on the law of mass action; Stumm and Morgan 1996)

$$K_{Cu} = \frac{(S—OCu^+)\,[H^+]}{(S—OH)\,[Cu^{2+}]} \tag{5.67}$$

Note that this reaction results in a surface-charge imbalance, with the surface having a net positive charge. A bidentate surface complex, as described by the reaction below,

$$2S—OH + Cu^{2+} = (S—O)_2Cu + 2H^+ \tag{5.68}$$

may occur simultaneously. This complex results in charge neutrality at the surface.

A good example of mineral surface hydroxyl groups is that of Fe(III) oxide; when acting as the central ion of a mineral surface, it results in the hydroxyl group $\equiv Fe—OH$, which acts as a Lewis acid and exchanges its structural OH against other ligands.

Because of the surface-charge imbalance created by the reaction given in equation 5.66, sorption of cations is affected by Coulombic forces that repel additional cations from the charged surface, and the equilibrium constant, K, must be corrected. This is accomplished by converting the law-of-mass-action equation first into a Langmuir-type equation, and then into a Frumpkin-type equation (Stumm and Morgan 1996) to correct for electrostatic effects. Consider again a reaction involving dissolved copper ions, such that

$$SH + Cu^{2+} \rightleftharpoons SCu^+ + H^+ \tag{5.69}$$

Then (by the law of mass action),

$$\frac{K_{Cu}}{[H^+]} = \frac{[SCu^+]}{[SH]\,[Cu^{2+}]} \tag{5.70}$$

and

$$[SH] + [SCu^+] + [S^-] = S_T \tag{5.71}$$

where S_T is the maximum concentration of surface sites, S represents the adsorption sites on the solid surface occupied by hydrogen ions, and SCu^+ represents the surface sites occupied by copper ions, and S^- are the sites where hydrogen ions have been removed through the process of mass action (i.e., exchange). The brackets, [ ], denote concentration, which for surface concentrations, may be expressed as moles/L solution, per kg solid; per m^2 of solid surface; or per mole of solid.

In most natural systems, either [SH] or [S^-] dominates, dependent on the nature of the surface and the pH. For surfaces dominated by iron and aluminum hydrous oxides, the concentration of SH_2^+ may be significant at low pH. Such sites are especially important in anion sorption; however, for most oxides in the normal pH range found in soils, [S^-] << [SH]. Consequently,

$$[SH] + [SCu^+] \approx S_T \tag{5.72}$$

S_T may be used to formulate a Langmuir type equation for

$$[SCu^+] = \frac{S_T K_{Cu}\,[H^+]\,[Cu^{2+}]}{1 + K_{Cu}\,[H^+]\,[Cu^{2+}]} \tag{5.73}$$

By definition, Γ_{Cu} is the surface concentration of copper which is [SCu$^+$]/mass of adsorbent and Γ_{max} is S_T/mass of adsorbent; consequently, one may write

$$\Gamma_{Cu} = \frac{\Gamma_{max} K_{Cu}[H^+][Cu^{2+}]}{1 + \dfrac{K_{Cu}}{[H^+]}[Cu^{2+}]} \tag{5.74}$$

Thus, K_{Cu} is now the surface (intrinsic) complex formation constant, and when divided by [H$^+$] it is the adsorption constant for that [H$^+$]. This intrinsic equilibrium needs to be corrected for Coulombic effects, resulting in the apparent equilibrium constant, K_{ap}, as a Frumpkin-type equation

$$K_{ap} = K_{it} \exp\left(-\frac{\Delta_c F\psi}{RT}\right) \tag{5.75}$$

where F is the Faraday constant (96,490 C mol^{-1}), ψ is the surface potential (V), and Δ_C is the change in ionic valence of the surface species. For our example of the complexation of copper ions, the equation becomes

$$\frac{[SCu^+]}{[SH][Cu^{2+}]} = \frac{K_{Cu(ap)}}{[H^+]} = \frac{K_{Cu}}{[H^+]} \exp\left(-\frac{F\psi}{RT}\right) \tag{5.76}$$

The surface charge (ψ) cannot be measured experimentally. A potential candidate, the zeta potential (ζ) can be determined from electrophoretic measurements, but is smaller than ψ. However, ψ may be estimated from the surface charge of the medium, based on the constant capacitance model, $\psi = \sigma/C$, where σ is surface charge per unit area and C is surface capacity or capacitance per unit area. Surface charge is given (if [SCu$^+$] is in units of moles/kg of absorbent,) by the equation $\sigma = F[SCu^+]/s$, where s is surface area of absorbent in m^2/kg. Surface capacity is given by the equation $C = \varepsilon\kappa/4\pi$ (Adamson 1990), where ε is permittivity of the fluid and κ is the Debye kappa, which is dependent on the concentration (as well as the valence) of ions in solution (see chapter 3). Thus, $\psi = (4\pi F[SCu^+])/s\varepsilon\kappa$. Substitution into equation 5.76 gives (if [SCu$^+$] is written as θS_T)

$$\frac{[SCu^+]}{[SH][Cu^{2+}]} = \frac{K_{Cu(ap)}}{[H^+]} = \frac{K_{Cu}}{[H^+]} \exp\left(-\frac{4\pi F^2 \theta S_T}{\varepsilon\kappa RT(s)}\right) \tag{5.77}$$

5.9   DIFFUSSION

Two types of diffusion are discussed in this text. The first type is the diffusion of solutions which will be discussed in brief here; the second type is the diffusion of particles (more accurately termed surface mobility). In the general process of diffusion that is commonly referred to as the molecular diffusion of solutions, ionic constituents are forced to move in the direction of a concentration gradient due to kinetic activity. Typically, the relative contribution of diffusion to transport is greatest in slow-flow conditions. Diffusion of a substance will continue to equilibrium (i.e., where no concentration gradients exist and there is no driving force). The transport of a molecule from one region to another involves work. The work of transporting a mole of solute from a zone where its chemical potential is $\mu(1)$ to a zone where its chemical potential is $\mu(2)$, is equal to $\mu(2) - \mu(1)$. Normally, the chemical potential of a substance in the vadose zone is a function of distance in the system; the work required to transfer a mole from x to $x + dx$ may be expressed as

$$d\mu = \mu(x + dx) - \mu(x) = [\mu(x) + (d\mu/dx)\,dx] - \mu(x)$$
$$= (d\mu/dx)\,dx \tag{5.78}$$

Because the work is equal to the negative of force $\times$ distance, the negative gradient of the chemical potential is also a force. For an ideal solution, $\mu = \mu^0 + RT \ln c$. Thus, the force is given by

$$F = \frac{d\mu}{dx} = -\frac{RT}{c}\frac{dc}{dx} \tag{5.79}$$

where R is the gas constant, T is temperature in Kelvin, c is concentration, and dc/dx is the concentration gradient. The force opposing the diffusion of a molecule is the velocity, v, times the frictional coefficient, f. By setting these forces for a molecule equal to each other, one obtains

$$vc = -\frac{RT}{N_A f}\frac{dc}{dx}; \qquad fv = -\frac{RT}{N_A c}\frac{dc}{dx} \tag{5.80}$$

where N_A is Avogadro's number. The relation expressed in these two equations corresponds to Fick's first law:

$$J = -D\frac{dc}{dx} \tag{5.81}$$

where J is the flux written in terms of mass of solute per unit area per unit time $[M/L^2 T]$, D is the diffusion coefficient $[L^2/T]$, and c is the solute concentration $[M/L^3]$; dc/dx is the concentration gradient. There is a relation here with Fick's second law (the continuity equation) and electric mobility, which will be discussed in greater detail in chapter 11.

5.10 SOLUBILITY

Water solubility is one of the most important parameters affecting the fate and transport of organic chemicals in the environment. The greater the solubility of the chemical, the more rapidly it will be dispersed in the hydrologic cycle. Generally, the higher the compound's solubility, the lower the adsorption coefficient, thus, the more easily it is degraded by soil microorganisms. Also, compounds that are sorbed to the particle surface are more likely to be completely degraded because the residence time of the compound in the soil is longer. But the higher the solubility, generally the weaker the attachment of the compound to the particle, which makes it a readily usable form of energy for microbes. Solubility is essentially the maximum amount of a chemical that will dissolve in pure water at a specific temperature. If the organic chemical is a solid or liquid at the specified temperature, two phases will exist at equilibrium: a saturated aqueous solution, and a solid or liquid organic phase.

The concentration of an aqueous solution is normally expressed on a mass-per-mass basis (such as weight percent, g/kg, et cetera), or mass-per-volume basis (mg L^{-1}, et cetera). Most organic chemicals are soluble in water to a certain extent; solubility may range from extremely low concentration to 100,000 mg L^{-1} and higher. Thus, many orders of magnitude in variation may be involved, depending on the chemical being measured. Several methods may be used to determine the solubility coefficient, S, of a given chemical. These methods include regression equations which are based on fitting equations of a specified form to experimentally measured solubilities, addition of atomic fragments to the solution, and theoretical equations using estimated activity coefficients. Only the latter method allows calculation of solubility at any temperature. The other methods yield a suitable answer only if the data have been collected over a range of temperatures, preferably ranging from about 5–35 °C. No one method can be recommended as "best" for determination of S, due to the special problems that arise when estimating S for hydrophobic compounds, various methods may yield different values, and many organic compounds become more soluble as temperature increases.

Caution must be used when determining S; for example, the magnitude of the equilibrium constant is not a good indication of one chemical's solubility in water compared to that of another chemical. This is because, in the equilibrium relation, the activity of the molecule is raised to the power of the number of moles in the dissociation expression. Additionally, great differences in solubility between minerals are commonly found in groundwater and organic chemicals. As an example, gypsum has a solubility of 2200 mg L^{-1} in water, and is widely used to flocculate soils to enhance greater root growth and improve drainage without altering their pH. In comparison, the pesticides parathion, atrazine, and methoxychlor (an organochlorine), have solubilities of 24, 33, and 0.003 mg L^{-1}, respectively.

Factors affecting solubility of organic chemicals include temperature, salinity, dissolved organic matter, chemical of the organic compound, and pH. Solubility will either increase or decrease at higher temperatures depending on the nature of the chemical being measured. The presence of dissolved salts normally leads to moderate decreases in S. The general relation between salinity and solubility may be expressed as

$$\log \frac{S^0}{S_1} = K_S C_S \tag{5.82}$$

where S^0 is the molar solubility (mg/L) in pure water, S^1 is the molar solubility (mg/L) in salt solution, K_S is the salting parameter (generally ranging from ~ 0.04 to 0.4), and C_S is the molar salt concentration (M/L^3). The presence of dissolved organic material such as humic and fulvic acids may lead to an increase in solubility for many organic chemicals. In general, solubility of organic acids tends to increase with pH, but the solubility of organic bases tends to decrease.

Determination of solubility for a solid in a liquid solvent is different than the method previously discussed. Not only is solubility a function of temperature, but the heat of fusion of the solute also must be considered, since energy is required to overcome intermolecular forces of molecules within the solid as it dissolves. In this case, it is fairly typical for the chemical with the high heat of fusion to have a lower solubility. Assuming one knows the heat of fusion ΔH_f, the melting point, T_m (in Kelvin), and the activity coefficient γ_1 (as a function of composition), the solubility denoted as x_1 (in mole fraction) may be obtained from

$$\ln \gamma_1 x_1 = \frac{\Delta H_f}{RT}\left(\frac{T}{T_m} - 1\right) \tag{5.83}$$

where T is the system temperature (in kelvin) and R is the gas constant. Equation 5.83 neglects certain correction terms which are proportional to Δc_p (the specific heat difference between liquid and solid) because these data are likely unavailable. However, the uncertainties and errors associated with these omissions will likely be small. The solution of equation 5.83 for x_1 must be by trial-and-error because γ_1 is a function of x_1. The estimation procedure involves: (1) the estimation of γ_1 for various values of x_1 to obtain a plot; (2) calculating the value of the right-hand side of equation 5.83; and (3) using calculated and interpolated values of γ_1 (step 1) to determine $\ln \gamma_1 x_1$ that will match the value from step 2. If no value is found that matches the right-hand side, the two chemicals are completely miscible at the specific temperature. Set $\gamma_1 = 1$ as a first estimate of x_1.

QUESTION 5.10

Using the regression equation $\log S = -0.922 \log K_{ow} + 4.184$ (good for many pesticides), what is the solubility of atrazine?

QUESTION 5.11

What is an estimation of the mole fraction solubility of naphthalene in 1-butanol at 40 °C? Assume $T_m = 353.4$ K, $\Delta H_f = 4494$ cal mol^{-1}, $\gamma_1 = 3.85$ mole fraction, and $x_1 = 0.111$ mole fraction.

5.11   CHEMICAL SATURATION

To determine whether a solution is saturated, one must ascertain the free energy of dissolution; that is, whether or not the dissolution of the solid phase is negative, zero, or positive. As mentioned in section 5.4, the free energy of dissolution is given by $-\Delta G = RT \ln Q/K$. The extent of saturation may be expressed by a saturation index, S_{id}, which is expressed as

$$S_{id} = \frac{Q}{K_{eq}} \tag{5.84}$$

If S_{id} (the ratio of Q/K) is positive ($Q/K > 1$), the solution is oversaturated and precipitation will occur; if $Q/K < 1$, the solid will continue to dissolve; and if $Q/K = 1$, the solution is at equilibrium.

For reactions that involve a solid phase, one can compare Q (ion activity product: IAP; also expressed as the solubility product, K_{sp}) with K_{eq} and can thus define the state of saturation. An example of this would be

$$CaCO_3 \text{ (solid)} + H^+ = Ca^{2+} + HCO_3^-$$

we can then write ΔG or the saturation index, S_{id} as

$$-AG = S_{id} = RT \ln \frac{[Ca^{2+}]_{act}[HCO_3^-]_{act}[H^+]_{eq}}{[H^+]_{act}[Ca^{2+}]_{eq}[HCO_3^-]_{eq}} = \frac{[Ca^{2+}]_a[HCO_3^-]_{act}}{[H^+]_{act}\ K_{eq}} \tag{5.85}$$

where the subscript *act* refers to actual activity and *eq* to activity at equilibrium. A simple test would be to compare either the actual concentration (or activity) of an individual reaction component such as H^+ with the concentration if it were in theoretical solubility equilibrium. Using this method, the state of saturation for calcite ($CaCO_3$) would be written in terms of $pH_{act} - pH_{eq}$. The resulting positive, negative, or zero value would have the same effect as discussed above: precipitation for a positive value; continued dissolution for a negative value; and equilibrium of solution for a zero value. Thus effectually, one may determine S_{id} by measuring the pH change upon addition of solid calcite.

5.12   OXIDATION AND REDOX REACTIONS

Oxidants and reductants are defined as electron, donors and electron acceptors. For purposes of our discussion, it should be remembered that an electron (e) has a negative charge and that there are no free electrons; thus, every oxidation reaction is accompanied by a reduction. The oxidation state of a species represents the hypothetical charge of the atom upon dissociation of the ion. The rules for assigning oxidation states can be found in any general college inorganic chemistry text; the student may wish to review these before proceeding further. Since every oxidation is accompanied by a subsequent reduction, there is always a balance of electrons. Using iron as an example, the oxidation state (written in half-reactions) is expressed as

$$4Fe^{2+} = 4Fe^{3+} + 4e$$

with the reduction state written as

$$O_2 + 4H^+ + 4e = 2H_2O$$

thus, the balanced, or redox, reaction is

$$O_2 + 4Fe^{2+} + 4H^+ = 4Fe^{3+} + 2H_2O$$

Most redox reactions that occur in ground water incur the use of oxygen. However, hydrogen is involved in many redox reactions as H^+, and other elements such as C, S, N, Mn, and Fe are involved in electron transfer. While surface water generally exhibits oxidizing conditions due to mixing with oxygen at its surface, water within soil pores often has reduced conditions because of decreasing oxygen content within the vadose zone. The oxygen that is present is normally consumed by biochemical and hydrochemical reactions.

As water content within the unsaturated zone increases, the availability of oxygen as an electron acceptor is severely restricted. This is particularly true in arid regions which have saline or sodic soils and low hydraulic conductivity. The increased wetting of a soil where seepage discharge is occurring and which contains decomposable organic matter, will cause an onset of anaerobic conditions because soil microorganisms that decompose the organic matter consume the available free oxygen in soil/ground water more rapidly than additional oxygen can diffuse from the land's surface into the wet soil. Typically, if microorganisms are respiring aerobically, the final electron sink is oxygen, which accepts electrons and converges with hydrogen to yield water. Thus, aerobic respiration involves the reduction of O_2 to water. If oxygen is deficient within the vadose zone, other substances can accept electrons and become reduced. Examples of this would be nitrate (NO_3^-) to nitrite (NO_2^-), sulfate to sulfide: $SO_4 + 8e + 10H^+ = H_2S + 4H_2O$; and H to H_2: $2H^+ + 2e = H_2$.

In the event of an oxygen deficiency in soil containing organic matter, the organic compounds present are no longer fully oxidized to carbon dioxide and water, but rather to intermediate products. These products are normally polycarboxylic acids, alcohols, simple fatty acids, et cetera. Further decomposition of these compounds will continue as methane and other hydrocarbons are produced in the soil. As a result, very wet and flooded soils may contain inorganic and organic compounds in reduced form.

The tendency of a solution to exchange electrons can be measured by its redox potential, E_h (measured in volts). The transfer of electrons is an electric current, and the more strongly reducing a substance, the lower its electric potential. As discussed earlier, the sign of the potential is positive in oxidizing reactions, and negative in reducing reactions. The standard potential, E^0, has been measured for many reactions at standard temperature and pressure. The total oxidation potential of a reaction is given by the Nernst equation, expressed as

$$E_h = E^0 + \frac{RT}{z\text{F}} \ln K_{eq} \tag{5.86}$$

where F is the Faraday constant (96,790 C/mol), z is valence, and K_{eq} is the equilibrium constant; in some references, K_{eq} is also termed the solubility product of the mineral, K_{sp}. The oxidation potential can be measured quickly and easily with a standard specific-ion electrode meter. The E_h of a specific solution is the difference in voltage between a platinum electrode and a hydrogen reference electrode; however, for convenience a calomel reference electrode is usually used to obtain this difference. Since the voltage difference between a calomel and a hydrogen electrode is 0.248 V, $E_h = E$ measured + 0.248 V. The transfer of electrons in a solution usually induces a transfer of protons as well, thus the redox potential depends on solution pH. As a result, E_h in well-aerated solutions decreases linearly by 59 mV for each pH-unit rise in solution. This relation usually holds well for aerated soil.

Numerous reduction reactions can occur within the vadose zone when it becomes anaerobic; these reactions depend on the redox potential at which the compound of interest is most strongly held. Compounds held at high redox potential, with their reactions going to completion before similar compounds held at a lower redox potential, can be important. The primary inorganic reactions (reductions) that occur in the unsaturated zone, when it becomes anaerobic, are: ferric hydroxide to ferrous ions; hydrogen ions to hydrogen gas; nitrate to nitrite; sulfate to sulfite to sulfide; and manganic salts or manganese dioxide to manganous ions. The reactions for both iron and manganese are reversible in the soil system. Once nitrate has been reduced to nitrite, it can easily be reduced further, and lost from the system as nitrous oxide gas. Likewise, sulfite is more easily reduced than sulfate. The reduction of sulfite to sulfide, in the absence of ferrous iron, can be lost as hydrogen sulfide gas. As a result, nitrate (always) and sulfate (sometimes) reductions are irreversible processes within the unsaturated zone.

The reduction of organic components within the unsaturated zone is much more complex than the reduction of inorganic compounds due to the wide sequence of reduction products normally involved. As the unsaturated zone first becomes very wet or flooded, decomposition will begin and a number of gases will be given off; typically, these will be nitrous oxide, hydrogen, and a range of low-molecular-weight hydrocarbons such as methane, ethane, ethylene (the latter having a marked effect on root growth), propane, and so on. However, production of these hydrocarbons will occur only in the initial stages of flooding and decreases rapidly until production ends within several days. Experimental results by Tindall, Petrusak, and McMahon (1995) comparing clay and sandy soils indicate that gases such as nitrous oxide are given off in larger quantities under mild reducing conditions than compared to stronger reducing conditions. Soils high in nitrate, or soils to which nitrate fertilizers have been added, maintain a redox potential of about $+400$ to $+200$ mV until all nitrate has been reduced and then the potential may fall rapidly; this may only take a few weeks (Couto, Sunzonowicz, and De O. Barcellos 1985). It is important to note that microorganisms are responsible for catalyzing most of the important redox reactions that occur within the unsaturated zone. Also, the microorganisms help to decompose the organic matter which may be present in the unsaturated zone.

In order to achieve reduction of inorganic constituents, other compounds must be oxidized; these other compounds are usually organic matter. The oxidation of organic matter is catalyzed by enzymes, soil bacteria, and other microorganisms deriving their energy by expediting the electron transfer. The dissolved oxygen in the unsaturated zone often decreases with depth, and the oxidation of even a small amount of organic matter can completely consume all of the oxygen present. Additionally, almost all of the reduction processes discussed regarding nitrate, sulfide, et cetera, consume oxygen and produce hydrogen gas, with little change in pH. In the root zone or near-surface area of the unsaturated zone, infiltrating water is in perpetual contact with organic matter, causing continual consumption of oxygen and production of carbon dioxide. The production of carbon dioxide will yield carbonate and bicarbonate.

5.13   MICROBIAL MEDIATION AND pH

Microorganisms act primarily as redox catalysts; they cannot carry out specific reactions that are not thermodynamically possible. Consequently, these organisms do not oxidize or reduce, they simply mediate electron transfer. For example, many species of bacteria found in soils are capable of reducing nitrates to nitrogenous gases (NO, N_2O, N_2), which are then released into the atmosphere. This dissimilatory reduction process is known as denitrification. The denitrifying bacteria that bring about these reactions are obligately aerobic, except for the ability to utilize nitrate in the absence of oxygen. Nitrate acts in lieu of oxygen as a terminal acceptor of electrons produced during anaerobic respiration. Denitrification is promoted by high soil-moisture conditions, neutral soil pH, high soil temperatures, a low oxygen-diffusion rate, the presence of organic matter, and nitrate. The amount of nitrogen gas released from soils under normal aerobic conditions found within the upper horizon of the unsaturated zone can vary widely, 5–50% of applied nitrate (Mengel and Kirkby 1982). The release of nitrogen gases is generally lower on well-aerated sandy soils than in clay soils; however, physical parameters and soil characteristics can change this. Lower rates of release of nitrogen gases generally indicates a lack of efficient biological activity or mediation within the soil due to reduced water content, unavailable nutrient source for microorganisms, occluded pore spaces which reduce microbial activity, and other factors (Tindall, Petrusak, and McMahon 1995). As a result, denitrification must occur indirectly such as the reduction of nitrate to nitrite, with a subsequent reaction of nitrite with ammonium to produce nitrogen gas and water. The conversion of nitrate to nitrogen gas is nonreversible, thus NO_3^{-2}—N_2 cannot be used as a reliable redox indicator.

The dominant types of microorganisms present are dependent on environmental conditions. As with most living organisms, they compete with each other in all types of conditions. An example of this is the predominance of fungi rather than bacteria at lower pH. It is not that the fungi particularly thrive more; the fundamental reason is that there is less competition from bacteria at the lower pH. The environmental factors of temperature, pH, ionic strength, and oxygen concentration influence the rate of biologically mediated transformations and also dictate whether these processes operate within the time frame of interest.

The effect of pH on microbial activity is very important. For example, organic matter will frequently develop on the surface of strongly acid soils that result from the acidifying effects of fertilizers or rainfall (rainfall produces a weak carbonic acid whose pH can decrease with time). In the absence of microorganisms and soil fauna such as earthworms, organic debris accumulates on the surface layer; in the presence of these organisms, the organic matter is distributed throughout the upper horizon depending on adequate oxygen and other environmental parameters. There is substantial evidence that fresh organic matter initially decomposes more slowly in strongly acid soils (pH 3.0 to 4.0) than in soils with pH values greater than 5.0. This is due primarily to a reduced microbial population. The same is true of certain chemical transformations within the unsaturated zone. Evidence suggests that a general pH range of 5.5 to 8.0 is the best range for the occurrence of nitrogen transformations (Mengel and Kirkby 1982). Because nitrogen transformation processes are mediated by bacteria, this general range is the same as that at which there seems to be the greatest microbial activity. Hence, degradation, reduction, oxidation, and transformation processes can be severely limited by pH.

SUMMARY

In this chapter, we discussed how the chemical properties of water become as important as the physical properties and quantity. Also, though it is commonly assumed that physical processes within a soil–water system are the dominant influences affecting change and equilibrium, it has often been found that chemical processes in ground and soil water are the controlling influences (Freeze and Cherry 1979).

We discussed organic compounds, along with humic and fulvic acids, all of which make up a large component of organics found in soil systems (see chapter 3). Although little is known about these compounds, they are not of great concern as contaminants within the unsaturated zone, but may be a source of contamination by complexation of heavy metals. The greatest concern for contamination in the unsaturated zone comes from man-made organic chemicals in the form of pesticides, volatile organic compounds such as solvents or petroleum products, or semivolatile organic compounds such as creosote or phenolic compounds. We also explained mechanisms such as adsorption, biodegradation, and so on, which help prevent transport of organics through the unsaturated zone.

We covered spontaneous reactions that usually proceed until a dynamic equilibrium is attained, as well as the relation that governs the relative proportions of compounds of chemical reactions and their products. Because very few solutions in nature are ideal, we illustrated how the activity coefficient related to a compound's concentration is a correction factor which compensates for nonideal behavior. (Ideal behavior for a solution is defined as a lack of interactions between the molecules of a liquid; that is, the solution would behave as if any molecular interactions present had a negligible effect on the intensive properties—Gibbs energy, entropy, enthalpy, and volume—of the solution.)

We also explained how the equilibrium of a system can be described by the entropy of a system and that, for the second law of thermodynamics at constant temperature and pressure, the Gibbs free energy, G, provides a more convenient measure of the thermodynamic property than the entropy. Also, the Gibbs free energy of any chemical reaction represents the driving force of the reaction (the change in internal energy per unit mass), which yields a measure of the system's ability to perform nonmechanical work.

We described the law of mass action, which illustrates equilibrium conditions between acids and bases, as well as the underlying principle of electroneutrality, which shows that there is generally an equal number of positive and negative charges so that a neutral charge condition exists in a solution. We discussed how chemical properties that determine acid–base interactions between a chemical and the soil matrix exert a major influence on partitioning between gaseous, solid, and solution components; and how hydrolysis, one of the most important reactions of organic compounds in aqueous solutions, is the chemical transformation process in which an organic molecule, R—X, reacts with water to form a new carbon–oxygen bond.

This chapter described how ion complexes may form, depending on the constituents present, and that in the unsaturated zone, the focus is generally more on metal-ion complexes which may be the result of agriculture and industry, such as copper and iron-hydrous oxides.

Additionally, we explained diffusion and its relative contribution to transport, which is greatest in slow-flow conditions (diffusion of a substance will continue to equilibrium; that is, where no concentration gradients exist and there is no driving force). Following this, we described water solubility, which substantially affects the fate and transport of organic chemicals in the environment, and described how the greater the solubility of the chemical, the more rapidly it will be dispersed in the hydrologic cycle. We also discussed was solution saturation, where the extent of saturation is expressed by a saturation index, S_{id}; that discussion was followed by oxidation and redox reactions, each being defined as electron donors and electron acceptors. Finally, we ended the chapter with a discussion of microbial remediation and pH: how microorganisms act primarily as redox catalysts, which cannot carry out specific reactions that are not thermodynamically possible. (Thus, microorganisms do not oxidize or reduce, they mediate electron transfer.) Ultimately, this chapter serves as a precursor to the following chapters discussing various principles of water flow in soils.

ANSWERS TO QUESTIONS

5.1. Equation 5.30 is the same as 5.17. Heat, expressed as q, may be obtained by taking the integral

$$q = \int_{300}^{1000} (26.984 + 5.910 \times 10^{-3}T - 3.77 \times 10^{-7}T^2)\, dT$$

$$q = 26.984(1000 - 300) + \frac{1}{2}(5.910 \times 10^{-3})(1000^2 - 300^2)$$

$$-\frac{1}{3}(3.377 \times 10^{-7}(1000^3 - 300^3) = 21.468 \text{ kJ mol}^{-1}$$

5.2. At 25 °C, A = 0.5085 and B = 0.3281 Å^{-1}; thus, using equation 5.30 log γ_i = $-[0.5085(2)^2 \times (0.0281)^{1/2}]/[1 + (6)(0.3281)(0.0281)^{1/2}]$ = -0.2564 and taking the antilog = 0.554. As a result the chemical activity α = (0.554)(0.00633) = 0.00351.

5.3. By sketching the structure of benzene, N_1 = 6, and for the structure of water, N_2 = 0. However, the B_2 term is not infinite because of modification. The correlation coefficients for alkylbenzenes in water are $A_{1,2}$ = 3.554, B_2 = 0.622, and C_1 = -0.466 (see table 5.1). Equation 5.36 must be modified and now takes the form

$$\log \gamma_1^\infty = A_{1,2} + B_2 (N_1 - 6) + C_1 \left[\frac{1}{N_1 - 4} \right]$$

For the typical modifications and additional information on equation 5.36, the student is referred to Reid, Pravsnitz, and Sherwood (1977). Thus, log γ_1^∞ = 3.554 + 0.622(6 − 6) − 0.466[1/6 − 4] = 3.321. Taking the antilog, we obtain 2,094. Repeat this sequence to obtain the infinite dilution activity coefficient for water as the solute and benzene as the solvent using correlation coefficients of $A_{1,2}$ = 3.04 and F_2 = -3.14. (Note that equation 5.36 must be modified for these parameters also.) For this case, B_2 = C_1 = D = 0. Thus, log γ_2^∞ = 3.04 − 3.14/6 = 2.517; take antilog = 329. The student may wish to estimate the solubility of benzene in water and water in benzene, as well as determine the Henry's-law constant for benzene in water. For solubility, x_2 = 3.5 × 10^{-3} mole fraction, and for the Henry's-law constant, H = 0.126/26.1 = 4.7 × 10^{-3} atm m^3/mole.

5.4. First, solve equation 5.41 for K. Thus, K = exp $(-77,772/8.3144 \times 298)$ = 2.33 × 10^{-14} = $[(P_{NO}/P^0)/(P_{N_2}/P^0)^{1/2}(P_{O_2}/P^0)^{1/2}]$ = $[(P_{NO}/P^0)/(0.80)^{1/2}(0.20)^{1/2}]$. Consequently, P_{NO} = 9.319 × 10^{-15} bar or 9.319 × 10^{-13} kPa. The partial pressure of NO in the denominator is ignored because it is very small.

5.5. Using equation 5.43 we obtain Δ = 147.25 − (298.15)[(354.4 − {5 × 5.74} − {6 × 130.68})/1000] = -10.6 kJ mol^{-1}. The experimental value for this formation is 8.3 kJ mol^{-1}. The equilibrium mole fraction is equal to the equilibrium pressure fraction. The equilibrium constant for the formation reaction is given by K_i = $O_i/P_{H_2}^6$ = exp $(\Delta G^0/RT)$.

5.6. By substituting the given values into the top part of equation 5.58, we have K_a^x = (1.46 × 10^{-2})10$^{(0.268)(0.755)}$ = 2.33 × 10^{-2}. Note that this is about 300 percent higher than an experimentally measured value of 0.56 × 10^{-2}; consequently, one of the special purpose equations may yield better results in this case.

5.7. First, simplify the Hammett correlation, equation 5.58, such that log k = $\rho\sigma$ + log k_o. Thus, log k_H = (0.11)(0.778) − 7.0 = -6.92. Hence, k_H = 1.2 × 10^{-7} M^{-1} s^{-1}. Thus, k_T = (1.2 × 10^{-7} M^{-1} s^{-1})(10^{-6}) = 1.2 × 10^{-13} s^{-1}. Using equation 5.63, the hydrolysis half-life $t_{1/2}$ = 0.693/(1.2 × 10^{-13}) = 5.8 × 10^{12} s^{-1}.

5.8. Using a simplification of the Hammett correlation, as in question 5.7, log k_{OH} = $\rho\sigma$ + log k_{OH}^0, the log k_{OH} = (2.38)(0.778) + log(7.2 × 10^{-3}) = -0.29. Hence, k_{OH} = 5.1 × 10^{-1} M^{-1} s^{-1}. You may wish to compare this to the k_{OH} for water.

5.9. Again, using a simplification of the Hammett correlation, log k_o = $(-1.31)(-0.17)$ − 5.21 = -4.99. Thus, k_o = 10^{-5} s^{-1}. This is for 25 °C; the student may wish to recalculate the value for 30 °C using the correction factor given by equation 5.65. The literature value at this temperature is k_o = 2.97 × 10^{-3} s^{-1}. How close are you to this upon recalculation? You may wish to contemplate the reason(s) for any differences.

5.10. The log K_{ow} for atrazine is 2.68, thus log S = $-0.922(2.68)$ + 4.184; log S = 1.713 and S = 51.642 mg/L.

5.11. Using equation 5.83 we have 4494/[(1.987)(313.2)] × [(313.2/353.4) −1] = -0.8214 (no units).

ADDITIONAL QUESTIONS

5.12. A contaminant spill has occurred; it consists of a mixture of ethanol (1)-n-hexane (2). The structure of ethanol is CH_3CH_2OH and hexane is $CH_3(CH_2)_4CH_3$. What is the infinite-dilution activity coefficient? In this case, N_1 = 2 and N_2 = 6; this is a mixture of an alcohol in paraffin. Assume a temperature of 25 °C. Also, $A_{1,2}$ = 1.96, C_1 = 0.475, and D = -4.9 × 10^{-4}.

5.13. Estimate the infinite-dilution activity coefficient of aniline in water at 25 °C. Assume the correlation factor for NH_2 on a benzene ring is -1.35.

5.14. Estimate the dissociation constant for p-tert butyl ($C(CH_3)_3$) benzoic acid. Assume K_a^o for benzoic acid is 6.26×10^{-5}, $\rho = 1.0$, $\sigma = -0.197$.

5.15. Estimate the dissociation constant for 3,4-dimethylanaline. Assume $\rho = 2.77$, $pK_a^o = 4.60$ and $\sigma = -0.303$.

5.16. What is k_{OH} for diisobutyl phthalate? Assume 25 °C and $\log k_{T2} = -3.03^{-1}$.

5.17. Using the regression equation given in question 5.11, calculate s for DDT and Carbofuran. Assume $\log k_{ow}$ is 5.98 and 1.60, respectively.

5.18. What is the solubility of 1,4-diiodobenzene in water? Assume 25 °C, $T_m = 402.6$ K, $\Delta Hf = 5340$ cal/mol and $\gamma_1^\infty = 1.66 \times 10^6$.

5.19. Estimate the solubility of 4-chloro-1,3 dinitrobenzene in water. Assume 50 °C, $T_m = 328$ K, and $\gamma_1^\infty = 27,500$; also assume that the value for ΔHf is not available.

5.20. Using the regression equation given in question 5.11, estimate s for 2-isopropoxyphenyl-N-methylcarbamate (mw = 209.2 g/mole). Assume $t_m = 91$ °C and $\log k_{ow} = 1.55$.

5.21. To determine s for benzene and benzene derivatives (aromatics), the following equation has been found to work well:

$$\log 1/s = 0.996 \log k_{ow} - 0.339$$

Using this equation, estimate s for trichloroethylene and 1,2,4-tribromobenzene (spilled on a nearby landfill, say). Assume $\log k_{ow} = 2.42$ and 4.98 and $t_m = 25$ and 44, respectively.

6

Principles of Water Flow in Soil

INTRODUCTION

By definition, a fluid is a substance that is capable of flow. In soils, this includes both the liquid and gas phases. In the gas phase molecules are spaced farther apart, while in the liquid phase, molecules are more closely bound. The intermolecular cohesive forces are considerably smaller in a gas because of the separated distance of molecules, compared to those of a liquid. Normally, fluids always possess elastic properties while under compression due to their inability to resist shear stress. Consequently, fluids can alter their shape and flow characteristics, depending on the physical and chemical characteristics of the medium.

Physically, fluid characteristics may be expressed as density, specific gravity, specific volume, and specific weight. Density is the mass of the fluid per unit volume, specific gravity is the

TABLE 6.1　Physical Properties of Water (Liquid Phase)

Temperature (°C)	Density, ρ (g cm^{-3})	Heat capacity, C_p (J g^{-1} K^{-1})[†]	Surface tension, γ (10^{-3} N m^{-1})	Thermal conductivity (10^{-3} W K^{-1} m^{-1})[‡]	Viscosity, η (Pa s)[§]
0	0.99984	4.2161	75.6	561.0	0.001793
4	1.00000	4.2077	75.0	569.4	0.001567
5	0.99999	4.2035	74.8	573.6	0.001519
10	0.99970	4.1910	74.2	586.2	0.001307
15	0.99913	4.1868	73.4	594.5	0.001139
20	0.99821	4.1826	72.7	602.9	0.001002
25	0.99708	4.1784	71.9	611.3	0.000890
30	0.99565	4.1784	71.1	619.6	0.000798
35	0.99406	4.1784	70.3	628.0	0.000719
40	0.99222	4.1784	69.5	632.2	0.000653
45	0.99024	4.1784	68.7	640.6	0.000596
50	0.98803	4.1826	67.9	644.8	0.000547
60	0.98320	4.1843	66.2	654.3	0.000466
70	0.97778	4.1895	64.5	663.1	0.000404
80	0.97182	4.1963	62.7	670.0	0.000354

Source: Data compiled from Lide (1992)

[†] To convert to (cal g^{-1} deg^{-1}), divide by 4.1868

[‡] To convert to (cal cm^{-1} sec^{-1} deg^{-1}) $\times$ 10^{-3}, divide by 418.68

[§] Dynamic viscosity: to convert to (g cm^{-1} sec^{-1}) $\times$ 10^{-2}, divide by 1000. To obtain kinematic viscosity, divide dynamic viscosity by fluid density.

ratio of fluid density to the density of pure water (which is dimensionless), specific volume is the volume per unit mass, and specific weight is the weight per unit volume of the fluid. Some of the physical properties of water are given in table 6.1. As shown in this table, the density, surface tension, and viscosity decrease with an increase in temperature.

The following is assumed of a perfect fluid: it lacks viscosity, it is resistance free, it is incompressible (i.e., it has constant density), and it has no irrotational flow. Fluid flow is irrotational when there is no angular momentum of the fluid about any object or point, that is, a small wheel with a fixed rotational point in the center of its mass will not rotate about that point if the wheel is placed (or submerged) in the fluid's path. The assumption of these characteristics implies that there will be no friction between various layers in the fluid as well as no friction between the fluid and the boundary wall. As a result, the fluid is like an aggregation of small particles that support pressure normal to the particle surface, but glide over other particles without resistance.

6.1   BERNOULLI'S EQUATION

The process of fluid transport always obeys the law of conservation of matter and energy. Simply defined, inflow = outflow ± change in storage. This principle applies equally to flow into a lake, the cross-section of an aquifer, or a specific volume of soil. Mathematical formulations of transport processes through soil must reflect the law of conservation, but because soil-water flow problems are generally considered to be isothermal, the law of conservation of energy can be omitted; however, the law of conservation of matter cannot. This law is expressed in the continuity equation, which for one-dimensional flow is $\partial\theta/\partial t = -\partial q/\partial x$, where θ is the volume fraction of water.

In figure 6.1, we examine fluid flow through a pipe, labeled A at one end and B at the other. For a specific time interval Δt, the fluid at point A (cross-sectional area of pipe) moves a distance $\Delta x_1 = v_1 \Delta t$. Considering cross-sectional area A, the mass within this area is $\Delta m_1 = \rho_1 A \Delta x_1 = \rho_1 A v_1 \Delta t$, where p and v refer to pressure and velocity; likewise, the mass of fluid moving through the upper end of the pipe may be expressed as $\Delta m_2 = \rho_2 B \Delta x_2 = \rho_2 A v_2 \Delta t$. Since the flow is steady and mass is conserved, the mass which crosses A during time Δt must also equal the mass which crosses B during this same interval, resulting in $\Delta m_1 = \Delta m_2$ and $\rho_1 A v_1 = \rho_2 B v_2$; this expression is termed the equation of continuity. The equation of continuity simply implies that the product of the area and the velocity of the fluid at all points through the pipe is constant—therefore, the pipe does not have to be the same diameter at each end.

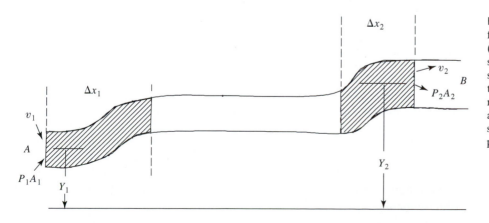

Figure 6.1   Incompressible fluid flowing through a constricted pipe (steady flow). The fluid in the cross-sectional length, Δx_1, moves to section Δx_2. The volume is equal in the two sections. Y is height above reference, v is velocity, x is length, and P is pressure. A and B are cross-sectional areas at each end of the pipe.

Assuming that fluid moves at a steady rate through a pipe of varying cross-sectional area and that the pipe varies in elevation, the fluid pressure along this pipe will change. In 1738, the Swiss physicist Daniel Bernoulli (1700–1782) derived an expression relating pressure to fluid velocity, elevation, and conservation of energy, when applied to a perfect fluid. The state of a fluid at any point may be characterized by the following four quantities: **(1)** pressure, P; **(2)** velocity, v; **(3)** elevation, h; and **(4)** density, ρ.

Consider a pipe with varying elevation and cross sectioned area, as in figure 6.1; flow velocity is nonuniform at any point of interest, Δx. The force, F, at the lowest end of the pipe in the figure is $P_1 A_1$, where P is pressure and A is cross-sectional area. The work done by the represented force is $W_1 = F_1 \Delta x_1 = P_1 A_1 \Delta_1 = P_1 \Delta V$, where ΔV represents the volume of the lower cross-hatched region. The work done at the upper end of the pipe would follow the same sequence (replacing all subscripts 1 by subscripts 2), except that the fluid force would be negative since the fluid force opposes the displacement, that is, the sign would be negative. In addition, the fluid volume passing through 1 in time Δt is equal to the fluid volume passing through 2 during the same Δt.

As a result, the net work performed by these forces during time Δt may be expressed as

$$W = (P_1 - P_2)\,\Delta V \tag{6.1}$$

The work will be divided into two parts: one part that changes the kinetic energy of the fluid and another part that changes the gravitational potential energy. Assuming that Δm is the mass passing through the pipe during Δt, the change in kinetic energy, ΔK, can be expressed as

$$\Delta K = \frac{1}{2}(\Delta m)v_2^2 - \frac{1}{2}(\Delta m)v_1^2 \tag{6.2}$$

The change in potential energy may be written as

$$\Delta U = \Delta m g y_2 - \Delta m g y_1 \tag{6.3}$$

where g is the acceleration of gravity. By applying the work–energy theorem ($W = \Delta K + \Delta U$) to the fluid volume we obtain

$$(P_1 - P_2)\Delta V = \frac{1}{2}(\Delta m)v_2^2 - \frac{1}{2}(\Delta m)v_1^2 + \Delta m g y_2 - \Delta m g y_1 \tag{6.4}$$

Since $\rho = \Delta m/\Delta V$, one may divide by ΔV; with some rearrangement, equation 6.4 reduces to

$$P_1 + \frac{1}{2}\rho v_1^2 + \rho g y_1 = P_2 + \frac{1}{2}\rho v_2^2 + \rho g y_2 \tag{6.5}$$

Equation 6.5 is Bernoulli's equation as typically applied to a nonviscous, incompressible fluid during steady flow. It is often expressed as

$$P + \frac{1}{2}\rho v^2 + \rho g y = C \tag{6.6}$$

This states that the sum of the pressure (P), the kinetic energy per unit volume of fluid ($\frac{1}{2}\rho v^2$), and the potential energy per unit volume ($\rho g y$), has the same value at all points along the streamline of flow. For a fluid at rest, this may be simply expressed as

$$\Delta P = \rho g(y_2 - y_1) = \rho g h \tag{6.7}$$

Some interesting applications of Bernoulli's equation include: the Venturi tube; streamline flow around an airplane wing; atomizers such as those used in perfume bottles and paint-sprayers; and vascular flutter associated with arterial blood flow (related to blood vessels and heart valves within humans).

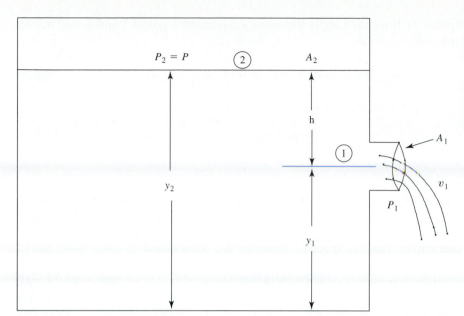

Figure 6.2   Efflux, v_1, from hole in side of container; $v_1 = (2gh)^{1/2}$. A is cross-sectional area of the exit hole (subscript 1) and container (subscript 2), y is fluid level to efflux hole (subscript 1) and container (subscript 2), P is pressure in container (point 2), and P_1 is atmospheric pressure or pressure at outlet (point 1).

QUESTION 6.1

A large tank is filled with water; it develops a hole in its side 20 m below the water level (shown in figure 6.2). If the rate of flow from the hole is 4.2×10^{-5} m³/s: **(a)** What is the speed at which water leaves the hole? **(b)** What is the hole diameter?

QUESTION 6.2

Geyser surges result from water becoming superheated and flashing to steam as pressure is first released. Using our imagination to idealize the famous geyser "Old Faithful" in Yellowstone National Park in Wyoming as a steady water spout, we can apply Bernoulli's equation to determine the velocity of the water as the geyser erupts, as well as determining the pressure in the heated chamber below ground. The height of the eruption typically reaches 40 m above ground surface. **(a)** What is the velocity of the water as it leaves the ground? **(b)** What is the pressure (above atmospheric) in the heated underground chamber?

6.2 TORRICELLI'S LAW

Because all fluids have mass, an unbalanced force that acts on the particles of a fluid will cause an acceleration of those particles, according to Newton's law of motion. An example of this can be seen in any water-supply system; this is especially true of those systems that use large tanks for storage. Such tanks can either be pressurized, or be placed at a higher elevation than the discharge point (e.g., a gravity system). Figure 6.2 shows a tank containing a fluid of density ρ with a hole in its side at distance y_1 from the bottom; here we assume the tank is not open to the atmosphere, so that the air space above the fluid level in the tank is maintained at a pressure P. By also assuming that the cross-sectional area of the tank is large compared to the cross-sectional area of the hole in its side ($A_2 \gg A_1$), the fluid will be at rest at the top (point 2). If we then apply Bernoulli's equation to points 1 and 2, and noting that $P = P_1$ at the hole, we obtain

$$P_1 + \frac{1}{2}\rho v_1^2 + \rho g y_1 = P_2 + \rho g y_2 \tag{6.8}$$

however, since $y_2 - y_1 = h$, equation 6.8 can be rewritten as

$$v_1 = \sqrt{\frac{2(P_2 - P_1)}{\rho} + 2gh} \tag{6.9}$$

The flow rate from the hole can be obtained by multiplying the cross-sectional area of the hole times its velocity, $A_1 v_1$. If P is large (in a pressurized system) compared to atmospheric pressure, the term $2gh$ in equation 6.9 can be neglected; the speed of efflux in this instance is primarily a function of P. For systems that are open to the atmosphere, $P_2 = P_1$ and $v_1 = (2gh)^{1/2}$. This implies that the speed of efflux for the open system is equal to the speed gained by a free-falling body through a vertical distance h; this is known as Torricelli's law.

Flow through the hole described above is different than if that same flow were through a pipe, because the confining walls of the pipe offer resistance to flow (assuming the fluid has viscosity). Because of viscosity, head losses occur in real systems due to frictional losses. The magnitude of these frictional losses depends upon whether flow is laminar or turbulent as defined by the Reynold's number. The flow system in figure 6.3 shows three equally spaced manometers at points C_1–C_3 along a pipe. The reservoir is held at a constant level and velocity of flow through the pipe is controlled by the valve at point C_4, such that constant pressure and steady flow are maintained. If the valve is closed, the fluid level will be equal in all manometers (point AB), hence h (AC_1) will indicate an equal pressure at all points C_1–C_3 along the pipe. However, once the valve C_4 is opened to a certain setting, steady flow will be achieved through the pipe and the height of fluid in each manometer will be at different levels, as indicated by points 1, 2, and 3. The greater the flow (i.e., the more open the valve), the greater the drop of the manometers wll be. As discussed in chapter 4, we know that the height of water in each manometer is a measurement of the pressure at points C_1–C_3 in the pipe. Because the system is at constant pressure and steady flow, the straight line along points 1–3 in figure 6.3 indicates a uniform pressure drop in the manometers. Now recall that the confining walls of the pipe offer resistance to flow, thus, the drop in pressure indicated by h_f in figure 6.3 is due to fluid friction, termed the friction head. For varying flow velocities

$$h_f = Kv^2 \tag{6.10}$$

where K is the proportionality constant (to be discussed in chapter 7) and v is fluid velocity in the pipe (m s^{-1}). Equation 6.10 depends on the Reynolds number (Re); for Re < 2,000, $h_f = Kv$.

Figure 6.3 shows an immediate drop in fluid level from point A of the reservoir to point 1 in the first manometer, which is indicated by h_v. Since equation 6.10 implies that friction head is proportional to the velocity squared, then from Torricelli's law, the drop in potential energy is due to the drop in fluid level from A to C_1. This drop must be converted into kinetic energy in the fluid, and since $v_1 = (2gh)^{1/2}$, then

$$h_v = \frac{v^2}{2g} \tag{6.11}$$

where h_v is the distance from point A to point 1, termed the velocity head in units of length (m). The pressure at any point from C_1–C_3 is determined from the height of fluid in the

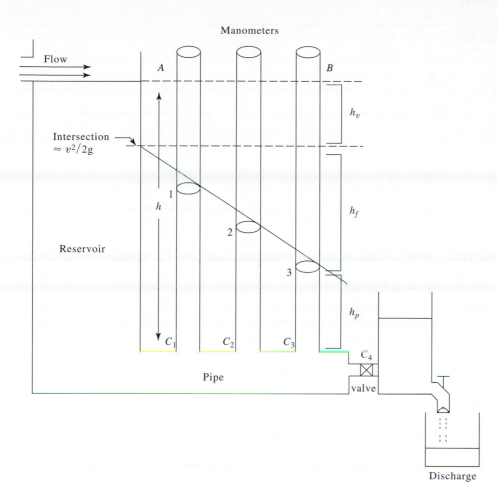

Figure 6.3   Illustration of manometer levels attached to pipe exiting reservoir. Points 1, 2, and 3 represent fluid levels in manometers located at points C_1, C_2, and C_3, while h_v, h_f, h_p refer to velocity head, pressure drop, and pressure head at point B; h is fluid level at point A.

manometer at the respective point such that

$$h_p = h - (h_v + h_f) \tag{6.12}$$

where h_p is the pressure head (m) and h is total head (m). An important point to remember here is that when fluid velocity increases, pressure decreases; and as fluid velocity decreases, pressure increases. In nature, an example of this would be when a wide, gently flowing river passes through a narrow canyon—the velocity of flow increases, but the pressure in the river decreases.

QUESTION 6.3

Calculate the pressure at a depth of 500 m beneath a lake's surface. Assume the density of water is 1.0×10^3 kg/m^3 and that atmospheric pressure, P_a, equals 1.01×10^5 Pa.

QUESTION 6.4

Water behind a dam of width w is filled to a height H. What is the resultant force on the dam? Give a general solution, not a numerical one.

QUESTION 6.5

Using figure 6.2 and the application of Bernoulli's equation to Torricelli's law (as given in equations 6.8 and 6.9), determine the velocity at which fluid will exit the small hole (point 1) on the right side of the tank when the fluid level is a distance h above the hole. In this case you are seeking a general solution, not a numerical one.

6.3   POISEUILLE'S LAW

Various models have been developed to investigate the effects of porosity and pore-size distribution on fluid flow. The best known is perhaps Poiseuille's law, which describes the laminar flow of a fluid in a small cylindrical tube. The tube radius in this case is that of a capillary tube, analogous to pore radius within soils and geologic material. Poiseuille's law states that the discharge rate, Q, of a fluid in a cylindrical tube of small, fixed radius, R, is dependent upon the driving force acting on the fluid as well as upon the internal friction forces between molecules within the fluid, characterized by the fluid viscosity, η. On a volume basis, the gradient of the hydraulic potential, $-dP_h/dx$, is the driving force, considered constant for this discussion.

Fluid viscosity may be explained with the aid of figure 6.4, which shows a fluid layer trapped between two parallel plates of solid substance. If one moves the lower plate at a constant velocity, v, relative to the upper plate, the velocity of the fluid at the boundary with the upper plate will be zero, but also equal to v at the boundary of the lower plate, because of adhesive forces. Assuming the steady velocity between the two plates, $v(y)$ will increase linearly with the distance y to the lower plate. The internal friction force per unit area within the fluid, τ_f, tends to retard the movement of the plate. To maintain the velocity, v, of the plate, the force per unit area applied to the plate must be equal and opposite to τ_f. The force is inversely proportional to the distance, h, between the two plates, also implying that the frictional force per unit area, τ_f, is proportional to the velocity gradient dv/dy. In this case, the proportionality factor is the fluid viscosity, η, which may be expressed such that

$$\tau_f = \eta \frac{dv}{dy} \tag{6.13}$$

As shown in figure 6.5, the rate of discharge from a small tube can be determined when the velocity distribution in the tube is known as a function of r. Since we assume laminar flow, the velocity of the fluid at the wall of the tube is zero, and has a maximum value in the center of the tube. Velocities from zero to maximum will depend on the radial distance from the center of the tube, r; equal velocities are present in concentric rings around the center. Thus, according to Poiseuille's law, the rate of discharge of a small cylindrical tube may be expressed as

$$Q = -\frac{\pi r^4}{8\eta}\left[\frac{dP_h}{dx}\right] \quad \text{or} \quad -\frac{\pi r^4}{8\eta}\left[\frac{\Delta P_h}{h}\right] \tag{6.14}$$

where Q is the rate of volume flow ($m^3\ s^{-1}$), η is the coefficient of viscosity of the fluid ($\mu Pa\ s^{-1}$), r is the tube radius (m), h is the tube length (m), and P is the pressure (Pa). Note here that $P = \rho g h$; often P is expressed as ΔH. If ΔH is used, the numerator on the right-hand side of the equation (within the brackets) must be multiplied by ρg. Normally, the pressure units would then be in the *cgs* system expressed as dynes cm^{-2}, and the resulting viscosity would have units of poise or dyne-sec cm^{-2}; all length units would be expressed in terms of cm rather than m. Equation 6.14 states that the rate of fluid flow through a cylindrical tube is

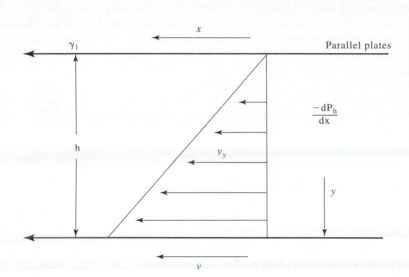

Figure 6.4   Flow between two parallel plates to illustrate viscosity, assuming a no-slip condition along plates. x is distance, v is velocity, γ_f is shear stress ($\gamma_f = \eta\, dv/dx$) exerted in direction x on fluid surface, y is distance of v_y above reference, and I is distance between plates (L).

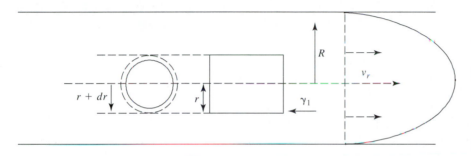

Figure 6.5   Laminar flow of a liquid in a tube illustrating Poiseuille's law. The rate of discharge can be calculated when the velocity distribution in the tube is known as a function of r. Due to friction between adjoining liquid layers and considering a liquid cylinder in the tube, r is radial distance from center, γ_1 is shear stress (as defined in figure 6.4), R is tube radius, and v_r is the velocity of fluid at radial distance r.

directly proportional to the fourth power of the radius of the tube and also to the pressure difference, but is inversely proportional to the viscosity of the fluid and the length of the tube. Poiseuille's law is true only when: **(1)** flow is steady and laminar; **(2)** the pressure is constant over every cross-section (no radial flow); and **(3)** fluid in contact with the tube wall is stationary.

QUESTION 6.6

An experiment to measure water flow is being conducted in a vertical capillary tube. At the bottom of the tube the water pressure, P_h (potential), is 7.0 kPa; at the top of the tube water pressure is 1.0 kPa. Assuming that $\eta = 10^{-6}$ kPa s, $l = 0.5$ m, and the diameter of the capillary tube is 3.6×10^{-4} m, what is the rate of flow of water, Q, through this capillary tube?

6.4   FLOW CHARACTERISTICS: LAMINAR AND TURBULENT FLOW

The movement of a fluid can be characterized as either laminar or turbulent. If each particle of the fluid flows along a smooth path and the paths of each particle do not cross each other, the flow is termed laminar. As a result, the velocity of the fluid at any point along its flow-path remains constant in time. However, above a certain critical speed, fluid flow becomes turbulent. Turbulent flow is an irregular flow often characterized by small whirlpool-like regions. A familiar example of this would be the flow of a stream around a rock which projects above the stream's surface. Upon close observation, one would see the small whirlpool-like regions and eddy irregularities around the rock.

The flow path taken by a fluid particle in laminar flow conditions is termed a streamline (see figure 6.6). For laminar flow, no two streamlines may cross each other; if they do, a fluid particle could move either way at the crossover point, and the flow would be termed turbulent.

Flow of water in soils and other geologic material can occur in three dimensions, as determined by the potential gradient of the system. If one assumes steady, one-dimensional laminar flow of a fluid (incompressible) along a solid plane surface, the velocity profile of this flow would be as illustrated in figure 6.7. Since the distance to the wall, x, is at a right-angle to the velocity, then at $x = 0$, the velocity $v = 0$, and v increases with distance from the wall but at a decreasing rate; at a certain distance from the wall, the fluid velocity will reach a maximum. Considering phases 1 and 2 in figure 6.7, which are at a distance Δx apart, the velocities along the phases will be v_1 and v_2; if $v_2 > v_1$ then $\Delta v = v_2 - v_1$. Hence, the velocity gradient

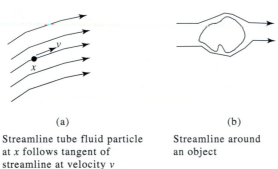

(a)

Streamline tube fluid particle at x follows tangent of streamline at velocity v

(b)

Streamline around an object

Figure 6.6 Illustration of streamlines.

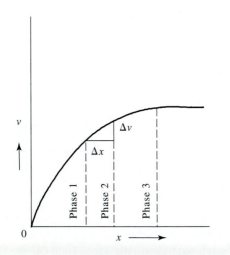

Distance from wall velocity profile

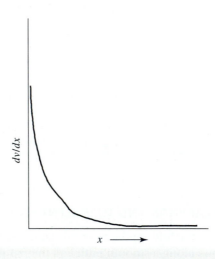

Velocity gradient

Figure 6.7 Illustration of velocity profile (left) and velocity gradient (right); v is velocity, x is distance, and dv/dx is velocity gradient, that is, change in velocity with change in distance.

dv/dx may be expressed as

$$\frac{dv}{dx} = \lim_{\Delta x \to 0} \frac{\Delta v}{\Delta x} \qquad (6.15)$$

The velocity gradient as illustrated in figure 6.7 is the reciprocal of the slope of the velocity profile. Since x is the measured distance perpendicular to the direction of flow, and from the definition of velocity we may state that

$$\frac{dv}{dx} = \frac{d\left(\dfrac{dy}{dt}\right)}{dx} = \frac{d\left(\dfrac{dy}{dx}\right)}{dt} \qquad (6.16)$$

where dy/dx is the shear at phase 2. Foregoing significant detail, the velocity gradient is the time rate of shear. As a consequence, when $dy = 0$, the shear vanishes and the velocity gradient also vanishes. Because real fluids resist shear, shear forces must always exist whenever there is a time-rate of shear. For more in-depth knowledge of shear and shear stress, the reader is referred to the literature discussing the science of rheology and related behavior (see the suggested readings section at the end of the text).

Osborne Reynolds was the first to demonstrate the difference between laminar flow and turbulent flow in 1883. Named in his honor, the Reynolds number, R_e, expresses the ratio of inertial forces to viscous forces during flow and is widely used to differentiate between laminar flow (low velocities) and turbulent flow (high velocities). For example, for a Reynolds number of value less than about 5, a condition of linear laminar flow exists; for values of about 5 to 100 the flow is termed nonlinear laminar, and for values greater than 100, flow is turbulent (Bear 1972). These general classifications assume that flow is occurring through a medium, and not a pipe or open channel. For Reynolds numbers less than about 5, viscous forces dominate; for values greater than 5 (up to about 100), inertial forces dominate. At values greater than 100, laminar flow is assumed to give way to turbulent flow. The Reynold's number is often used to determine the magnitude of friction loss within a system. For flow through soil, the Reynolds number is expressed as

$$R_e = \frac{\rho q l}{\eta} \qquad (6.17)$$

where ρ is the fluid density (kg m^{-3}), q is the flux density (some hydrologists refer to this as the specific discharge; m s^{-1}), l is a representative length-dimension of the medium in question (usually taken as the pore diameter or mean-particle diameter), and η is the viscosity (kg m^{-1} s^{-1}; converted from Pa s^{-1} listed in Table 6.1 for convenience of calculation). If R_e remains constant, fluid flow will be steady. For a detailed discussion of the Reynolds number and its application to flow through soils, the reader is referred to Bear (1972).

QUESTION 6.7

Ten cubic cm of water at 25 °C is passed through a steel capillary tube 30 cm length and 1.5 mm diameter in 4 seconds. What is the pressure required, and the Reynold's number?

SUMMARY

In this chapter we discussed the basic principles of water flow in soils, and how Bernoulli's expression relates pressure to fluid velocity, elevation, and conservation of energy when applied to a perfect fluid. In addition, the state of a fluid at any point may be characterized by four quantities: **(1)** pressure, P; **(2)** velocity, v; **(3)** elevation, h; and **(4)** density, ρ. Also described was Torricelli's law, which states that the speed of efflux for an open system is equal to the speed gained by a free-falling body through a vertical distance h. Additionally, Poiseuille's law, describing the laminar flow of a fluid in a small cylindrical tube, was discussed. Poiseuille's law states that the discharge rate, Q, of a fluid in a cylindrical tube of small, fixed radius, R, is dependent upon the driving force acting on the fluid as well as upon the internal friction forces between molecules within the fluid, characterized by the fluid viscosity, η. Finally, it was discussed how the movement of a fluid can be characterized as either laminar or turbulent. If a particle of fluid flows along a smooth path and the path of this particle and others do not cross, the flow is termed laminar, and the velocity of the fluid at any point along its flow-path remains constant in time. Turbulent flow was defined as an irregular flow, often characterized by small whirlpool-like regions.

ANSWERS TO QUESTIONS

6.1. Assume A_1 is the cross-sectional area of the hole and v_1 is the velocity of fluid exiting the hole, then $A_2 \gg A_1$ and $v_2 \ll v_1$. Also assume $v_2 \approx 0$ and $P_1 = P_2 = P_a$. Thus,

(a)

$$P_1 + \frac{\rho v_1^2}{2} + \rho g y_1 = P_2 + \frac{\rho v_2^2}{2} + \rho g y_2$$

$v_1 = [2g(y_2 - y_1)]^{1/2} = [2(9.80)(20)]^{1/2} = 19.8$ m/s.

(b) The flow rate is $A_1 v_1 = (\pi d^2 / 4)(19.8) = 4.2 \times 10^{-5}$ m^3/s. Solve for d to obtain 1.64×10^{-3} m or 1.64 mm.

6.2. By utilizing Bernoulli's equation, the pressure is converted entirely to kinetic energy, which is converted into gravitational potential energy. Thus,

$$\Delta P \rightarrow \frac{1}{2} \rho v^2 \rightarrow \rho g h$$

where $\rho = 1000$ kg/m^3. Hence, **(a)** $\rho g y = (10^3)(9.80)(40$ m$) = 1/2\ \rho v^2$; solving for v, we obtain $v = 28$ m/s. **(b)** $\Delta P = (10^3)(9.80)(40$ m$) = 3.92 \times 10^5$ Pa or 3.87 atm.

6.3. To obtain the solution we may use the formula $P = P_a + \rho g h$. Thus, $P = (1.01 \times 10^5$ Pa$) + (1.0 \times 10^3$ kg/m$^3)(9.80$ m/s$^2)(500$ m$) = 5.0 \times 10^6$ Pa. This is roughly 50 times greater than atmospheric pressure.

6.4. For the solution to this problem, the equation used in question 6.1 becomes $P = \rho g h = \rho g (H - y)$, and to find the force exerted by the fluid over a specific surface area ΔA we may use

$$P = \lim_{\Delta A \to 0} \frac{\Delta F}{\Delta A} = \frac{dF}{dA}$$

Consequently, the force is given by $dF = P\, dA = \rho g (H - y) w\, dy$ and the total force on the dam is

$$F = \int P\, dA = \int_0^H \rho g (H - y) w\, dy = \frac{1}{2} \rho g w H^2$$

Hint: draw a diagram of the dam's surface, shade to the height of a chosen water level, then select a small strip across the width of the shaded face to calculate pressure. Remember, the total force on the dam must be obtained from the expression $F = \int P dA$, where dA is the area of the small strip. As a result: w is the dam width; H is the total height of water; h is the height of water above the small strip; y is the depth of water below the small strip; and, of course, dy is the thickness of the small strip.

6.5. This is a classic case of Torricelli's law. First, assume the tank is large in cross-sectional area compared to the exit hole, that is, $A_2 \gg A_1$. Thus, the fluid will be relatively at rest at the top of the tank (point 2). Now, applying Bernoulli's equation to points 1 and 2 and at the exit hole $P = P_a$, we obtain

$$P_a + \frac{1}{2}\rho v_1^2 + \rho g y_1 = P + \rho g y_2$$

however, since $y_2 - y_1 = h$, this will reduce to

$$v_1 = \sqrt{\frac{2(P - P_a)}{\rho} + 2gh}$$

Since A_1 is the area of the exit hole, the flow rate from this hole is $A_1 v_1$. When the pressure P is large compared to atmospheric pressure, the term $2gh$ can be neglected and the speed of efflux is primarily a function of P. Also, if the tank is open to atmospheric pressure, $P = P_a$ and $v_1 = (2gh)^{1/2}$.

6.6. Using Poiseulle's law (equation 6.14), we find that $Q = [\pi(1.8 \times 10^{-4}\,m)^4/(8)(10^{-6}\,kPa \cdot s)] \times [(7.0\,kPa - 1.0\,kPa)/0.5\,m] = 4.947 \times 10^{-9}\,m^3/s$, or 0.297 mL/min.

6.7. The pressure, P, required to force the fluid through the capillary tube is

$$P = (8Vl\eta)/(\pi r^4 t)$$

$$= [(8)(10 \times 10^{-6}\,m^3)(0.30\,m)(8.90 \times 10^{-4}\,kg\,m^{-1}\,s^{-1})]/[\pi(7.5 \times 10^{-4}\,m)^4\,(4\,s)]$$

$$= 5373\,N\,m^{-2}.$$

To calculate the Reynolds number, $R_e = \rho q l/\eta$, we must first obtain q, thus

$$q = [(10 \times 10^{-6}\,m^3)/(4\,s)]/\pi(7.5 \times 10^{-4}\,m)^2 = 1.4147\,m\,s^{-1}.$$

$$R_e = (1.5 \times 10^{-3}\,m)(1.4147\,m\,s^{-1})(0.99708 \times 10^3\,kg\,m^{-3})/(8.90 \times 10^{-4}\,kg\,m^{-1}\,s^{-1})$$

$$= 2377.1.$$

A value of this magnitude would indicate turbulent flow.

ADDITIONAL QUESTIONS

6.8. You have a field-study site near the western coast of the United States. An oceanographer asks your assistance in calculating the pressure 1000 m beneath the ocean's surface. Assume water density is 1.0×10^3 Kg/m^3 and $P_a = 1.01 \times 10^5$ Pa.

6.9. At what depth in a lake is the absolute pressure three times the atmospheric pressure?

6.10. In Greenland, the ice sheet is 1 km thick. What is the pressure on the ground beneath the ice? Assume $\rho_{ice} = 920$ kg/m^3.

6.11. You are calibrating pressure transducers in the laboratory with a u-shaped tube (see figure below). What is the absolute pressure, P, on the left side if $h = 20$ cm? What is the gauge pressure?

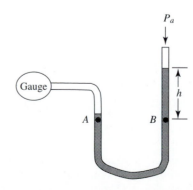

6.12. The rate of flow through a horizontal pipe is 1.5 m^3/min. What is the velocity of flow at a point where the pipe is **(a)** 5 cm and **(b)** 2 cm?

6.13. Water flows through a 6.35-cm diameter hose at a rate of 0.012 m^2/s. At what velocity does water exit the nozzle at the end of the hose?

6.14. A Venturi tube can be used as a fluid flow meter. If $P_1 - P_2 = 21 \times 10^3$ Pa ($\approx$ 3 lb/in^2), what is the flow rate (m^3/s) if the outlet radius is 1 cm, and inlet radius of the tube is 2 cm? Assume fluid density is 700 kg/m^3.

6.15. You are working with an above-ground storage tank (AST) which is filled to a height h_o. If this tank is punctured at a height h from the bottom of the tank, how far from the tank will the stream land? Assume $h_o = 5$ m and $h = 2$ m.

6.16. What is the Reynolds number for flow of a liquid in a 1.2-cm diameter soil pore? The fluid is from an oil spill: $\rho = 850$ kg/m^3; viscosity is 0.3 Pa $\cdot$ s; and velocity is 3.0 $\times$ 10^{-5} m/s.

6.17. You determine a fluid viscosity at 40 °C by measuring flow rate through a capillary tube at a known pressure difference between the ends. The capillary radius is 0.70 mm, length is 1.5 m. When a pressure difference of 1/20 atm is applied, a volume of 292 cm^3 was collected in 10 minutes. What is the viscosity of the fluid? Identify the fluid.

7

SATURATED WATER FLOW IN SOIL

INTRODUCTION

This chapter will serve as a brief review of fluid flow in saturated media, for some; for others, it is new material to learn before studying the more complex factors involved in unsaturated fluid flow. There are essentially three types of fluid flow: (1) saturated flow; (2) unsaturated flow; and (3) vapor flow. This chapter focuses on saturated flow; for types 2 and 3, the reader is referred to chapter 8 (unsaturated water flow) and chapter 9 (gaseous diffusion). The driving forces that cause saturated flow are gravitational and pressure-potential gradients. For steady flow in a saturated medium, the change in volumetric water content with respect to time is zero, that is, the water content is equal to the porosity of the medium.

7.1 DARCY'S LAW

The flow of water through water-saturated sand was studied by the French scientist Henry Darcy (1803–1858) in Dijon, France (see Philip 1995). In 1856, in one of the most famous hydrology experiments performed so far, Darcy showed that the volume of water that passes through a bed of sand per unit time is dependent on four things: (1) cross-sectional area of the bed; (2) the bed thickness; (3) the depth of ponded water atop the bed; and (4) on K, the hydraulic conductivity. This is now known as Darcy's law, expressed mathematically as

$$Q = K\frac{(A\,\Delta H)}{L} \tag{7.1}$$

where Q is the volume of water that passes through the bed (or column), cm^3 per unit time; K is the hydraulic conductivity (also called the proportionality constant), in cm per second; A is the cross-sectional area of the column (cm^2); and ΔH is the difference between the head at the inlet boundary and the head at the outlet boundary; these two boundary heads will be discussed later in this chapter. (It should be noted that H is equal to $h + L$, L representing the length of the column or bed thickness, cm.) Since $q = Q/A = V/At$, then equation 7.1 is also expressed as

$$q = \frac{V}{At} = K\frac{(\Delta H)}{L} \tag{7.2}$$

or

$$K = \frac{VL}{At\,\Delta H} \tag{7.3}$$

where V is the volume of water, cm^3, and t is time, in seconds.

The flux density, q, is the rate of water movement through a medium. To be precise, flux density is the volume of water that passes through a plane perpendicular to the direction of flow, per unit time. Although often referred to as flux, this is incorrect; flux is the volume of water flowing through the medium per unit time, expressed as $Q = V/t$. Flux density (as expressed by Darcy's Law) is dependent on the hydraulic gradient as well as on the type of medium involved. The term filter velocity is sometimes mentioned in conjunction with flux density when the unit cross-sectional area of a specific soil volume is referred to, usually as the mean speed of the soil water through the pore space, described as $q' = q/\theta$, where θ is volumetric water content.

7.2   HYDRAULIC CONDUCTIVITY AND PERMEABILITY

In the previous equations we showed one proportionality constant, K. This is often termed "big" K to represent hydraulic conductivity, which depends on properties of both the fluid and the medium; there is also a "small" k. A difference exists between the two, that we attempt to clarify here, to eliminate any confusion about Darcy's law and the proportionality coefficients associated with it (i.e., big K and small k). If we express Darcy's law as a flux density equation, then

$$q = -K\frac{\partial P_h}{\partial s} = -K\left(\frac{\partial P_t}{\partial s} + \rho_l g\frac{\partial z}{\partial s}\right) \tag{7.4}$$

or

$$q = \frac{k}{\eta}\left(\frac{dP_h}{ds}\right) \tag{7.5}$$

where K is the hydraulic conductivity (m s^{-1}) and k/η is the mobility. Mobility has the units of m^2 Pa^{-1} s^{-1}, and provides the Darcy flux (in cm/s) when multiplied by the gradient in the pressure potential; the hydraulic conductivity (K) provides the Darcy flux when multiplied by the hydraulic head gradient. Mobility can be determined from K by dividing by ρg, noting that s is distance. Thus, units for small k are m^2.

Hydraulic conductivity ($K = k\rho g/\eta$) can be separated into two factors: fluidity (defined as $\eta/\rho g$), and intrinsic permeability. The intrinsic permeability (k) of a medium is a function of pore structure and geometry. We can gain further understanding of the concept of permeability if we optimize the pore space by using capillary-tube models, starting with a single, smooth, straight capillary. For this assumption, Poiseuille's law shows that the rate of discharge for the tube is expressed as

$$Q = -\frac{\pi r^4}{8\eta}\frac{\partial P_h}{\partial s} \tag{7.6}$$

where r is the radius and η is the viscosity of the fluid; all other parameters are as previously discussed. Since $q = Q/A$, and $A = \pi r^2$, then

$$q = -\frac{r^2}{8\eta}\frac{\partial P_h}{\partial s} \tag{7.7}$$

and, by comparison with equation 7.6, $k = r^2/8$.

If the column contains a bundle of capillary tubes or parallel pores of only one size with a certain number of pores, n (perpendicular to flow), then $q = nQ/A$. However, not all pores within a medium are of the same diameter. As a result, n_i represents the number of pores within a system in the ith class with a radius r_i, and using this approach, the flux density is expressed as

$$q = \sum \frac{n_i Q_i}{\pi r^2} = -\frac{1}{8\eta}\frac{\partial P_h}{\partial s}\sum_{i=1}^{n} n_i r_i^2 \tag{7.8}$$

Equation 7.8 is invalid since all pores are not straight, smooth, and parallel to each other. Pore water does not flow in a straight line, but instead travels around individual particles and through varying pore sizes within the medium, resulting in a much longer path than a straight line. The effect of this meandering flow path on the permeability of the medium can be accounted for by tortuosity τ. This is defined as the square of the ratio of the fluid flow path to L, the length of path over which the pressure gradient is effective.

Bear (1972) states that the ratio of $(L/L_e)^2$ (where $L_e > L$) is the correct form to use for tortuosity, rather than L/L_e, as presented in some texts. This is because the effective flow path through a porous medium affects velocity and driving force (which is the hydraulic gradient in cases that involve solution, and pressure in cases that involve air). If we assume a straight path of length L for fluid flow through a soil column (versus the effective length, L_e, which is a meandering path), we can project the flow direction, x, of L_e onto L to obtain an average velocity, $\bar{v}$, in a direction tangential to the axis of the soil capillary or tube of interest. We define this velocity as v_s. Because of the meandering path, even if $|v|$ is constant, v_s can vary. Based on analysis presented by Carman (1937), $\bar{v}$ is defined as the magnitude of the average tangential velocity, and the mean value of v_s is defined as $\bar{v}_s = v(L/L_e)$, which is the velocity component. The absolute value of the mean hydraulic gradient, $|\bar{v}\phi|_s$ (where $\phi = dh/dl$), acts as the driving force in the porous medium; thus, $|\bar{v}\phi|_s = (\Delta\phi/L_e)/(L_e/L)$ and, extending Poiseuille's law to flow in a noncircular tube, $v = (R^2\rho g/m\eta)/(\Delta\phi/L_e)$, where R is the hydraulic radius of the tube and m is a shape factor accounting for the noncircular shape of the tube. Thus,

$$\bar{v}_s = (R^2\rho g(L/L_e)^2|\bar{\nabla}\phi|_x/m\eta = [(R^2\rho g/m)(L/L_e)^2|\bar{\nabla}\phi|_x]/\eta,$$

where $(L/L_e)^2 < 1$ is called the tortuosity factor. The erroneous presentation of L/L_e as the tortuosity factor by some authors arises from failure to account for the effects of L_e on both velocity and the driving force for flow. Values of L/L_e mentioned in the literature vary from 0.56 to 0.8 (Bear 1972).

Because $n_i r_i^2$ is the contribution of $(\Delta\theta)_i$ of the pores with radius, r_i, to the total volume fraction of water (θ), the flux density is also

$$q = -\frac{1}{8\eta\tau}\frac{\partial P_h}{\partial s}\sum_{i=1}^{n}(\Delta\theta)_i r_i^2 \qquad (7.9)$$

where $(\Delta\theta)_i = n_i r_i^2$ is the contribution of pores of radius class i to the total volume of water within this fraction. We can compare equation 7.9 to equation 7.4 to obtain the hydraulic conductivity, such that

$$K = \frac{\rho g}{8\eta\tau}\sum(\Delta\theta)_i r_i^2 \qquad (7.10)$$

This equation permits an estimation of either hydraulic conductivity or mobility, given the size distribution of pores filled with water.

The most concise way to express tortuosity is $\tau = (L/L_e)^2$, where L_e is effective length (or actual path length) through which a molecule of water must travel through the length (L) of a column. Tortuosity can also be evaluated by measuring the electrical resistivity of a medium, E_s, which assumes the pore space is filled with fluid of known resistivity E and saturated conditions. Then

$$\left(\frac{L}{L_e}\right)^2 = \frac{E}{\phi E_s} \qquad (7.11)$$

where ϕ is the porosity (i.e., the area of conducting pore per unit cross-sectional area).

QUESTION 7.1

What is the intrinsic permeability of a medium in which $K = 4.15 \times 10^{-7}$ m^2 Pa^{-1} s^{-1} ? Assume $\eta = 10^{-3}$ Pa s.

QUESTION 7.2

Poiseuille's law as listed in equation 7.7 is a special case of Darcy's law. In Poiseuille's law, what terms correspond to K in Darcy's law?

QUESTION 7.3

Express the hydraulic potential on a per-volume, per-mass, and per-weight basis.

QUESTION 7.4

Express the units for K and k.

QUESTION 7.5

How would you define the volume basis for flux density, for water flow in soil?

7.3   HYDRAULIC CONDUCTIVITY VALUES OF REPRESENTATIVE SOILS

The previous sections have outlined Darcy's law and given the basic equations for water flux. The constant of proportionality of Darcy's law (K) has been termed the hydraulic conductivity, and is a function of both the properties of the medium and the fluid. The hydraulic

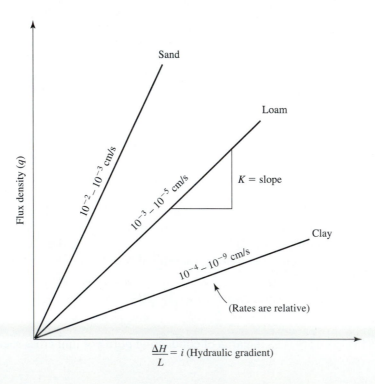

Figure 7.1 Plot of flux density versus the hydraulic gradient for various soil types

conductivity of soils has a wide range; generally, K can range from 10^{-9} cm/s for clay to 1 cm/s for clean sand. Lower values of K for a clay medium (with smaller pore sizes) are likely due to the drag exerted on the viscous fluid by the walls of the pores. A medium with a wide range of pore sizes conducts fluid much more rapidly than a medium with a narrow range of pore sizes; this is especially true if pores, preferential flowpaths, or macropores form continuous paths through the body of the soil.

Particles of smaller-sized individual grains (such as clays compared to sands) have a larger specific surface area, increasing the drag on water molecules that flow through the medium. Thus the result is a reduced intrinsic permeability and K (see figure 7.1).

7.4   FACTORS AFFECTING PERMEABILITY AND HYDRAULIC CONDUCTIVITY

Intrinsic permeability is a property of the medium alone, as the units (L^2) suggest. For soils that do not interact with fluids and change fluid properties or vice versa, this indicates that the same value for k will be obtained for different fluids. However, if fluid–media interactions alter the medium structure, the intrinsic permeability can be altered greatly. For example, a soil with a different composition than that of the native (or residual) water may react with clay minerals in the medium. Such interactions can cause swelling of the clay lattice, thereby reducing the pore space available for flow, or can result in clay disaggregation, allowing clay platelets to migrate and block pore throats.

McNeal and Coleman (1966) evaluated seven soils of varying clay mineralogy and found that decreases in permeability were more pronounced for soils high in 2:1-layer silicates, and that the most labile permeability was exhibited by soils that contained the most montmorillonite. They also found that soils containing considerable amorphous material were more stable than the average soil, and that a soil high in kaolinite and sequioxides was basically insensitive to variation in solution composition.

Petroleum engineers are also concerned with the effects of water composition on permeability, since they often inject water foreign to the formation in order to enhance oil recovery. In laboratory studies, Johnston and Beeson (1945) determined that the permeability of clay-rich (i.e., water-sensitive) formation samples to distilled water can be thousands of times less than to salt water. Originally, petroleum engineers had many problems with water-floods of petroleum reservoirs, until they learned to use water compatible with the geologic formation.

Permeability can also be affected by the kinetic energy associated with rainfall and irrigation events since this energy disperses particles on the particle surface, causing mechanical crusting that results in lower values of K at the surface. In addition to mechanical crusting, water quality can be a factor in crusting during precipitation, as the nearly distilled rain water contacts the surface sediments and causes the clays to disperse due to double-layer effects (see chapter 3).

A third cause of permeability reduction occurs due to entrapped air. As a previously unsaturated medium is saturated with water, some air is trapped within the system. This trapped air can form bubbles of varying size within the soil, depending on the amount of air present and pore size; such bubbles can obstruct the flow of water.

An additional cause of hydraulic conductivity reduction involves microbial processes. The bacteria themselves may clog the medium, or they may induce chemical reactions that produce clogging slimes. For soils subjected to continuous ponding or submergence, K will decrease initially, due to air entrapment and the leaching of electrolytes, causing dispersion of clay aggregates that seal some pores. Typically, after the initial decrease, an increase in K is seen due to the gradual dissolution of entrapped air caused by water movement. Finally, a reduction of K follows because microbial sealing (the growth of bacterial colonies that clog soil pores) exceeds the rate of increase of K caused by the removal of entrapped air; this results in a gradual, yet continual decrease in K. The reduction of K due to microbial sealing can be significant; it is a major reason why artificial recharge basins are rotated for use.

7.5   LIMITS OF DARCY'S LAW

One of the assumptions of Stokes' law (as discussed in chapter 2) was that the velocity of the particle falling through the solution must be small enough to ensure that flow is always laminar. Similarly, for small-flow velocities such as flux densities or specific rate of discharge (as q is often referred to), the drag forces between the solid surfaces of the medium and the water molecules is proportional to q. Considering the specific surface area of a medium, the area of contact between solid particles and water can be quite large. This means that the force of drag divided by the mass of water increases rapidly with an increase in flow velocity; however, the driving forces acting on the water are not very large. Therefore, the drag force acting on the water reaches a magnitude equal to the driving force. When this occurs, the net force is zero, because each of these forces acts in an opposite direction. As a result, acceleration ceases and a steady flow is achieved until the balance of forces is disturbed; in this situation, inertial forces are insignificant.

As shown by the departure of the "observed" curve from the predicted curve in figure 7.2, Darcy's law does not apply at large enough flow velocities, that inertial forces become significant relative to viscous forces. Darcy's law generally applies when the Reynolds number, defined as the ratio of the inertial forces (as given by the product of fluid density, flux density, and median pore size), to the viscous forces (as given by the fluid viscosity) is less than about 1 (Bear 1972). Darcian flow rates are usually not exceeded in granular media or nonindurated rocks, but flow rates that do exceed the upper limit of Darcy's law (as shown in figure 7.2) are common in karstic limestones and dolomites, as well as in cavernous volcanics.

For soils with low permeability such as clays (see figure 7.2, lower curve), deviations from Darcy's law have been observed, especially at low gradients. Low hydraulic gradients

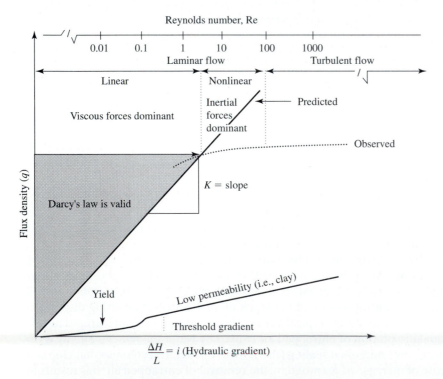

Figure 7.2   A schematic classification of flow through soil, illustrating the Reynolds number, flow conditions under which Darcy's law is valid, and predicted versus observed plots for K.

can cause no-flow conditions, or flow rates that are so low that they are less than proportional to the hydraulic gradient. This likely results from water molecules that are subjected to adsorptive forces near the solid particles, which means the water near the particle is more "rigid" than bulk water within the medium. Such water exhibits properties associated with a "Bingham liquid" (which has a yield value), rather than a "Newtonian liquid." A good example of bound (or rigid) water is hygroscopic water, which develops a structure similar to that of ice. Certain media have a threshold gradient (figure 7.2, bottom curve), below which the flux density is lower than that predicted by Darcy's law, and usually very near zero. In these cases, the flux density is proportional to the hydraulic gradient only when the gradient value exceeds the threshold value. Olsen (1966) showed that (except for montmorillonite clays), deviations from Darcy's law at the low-range are generally experimental artifacts; also, the likely cause of non-Darcy flow is the quasi-crystalline structure of water near the clay surface, and reversible clay fabric changes induced by seepage forces. The two characteristics governing the extent to which these mechanisms cause deviation from Darcy's law are pore size and rigidity of the clay fabric.

Considering the previous discussion, a more general form of Darcy's law can be written, such that

$$q = -K\left(\frac{\Delta H}{\Delta L}\right)^m \tag{7.12}$$

where m would equal 1 for most flow situations in which Darcy's law could be applied (Ghildyal and Tripathi 1987). If m is greater than 1, laminar flow does not apply, and Darcy's law should not be used (see fig. 7.2).

7.6 DARCY'S LAW AND WATER FLOW THROUGH SOIL COLUMNS

A common practice has been to utilize the principles associated with Darcy's law on many small-scale studies, especially those laboratory soil columns that allow scientists to determine hydraulic conductivity in an economic, efficient, and controlled manner. While using manometers for measuring H in soil columns and in-situ has been a common practice for several decades, recent technological advances allow the use of pressure transducers for more accuracy and automation of sequential measurements.

Flow within a soil profile (or system) can occur in any direction, whether vertical or horizontal. When the hydraulic heads H_1 and H_2 are measured by manometers labeled 1 and 2 (see figure 7.3), Darcy's equation is expressed as

$$q = K\frac{(H_1 - H_2)}{L} \qquad \text{or} \qquad q = K\frac{(h_1 - h_2) + (z_1 - z_2)}{L} \tag{7.13}$$

If the manometers in figure 7.3 are replaced by pressure transducers, the pressure potential, P, for the two transducers will be $P_1 = P_1 + \rho g z_1, P_2 = P_2 + \rho g z_2$, and

$$\Delta P = [P_1 - P_2 + \rho g(z_1 - z_2)] \qquad \text{or} \qquad \Delta H = \left[\frac{(P_1 - P_2)}{\rho g}\right] + (z_1 - z_2) \tag{7.14}$$

QUESTION 7.6

Using the soil column shown in figure 7.3, what is the volume of water, V, that will flow through the soil in 30 minutes? Assume $K = 3.52$ cm/hr, $L = 70$ cm, cross-sectional area $= 7.5$ cm^2, $z_2 = 0, h_1 = 10$ cm, $h_2 = 5$ cm, and the angle of incline for the column is 45 degrees.

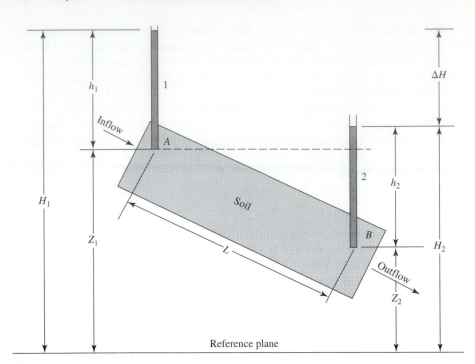

Figure 7.3   Water flow through an inclined column

7.7   HOMOGENEITY, HETEROGENEITY, ISOTROPY, AND ANISOTROPY OF SOILS

Models of water flow that are based on analytical equations as solutions to the Laplace or diffusion equation treat a medium as homogenous—that is, the same spatial properties exist throughout the medium. In this case, K would be independent of position; however, due to spatial variability, K is dependent on physical properties at any given position within a medium, and as such, the medium is considered heterogeneous.

Because K values often vary by more than two orders of magnitude within the same soil profile or hydrogeologic unit, an arithmetic mean appears to give more emphasis to the higher values of K. A more representative description of the average K value of the system is the geometric mean. We determine the geometric mean by taking the natural log of each value, find the mean of the natural logs, and then obtain the exponential (e^x) of that value (Fetter 1994; Freeze and Cherry 1979). Statistical methods are usually performed on media parameters in order to obtain more information about them than either arithmetic or geometric means provide.

In general terms, statistical distributions can provide quantitative descriptions of a medium and therefore, the degree of heterogeneity *for* that medium. A well-known fact is that the probability density function, *pdf*, for hydraulic conductivity (K) often fits a log–normal function, for which log K has a normal distribution. Freeze (1975) showed that the standard deviation of log K (independent of the units of measurement) ranges from approximately 0.5 to 1.5. Additionally, Greenkorn and Kessler (1969) state that a homogenous formation has a *pdf* of K that is monomodal, while for a heterogeneous medium, the *pdf* of K is multimodal; and for the homogenous medium, K varies only slightly in spatial terms, but has a constant mean K throughout the medium.

In most natural-aquifer systems comprised of unconsolidated sediments or sedimentary rocks, the geometry of the void space varies with direction, due to the preferred orientation of plate-shaped clay particles within the medium. Because of these pore-geometry variations, the hydraulic conductivity of the medium varies with direction, and is termed "anisotropic." Permeability is usually greatest when parallel, and least when perpendicular, to the plate orientation. For systems in which flow is three-dimensional, the medium exhibits three permeability axes, including those where permeability is greatest and least (the principal axes), and a third that is orthogonal to the principal axes.

By generalizing the one-dimensional form of Darcy's law for a medium that is anisotropic, we can describe flow in three dimensions. For three-dimensional flow, the velocity (v) is a vector with components in the x, y, and z directions. The velocity for each direction can be expressed by

$$v_x = -K_x \frac{\partial h}{\partial x}$$

$$v_y = -K_y \frac{\partial h}{\partial y} \qquad (7.15)$$

$$v_z = -K_z \frac{\partial h}{\partial z}$$

where K_x is the hydraulic conductivity in the x direction. Note that we use partial derivatives here, since h is a function of x, y, and z. For generalized flow in three dimensions for the x direction (Freeze and Cherry 1979), we may write

$$v_x = -K_{xx} \frac{\partial h}{\partial x} - K_{xy} \frac{\partial h}{\partial y} - K_{xz} \frac{\partial h}{\partial z}$$

$$v_y = -K_{yx} \frac{\partial h}{\partial x} - K_{yy} \frac{\partial h}{\partial y} - K_{yz} \frac{\partial h}{\partial z} \qquad (7.16)$$

$$v_z = -K_{zx} \frac{\partial h}{\partial x} - K_{zy} \frac{\partial h}{\partial y} - K_{zz} \frac{\partial h}{\partial z}$$

Thus, we obtain nine components of K for the most general case. In matrix form, these three equations form a second-rank tensor, the hydraulic conductivity tensor (Bear 1972). If the principal directions of anisotropy coincide with the x-, y-, and z-coordinate axes, equation 7.15 can be used instead of equation 7.16. In many (but not all) situations, it is possible to choose the coordinate system that will satisfy this condition. A situation in which it may not be possible is a heterogenous anisotropic system, where the direction of anisotropy varies from one location to the next.

7.8   SATURATED FLOW IN LAYERED MEDIA

For heterogeneous media, K varies in spatial terms and is commonly modeled to predict water movement through a layered medium, assuming that each layer is homogenous but the medium itself is heterogeneous. The average K depends on the direction of flow: for horizontal flow, the average K is the arithmetic mean; the vertical K is defined as the harmonic mean, generally smaller. Thus, a heterogeneous, layered medium can be modeled as an anisotropic homogeneous medium in which $K_H = \Sigma K_i b_i / \Sigma b_i$ and K_z = eq. 7.17. Considering the fact that many geologic media such as sedimentary rocks and marine deposits contain depositional layers, this approach is reasonable. Each layer will have a different hydraulic conductivity denoted as K_1, K_2, etc.

Soils are often comprised of several layers of thickness d. Each layer is generally considered homogeneous and will have a varying hydraulic conductivity, as illustrated in

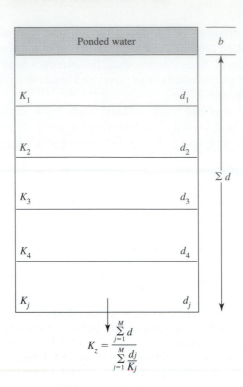

Figure 7.4 Illustration of calculation of K_z through a layered-soil profile (see equation 7.14); K is the hydraulic conductivity for layer $1, \ldots, j$; d is layer thickness for layers $1, \ldots, j$; and K_z is the effective hydraulic conductivity through the entire profile

figure 7.4. If the medium were composed of one layer, the vertical hydraulic conductivity could be simply calculated using equation 7.3. However, for a layered medium the effective vertical K will depend on resistance to flow through each layer. As a result, one obtains a hydraulic resistance, R (derived from Ohm's law), such that $R = d_j/K_j$, where d_j is the thickness of each layer and K_j is the hydraulic conductivity of each layer. Consequently, we can write an expression for K as a sum of all layers and obtain the effective vertical hydraulic conductivity such that

$$K_z = \frac{\sum\limits_{j=1}^{m} d}{\sum\limits_{j=1}^{m} \left(\dfrac{d_j}{K_j} \right)} \tag{7.17}$$

where K_z is the effective hydraulic conductivity; the summation of d in the numerator is the thickness of the entire medium; and the hydraulic resistance is the denominator, summed for each layer assuming perpendicular flow.

Use the following steps to obtain the effective hydraulic conductivity and the flux density, q, for a layered profile: (1) define a reference point (elevation); (2) determine the effective hydraulic conductivity using equation 7.17; (3) using the reference elevation, select two points (1 and 2) where the hydraulic head (H) is known, and determine the gradient, $\Delta H/\Delta d$, then solve for $q = -K[\Delta H/\Delta d]$; and (4) substitute the values obtained for q and apply Darcy's law across each layer to calculate the pressure at the layer interfaces. As previously explained (and assuming K is constant), we should expect the pressure to change linearly within each layer.

QUESTION 7.7

You are working in the Florida Everglades, which has just been ravaged by a major hurricane. A large area around a municipal wastewater treatment plant in a small town has 100 cm of ponded water covering the land surface. State health officials fear that unless the water drains rapidly, a severe health hazard will be created for the nearby town; they want you to tell them how long it will take the water to subside. Additionally, they need you to calculate the flux density, total potential, and pressure potential (weight basis) at the interface between each layer, in the soil report they provided to you. Based on soil reports of the area, there is a surficial aquifer about 1 m below the surface which is comprised of two 50-cm layers of soil with different hydraulic conductivities. Therefore, you must consider two cases of saturated flow down through a two-layer medium. The conductivity of layer 1 is 10^{-4} cm/s, and K for layer 2 is 10^{-5} cm/s. These parameters are true for only 60% of the submerged area; for the remaining 40%, K for each layer is reversed (i.e., the less conductive medium overlies the more conductive medium).

QUESTION 7.8

In question 7.7, what causes the difference in sign of the interface pressure potentials?

QUESTION 7.9

A soil column has a 2-mm thick crust at the top and a total length of 20 cm, thus the remaining length of column is 19.8 cm. Water is kept ponded on the surface to a depth of 1 cm and steady-state flow of water is occurring through the column; the column is open to the atmosphere at the bottom. The saturated K of the crust is 0.001 cm/hr and the conductivity of the underlying soil layer is 5 cm/hr. **(a)** What is the effective hydraulic conductivity of the column? **(b)** What is the flux density? **(c)** What is the pressure pontenial on a weight basis at the interface between the crust and the underlying soil layer?

7.9   LAPLACE'S EQUATION

Steady-state saturated flow in three dimensions is usually described by the Laplace equation. Steady-state flow occurs when the magnitude and direction of the flow velocity, at any point in a flow field, are constant with time. As previously mentioned, for steady-state flow to occur through a medium, the law of conservation of mass must apply. This implies that, when considering a specific volume element of interest, the rate of fluid flow into the volume must equal the rate of fluid flow out of the volume element; simply stated, inflow − outflow = 0. The continuity equation relates this law in mathematical notation such that

$$-\frac{\partial(\rho v_x)}{\partial x} - \frac{\partial(\rho v_y)}{\partial y} - \frac{\partial(\rho v_z)}{\partial z} = 0 \tag{7.18}$$

The ρv terms have dimensions of mass rate of flow per unit cross-sectional area. For constant density fluid, the expression can be further simplified to

$$-\frac{\partial(v_x)}{\partial x} - \frac{\partial(v_y)}{\partial y} - \frac{\partial(v_z)}{\partial z} = 0 \tag{7.19}$$

A simplified expression of Darcy's law in differential notation is $v = -K \, dh/dl$; substituting Darcy's law for v in the above equation for all x, y, and z, we obtain

$$\frac{\partial}{\partial x}\left(K_x \frac{\partial h}{\partial x}\right) + \frac{\partial}{\partial y}\left(K_y \frac{\partial h}{\partial y}\right) + \frac{\partial}{\partial z}\left(K_z \frac{\partial h}{\partial z}\right) = 0 \tag{7.20}$$

This equation applies to conditions of steady-state flow through saturated anisotropic media under the condition that the principle axes of flow coincide with those of the K tensor. If we assume a homogeneous, isotropic medium, the equation can be simplified to

$$\frac{\partial^2(h)}{\partial x^2} + \frac{\partial^2(h)}{\partial y^2} + \frac{\partial^2(h)}{\partial z^2} = 0 \tag{7.21}$$

This last equation is termed Laplace's equation, the solution of which is a function $h(x,y,z)$ that gives the value of the hydraulic head, h, at any point within a three-dimensional volume, subject to specified boundary conditions. Solutions of Laplace's equation allow us to plot flow nets, flowlines, and equipotential maps of h. The Laplace equation usually assumes that all flow is from water stored in the aquifer; however, field research often shows significant flow generated from leakage into the aquifer through overlying or underlying confining layers.

The boundary conditions specified in equation 7.21 enable us to obtain a numerical solution to a flow problem. If steady horizontal flow through a homogeneous, isotropic porous medium is assumed, Laplace's equation can be written as

$$\frac{\partial^2(h)}{\partial x^2} + \frac{\partial^2(h)}{\partial y^2} = 0 \tag{7.22}$$

Equation 7.22 is referred to as the equation of flow for steady-state saturated in flow in the xy plane. Consider the simple problem of flow through a rectangle bounded by equipotentials at $x = 0$ and $x = x_L$, and by impermeable boundaries at $y = 0$ and $y = y_L$. For this simplified problem, the mathematical statement of the boundary condition, given by Freeze and Cherry (1979), is

$$\frac{\partial h}{\partial y} = 0 \quad \text{on} \quad y = 0; \quad y = y_L$$

$$h = h_0 \quad \text{on} \quad x = 0 \tag{7.23}$$

$$h = h_1 \quad \text{on} \quad x = x_L$$

where $h = h_0$ at $x = 0$ and $h = h_1$ at $x = x_L$ and x is the direction of flow; the y direction represents plane thickness that would extend from $y = 0$ at $x = 0$ to $y = y_L$. When boundary conditions are specified, the problem becomes a boundary-value problem, in actuality a mathematical model. The mathematical model has a four-step process of analysis: (1) examine the physical problem; (2) specify boundary conditions and replace the physical problem with a mathematical statement; (3) solve the mathematical statement; and (4) interpret the results in terms of the original physical problem. For any problem that involves steady flow through a confined or water-table aquifer, specified boundary conditions allow us to obtain a solution to the problem via the four basic steps listed above. A more detailed discussion of boundary conditions and boundary value problems is given in Bear (1972), Freeze and Cherry (1979), Bear and Bachmat (1991), and Smith (1985).

The water table is defined as the phreatic (or free) surface in an unconfined aquifer or confining bed, at which pore-water pressure is atmospheric. The pressure (pressure head) at the water table is zero, thus at any given point along the water table, the hydraulic head is equal to height of elevation of that point from the reference plane; see figure 7.5. The slope of the water table is equal to $\Delta H/S$ or $\tan \theta$ (see figure 7.5); the hydraulic gradient is given by $\Delta H/L$ or $\sin \theta$. Also note in figure 7.5 that the saturated zone extends to the top of the water table, and that above the saturated zone is the capillary fringe, the thickness of which depends on the pore-size distribution of the media—that is, the finer the medium, the greater the capillary fringe's extent.

In a water-table situation, flow is in fact three-dimensional. Suppose we have a water-table aquifer with a horizontal impermeable base that discharges to a full penetrating stream, as seen in figure 7.6. Equipotentials are perpendicular to both the impermeable base

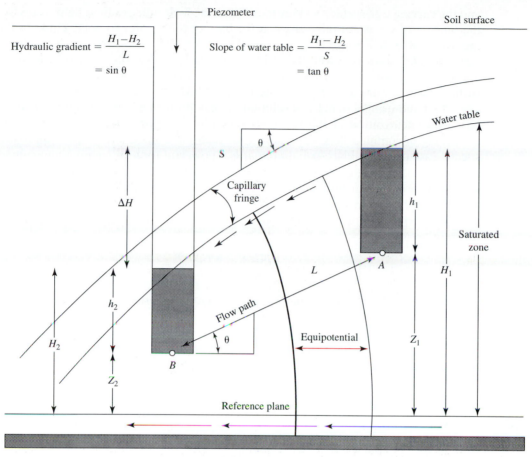

Figure 7.5   Illustration of hydraulic gradient showing differences between gradient and slope of water table

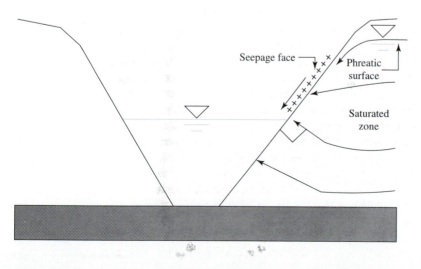

Figure 7.6   Water-table aquifer with horizontal impermeable base discharging to full penetrating stream (after Bear and Bachmat 1991)

and the curving free surface, so they must bend; the hydraulic gradient then varies from being the tangent of the water-table slope at the base of the aquifer to being equal to the sine at the water table. A seepage face is required wherever the water table discharges to a stream, ocean or lake shore, or a well. Typically, the phreatic surface represents a nonlinear boundary whose location is a priori unknown. Therefore, in most cases analytical solutions to free-surface problems are difficult, so numerical methods are often employed.

To avoid using a numerical solution, we can derive analytically approximate solutions based on linearization of the boundary conditions and/or nonlinear continuity equations describing unconfined flow. An example of such methods is the Dupuit approximation, a special case of the Laplace equation that is a powerful tool for treating unconfined flow. It assumes two things: (1) the equipotential extends vertically down from the point of intersection with the water table and the streamlines are horizontal (for small water-table gradients); and (2) the hydraulic gradient is equal to the slope of the water table. The resulting continuity equation for an isotropic, homogeneous water-table aquifer based on the Dupuit assumption is:

$$\frac{\partial(h^2)}{\partial x^2} + \frac{\partial^2(h^2)}{\partial y^2} = 0 \tag{7.24}$$

Although the Dupuit assumptions represent an approximate solution, results are usually accurate if the water-table slope is small. In addition, the Dupuit assumption does not allow a seepage face, and will give erroneous results in the immediate vicinity of such a boundary. The end result is that a three-dimensional problem can be treated approximately as a two-dimensional one, with no complicated seepage face boundary.

7.10 DIFFUSION EQUATION

The diffusion equation describes transient saturated flow. The equation of continuity is no longer equal to zero as in the case with steady-state saturated flow, but takes the form

$$-\frac{\partial(\rho v_x)}{\partial x} - \frac{\partial(\rho v_y)}{\partial y} - \frac{\partial(\rho v_z)}{\partial z} = \frac{\partial(\rho\phi)}{\partial t} \tag{7.25}$$

which, by expansion of the right-hand side, yields

$$-\frac{\partial(\rho v_x)}{\partial x} - \frac{\partial(\rho v_y)}{\partial y} - \frac{\partial(\rho v_z)}{\partial z} = \phi\frac{\partial\rho}{\partial t} + \rho\frac{\partial\phi}{\partial t} \tag{7.26}$$

where ρ and ϕ refer to water density and porosity of the media, respectively. The first term on the right-hand side of the equation is the mass rate of water due to expansion caused by a change in the water density; the second term is the mass rate of water produced by compaction of the soil as a result of change in porosity. Consequently, the first term is due to fluid compressibility and the second is due to aquifer compressibility. A change in both density and porosity results from a change in head, h. The volume of water produced (per unit volume and unit-head change) is termed the specific storage, S_s, where $S_s = \rho g(\alpha + \phi\beta)$, in which α is the aquifer compressibility and β is the fluid compressibility. The mass rate of water produced (time rate of change of fluid mass storage) is $\rho S_s \partial h/\partial t$. Using this relation, we apply the chain rule of calculus and insert Darcy's law, and obtain

$$\frac{\partial}{\partial x}\left(K_x\frac{\partial h}{\partial x}\right) + \frac{\partial}{\partial y}\left(K_y\frac{\partial h}{\partial y}\right) + \frac{\partial}{\partial z}\left(K_z\frac{\partial h}{\partial z}\right) = S_s\frac{\partial h}{\partial t} \tag{7.27}$$

which is the equation that describes transient flow through saturated anisotropic media. For transient flow through saturated, homogeneous, and isotropic media, this equation may be reduced to

$$\frac{\partial^2(h)}{\partial x^2} + \frac{\partial^2(h)}{\partial y^2} + \frac{\partial^2(h)}{\partial z^2} = \frac{S_s}{K}\frac{\partial h}{\partial t} \qquad (7.28)$$

This is the diffusion equation, first developed by Jacob (1940). The solution $h(x,y,z,t)$ describes the value of h for any point in the flow field at time t. In addition to specifying boundary conditions (as in the Laplace equation), we must also specify initial conditions to obtain a solution for the diffusion equation. The following parameters must also be known: fluid density, ρ; fluid compressibility, β (fluid parameters); hydraulic conductivity, K; aquifer compressibility, α; and porosity, ϕ (aquifer parameters). This equation takes on different forms for confined aquifers, unconfined aquifers, and other situations.

SUMMARY

In this chapter, we discussed Darcy's law and its application to saturated soils and how a hydraulic gradient is defined as a change in total head (H), with a change in distance through the medium in the direction that yields a maximum rate of decrease in head. We also demonstrated that intrinsic permeability is a property of the medium alone; this indicated that, for soils not interacting with fluids and change fluid properties or vice versa, the same value for k is obtained for different fluids. We defined flux density as the amount of water passing through a plane perpendicular to the direction of flow per unit time. We also discussed how Darcy's law does not apply at flow velocities large enough that inertial forces become significant relative to viscous forces (i.e., the limits of Darcy's law).

Additionally discussed was how Darcy's law applied to laboratory work with small soil columns. Definitions of homogeneity, heterogeneity, isotropy, and anisotropy of soil were presented, as well as generalized flow forms of Darcy's law in three-dimensions. We also explained how K varied in spatial terms, commonly modeled to predict water movement through layered soils, using the assumption that each layer was homogeneous. We ended the chapter with discussions of Laplace's equation that described steady-state saturated flow in three dimensions as well as the diffusion equation, that described transient saturated flow. The next chapter presents the fundamentals of water flow in unsaturated soils.

ANSWERS TO QUESTIONS

7.1. Using equation 7.12, $k_i = \eta K = 4.15 \times 10^{-10}$ m^2. Of course, this is only a rough estimate, without knowing the tortuosity.

7.2. We simply need to express each law in terms of Q and remember several basics: $A = \pi r^2$ (we use the term P for pressure here, instead of ∂h); $P = \rho g h$; and that ρ and g are constant. For Darcy's law, $Q = K[(A\Delta H)/L]$, and since $A = \pi r^2$ and $P = \rho g h$, we can insert these values into equation 7.7. Hence, for Poiseuille's law, $Q = Ar^2\Delta\rho g h/8\eta L$, which is also expressed as $Q = Ar^2\rho g\Delta H/8\eta L$ (because ρ and g are constant). Rewriting, $Q = [(A\Delta H)/L][(\rho g r^2)/8\eta]$; consequently, the Darcy's law $K = [(\rho g r^2)/8\eta]$ of Poiseuille's law.

7.3. On the basis of volume, the hydraulic potential is expressed as J m^{-3} m^{-1} = N m^{-3}, which is the force divided by volume. The hydraulic potential per unit mass is J kg^{-1} m^{-1} = N kg^{-1}, which is force divided by mass. The hydraulic potential per unit weight is $J\ N^{-1}\ m^{-1} = N\ N^{-1}$, or $m\ m^{-1}$ (the hydraulic head gradient).

7.4.

$$K = \left(\frac{q}{\frac{\partial h}{\partial x}}\right) = \frac{ms^{-1}}{mm^{-1}} = ms^{-1}$$

$$\frac{k}{\eta} = \left(\frac{q}{\frac{\partial P_h}{\partial x}}\right) = \frac{ms^{-1}}{J\ m^{-2}\ m^{-1}} = \frac{ms^{-1}}{Pa\ m^{-1}} = m^2\ Pa^{-1}\ s^{-1}$$

7.5. On a volume basis, the flux density (q) is simply the volume of water flowing through a cross-sectional area of interest perpendicular to the direction of flow per unit time. This is expressed in the units for the last equation, for "little" k in answer 7.2.

7.6. One problem we have is that we do not know z_1; however, we do know that the angle of the column is 45 degrees, thus z_1 is equal to the length of the column multiplied by the sine of the angle. Hence, $z_1 = 70$ cm $\times \sin 45 = 49.5$ cm. Since we are trying to determine V, then we must assume that $q = V/At$. As a result we can set up the problem such that $V/At = K[\{(h_1 + z_1) - (h_2 + z_2)\}/L]$. Inserting both the given and calculated values, we have $V/(7.5\ cm^2)(0.5\ hr) = 3.52$ cm/hr$[\{(10 + 49.5) - (5 + 0)\}/70]$; $V/3.75\ cm^2$ hr $= 2.74$ cm/hr. Thus, $V = 10.28\ cm^3$. There are many ways to calculate q, V, and K. What formula would you obtain if you solved for K? Since the numerator of the calculation above is really ΔH, we obtain equation 7.3. If we use pressure transducers instead of manometers, the solution is simpler.

7.7. To begin with, we need to set up two equations: one for flow through the top layer, and another for flow through the bottom layer; the pressure potential (h) at the interfaces is an unknown in both equations. Discharge through the top layer must equal discharge through the bottom layer. Thus, for Case 1 we have

$$q = K\left[\frac{(h_1 + z_1) - (h_2 + z_2)}{L}\right]$$

$$10^{-4}\ cm\ s^{-1}\left[\frac{(100 + 100) - (h_2 + 50)}{50}\right] = 10^{-5}\ cm\ s^{-1}\left[\frac{(h_1 + 50) - (0 + 0)}{50}\right]$$

$$50 \times 10^{-4}(150 - h_2) = 50 \times 10^{-5}(h_1 + 50)$$

$$\frac{0.725}{0.0055} = h = 131.82\ cm$$

$$q = 10^{-4}\left[\frac{(200 - 132.82 - 50)}{50}\right] = 3.64 \times 10^{-5}\ cm\ s^{-1}$$

Notice that we list h as both h_1 and h_2; this is to keep them initially separate, to avoid confusion. The total potential in this case is simply $h + 50$ cm $= 181.82$ cm. To solve for Case 2, put 10^{-5} on the left side of the equations above and 10^{-4} on the right side (i.e., reverse their order), and perform the calculations in the same order to find $h = -31.82$ cm with a total potential of 18.2 cm. For Case 2, $q = 10^{-5}\ [(150) - (-31.82)]/50] = 3.64 \times 10^{-5}$ cm/s, which is the same as for Case 1. Now, using the flux density for determination of water subsidence, we obtain 3.14 cm/d, neglecting loss by evaporation. Thus, the 1-m depth of water will take 31.8 days to subside, causing significant health hazards.

7.8. The difference between the sign in the pressure heads in question 7.7 is simply caused by a difference in K between the heads.

7.9. **(a)** We can use equation 7.17 to obtain the effective hydraulic conductivity:

$$K_z = \frac{\sum\limits_{j=1}^{m} d}{\sum\limits_{j=1}^{m} \left(\dfrac{d_j}{K_j}\right)} = \frac{20}{\dfrac{0.2}{0.001} + \dfrac{19.8}{5}} = 0.098 \text{ cm/hr}$$

(b) The flux density q may be obtained by:

$$q = -K\frac{\Delta H}{L} = -0.098\left(\frac{(h_i + z_i) - (h_o + z_o)}{L}\right)$$

$$= -0.098\left(\frac{(1 + 20) - (0 + 0)}{20}\right) = -0.103 \text{ cm/hr}$$

where subscripts i and o represent parameters at the inlet and outlet boundaries.

(c) The pressure potential we denote as P_3 may be calculated in the same manner as in part **(b)**, but using the saturated hydraulic conductivity of the underlying layer such that

$$q = K_s\left(\frac{(H_3 - 0)}{L}\right) = -0.103 = -5\left(\frac{(P_3 + 19.8)}{19.8}\right)$$

we must now solve for P_3, which yields $P_3 = -19.39$ cm.

ADDITIONAL QUESTIONS

7.10. The saturated hydraulic conductivity of a soil, K_{sat}, has been measured at 4.3×10^{-8} m²/Pa · s. What is the intrinsic permeability, K_i, for this soil? Assume $\eta = 10^{-3}$ Pa · s.

7.11. What is K_i for a poorly conductive soil?

7.12. You have a soil column in the lab with a soil depth (z), equal to 0.80 m, and a ponded head (H) equal to 0.10 m. What is K_{sat} for this column? Assume a discharge rate of 1.5 cm³/min and a cross-sectional area of 20 cm².

7.13. You are investigating saturated flow within a soil profile with layers of varying hydraulic conductivity; the flux density is the same in all layers. Why?

7.14. For question 7.3, $K_{sat} = -1.185 \times 10^{-5}$ m/s. Assume $z = 0$, $H = 0$, $z = 0.2$ m at $H = 0.6$ m, and $\Delta H = 0.6$ m. What is the flux density?

7.15. You are working with a sugarcane producer who has just harvested his crops and says he must maintain saturation of this soil for 14 days to control root-knot nematodes. Using the answer in question 7.6 (5.92×10^{-6} m/s or 5.12×10^{-1} m/day), how much water will be required over a 14-day period to satisfy this requirement?

7.16. You have a soil profile consisting of two layers. Layer 2 is 0.2 m thick and the bottom layer overlies evenly spaced drains. Since the drainage system provides atmospheric pressure, what is the flux density for each layer? Assume $h = 0$ at $z = 0.2$ m for layer 2; for layer 1, $h = 0.7$ m at $z = 0.6$ m; $K_1 = 1.18 \times 10^{-5}$ m/s and $K_2 = 1/6 \, K_1$.

7.17. Suppose you have a two-layered soil with hydraulic characteristics similar to those listed in question 7.6, with a flux density of 5.92×10^{-6} m/s, and this value reduces to -1.97×10^{-6} m/s (q_2 in question 7.7). Using the drainage set-up from question 7.7, will the drainage system be effective in reducing percolation loss through layer 2?

7.18. You have set up a field project that has a two-layered soil profile. The upper layer is 0.10 m and the lower layer is 0.9 m. The lower layer overlies a bed of coarse gravel and K_{sat} of the lower layer is 4.0 times greater than that of the upper layer. Assuming $K_1 = 1.1 \times 10^{-5}$ m/s and $dH_2/dz_2 = 0.875$, what is the flux density to the gravel layer?

7.19. In the text, we showed that at a constant static force (F^s) we have

$$\frac{F^s}{m} = -\lim_{\Delta s \to 0} \frac{\Delta \psi}{\Delta s} = -\frac{d\psi}{ds}$$

We also discussed that the corresponding potential in the case of water flow in soil is ψ_h and thus,

$$\frac{\sum F^s}{m} = -\frac{\partial \psi_h}{\partial s}$$

In addition, we showed that for the water flow in soil, the driving forces as divided by volume and by weight (respectively) were $-\partial p_h/\partial s$ and $-\partial H/\partial s$. For all normal water flow in soil, the average terminal velocity is much smaller than the value at which the drag force is no longer proportional to the velocity. Thus, when $\sum F^d = -\sum F^s$, the terminal velocity is constant, and proportional to $-\sum F^d/m = \sum F^s/m$. Using this information and the fact that the flux density (q) is proportional to terminal velocity: **(a)** define the flux density on a volume basis for the water flow in soil and state its dimension (unit); and **(b)** obtain (i.e., derive) the flux density equations (also called Darcy's law) on a weight and volume basis.

8

Unsaturated Water Flow in Soil

INTRODUCTION

So far in this text, we have discussed the physical properties and characteristics of soil, and why these properties are important in chemical transport. We also discussed the behavior of clay–water systems as well as the importance of electrokinetic properties, ion exchange, swelling, dispersion and other parameters associated with clay behavior. We then discussed the aspects of potential and thermodynamics of soil water, including capillary theory, chemical and matric potential, and hysteresis. A brief review of the chemical properties associated with soil water and the basic principles involved in water flow followed, then a very brief review of saturated water flow, including various flow cases that adhere to (or can be explained by) Darcy's Law. Although water flow in saturated soils and other media is important, soils are usually not saturated with fluid in typical vadose-zone investigations. As a consequence, the principles previously discussed are necessary in order to understand the concepts involved in water flow and the transport of chemicals in the unsaturated zone.

In chapter 7, we discussed how saturated flow depends on a positive hydraulic gradient, that is, a combination of the gravitational (elevation) and pressure potentials of the fluid that will cause water flow from high potential to low potential. Within the unsaturated zone the pore spaces are not completely filled with fluid, thus, the effective conducting pore space is much smaller than if the medium were saturated and the pore space was normally filled with both liquid and gas phases. Water flow under these conditions is termed unsaturated flow.

The unsaturated zone itself can be either very shallow or very deep, depending on geographic location. In the eastern United States, it is typical to find shallow water tables in which the root-zone portion of the unsaturated zone may be of major concern, especially since the presence of plants and greater populations of bacteria and other microbes make this shallow zone (usually about 1 m) much more dynamic. In contrast, the vadose zone in the western United States is deeper, anywhere from many meters to hundreds of meters thick. A schematic of the unsaturated zone is shown in figure 8.1. Regardless of depth the principles involved are the same, although deeper vadose zones generally require more expense in instrumentation and data collection.

Because soil pores are generally filled with both liquid and gas within the unsaturated zone, the degree of saturation refers to the portion of pore volume filled with water. Also, since the largest water-filled pores empty out first, the unsaturated hydraulic conductivity decreases rapidly as the volumetric water content decreases. This is due to the fluid that is

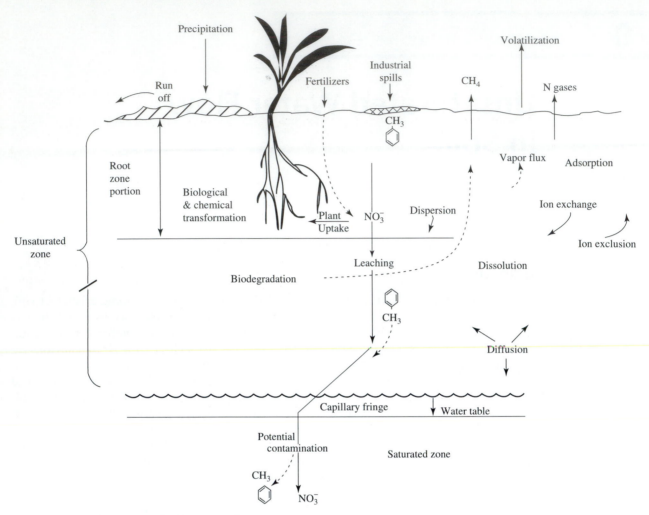

Figure 8.1 Simplified schematic of unsaturated zone processes

constrained to smaller flow channels as water content decreases. The channels not only become narrower, but the flow path becomes more tortuous and fluid can literally flow as a hydration film from one position to the next. Fluid flow in the unsaturated zone primarily is due to water content, matric potential gradient (also known as soil pressure or capillary potential), and gravitational potential. As opposed to a positive head (as discussed in saturated fluid flow), flow in the unsaturated zone is usually by a negative gradient, as well as the other parameters mentioned above. (See chapter 4 for a review of potential.)

In this chapter we discuss the validity of Darcy's Law for unsaturated conditions; factors affecting unsaturated conductivity; basic flow equations for unsaturated flow; and some fundamental mathematics that we use to solve the general unsaturated flow equation. The general continuity equation (as derived in chapter 10, question 10.2) must be altered to describe unsaturated flow fully; this alteration results in a nonlinear differential equation. A nonlinear differential equation is one in which the sum of the independent solutions, each multiplied by an arbitrary constant, does not bring about a solution to the equation. Due to this nonlinearity, the mathematics of unsaturated flow are highly complex. An additional major concern is that Darcy's Law may not apply for all unsaturated flow conditions.

8.1 VALIDITY OF DARCY'S LAW FOR UNSATURATED CONDITIONS

As discussed in chapter 7, we express Darcy's Law as

$$q = -K(\theta)i \tag{8.1}$$

where q is the flux density (volume of water flowing across a unit area per unit time), $K(\theta)$ is the conductivity (and in this case is termed the capillary conductivity to differentiate it from saturated hydraulic conductivity), and i is the hydraulic gradient (typically expressed as $\Delta H / \Delta L$). Equation 8.1 implies that if $K(\theta) \neq 0$ and a hydraulic gradient exists, flow will occur at a rate proportional to $K(\theta)$, assuming i is constant. If a gradient (no matter the size) is applied to fluid in an unsaturated medium and the water does not move, or the velocity (v) is not a straight-line function of the hydraulic gradient, or K does not vary as θ varies, then Darcy's Law will not apply and the fluid is termed non-Newtonian in behavior. However, fluids *not* conforming to Darcy's law are not necessarily non-Newtonian (Bear 1972). Often, the fluid simply does not flow until the hydraulic gradient reaches a certain value. In this case, the value the gradient reaches when flow begins is called the threshold gradient.

We do not understand fully the extent to which non-Darcy flow exists in soil, but we do know that many forces affect the flow of water within a medium. Among these is the hydrogen and covalent bonding between the water and clay particles, that in some instances is sufficient to prevent flow (in the presence of a very small hydraulic gradient). Additional forces can include: ion adsorption; double-layer thickness; and streaming potential (all discussed in chapter three). Soil structure factors such as the presence of macropores (cracks, worm holes, etc.); preferential flowpaths (areas with lower bulk density than the surrounding soil matrix); fingering; very structured and aggregated media; sand lenses; and other heterogeneities, may also lead to non-Darcy flow. However, understanding is still limited on the effects of these conditions on unsaturated flow and transport, although under investigation for more than a decade.

The continuity equation, while commonly used to simulate unsaturated flow and transport, does not consider the validity of Darcian flow. As a result, the continuity equation (in three dimensions)

$$-\left(\frac{\partial q_x}{\partial x} + \frac{\partial q_y}{\partial y} + \frac{\partial q_z}{\partial z}\right) = \frac{\partial \theta}{\partial t} \tag{8.2}$$

can be used without having to consider the limitations placed on Darcy's Law. However, when non-Darcy flow occurs (whether due to fluid property or structural effects), models based on the continuity equation are quite poor in their prediction of water and solute transport. When this and other complex relations (between matric potential, volumetric water content, and unsaturated hydraulic conductivity) are taken into account, vadose-zone experimentation frequently involves extensive instrumentation, rigorous mathematics, and methods of analysis that typically include numerical approximations for solution of a problem.

8.2 FACTORS AFFECTING UNSATURATED HYDRAULIC CONDUCTIVITY

We discussed several flow cases for saturated-flow problems in chapter 7. To show factors affecting unsaturated water flow, we begin this chapter discussing an unsaturated flow system where the pores are filled with a mixture of air and water (i.e., the pores are only partially water filled) and water flows under the influence of a negative suction or matric potential. A schematic of the flow system is shown in figure 8.2. For this case, the hydraulic head H at the inflow (point A) is $-h_1 + z_1$, where $-h$ is the matric potential and z is the gravitational

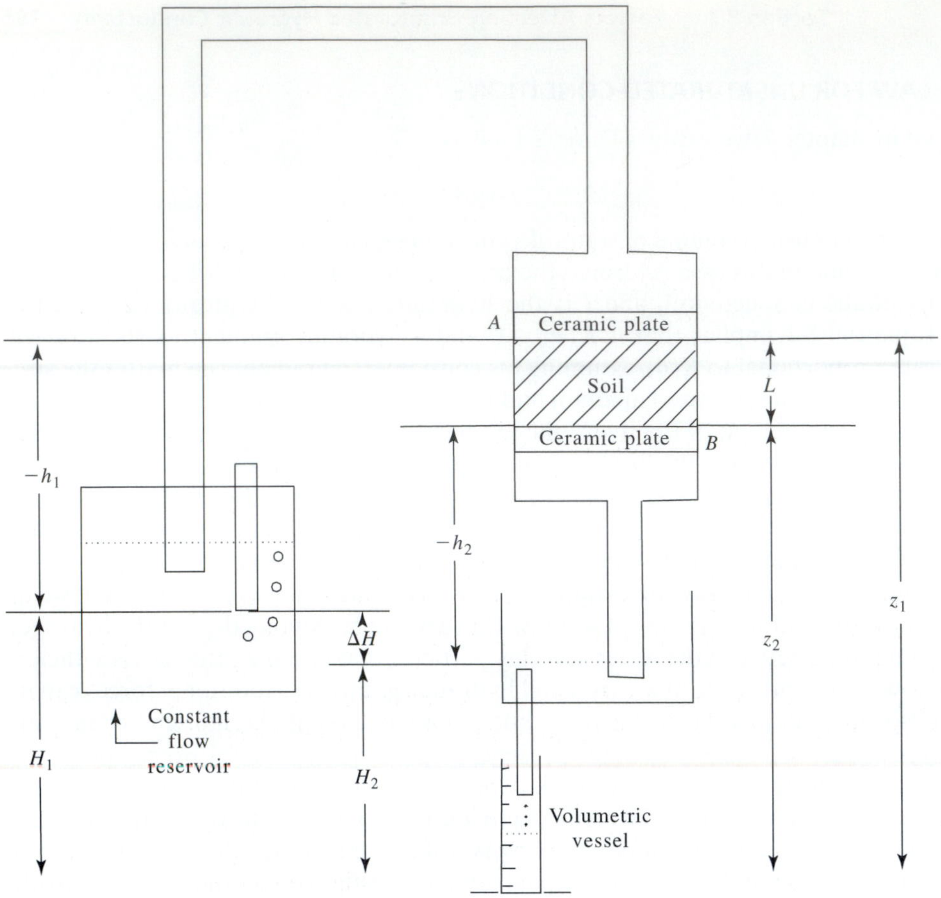

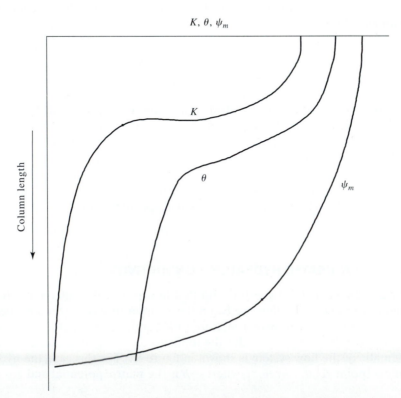

Figure 8.2 Illustration of flow system under the influence of matric potential (i.e., negative suction)

A Ceramic plate

Soil

Ceramic plate B

$-h_1$

$-h_2$

ΔH

z_1

z_2

L

Constant flow reservoir

H_1

H_2

Volumetric vessel

K, θ, ψ_m

Figure 8.3 Illustration demonstrating non-uniform wetness and thus, variation in unsaturated hydraulic conductivity, K, volumetric water content, θ, and matric potential, ψ_m (after Hillel 1971)

K

θ

ψ_m

Column length

potential; the hydraulic head H at the outflow (point B) is $-h_2 + z_2$. Subtraction of the two heads will give a change of hydraulic head: $\Delta H = H_1 - H_2 = -(h_1 - h_2) + (z_1 - z_2)$.

Assuming a constant matric potential along the column, we surmise that flow is steady, since the column is short and the flux through the column follows Darcy's law (equation 8.1). However, for an unsaturated medium (even in a short column but especially for longer soil columns), it is unlikely that the gradient is constant along the column length. As a rule, this is a result of non-uniform wetness along the column. Consequently, the volumetric water content, matric potential, and unsaturated hydraulic conductivity all vary with distance in the column. Figure 8.3 represents this graphically, and also illustrates that as matric potential decreases (increased suction), the suction gradient usually increases with a subsequent decrease in capillary conductivity along the column. Since the hydraulic gradient along the column length is not constant, we do not obtain the capillary conductivity by dividing the flux by the gradient ($\Delta H/\Delta L$), as we do in a saturated system. Instead, the flux has to be divided by the specific gradient at a given location; thus, we have to obtain a flow solution iteratively.

We have discussed a soil column that is a single layer, basically of one soil type up to this point. We now consider what happens to matric potential and moisture content in a stratified column, which is more applicable to field situations in which there is a layered profile. Using a layered column with a coarse media such as sand for layer one and a fine medium like clay for layer two (in contact with a water table), and assuming a steady downward flow, we find a distinct discontinuity in moisture content at the interface of the two layers, although the flux rate is constant throughout the column (see figure 8.4a). The flux rate is accommodated by a

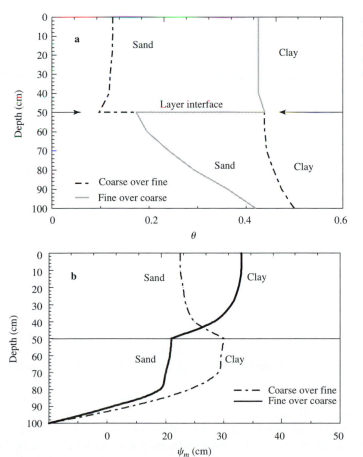

Figure 8.4   Discontinuity of θ at soil interface in a layered column with two different soil textures (a), and relation of θ to ψ_m using same soils (b)

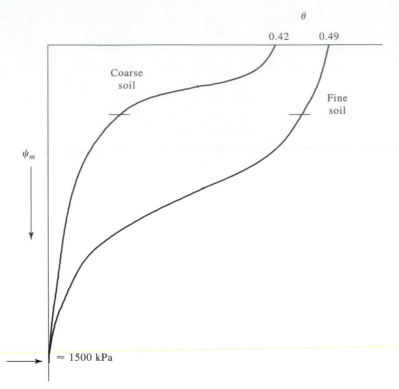

Figure 8.5 Soil moisture curves for soils shown in figure 8.4

greater tension in the fine soil (figure 8.4b), since enough pores remain saturated at that tension to transport the flux by a gravity gradient; whereas in the coarse soil, fewer pores are available at the same tension. Thus, even though the coarse soil has a higher saturated hydraulic conductivity, both soil water content and unsaturated hydraulic conductivity decrease more rapidly with increasing tension than for the fine soil (figure 8.4a). When the fine soil overlies the coarse soil (figure 8.4b), the soil suction decreases at the interface of the two layers, since water enters the coarse soil only when soil suction is reduced to that of the pores in the coarse soil. Typically, a ponded head or capillary fringe must build a positive pressure to allow water to flow from the fine to the coarse layer, meaning that volumetric water content increases and matric potential, ψ_m, decreases. Comparing figure 8.4a to 8.4b, we see that when θ decreases, matric potential decreases—that is, it becomes more negative (has a greater suction). When θ increases, matric potential increases (i.e., approaches zero). We show the soil moisture characteristic curves for these two media in figure 8.5.

A general relation for capillary conductivity versus matric potential for three different (non-layered) classes of soil is shown in figure 8.6. Initially, the coarser-textured soil has a higher $K(\theta)$; however, as suction increases the finer materials have the higher $K(\theta)$.

A medium such as sand generally has a uniform distribution of pore sizes (i.e., the pores are fairly uniform in size throughout the medium), which yields a fairly uniform mixture of soil-pore water and air. A well-structured soil that has large aggregates, cracks, macropores, and other types of voids is in direct contrast to this. In the well-structured medium, there are more phases of desaturation; desaturation is not uniform with increased suction. For example, much of the initial water flow is through larger macropores during significant recharge events. Although water can move through the aggregates, primarily it moves through the larger macropores, the size and distribution of which determines the capillary conductivity of the medium. However, with only a small suction these larger pores empty

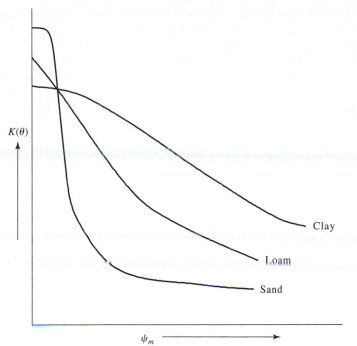

Figure 8.6   General relation for capillary conductivity ($K(\theta)$) versus matric potential (ψ_m) for three different soil classes

$K(\theta)$

Clay

Loam

Sand

ψ_m

very rapidly, leaving interped and inter-aggregate pores still saturated. It is at this phase that the water within the medium begins to move as hydrated-water films since the area of conductivity (i.e., the number of pores conducting water) rapidly decreases. Thus, even though the aggregates are saturated, the overall conductivity of the entire system is very low. As we discussed earlier, it is unlikely that full saturation is achieved due to air entrapment (likely to occur during significant recharge events). This discussion shows that Darcy's law is nonlinear under unsaturated conditions.

Unsaturated hydraulic conductivity is affected by the presence of swelling clay soil. As water content decreases with increasing suction, the soil shrinks, reducing pore size and then K. Good examples of this are well-structured vertisols, that have a high K during a state of low-moisture content but due to swelling after a significant recharge event, are reduced dramatically, to almost zero. Therefore, in general terms, the factors that can cause distinct reductions in unsaturated conductivity with an increase in soil suction are: reduction in pore conductivity due to water loss from the largest pores; reduction in effective porosity; increased tortuosity; as well as the fact that water near clay surfaces can have a viscosity four times higher than in bulk solution (see chapter 3). Temperature effects, more pronounced at water contents from about 0 to -30 kPa, result in higher K at higher temperatures, due to decreased viscosity and surface tension. Because of clay–water interactions as discussed briefly above (and in chapter 3), the effect of temperature is less dominant at low-water content.

8.3   DEVELOPMENT OF UNSATURATED FLOW EQUATIONS

The conventional form for Darcy's law does not describe adequately water flow in the vadose zone, due to the rapid decrease of both hydraulic conductivity with decreasing water content and total cross-sectional area available for water flow in an unsaturated medium. Consequently, it has been extended; this extension assumes that K is a function of matric potential and/or volumetric water content. However, which of these to use is still somewhat controversial; the typical form extends Darcy's law as a function of water content. This is due to the difficulty obtaining a specific value of θ for a particular value of ψ_m, unless we know a great deal about the hysteresis of the medium in question. Therefore, extending Darcy's law and writing hydraulic conductivity as a function of theta $[K(\theta)]$, makes the problem of hysteresis avoidable, primarily because $K(\theta)$ is less hysteretic than $K(\psi_m)$. Thus, in tensor form, the extension of Darcy's law is written as

$$q = -K(\theta)\nabla H \tag{8.3}$$

where ∇H is the hydraulic gradient and all other parameters are as previously discussed. It is important to remember that $H = h + z$, where h can be expressed as the suction or matric potential (ψ_m). As a result, we rewrite equation (8.3) as

$$q = -K(\theta)\nabla(h + z) \tag{8.4}$$

For a one-dimensional vertical coordinate system, equation (8.4) is rewritten as

$$q = -K(\theta)\left(\frac{\partial h}{\partial z} + 1\right) \tag{8.5}$$

Also, should we choose to use the matric potential rather than theta, then we obtain

$$q = -K(\theta)\left(1 - \frac{\partial \psi_m}{\partial z}\right) \tag{8.6}$$

Equation 8.6, assuming ψ is a single-valued function of θ, $\partial \psi_m/\partial z$ is rewritten as follows, using the chain rule of calculus:

$$\frac{\partial \psi_m}{\partial z} = \frac{d\psi_m}{d\theta} \frac{\partial \theta}{\partial z} \tag{8.7}$$

The first term on the right-hand side of equation 8.7 is the inverse of the specific water capacity, that is, the reciprocal of change in water content per unit change in soil suction or matric potential. The second term is the water-content gradient, with respect to depth.

By substituting equation 8.7 into equation 8.6, we have

$$q = -K(\theta)\left(1 - \frac{\partial \psi_m}{\partial \theta} \frac{\partial \theta}{\partial z}\right) \tag{8.8}$$

which is written as

$$q = -K(\theta) - K(\theta)\left(-\frac{\partial \psi_m}{\partial \theta}\right)\frac{\partial \theta}{\partial z} \tag{8.9}$$

Equation 8.9 is Richards' equation (Richards 1931), and is written so that the hydraulic diffusivity term is readily introduced into the basic flow equation, as described below.

According to Poiseuille's law (discussed in chapter 6), flow is related to pore radius. Consequently, if the conducting pore size is reduced by one-half, the capillary conductivity will be reduced by one-quarter. As more water is removed so that transport is by hydrated films, the effective path length over which the fluid travels is lengthened, that is, the fluid cannot go straight through the medium but must meander around individual particles that are covered with the hydrated-water films. This results in a tortuous path of flow that decreases conductivity further. As continuity fails within the pore system, no fluid flow occurs. At this point, only vapor transport occurs within the system; however, vapor flow is usually minimal unless there are significant temperature gradients.

By considering a bundle of capillary tubes similar to those of Poiseuille's law, but taking into account the tortuosity of the medium, we rewrite a generalized form of Kozeny's equation to determine K, such that

$$K = \frac{\rho_w g}{k' \eta s^2} \left(\frac{L}{L_e}\right)^2 \left[\frac{\phi_e^3}{(1 - \phi_e)^2}\right] \tag{8.10}$$

where ρ_w is density of water (g cm^{-3}), g is the gravitational constant, k' is the pore-shape factor (ranging from 2 to 2.5), η is the viscosity of water (Pa · s), s is the specific-surface area (ratio of total surface area of solid to volume of same solid; cm^2 cm^{-3}), ϕ_e is the effective porosity (area of conducting pore or channel per unit area of cross section—sometimes expressed as $\theta - \phi_r$; where ϕ_r is the residual moisture content), L is length of column or profile (L), and L_e is the effective length (L). The ratio of length to effective length may be determined by

$$\left(\frac{L}{L_e}\right)^2 = \frac{E_o}{\phi_e E_s} \tag{8.11}$$

where E_s is a measurement of the electrical resistivity of the soil (S m^{-1}—Siemens per meter) when the pore space is filled with a liquid of known electrical resistivity, E_o. Using equation 8.10, we can account for both the pore-shape factor and tortuosity, when determining K. The generalized Kozeny's equation 8.10 is a semiempirical equation relating K to θ and to ψ and is based on the assumption that tortuosity increases as a power of $1/S_e$, where $S_e = \theta - \theta_r/\phi - \theta_r$; see equation 8.12f.

Since the hydraulic conductivity depends on the volumetric water content and/or matric potential, we have modified Darcy's law for unsaturated media, with the more general forms as expressed in equations 8.3 and 8.4 (where K is written as a function of θ). Since no universal relations are available for capillary conductivity versus soil suction or water content, several empirical relations are proposed; these relations follow.

$$K(\psi_m) = \frac{a}{\psi_m} \qquad \text{(Baver, Gardner, and Gardner 1972)} \tag{8.12a}$$

$$K(\psi_m) = a(b + \psi_m^n)^{-1} \qquad \text{(Childs and Collis-George 1950a)} \tag{8.12b}$$

$$K(\psi_m) = \frac{K_s}{\left[1 + \left(\dfrac{\psi_m}{\psi_c}\right)^n\right]} \qquad \text{(Gardner 1958)} \tag{8.12c}$$

$$K(\psi_m) = \frac{K_s}{b + \psi_m^n} \qquad \text{(Childs and Collis-George 1950b)} \tag{8.12d}$$

$$K(\theta) = a(\theta)^n \qquad \text{(Marshall and Holmes 1979)} \tag{8.12e}$$

$$K(\theta) = K_s\left(\frac{\theta - \theta_r}{\phi - \theta_r}\right)^n \qquad \text{(Brooks and Corey 1966)} \tag{8.12f}$$

$$K(\theta) = K_s \exp(a\psi_m) \qquad \text{(Mualem 1976)} \tag{8.12g}$$

$$K(\theta) = K_s \sqrt{\frac{\theta - \theta_r}{\phi - \theta_r}} \left[1 - \left(1 - \left(\frac{\theta - \theta_r}{\phi - \theta_r}\right)^{1/m}\right)^m\right]^2 \qquad \text{(Van Genuchten 1980)} \tag{8.12h}$$

where $m = 1 - 1/n$; $K(\theta)$ is the unsaturated hydraulic conductivity; K_s is the saturated hydraulic conductivity for the same medium; a, b, m, and n are empirical constants (for fine textured media $n = 1$–2 and can be up to 4 or more for coarse media); ψ_c is the matric potential for which $K = 1/2(K_s)$; and θ_r is the residual saturation (where the moisture characteristic curve goes vertical; i.e., the water content will not get much lower). Note that equation 8.12a is not used for soil near saturation, that is, where the matric potential is near zero. Also, the relation of conductivity to matric potential depends on hysteresis, which has to be considered in the rigorous analysis of unsaturated-flow problems. For further information about equations 8.12a to 8.12h, we refer the reader to the respective references for each equation. Also, equations 8.12a to 8.12h are of different categories based on the information required to determine their coefficients (those requiring empirically developed values of K versus θ or ψ) and those based on mathematical analysis of the moisture-characteristic curve.

QUESTION 8.1

You conducted an experiment using some large sandy soil cores (97% sand) in the laboratory. The cores were set up with a matric suction (-20 kPa) applied to the bottom, across a porous ceramic plate. You assume that the suction applied was equal throughout the core. From soil samples taken at the site (where the cores were extracted), you run standard texture, soil moisture, organic carbon tests, et cetera. Using data from the soil-moisture characteristic curve in a computer program [after principles developed by Millington and Quirk (1961)] to obtain an unsaturated hydraulic conductivity curve, you determine the following: $\psi_c = 0.55$ m, $K_s = 16.72$ m/day, and the empirical constant for this soil, $n = 4.0$. What is $K(\psi)$?

8.4 HYDRAULIC DIFFUSIVITY

Working from the previous discussion, we now introduce the hydraulic diffusivity, D, often referred to as the soil-water diffusivity. Because both K and ψ_m are assumed to be single-valued functions of θ, D is expressed as

$$D(\theta) = K(\theta)\left(\frac{d\psi_m}{d\theta}\right) \tag{8.13}$$

where D is expressed in units $L^2\,T^{-1}$. Equation 8.13 is rewritten as

$$D(\theta) = \frac{K(\theta)}{\left(\dfrac{d\theta}{d\psi_m}\right)} \tag{8.14}$$

From equation 8.14 we see that D is the ratio of the unsaturated hydraulic conductivity to the specific moisture capacity (the denominator of equation 8.14), that is, a change in volumetric water content per unit change in matric potential. The denominator sometimes is expressed mathematically in "shorthand" as $c(\theta)$. The specific moisture capacity (hydraulic diffusivity) is also considered the reciprocal of the slope of the moisture characteristic curve at the same water content.

Substitution of equation 8.13 into equation 8.9 gives

$$q = D(\theta)\frac{\partial\theta}{\partial z} - K(\theta) \tag{8.15}$$

However, if we were to express D for horizontal flow in which the effects of gravity can be neglected, then equation 8.15 is rewritten as

$$q = -D(\theta)\frac{\partial\theta}{\partial x} \tag{8.16}$$

which expresses a measure of moisture flow due to a moisture or specific water-content gradient. Equation 8.16 is analogous to Fick's law of diffusion or Fourier's law of heat flow. However, we have to remember that equation 8.16 is written assuming that K, ψ_m, and $(d\psi_m/d\theta)$

are unique or single-valued functions of θ—not exactly true for unsaturated flow because of hysteresis in each of these parameters (i.e., the wetting and drying history of the medium; entrapped air; overburden pressure; water-contact angle, and so on). Essentially, the value of D for a drying medium is likely different than that for the same medium upon wetting.

The introduction of $D(\theta)$ allows us more readily to solve unsaturated flow problems because the flow equations are of a form similar to the diffusion equation, for which analytical solutions are available. Also, we can readily substitute equation 8.16 in the solute transport equation (see chapter 10). By introducing D into the differential flow equation (again, see chapter 10) and assuming only horizontal flow, we have

$$\frac{\partial \theta}{\partial t} = \frac{\partial \left(\dfrac{D\, \partial \theta}{\partial x} \right)}{\partial x} \tag{8.17}$$

Three general approaches are available for solving for D. The first assumes that D is a unique function of θ; in this case equation 8.17 remains in its current form. A second approach assumes steady flow (i.e., $(\partial \theta/\partial t) = 0$), in which case equation 8.17 is rewritten as

$$0 = \frac{\partial \left(\dfrac{D\, \partial \theta}{\partial x} \right)}{\partial x} \tag{8.18}$$

The third approach assumes that D is constant, and rewrites equation 8.17 as

$$\frac{\partial \theta}{\partial t} = D\left(\frac{\partial^2 \theta}{\partial x^2} \right) \tag{8.19}$$

Equation 8.19 is the best-known form of the diffusion equation, and is similar to the equation for heat flow or electron flow.

The most convenient way to measure D is in the laboratory, for which there are two commonly preferred methods. These include the pressure-plate outflow method (Gardner 1956), and the horizontal infiltration method (Bruce and Klute 1956), which uses the Boltzmann transformation and is discussed in the following section. In this section, we focus our discussion on the pressure-plate outflow method.

By placing a sample of a given soil in a pressure-plate apparatus, we subject it to a specific gas–phase pressure, for which the volume of water released from the sample may be recorded for each increment in pressure increase as a function of time. The pressure-plate apparatus typically is attached to a burette (used for outflow collection), and at time zero there is an initial gage pressure, P_i, for which the pressure in the apparatus and the water in the medium sample is in hydraulic equilibrium. Also at time zero, we apply a specific pressure ΔP, such that the final pressure in the apparatus, P_f, can be represented as $P_f = P_i + \Delta P$. The outflow volume collected in the burette is measured as a function of time. The smaller the pressure increments, ΔP, the more accurate the measurement of D and/or K. Remembering that $H = \psi_m + z$, then,

$$H = \frac{P}{\rho_f g} + z \tag{8.20}$$

where ρ_f is the fluid density (kg m^{-3}) and g is the gravitational constant. Since $P/\rho_f g \gg z$, z is safely neglected which results in the following non-linear differential equation:

$$\frac{\partial \theta}{\partial t} = \frac{1}{\rho_f g} \frac{\partial}{\partial z}\left[K(\theta) \frac{\partial P}{\partial z} \right] \tag{8.21}$$

Equation 8.21 is difficult to solve analytically, but if we linearize it, it is readily solved. We do this by assuming that the pressure increment, ΔP, is very small, yet large enough to obtain a measurable volume of outflow. In so doing, K is constant during outflow and is moved to the outside of the differential operator. Additionally, by using this same process we assume that the water content and pressure have a linear relation during outflow, and that both fluid density and the gravitational factors are constant also. This allows us to write (skipping several steps in the math):

$$K(\theta)\left(-\frac{d\psi_m}{d\theta}\right) = \frac{K(\theta)}{\rho_f gb} = D(\theta) \tag{8.22}$$

where b represents the reciprocal of $-d\psi_m/d\theta$, assumed constant. Assuming D is constant during each pressure increment, we calculate D from the quantity of water released from a sample. Skipping about 25 mathematical steps, we can plot $\ln[Q_o - Q(t)]$ as a function of time. In this plot, Q_o represents the total amount of water released from the sample over a pressure increment ΔP, and $Q(t)$ is the quantity of outflow from the sample at time, t. Total outflow, Q_o, is expressed as $Q_o = bV\Delta P$, where V is the volume of the sample. Thus, in order to find D from the quantity of water collected from a sample, we use

$$\ln[Q_o - Q(t)] = \left[\ln\left(\frac{8Q_o}{\pi^2}\right)\right] - \left(\frac{\pi}{2L}\right)^2 Dt \tag{8.23}$$

where L is length of soil sample. The plot of $\ln[Q_o - Q(t)]$ versus time should yield a straight line with an intercept $\ln(8Q_o/\pi^2)$; S is slope of this line per unit time (see figure 8.7), given by

$$S = -\frac{\pi^2}{4L^2}D \tag{8.24}$$

which we rewrite to obtain

$$D = S\left(\frac{4L^2}{\pi^2}\right) \tag{8.25}$$

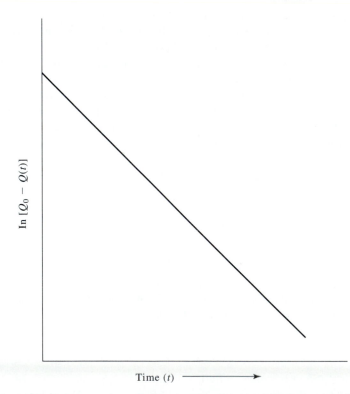

Figure 8.7   Plot of $\ln[Q_o - Q(t)]$ versus time, t, to find D; where $\ln[Q_o - Q(t)] = [\ln(8Q_o/\pi^2)] - (\pi/2L)^2 Dt$ (after Gardner 1956)

For our theory, the plot is a straight line; however, experimental data often results in "tailing," as with a break-through curve (see chapter 10). Such behavior shows that the slope S and consequently, $D(\theta)$, is not constant over the range of pressure change ΔP; it does not mean that D or K is no longer a function of theta, since they depend on the total quantity of water outflow, Q_o, from the sample. Thus, the expression for capillary conductivity is written as

$$K_c = \frac{\rho_f g Q_o 4 S L^2}{V \Delta P \pi^2} \tag{8.26}$$

where V is the volume of the sample. This is equation 16 of Gardner (1956), and allows us to solve for both K and D simultaneously from the same set of data, where the initially sought diffusivity D is given by equation 8.25. The constant, b, used in equation 8.24 is determined from the relation $b = (Q_o/V \Delta P)$.

It is of interest to note that this method was originally developed by Terzaghi (1943) to determine the diffusivity of soil samples during consolidation. Water is derived by compression of the sample due to the imposition of an incremental load, and the samples are saturated, but the theory and approach are the same.

8.5   BOLTZMANN'S TRANSFORMATION

The horizontal diffusivity equation is written as

$$\frac{\partial \theta}{\partial t} = \frac{\partial}{\partial x}\left[D(\theta)\left(\frac{\partial \theta}{\partial x}\right)\right] \tag{8.27}$$

However, equation 8.27 is a non-linear partial differential equation and cannot be solved by the usual methods. Assuming that θ is a single-valued function of a variable B (also a function of both time and distance), equation 8.27 is transformed to an ordinary differential equation through the use of Boltzmann's transformation (Boltzmann 1894; Crank 1956). The Boltzmann transformation is expressed as

$$B = \frac{x}{\sqrt{t}} \tag{8.28}$$

where x is distance (L) and t is time (T). The boundary conditions are

$$\theta = \theta_i, \quad \text{for} \quad B = \infty \quad (B \to \infty) \tag{8.29}$$

and

$$\theta = \theta_s, \quad \text{for} \quad B = 0 \tag{8.30}$$

These boundary conditions actually create a third condition because θ must be both continuous and differentiable (vary smoothly) with x, t, and also B, thus

$$\frac{d\theta}{dB} = 0, \quad \text{for} \quad \theta = \theta_i \tag{8.31}$$

These relations are seen in figure 8.8 using reduced measurements after Nielsen and Biggar (1962). By manipulating the above equations we obtain the differential equation of the curve in figure 8.8, and integrate it to find that

$$D(\theta_x) = \left(-\frac{1}{2\left(\dfrac{d\theta}{dB}\right)_{\theta_x}}\right)\int_{\theta_i}^{\theta_x} B\, d\theta \tag{8.32}$$

where θ_x is the volumetric water content at distance x, B can be obtained from equation 8.28, and θ_i is the initial volumetric-water content (preferably air-dry). One method to obtain $D(\theta_x)$ is to use a computer program. However, if you do not have such a program, the basics steps in obtaining $D(\theta_x)$ is to: (1) Plot θ versus x as in figure 8.8; (2) From this plot, evaluate $(d\theta/dB)_{\theta_x}$ and the integral of equation 8.32 at various values of θ_x. The value of the derivative $(d\theta/dB)$ is found by drawing tangents at various points along the curve ("eyeball it") and finding the slope, or semianalytically from the raw data. To evaluate the integral, simply divide the area under the curve of θ versus x into a finite number of different strips and find the approximate cumulative sum (area) of these strips; and (3) Determine D at the value of θ_x used in step 2 to obtain $D(\theta)$.

Since $D(\theta_x)$ only appears on the left side of equation 8.32, it is important to remember its relation to θ_x. Because θ_i is the lower limit in the integral, we have to integrate completely to the end of the tail in figure 8.8, that is, where the curve joins the x axis or the value of B as given in equation 8.28. As a result of this, the horizontal tube must be long enough so that the diffusing fluid front does not reach the end of the tube.

From the discussion above and the original experiment of Bruce and Klute (1956, 1962), we perceive that the Boltzmann transformation is valid for fluid movement in unsaturated soils for cases in which equation 8.27 is valid and where the following boundary conditions exist: $\theta(x,t) = \theta_i$ for $x > 0$, $t = 0$ and also $\theta(x,t) = \theta_s$ for $x = 0$, $t > 0$. Additionally, since $\theta = f(B)$, then equation 8.28 is expressed as

$$B(\theta) = \frac{x}{\sqrt{t}} \tag{8.33}$$

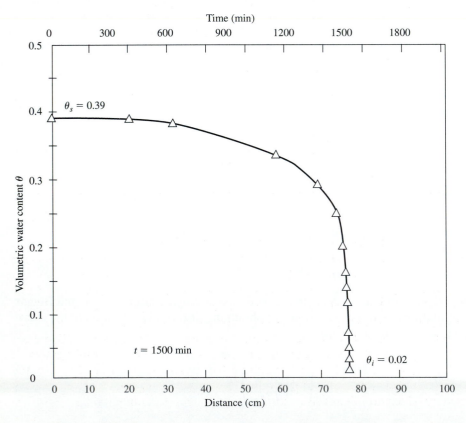

Figure 8.8   Plot of θ versus distance, x, to assist in obtaining diffusivity, $D(\theta_x)$, as described in equation 8.32 (data from Nielsen and Biggar 1962)

Simply, this implies that the quantity B as a function of θ is equal to the product on the right-hand side of equation 8.33. It should be noted that $B(\theta)$ versus θ will be different for different soils. Consequently, for positions along the horizontal column that have the same value of θ, then, despite distance x and time t, the product on the right-hand side of equation 8.33 is constant. If it is determined that θ at the wetting-front is constant, then we can plot distance of the front versus $\sqrt{t}$ at the front. The result should be a straight-line relation. However, if a straight line is not obtained, it usually indicates that the researcher failed to measure the portion of the wetting front that is at constant θ, not that the diffusion theory is invalid for the medium.

QUESTION 8.2

Volumetric-moisture content, θ, versus distance for a horizontal soil column of a sandy loam soil, for a time of 1500 minutes after water was applied at the input end ($x = 0$), is given in table 8.1. Utilize equation 8.33 to plot the Boltzmann transformation versus θ for this data.

Table 8.1

θ	Distance (cm)	θ	Distance (cm)
0.02	76.00	0.21	74.45
0.03	75.90	0.22	74.15
0.04	75.85	0.23	73.80
0.05	75.75	0.24	73.40
0.06	75.71	0.25	72.80
0.07	75.67	0.26	72.08
0.08	75.63	0.27	71.20
0.09	75.58	0.28	69.90
0.10	75.50	0.29	68.05
0.11	75.45	0.30	66.05
0.12	75.40	0.31	63.75
0.13	75.35	0.32	61.10
0.14	75.30	0.33	58.15
0.15	75.25	0.34	54.80
0.16	75.18	0.35	51.20
0.17	75.05	0.36	47.10
0.18	74.90	0.37	41.80
0.19	74.75	0.38	32.50
0.20	74.65	0.39	0

Source: Nielsen et al., 1962

SUMMARY

In this chapter we discussed the basic principles of water flow within an unsaturated medium; how, within the unsaturated zone the degree of saturation refers to the portion of pore volume filled with water, and that the unsaturated hydraulic conductivity decreases rapidly as volumetric water content decreases. Also, rather than a positive pressure head (as discussed in saturated fluid flow), flow in the unsaturated zone is primarily by a negative suction or matric potential. We also discussed the validity of Darcy's law for use in unsaturated flow. If a non-zero hydraulic gradient is applied to fluid in an unsaturated medium (no matter how small) and the water does not move or the velocity, v, is not a straight-line function of the hydraulic gradient, Darcy's law will not apply and the fluid is termed non-Newtonian in behavior. However, for most cases, Darcy's law is valid for describing water movement through unsaturated porous media. The extent to which non-Darcy flow exists in a soil is not fully understood.

We also discussed various physical and other flow factors such as stratified medium, the dependence of K on θ, and so on, which affect unsaturated hydraulic conductivity, and we developed the general flow equations for hydraulic conductivity and diffusivity. It was also discussed how to linearize the flow equation 8.27 for diffusivity by use of the Boltzmann transformation. This chapter is to serve as a beginning point for understanding the unsaturated flow concepts involving gaseous diffusion, contaminant migration, and moisture redistribution—all of which will be discussed in chapters 9, 10, and 11, respectively.

ANSWERS TO QUESTIONS

8.1. Using 8.12c, we obtain (remembering to convert from kPa to m for ψ): $K(\psi) = (16.72$ m/day$)/ [1 + (2.044$ m$/0.55$ m$)^4] = 0.087$ m/day.

8.2. Using a spreadsheet program, we obtain Table 8.2, as well as the corresponding plot for the Boltzmann transformation.

Table 8.2

Moisture content (cm³/cm³)	Distance (cm)	Boltzmann[1] (cm/min$^{0.5}$)	Area at Point[2]	Cumulative area under curve[3]	Slope of curve at Point[4]	Diffusivity (cm²/min)[5]
0.02	76.00	1.9263	–	–	–	–
0.03	75.90	1.9597	0.0194	0.0194	0.2991	–0.0325
0.04	75.85	1.9584	0.0196	0.0390	–7.7460	0.0025
0.05	75.75	1.9559	0.0196	0.0586	–3.8730	0.0076
0.06	75.71	1.9548	0.0196	0.0781	–9.6825	0.0040
0.07	75.67	1.9538	0.0195	0.0977	–9.6825	0.0050
0.08	75.63	1.9528	0.0195	0.1172	–9.6825	0.0061
0.09	75.58	1.9515	0.0195	0.1367	–7.7460	0.0088
0.10	75.50	1.9494	0.0195	0.1562	–4.8412	0.0161
0.11	75.45	1.9481	0.0195	0.1757	–7.7460	0.0113
0.12	75.40	1.9468	0.0195	0.1952	–7.7460	0.0126
0.13	75.35	1.9455	0.0195	0.2147	–7.7460	0.0139
0.14	75.30	1.9442	0.0194	0.2341	–7.7460	0.0151
0.15	75.25	1.9429	0.0194	0.2536	–7.7460	0.0164
0.16	75.18	1.9411	0.0194	0.2730	–5.5328	0.0247
0.17	75.05	1.9378	0.0194	0.2924	–2.9792	0.0491
0.18	74.90	1.9339	0.0194	0.3117	–2.5820	0.0604
0.19	74.75	1.9300	0.0193	0.3310	–2.5820	0.0641
0.20	74.65	1.9275	0.0193	0.3503	–3.8730	0.0452
0.21	74.45	1.9223	0.0192	0.3696	–1.9365	0.0954
0.22	74.15	1.9145	0.0192	0.3888	–1.2910	0.1506
0.23	73.80	1.9055	0.0191	0.4079	–1.1066	0.1843
0.24	73.40	1.8952	0.0190	0.4269	–0.9682	0.2204
0.25	72.80	1.8797	0.0189	0.4457	–0.6455	0.3453
0.26	72.08	1.8611	0.0187	0.4645	–0.5379	0.4317
0.27	71.20	1.8384	0.0185	0.4829	–0.4401	0.5487
0.28	69.90	1.8048	0.0182	0.5012	–0.2979	0.8411
0.29	68.05	1.7570	0.0178	0.5190	–0.2094	1.2395
0.30	66.05	1.7054	0.0173	0.5363	–0.1936	1.3847
0.31	63.75	1.6460	0.0168	0.5530	–0.1684	1.6421
0.32	61.10	1.5776	0.0161	0.5692	–0.1462	1.9472
0.33	58.15	1.5014	0.0154	0.5846	–0.1313	2.2262
0.34	54.80	1.4149	0.0146	0.5991	–0.1156	2.5912
0.35	51.20	1.3220	0.0137	0.6128	–0.1076	2.8481
0.36	47.10	1.2161	0.0127	0.6255	–0.0945	3.3109

(*contd.*)

Table 8.2 (*continued*)

Moisture content (cm³/cm³)	Distance (cm)	Boltzmann[1] cm/min^0.5	Area[2] at Point	Cumulative[3] area under curve	Slope[4] of curve at Point	Diffusivity[5] cm²/min
0.37	41.80	1.0793	0.0115	0.6370	−0.0731	4.3585
0.38	32.50	0.8391	0.0096	0.6466	−0.0416	7.7630
0.39	0.00	0.0000	0.0042	0.6508	−0.0119	27.3049

[1]Distance ÷ √t, where t = 1500 min
[2]0.01 ((1.9263 + 1.9597)/2)
[3]Cumulative sum
[4]Current θ − Prev. θ/Boltzmann, that is, (0.03 − 0.02)/(1.9597 − 1.9263)
[5](1/slope)*(−0.5)*Cumulative Area; here = 0.0194

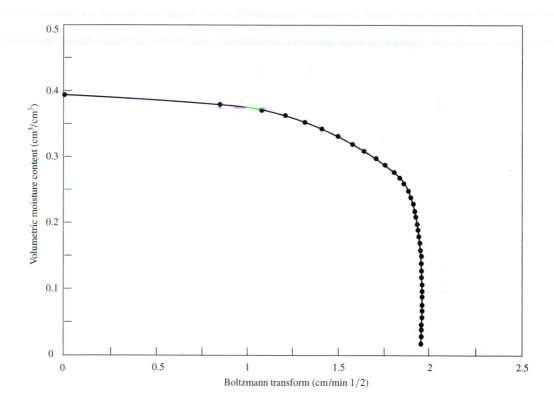

ADDITIONAL QUESTIONS

8.3. You are given two soils, a loam and a sand. At the same θ, the loam has a lower hydraulic conductivity than the sand. Explain this.

8.4. Why can a soil have different K values for a specific h_m?

8.5. Suppose $\partial H/\partial z$ is negative, positive, and zero. What is the direction of flow for each case?

8.6. You have a soil with the condition $h < 0$ at the surface. Would you expect a water layer to be present on the surface?

8.7. We can express $D = K / (d\theta/dP_m) = K (dP_m/d\theta)$, where P_m is the pressure equivalent of the matric potential. With this in mind, is D a function of the volume fraction of water?

9

Transport of Heat and Gas in Soil and at the Surface

INTRODUCTION

Plant and animal life depend upon the physical processes that govern soil heating, spatial distribution of water, and gaseous exchange between the soil and atmosphere. These interrelated processes are responsible for determining the following: plant and microbial growth rates; rates of decomposition; soil temperature and moisture content; temperature and moisture distribution; evaporation; transpiration; and the gaseous composition of both the soil-air and that of the overlying atmosphere. While soil heating and cooling affect the distribution and movement of water in the unsaturated zone through the processes of condensation and evaporation, soil temperature and moisture can indirectly determine the rate of gaseous exchange between the soil and the atmosphere, thereby influencing the gaseous composition of both soil and atmosphere.

Soil temperature, moisture, and gaseous composition have a tremendous impact on agricultural operations. The rate of soil heating, as well as both the spatial distribution and temporal trends of soil temperature, determines planting and subsequent agronomic schedules. Planting too early (when soils are cool) can delay germination, thereby resulting in a higher potential of seed rot, while planting much later than normal can lead to a shortened growing season, likely decreasing yields. The composition of soil-air can influence biological activity, including plant and root growth. For example, productivity of agricultural crops and forests depends upon soil aeration, that is, the movement of oxygen (O_2) into the soil and subsequent release of carbon dioxide (CO_2) from the soil into the atmosphere. Respiration by plant roots and soil heterotrophs (the microbiological community living within the soil) consumes O_2 and generates CO_2. Typically, plant productivity and root growth are reduced when the volume fraction of O_2 is less than 15 percent and the volume fraction of CO_2 is greater than 5 percent or 50,000 ppm (Rosenberg, Blad, and Verma 1983). In fact, it appears that for well-watered and fertilized crops, the major limitation to attaining optimal growth rates is lack of sufficient soil aeration (Hillel 1982).

"Greenhouse gases" in the atmosphere prevent some of the heat that is radiated upward by the Earth's surface from escaping to space by absorbing it, and then re-radiating some of this energy back toward the Earth's surface. As a result, with solar radiation nearly constant, the total amount of radiant energy received at the surface increases, causing increased surface temperature. Biophysical processes within soil also play an important role in determining climate, by affecting atmospheric concentrations of most greenhouse gases, either through consumption or generation of these gases. While evaporation of soil water can affect water availability for plant growth, it also is an important component of the hydrologic cycle. Water vapor released into the atmosphere during evaporation from soils can affect climate as a greenhouse gas and by determining the amount of water vapor available for cloud formation and precipitation. Other important greenhouse gases include CO_2, methane (CH_4), and nitrous oxides (NO, N_2O). Although both the burning of fossil fuels and forest clearing result in the greatest proportion of CO_2 in the atmosphere, warm, fertile soils in agricultural use also release significant amounts of CO_2 to the atmosphere (Houghton 1995). It is well documented that saturated soils, such as those found in rice paddies, beaver ponds, and areas seasonally flooded, host production of globally significant amounts of CH_4 (Conrad 1989). While arable lands consume substantial amounts of CH_4, the conversion of forest and pasture lands to intensive agriculture often decreases the strength of this sink (Goulding et al. 1995). Additionally, conversion of forest to agriculture usually results in a several-fold increase in the release of nitrous oxides; emissions of these gases are enhanced following the application of N-based fertilizers (Firestone and Davidson 1989).

As a final point of introduction, radon gas is produced by radioactive-decay reactions in a number of uranium-bearing geologic environments. This gas can diffuse through rock fractures or soil pores, and can enter dwellings via cracks in building foundations. Radon is a suspected carcinogen at low concentrations, and hence poses a public-health risk. Understanding its transport has led to building methods that divert the gas away from buildings, thereby lowering gas concentrations in occupied spaces.

This chapter introduces basic physical principles governing soil heat and gaseous transport, and their mathematical descriptions. Soil heating, evaporation, and gaseous transport are presented together for these reasons: (1) temperature and moisture are linked thermodynamically in both the soil-air and the atmosphere; and (2) soil–gas transport, as well as the generation (or consumption) of soil gases by biological activity, is affected by both soil temperature and moisture distribution. This chapter initially begins with an examination of the modes of energy transfer and the surface-energy budget of the Earth's surface, driven by solar radiation. This radiant energy is responsible for heating the soil, and determines both the spatial and temporal distribution of temperature and water in soil, as well as surface evaporation rates. Its influence is also important in determining the types of vegetation supported by soils. In this vein, the mechanisms responsible for heat and gaseous transport (as well as evaporation) are detailed; this includes an introduction to some of the models used to quantify heat and mass transport in soil. We also examine the role of vegetation in determining soil temperature and moisture. The chapter then proceeds with a discussion of the role of biology (plants and microbes) in the development of soil-air composition, describing the instrumentation and methodologies that help us quantify temperature distribution as well as the fluxes of heat, water vapor, and other gases.

9.1 BASIC CONCEPTS AND DEFINITIONS

Energy Transfer and Heat Content

Energy may be transferred from place to place by some combination of the following three processes:

1. **Conduction**—kinetic energy is transferred from molecule to molecule through physical contact;

2. **Convection**—a mass having some level of kinetic energy is transported from one place to another;

3. **Radiation**—electromagnetic energy is emitted from one object, transmitted through space, and is then received by another.

All three of these processes are relevant to the study of heat transport in soil. The Sun's radiant energy heats the soil's surface; surface heat is transferred to the subsurface by conduction, and to the air by convection.

Any mass composed of molecules that have kinetic energy is considered to have heat. If an object with an initially warmer temperature (T_i) is placed in contact with an object that has a cooler temperature (T_o), heat will flow from the warmer object to the cooler object until an equilibrium is reached, such that both objects are at the same temperature, T_f (where $T_i > T_f > T_o$). The initially warmer object loses a quantity of heat, ΔH:

$$\Delta H = -C(T_f - T_i) \tag{9.1}$$

where C is the heat capacity (J K^{-1}). C is considered either on a per-unit-mass or per-unit-volume basis; that is, mass (J kg^{-1} K^{-1}) or volume-specific heat capacity (J m^{-3} K^{-1}). Specific heats of common materials and soils are given in table 9.1.

Evaporation and Condensation

The term evaporation generally refers to the volatilization of water, in which the phase change from liquid to vapor occurs. For example, consider a body of water consisting of molecules in a constant state of random motion. These molecules are bound by a weak physical attraction between them that diminishes with the sixth power of the distance known as Van der Waal's forces. At any given moment, the kinetic energy of these water molecules results in a certain number of molecules gaining enough energy to leave the surface of the body of water. In order to do so, they must gain energy equivalent to their mass times the latent heat of vaporization (L_v) from the water. These molecules are now in a gaseous state, called water vapor; the pressure exerted by this gas is called vapor pressure. At the same time, a certain number of water molecules that are in the air are driven to the water's surface when they collide with other air molecules; subsequently, they lose their energy to liquid water molecules when they release a quantity of heat (mass times L_v). If the number of molecules that leave the water equals the number of molecules that return, the system is in equilibrium, and the overlying air is considered saturated with water vapor at that temperature. The pressure exerted by water vapor at equilibrium is the saturated vapor pressure. The evaporation rate is zero in this case, but would be nonzero if more molecules left the water than returned. Condensation is the reverse of evaporation—more molecules leave the gaseous state than enter it. Thus, both evaporation and condensation are continuous processes.

TABLE 9.1   Thermal Properties of Common Materials and Soils

Substance	Density (kg m^{-3})	Specific heat capacity (J kg^{-1} K^{-1})	Volumetric heat capacity (J m^{-3} K^{-1})	Thermal conductivity (W m^{-1} K^{-1})
Air (20 °C)	1.2	1.0×10^3	1.2×10^3	0.025
Water (20 °C)	1.0×10^3	4.2×10^3	4.2×10^6	0.58
Ice (0 °C)	9.2×10^3	2.1×10^3	1.9×10^6	2.2
Quartz	2.66×10^3	8.0×10^2	2.0×10^6	8.8
Mineral clay	2.65×10^3	8.0×10^2	2.0×10^6	2.9
Soil organic matter	1.3×10^3	2.5×10^3	2.7×10^6	0.25
Light soil with roots	4.0×10^2	1.3×10^3	5.0×10^5	0.11
Wet sand ($\theta = .4$)	1.6×10^3	1.7×10^3	2.7×10^6	1.8

Source: Data from de Vries (1963) and Rosenberg, Blad, and Verma (1983).

There are four requirements that must be met for soil evaporation to occur: (1) a supply of water; (2) a supply of heat to enable the phase change of water to vapor; (3) a vapor-pressure gradient such that the overlying air contains fewer vapor molecules than if the air were saturated; and (4) turbulence in the overlying atmosphere that carries vapor away from the surface, thus maintaining a vapor-pressure gradient between the soil and the atmosphere.

At times, in both temperate and arctic soils, water can be found in all three phases. As previously described, for water to become water vapor, a quantity of energy (mass of water times L_v) must be supplied to the water body ($L_v = 2.5$ [MJ kg^{-1}] $- 2.37 \times 10^{-3}$ T[MJ · kg^{-1} °C^{-1}]; where T is temperature (°C)). This heat is obtained from the water and its surroundings, thereby cooling them. For ice to melt, energy equivalent to the latent heat of fusion ($L_m = 0.34$ MJ kg^{-1}) is needed; however, water molecules in the solid phase do not necessarily have to make the transition to liquid, to become vapor. At temperatures below freezing, solid-phase water molecules can obtain sufficient energy to reach the vapor phase, but in order to do this they have to obtain the latent heat of sublimation ($L_s = L_m + L_v$), which at 0 °C is 2.84 MJ kg^{-1}. As shown in a phase diagram such as figure 9.1, at 273 K and a vapor pressure of 0.611 kPa (the "triple point," point A in the diagram), all three phases of water can exist at the same time. Water can exist as a liquid (if it is pure) at temperatures as low as −40 °C, provided it does not come into contact with a solid of a crystalline structure similar to ice; below about −40 °C, water undergoes spontaneous self-crystallization to ice. Curve segment A–B (see figure 9.1, inset) shows the distinction between saturation vapor pressure over water and over ice. Note that below freezing, the saturation vapor pressure over water (solid line) is greater than that over ice (dotted line). Above 273 K, curve segment A–C becomes the saturation-vapor pressure curve discussed earlier. At the boiling point of water (100 °C at sea level), saturation vapor pressure is 1 atmosphere (101.325 kPa). By increasing the pressure of an enclosed system, equilibrium saturation vapor pressure and boiling point may be increased. Curve segment A–D is the latent heat of fusion (melting/freezing), showing the dependence of water's phase on pressure.

The phase diagram of figure 9.1 helps us to realize the significance of phase changes (determined by temperature) as an important variable in controlling the movement of water in soil. Consider the following: Due to the net loss of longwave radiation at night, the soil's surface cools to the dew-point temperature (T_{DP}), the temperature at which air in contact

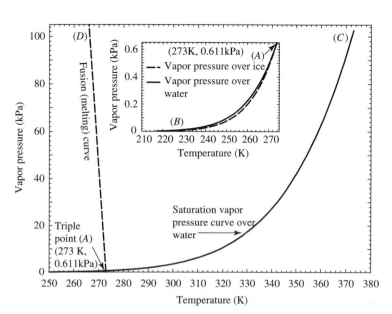

Figure 9.1 Phase diagram for water at standard pressure (101.325 kPa). Saturation vapor pressure curve (segment A–C) extends from the "triple point" (273 K) to the boiling point (373 K), and theoretically extends below freezing (segment A–B) to 0 K. The saturation vapor pressure curve over ice (segment A–B, dotted line) extends below that for vapor, along which sublimation occurs. Melting (or freezing) occurs along curve segment A–D.

with the soil cools to the saturation vapor pressure. As the soil continues to cool, water vapor from the atmosphere condenses onto the soil, forming dew and releasing heat to the soil; soil-air also has a dew-point temperature that can be different from that of the atmosphere. Given sufficient permeability of the soil crust, water vapor from within the soil can condense onto the subsurface of the crust, or even further below the crust if the soil temperature becomes cool enough. If the soil's surface cools enough to freeze, water tends to migrate by vapor-phase transport from the liquid water mass (below the ice) to the ice at the surface, because the saturation vapor pressure over ice is lower than that over water. However, it should be noted that, in contrast to the liquid phase, water vapor can move just as easily in any direction—vertical or horizontal—and that the direction of its movement is determined solely by the gradient in vapor pressure.

To summarize, water and heat are almost always in a state of transition. They are thermodynamically linked via water's ability to change phase through some combination of the following: evaporation; condensation; melting; freezing; or sublimation. Each change of phase involves a release or absorption of energy between water and its surroundings. For instance, energy needed for evaporation within soil is derived from the soil mass, thereby cooling it. Finally, we stress that the movement of water vapor is driven by gradients of both humidity and temperature.

QUESTION 9.1

In a closed soil column having a temperature of $-10\,°C$ at one end and $5\,°C$ at the other, soil-water content (in the form of ice) was observed to increase at one end while moisture content, at the other end, decreased. Explain why.

Quantifying Water Content in Air

So far, we have described water vapor qualitatively. Using the terms discussed in the following text, we obtain the means for quantifying water vapor.

Water vapor is a component of ambient air; we speak of its partial pressure as vapor pressure. We can write Dalton's Law of partial pressures (discussed in detail in section 9.7) as:

$$P_{total} = P_{dryair} + e \qquad (9.2)$$

Vapor pressure (e) is small compared to pressure of dry air (P_{dryair}). At sea level, P_{total} averages 101.325 kPa, while e usually has a range of 0.5–3.0 kPa.

Water vapor equilibrates with water in a steady-state closed system at a vapor pressure; this is referred to as saturation vapor pressure (e_s). There is a direct physical dependence of e_s on temperature (see figure 9.1), which can be derived from basic laws of thermodynamics by invoking the concept of entropy. The rate of change of e_s with temperature (T) is quantified in the Clausius–Clapeyron equation:

$$\frac{de_s}{dT} = \frac{L_x}{T(a_2 - a_1)} \qquad (9.3)$$

where L_x is a latent heat quantity determined by the phase change that is occurring—vaporization (v) or sublimation (s)—and a is specific volume (inverse of density), with subscripts identifying phase (or state) of matter. If we integrate equation 9.3, inserting the constants at the triple point of water (the temperature and vapor pressure at which all three phases of water exist), we derive the dependence of saturation vapor pressure on temperature:

$$\ln\left(\frac{e_s}{0.6108[kPa]}\right) = \frac{L_x}{R_v}\left(\frac{1}{273} - \frac{1}{T[K]}\right) \qquad (9.4)$$

where R_v is the gas constant for water vapor (461.5 J kg^{-1} K^{-1}), derived by dividing the universal gas constant ($R^* = 8.3144$ J mol^{-1} K^{-1}) by the molecular weight of water ($m_v = 0.018016$ kg mol^{-1}). The solution for e_s in equation 9.4 may be approximated (above 80 kPa total pressure) with Teten's empirical formula (1930):

$$e_s[\text{kPa}] = 0.6108 \times 10^{7.5T[°C]/(T[°C] + 237.3)} \tag{9.5}$$

We find somewhat better accuracy for a range of temperatures (-50 to $+50$ °C) using Buck's relation (1981):

$$e_s[\text{kPa}] = 0.61365 \, exp \, (17502T[°C]/(240.97 + T[°C]) \tag{9.6}$$

Where exp is the exponential function.

The consideration of saturated conditions gives us a relative basis for gauging humidity or water content in air. The ratio of vapor pressure in air to saturated vapor pressure at the same temperature is known as relative humidity (rH):

$$\text{rH}(\%) = \frac{e}{e_s} \times 100 \tag{9.7}$$

Note that for any fixed rH there exists a range of water contents (or vapor pressures), if the temperature is allowed to vary (see figure 9.2). That is, relative humidity is a function of vapor pressure and temperature; it does not quantify the amount (mass) of moisture in air. One way of expressing the amount of water-vapor mass in air is by using absolute humidity

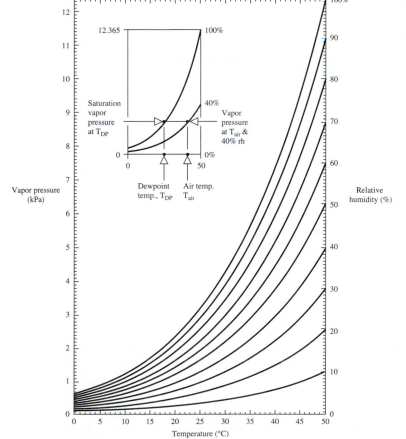

Figure 9.2   Psychrometric chart depicting the relation between relative humidity, vapor pressure, temperature, and dew-point temperature

or vapor density (kg m^{-3}):

$$\rho_v = \frac{e}{R_V T} = \frac{m_v e}{R^* T} = \frac{\varepsilon e}{RT} \tag{9.8}$$

where

$$\varepsilon = \frac{m_v}{m_d} \tag{9.9}$$

R (R^*/m_d) is the dry-air gas constant and m_d is the molecular weight of dry air (0.028996 kg mol^{-1}). If vapor pressure is known, absolute humidity can be calculated from

$$\rho_v[\text{kg m}^{-3}] = \frac{e[\text{Pa}]}{461.5[\text{J kg}^{-1}\,\text{K}^{-1}] \times (T[\text{K}])} \tag{9.10}$$

A related quantity, specific humidity (unitless), is given by this ratio:

$$q = \frac{\rho_v}{\rho} \tag{9.11}$$

where

$$\rho = \frac{m P_{\text{total}}}{R^* T} \tag{9.12}$$

and m is the molecular weight of moist air. Specific humidity is approximately equal to the mixing ratio of vapor in air, and is identical to the mixing ratio if we replace ρ by the dry-air density in equation 9.11.

Vapor pressure can be calculated by inserting the dew-point temperature (T_{DP}) into either Teten's (equation 9.5) or Buck's equation (equation 9.6), since dew-point temperature is the temperature at which air becomes saturated (i.e., saturation vapor pressure). Alternately, vapor pressure can be calculated from wet-bulb temperature (T_w), the temperature to which a wetted object cools due to the vaporization of water from its surface:

$$e = e_{\text{sw}} - \gamma_T P(T - T_w) \tag{9.13}$$

Equation 9.13 is known as the psychrometric equation, where: e_{sw} is the saturation vapor pressure at T_w; T and T_w are in °C; P is ambient air pressure; γ_T is the thermodynamic psychrometric constant ($\gamma_T = C_p/(\varepsilon L_v)$, and C_p is specific heat of air at constant pressure (1.005 kJ kg^{-1} K^{-1}). The product, $\gamma_T P$, is referred to as the psychrometric constant (γ). A handy approximation for γ_T is given by Gay (1972):

$$\gamma_T = 6.97 \times 10^{-4}(1 + 0.00115 T_w) \tag{9.14}$$

The term "wet bulb" refers to humidity measurements that are based on wet- and dry-bulb psychrometers. This instrument is composed of a housing that contains two aspirated thermometers, referred to as "bulbs." One thermometer (wet bulb) is wrapped in a hygroscopic, wetted cloth, while the other is kept dry, to measure air temperature. By using these readings and equations 9.6, 9.13, and 9.14, vapor pressure can be estimated. Wet-bulb temperature varies with air flow over the wetted surface at low-flow speeds, hence equations 9.13 and 9.14 apply to psychrometers aspirated with an air flow of 4–10 m s^{-1}, over which a negligible change in T_w occurs. Wet-bulb temperature should not be confused with dew-point temperature (T_{DP}), the temperature below which air has to be cooled in order for moisture to condense out of it.

Unsaturated soil usually contains both water vapor and some quantity of liquid water. Water-vapor content can be calculated if both water potential and soil temperature are known. Vapor density can be calculated directly from:

$$\rho_v = \rho_{vs} exp(\psi/R_v T) \tag{9.15}$$

where ρ_{vs} is saturation vapor density, and ψ is water potential (J kg^{-1}). More often we want to know ψ, having some measure of humidity. Solving equation 9.15 for ψ yields the Kelvin equation:

$$\psi = R_v T \ln\left(\frac{e}{e_s}\right) \tag{9.16}$$

A deep, uniform soil that is initially wetted and allowed to drain for several days typically has a ψ of about -30 J kg^{-1}. The permanent wilting point of many plants is typically below $-2,000$ J kg^{-1} (equivalent to $-2,000$ kPa or -2 MPa), and is about $-1,500$ J kg^{-1} for agricultural crops. Even at this point, soil air is humid compared to that usually found in the atmosphere; at the permanent wilting point, soil air at 25 °C has a relative humidity of about 99 percent. Relative humidities sometimes fall below 95 percent in soils, but only in arid regions (Loskot, Rousseau, and Kurzmack 1994). Therefore, over typical soil-moisture ranges, equation 9.16 can be closely approximated by:

$$\psi[\text{J kg}^{-1}] = 461.5 \times T[\text{K}]\left(\frac{e}{e_s} - 1\right) \tag{9.17}$$

QUESTION 9.2

For a fixed amount of moisture in the air, how does rH change with increasing T?

Measurement of soil-air temperature Temperature is often measured with either a thermistor, RTD (resistance-temperature device), or thermocouple. In choosing a sensor, there are a number of factors to consider: sensor stability (i.e., minimal drift in output or calibration); resolution; accuracy; and cost.

Thermistors are specialized variable resistors whose electrical resistance varies strongly with temperature, in a predictable way; voltage drop across the thermistor is calibrated to temperature. Thermistors have an advantage over RTDs or thermocouples because of their adjustable-voltage output and low cost. A brief description of these sensors is given below, details are given in Fritschen and Gay (1979).

RTDs are of two types: thin metal film, or wire-wound (usually platinum wire). Resistance of RTDs strongly varies with temperature; therefore, the principle of temperature measurement with RTDs is similar to that of thermistors. However, in contrast to thermistors, RTDs are quite stable, generally offer a larger temperature measurement range, but are somewhat more expensive.

Thermocouples are welded junctions of dissimilar metals, across which a weak electrical potential develops as a function of temperature (see tables 9.2 and 9.3). Thermocouple

TABLE 9.2   Thermocouple Sensitivity to Temperature

Thermocouple	ISA symbol	Sensitivity (μV °C^{-1})
Chromel (P)–constantan	E	6.32
Iron–constantan	J	5.27
Chromel (P)–alumel	K	4.10
Copper–constantan	T	4.28

TABLE 9.3   Elemental Composition of Metals Used in Thermocouples

Alumel	95% nickel, 2% manganese, 2% aluminum
Chromel	90% nickel, 10% chromium
Constantan	55% copper, 45% nickel
Evanohm	75% nickel, 20% chromium, 2.5% copper, 2.5% aluminum

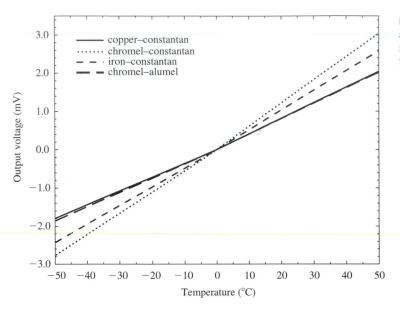

Figure 9.3   Voltage output as a function of temperature for commonly used thermocouples

junctions can be built using a variety of metals whose sensitivities to temperature differ (see figure 9.3). To measure temperature, two thermocouples are connected in series. A reference junction is maintained at 0 °C, to yield a constant voltage. Voltage across the sample-sensing junction increases with temperature when measured with respect to the reference junction in the same circuit. More often, a precision microvolt source is used rather than a reference junction, to produce the required reference voltage. Due to very weak thermocouple output (on the order of microamps and microvolts), a high-impedance, high-precision, voltmeter is needed to measure output accurately. In addition, the sampling junction must be electrically isolated from soil (usually encased in a Teflon shield), to prevent electrical-ground loops. While most thermocouples are inexpensive and can be built with minimal equipment costs, they also have the potential to provide measurements with high accuracy and resolution.

Measurement of humidity (water potential) Soil-air humidity is difficult to measure accurately in situ, yet its determination is often essential in interpreting water potential in unsaturated soil; this assumes the vapor phase of moisture is in equilibrium with the liquid phase. Frequently, a soil psychrometer, as shown in figure 9.4 (see Spanner 1951; Rawlins and Campbell 1986), or a soil dew-point hygrometer (Neumann and Thurtell 1972) is used to obtain water potential. Both instruments employ thermocouples and are unaspirated; however, their procedures for obtaining water potential differ.

Calculating water potential from a soil psychrometer involves the measurement of both dry- and wet-bulb temperatures. Typically, one of the psychrometer's thermocouple junctions is cooled by the Peltier effect (i.e., the current flow through a thermocouple junction is reversed), thereby water condenses onto it; temperature of the wetted thermocouple is the wet-bulb temperature. However, the evaporation of water droplets from the unaspirated pyschrometer's wet bulb is a function of both droplet shape and size, and in turn somewhat

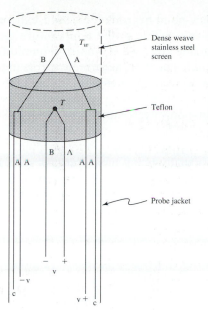

Figure 9.4   Diagram of a soil psychrometer for measurement of soil-air humidity. Thermocouples are used to measure air temperature (dry bulb, T), and wet-bulb (T_w) or dew-point temperature (T_{DP}), from humidity is calculated. Wires A (usually copper) and B differ in elemental composition. The screen is of high density, to prevent the entrance of water; the Teflon block helps to eliminate temperature differences between dry junctions. Switchable voltage is applied to the wet bulb thermocouple loop at points C so that current runs opposite to that of the thermocouple circuit, cooling the T_w junction and condensing water onto it. Voltage is read off of loops at points $v(+, -)$.

dependent upon soil-air humidity. Thus, wet-bulb depression ($T - T_w$) cannot readily be used to determine humidity and water potential when using the combination of equations 9.13 and 9.16. Instead, a strictly empirical approach is usually employed.

Following the procedure of Loskot, Rousseau, and Kurzmack (1994), the psychrometer probe (with both dry- and wet-bulb thermocouples) is outfitted with a tightly woven metal screen that prevents water seepage into the thermocouples. Since saturation vapor pressure varies as a function of osmolality and temperature, normally this instrument is calibrated by immersing it in a series of salt solutions with varying molality. Alternately, the probe can be enclosed in a small, sealed chamber that contains the forementioned solutions, and the procedure is duplicated at various temperatures. The final product is a family of calibration curves spanning a range of humidities that encompass the anticipated range of environmental conditions; humidity is found by interpolating between these calibration curves with psychrometer voltage. A procedure outlined in Rawlins and Campbell (1986) substantially reduced the number of humidity calibrations required when using the empirical approach described above. This reduction is achieved by using equation 9.16 as well as a semi-empirical equation based on the conventional psychrometer equation. When this approach is followed, water potential can be calculated directly from the psychrometer voltage without the necessity for interpolation between calibration curves. Details of some salt-solution preparations can be found in Brown and Van Haveren (1971), Rawlins and Campbell (1986), or Loskot, Rousseau, and Kurzmack (1994).

Neumann and Thurtell (1972) introduced a modification of the wet-bulb psychrometer (called a thermocouple dew-point hygrometer), that improved the accuracy of humidity measurements. This instrument measures dew-point temperature in an iterative process. First, a thermocouple is cooled until moisture condenses onto it. Next (as with the psychrometer), the wetted thermocouple's temperature is measured until all the water on it is reevaporated. During several more cycles, the junction is allowed to cool and warm repeatedly, but in each iteration cooling and heating cycles are shortened, thereby approaching the precise temperature of condensation, the dew point. The dew-point hygrometer has the advantage of negligibly disturbing the natural soil-moisture distribution, and is 2 to 5 times more accurate than soil psychrometers (Brunini and Thurtell 1982; Rawlins and Campbell 1986).

Whether a soil psychrometer or dew-point hygrometer is used, extra care is essential to measure temperature accurately and avoid within-probe temperature gradients, since humidity measurements are frequently made over a very narrow humidity range (99% to 100%). To resolve a 10 J kg^{-1} water potential at 20 °C accurately, errors in temperature measurement must be < 0.001 °C. This can be accomplished with proper selection of materials and probe design (Rawlins and Campbell 1986). Some guidelines for assessing errors involved with soil psychrometers and dew-point hygrometers are given in Savage, Cases, and de Jager (1983), and Savage and Wiebe (1987).

When measuring soil moisture with psychrometers, another point to keep in mind is that it is the osmotic potential that is actually being measured. As such, humidity will decrease with increasing solute concentration in soil water. Alternately, a soil containing water with a low solute concentration will have a higher measured water content than one with an identical amount of water but with higher solute concentration.

9.2   ENERGY EXCHANGES AT THE SURFACE

Heat and water are continually exchanged between the Earth's surface and the overlying atmosphere; each influences the heat and moisture content of the other (Entekhabi, Rodriguez-Iturbe, and Castelli 1996). Solar energy heats the Earth's surface, including the soil. Heat, water vapor (from evaporation of surface and subsurface water), and other gases are transported to (or from) the surface by a variety of air motions. Transport from the surface occurs across two distinct air layers. Adjacent to every surface is a layer (a few millimeters thick) called the laminar sublayer in which heat, water vapor, and other gases are carried to and from the surface by molecular motions. Laminar flow is well organized, but both heat and mass transfer within it are slow compared to the turbulent sublayer immediately above (see figure 9.5(a)), that is dominated by turbulent motions (eddies). Together, the laminar

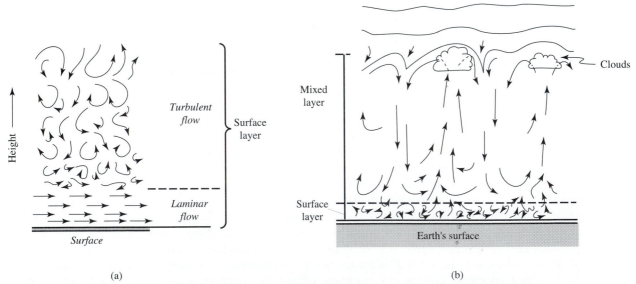

(a)   (b)

Figure 9.5   Conceptual diagram of the surface layer (a), divided into laminar and turbulent sublayers. Arrows indicate characteristics and direction of air movement. The laminar sublayer is characterized by molecular transport of heat and gas, and is only millimeters thick. The surface layer is of the order of tens of meters deep, and the atmospheric-boundary layer (ABL) or mixed layer (b) is of the order of a kilometer deep. Motion in the ABL is dominated by large-scale convection that carries surface energy and moisture through the atmosphere.

and turbulent sublayers comprise the surface layer, typically on the order of tens of meters thick. Above the surface layer is the mixed layer (see figure 9.5(b)), dominated by large-scale convection. The top of the mixed layer defines the top of the atmospheric-boundary layer, approximately one or more kilometers deep.

Solar energy is absorbed differentially by the mosaic of surfaces covering the Earth. Thus, the atmosphere is heated and humidified differentially by these surfaces. Consequently, density variations occur in the atmosphere, from which pressure patterns develop. Since it tends to flow down pressure gradients, air flow (wind) over the mosaic of Earth's surfaces carries warm air over cold surfaces, and vice versa. Additionally, storms embedded in the atmospheric circulation carry moisture from one area to another.

Near-surface soils are heated by a combination of solar radiation, sensible (i.e., thermal energy) and latent heat exchanges (described later in this chapter), and to a lesser extent, geothermal sources. Soils are cooled by the loss of radiant energy at night (see the next section) or by contact with cold air advected from another region of the Earth's surface by the atmosphere. Averaged spatially and temporally on a global basis, solar radiation is by far the greatest source of heat, supplying more than 99 percent of the energy that heats soil. Some tens of meters below the surface, geothermal heating from the Earth's core (in excess of 3,000 °C) begins to determine temperature substantially, with rate-of-increase with depth of about 2–4 °C/100 m (Foster 1969; Sorey 1971; Sass et al. 1988). It is believed that core heat is due to residual heat of creation and radioactive decay.

An aside of historical interest regarding soil heating: in the 1800s Lord Kelvin of England used measurements of the rate-of-heat flow to the Earth's surface to estimate its age, assuming the Earth was molten at the time of its formation and had been cooling ever since. Kelvin's assumptions led him to estimate the Earth's age at less than 100 million years. Geologists immediately dismissed the estimate based upon their understanding of the rates of sedimentation; scientists of Kelvin's time were unaware of radioactive decay.

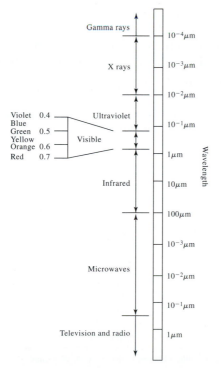

Figure 9.6   The electromagnetic spectrum

Radiation Balance

Our examination of radiation at the Earth's surface concerns just a small fraction of the electromagnetic spectrum (shown in figure 9.6), ranging in wavelength from ultraviolet (0.01 μm) through the far-infrared (100 μm); the range encompasses almost all radiation emitted as solar and terrestrial radiation. The discussion following concerns this region but first, let us consider some basic concepts.

Wavelength (λ) is related to frequency (ν) by:

$$\lambda = \frac{c}{\nu} \qquad (9.18)$$

where c is the speed of light (3.0×10^8 m s^{-1}). A portion of radiation, of a certain wavelength that is incident upon an object, can be absorbed while passing through it. The ratio of the amount absorbed to that incident is the monochromatic (single wavelength) absorptivity, $\alpha(\lambda)$. Radiation not absorbed can be reflected or transmitted, or both. When described with respect to incident radiation, these terms are called reflectivity (r) and transmissivity (τ), respectively. Summing all three terms, we account for the fate of all monochromatic radiation:

$$\alpha(\lambda) + r(\lambda) + \tau(\lambda) = 1 \qquad (9.19)$$

All bodies at temperatures above absolute zero (0 K) possess molecular kinetic energy and emit radiation in the thermal or far-infrared wavebands, due to specific molecular vibrations (Williamson 1973). Bodies that emit radiation at shorter wavelengths (higher frequencies) have to derive this energy from electron activity at the atomic level.

In the thermal waveband, Kirchhoff's law states that absorptivity of a mass (α, ratio of energy absorbed to that incident) is equal to its emissivity (ε) at that wavelength:

$$\alpha(\lambda) = \varepsilon(\lambda) \qquad (9.20)$$

A mass that is both a perfect emitter and absorber of radiation is called a blackbody. For its surface, $\alpha(\lambda) = 1$ and $r(\lambda) = \tau(\lambda) = 0$. A perfect blackbody does not exist in nature, although many natural objects come quite close in the far-infrared wavelengths. Objects that are not blackbodies are referred to as greybodies if their emissivity is less than 1.0, or whitebodies if their emissivity is 0. Soils are greybodies, with typical emissivities between 0.90–0.98 (see table 9.4).

TABLE 9.4   Emissivity and Albedo of Some Common Objects

Substance	Emissivity	Conditions	Albedo
Snow	0.99	Fresh	0.80–0.95
		Old	0.42–0.70
Water body	0.97	Calm	0.07–0.08
		Windy	0.12–0.14
Crops	0.96–0.98		0.20–0.26
Forests	0.97	Deciduous	0.15–0.20
		Coniferous	0.10–0.15
Soil: sandy	0.949	Wet	0.24
		Dry	0.37
Soil: silty clay	0.966	Wet	0.12
		Dry	0.21
Soil: loam	0.967	Wet	0.16
		Dry	0.23

Source: Data from Rosenberg, Blad, and Verma (1983), Davies (1979), and Idso (1969).

Any body will emit energy at an intensity (I) as a function of its surface temperature (T), and emissivity as given by Stefan's law:

$$I(\lambda) = \varepsilon(\lambda)\sigma T^4 \tag{9.21}$$

where σ is the Stefan–Boltzmann constant (5.67×10^{-8} W m^{-2} K^{-4}) and I is in W m^{-2}. Stefan's law also applies to emissions from liquid and gas bodies. However, since there is no definable surface of a region or parcel of gas (such as the atmosphere), an effective temperature is used that is determined from a density and composition profile of the gas parcel (Fleagle and Businger 1963).

The radiation budget at the Earth's surface (see figure 9.7) may be partitioned into two components: incoming (received by the surface) and outgoing (leaving the surface); net radiation is the sum of these components. Incoming solar radiation of short wavelengths (0.15–3.0 μm) dominates during the day, heating the Earth; outgoing terrestrial radiation at long wavelengths (3–100 μm) dominates at night, cooling the Earth.

Solar radiation that reaches the top of the atmosphere has a magnitude of 1367 W m^{-2} as projected on a flat surface. The annual average over the curved surface of the Earth is 339 W m^{-2}. Approximately 47 percent of solar radiation is in the visible portion (0.36–0.75 μm) of the electromagnetic spectrum and 48 percent in the near infrared (0.75–3.0 μm). Of the remaining 5 percent of solar radiation, about 3 percent occurs in ultraviolet ($> 0.3 \mu$m) and 2 percent in far-infrared ($> 3 \mu$m) spectrums. At the Earth's surface, the 50:50 clear-sky ratio of visible to infrared radiation becomes a 60:40 ratio under overcast conditions.

Some portion of solar radiation received at the surface is reflected; the ratio of reflected to incoming solar radiation (within visible or infrared wavebands) is called the albedo. The distinction between albedo and reflectivity of a surface is that reflectivity is wavelength-specific, while albedo pertains to wavebands. Materials that compose the Earth's surface have a wide range of albedos (see table 9.4). The albedo of some objects can differ for different wavebands. For example, snow has a high, visible albedo that exceeds its near-infrared value.

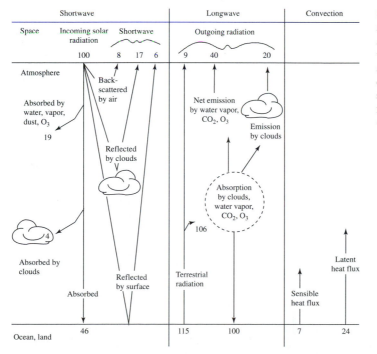

Figure 9.7   Earth's energy balance (from an annual global average), depicting the fate of solar and terrestrial radiation at the surface and in the atmosphere. All values expressed as percent of incoming solar radiation (339 W m^{-2}); data from MacCracken and Luther (1985)

Most soils exhibit a range of albedos depending upon mineral, organic, or water content of the crust; values range from about 5–45 percent. For the same soil, an increase in soil water content decreases albedo. For example: Idso, Asae, and Jackson (1975) provided the following relation between albedo (a) and fractional water content (w/w_s; w is water content and w_s is field capacity) of a soil:

$$a = 0.31 - 0.34\left(\frac{w}{w_s}\right); \quad \frac{w}{w_s} \leq 0.5 \tag{9.22a}$$

$$a = 0.14; \quad \frac{w}{w_s} > 0.5 \tag{9.22b}$$

Thus, as near-surface soil moisture can vary with time, so can albedos, for the same surface.

Terrestrial radiation is also referred to as thermal or longwave radiation, and occurs in the far-infrared region (3–100 μm). It is emitted by the Earth's surface at an intensity determined by temperature (see equation 9.21). Under clear skies, surface temperature—and hence outgoing terrestrial radiation—usually reaches a maximum just after midday, and a minimum shortly before sunrise. As shown in figure 9.7, about 80 percent of the upward-component of terrestrial radiation from the surface is absorbed by the atmosphere then returned to the surface. Returned energy is emitted at an intensity that is determined by the effective temperature of the atmosphere, and specifically related to its radiating components: clouds, aerosols, water vapor, and other greenhouse gases (these are listed further on, in table 9.8). Increasing atmospheric concentrations of these important radiative atmospheric constituents also increases the amount of energy re-radiated back to the Earth's surface, which in turn, can cause increases in surface temperature. We often experience the effect of increased terrestrial re-radiation at night, under calm winds and changing sky conditions. Under a clear sky, the air cools rapidly as net radiation typically reaches about −40 to −60 W m^{-2}. As cloud cover appears, the rate of cooling decreases rapidly, due to increased re-radiation from the atmosphere; net radiation often approaches 0 under these circumstances.

Greenhouse gases also play an important role in thermal regulation of the Earth's surface. If greenhouse gases did not exist in the atmosphere, the average global-surface temperature would more likely be −18 °C, rather than our present 15 °C (Lindzen, 1990). Surface

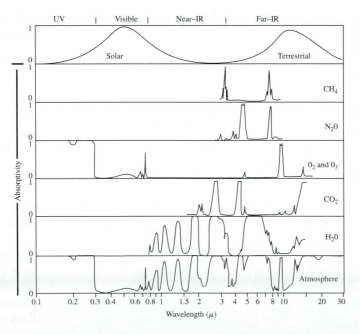

Figure 9.8   Relative emission spectra for the Sun and Earth (top), and absorption spectra for major atmospheric gases. Lowest panel is the summation of absorption by all atmospheric gases listed in panels above. Note that the atmosphere is relatively transparent near peak solar radiation levels, while it is opaque over much of the terrestrial radiation curve, with the exception of 8.5–9.5 and 9.7–12 μm, the atmospheric window. Data from Fleagle and Businger 1963

temperatures could even be warmer if not for the "atmospheric window"—a split region in the far-infrared (8.5–9.5 and 9.7–12 μm)—where absorption by greenhouse gases is markedly minimal (see figure 9.8); about 8 percent of terrestrial radiation escapes to space through these wavebands. The split in the atmospheric window around 9.6 μm is due to a strong absorption peak by O_2 and O_3. Strong absorption of far-infrared radiation by water vapor (at shorter wavelengths), and by CO_2 (at longer wavelengths) borders this window. Among greenhouse gases, water vapor usually occurs in the greatest concentration in the atmosphere, and accounts for the largest portion of reradiated energy.

Incoming energy not reflected or emitted by the surface is absorbed by the soil and ground cover. Here it is converted into kinetic energy, then released into the atmosphere as sensible (kinetic energy) and latent heat. On an annual global basis, these two convective terms account for 31 percent of solar energy. Note that no net energy goes into conduction; it is assumed that the Earth is not warming measurably, so the net flux of heat into the solid Earth (soil-heat flux) is negligible. We also note that incoming solar energy is balanced by that outgoing, which prevents the Earth from warming or cooling.

Net radiation exchange at the soil's surface is complicated by the presence of vegetation. Vegetation type, vigor, age, height, and density can all be significant factors in determining shortwave absorption and longwave emission, thereby affecting net radiation; vegetation can also reduce net radiation at the surface substantially. Baldocchi et al. (1984) report that net radiation at the floor of a closed-canopy, deciduous forest is only a few percent of that above-canopy during the growing season, but during winter months, floor values are much closer to above-canopy values. Methods for calculating and measuring net radiation at the soil's surface, at the floor of a stand of vegetation, can be found elsewhere (Baldocchi et al. 1984; Norman 1979).

QUESTION 9.3

Examining equation 9.21, a surface temperature increase from 20 °C to 22 °C leads to what size increase in I for a wet, sandy soil?

QUESTION 9.4

Referring to the value at the bottom of the center panel (longwave) of figure 9.7, terrestrial radiation is 115 percent of incoming solar energy. Explain how terrestrial radiation leaving the Earth's surface can exceed the value of incoming solar radiation (100 percent).

Energy Balance

Bare soil The net radiation (R_n) received by a bare-soil surface is partitioned into heating the soil and atmosphere, and evaporating water. Under steady-state conditions (i.e., time-invariant, horizontally homogeneous), the energy budget of the surface can be written as:

$$R_n = H + L_v E + G \qquad (9.23)$$

where H is sensible-heat flux, G is soil-heat flux, and $L_v E$ is latent-heat flux—a product of the latent heat of vaporization (L_v) and mass flux of water vapor (E). The term 'flux' as used here refers to the amount of energy (or matter) that passes through a unit area per unit time and is alternately known as flux density. Sign convention for fluxes in equation 9.23 varies in the literature, although frequently (as here), the favored convention considers fluxes toward the surface as positive. Fluxes are commonly expressed in W m^{-2} units; therefore, the units of E are typically [kg m^{-2} s^{-1}], and L_v units are [MJ kg^{-1}]. The partitioning of energy derived from the radiation balance is shown in figure 9.9. Parts (a) and (b) of this figure show the energy-

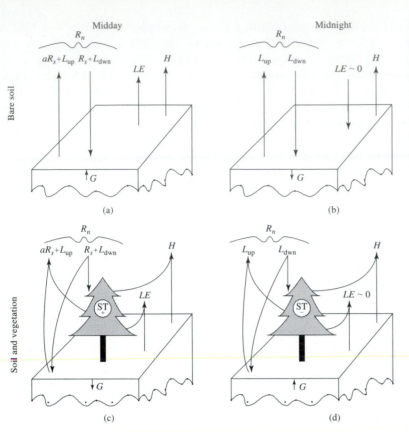

Figure 9.9 Energy-budget components for bare soil and soil with vegetation for both daytime and night-time conditions under clear skies. Radiation is decomposed into shortwave (R_s), reflected shortwave (aR_s), and longwave (L) components. H, LE, and G are sensible, latent, and soil-heat fluxes, respectively. Storage (ST) is positive during the day before solar noon, as the canopy vegetation (including air within it) lags ambient temperature. ST becomes negative late in the day and at night as the canopy releases stored energy.

balance terms of bare soils and how these change as time-of-day functions. Energy gain during the day at the surface from solar radiation (figure 9.9(a)) is used to heat the air through sensible heat (H), in evaporation as the latent heat of evaporation ($L_v E$), and in heating the soil (G). At night, the soil's surface cools because net radiation (incoming minus outgoing longwave) is negative; heat flow in the soil is upward since the subsurface soil layers are warmest.

The energy derived from net radiation and from that of soil that heats the air (through sensible and latent heat) are sometimes lumped together as available energy (A):

$$A = (R_n - G) = H + L_v E \tag{9.24}$$

In order to appreciate the partitioning presented in equation 9.24 more fully, we should examine data gathered on a warm, clear day (figure 9.10). Environmental measurements were taken hourly over a dry desert lake bed, where $L_v E \sim 0$. As indicated in the figure, soil-heat flux roughly tracked net radiation in the early morning hours. Sensible-heat flux lagged behind that of soil, since it depended upon atmospheric turbulence to transport heat away from the surface. During these hours, mechanical transport by turbulence was presumably weak, since wind speed was light. As a result, surface temperatures increased rapidly, which led to strong soil-heat flux to subsurface layers; by mid-morning, soil-heat flux had peaked. Past this time, the trend to decrease was likely due to increased wind-generated turbulent transport of heat (H) from the surface to the atmosphere. If the surface had sufficient moisture, $L_v E$ would also have been expected to increase. Net radiation reached a maximum near noon, or near the time that peak solar radiation was expected. With increased wind speed, sensible-heat flux from the soil to the atmosphere continued to increase; its peak followed that of net radiation by two hours. Then, with declining input of solar radiation into the soil-

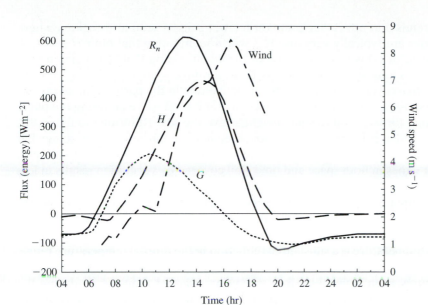

Figure 9.10   Principle energy-budget components of a dry, bare soil (e.g., dry desert lake bed), El Mirage, California, June 10–11, 1950 (data from Vehrencamp 1953)

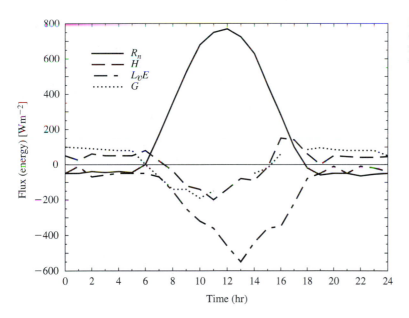

Figure 9.11   Diurnal trends of energy-balance components over a wet soil, Phoenix, Arizona (data from Fritschen and van Bavel 1962). Note sign convention; fluxes to the surface are positive.

atmosphere system, sensible-heat flux steadily declined, despite relatively strong winds. From 1600–2000 hours, sensible-heat flux exceeded net radiation, as it derived heat from the soil (note that soil-heat flux became upward, or negative). After 1800 hours, net radiation became negative since incoming solar radiation was low, and emission by the hot-soil surface that radiated in the far-infrared was high. After 2000 hours, net radiation remained a fairly constant negative value as the surface continued to lose heat, deriving energy from subsurface soil, and thereby lowering soil temperatures.

In the case of a moist soil as shown in figure 9.11, energy can be partitioned into soil-, sensible-, and latent-heat fluxes. The net radiation curve shown again depicts a smooth parabolic shape, indicative of a clear day. Latent-heat flux peaks near 1300 hours, at which time it consumes 80 percent of net radiation. Evaporative demand is so large that after 1500 hours, the energy that drives evaporation comes from net radiation, soil-, and sensible-heat fluxes.

This figure depicts trends that are rather common for wet soils in semi-arid regions, where atmospheric humidities are typically very low. It is worth mentioning that both H and L_vE increase (at the expense of G) with the increase of turbulent mixing that is associated with increasing wind speed, as long as available energy is not limiting.

Soil with vegetative cover   Vegetation plays an active role in the partitioning of incoming solar energy into sensible- and latent-heat fluxes, in that it is effective in raising the plane of interaction from that for bare soil to some height above the soil. Returning to figure 9.9, we note that with the added presence of vegetation, the energy balance becomes somewhat more complex (panels c and d). Energy can be transferred between soil and vegetation in such a manner that it varies in both space and time. Soil covered by vegetation results in less soil heating (see figure 9.12); by intercepting solar radiation during the day, vegetative cover reduces the radiative input to soil. At night, the vegetative canopy becomes the effective radiative surface, losing energy to space while reducing the soil's radiative losses. Thus, the canopy replaces the soil's surface as the plane of radiative interaction. The net effect of canopy cover on soil temperature is a substantial reduction in the diurnal temperature range, in contrast to soils not covered by a canopy. This can be seen by comparing in-row versus between-row temperatures, as in figure 9.13. Tree crowns are warmer than the air during the day, thereby heating the air. They are cooler at night due to radiative heat loss, cooling the air.

Vertical profiles of temperature and humidity as measured in a forest (see figure 9.14) indicate that conditions where foliage is highest in density are both the warmest and moistest. Due to the frictional resistance foliage imparts on air flow, wind velocities are lightest (with the exception of near-ground) at this location as well. Because vegetation is elevated above soil and is aerodynamically "rougher," it experiences higher wind velocities and greater turbulence than the soil's surface. Consequently, vegetation exchanges heat and moisture with the atmosphere more efficiently, since air flows freely on either side of leaves

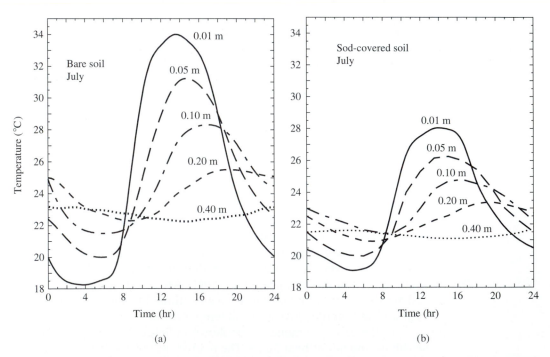

(a)   (b)

Figure 9.12   Typical diurnal patterns of soil temperature at various depths under clear skies for bare-soil surface (left) and sod-covered soil surfaces (right). The effect of lowering soil temperature in the presence of vegetation decreases with increasing soil depth (data from Backer 1965)

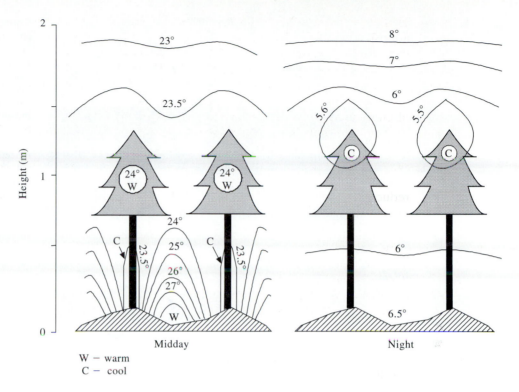

Figure 9.13   Distribution of temperature in an orchard (data from Geiger 1965) for typical conditions at mid-day, and at night under clear skies. Windy conditions tend to diminish temperature differences during the day or at night.

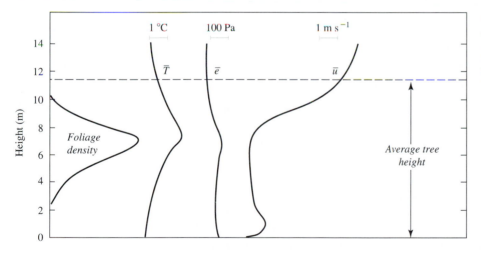

Figure 9.14   Typical profiles of relative foliage density (function of the amount of leaf area), air temperature, vapor pressure, and wind speed in a dense forest (data from Jarvis, James, and Landsberg 1976)

and other plant elements. Although much of our discussion involves tall canopies, the same statements generally apply to short canopies (such as grass) as well. In either case, under vegetative cover, the near-surface soil temperature range decreases and evaporation of soil moisture at the surface is reduced through a combination of decreased ventilation (lower wind speed) and lower available energy at the surface.

Referring once again to figure 9.9, we see that heat can be stored or released from vegetation (St) as a combination of sensible and latent heat. Some energy is consumed in

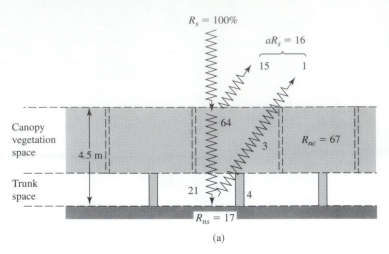

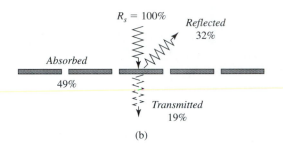

(b)

Figure 9.15 (a) Fate of radiant energy falling upon an orange grove (data from Kalma 1970). All values are expressed as a percent of incident radiation: R_s is shortwave (solar) radiation; R_{ns} is net shortwave radiant flux at the soil; and R_{nc} is net shortwave flux in the canopy. (b) Percentages of reflected, absorbed, or transmitted shortwave radiation for a single layer of orange-tree leaves. These numbers differ from those of a grove due to interactions between leaves, and varying leaf orientations within a grove.

transpiration from plant leaves (details in section 9.5) and a small amount of energy is consumed by photosynthetic and metabolic activity in plants (P). Evaporative flux from soils (E) can be combined with that involved in transpiration (T), in a term called evapotranspiration (ET). Latent heat of vaporization times the combined moisture flux comprising ET is included in the $L_v E$ flux term. With the presence of vegetation, equation 9.23 becomes

$$R_n = H + L_v E + G + St + P \tag{9.25}$$

The effects of vegetation on the magnitudes of energy-budget terms vary, but data gathered from an orange grove (Kalma 1970) provides us with some typical values (see figure 9.15): About 67 percent of incident radiation was absorbed by the canopy and partitioned into H, LE, St, and P; and about 17 percent was absorbed by the soil's surface. Over the course of a day, the sum of St is negligible and over a year, G typically sums near 0 as well; term P is also negligible, even for active vegetation. During the day P is about 6–16 W m^{-2} as light is utilized for photosynthesis, and becomes -3 W m^{-2} at night due to release of energy during nocturnal respiration (Oke 1987). With respect to the shortwave radiation budget (equation 9.19), orange leaves absorbed 49 percent of incoming radiation, reflected 32 percent, and transmitted 19 percent.

The partitioning of R_n into H and LE over vegetation can vary in time due to changes in plant physiology, environmental conditions (notably wind speed), and availability of soil water. Van Bavel (1967) shows the influence of water availability in a semi-arid region on an alfalfa crop that was flood irrigated at the beginning of the month, receiving no water thereafter. Note how energy partitioning changes as soil-water availability declines during the month (see figure 9.16). For the first several days, H is negative since latent heat loss due to transpiration by plant leaves (the active surface), enables them to become cooler than the air. Later, $L_v E$

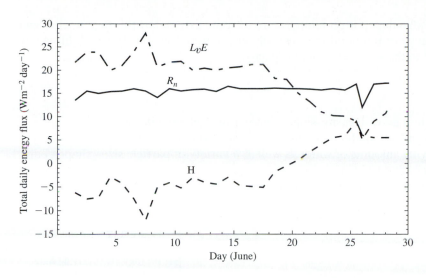

Figure 9.16 Daily sums of key energy-budget components measured over an alfalfa crop near Phoenix, Arizona, June, 1964. The crop was flooded in late May, and not watered in June (data from van Bavel 1967).

declines as soil water availability decreases, slowing transpiration. With decreased transpiration, latent heat loss declines, resulting in more energy to be partitioned into H.

9.3   SOIL-HEAT TRANSFER

Heat Capacity, Conductivity, and Diffusivity

The conduction of heat through soil depends on several factors: the soil's water content; texture; mineral composition; organic content; and compaction. These soil characteristics can affect heat transfer by changing the soil's heat capacity and thermal conductivity, and consequently, thermal diffusivity. Many of the factors that characterize a soil are interrelated. For instance, coarse-textured sandy soil usually has a high air-filled porosity (typically about 25 percent), and dry quickly since they have a high nonhygroscopic mineral content. When wet they conduct heat well, since particle sizes are large and water is a good conductor of heat.

Calculating heat capacity Heat capacity (c) was introduced on a per-mass basis earlier in this chapter. It is often more convenient to express it on a volumetric basis

$$C_i = \rho_i c_i \tag{9.26}$$

where ρ is density of soil constituent (i). Volumetric heat capacity for a soil (C_{soil}) can be expressed as

$$C_{soil} = \sum (f_i C_i)_s + \sum (f_j C_j)_w + (fC)_a \tag{9.27}$$

where f is the volume fraction and subscripts s, w, a refer to solids, water, and air, respectively. Solids can be divided into mineral and organic matter categories (subscript i), and water into liquid or ice (subscript j). For practical applications, it is not usually necessary to differentiate among minerals, since their heat capacity varies little when compared to the differences seen between minerals and organic material (see table 9.1). The heat capacity of water is almost twice as large as that of ice and should therefore be treated separately, as in equation 9.27.

Conductivity and diffusivity To predict heat flow and temperature changes in soil, the thermal conductivity and diffusivity of soil must be known. Soil thermal conductivity (κ_c) is simply the ability of a soil to conduct a quantity of heat in the presence of a temperature gradient.

$$\kappa_c = \frac{G}{\left(\dfrac{\partial T}{\partial x}\right)} \tag{9.28}$$

κ_c [W m^{-1} K^{-1}] of a solid depends on temperature to some extent, but this dependence can be ignored for the range of temperatures typically encountered in soil. Thermal diffusivity (D_h) is the ratio of κ_c to the volumetric heat capacity,

$$D_h = \frac{\kappa_c}{\rho c} = \frac{\kappa_c}{C} \tag{9.29}$$

D_h [m^2 s^{-1}] governs the rate-of-transmission of temperature change in soil.

Soil factors that affect conductivity and diffusivity Soils, typically complex in nature, are composed of varying amounts of water as well as a variety of particle sizes, shapes, composition, and densities (due to compaction, activity of resident flora or fauna, or other disturbance); the end result is an array of conductivities. Therefore, thermal conductivity and

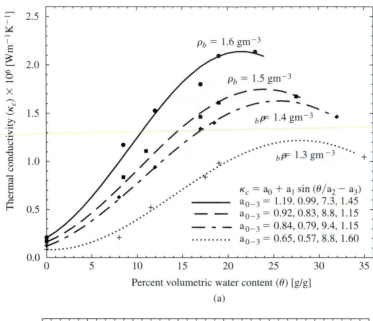

(a)

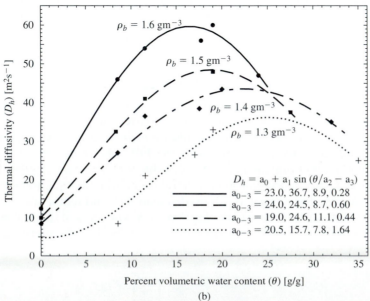

(b)

Figure 9.17   Thermal conductivity (a) and diffusivity (b) for soils of varying water content and bulk densities (values given on the curves). Data from Ghildyal and Tripathi 1987

thermal diffusivity can vary both spatially and temporally. κ_c and D_h depend greatly on soil-water content (see figure 9.17), and generally increase with it. This stems largely from water's greater thermal conductivity, 23 times larger than that of air (refer to table 9.1). Interestingly, the influence of water on conductivity and diffusivity is not linear, as depicted in the figure. An explanation for this is that at soil-water contents below ≈ 7 percent, a water film exists that merely coats soil particles. When soil-water content reaches between 7–20 percent, enough water fills soil pores to bridge the air-filled gaps between soil solids, thereby sharply increasing conductivity. Above ≈ 20 percent, further increases in soil-water content have little effect on increasing conductivity.

Thermal conductivity is higher for soils of coarser texture: sand > loam > clay. This can be due to a reduced area of contact between particles (Nakshabandi and Kohnke 1965), but thermal conductivities of these soils' dominant minerals can also be a factor. Increasing amounts of organic matter in the soil tend to decrease soil conductivity; organic matter's thermal conductivity is less than that of mineral constituents and it tends to have a loose, friable structure that promotes granulation.

Finally, κ_c and D_h also increase with soil compaction or bulk density, as either air or water-filled porosity decreases (see figure 9.17). Note that the thermal conductivity of minerals is substantially higher than that of air or water, but not substantially different from that of ice.

Calculating conductivity Considering the variability of κ_c and the variety of soil constituents, de Vries (1975) and Hillel (1982) proposed the following in calculating it:

$$\kappa_c = \frac{\sum\limits_{i=1}^{n} k_i f_i \kappa_i}{\sum\limits_{i=1}^{n} k_i f_i} \tag{9.30}$$

where f is the volume fraction of particle class (i) having a conductivity κ_i, and k_i is the ratio of the average temperature gradient in particle class (i) versus the temperature gradient of the fluid medium (air or water; m) surrounding the particles, as defined by (de Vries 1975)

$$k = \frac{(dT/dz)_i}{(dT/dz)_m} \tag{9.31}$$

A more practical form of equation 9.30 is found by segregating the soil constituents into air (a), water (w), and solid (s) components (after Hillel 1982):

$$\kappa_c = \frac{k_a f_a \kappa_a + f_w \kappa_w + k_s f_s \kappa_s}{k_a f_a + f_w + k_s f_s} \tag{9.32}$$

where the conductive-media reference is water.

Soil-Heat Flux

First law of heat conduction (Fourier's law) For steady-state conditions (i.e., constant temperature gradient), the rate of heat flow (heat flux) through soil is described by the first law of heat conduction:

$$G = -\kappa_c \nabla T \tag{9.33}$$

or, for the vertical dimension only:

$$G = -\kappa_c \frac{\partial T}{\partial z} \tag{9.34}$$

where κ_c is considered constant. The flux is in the opposite direction of the gradient, hence the right side of the equation carries a minus sign. Equations 9.33 and 9.34 are commonly referred to as Fourier's Law, in honor of the French mathematician (1768–1830) who pioneered the analysis of heat conduction in solids, in 1822. These equations are analogous to Darcy's equation for the conduction of fluids in soils and to Fick's law for diffusion in fluids, which is discussed later in this chapter (see section 9.7).

When considering a specific volume of soil, we employ a continuity equation for heat similar to that for water (see equation 7.18), enforcing conservation-of-energy principles. As such, all energy is accounted for in a closed-volume of a conducting medium. The difference in the local time–rate-of-change in heat equals the divergence (or rate of change with distance), plus the contribution by sources and sinks (S) within the volume:

$$\rho c \frac{\partial T}{\partial t} = -\nabla \cdot G \pm S(x,y,z,t) \tag{9.35}$$

or (again), for the vertical direction only:

$$\rho c \frac{\partial T}{\partial t} = -\frac{\partial G}{\partial z} \pm S(z,t) \tag{9.36}$$

where ρc is the volumetric-heat capacity.

Combining the Fourier equation (equation 9.34) with the continuity equation for heat (9.36), we get an expression for the time-dependent change in heat flow, known as the second law of heat conduction:

$$\frac{\partial T}{\partial t} = \frac{\partial}{\partial z}\left(D_h \frac{\partial T}{\partial z}\right) \pm \frac{S(z,t)}{\rho c} \tag{9.37}$$

The source (or sink) term is often omitted for the sake of simplicity, yet it can be significant. Concurrent with the movement of kinetic energy as heat, phase changes of water can result in substantial transport of energy in the form of latent heat. Thus, evaporation and condensation can appear as substantial positive or negative S values; this phenomenon is examined in greater detail, later in this chapter. Oxidation of organic materials mediated by microbial activity can also be a source of energy in equation 9.37.

Sinusoidal Solution to the heat transport equation Initial and boundary values are chosen to represent steady-state, periodic conditions in the solution of equation 9.37. In approximation, the average annual and diurnal cycles of solar radiation that force surface temperature and soil heating are sinusoidal in character. We can assume: **(1)** an initial ($t = 0$) average temperature T_a at the surface with an isothermal profile such that the average soil temperature throughout the profile is the same (T_a); **(2)** a maximum soil-surface temperature that occurs just after solar noon ($\approx$13:00 solar time); **(3)** a period (Π) of the temperature wave of 24 hours; and **(4)** a trend in surface temperature at the surface over the diurnal period given by:

$$T(z,t) = T_a + A_o \sin \omega t \quad \text{for } z = 0, t > 0 \tag{9.38}$$

where A_o is the temperature amplitude and ω is angular frequency given by

$$\omega = \frac{2\pi}{\Pi} \tag{9.39}$$

At $t = 13$ at the surface, equation 9.38 reduces to

$$T(0, 13) = T_a + A_o \tag{9.40}$$

The solution to equation 9.37 (for simplicity, we ignore the term S), is a linear and homogeneous differential equation with constant coefficients, and has the form (details in Ghildyal and Tripathi 1987):

$$T(z, t) = ae^{(bt+cz)} \tag{9.41}$$

where b and c are constants. Substituting equation 9.41 into equation 9.37 (neglecting term S) and solving for $T(z, t)$, yields

$$T(z, t) = ae^{bt \pm z(b/D_h)^{1/2}} \tag{9.42}$$

where we require that $D_h = b/c^2$. In meeting the requirements of the boundary condition (a sine function must exist in t at $z = 0$), b must be defined as a complex value ($\pm i\beta$). Continuing, we apply boundary conditions in equation 9.42, to find

$$T(z, t) = b \exp\left[\pm i\beta t \pm z\left(\frac{\beta}{2D_h}\right)^{1/2}(1 \pm i) \right] \tag{9.43}$$

which can be decomposed into four solutions; of these four, the one realistic solution is

$$T(z, t) = T_a + A_o \exp\left[-z\left(\frac{\omega}{2D_h}\right)^{1/2} \right] \sin\left[\omega t - z\left(\frac{\omega}{2D_h}\right)^{1/2} \right] \tag{9.44}$$

Examining equation 9.44, we see that the amplitude of the diurnal temperature wave at the surface decreases with depth by a factor of $\exp(-z(\omega/2D_h)^{1/2})$; at that depth, it is out-of-phase, lagging surface temperature by the factor $-z(\omega/2D_h)^{1/2}$. The reciprocal of this factor, $(2D_h/\omega)^{1/2}$, is a constant (independent of z) known as the damping depth (Z_D), or the depth that the diurnal or annual temperature wave's amplitude decreases to, $1/e$ (~37 percent) of A_o. The amplitude decrease (with depth of the propagating temperature wave) is due to cumulative loss (or gain) of energy from the wave to a temperature-contrast in the soil, within which it penetrates. That is, a wave of warmer (or cooler) temperature loses (or gains) heat to the cooler (or warmer) soil it penetrates. Damping-depth increases with the period of the wave, such that the annual wave's penetration is about $(365)^{1/2} \sim 19$ times deeper than that of the diurnal wave. Diurnal and seasonal waves as seen in field data are examined in the next section.

We may rewrite equation 9.44:

$$T(z, t) = T_a + A_o \exp\left(-\frac{z}{Z_D}\right) \sin\left(\omega t - \frac{z}{Z_D}\right) \tag{9.45}$$

To satisfy condition (2) (maximum surface temperature occurring at $t_o = 13$ hours), a phase adjustment (Φ, a constant) is needed:

$$T(z, t) = T_a A_o \exp\left(-\frac{z}{Z_D}\right) \sin\left(\omega t - \frac{z}{Z_D} + \Phi\right) \tag{9.46}$$

We can solve for (Φ) by applying the boundary condition (equation 9.40; $T(0, 13)$) in equation 9.46

$$T_a + A_o = T_a A_o \sin\left(\frac{13\pi}{12} + \Phi\right) \tag{9.47}$$

which requires $\Phi = -7\pi/12$. Equation 9.45 can be extended to include prediction of the ensemble temperature-wave in soil arising from diurnal and annual forcing (Hillel 1982):

$$T(z, t) = T_a + A_d \exp\left(-\frac{z}{Z_{Dd}}\right) \sin\left(\omega t_d - \frac{z}{Z_{Dd}} + \Phi_d\right)$$

$$+ A_y \exp\left(-\frac{z}{Z_{Dy}}\right) \sin\left(\omega t_y - \frac{z}{Z_{Dy}} + \Phi_y\right) \tag{9.48}$$

Terms related to diurnal and annual periods are denoted by subscripts (d) and (y), respectively.

From equation 9.45, we can determine an expected temperature range (maximum–minimum) at a given level (z):

$$\text{Range}(z) = \text{Range}(0) \exp\left[-\frac{z}{Z_D}\right] \quad (9.49)$$

We can also predict when maxima and minima occur relative to their occurrence at some other level

$$t(z_2) = t(z_1) + \frac{z_2 - z_1}{2}\left(\frac{\pi}{D_h \Pi}\right)^{1/2} \quad (9.50)$$

where levels are denoted by subscripts (2) and (1), and where D_h and Π must be in the same units of time.

Semi-infinite slab solution to the heat transport equation   Consider a soil column at a uniform initial temperature (T_o), heated by a source at the surface of a constant temperature (T_a). Initial conditions and boundary values are

$$T(z, 0) = T_o \quad \text{for } 0 \leq z \leq \infty \quad (9.51a)$$

$$T(0, t) = T_a \quad \text{for } t > 0 \quad (9.51b)$$

$$T(z, t) = T_a \quad \text{for } z \rightarrow \infty, t > 0 \quad (9.51c)$$

Details of the solution to equation 9.37 (second law of heat flow) with respect to the above conditions is given in Ghildyal and Tripathi (1987). Key steps include multiplying both sides of equation 9.37 (sink/source term S dropped for simplicity) by e^{-st} (s is a dummy parameter), and integrating over time:

$$\int_0^\infty e^{-st} \frac{\partial T}{\partial t}\, dt = D_h \int_0^\infty e^{-st} \frac{\partial^2 T}{\partial z^2}\, dt \quad (9.52)$$

If we apply the Laplace transformation to solve the integral, it yields a linear nonhomogeneous equation. After some algebra and application of boundary conditions, the following result can be obtained:

$$T = T_a + [T_o - T_a]\left[1 - \text{erf}\left(\frac{z}{2(D_h t)^{1/2}}\right)\right] \quad (9.53)$$

where erf is the error function (see appendix 3). Equation 9.53 can be used to determine temperature change with depth of confined-soil columns, or to determine thermal diffusivity of columns in a controlled (laboratory) environment.

Temperature Distribution in Soil

Temperature variability and its relevance   The distribution of soil temperature and its temporal variability have important agronomic implications, particularly with respect to seed germination and plant development (Rosenberg, Blad, and Verma 1983). The warming of cool soil can accelerate seed germination and substantially increase root uptake of nitrogen, potassium, and phosphorous—all essential nutrients for plant growth. Kaspar, Wooley, and Taylor (1981) noted much deeper root penetration of soybeans in warm soils as opposed to that in cold soils, where the roots tend to spread laterally, almost paralleling the soil's surface. This tendency appears to be a trait characteristic of root-adaption to soils of cold climates (Kramer and Boyer 1995).

Earlier in this section, equations that described the penetration of diurnal and annual temperature waves were determined; these waves are identified easily in field data. For example, diurnal patterns in soil temperature (with discernible amplitude dampening and phase shift) are found in a sandy soil (figure 9.18) beneath a boreal Jack pine forest in central Canada, just south of permafrost (data courtesy of D. I. Stannard, U.S. Geological Survey, Denver, Colorado). The solar-forcing responsible for the amplitude of the waves varies, in turn affecting soil temperature. Clear skies dominate the first two days (September 24 and 25), followed by two increasingly cloudy days (September 26 and 27). The character of the temperature-profile over the course of a 24-hour period for the same site on a clear day (see figure 9.19) indicates the speed and magnitude of the penetrating temperature wave through soil. The figure also shows null-points (zero heat-flux) for two of the profiles.

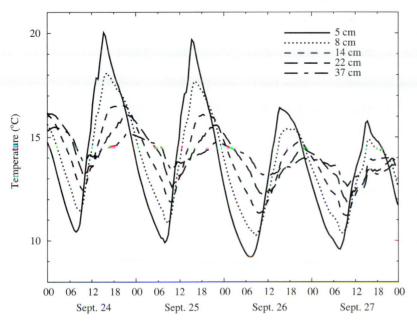

Figure 9.18 Diurnal variation of temperature in a sandy soil beneath a central Canadian forest. Various depths are given, illustrating both the dampening of the diurnal temperature wave with depth, and its increasing phase lag with respect to near-surface temperatures.

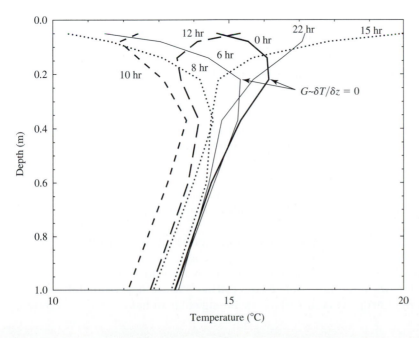

Figure 9.19 Soil-temperature profiles for various times of day, measured in the same soil as that of figure 9.18. Note location of soil-heat flux null-points (zero flux).

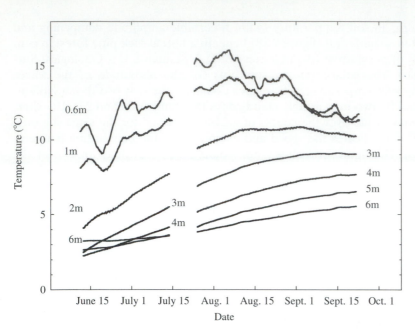

Figure 9.20   Seasonal variation of soil tempera-ture at various depths, measured in the same soil as that of figures 9.18 and 9.19. Upper layers (0.6 m and 1 m) indicate peak soil temperatures in early August. The penetration of the annual temperature wave (winter temperatures) appears to reach the 6 m level in late June. Dips in the seasonal pattern at 0.6 m and 1 m in late June and mid-August are due to cloudy conditions and infiltration of cool rainwater. Data were missing from June 16–20 at all levels.

Strong seasonal variation of soil-temperature is also observed in this soil (see fig-ure 9.20). Note that the annual wave responsible for the minimum in soil temperature at 6 m arrives in early July, following the minimum air temperatures of late January. Precipitation influences the amplitude of both the diurnal and annual patterns (altering heat capacity, con-ductivity, and diffusivity of soil), as well as other short-term climatic events (increased cloudi-ness, departures from seasonal temperatures). In mid- to late June, a prolonged cloudy pe-riod of nearly two weeks results in a deviation from the increasing seasonal soil temperature pattern, at 0.6 m and at 1 m. Note that the annual wave's period is not the longest wave to penetrate soil; boreholes of 100 m and greater in the Canadian arctic yielded temperature waves (forced by climate change) with periods in lengths varying from decades to 1,000 years (Wang 1992).

QUESTION 9.5

Find the damping depth of the diurnal temperature wave and average thermal diffusivity of the sandy soil in central Canada mentioned in this section (refer to figure 9.18, using the 5- and 37-cm depths).

Effect of latitude, slope, aspect, and mulches on soil temperature   On a global scale, temperature distribution varies spatially, as it is affected by the magnitude of incident solar radiation (latitude). At the local or regional scale, soil properties—including specific heat, conductivity, water content, bulk density and soil cover (vegetation, decaying matter, and mulches)—determine temperature distribution. Temporal variability itself is due to the vari-ability in solar-forcing, air temperature, and the infiltration of precipitation (Ghildyal and Tripathi 1987).

Now, we need to examine how latitude determines temperature distribution. Passing through a plane parallel to the Earth's surface are surface-energy fluxes (Rn, H, L_vE, G); the magnitude and sign of net radiation is determined largely by astronomical and geometrical considerations. While longwave radiation leaves the Earth's surface approximately normal to a plane tangent to the Earth's curved surface, incoming solar radiation is received on a

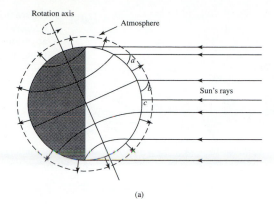

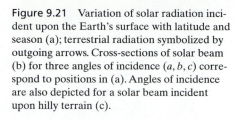

Figure 9.21   Variation of solar radiation incident upon the Earth's surface with latitude and season (a); terrestrial radiation symbolized by outgoing arrows. Cross-sections of solar beam (b) for three angles of incidence (a, b, c) correspond to positions in (a). Angles of incidence are also depicted for a solar beam incident upon hilly terrain (c).

(a)

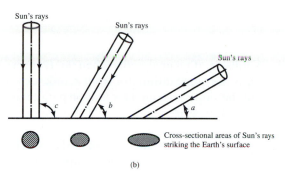

(b)

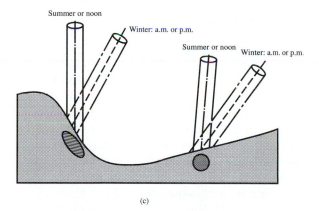

(c)

plane normal to Earth's orbital plane (see figure 9.21a). Thus, the solar radiation received (per unit surface) generally decreases with the cosine of latitude (figure 9.21b and c), while longwave is not directly latitude-dependent. As a result, annual net radiation is greater in the tropics than in polar regions. The latitudinal difference is enhanced by a greenhouse-effect involving water vapor. Water vapor is a strong absorber of thermal longwave radiation. On average, it reradiates a greater percentage of outgoing radiation back to Earth at lower latitudes, than at higher latitudes, because it occurs at a higher density at lower latitudes. Water vapor occurs in greater density at lower latitudes largely because of a higher-saturation vapor pressure (higher temperature), and greater abundance of surface water.

    With respect to slope, the solar radiation received on a hillside increases as the sine of the angle between the hills' slope and the Sun's rays increases (figure 9.21, c). In instances

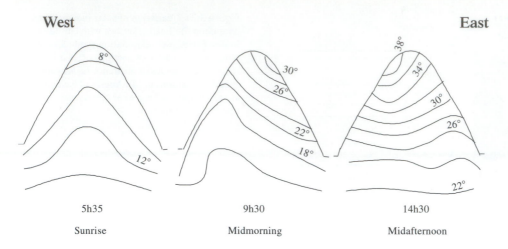

West   **East**   Figure 9.22 Temperatures (Celsius) of bare-soil furrows at various times of day (data from Geiger 1965)

5h35   9h30   14h30

Sunrise   Midmorning   Midafternoon

when the solar beam strikes a hillside at an angle other than 90° (zenith), solar radiation received is expected to decrease, as the angle difference from zenith increases. Thus, soil heating varies with both the angle of the slope (maximal for a slope parallel to the Sun's rays) and aspect (cardinal direction), as field measurements show. The temperatures of furrows in a plowed field shown in figure 9.22 provide an example of this effect. To go beyond this example, we note that differences in received energy due to slope and aspect increase with latitude (being least-important nearest the equator), and are more important in summer, when solar elevation is highest. Higher solar elevation reduces the optical path-length of solar radiation through the atmosphere, thereby increasing the ratio of direct beam versus diffuse (also referred to as scattered, isotropic, or nondirectional) radiation. Due to its isotropic nature, diffuse radiation more evenly irradiates a non-uniform surface.

It is widely known that changing the surface characteristics of a soil can modify the soil's temperature (Rosenberg, Blad, and Verma 1983). This can be accomplished with the application of a mulch, to the surface of the soil. The agronomic incentive in applying a mulch is either to increase or decrease soil temperature, and decrease soil evaporation. A dark mulch can decrease albedo, which increases absorption of solar energy and generally increases soil temperature. Most organic mulches (including crop residues) have low heat capacity and low thermal conductivity. As such, they tend to dampen extremes of soil temperature that might otherwise have been experienced (van Doren and Allmaras 1978). Mulches that are hydrophobic with a high-porosity and low heat-capacity can decrease soil evaporation. Waggoner, Miller, and DeRoe (1960) and Rosenberg, Blad, and Verma (1983) present a comprehensive review of mulches and their applications.)

9.4 SOIL MOISTURE EVAPORATION AND THE STAGES OF SOIL DRYING

The evaporation of soil moisture is a much-studied process, considered to proceed along three stages of drying (Hillel 1980; 1982), identifiable in figure 9.23. Let us consider a scenario with soil moisture at field capacity, initially. At this point, evaporation proceeds at a rapid and constant rate, near that of potential evaporation, since the system approximates that of a free-water surface. Evaporation rate is determined by atmospheric transport in the boundary-layer above soil, and by available energy (A). The rate of evaporation increases as any of the following increase: wind speed, surface roughness (which increases turbulent transport), available energy at the surface, and vapor-pressure deficit ($e_s - e_a$). In some cases, the evaporation rate can exceed that of a free-water surface. This situation occurs more often over an

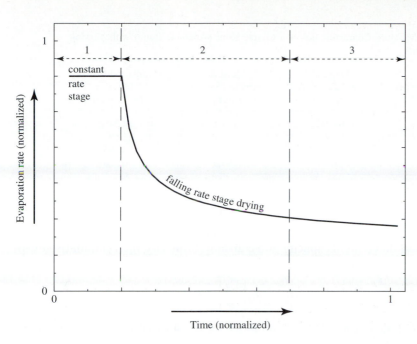

Figure 9.23   Conceptual diagram illustrating the three stages of soil drying over time

aerodynamically rough surface that increases the turbulent mixing of air. A rough, irregular surface can also have a greater amount of surface area (per horizontal distance) available for evaporation. The flow equation for a uniformly wet, isothermal soil-column of depth (Z) is:

$$\frac{\partial \theta}{\partial t} = \frac{\partial}{\partial z}\left[D(\theta)\frac{\partial \theta}{\partial z}\right] \tag{9.54}$$

where θ is water content [$cm^3\ cm^{-3}$], and D ($K(\theta)/(d\theta/d\psi)$) is soil-moisture (liquid) diffusivity [$cm^2\ s^{-1}$]. Conceptually, D is the ratio of hydraulic-conductivity to water-capacity of soil, as previously defined in chapter 8. We can derive a solution for the mean water-content in the column at the end of the first stage of drying, after successive integrations of equation 9.54 (Hillel 1980; Ghildyal and Tripathi 1987):

$$\bar{\theta} = \theta_a + \frac{1}{\beta}\ln\left(1 + \frac{\beta E z}{2D_a}\right) \tag{9.55}$$

where θ_a is soil-water content at air dryness (i.e., liquid water in equilibrium with air) and β is a soil-specific, dimensionless constant. D_a is the soil's hydraulic diffusivity at air dryness, related to $D(\theta)$ as: $D(\theta) = D_a \exp\{\beta(\theta - \theta_a)\}$. We assume: the supply of water required to meet the rate of evaporation (E) at the surface is constant with depth; there is no water flux out of the bottom ($d\theta/dz = 0$); and that evaporativity is determined by available energy. (Gardner and Hillel (1962) argue that θ varies little within the soil column, except at the surface.)

Transition to the second stage of drying occurs when the soil no longer can transmit enough water to the surface to meet evaporative demand; thus, the primary control on evaporation becomes the soil's hydraulic conductivity. As the soil dries, the plane of evaporation (i.e., source of water for evaporation) continues to move downward from the surface, and the water content of the surface layer eventually reduces to that of air (θ_a). Consequently, cumulative evaporation decreases (approximately) as the square root of the elapsed time (Gardner 1959), until it is about one-third the initial rate. Cumulative evaporation is derived from the flow equation (9.54), using the Boltzmann transformation (Kirkham and Powers

TABLE 9.5   Moisture Characteristics of a Coarse Sand (Adelaide dune) with Varying Water Content

Water content (θ) [cm³/cm³]	Hydraulic conductivity (K) [cm/min]	Moisture characteristic curve ($d\psi/d\theta$) [cm]	Capillary diffusivity of water (D) [cm²/min]
0.293	0.0417	104.1	4.34
0.259	0.0181	90.1	1.65
0.150	0.00358	74.0	0.265
0.121	0.00257	84.0	0.216

Source: Data from Elrick(1963).

1972; Hillel 1980; Ghildyal and Tripathi 1987). If we assume a semi-infinite column length, a water content about half that of field capacity, and constant diffusivity, then cumulative evaporation is given by:

$$E_{\text{cum}} = 2(\theta_o - \theta_a)\left[\frac{\overline{D}t}{\pi}\right]^{1/2} \tag{9.56}$$

where $\overline{D}$ is the weighted-mean diffusivity of soil moisture (comprised of liquid and vapor), t is time, and subscript (o) refers to the initial conditions. Although Hillel (1980) presents a numerical method of calculating $\overline{D}$, empirical data are relied upon to set soil-specific constants. A more practical means of finding $\overline{D}$ is to solve for it using equation 9.56, expressing it in terms of slope S ($E_{\text{cum}}/t^{1/2}$). A rough estimate of S can be obtained in the literature (Jury, Gardner, and Gardner 1991; Tripathi and Ghildyal 1975). For instances requiring more precision and reliability, it is necessary to conduct laboratory or field measurements with the specific soil under study, to find S (Hillel 1980, 1982).

Historically, the soil-drying process was divided into constant-rate and falling-rate stages. Kimball and Jackson (1971) further divided the falling-rate stage into two phases. The first phase is controlled by liquid-water diffusivity ($D_l(\theta)$), which decreases exponentially with decreasing θ (Gardner 1959; Hillel 1980); $D_l(\theta)$ is defined (as above) by hydraulic conductivity. The second phase is determined increasingly by the diffusivity of vapor ($D_v(\theta)$), as θ decreases. This phase begins essentially when the surface layer of soil becomes so dry that liquid water no longer can be conducted through it. The transition from one phase to the other is not discrete but gradual, due to the coexistence of both diffusivities in soil ($D(\theta) \approx D_l(\theta) + D_v(\theta)$). The transport of moisture by $D_v(\theta)$ is rather complex, since it is affected by gaseous transport and the absorptive forces of soil particles. We emphasize that $D_v(\theta)$ controls the gaseous diffusion of vapor in soil, and is not a function of hydraulic conductivity.

Much of our discussion to this point is based on laboratory work. Field studies, such as that of Brutsaert and Chen (1995), indicate that the transition from constant-rate to falling-rate stages of drying is not as discrete as presumed in figure 9.23; they appear much smoother, and occasionally last from several days to a week or longer. Also, transition periods appear to lengthen with lower net radiation and also when soil water content well below the surface is high.

Another point to keep in mind with regard to figure 9.23 is that the curve shape is expected to vary among soil types. Much of the variability is due to hydraulic properties of soils, as defined by their physical structure and chemistry. Fine-textured soils (composed largely of clays and loams) tend to lose more water to evaporation than sandy soils, which drain faster. Consequently, sandy soils progress through the constant- and falling-rate stages of drying quickly, which results in lower cumulative evaporation. On the other hand, grummusol and vertisol soils contain smectite clay, which shrinks during drying. Dessication cracks in

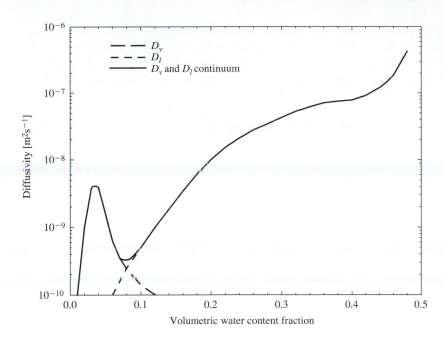

Figure 9.24   Relation between soil-moisture diffusivity (combination of liquid, l, and vapor, v) and water content for a Yolo light clay. Dashed lines illustrate extrapolations from liquid and vapor curves if only one phase of water is present in soil (data from Philip 1974; Hillel 1980)

these soils sometimes can reach several centimeters in width and a meter in depth. As these pores widen and deepen, a secondary plane of evaporation is created. In this case, soil drying can increase to 3 to 4 times that of a noncracking soil (Adams and Hanks 1964; Hillel 1980).

Soil evaporation is a dynamic, interactive process which we can appreciate in context of the material presented thus far, including the surface-energy balance. During stage-one drying, soil temperature varies little diurnally, due to the high heat capacity of soil with a high-water content. Low surface albedo of wet soil promotes absorption of radiant energy, which is partitioned into heating the surface, vaporizing water (latent-heat flux), and, in turn, humidifying the atmosphere. Following stage-one drying, the soil's albedo increases with progressive surface drying, lowering the soil's shortwave absorbence. However, the diurnal extremes of surface temperature increase due to the decrease in both heat capacity and thermal conductivity. This allows increasingly larger thermal gradients with decreasing water content to be established. The intensity of longwave radiation is expected to decrease slightly with decreasing soil-water content, since water's emissivity is slightly greater than that of minerals. However, the amplitude of the diurnal pattern of emitted longwave increases, with an increasing diurnal-temperature range of the surface. Recall that the intensity of this emitted longwave radiation increases with the fourth power of temperature. In turn, diurnal heating and cooling of near surface soil affects vapor transport by alternately vaporizing and condensing soil moisture within the soil column. A condition of soil-moisture hysteresis can develop that affects soil-moisture distribution, since sorption and desorption of soil moisture proceed at different rates.

QUESTION 9.6

Derive an equation for the evaporative flux using equation 9.56. Also using this equation, find $\overline{D}$ for a soil where $\theta_o = .35$ and a linear regression of E_{cum} versus $(\text{time})^{1/2}$ yields: $E_{\text{cum}}/t^{1/2} = 2.5 \text{ cm/day}^{1/2}$.

Evaporation from a shallow water table   Soils overlying a shallow water table rarely experience the latter stages of soil drying. Instead, their evaporation usually proceeds at the constant-rate stage. To meet evaporative demand or suction at the surface, water is conducted upward from the water table initially by capillary action through the capillary fringe, immediately above the water table and within the unsaturated zone. Above this zone, hydraulic conductivity controls the upward movement of water to the surface, as defined by Darcy's equation for vertical flow

$$E = -K \frac{d}{dz}(H - z) \tag{9.57}$$

which can be rewritten as

$$E = K(\psi)\left(\frac{d\psi}{dz} + 1\right) \tag{9.58}$$

where $K(\psi)$ is hydraulic conductivity, H is head pressure and z is distance downward from the soil's surface. H is a negative value in the unsaturated zone, expressed as suction head (ψ) in equation 9.58. Ripple et al. (1972), following Gardner (1958), define $K(\psi)$ in this instance as

$$K(\psi) = \frac{K_{sat}}{(\psi/\psi_{1/2})^n + 1} \tag{9.59}$$

where K_{sat} is the hydraulic conductivity of saturated soil [m s^{-1}], $\psi_{1/2}$ [m s^{-1}] is a constant representing ψ when $K(\psi) = 1/2\,K_{sat}$, and n is an integer ranging from 2 (for clays) to 5 (for sands). Procedures for calculating evaporation from a soil column above a shallow water table are given in Ripple et al. (1972) where equations 9.58 and 9.59 are applied to describe soil's ability to meet evaporative demand.

Evaporative potential is determined by available energy, vapor pressure deficit, and turbulent transport at the surface. Soil-evaporation rates can reach that of potential evaporation at times (defined in the following section) but are typically less, since they are limited by the rate at which a soil can conduct water upward (hydraulic conductivity), through the unsaturated zone and to the surface. Ripple et al. (1972) derive the soil-limited evaporation rate (E_{lim}):

$$E_{lim} = K_{sat}\left[\frac{\psi_{1/2}}{Z}\right]^n \left[\frac{\pi}{n \sin(\pi/n)}\right]^n \tag{9.60}$$

where Z is the depth from the surface to the water table, and n is an integer value; it is assumed that $E_{lim} \ll K_{sat}$. We can see from equation 9.60 that E_{lim} decreases linearly with decreasing K_{sat}, but decreases nonlinearly with increasing water-table depth.

9.5   EVAPOTRANSPIRATION AND THE INFLUENCE OF VEGETATION ON SOIL MOISTURE

Plants play an important hydrologic role (see figure 9.25), in that they provide a rapid means of transporting large amounts of soil-water to the atmosphere. Plants have a biological need for water; most (terrestrial) plants obtain this water from unsaturated soil near their roots. Roots can access water over a wide and deep volume of soil, depleting water at rates far in excess of that by soil evaporation alone—especially in comparison with that of soil drying's latter stages. Tap roots can extend several meters in depth, at times reaching the water table.

Plant Physiology

To understand the spatial and temporal character of plants' demand on soil moisture more completely, we need to briefly examine some basic concepts of plant physiology (a com-

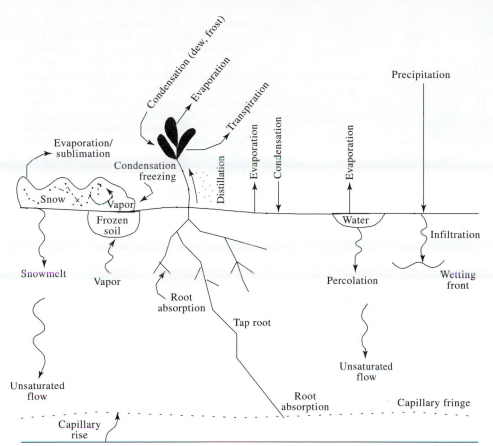

Figure 9.25 Fate of water in the soil–plant–atmosphere continuum, indicating the hydrologic importance of plants

prehensive discussion is given in Salisbury and Ross 1978). Plant growth and maintenance both depend on photosynthesis—a process in which CO_2 and water are combined (in the presence of light and chlorophyll) to produce simple sugars. The overall reaction of photosynthesis is

$$6CO_2 + 12H_2O + Light \rightarrow C_6H_{12}O_6 + 6O_2 + 6H_2O \tag{9.61}$$

Soil moisture supplies the water needed in the reaction, as well as the O_2 that is released to the atmosphere (as determined by Ruben, Randall, and Hyde 1941, using oxygen isotopes). CO_2 is largely sequestered from the atmosphere, although a small percentage can be absorbed from soils (Higuchi, Yoda, and Tensho 1984; Wium-Andersen 1971). The energy needed for the reaction is derived from photon absorption by chlorophyll molecules, largely held in mesophyll cells (see figure 9.26, top) by chloroplasts.

Appearing among the leaf's epidermal cells are pores (openings) in the leaf interior called stomata. The size and shape of stomata are defined by guard cells that respond to light intensity, temperature, and leaf-water potential. The size of the stomatal opening directly affects the rate of diffusion of water vapor and CO_2 into (or out of) the stomatal cavity within the leaf.

In the soil–plant–atmosphere continuum, water exists at a variety of potentials (see figure 9.26, bottom). Vascular plants acquire water in the root zone, where it follows a gradient in chemical-energy potential from the soil into the roots through pores in the epidermis, and in the long peripheral cells on young roots (called root hairs). Water is absorbed along roots

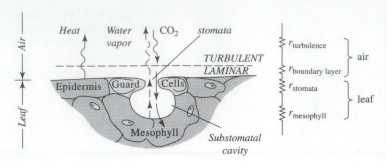

Figure 9.26   Resistances to water flow through a plant (after Kramer and Boyer 1995). Near-surface cross-section of a leaf relating physiologically significant tissue and related resistances to water flow (top). Overall-plant-mediated transport of water from the soil to the atmosphere (bottom). Distillation occurs as evaporated soil water condenses onto plant surfaces. Guttation is an exudation of plant water, due to root pressure occurring from hydathodes (stomate-like pores in the epidermis, which are the terminus of small leaf veins).

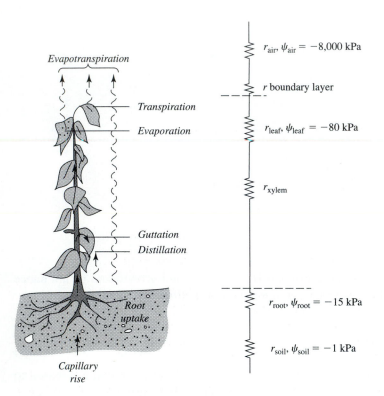

differentially, at rates that depend on cell structure and development (figure 9.27). Most of the water is absorbed where full root-cell structure develops in young cells; little water is absorbed at the root tips (for details, see Kramer and Boyer 1995). Once again, following the water-potential gradient (see figure 9.26, bottom), water is absorbed by roots and eventually reaches the root xylem; these are the fine, tubular structures near the center of roots and stems. Water in the xylem responds to water potential and consequently, it moves up the stems into the leaves. Continuing along the gradient, water passes through mesophyll cells and into the stomatal cavities, where it diffuses through stomata and into the atmosphere.

Transpiration is the process whereby water within plants is evaporated—either within stomata or at plant surfaces—and released to the atmosphere. Evapotranspiration is the combination of evaporation (including that from soil, soil cover, and plant surfaces) and transpiration. As a general rule, water loss to the atmosphere by transpiration is substantially

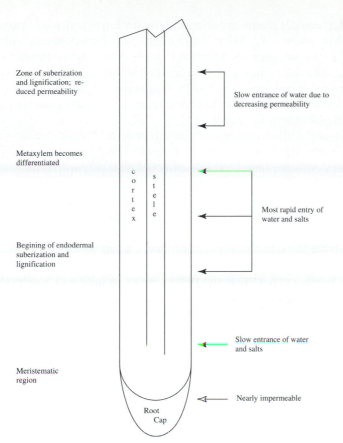

Zone of suberization
and lignification; re-
duced permeability

Slow entrance of water due to
decreasing permeability

Metaxylem becomes
differentiated

Most rapid entry of
water and salts

Begining of endodermal
suberization and
lignification

Slow entrance of water
and salts

Meristematic
region

Nearly impermeable

Root
Cap

Figure 9.27   Diagram of a root cross-section indicating areas of water and solute absorption related to cell structure and development (after Kramer and Boyer 1995). Greatest absorption occurs where root xylem becomes fully functional, behind the meristematic region but before the older endodermal region, where suberization and lignification (corky and woody tissue) reduce permeability of water.

larger than that of soil evaporation alone. At the height of the growing season, crops typically extract $\simeq 8$ mm of water per day from soils, transpiring up to 5–10 times the amount of water they contain at any one time (Rosenberg, Blad, and Verma 1983). Over the course of a growing season, these rates account for substantial cumulative totals; for example, up to 1,000 kg of water is used in evapotranspiration to produce 1 kg of wheat. Citing corn growing in a semi-arid area, Salisbury and Ross (1978) found soil evaporation to be only one-fourth of the cumulative evapotranspiration over the growing season.

Among a variety of other factors (e.g., soil-moisture levels and plant physiology), the transpiration rate depends upon the number and size of stomata, their function, and the amount of leaf area (for details, see Gates 1980). Stomatal openings occupy only $\simeq 1$–3 percent of leaf surface area, yet about 90 percent of transpiration occurs through them. The remaining 10 percent of transpiration occurs by the diffusion of water through leaf epidermal cell walls and the overlying cuticle (a waxy covering). Resistance to diffusion through stomata normally is in the range of several hundred seconds/meter, while it is more than 1 to 2 orders of magnitude greater across the cuticle. Stomata tend to close under the following conditions: low-light intensity; temperature extremes; mechanical disturbance (such as wind); exposure to certain pollutants; and the lack of either sufficient soil moisture in the root zone or the ability of roots to uptake water sufficient to meet transpiration rates. When stomata close, the only source of transpiration is that of diffusion through the leaf cuticle (Gates 1980).

While stomatal function is the primary means for controlling transpiration and the subsequent movement of water from the root zone (rhizosphere) into the plant, it is not the only control. Most (terrestrial) plants experience dehydration if soil-water potential (osmotic + matric) falls below about −1.5 MPa. Many plants also experience dehydration if soils become saturated; in this instance, O_2 diffusion to roots is too low to support metabolic activity, thus leading to a decrease in their ability to conduct water (Kramer and Boyer 1995). If water loss by transpiration continues, guard cells that sense declining leaf water-potential (i.e., larger negative values) close stomata to conserve water, thereby shutting down transpiration and, subsequently, photosynthesis as well.

Although transpiration occurs as a consequence of a plant's need to obtain CO_2 from the atmosphere, the resulting water-potential gradients help to transport dissolved soil-water nutrients to leaves. Another beneficial side-effect of transpiration is that it helps to cool plant leaves, when humidity is low and radiation load is high, sometimes by several degrees (Rosenberg, Blad, and Verma 1983).

Plant-root uptake of soil water can change soil-moisture distribution substantially in the rhizosphere and beyond. Data gathered in the vicinity of onion roots by Dunham and Nye (1973) illustrate the effect of root absorption on soil water and its distribution (see figure 9.28). The effect on soil-moisture distribution becomes more substantial at lower soil-moisture levels. We can expect that the plant's water uptake will accelerate the rate of soil drying, more than through evaporation alone.

Model Estimates of Evapotranspiration

Potential ET (ET_p) is the evapotranspiration rate of short, actively transpiring vegetation (e.g., grass) that: completely covers the ground; is well-supplied with water; and exerts negligible resistance to water movement through the plant. In arid to semi-arid regions, ET_p may exceed the free-water evaporation rate, E_o. Equilibrium ET (ET_{eq}) defines the minimum possible evaporation rate from a moist surface, and is quantified as (Slatyer and McIlroy 1961):

$$ET_{eq} = \left[\frac{s}{s+y}\right](R_n + G) \tag{9.62}$$

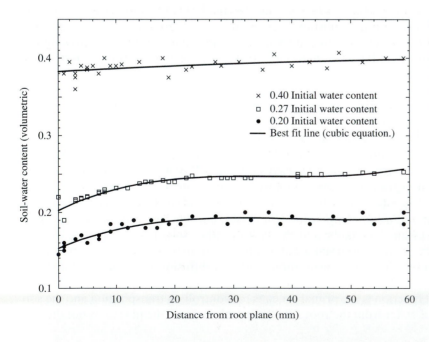

Figure 9.28   Soil-moisture distribution of a sandy soil in the vicinity of an onion root mat attached to live shoots after 2.5 days. Water content was 42–45 percent (top), 27 percent (middle), and 20 percent (bottom). Data represent measurements from either side of the root mat (data from Dunham and Nye 1973)

where s is the slope of the saturation–vapor-pressure curve at the wet-bulb temperature and γ is the psychrometric constant ($C_p P / \varepsilon L_v$).

A variety of approaches are available for estimating ET, many of which are described in Rosenberg, Bard, and Verma (1983). This chapter focuses on a number of physically derived and semi-empirical models that have global applicability. Some regression (and other, strictly empirical) approaches of ET estimation based on the use of routinely measured climate variables are covered in chapter 12.

Penman model A model proposed by Penman (1948) was derived from energy-balance and aerodynamic equations. This method has been widely accepted in agriculture and hydrology, since few measurements are needed and it usually produces reasonable estimates. Originally derived for open water, evaporation is given by:

$$E = \frac{s R_{nw} + \gamma E_a}{s + \gamma} \tag{9.63}$$

where R_{nw} is net radiation over water. A variety of semi-empirical equations have been used to specify E_a (see Rosenberg, Bard, and Verma 1983), which is a function of wind speed and the vapor-pressure deficit. Penman (1948) multiplied equation 9.63 by an empirically derived factor to estimate ET, but Thom and Oliver (1977) found a better global application of equation 9.63 by replacing R_{nw} with ($R_n + G$), without the empirical coefficient.

Priestly–Taylor A semi-empirical formula for ET_p was proposed by Priestly and Taylor (1972):

$$ET_p = \alpha \frac{s}{s + \gamma}(R_n + G) \tag{9.64}$$

where α is an empirically derived constant (conceptually equivalent to ET_p / ET_{eq}), with values ranging from 1.08–1.34 and a mean of 1.26 (Priestly and Taylor 1972). α appears sensitive to surface soil-moisture levels ($\alpha = 1.26$ for wet surfaces). The relation for non-ET_p conditions appears to be inconsistent (Williams et al. 1978; Rosenberg et al. 1983).

Penman–Monteith Monteith (1965) expanded the Penman model to improve simulation of vegetation feedbacks on ET by inclusion of resistance terms (units of s m^{-1}) for the aerial transport of water vapor (r_a) from the canopy, and the resistance to vapor transport through stomata (r_c). This was an important development, since for any land surface covered with vegetation, r_c largely controls the transpiration rate and, consequently, much of $L_v E$. In contrast to previous models, this model predicts actual ET (as $L_v E$; energy) in contrast to ET_p:

$$L_v E = \frac{s(R_n + G) + \rho_a C_p [(e_s - e_a)/r_a]}{(s + \gamma)[(r_a + r_c)/r_a]} \tag{9.65}$$

Difficulties in the practical application of equation 9.65 include obtaining stomatal-resistance data (either through direct measurements or a surrogate). Wind-speed measurements are required for r_a, which varies with vegetation type. Details on the application of the Penman–Monteith and closely related models—some more sensitive to sparse vegetation canopies than equation 9.65—can be found in Jensen, Burman, and Allen (1990) and Stannard (1993). An important finding by Stannard was that, when r_c must be estimated (i.e., when measurements are not available), the Penman model predicted measured-ET better than the Penman–Monteith model.

Water-budget ET is one of the key elements of the hydrologic cycle (or water budget) of a land surface (Garratt 1992):

$$\rho_w \left(P_R - d \frac{\partial \theta}{\partial t} \right) = ET + D + R \tag{9.66}$$

where all terms are kg m^{-2} s^{-1}. Terms on the left side of equation 9.66 account for water in a soil layer of depth (d). P_R is the precipitation rate, m s^{-1}, and θ is soil-water content. These terms are balanced by the "loss" terms on the right, where D is drainage (or water loss) through the bottom of the soil layer and R is surface run-off. Assuming the budget is closed, ET can be estimated as a residual term in equation 9.66, provided all other terms are measured accurately.

9.6   MEASURING ENERGY BUDGET TERMS

Soil heat flux   Knowing a temperature profile and thermal conductivity, soil-heat flux can be calculated by applying equation 9.34. In practice, conductivity is a difficult parameter to obtain in the field, made more complex by the temporal variability of water content and soil compaction. Soil-heat flux plates (see figure 9.29) allow flux to be measured directly. Heat-flux plates consist of differential thermopiles (or multiple thermocouple junctions) embedded in the top and bottom of a substance with a known thermal conductance (usually $\simeq 1.2$ W m^{-2}); chosen to be similar to that of typical soil, so that the naturally occurring thermal gradient in their vicinity is minimally distorted. They are usually made of a material with low heat-capacity and kept small to enable them to respond quickly to temperature changes. The small size of the soil-heat flux plates also minimizes latent-heat exchanges, due to the condensation or vaporization of water caused by differential heating or cooling that results from their presence. Although the plate's thermal conductivity is representative of that of many soils, it can differ substantially from that of the soil at the measurement site. Thus, the plates cannot be positioned at or near the surface, since their presence can substantially alter heat flux in this region of very strong temperature gradients. Typically, they are instead placed at a depth of 5 to 10 cm, where thermal gradients are smaller and the distortion of these gradients is less. However, short-term fluctuations in heat flux cannot be measured with this placement, because they are damped-out in the top few centimeters of soil.

The combination method (Stannard et al. 1994) offers a means by which soil-heat flux (at some depth) can be measured with heat-flux plates, and combined with a measurement of change in heat storage in soil above the plate:

$$G_{\text{Tot}} = G_x + G_{\text{stor}} \tag{9.67}$$

G_x is the heat flux measured by a heat-flux plate; G_{stor} is calculated by measuring soil temperature at a number of depths, averaging the temperatures, and then multiplying the result by the specific-heat capacity of the soil layer above the plate, determined from specific mea-

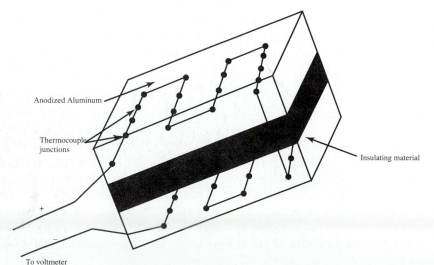

Anodized Aluminum

Thermocouple junctions

Insulating material

+

To voltmeter

Figure 9.29   Schematic diagram of a soil-heat flux plate. The model consists of a differential thermopile with thermocouples embedded in substrates near the top and bottom surfaces of the instrument. Magnitude of the difference between thermocouple (in series) on the top of the plate versus those (in series) on the bottom is proportional to heat flux (plate thickness exaggerated to show detail).

surements of soil density, composition and moisture. Temperature can be measured at equidistant points above the heat-flux plate, or if the depth is greater than several centimeters, at logarithmic spacing (densest sampling near the surface), to capture the non-linear temperature profile. Another problem that may need to be addressed is properly accounting for latent-heat transport, related to the movement of water vapor. If significant, the soil-moisture profile would need to be measured concurrently.

Some practical problems with field measurements of soil-heat flux include: disruption of the natural soil profile (its structure and moisture) when installing sensors; installing enough sensors for measuring the soil-temperature profile; and sampling heat flux at enough locations to avoid spatial bias due to soil heterogeneities—either inherent in the soil, slope, or with vegetative cover.

Net radiation As shown in section 9.2, net radiation is the difference of incoming radiation minus that of outgoing. Net radiometers are identical, in the principle of operation, to soil heat-flux plates, where thermopiles are embedded in both upper and lower portions of a substrate of known thermal properties. The upward- and downward-facing surfaces that intercept radiation are painted with a high-emissivity black paint, simulating blackbody emissivity. Transparent domes cover the radiometers' sensing surfaces to shield them from dust and moisture, which could modify radiative characteristics (absorptivity and emissivity) of these surfaces (see Rosenberg, Bard, and Verma 1983 for more details).

Sensible and latent-heat fluxes These surface fluxes are typically measured with micrometeorological methods: gradient; aerodynamic; Bowen-ratio-energy-balance (BREB); or eddy covariance. Ideally, all micrometeorological methods require a homogeneous (in terms of surface roughness and flux), flat surface area upwind of the point of measurement. This surface area is referred to as the flux footprint, and the linear distance of its upwind leading edge to the instruments is called the fetch. The depth of the atmosphere that is in equilibrium with the surface area (the equilibrium boundary layer, lying at the bottom of the surface layer) increases at about 1 m per 100 m of fetch. Considering limitations of current instrumentation for flux measurement and characteristics of atmospheric turbulence over vegetation it is generally desirable to install instruments 1–2 m above ground level over short vegetation (e.g., grass, wheat), and at least twice the plant height over tall vegetation such as forests. Thus, the decision regarding site selection for micrometeorological flux measurements primarily must weigh fetch (being aware of both wind direction and speed variability), instrumentation limits, and surface characteristics. For small areas or plots, latent-heat flux can be derived from weighing lysimeters that directly measure the amount of water lost to evaporation. Details of these methods may be found in Rosenberg, Bard, and Verma (1983), Verma (1990), and Daamen et al. (1993). For the sake of brevity, our present review of methods will focus on the BREB and eddy-covariance methods, which are used most often in the field for flux estimates from large areas (on the order of several hectares or more).

BREB The BREB method provides a robust and low-cost means of obtaining H and $L_v E$. This method relies on the assumptions of energy-budget closure and similarity in the turbulent transport of sensible and latent heats. Energy-budget closure is expressed as:

$$R_n + G + H + L_v E = 0 \qquad (9.68)$$

In applying equation 9.68, we assume that there is no energy storage above the soil surface (such as within a vegetation canopy). Expressions for H and $L_v E$ in terms of thermal and vapor gradients in the atmosphere's turbulent surface layer (see figure 9.5) are:

$$H = \rho C_p K_h \frac{\partial \overline{T}}{\partial z} \qquad (9.69)$$

$$L_v E = L_v \rho \frac{\varepsilon}{P} K_v \frac{\partial \overline{e}}{\partial z} = L_v K_v \frac{\partial \overline{\rho_v}}{\partial z} \qquad (9.70)$$

where C_p is specific heat of air at constant pressure ($\sim$1.005 kJ kg^{-1} K^{-1}), ε is the ratio of molecular weights between water and dry air (0.622), P is atmospheric pressure [kPa], z is height above ground, and K is the turbulent-exchange coefficient, or eddy diffusivity [m^2 s^{-1}]. Subscripts (h and v) refer to heat and water vapor, respectively; overbars indicate a time average. The BREB method assumes that $K_v = K_h$. While this can be true under neutral thermal stability (windy daytime conditions in the presence of small ($\sim$0) air-to-surface temperature gradients), it is only an approximation under other conditions. Substituting equations 9.69 and 9.70 into 9.68, solving for K, and assuming eddy diffusivities are equivalent, we find

$$K_h = K_v = -\frac{(R_n + G)}{\rho\left(C_p\dfrac{\partial \overline{T}}{\partial z} + L_v\left(\dfrac{\varepsilon}{P}\dfrac{\partial \overline{e}}{\partial z}\right)\right)} \tag{9.71}$$

By substituting equation 9.71 into 9.70, we obtain:

$$L_v E = -\frac{R_n + G}{1 + \beta} \tag{9.72}$$

where β is the Bowen ratio $H/L_v E = \gamma(d\overline{T}/d\overline{e})$.

The beauty of the BREB method is that no wind-speed measurements are needed, and the equipment required for taking the measurements is robust and relatively inexpensive. Typically, only an aspirated, wet- and dry-bulb psychrometer system is needed, in addition to radiation and soil-heat flux measurements. However, we need to remember that BREB measurements must be conducted at sites where the inherent assumptions in method development closely approximate reality.

Eddy covariance Increasing in popularity because of its lowered instrument costs, more robust instrumentation, and ease of use, the eddy-covariance (or eddy-correlation) method offers a direct means to measure H and $L_v E$. The principle of the measurement is to obtain the covariance of the product of fluctuations in vertical velocity and the scalar variable of interest (e.g., temperature, T' or water-vapor density fluctuations, ρ_v'), measured at a distance above the surface

$$H = -\rho C_p\,\overline{w'T'} \tag{9.73}$$

$$L_v E = -L_v\overline{w'\rho_v'} \tag{9.74}$$

where the deviations (primes) from the means (overbars) in vertical velocity and the entity of interest are defined by: $w = \overline{w} + w'$; $T = \overline{T} + T'$; $\rho_v = \overline{\rho_v} + \rho_v'$. Because of fluctuations in air density, equation 9.74 (actually a simplified result) must be adjusted. Following Webb, Pearman, and Leuning (1980), equation 9.74 becomes:

$$L_v E = L_v\overline{w'\rho_v'} + (1 + 1.61\overline{q})\frac{\overline{\rho_v}}{\overline{T}}\,\overline{w'T'} \tag{9.75}$$

where q is specific water vapor (defined in equation 9.10).

Key to the success of eddy-covariance measurements are fast-response instruments. A basic deployment of some often-used eddy-covariance instruments is depicted in figure 9.30. Most eddy-covariance sensors are capable of making accurate measurements at a frequency faster than 10 Hz. Sonic anemometers are often chosen for the wind-velocity measurement and some can derive temperature fluctuations from speed-of-sound measurements as well. However, in this case, temperature is virtual temperature, (temperature that dry air would have if it was of the same density as air containing water vapor) and must be converted to kinetic temperature to calculate sensible-heat flux. Other sonic anemometers are outfitted with a fine-wire ($\sim$.02 mm) thermocouple for kinetic-temperature measurement. Humidity often

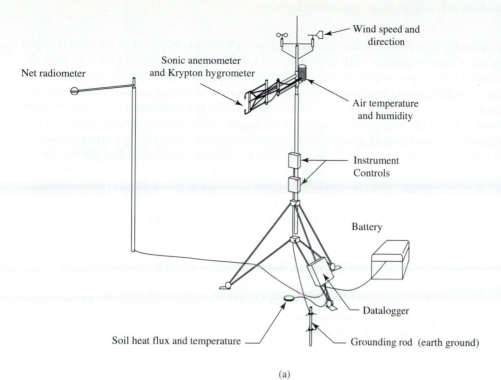

Wind speed and direction

Sonic anemometer and Krypton hygrometer

Net radiometer

Air temperature and humidity

Instrument Controls

Battery

Datalogger

Soil heat flux and temperature

Grounding rod (earth ground)

(a)

Figure 9.30 A basic eddy-co-variance deployment, providing sensible- and latent-heat flux measurements. A net radiometer and soil-heat flux plates supple-ment the array to provide mea-surements of major energy-budget terms (a). Single-axis sonic anemometer (vertical ve-locity sensing only) outfitted with a fine-wire thermocouple and krypton hygrometer (to sense water-vapor density) is de-picted in (b). Tri-axis sonic anemometers (c) provide com-plete velocity measurement and virtual temperature. Panels (a) and (b) courtesy of Campbell Scientific, Logan, UT; panel (c) courtesy of Applied Technolo-gies, Boulder, CO.

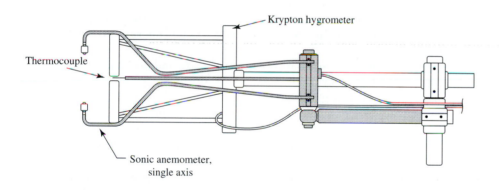

Krypton hygrometer

Thermocouple

Sonic anemometer, single axis

(b)

Sonic anemometer, tri-axis

(c)

is sensed optically by measuring the amount of monochromatic infrared or ultraviolet light remaining after passing along a path from source to detector, through ambient (sample) air. The natural log of voltage output from the detector, is inversely related to the number (density) of vapor molecules in the sensing path. Hygrometers that use ultra-violet light typically use the Lyman–alpha wavelength (~121 nm), which is affected somewhat by O_2 absorption of UV. Those who use infrared generally use wavelengths near 2.6 μm, where absorption by water molecules is substantially greater than that due to other infrared-absorbing gases.

The eddy-covariance method is the only means currently available that allows for direct measurement of H and L_vE. It is also a more labor-intensive and expensive approach than other micrometeorological methods. The same cautions of site-selection mentioned for the BREB-method apply to this method as well. Instrumentation is delicate and a data logger is needed capable of recording high-frequency data for computation of fluxes. A variety of adjustments to the raw measured flux need to be made that take instrument characteristics and their deployment into account (Moore 1986). The reader is referred to Verma (1991) for details regarding eddy-covariance instrumentation and application of the method.

Measuring transpiration   Transpiration is a difficult quantity to determine from L_vE or ET measurements; Rosenberg, Bard, and Verma (1983) list models used to do so. Measurement of transpiration is accomplished primarily with either sap-flow measurements in stems, plant chambers or enclosures, or isotopic analyses.

Sap-flow or sap-velocity measurements (Dugas 1990; Kramer and Boyer 1995) are used to estimate the amount of water traveling through stems. The operating assumption is that the amount of water measured by sap flow is proportional to the amount transpired; however, plants can store (or release) water in sufficient quantity at times, as to invalidate the assumption. Rates of storage (or release) can vary over the course of a day, even hourly. Thus, we need to take plant water storage into account when using sap-flow measurements to estimate transpiration.

Chambers used to measure transpiration can vary in size and in construction but all are based upon a similar concept—transpiration rate is proportional to the rate of increase in water-vapor density in an enclosure containing a known area of plant leaves. Chambers range in size from that which is inclusive of only a plant leaf (including porometer measurements), to branches (Saugier et al. 1997), and to entire plants or trees (Garcia, Norman, and McDermitt 1990; Cienciala and Lindroth 1995). Leaf and branch measurements are relatively simple to perform and are more frequently conducted. However, these measurements must carefully be scaled-up (Baldocchi 1993) to the whole plant or canopy level to understand the significance of transpiration on soil moisture. Successful use of this method requires a well-planned temporal and spatial sampling strategy.

Recently, the stable isotopes deuterium (^2H) and oxygen-18 (^{18}O) have been used in attempts to quantify transpiration and the movement of soil-water in the soil–plant--atmosphere continuum (Walker and Brunel 1990; Thorburn, Hatton, and Walker 1993). There is a naturally occurring variability in the amount of these stable isotopes found in the air, foliage, soil, and ground water. The isotopic value of soil water is initially determined by precipitation. At the soil's surface and below, this water can become enriched with ^2H and ^{18}O during the processes of evaporation and vapor diffusion. Water molecules consisting of the combination of these isotopes are heavier than ordinary water molecules, and thereby require more energy to reach the vapor phase or to be moved during diffusion. Provided that little isotope-enrichment or depletion occurs during root-absorption of water, the isotopic concentration of transpired water is expected to be similar to that of the xylem, and similar to the soil source of this water. Thus, transpired water can be expected to have different concentrations of stable isotopes than water evaporated from plant and soil surfaces. By evaluating the isotopic ratio of air above canopy vegetation, the relative strength of transpiration to measured evapotranspiration is obtained for a plant canopy. If ET is concurrently measured, then transpiration can be quantified. The stable-isotope method offers an attractive potential as a means of quantifying transpiration and determining the source of transpired water. However, at this time, it is still undergoing experimental development and field verification.

9.7   SOIL-GAS TRANSPORT

Soil gases move in response to pressure gradients; that is, air moves from a higher pressure potential to a lower one. Additionally, since air is made up of a number of gases, each species can have different concentration gradient that drive individual gases in different directions. Thus, at any time, gas transport can occur in response to the total pressure gradient of air, as well as that of the partial pressure gradients of individual gas species in the air. The objective of this section is to develop an understanding of these and other gas-transport mechanisms operating in soil. Before doing so, a review of basic properties of gases is provided.

General Properties of Gases

Variables of state and ideal gases If we consider a gas as consisting of a number of individual molecules, we could describe it as having a chaotic system of motions and collisions. However, when considered en masse, this same population of chaotic molecules could instead be described by nonchaotic variables of state (the basis of the continuum hypothesis): mass (M), volume (V), pressure (P), temperature (T) and composition. For now, we focus on ideal gases that are stable in time (nonreactive, nondecaying) and do not exhibit a change in phase with a change in temperature. Variables of state can be combined to form specific variables, such as density ($\rho = M/V$) and specific volume ($a = V/M$). In this review, we refer to standard conditions for a gas: pressure ($P_o = 101.3$ kPa), temperature ($T_o = 273$ K), and mole-volume (mole = 22.4 liters).

Charles' law In the late 1700s, Jacques Charles and Joseph Gay-Lussac found that an ideal gas, contained in sealed insulated vessels and uniform pressure, undergo a change in temperature corresponding to a change in volume. Thus, at constant pressure (P_o) a gas initially at temperature (T_o) and specific volume (a_o), will have a new temperature (T) when its specific volume has been changed to (a):

$$\frac{a(T, P_o)}{T} = \frac{a_o(T_o, P_o)}{T_o} \tag{9.76}$$

Boyle's law In the 1660s, Robert Boyle adjusted pressure and volume at a constant temperature, and found that pressure is inversely proportional to specific volume

$$Pa = C(T) \tag{9.77}$$

where C is a temperature-dependent constant.

Ideal gas law It was later found that Charles' Law compliments Boyle's Law by providing a means of relating P, T, and a in a powerful relation we call the ideal gas law. From Boyle's Law, we can write (at a fixed temperature)

$$Pa(T, P) = P_o a(T, P_o) \tag{9.78}$$

Using Charles' Law (equation 9.76), we substitute for $a(T, P_o)$ in equation 9.78 to obtain:

$$Pa(T, P) = \frac{P_o a(T_o, P_o)}{T_o} T = RT \tag{9.79}$$

where R is the specific gas constant. Equation 9.79 is the ideal gas law, which can be written in terms of volume if we consider a mass of gas equal to its molecular weight (i.e., a mole of gas)

$$PV = mRT \tag{9.80}$$

where m is molecular weight, and volume is $V = ma$. Equation 9.80, Avogadro's Law states that a mole of ideal gas will occupy the same volume as a mole of any other ideal gas, at the same temperature and pressure. Therefore, a more general form of equation 9.80 is

$$\frac{PV}{T} = mR = R^* \tag{9.81}$$

where R^* is the universal gas constant ($8.3144 \text{ J mol}^{-1} \text{ K}^{-1}$).

Dalton's law The soil-air and atmosphere are a mixture of a number of gases. Dalton's Law of Partial Pressures states that the total pressure (P_{total}) of a gas mixture is equal to the sum of pressures of each gas:

$$P_{total} = \sum_{i=1}^{n} P_i \tag{9.82}$$

For each ideal gas

$$P_i = \frac{R^*T}{m_i a_i} = \frac{R^* M_i T}{m_i V} \tag{9.83}$$

where M_i is the mass of gas (i). For the gas mixture, we write

$$P_{total} = \frac{R^*T}{V} \sum \frac{M_i}{m_i} \tag{9.84}$$

Since $V = Ma$, equation 9.84 can be rewritten as:

$$Pa = R^*T \frac{\sum \dfrac{M_i}{m_i}}{\sum M_i} = \frac{R^*T}{\overline{m}} \tag{9.85}$$

Thus, the specific-gas constant for dry air ($R = R^*/m_d = 287.04 \text{ J mol}^{-1} \text{ K}^{-1}$) can be calculated once the mean molecular weight of dry air is found. Weight, $\overline{m}$, is defined more specifically as the mass-weighted harmonic mean of the molecular components that comprise dry air.

Porosity and Permeability

Porosity ϕ is defined as the volume ratio of fluids (gas and liquid) in soil, to the total soil volume (solid and fluid components):

$$\phi = \frac{V_a + V_l}{V_a + V_l + V_s} \tag{9.86}$$

Coarse-textured, sandy soils tend to have the greatest porosity, typically about 35 percent, and have a high V_a (field-air capacity) fraction. Water-filled pores drain quickly, since these soils usually consist of siliconitious, nonhygroscopic particles, and large pores. In contrast, fine-textured soils, which may have a high content of loam or clay, drain slowly and typically, are easily compacted. They tend to be poorly aerated and can have a field-air capacity of only 5 percent. These soils often contain soil macropores and soil aggregates which can exceed 5 mm in diameter. Thus, the soil porosity in these soils can vary greatly in space and time.

In contrast to porosity, air permeability defines the ability of a soil to conduct fluids (air and water) through it via connected interstitial spaces, or pores (Katz et al. 1959). Permeability is determined solely by structure, and hence is also referred to as intrinsic permeability. Changes in permeability occur if the soil structure is altered by water (Marshall and

Holmes 1979), or by compaction. Permeability [Darcys, L^2], and can be calculated from the equation for flux

$$k = -\eta q \left[\frac{dx}{dP} \right] \qquad (9.87)$$

where q is flux density (a volume flow-rate per unit area), η is dynamic viscosity [Pa s], x is distance over which a pressure (P) differential [Pa] is measured. Hydraulic conductivity (K), on the other hand, is a function of the fluid properties of viscosity and density, as well as the properties of the medium. The following expresses its relation to permeability:

$$K = \frac{gk}{\nu} \qquad (9.88)$$

where g is gravitational acceleration and $\nu (= \eta/\rho)$ [m^2s^{-1}] is kinematic viscosity.

QUESTION 9.7

Why would an increase in field air capacity not necessarily show a similar increase in air permeability?

Physical Mechanisms Responsible for Soil-gas Transport

There are three modes of gas transport in the vadose zone that usually operate simultaneously and in parallel (see figure 9.31). Viscous flow (also known as mass, advective, or bulk flow) is caused by a pressure (force per unit area) gradient in a gas mixture that results in mass movement of gas molecules down the pressure gradient. Gas movement by diffusion occurs due to molecular interactions. When a gas is concentrated in one region of a mixture more than another, it is a probabilistic outcome of the random motions of gas molecules that this gas spreads or diffuses into other regions. The molecules of the diffusing gas possess an overall velocity differing from that of the mixture; this is the diffusive flux. Thus, a diffusive flux occurs as a result of a concentration or partial-pressure gradient of individual gases. Generally, viscous and diffusive fluxes are the primary means by which gases move through soil. The third mode of gas transport is surface flow, occurring by nonreactive absorption of gas onto soil surfaces. It is a complex phenomenon, typically of secondary significance (Mason and Milanausksa 1983), and is beyond the scope of this text. The relative significance of the differing modes of transport depends upon the total- and partial-pressure gradients, molar concentration of the gases and their viscosities, and the physical properties of the soil. Small environmental changes can determine the relative significance of transport mechanisms. For instance, Thorstenson and Pollock (1989b) found that a pressure gradient of only 1 Pa m^{-1} in a sandy soil can lead to the development of a viscous flow that exceeds most diffusive flows under most circumstances.

Mechanisms inducing viscous (mass) flow   Viscous flow is facilitated by the presence of a pressure gradient or potential. To equalize pressure, a mass of air travels from a region of higher pressure and density to a lower one. The air flowing from region to region through soil experiences a resistance to movement from the rough surfaces of pore walls; specifically, momentum of the moving gas molecules is lost to the stagnant pore walls. We call this resistance friction. The efficiency of momentum-transfer from the gas to the walls grows with increasing viscosity of the gas, and is much higher if flow is turbulent. Viscous transport within soil pores is ordinarily well-organized, (laminar) because flow velocities are low and soil-pore sizes are small. In contrast, turbulent flow, dominating atmospheric transport, is characterized by random motions of air.

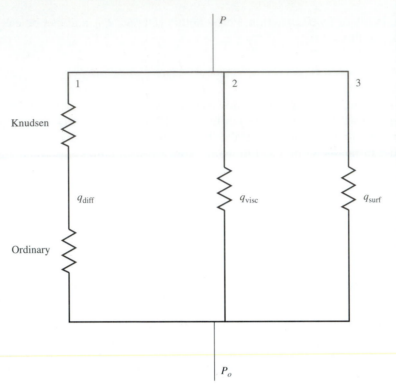

Figure 9.31 Conceptual diagram using an electrical analog to illustrate the relations between fluxes (q) and their associated resistances (jagged symbol) to the transport of gases in soil. Diffusive, viscous, and surface fluxes operate in parallel (independently), responding to the difference ($P_o - P$). Their sum is the total flux. For simplicity, nonequimolar flux is not shown, since it can be a positive or negative contribution to total diffusive flux. If it were included, it would appear in some unique form in branch 1.

1: Diffusive flow branch
2: Viscous flow branch
3: Surface flow branch

The nondimensional Reynolds number provides us with a convenient means for determining whether flow is laminar or turbulent, based on basic properties of the fluid and the medium containing it:

$$Re = \frac{Uk^{1/2}}{\nu} \qquad (9.89)$$

This nondimensional quantity is applicable to any fluid (gas or liquid) in soil, and is comprised of the mean-flow velocity of the fluid (U) [m s^{-1}], gas permeability (k) [m^2], and kinematic viscosity of the gas (ν), which is a function of the molecular properties and density of the gas. Equation 9.89 has been adapted from classical fluid-dynamic studies of pipe flow, where a characteristic length (such as diameter) is used in place of $k^{1/2}$; Ward (1964) shows how $k^{1/2}$ is related to characteristic length. In soil, fluid flows with Reynolds' numbers below about 1 are laminar. They are transitional in the 1–200 range and fully turbulent above 200 (Ward 1964). Equation 9.89 is similar to equation 6.17, except that the latter assumes an equivalence between grain and pore size for the characteristic-length dimension.

Mechanisms initiating mass flow are varied and can be linked to either density (thermal) or pressure-induced changes occurring above or within soil. Differential soil heating creates thermal gradients and consequently, density-driven mass (viscous) flow. However, it only accounts for ≈0.1–0.5 percent of soil-gas exchange (Romell 1922). In contrast, pressure fluctuations associated with large-scale weather patterns that occur with periods from many hours to many days, can penetrate well into soil and can enhance movement of gases significantly, particularly if the unsaturated zone is deep (Buckingham 1904; Nilson et al. 1991).

This mechanism is referred to as "barometric-pumping." The depth of influence and impact on soil-gas exchange of this phenomenon varies with the amplitude of the pressure change at the surface, its period, soil-water content of the unsaturated zone and its depth, and with soil porosity. Over a 10-hour period Weeks (1978) found a change of 20 Pa in soil-air pressure at a 32-m depth in soil due to an atmospheric-pressure change of $\simeq 100$ Pa at the surface. Clements and Wilkening (1974) studied ^{222}Rn efflux in an arid region, and found a 20–60 percent increase in gas-exchange rates with changing synoptic weather patterns that were responsible for a 100–200 Pa surface pressure change in a day. In contrast to deep, dry soils, barometric pressure changes above shallow soils (≤ 15 m above a high-water table) have little effect on total soil-gas exchange (Kimball 1983).

Soil-gas exchanges are also induced by pressure fluctuations of a more-minute scale (on the order of several Pa in soil), that occur as turbulent wind gusts pass over a rough surface. This mechanism is referred to as "wind-pumping." The associated pressure fluctuations have a duration on the order of fractions-of-a-second to minutes. As a consequence of their high frequency and viscous damping in soil pores, they have a much shallower penetration into soil than those related to barometric-pumping. Kimball and Lemon (1971) found that wind-pumping strongly diminishes from the soil-atmosphere interface downward, becoming negligible below several centimeters, even in coarse, sandy soils. Wind-pumping occurs in two parts. Ahead of an approaching wind gust, air pressure is slightly higher than the time-average and some air is forced downward, into the soil. With the gust's passage, pressure decreases and air moves upward, out of the soil. Shaw et al. (1990) document the creation of coherent, well-defined, turbulent structures penetrating from the top of a forest to the forest floor, thereby inducing pressure fluctuations. Baldocchi et al. (1991) found that these pressure fluctuations at the soil's surface measurably increased flux of CO_2 from the forest floor (litter and soil). Although there is typically little net effect on gaseous exchanges in the entire soil-air volume (Romell (1922) estimates about 0.1 percent of soil-gas is exchanged this way), the rate of soil-atmosphere exchanges in the uppermost portion of soils can be affected greatly (Farrell, Greacen, and Gurr 1966; Kimball and Lemon 1971). Wind-pumping can substantially enhance the local rate-of-drying of the top layer of soil (Baldocchi and Meyers 1991), which includes mulches and humus. Kimball and Lemon (1971) report that soil evaporation in a 2-cm layer of coarse sand proceeded at a rate about double that attributable to diffusion alone. With increasing soil-water content, soil-gas exchanges initiated by wind-pumping rapidly diminish.

Wind flow over topography (mountains), or topographic flow, can produce pressure gradients between upwind and downwind exposures that can induce mass flow through the unsaturated zone. Weeks (1993) documented an example of this flow through highly fractured rock, deep within Yucca Mountain (a prospective nuclear-waste repository). Flow was in response to a combination of atmospheric-pressure changes and wind-induced pressure patterns over the mountain.

Pressure-induced viscous flow also occurs with the infiltration of rainfall and snowmelt that initiate mass flow by displacing air-filled pore space ahead of the wetting front. As water moves downward, it draws air from the surface into the soil above the drying front. The amount of soil-gas exchange due to infiltration by water can vary widely due to soil porosity, frequency, and intensity of wetting events. Romell (1922) estimated that overall, about 7–9 percent of soil–atmosphere gas-exchange was attributable to this mechanism.

QUESTION 9.8

Determine whether viscous flow in a fine, sandy soil is laminar or turbulent. We know that $\nu \sim 10^{-6}$ m^2 s^{-1}, $U \sim 10^{-4}$ m s^{-1}, $K \sim 10^{-5}$ m s^{-1}.

Mathematical Description of Gas Transport

Equations describing physical properties and transport of soil gases will be introduced now, including both viscous and diffusive modes. The examination of diffusive transport will begin with Fickian diffusion but later considers non-Fickian modes as well. A model incorporating viscous and diffusive modes of transport is presented last.

Vicous flow and the pressure field Viscous flow responds to pressure potential of air while diffusion occurs in response to concentration gradients or partial pressures of gases in the air mixture. Dynamic pressure originates from varying atmospheric pressure, infiltrating water, a change in water table depth, or pressure changes created by chemical or bio-physical activities within soil. Examples of the latter include biological generation of gas through decomposition, phase changes of water, or, thermal expansion or contraction of soil-air. The flux density of air due to dynamic pressure is:

$$q = \frac{k}{\eta} \nabla P \qquad (9.90)$$

Although we are presently concerned with the movement of air, this equation is equally applicable to the flux of water (eqn. 7.5). Soil permeability to air (k) is a function of soil water content, porosity, and the connectivity and tortuosity of pores (related to soil structure). To solve for q, gas permeability (k) must be determined. Ordinarily this is done under stable, steady state conditions such as in an environmentally controlled chamber. If a liquid is used in the determination of k, Corey (1986) points out that while the intrinsic permeability of soil under saturated conditions should be equivalent to k, many times it may not. Usually k is underestimated if a liquid such as water is used in its determination. It appears that air as a fluid behaves differently in soil than a liquid due to slippage of flow along pore surfaces (the Klinkenberg effect, discussed in more detail later in this section). Also because of interactions between water and soil solids (particularly clayey soils) permeability may be further reduced from what would be representative of air.

Geopotential pressure arises from gravitational pull on an air-mass, much the same as defined for water (eqn. 7.5). The combination of dynamic and geopotential pressures is total pressure (P_{total}):

$$P_{total} = P + \rho g \Delta z \qquad (9.91)$$

where the geopotential terms are density (ρ), gravitational acceleration (g), and height ($\Delta z = z - z_0$) above a chosen reference (z_0).

Geopotential pressure differences increase with increasing depth from a reference or datum. For steady-state, isothermal conditions, pressure change with depth is given by the hydrostatic equation:

$$\frac{dP}{dz} = -\rho g \qquad (9.92)$$

Applying a finite difference form of equation 9.92, pressure at a 1 m depth is found to be 12 Pa higher than at the surface, at sea level ($\rho = 1.21$ kg m^{-3}, $T = 25$ °C). If pressure is measured over a thick, unsaturated layer, the geopotential component to equation 9.90 should include a density that is a function of depth and temperature. The derivation of an expression for static pressure in terms of temperature may begin with the substitution of the ideal gas law

$$\rho = \frac{P}{RT} \qquad (9.93)$$

into equation 9.91; we have, after rearranging terms

$$\frac{dP}{P} = -\frac{g}{RT}\,dz \tag{9.94}$$

The change in temperature with depth in the soil profile may be written:

$$T = T_R - \zeta z \tag{9.95}$$

Substitution of this into equation 9.91, where ζ is the rate of temperature change with depth, and T_R is temperature at the reference point, $z = 0$, yields

$$\frac{dP}{P} = -\frac{g}{R}\frac{dz}{(T_R - \zeta z)} \tag{9.96}$$

Integrating from the reference point (having P_R, T_R) down, we use

$$\int_{P_R}^{P} \frac{dP}{P} = -\frac{g}{R}\int_{T_R}^{z} \frac{dz}{(T_R - \zeta z)} \tag{9.97}$$

to obtain

$$\ln\left(\frac{P}{P_R}\right) = \frac{g}{R\zeta}\ln\frac{(T_R - \zeta z)}{T_R} \tag{9.98}$$

After taking the antilog of both sides, we find

$$P = P_R\left[\frac{(T_R - \zeta z)}{T_R}\right]^{g/R\zeta} \tag{9.99}$$

If the soil is isothermal, equation 9.99 reduces to

$$\frac{dP}{P} = -\frac{g}{R}\frac{dz}{T} \tag{9.100}$$

with the result

$$P = P_R\exp^{-(gz/RT)} \tag{9.101}$$

Using equation 9.99 or 9.101, geopotential pressure can be subtracted from total pressure to arrive at a dynamic pressure which could drive a viscous flux.

Fluid (gas) flow Theoretical development and mathematical description of gas flow in soil is facilitated with simplifying assumptions regarding soil gases. We assume soil gases behave like ideal gases and that the continuity equation, also known as the conservation-of-mass equation, for a gas (concentration, C) applies:

$$\frac{\partial C}{\partial t} = -\left(u\frac{\partial C_x}{\partial x} + v\frac{\partial C_y}{\partial y} + w\frac{\partial C_z}{\partial z}\right) = -C\nabla \cdot \mathbf{U} \tag{9.102}$$

where u, v, and w are unit velocities along the x, y, and z coordinates, respectively, and $\mathbf{U}$ is the total component velocity vector. Equation 9.102 is analogous to that for temperature (section 9.4) and for water (equation 7.18).

Viscous flow The movement of air (or any other fluid in free space) is described by the equations of motion, also known as the Navier–Stokes equations, established in the 19th century. The relevant terms of the mean-flow equations for a viscous fluid in soil are:

$$\rho\left(\frac{\partial u_i}{\partial t} + u_j\frac{\partial u_i}{\partial x_j}\right) = -\frac{\partial P}{\partial x_i} + \rho g + \eta\frac{\partial^2 u_i}{\partial x_j\partial x_j} \tag{9.103}$$

where η is dynamic viscosity ($L^2 t^{-1}$) and subscripts (i and $j, i \neq j$) refer to Cartesian coordinates (x, y, z). The left side of the equation contains the inertia terms that describe the rate of change in momentum of the moving air. This includes the local-time rate of change in fluid motion and the advection of fluid, respectively. The first two terms on the right side are force terms that drive momentum, consisting of the dynamic-pressure gradient and gravimetric- pressure potential. The last group on the right side holds the viscous terms through which momentum is lost to internal friction. In soils, the inertial terms are dropped, since velocities are very low. Under steady-state conditions and hydrostatic equilibrium, equation 9.103 reduces to

$$\frac{\partial P}{\partial x_i} - \rho g = \eta \frac{\partial^2 u_i}{\partial x_j \partial x_j} \tag{9.104}$$

Assuming density is a constant, the force terms on the left side of equation 9.104 can be combined (for details, see Corey 1977), and referred to as total (or piezometric) pressure. As mentioned earlier, the contribution of the hydrostatic term is only effective in the vertical dimension, and is very small over short vertical distances. Solutions of equation 9.104 are given in Corey (1977).

We generally consider pore space in soils to be a three-dimensional maze of interconnected channels. Conceptually, we can think of the tortuous flow-path through pores as having an effective (or actual) length (l_e), which corresponds to a straight-line distance (l) (see figure 9.32). The square of the ratio of the effective length of the pathway traveled by air to the straight-line distance traveled (l) is the tortuosity (details in Chapter 7):

$$\tau = \left[\frac{l}{l_e}\right]^2 \tag{9.105}$$

Bear (1972) notes that $\tau < 1$ and is typically about $(0.4)^{1/2}$, or about $2/\pi$, with a usual range of 0.5 to 0.8. We expect τ to decrease as liquid soil-water content increases. Tortuosity reduces

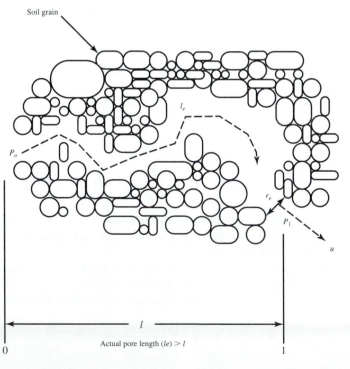

Soil grain

Figure 9.32   Conceptual diagram of tortuosity and its effect on gas movement in dry soil. Flow velocity (u) through the pore of radius (r_e) is in response to a pressure differential ($P_0 - P_1$) between points of reference, (0) and (1). Actual pore length (l_e) is longer than the straight-line distance (l) between the two reference points.

Actual pore length (le) $> l$

mean-mass flow velocity through soil as shown by Corey's (1977) derivation of the mean velocity for a fluid passing through soil:

$$u_i = -\frac{r_e^2}{s_k \tau \eta} \frac{\partial P}{\partial x_i}$$

(9.106)

where u_i is velocity of flow in the direction of a unit vector ($\mathbf{i}$) through a channel of effective radius (r_e), and s_k is a nondimensional geometrical-shape factor. The average (or effective) hydraulic radius of the channel carrying the air is defined as the internal volume of the channel divided by its internal surface area. Note, too, that flow velocity decreases with increasing viscosity.

Fickian diffusion The focus of our discussion now turns to gaseous diffusion—an omnipresent phenomenon, regardless of the mechanism that forces gas transport. Adolf Fick, following work on diffusion by Thomas Graham in the late 1820s, first formulated a law of diffusion in 1855, establishing that the rate of diffusion of a solute in any direction is proportional to the concentration gradient in that direction. Similar to gradients in potential that drive fluid flow (as described by Darcy's law), and analogous to Fourier's law of heat conduction, Fick's law defines a gaseous flux (q) driven by a concentration gradient:

$$q = -D_{ij}(x, y, z)\,\nabla C$$

(9.107)

where C is gas concentration (in density units, when a mass flux is desired). D_{ij} [m^2 s^{-1}] is the binary gas diffusion coefficient (for gas (i) in gas (j)), that ordinarily varies in both horizontal and vertical directions due to variable soil composition, presence of water, and soil compaction. Since, in most cases, vertical transfer of gas is dominant, equation 9.107 reduces to

$$q = -D_{ij}\frac{\partial C}{\partial z}$$

(9.108)

where D_{ij} is assumed constant. The binary diffusion coefficient controls diffusive flow in a system composed of two gases. Diffusive flow of a gas in one direction (due to a partial pressure gradient) must be balanced by the flow of the other gas in the opposite direction in response to the development of a pressure potential, thereby reattaining an equilibrium. The value of D_{ij} has to be determined experimentally; it varies with temperature and pressure as described below

$$D_{ij}(T, P) = D_{ij}(T_o, P_o)\left(\frac{T}{T_o}\right)^n\left(\frac{P_o}{P}\right)$$

(9.109)

where subscript (o) refers to standard conditions. Values of n vary, with $n = 1.823$ for nonpolar gas pairs (Bird, Stewart, and Lightfoot 1960). D_{ij} in soils is affected by soil porosity and soil moisture. It may be estimated from measured binary diffusivity in free air, D_a (see table 9.6 for selected gases):

$$D_{ij} = \tau_f D_a$$

(9.110)

The gas tortuosity factor (τ_f) is specific to a soil with known porosity, volumetric air (θ_a) and water content. An often-used relation expressing τ_f in terms of θ_a and porosity (ϕ) was derived by Millington and Quirk (1961):

$$\tau_f = \frac{\theta_a^{10/3}}{\phi^2}$$

(9.111)

A few other models of τ_f can be found in Jury, Letey, and Collins 1991.

We emphasize the distinction between τ and τ_f. Tortuosity is typically applied to viscous-flow problems, since fluid velocity through a soil column decreases as τ increases. While it can include some effects of tortuosity, τ_f is usually determined empirically from diffusion experiments where a number of transport processes can be simultaneously involved in gas transport.

Fickian–diffusive gas transport Fick's law describes a flux due to diffusion of a gaseous constituent, in response to a partial-pressure or concentration-gradient. If we require conservation of mass in a soil volume, equation 9.108 can be inserted into equation 9.102 to obtain an equation that describes gas transport due to diffusion (Fick's second law):

$$\alpha \frac{\partial C}{\partial t} - D_{ij} \frac{\partial^2 C}{\partial z^2} - S(z, t) = 0 \tag{9.112}$$

where, again, D_{ij} is assumed constant, α is a volumetric-air constant, and a source/sink term was added for completeness. D_{ij} can be calculated, under steady-state, homogeneous conditions, if gas flux and the concentration profile is known. Tindall, Petrusak, and McMahon (1995) discuss laboratory measurements of soil-gas flux and concentration gradients from confined soil columns, from which D_{ij} is determined.

QUESTION 9.9

Calculate the concentration profile of O_2 through the rhizosphere (the layer of soil containing roots) from the surface ($z = 0$) to maximum rooting depth ($z = \delta$). Assume that there is no source or sink of O_2 below the roots, that there is a steady rate of consumption of O_2 in the rhizosphere with depth, and that O_2 concentration at the surface remains constant.

Diffusion involving point and line sources (or sinks) Thus far, only planar transport has been considered, in which diffusion has been expressed in Cartesian coordinates. In natural settings, gaseous diffusion often involves point or line sources and sinks. In these instances, the problems are better-posed in polar coordinates. Roots, with associated microbes, can be considered line sources of CO_2 and sinks for O_2. The governing equation for diffusion to (or from) roots is described in cylindrical coordinates:

$$\alpha \frac{\partial C}{\partial t} = \frac{1}{r} \left[\frac{\partial}{\partial r} \left(r D_{ij} \frac{\partial C}{\partial r} \right) + \frac{\partial}{\partial \theta} \left(\frac{D_{ij}}{r} \frac{\partial C}{\partial \theta} \right) + \frac{\partial}{\partial z} \left(r D_{ij} \frac{\partial C}{\partial z} \right) \right] \tag{9.113}$$

where the coordinates are radius (r), length (z), and arc (θ). Cylindrical coordinates are related to Cartesian coordinates in the following way: $x = r \cos \theta, y = r \sin \theta, z = z$, for a vertically oriented cylinder.

Diffusion to (or from) a point or sphere (perhaps a soil aggregate or microsite), expressed in spherical coordinates, is

$$\alpha \frac{\partial C}{\partial t} = \frac{1}{r^2} \left[\frac{\partial}{\partial r} \left(r^2 D_{ij} \frac{\partial C}{\partial r} \right) + \frac{1}{\sin \theta} \frac{\partial}{\partial \theta} \left(D_{ij} \sin \theta \frac{\partial C}{\partial \theta} \right) + \frac{\partial}{\partial \theta} \left(\frac{D_{ij}}{\sin^2 \theta} \frac{\partial^2 C}{\partial \phi^2} \right) \right] \tag{9.114}$$

where ϕ is azimuth angle. Expressions relating spherical coordinates to Cartesian coordinates are: $x = r \sin \theta \cos \phi, y = r \sin \theta \sin \phi, z = r \cos \theta$.

Analytical solutions and suggestions for numerical methods to solve equations 9.113 and 9.114 are given in Carslaw and Jaeger (1959). Although this reference concerns itself with heat flow, the solutions for heat and gas transport are analogous, other than a change of coefficients.

Non-Fickian diffusion: non-equimolar and Knudsen Historically, the Fickian model has been assumed to represent diffusion problems in soil adequately. However, Thorstenson and Pollock (1989a, b), Baehr and Bruell (1990), Abu-El-Sha'r and Abriola (1997) have shown that this model can grossly underestimate fluxes. Baehr and Bruell (1990) document that the underestimates can be of significant practical importance when modeling transport of hydrocarbon vapors. Fickian diffusion considers only the movement of gas in response to the partial-pressure potential driving it, and the resistance to this movement by intermolecular collisions of gas molecules. Actually, there are four different types of diffusion (including Fickian), that can simultaneously occur (at differing levels of significance) in soil:

1. Ordinary molecular
2. Knudsen
3. Nonequimolar
4. Surface flow

Ordinary diffusion, most closely represented by Fickian models, refers to the movement of molecules from a region of higher, to lower concentration. Resistance to diffusion comes from momentum loss due to intermolecular collisions and, to a lesser extent, losses to collisions with pore walls. In this instance, pore dimensions are much larger than the mean free path of travel—the distance a gas molecule travels before colliding with another gas molecule. The molecular free path decreases with increasing gas density, as the probability of a gas molecule colliding with another molecule (within a given travel distance) increases.

If, in contrast to the conditions above, the molecular free path of diffusing gas molecules was equal to or greater than the dimensions of pores, Knudsen diffusion would occur. In this instance the only collisions are those of gas molecules and pore walls.

Nonequimolar diffusion occurs when gases of differing molecular weight diffuse into one another. A lighter molecular-weight gas will diffuse into a heavier one at a higher rate than in the reverse process (see Mason and Malinauskas (1983) for details). Gas flux resulting from non-equimolar diffusion can cause a compensating viscous flow to develop that can move gas in the opposite direction to the diffusive flux. To explain this event conceptually, consider a two-bulb experiment involving two initially pure gases under isothermal and isobaric (constant pressure) conditions (see figure 9.33). Gas in bulb A has a lighter molecular weight and higher thermal energy than that in bulb B. Consequently, the lighter gas in bulb A diffuses into bulb B faster than the gas in bulb B diffuses into bulb A, causing an initially higher total pressure to build in bulb B. This pressure differential leads to a viscous flow from bulb B to bulb A. Following the diagram, for the system to reach equilibrium, the piston shown in the diagram must move toward bulb A.

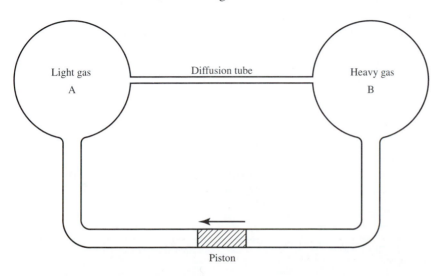

Figure 9.33 A two-bulb experiment illustrating nonequimolar flux due to differential diffusion of a lighter gas into a heavier one. Lighter gas flux into bulb B results in a higher pressure in the gas mixture in bulb B. Viscous flow attempts to equilibrate pressures, moving the piston toward bulb A. Reproduced from Thorstenson and Pollock 1989b, published by the American Geophysical Union

Diffusion through pores having sizes of the same order of magnitude as the molecular free path may include ordinary, Knudsen, and non-equimolar diffusion. Under the conditions of this transitional regime and with the presence of a bulk-pressure gradient, viscous flow occurs, as is expected. However, Klinkenberg (1941), found that measured flux is greater than that predicted by permeability considerations alone, due to viscous slip of gas through pore space. Now known as the "Klinkenberg effect," this property of flow is used to determine Knudsen-diffusion coefficients, needed to predict diffusive flow (Thorstenson and Pollock, 1989a).

The fourth type of diffusion—surface-flow—refers to the non-reactive absorbence of gas molecules onto solid-soil surfaces and their subsequent transport down the sorbed phase concentration gradient (Cunningham and Williams, 1980). It is distinctly different from the other forms of diffusion, therefore it is listed as another mode of transport. Ordinarily, it is of secondary importance with respect to other modes of transport, and is beyond the scope of this chapter.

The dusty-gas model for combined gas transport   The dusty-gas model, first described by Mason, Malinauskas, and Evans (1967), combines viscous and diffusive gas-transport mechanisms. Conceptually, the model considers the solid components of soil to be giant dust molecules. Consequently, one can think of the model as applying to a gas mixture that includes dust particles as a component of the mixture. The governing equations for transport of gas component (i) in a mixture of (l) number of gases are:

$$\sum_{j=1;\,j\neq i,\,p}^{l} \frac{X_i F_j - X_j Q_i}{D_{ij}} + \frac{Q_i}{D_i^k} = \frac{\nabla P_i}{RT} + \frac{k P_i \nabla P}{D_i^k \eta RT} \tag{9.115}$$

where X_i is the mole fraction of gas (i) having a partial pressure P_i, Q_i is its molar flux (mol $L^{-2}\,t^{-1}$), D_i^k is its Knudsen-diffusion coefficient, P is air pressure of the mixture, k is permeability (L^2) of the media to air, and η is the dynamic viscosity of the gas mixture ($mL^{-1}\,t^{-1}$ = Pa s). Dynamic viscosity of common gases are given in table 9.6 and binary-diffusion coefficients for common gas pair mixtures are listed in table 9.7. The first term in equations 9.115 represents molecular transfer among gas species (classical Fickian-type diffusion). By not summing over p (particles) in the first term, the system of equations are independent of one another. Proceeding through the equation, the second term represents Knudsen diffusion; the third term (first one right of the "equals" sign) is the partial-pressure contribution to diffusion; and the last term is the viscous-flow contribution to flux. Details of the model are given in Mason and Malinauskas (1983) and Cunningham and Williams (1980).

TABLE 9.6   Dynamic Viscosities of Some Common Soil Gases

Gas	Viscosity (Pa s) $\times 10^{-5}$ $T = 293$ K
N_2	1.75
O_2	2.03
H_2O	0.95
CO_2	1.46
CH_4	1.09
N_2O	1.47
Ar	2.21
H_2	0.87
He	1.94

Source: Data from CRC Handbook of Chemistry and Physics (1972), Chemical Rubber Co., Cleveland, OH.

TABLE 9.7   Binary Diffusion Coefficients for Some Common Gas Mixtures

Gas mixture (i, j)	D_{ij} (m^2 s^{-1}) $\times 10^{-4}$ $T = 293$ K
N_2—O_2	0.2083
N_2—CO_2	0.1649
N_2—CH_4	0.2137
N_2—Ar	0.1954
O_2—CO_2	0.1635
O_2—CH_4	0.2263
O_2—Ar	0.1928
CO_2—CH_4	0.1705
CO_2—Ar	0.1525
Ar—CH_4	0.2045

Source: Data from Thorstenson and Pollock, 1989a.

Thorstenson and Pollock (1989a, b) examined gas transport using a form of equation 9.115. They considered an example of CH_4 flux, constant with depth (0–10 m) and time, as ideally can occur in the soil overlying a landfill. They also assumed that the primary gases (N_2, O_2) were stagnant—that is, they were not being generated or consumed in the soil column. Soil-gas concentration gradients for the numerical experiment depict strong non-linearity with depth (see figure 9.34). Referring to the figure and with only Fickian-diffusion in mind, we can expect to find a downward flux of N_2, and no CH_4 flux below $\simeq 5$ m. However, an interesting story is told by examining components of total flux (figure 9.35), computed from equation 9.115. Ordinary or Fickian-type diffusive flux of CH_4 is upward. Because CH_4 is much lighter than N_2 and O_2, its non-equimolar flux becomes a large upward flux, nearly one-third that of ordinary diffusion. The flux due to Knudsen diffusion alone was not determined, but relative to ordinary diffusion, it is believed to be small. Thus, the combined total diffusive-flux is upward. Viscous flux of CH_4 is also upward, becoming an increasing component of total flux with increasing depth. For example, at 10 m it is responsible for all the flux, while at 1 m it is comparable in size with diffusive flux. Turning now to N_2, we find a diffusive flux of N_2 into the soil. However, without a sink for N_2 in the soil, why should a downward flux occur? On closer inspection, we find that net flux of N_2 is indeed zero (figure 9.35), but only because diffusive flux into the soil is precisely balanced by viscous and non-equimolar flux out of the soil. Thorstenson and Pollock (1989b) note that in this example, viscous flux accounts for 65–90 percent of total CH_4 flux, driven by a pressure gradient of only 24 Pa m^{-1}.

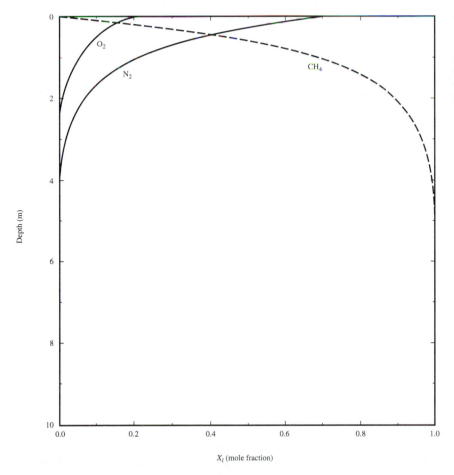

Figure 9.34 Hypothetical gas concentration profiles in soil; concentration is expressed as a molar fraction (X_i). These profiles were used in a dusty-gas model simulation of soil-gas flux. Reproduced from Thorstenson and Pollock, 1989a, copyright by the American Geophysical Union

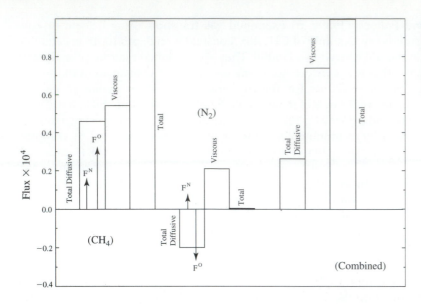

Figure 9.35   Relative magnitudes and directions of CH_4, N_2, and total combined gas fluxes. F^N is non-equimolar flux and F^O is ordinary diffusive flux. Reproduced from Thorstenson and Pollock, 1989a, copyright by the American Geophysical Union

The extent to which pressure fluctuations in soil derived from atmospheric weather affect soil-gas transfer in the dusty-gas model was examined by Massman and Farrier (1992). They note that atmospheric-pressure fluctuations on the order of 200 Pa in 24 hours can occur several times a year in mid-latitudes. Due to the intrusion of "fresh air" from the surface, a viscous flow is established with subsequent, coupled diffusive fluxes. This results in substantial vertical and horizontal (if soil permeability varies horizontally) fluxes to depths of several meters in deep, unsaturated soils. Of practical note to investigators measuring soil gases with probes, Massman and Farrier (1992) found enhanced movement of gas in the vicinity of probes was due, perhaps, to disturbance of soil structure.

9.8   COUPLED TRANSPORT OF WATER, HEAT, AND WATER VAPOR

A substantial amount of soil heat can be transferred by phase changes of a water-substance (liquid or gas) through vapor's movement in soil. Redistribution of heat affects thermal gradients, which, in turn, affect vapor flow and phase changes in water elsewhere in the soil. Evidence of vapor flow and phase changes have been reported by a number of researchers. Depicted in figure 9.36 are fluxes measured at two different soil depths in a soil three days after irrigation. Soil-moisture flux is always positive (upward) at the 10-mm level, but is occasionally downward at 90 mm. Negative values at 90 mm between 8–12 hours indicate condensation of vapor from upper levels and the concomitant downward transport of latent heat energy from the soil above. Latent-heat energy can also be transported by vapor that moves from evaporating to freezing-soil areas. Helping to drive vapor movement is the lower-saturation vapor pressure over ice versus that over water. By way of illustration, figure 9.37 depicts the drying of soil below ≈ 25 cm and increasing water content in frozen soil above, at the expense of the deeper layer.

Flux of water vapor, considering only Fickian-diffusion, can be written as

$$F_v = -D_v(\theta)\, \nabla \rho_v \qquad (9.116)$$

where $D_v(\theta)$ is molecular diffusivity of water vapor in soil. At first glance, we expect soil moisture to respond to a temperature gradient by moving from warmer to cooler areas. The basis of this conclusion is that, at saturation, ρ_v increases with temperature. We expect a

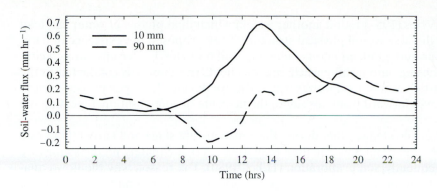

Figure 9.36   Soil-water flux measured at 10-mm and 90-mm depths, three days after irrigation (data from Jackson et al. 1973). Upward flux is positive.

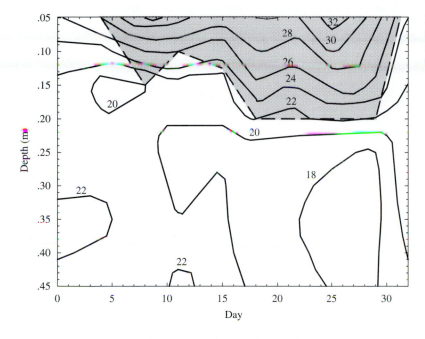

Figure 9.37   Soil-water content (percent volumetric) in a cold soil, with an intermittently frozen surface (data from Geiger 1965). Frozen soil is shaded.

vapor-pressure gradient in the direction from cool (low saturated vapor density) to warmer (high saturated vapor density) areas in soil and, therefore, for a flux (equation 9.116) of water to move down this gradient, from high (warm) to cool (low).

   With the presence of both liquid and vapor, the net-overall transport of water in a multi-phase problem becomes more complex. Gravitational effects aside (for the moment), due to the temperature dependence of matric potential, the net flow of liquid water is from warm to cool areas. However, citing the work of others using closed-horizontal soil columns, Marshall and Holmes (1979) point out that in the presence of liquid-vapor phase changes, just the opposite occurs; that is, net flow of water will be from cool to warm. The explanation is that, initially, vapor flux is in the direction of the temperature gradient (from warm to cool), resulting in the condensation of water in the cool area. Next, the increasing liquid-water content that results from the condensation of vapor flux leads to the development of a water-content gradient in the opposite direction to that just created by the thermal gradient. Thus, the matric-potential gradient from liquid water becomes greater than that caused by the thermal gradient. As a result of the difference in potentials, water moves from the cooler to warmer areas of soil in this instance.

Philip and deVries (1957) found another way in which dual phases of water can lead to enhanced-water flux due to soil physical structure. They proposed the existence of a migrating, sharp temperature gradient across individual soil particles. Under unsaturated conditions, water condensing onto a soil particle releases heat. This energy is conducted to the other side of the particle where it vaporizes water. Responding to the vapor-pressure gradient, water vapor moves forward from this site to the next site of condensation, repeating the scenario.

It is generally acknowledged that during the constant-rate stage and early falling-rate stages of drying, the isothermal flow equation (equation 9.54) adequately predicts moisture transport and, consequently, soil evaporation (Hillel, 1980). The reason why the flow equation holds for other than constant-rate stage drying (under isothermal conditions), is that the nonlinear form of the vapor-diffusivity equation is similar to that for liquid water (Hillel 1980). However, as the falling-rate stage proceeds, near-surface soils dry and temperature gradients become larger. Simultaneous equations for water and heat are needed to analyze soil-moisture transport under these conditions. Philip and deVries (1957) proposed the following equations for soil fluxes of heat (G_c) and moisture (Q_w):

$$Q_w = -\rho_w\left(D_T\frac{dT}{dz} - D\frac{d\theta}{dz} - K(z)\right) \tag{9.117a}$$

$$G_c = -\kappa_c\frac{dT}{dz} - \rho_w L_v D_v\frac{d\theta}{dz} \tag{9.117b}$$

where D is total-moisture diffusivity (liquid plus vapor); G_c is a combination of kinetic- (sensible) and latent-heat fluxes; $K(z)$ is the hydraulic conductivity in the vertical direction and D_T [m^2 s^{-1} °C^{-1}] is a diffusivity for thermally driven moisture flux. D_T is comprised of two components, individually related to the movement of liquid water and water vapor. D_v is similarly comprised of components related to liquid- and gas-phase movements. We can identify that the last two terms in equation 9.117a are derived from Richard's equation for unsaturated flow (see chapter 8). Governing equations based on equations 9.117 were formulated by Milly (1982; 1984) and written into a numerical model used by Scanlon and Milly (1994). Model simulations for combined water vapor, water, and heat transport in shallow desert soils were found to predict field conditions closely. Success was attributable to the robustness of thermal calculations. Main sources of uncertainty in simulations centered on estimated hydraulic conductivity.

9.9   MULTI-PHASE TRANSPORT OF VOLATILE COMPOUNDS IN SOIL

Volatile compounds can also exist in liquid and vapor phases in soil. Additionally, some solutes may be adsorbed onto surfaces of solids. Thus, concentration of a solute in soil may exist as a combination of absorbed solute (C_s), dissolved solute (C_l), or as a vapor (C_g). These concentrations may be expressed on a per-mass basis by multiplying by bulk density (ρ_b), volumetric water content (θ), or volumetric air content (θ_a), yielding $\rho_b C_s$, θC_l, and $\theta_a C_g$, respectively. The governing equation for multiphase transport of the solute within a specific volume is

$$\frac{\partial}{\partial t}(\rho_b C_s + \theta C_l + \theta_a C_g) - \frac{\partial}{\partial z}\left(D_g\frac{\partial C_g}{\partial z}\right) + \frac{\partial}{\partial z}\left(D_e\frac{\partial C_l}{\partial z}\right) - \frac{\partial Q_w C_l}{\partial z} - S(z,t) \tag{9.118a}$$

where the effective dispersion coefficient (D_e) is the combination of the liquid diffusion coefficient and the hydrodynamic dispersion coefficient (viz.; $D_e = D_l + D_{lh}$), D_g is diffusivity of the vapor phase, and Q_w is the liquid water flux. The term on the left describes the local time rate of change in total solute concentration due to the transport and production/decay terms

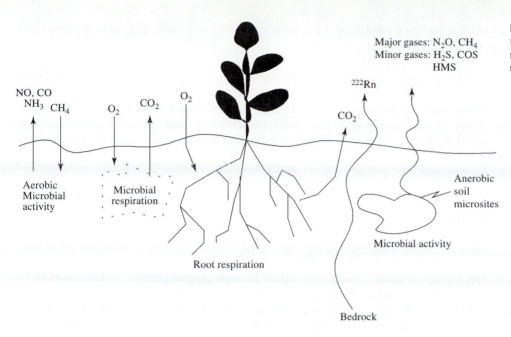

Figure 9.38 Schematic of trace gas fluxes in unsaturated soil and associated sources and sinks

Major gases: N₂O, CH₄
Minor gases: H₂S, COS
HMS

TABLE 9.8　Nondimensional Henry's Law and Distribution Coefficients of Selected Volatile Pesticides and Organic Compounds (at 20 °C)

Compound	K_H	K_S [cm³ g⁻¹]
Atrazine	2.5×10^{-7}	1.6
Benzene	1.77×10^{-1}	
Bromacil	3.7×10^{-8}	0.7
Carbon disulfide	5.82×10^{-1}	
Chloromethane	$3.9 \times 10^{1\,*}$	
Cis-1, 2-Dichloroethane	3.7×10^{-1}	
DBCP	8.3×10^{-3}	1.3
DDT	2.0×10^{-3}	2400
Lindane	1.3×10^{-4}	13
MTBE (methyl tert-butyl ether)	1.69×10^{2}	
Napthalene	1.74×10^{2}	
Chloroform (trichloromethane)	1.2×10^{-1}	
Toluene	2.28×10^{-1}	
Phorate	3.1×10^{-4}	6.6

Source: Data from Jury, Gardner, and Gardner (1991), Howard (1991), Mackay and Shui (1981), and Robbins (1993).
*This value determined at 25 °C

on the right: Fickian diffusive transport of the vapor phase (first term on the right), the convective—dispersive transport in the liquid phase (second term), mass liquid water flux, and the production/decay term (S), which is spatially and temporally variant.

To utilize equation 9.118a, a relation is needed reducing the number of unknowns (C_s, C_l, and C_g) to one. An approach used by Jury, Gardner, and Gardner is to describe the left side of 9.118a in terms of C_l. Applying Henry's law and the Ideal Gas law, a phase-partitioning law is derived:

$$C_g = k_H \frac{C_l}{\rho_w R_v T} = K_H C_l \qquad (9.118b)$$

where k_H is Henry's law constant, and K_H, its nondimensional form. The sorbed phase concentration may be expressed as

$$C_s = K_S C_l \tag{9.118c}$$

where K_S (known as the distribution coefficient) is the slope of the relation between the adsorbed phase concentration and the dissolved concentration, at multi-phase equilibrium. Values of K_S and K_H for several volatile pesticide and organic compounds are given in table 9.8.

9.10   COMPOSITION OF SOIL-AIR

Soils vary widely in their composition and structure, largely the result of their parent material, vegetation, hydrology, and climate. The composition of soil gases is also largely determined by soil's mineral composition, vegetation, hydrology, and climate, but resident flora and fauna and exchanges with the atmosphere affect it, too. A variety of important trace gases are either generated or consumed in the unsaturated zone (figure 9.38). Typical concentrations of the primary gaseous components of soil-air are given in table 9.9. The abundance of N_2 in soil is due to equilibration of soil-air with the atmosphere.

Some soil-gas concentrations differ from atmospheric levels. Concentrations can vary both horizontally and vertically within the soil, as determined by the distribution of their sources and sinks. Since the composition of soil-air is influenced by biological activity, concentrations vary temporally in response to changes in soil temperature, water content, nutrient availability, and soil aeration. O_2 and CO_2 are good examples of contrasts in gas concentrations between the atmosphere and within soils. O_2 is generated primarily by photosynthesis above-ground, while a large quantity of this gas is consumed by below-ground respiring soil organisms. As a result, in most ecosystems its concentration is higher in the atmosphere than in the soil. Just the opposite is true for CO_2, which is respired by roots and soil organisms but consumed in photosynthesis.

TABLE 9.9   Constituents of Clean Air, and Soil

Constituent	Atmosphere (percent volume or as indicated)	Conditions regarding soil environment	Soil-air (percent volume or as indicated)
Nitrogen	78.08		78
Oxygen	20.95	Variable due to respiration	19
Argon	0.93		0.93
Carbon Dioxide[G]	350 ppm$_v$	Variable due to respiration	400–10,000 ppm$_v$
Water vapor[G]	0–3	Usually near saturation	0–3
Methane[G]	1.8 ppm$_v$	Variable due to soil source/sinks	0–20 ppm$_v$
Nitrous oxide[G]	320 ppb$_v$	Variable due to soil source/sinks	0.3–134 ppm$_v$
Hydrogen	0.5 ppm$_v$	Variable due to sources (bedrock, biological)	0–0.5 ppm$_v$
Radon	0.1–1 pC L^{-1}	Range is function of bedrock composition (granitic is highest)	100–100,000 pC L^{-1}

[G]Indicates a greenhouse gas.
Source: Data compiled from Williamson (1972); Denmead (1991), Lal et al. (1995), Otton Gunderson, and Schumann (1993).

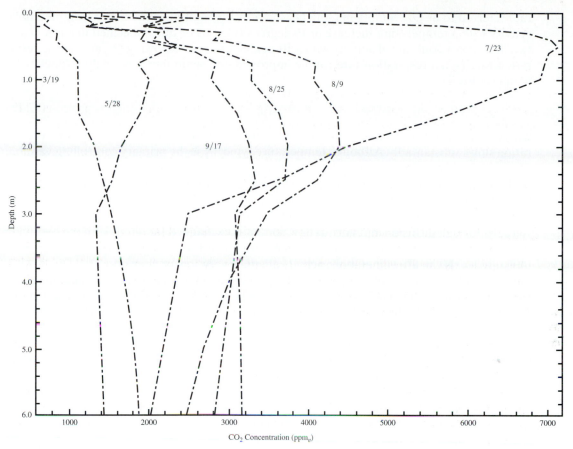

Figure 9.39 CO_2 concentration profiles measured in sandy soil beneath a boreal jack pine forest in central Canada. Month and day given alongside individual profiles (data from Wickland and Striegl, 1997)

Due to its agricultural and climatological significance, CO_2 is the most commonly studied of soil gases. CO_2 is generated by plant roots, fauna, and soil microbes (bacteria and fungi) as a result of respiration. This process involves the consumption of carbohydrates and oxygen, and the release of combustion energy, along with CO_2 and water vapor as by-products. Energy acquired through respiration is utilized to conduct metabolic activity, some of which is related to plant growth and decomposition. Both fauna and microbes are involved with decomposition. The former mechanically reduce organic structures, while microbes in both aerobic and anaerobic soils reduce complex organic-carbon molecules to simple carbohydrates. In turn, these carbohydrates can be utilized to attain energy through respiration.

It is difficult to determine the relative significance of plant-root and soil-microbial contributions to soil CO_2. Field studies by Bowden et al. (1993) indicate that for a temperate hardwood forest, roots contribute about one-third of the total respired flux of CO_2. Near-surface soil microbes, largely consuming litter-fall, also generate about one-third, as do deep-soil microbes. Studies of other ecosystems indicate a greater contribution (up to 50 percent) by roots (Behera, Joshi, and Pati 1990; Nakane et al. 1983).

Responding to daily and seasonal temperature- and moisture-patterns in soils, the rates of decomposition by soil microbes and respiration by plant roots and microbes change at a rate referred to as Q_{10}. Simply put, the Q_{10} factor is the multiplicative increase in metabolic activity with a temperature increase of 10 degrees (Celsius or other specified unit). A Q_{10} of two indicates a doubling of activity with a 10-degree increase. Knowing Q_{10} and a base respiration rate (r_b), a respiration rate can be approximated, assuming it is solely a function of temperature:

$$r = r_b Q_{10}^{(T-b)/10} \tag{9.119}$$

where b is an arbitrarily chosen base-temperature—typically, a temperature at which respiration slows substantially. Although temperature typically is the primary controlling variable of r, soil moisture, pH, and nutrient availability can be limiting factors at times. The influence of these latter factors can lead to erroneous predictions with the use of equation 9.119.

With increasing rates of metabolic activity, consumption of O_2 increases while production of CO_2 also increases, such that soil concentrations reach thousands-of-ppm CO_2. Largely through diffusion and viscous flow, soil-air is exchanged (aerated) with the overlying atmosphere, where the concentrations of CO_2 are much lower, and that of O_2 higher. Soil-moisture levels can affect metabolic activity at either very-low or near-saturated levels. Also, high soil-water content reduces gaseous transport or can block it entirely. Magnusson (1992) found occasions in thawing boreal forest soils in which a saturated soil layer (perched above frozen soil), capped a layer of active respiration, thereby preventing exchange with the atmosphere. On these occasions, CO_2 concentrations reached their highest levels of the year, while those of O_2 were the lowest.

Observations of soil CO_2 concentrations in boreal and temperate ecosystems reveal seasonal trends. Referring to the boreal forest example shown in figure 9.39 where the soil is a deep, coarse sand, we note a wide range of CO_2 concentrations at a depth near 1 m. CO_2 concentrations at this depth were highest in July, when the soil was moist and its warmest. Concentrations decrease below 1 m, since the majority of roots and microbes responsible for respiration are above this level. Below 3 m, CO_2 concentrations are the lowest of observed levels on May 28, likely the result of the penetration of the seasonal temperature wave. Temperatures at 3 m were the coldest of the entire profile in early June (see figure 9.20). In contrast to this soil, larger CO_2 concentrations are often found in agricultural soils (Rosenberg, Blad, and Verma 1983) or other settings where soils are rich in organic carbon and fixed nitrogen, that support greater rates of biological activity.

Other gases produced by biological activity in soils that have a substantial influence on the atmosphere's radiation balance (through the greenhouse effect) are nitrous oxide (N_2O), nitric oxide (NO) and methane (CH_4). Additionally, biological activity in soils may produce smaller amounts of other gases (see figure 9.38), including NH_3 (ammonia), CO (carbon monoxide), H_2S (hydrogen sulfide), COS (carbon monosulfide) and DMS (dimethyl sulfide) (Conrad 1995). All of these gases (CH_4, N_2O, H_2S, COS, DMS) are typically produced under anoxic conditions, which generally occur in saturated soils. In some instances, substantial amounts of certain of these gases, such as N_2O, have been measured from generally unsaturated soils (Rosswall 1989). It appears that saturated soil aggregates (microsites) within the drier soil media provide the needed anoxic environment for gas production.

N_2O and NO are released during nitrification and denitrification processes associated with reduction and oxidation reactions involving nitrate-reducing bacteria and, on occasion, by fungi in acidic forest soils (Rosswall, 1989). Nitrification is a major source of N_2O production in soils following the application of urea- or ammonium-based fertilizers. Mineral-N (nitrogen) availability increases with the application of N-based fertilizers or in the absence of plant N uptake. With the occasion of plant root die-off, the large increase in organic C availability intensifies denitrification, leading to increased N_2O production. In general, N_2O and NO production are highest under conditions of intermediate soil water content, high

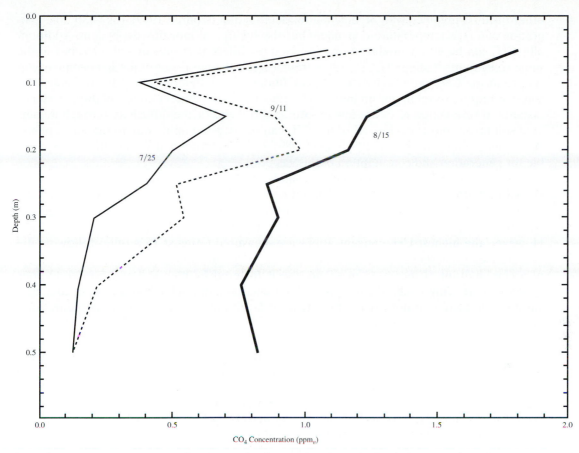

Figure 9.40　CH_4 concentration profiles measured at the same location as for data in figure 9.39 (data from Wickland and Striegl, 1997)

mineral-N availability, and warm temperatures (Rosswall 1989). Coincident soil characteristics are high fertility (making sufficient N available) and low O_2 levels, thereby resulting in anaerobic microsites. Generation and transport of NO and N_2O are largely controlled by soil texture. For instance, anaerobic metabolism resulting in NO and N_2O by-products is fostered by numerous dense-soil aggregates while gas-transport depends largely on diffusion through soil. Anaerobic conditions are more prevalent in saturated soils, but since diffusion of gas is more than 3 to 4 orders of magnitude slower in water than it is in air, there is an optimum soil-water content that facilitates anaerobic processes within saturated microsites while simultaneously facilitating gaseous transport through soil. Bouwman et al. (1993) present a general algorithm and model, simulating N_2O production for a variety of soils. Disturbed soils such as those put to agricultural uses exhibit strong seasonal emissions of N_2O following tillage, as well as strong consumption of CH_4 as a consequence of mechanical aeration (Mosier et al. 1997; Powlson et al. 1997).

　　CH_4 and CO_2 are produced by methanogenic bacteria under strictly anoxic conditions where organic matter (e.g., cellulose, hexose) is degraded into gaseous by-products

$$C_6H_{12}O_6 \rightarrow 3CO_2 + 3CH_4 \tag{9.120}$$

Equation 9.120 is a simplification of a number of reactions. Methanogenic bacteria are not solely responsible for the above reaction, and are the recipients of by-products from three other bacteria involved with organic-matter degradation (Conrad 1989). Like CO_2, the generation of CH_4 is temperature-dependent but also highly soil-moisture dependent. Additionally, CH_4 can be either produced or consumed by different groups of soil microbes in the same soil column. Generally, CH_4 efflux from a soil is the net result of methanogenic bacteria producing CH_4 under strictly anoxic conditions and methanotrophic bacteria in aerobic soil that require oxygen, consuming it. CH_4 efflux is also dependent upon soil diffusivity. For instance, if production rates are low in saturated soil and/or the diffusivity through an aerated soil-layer above it is low, all of the CH_4 can be oxidized on its way to the atmosphere. Accounts given in Andreae and Schimel (1989) detail soil microbial activity that is responsible for generation and consumption of this gas.

Profiles of CH_4 concentrations in a sandy soil beneath a boreal Jack pine forest are shown in figure 9.40. This soil was well-drained and the water-table depth was more than 10 m below the surface. Ambient air CH_4 concentrations were about 1.8 ppm$_v$. Soil concentrations are typically lower than that of air, indicating that the soil is an overall sink for CH_4. However, the highest CH_4 concentrations are measured (August 15 profile) following the occurrence of several rain events. These high CH_4 concentrations could have been the result of a combination of increased methanogenesis fostered by high soil-moisture conditions (possibly harboring anaerobic microsites) and decreased methane consumption, commensurate with decreased diffusivity of O_2 associated with increased soil-water content. Volumetric soil moisture averaged 5 percent July 15–25, 10 percent on Aug. 5–15 and 4 percent Sept. 1–11.

The soil air concentration of nitrogen (N_2) closely follows that of the atmosphere, since there are no major sources of production nor sinks of the gas in either realm; the same can be said of the noble gases: Neon, Argon, Krypton and Xenon (Ozima and Podosek 1983). Helium and radon are noble gases as well, but in areas underlain by uranium laden bedrock, their concentrations will be substantially higher in soil. Produced in some instances by anaerobic metabolism hydrogen concentrations in soil can exceed that of the atmosphere.

Water-vapor concentration varies (particularly in the upper soil layers) on daily to weekly time-scales due to variability in temperature and infiltrating water. Vapor concentrations are typically higher in soils than in the overlying atmosphere for the same location. This is because soils generally contain water (at some depth) and resistance to diffusive transport of vapor is large in contrast to transport in the atmosphere above the surface. Relative humidities in soils are seldom below 90 percent, even in arid regions. As a result, vapor density or concentration is strongly determined by soil temperature.

Naturally occurring radioisotopes Transport of radon (^{222}Rn) gas in soils—and from soils into buildings—has become a subject of practical importance to homeowners, building contractors, and to workers exposed to air in enclosed, poorly ventilated, subterranean environments. Radon is the product of successive disintegrations of uranium (^{238}U), and follows radium (^{226}Ra) in the decay sequence. Its occurrence and abundance is directly linked to area geology. Rocks having a relatively high uranium content include those that are either: of volcanic origin; of granitic composition; sedimentary rocks containing phosphates; or metamorphic rocks derived from rock types in this list. Radon decomposes into a series of unstable isotopes of polonium, bismuth, and lead; these are known as "radon daughters." Disintegration ends with the stable lead isotope (^{206}Pb). During the decay, a series of alpha and beta particles are emitted, sometimes coincident with gamma radiation. Radon gas contributes only about 5 percent of the combined alpha emissions from radon and radon 'daughters.'

Radon and its decay products present a public-health risk when inhaled, since its radioactive decay results in the release of alpha particles into lung tissue. The radioactive half-life (time for a substance to lose one-half of its radioactivity) of ^{222}Rn is 3.8 days. Typically, then, it is exhaled before large amounts of alpha particles are emitted. However, some radon-

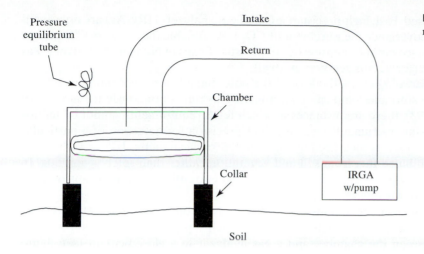

Pressure
equilibrium
tube

Intake

Return

Chamber

Collar

Soil

IRGA
w/pump

Figure 9.41   Schematic of chamber method of measurement of atmsophere–soil gas fluxes

daughter products posing substantial health risks have much shorter half-lifes. These atoms can also become embedded into dust, and after inhalation, accumulate in lung tissue. Radon daughter product, polonium, is that most linked to lung cancer (Miller and Dalzell, 1982).

The migration of radon from parent material occurs mostly by diffusion through rock fractures and soil pores in overlying soil. In addition to parent bedrock, radon can be transported from areas with decaying radium in the soil. In many regions, radium is found in concentrations on the order of picograms per gram of dry soil (Clements and Wilkening 1974). The flux of radon to the atmosphere varies by region from 0.1 to 2.5 atoms $cm^2 s^{-1}$, and averages 0.75 (Wilkening, Clements, and Stanley 1974). Only about 10–50 percent of the radon generated in subsurface soil and rock ever reaches the surface. A substantial amount of radon is driven into soil-mineral grains or water-filled pore space due to the recoil of particles when radium decays. Radon transport in dry soil can be as much as 3 m within its half-life, but only about 0.02 m in wet soil over the same interval. Radon transport may be enhanced by barometric-pumping by the atmosphere (Kraner, Schroeder, and Evans 1964).

Buildings that are situated on dry and highly permeable soils, fractured bedrock, hill-sides, bottoms of canyons, or on coarse glacial debris will have high-indoor radon levels unless precautions are taken during construction. Radon moves into buildings because of pressure differences that enable transport by mass flow and diffusion between soil and the interior of these buildings. Pathways for radon to enter a building are through unfinished basements or crawl-spaces, cracks in walls or floors and, around the subterranean entrances of water, sewer and electrical lines (Nazeroff et al. 1987). The simplest preventative measure against radon's intrusion into buildings is to back-fill loose stone and gravel during construction of concrete footings, foundation walls, and floors poured over disturbed soils. Back-fill tends to be much more permeable than the surrounding soil and rock, thereby facilitating gas transport around the foundation. Sometimes, ventilated porous tubing is embedded in the fill to enhance movement of gas away from footings and foundations (Otton, Gunderson, and Schumann 1993).

9.11   MEASURING SOIL-GAS FLUX

Soil–Atmosphere Exchanges

There are two categories of approaches used to measure gas flux between the soil and atmosphere: micrometeorological approaches, and surface chambers. We examined micrometeorological approaches in section 9.6, where means of measuring evapotranspiration and heat flux where discussed. Fast, high-precision infrared gas analyzers (IRGAs) are now available that allow eddy-covariance measurement of CO_2 flux. Tunable diode lasers (TDLs) are also being used in eddy-covariance approaches to measure fluxes of N_2O and CH_4 over areas where fluxes are large (for details, see Verma, 1990).

While micrometeorological methods offer the potential advantage of sampling a large area with negligible disturbance, and can be automated to run continuously for months to years (Wofsy et al. 1993), there are instances in which fetch requirements cannot be met or fast-response gas-sensing instruments lack sufficient precision for detection of a particular gas flux. Additionally, CO_2 flux measured by micrometeorological methods over viable vegetation provides a combined flux from soil and vegetation rather than solely a soil flux. In these situations, surface gas-flux chambers offer the only currently available means of measuring soil—atmosphere gas flux.

Two basic types of chamber systems are used: closed and open. A closed system is comprised of a single closed-end cylinder (the chamber), which is placed on the soil surface. Air can be circulated between the chamber and a gas analyzer in a closed-circuit (see figure 9.41), or be extracted from the chamber with syringes for measurement off-site. Gas flux is determined by first measuring the rate of change of concentration with time; flux can then be calculated from the following equation

$$q_i = \left(\frac{V}{A}\right)\frac{d\rho_i}{dt} = h\frac{d\rho_i}{dt} \qquad (9.121)$$

where ρ_i is the density of gas (i), V is volume of the air space enclosed by the chamber, A is the enclosed soil area, and h is the mean height of the top of the chamber above the soil's surface. Gas flux is expressed in moles (or mass) per unit area per unit time. The key to accurate flux measurement is to keep sampling duration as short as possible and to minimize soil disturbance when conducting the measurement. It is particularly important to strive for minimal disturbance if the gas has a significant near-surface source or sink, as is the case for CO_2. The closed-chamber method is intended for diffusive flux measurement only. It is criticized for creating an artificial environment in which atmospheric-turbulent transport is eliminated, and for allowing gas concentrations to increase above ambient levels. Increasing above soil concentrations diminishes the natural concentration gradient and can lead to flux underestimates. Methods to estimate flux missed by employing chambers are given in Nakayama (1990) and Healy et al. (1996). A variety of other effects can introduce bias errors in either the positive or the negative. Using CO_2 flux measurement as an example, placing a closed-chamber on the surface forces atmospheric air into the soil, which can decrease measured flux by reducing the near-surface CO_2 gradient. On the other hand, forcing the chamber wall into the soil compacts it, possibly enhancing flux. Also, forcing chamber walls into the soil can provide pathways for enhanced exchange, increasing flux. To help avoid some problems related to the effects of soil disturbance, a collar (see figure 9.41) may be placed into the soil's surface some time before measurements are to be made; a chamber may then be attached to the collar at the time of measurement.

Open-chamber systems have the advantage of continuous measurement with minimal soil disturbance, preservation of the natural air-soil concentration gradient, and allowing (for the most part) naturally occurring atmospheric-turbulent exchange. Flux is determined by measuring the gas concentration entering (ρ_{in}) and leaving (ρ_{out}) the chamber, and accurately measuring the flow rate (v) through the chamber. Applying the following formula, a flux is calculated:

$$q_i = \frac{v}{A}[\rho_{in} - \rho_{out}] \qquad (9.122)$$

Accurate measurements are made by keeping the concentration difference small by adjusting v and accurately measuring v. Open-chamber systems are much more difficult to operate than closed chambers however, and critics argue that pressure fluctuations associated with air flow through them can enhance flux estimates.

In the field, chamber measurements offer a comparatively low-cost means of measuring flux. Often, if the flux or the gas concentration is small, they offer the only means of measuring flux currently available. Beyond the disadvantages mentioned above, sampling density and frequency are often highly variable with site characteristics (physical and biological), potentially introducing some bias into measurements.

Gas Flux within Soil

It is very difficult to accurately determine gas flux from one region to another within soil. A typical approach would be to determine diffusivity of the specific gas in the soil and make gas concentration and water content measurements along the path of interest. Since gas permeability is much lower through water-filled than dry pores, permeability used in the determination of diffusivity must be carefully adjusted for a known soil moisture level. Using Fick's law (equation 9.107), we may attempt to estimate flux from:

$$q_i = -D_{ij}\left(\frac{\partial C}{\partial x}\right) + S(x, t) \tag{9.123}$$

Beyond soil measurement of gas concentrations and determination of diffusivity (D_{ij}), knowledge is needed of the spatial and temporal character of the source/sink term, $S(x, t)$. Often, investigators refer to laboratory measurements of D_{ij} and $S(x, t)$ in moisture and temperature controlled environments. Once a model has been generated from laboratory data, field measurements of soil moisture, temperature and other parameters relevant to the source or sink are taken to predict D_{ij} and $S(x, t)$. To obtain soil gas concentrations, many investigators insert a small, stainless steel tube (about 2 mm diameter) with a sealed insertion tip into soil. Small inlet ports are manufactured near the tip or sealed-end through which soil-air may be drawn for analysis. It is important to keep the extracted sample volumes as small as possible to avoid significant alteration of the natural concentration profile of the gas. Field application of this method is increasingly difficult with depth in drier, low porosity soils since tube insertion is difficult and new gas flow paths, particularly along the outside of the tube, are created.

SUMMARY

We have shown that the movement of heat, water vapor, and other gases through the soil, and between soil and the atmosphere is an important influence on life dependent on the soil for sustenance (both directly and indirectly), as well as an important influence on the world's climate. Incoming radiation was shown to influence not just surface heating of soil, but the heating of the entire soil column. Additionally, we showed that incoming radiation received at the Earth's surface differs, according to a variety of factors: season; latitude; slope and aspect of the surface; and vegetative cover. This energy is partitioned into heating the soil as well as the air above it, and evaporating water (at the surface, within soil, and within plants). Methods of quantifying these fluxes of energy and mass were examined, including models and instrumentation.

The composition of soil air was reviewed, as well as the importance of atmospheric exchanges and biological activity in determining its make-up. The generation of radon and its transport were also discussed as were the mechanisms responsible for the movement of soil gases and exchanges with the atmosphere. Key mathematical models used to quantify soil-gas transport were reviewed as well, and field methods to measure them were briefly described. Although—due to its complexity—coupled-transport of gases, heat, and water (liquid and vapor) is given only brief mention in this chapter, it is important to realize that understanding this phenomenon is the next step essential for fully understanding heat and gas transport in the unsaturated zone.

ANSWERS TO QUESTIONS*

9.1. Referring to figure 9.1, we find that the saturation vapor pressure over water is greater than that over ice. Water migrates down the potential gradient to the region of lower pressure (the cold end).

9.2. Using the definition of rH (see equation 9.7), $rH = e/e_s$ and the relation between e_s and temperature, we see that an increase in temperature leads to an increase in e_s, which decreases rH if e is constant.

9.3. At 20 °C, $I_{20} = (0.949)(5.67 \times 10^{-8} \text{ W m}^{-2} \text{ K}^{-4})(293 \text{ K})^4 = 396.6 \text{ W m}^{-2}$. $I_{22} = 407.5 \text{ W m}^{-2}$. The net difference is a 10.9 W m^{-2} increase.

9.4. Total energy received by the surface is 146 (100 from incoming longwave +46 absorbed short-wave). The surface loses 31 percent (of incoming solar energy) to sensible- and latent-heat fluxes. Therefore, 146 − 31 percent = 115 percent leftover energy to radiate upward.

9.5. First, let us determine the damping depth of the diurnal temperature wave. Referring to figure 9.18, we find the following:

Depth (m)	T max (°C)	T min (°C)
0.05	20.0	10.4
0.37	15.2	13.8

Eqn. 9.45 can be applied to find maximum and minimum temperature with depth.

$$T_{max}(z) = T_a + A_o \exp\left(-\frac{z}{Z_D}\right); \quad \sin\left(\omega t - \frac{z}{Z_D}\right) \rightarrow 1$$

$$T_{min}(z) = T_a - A_o \exp\left(-\frac{z}{Z_D}\right); \quad \sin\left(\omega t - \frac{z}{Z_D}\right) \rightarrow -1$$

At depths z_1 and z_2:

$$[T_{max}(z_1) - T_{min}(z_1)] = \Delta T(z_1) = 2A_o \exp\left(-\frac{z_1}{Z_D}\right)$$

$$\Delta T(z_2) = 2A_o \exp\left(-\frac{z_2}{Z_D}\right)$$

We can eliminate A_o and solve for Z_D using the ratio

$$\frac{\Delta T(z_1)}{\Delta T(z_2)} = \frac{\exp\left(-\frac{z_1}{Z_D}\right)}{\exp\left(-\frac{z_2}{Z_D}\right)} = \exp\left(\frac{z_2 - z_1}{Z_D}\right)$$

*Before answering these questions, the reader may wish to refer to the conversion factors in appendix 3.

and

$$Z_D = \frac{z_2 - z_1}{\ln\left[\dfrac{\Delta T(z_1)}{\Delta T(z_2)}\right]}$$

Plugging in values to calculate damping depth:

$$Z_D = \frac{0.37 - 0.05 \text{ m}}{\ln(9.6/1.4)} = 0.166 \text{ m}$$

We can solve for thermal diffusivity now that Z_D is known

$$Z_D = \left[\frac{2D_h}{\omega}\right]^{1/2}$$

For the diurnal damping depth ($\tau = 1$ day):

$$D_h = \frac{\pi Z_D^2}{\tau} = 1.0 \times 10^{-6}\,\text{m}^2\,\text{s}^{-1}$$

9.6 Taking the derivative of equation 9.56, we have

$$E = \frac{dE_{\text{cum}}}{dt} = (\theta_i - \theta_f)\left(\frac{\overline{D}}{\pi t}\right)^{1/2}$$

where E is the evaporative flux.

In the second part of this question, we are to find $\overline{D}$ given $\theta_o = 0.35$ and $E_{\text{cum}}/t^{1/2} = 2.5$ cm/day$^{1/2}$. Following the accompanying text we rearrange equation 9.56 and set $\theta_a = 0$:

$$\text{slope} = \frac{E_{\text{cum}}}{t^{1/2}} = 2\theta_o\left[\frac{\overline{D}}{\pi}\right]^{1/2}$$

$$\overline{D} = \pi\left[\frac{\text{Slope}}{2\theta_o}\right]^2 = \left(\frac{40.1 \text{ cm}^2}{\text{day}}\, \frac{1\text{m}^2}{10000 \text{ cm}^2}\right)\left(\frac{1 \text{ day}}{86400 \text{ s}}\right) = 4.6 \times 10^{-8}\,\text{m}^2\,\text{s}^{-1}$$

9.7 We are given U, v, and K (hydraulic conductivity). To determine whether a flow is turbulent or not we will use the Reynolds number. To do so we need k, which can be solved from equation 9.88:

$$k = \frac{Kv}{g} = \frac{1 \times 10^{-5}\, 1 \times 10^{-6}}{9.8} = 1 \times 10^{-12}\,\text{m}^2$$

Now using equation 9.89:

$$Re = \frac{(1 \times 10^{-4}\,\text{m s}^{-1})\,(1 \times 10^{-12}\,\text{m}^2)^{1/2}}{1 \times 10^{-6}} = 1 \times 10^{-4}$$

Since $Re \ll 1$, this flow is laminar.

9.8 Field-air capacity is a measure of the volume of air in soil (including entrapped air), while permeability only relates to the connectivity of air-filled pores.

9.9 We are given that, below rooting depth δ the flux of O_2 (q_{O_2}) = 0, that steady-state conditions exist ($dC/dt = 0$), the sink for O_2 is constant in time and depth ($S(z, t) = S$), and that the concentration at the surface (C_o) is constant. From equation 9.112, we have

$$D_{ij}\frac{d^2C}{dz^2} = \frac{dq_{O_2}}{dz} = -\overline{S}$$

Applying boundary conditions and integrating:

$$\frac{d^2C}{dz^2} = \frac{dq_{O_2}}{dz} = -\overline{S}$$

$$\int_0^{F_{O_2}} dq_{O_2} = -\overline{S}\int_{+\delta}^z dz$$

Solving, we find:

$$q_{O_2} = -\overline{S}(z - \delta)$$

Substituting for q_{O_2}:

$$-D_{ij} \frac{dC}{dz} = -\overline{S}(z - \delta)$$

Rearranging:

$$dC = -\frac{\overline{S}}{D_{ij}}(z - \delta)\, dz$$

and after integration

$$\int_{C_o}^{C} dC = \frac{\overline{S}}{D_{ij}} \int_{o}^{z} (z - \delta)\, dz$$

we find the profile equation

$$C - C_o = \frac{\overline{S}}{D_{ij}} \left(\frac{z^2}{2} - \delta z \right)$$

ADDITIONAL QUESTIONS

9.10. Confirm that a water potential measured as -2000 J kg^{-1} converts to -2.0 MPa.

9.11. Calculate the heat content of 1 kg of wet ($\theta = 0.4$), sandy soil at 20 °C.

9.12. Calculate the relative humidity in a soil where $\psi = -4000$ J kg^{-1}, $T = 10$ °C.

9.13. (a) Find the soil-heat flux at the surface for a soil having the following composition: 40 percent mineral solids, 15 percent organic solids, and 20 percent volumetric water content. A heat flux plate at a 5-cm depth sensed a heat flux of -30 W m^{-2} and thermocouples within the upper 5 cm of soil were used to calculate an average temperature increase of 2 °C in a hour.

    (b) What would be the heat flux if half of the soil moisture was ice? (Assume the ice is randomly distributed.)

9.14. (a) Find the total evaporation (over the course of a day) from a soil column with initial volumetric water content 15 percent and surface volumetric content of 5 percent. Assume a sandy soil (from question 9.2, use a mean value of $D = 0.265$ cm^2 min^{-1}), and a constant soil temperature.

    (b) Assuming that the only energy available for evaporation is derived from net radiation at the surface, calculate the average 24-hour net radiation needed.

9.15. (a) Convert 350 ppm$_v$ CO$_2$ to mg m^{-3} and mmol m^{-3} units when $T = 20$ °C and P $= 900$ mb.

    (b) Compare this result with that calculated for a -10 °C temperature.

9.16. (a) Steady-state soil respiration generates CO$_2$ uniformly from the surface ($z = 0$) to rooting depth ($z = \delta = 1$ m). Assuming that atmospheric concentration at the surface ($C_o = 350$ ppm$_v$) is constant and that there is no sink or source of CO$_2$ below the roots, derive the CO$_2$ profile from $z = 0$ to δ.

    (b) Assuming that the parameters of respiration rate equation are: $Q_{10} = 3$, $b = 15$, $r_b = 7$ $\mu \cdot$ mol m^{-3} s^{-1}, calculate soil CO$_2$ concentrations in ppmv at the 0.5-m depth when soil temperature is 20 °C, and also when it is 10 °C. Assume that soil temperature is uniformly distributed and is constant, and that the soil-air is a mixture of CO$_2$ and N$_2$. Use the Millington-Quirk method to find an equivalent diffusivity in soil with $a = 0.35$ and $\phi = 0.55$.

9.17. Calculate CO$_2$ flux from the CO$_2$ concentration measurements obtained from a closed-chamber (cylindrical shape: radius $= 15$ cm, height $= 20$ cm) placed on the soil surface. Initial concentration is 360 ppm$_v$; average ambient temperature is 20 °C and pressure is 90 kPa. Calculate flux in μmol m^{-2} s^{-1} and mg m^{-2} s^{-1} units. The IRGA used in the concentration measurements reported the amount of time between 1 ppm$_v$ increases in CO$_2$ concentration. Here are the times between the 1 ppm$_v$ increases, beginning immediately after chamber placement (in seconds): 1.0, 0.9, 1.3, 0.8, 1.6, 2.4, 3.4, 1.5, 3.0, 4.2.

10

Contaminant Transport

INTRODUCTION

Numerous environmental management problems involve the transport and reactions of dissolved chemicals that are either native to the soil, added deliberately to the soil surface, or are accidentally spilled. The design of optimum application rates and application timing of fertilizer, domestic waste (sewage sludge), wastewater irrigation, herbicide, and low-level radioactive waste (disposal) depends on methodologies that maximize the degradation or retention of these chemicals within the unsaturated zone, while minimizing their mobility. Increased public awareness of the contamination of ground water by agricultural, industrial, and municipal chemicals (Pye, Patrick, and Quarles 1983) has focused considerable attention on solute transport, creating a heightened awareness and increased research in this area (Van Genuchten and Jury, 1987).

Transport of chemicals through the unsaturated zone depends on many factors: ion exclusion; ion exchange; volatilization; dissolution and precipitation; chemical and biological transformation; biodegradation; adsorption; diffusion; dispersion and volumetric water content; unsaturated hydraulic conductivity; and the matric potential of the medium (see figure 10.1). The investigation of transport through the unsaturated zone evolved primarily within the domain of soil science (i.e., soil physics), but, because of its importance and its complexity, has enlarged to encompass the fields of hydrology; hydrogeology; geochemistry, agronomy, agricultural engineering, geology, and environmental science. Due to its diversity, research has expanded to include: areas of mathematical approaches to solve flow and transport equations; theoretical investigations concerning homogeneous media through laboratory studies; the effects of preferential pathways and macropores on flow; fractal flow of solutes; field experiments with inherent heterogeneity; laboratory experiments focusing on exchange chromatography; radiation-scanning tomography; and the influence of biological, hydrodynamic, and geochemical processes during unsaturated flow conditions. This is not a complete list, but serves to illustrate the broad scope of research currently being conducted in the area of contaminant transport within unsaturated soils.

10.1 PHYSICAL PROCESSES AND MOVEMENT OF SOLUTES

Many texts treat the unsaturated zone as if it were in static equilibrium. Because soil is a dynamic system, however, it is seldom in equilibrium. This is because conditions continually change due to physical, chemical, and atmospheric influences or disturbances. For

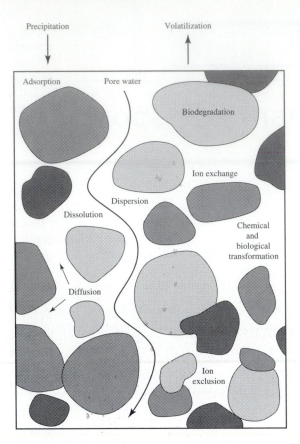

Precipitation

Volatilization

Figure 10.1 Schematic showing various processes
that occur in the unsaturated zone

convenience in modeling, flux, and calculations of transport, the assumption of equilibrium conditions allows easier determination of these parameters. Disturbances that take place within the unsaturated zone normally will quickly dissipate as a system moves toward equilibrium. Earlier, we discussed the potential and thermodynamics of water within the unsaturated zone, the chemical properties of that water, and unsaturated water flow (chapters 4, 5, and 8). This section briefly outlines the physical processes that concern the concepts of water and solute flow in the unsaturated zone.

The flow of a solute in soils involves both movement and accumulation. Solute movement may be described as the change of position of water (solute) over a given time within the unsaturated-soil matrix. The accumulation of a solute usually refers to the change in mass of solute at a given volume of the matrix for a specific time period. This is usually caused by recharge events such as rainfall or irrigation, spills, and intentional dumping. However, the volume at a specific location can decrease by evaporation, drainage, advective flow, and/or plant uptake. The driving force for advective flow is the hydraulic gradient. During equilibrium conditions no fluid movement takes place because the potential within the soil matrix is the same at each point (i.e., hydraulic heads are static), but concentrations must also be in equilibrium to prevent diffusive flow of mass. Thus, the sum of static forces, $\Sigma F^s = 0$. As equilibrium changes during disturbance or other influencing factors, the sum of static forces is not zero, and we can say that this becomes the driving force for advective transport. To give more detail, Newton's second law of motion describes force, F, as

$$F = ma \qquad (10.1)$$

where F is force in N (kg m s^{-2}), m is mass in kg, and a is acceleration in m s^{-2}. When more than one force is acting on the fluid, they can be added in the proper order so that the sum of all forces acting on the fluid simultaneously is the vector sum, or resultant force ΣF. If the vector sum is zero, acceleration of the fluid is zero and the body is at mechanical equilibrium. The sum of forces can be separated into two types of forces: drag forces (dynamic), F^d, and static forces, F^s, such that $\Sigma F = \Sigma F^s + \Sigma F^d$. Static forces are always present, but drag forces occur only when static forces cause a body to move. These are the reaction forces associated with movement and may be written as $\Sigma F^s = -\Sigma F^d$. The conditions for static equilibrium can also be expressed in terms of potential, ψ_h (called hydraulic potential). This is expressed mathematically as

$$\frac{\Sigma F^s}{m} = -\frac{\partial \psi_h}{\partial z} \tag{10.2}$$

where ψ_h is the sum of both the pressure potential, ψ_p, and gravitational potential, ψ_g (see chapter 4), and z is the vertical direction; the minus sign indicates that movement is taking place from higher to lower potential. The driving force (hydraulic gradient) can also be written in terms of force per-unit volume as $= -\partial p_h/\partial z$ or per-unit weight $= -\partial H/\partial z$ where ψ_h is in J kg^{-1}, p_h is in J m^{-3} = Pa, and H is in J N^{-1} = m.

The rate of water movement through the unsaturated zone is typically referred to as a flux density, q. Often, the term "flux" (the volume of water divided by time) will be used when "flux density" is actually meant. Flux density is the amount of water (solute) passing through a perpendicular plane of unit area in direction z during a specific time interval. This volume or mass is divided by the area of the plane and the magnitude of the time interval—that is, the mass of solute/(area * time). The flux density is normally expressed on a volume basis, but is also expressed on a mass and weight basis. This becomes more apparent later in this chapter, in the discussion of the continuity equation.

We have mentioned both water flux and solute flux. Water flux is aptly described by Darcy's law (discussed in chapter 7); solute flux entails the movement of solutes (or chemicals) of various kinds with water. To obtain the solute flux, we multiply the water flux or flux density q by the dissolved concentration C of solute in solution, expressed as mass of solute per volume of soil solution. While this gives the general case qC (hereafter referred to as J), it is only an approximation because q is averaged over many soil pores, and does not represent actual water-flow paths that must meander around individual soil particles and various air pathways. There is an extra motion involved, described as hydrodynamic dispersion, D_H, which must be added to J to describe the motion of the solute relative to the average motion, so that $J + D_H$ = total flux. This is a simplified equation for solute flux. A more-detailed discussion regarding hydrodynamic dispersion and other parameters that influence solute flux is given in the following pages.

10.2   TYPES OF FLUID FLOW

When fluid flow takes place within the unsaturated zone, we consider only the water or solute within the zone of interest. When investigating the flow of infiltrating water, there are always two fluids present within the system: the fluid introduced and the fluid that is already present. The fluid present will almost always have a different ion concentration and activity level than the introduced fluid. Thus, we can think in terms of dispersion and mixing of fluids. Generally, there are two types of flow that are possible when two or more fluids occupy the same space within the soil matrix; these are miscible and immiscible flow, commonly referred to as "miscible displacement" and "immiscible displacement." In the case of miscible displacement, the fluids are completely soluble in each other, indicating that interfacial tension between the fluids is zero and there is no distinct fluid–fluid interface; this is the case with hydrodynamic dispersion. For immiscible displacement, the fluids do not mix and there is both interfacial tension and a distinct interface between the fluids, indicating a pore- or capillary-pressure difference across the fluid–fluid interface.

For unsaturated-zone transport studies, the case of miscible displacement is the most common. As an example, if we investigate the hydraulic parameters of a field site, it is normal procedure to apply a tracer (usually conservative, such as a bromide) in recharge water. The infiltrating recharge water miscibly displaces the soil water already present. The two fluids initially can be separated by an abrupt interface, but due to hydrodynamic dispersion and diffusion, this interface immediately transforms into a transition zone between the two fluids. In comparison to the entire matrix domain of consideration, the transition zone is relatively small.

The most common occurrence of immiscible displacement in the unsaturated zone is when air fills the void space not occupied by water. This is a special case of the simultaneous flow of two immiscible fluids, with air being the non-wetting fluid. Various types of two-fluid flow occur in the engineering field. Examples of these are the flow of oil, water, and gas in oil reservoirs during production and secondary recovery operations (immiscible displacement), and injection of solvents during the secondary recovery process (miscible displacement).

10.3 BREAKTHROUGH CURVES, PISTON FLOW, AND HYDRODYNAMIC DISPERSION

A breakthrough curve is a graphical representation (or plot) of outflow concentration versus time or cumulative water drainage during an experiment. Much of the literature dealing with solute transport in the unsaturated zone reports the use of breakthrough curves in great detail—that is, mass-transfer studies in sorbing porous media (Van Genuchten and Wierenga 1976), miscible displacement in soils (Nielsen and Biggar 1962), and many others. The breakthrough curve indicates the relative tracer distribution of the effluent, with respect to the column or area of the soil matrix under consideration as it relates to either pore volume, time, or both. It is also a very useful way to illustrate the physical meaning of the advection–dispersion equation in one-dimensional form, as well as for comparing results from a model to data collected in the field or laboratory—that is, a plot of the modeled values versus experimental data. Examination of a breakthrough curve can indicate how aggregated the soil is, the presence of macropores or preferential flow paths, or presence of adsorption sites. A typical breakthrough curve is shown in figure 10.2. For an ideal medium, a C/C_0 would reach 0.5 at $V/V_0 = 1$. However, this rarely happens in normal conditions due to the effects of mechanical dispersion and molecular diffusion which cause spreading of the curve. Because of these effects, the tracer begins to appear in the effluent at the outflow end of the column at time t_1 (initial breakthrough), before the arrival of water traveling at velocity t_2 (average velocity; see figure 10.2).

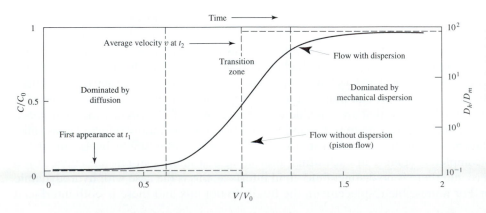

Figure 10.2   The longitudinal dispersion of a typical breakthrough curve, illustrating the effects of molecular diffusion and mechanical dispersion, velocities, and other parameters based upon both time and pore volumes of fluid passing through the column. t_1 represents the time of first appearance of the solute, and t_2 represents the time at which the solute would reach the midpoint at average velocity, discounting anion exclusion (i.e., by piston flow).

Additionally, the drier the soil, the greater the volume of effluent required for $C/C_0 = 1$ and the more tailing that occurs, but the sooner the effluent would break through. This happens due to discontinuity of the number of macropores as water content decreases (see figure 10.3). The overall number of macropores decreases, making less pore-space available for transport as water films around the particles decrease in thickness and length of continuous flow. As long as macropores are present (and because they fill fastest) effluent appears at an earlier breakthrough time and significant tailing occurs due to diffusion into areas of immobile (stagnant) water or intra-aggregates. Initially, these stagnant areas are bypassed, but as time proceeds, diffusion into them increases.

Piston Flow

Piston flow refers to total displacement of the original solution with the incoming solute or tracer, without mixing. This is a direct result of the "piston" (sharp wetting front) displacing the total amount of solution in the column (see figure 10.2). In a strict sense, piston flow is a special case of immiscible displacement in which the solute moves into an area and displaces not by mixing, but by pushing out the original solution and replacing it. This happens due to the properties of the medium and of the incoming and original solutions. These properties include temperature, viscosity, solubility, concentration, and other chemical and physical parameters. Since no mixing takes place during piston flow, the solute pushes out the original concentration, dependent on the total amount of incoming solute. As the total amount of incoming solute increases, more of the original solution is displaced; this is an additive effect. Also, because no mixing occurs, the displacement relies on advective velocity and no diffusion takes place. An example of this is the displacement of water by air, or oil by water. Since both components or liquids are immiscible and advective velocity overcomes any effects due to diffusion, we can expect a very steep front on the breakthrough curve during piston flow. However, since both the radius of the soil pores and the diameter of individual soil particles are not constant—and changes occur in the hydrodynamic dispersion, water content, and chemical-diffusion coefficient—it is highly unlikely that pure piston flow occurs under typical immiscible displacement or any other soil condition.

Hydrodynamic Dispersion

Hydrodynamic dispersion is a non-steady, irreversible process (i.e., the initial tracer distribution cannot be obtained by reversing the flow) in which the tracer mass mixes with the

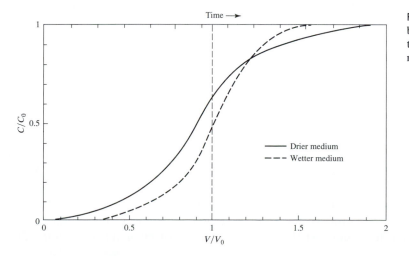

Figure 10.3 The shift of a breakthrough curve based on the water content of the medium

nonlabeled portion of the solution. This is due to the presence of flow through a complicated system of pathways within the soil matrix. Hydrodynamic dispersion consists of two parts: **(1)** mechanical dispersion (sometimes referred to as convection) and **(2)** molecular diffusion. These two conditions are usually artificially separated, but in actuality are totally inseparable in form because both occur together; the dependence of each on different parameters can vary due to changes in physical and chemical conditions, however. As an example: mechanical dispersion is more prominent at high-water content and greater-flow velocities because it is here that contaminant particles mix more freely with water within soil pores, as that water meanders around individual particles. Molecular diffusion predominates at low-water content and low-flow velocity since at this stage, chemical phenomena associated with a tracer or contaminant continue, even though mechanical dispersion due to water movement has ceased. Molecular diffusion alone takes place at the molecular level, in the absence of motion (both in a soil and/or solution); dispersion occurs at the pore level. Hydrodynamic dispersion is also generally associated with early breakthrough of the contaminant.

The causes of hydrodynamic dispersion are: **(1)** the range in pore size causes solutes to arrive at the end of a soil column (used for an example) at different times; **(2)** transverse diffusion into pores (especially stagnant areas) of some of the solute, while direct flow through other pores causes solutes to arrive at different times; and **(3)** molecular (chemical) diffusion ahead of the wetting front as it varies with time.

Soil structure affects hydrodynamic dispersion and the resultant shape of the breakthrough curve in numerous ways. As particle size increases or the soil is aggregated, the graphic representation of the breakthrough curve has more tailing due to diffusion into stagnant regions, but the effluent appears at the end of the column sooner due to larger pore size. However, to get $C/C_0 = 1$, a larger volume of effluent would be required, due to diffusion into both the stagnant areas and into the tortuous path of large aggregates. Smaller-particle sizes yield a more even distribution in pore size, which means there is less stagnant water, so the effluent appears at a later time, and requires less volume for $C/C_0 = 1$ than for a soil with large aggregates (see figure 10.4).

The symmetry of a breakthrough curve is due to the contribution of longitudinal dispersion and the narrow range in pore-water velocity distribution. This usually means the medium is more uniform in texture and pore geometry. In such circumstances, it can be assumed that no interaction takes place between the solute and the solid. However, due to tortuosity of path, volumetric water content, and other parameters previously mentioned, we do not normally expect a symmetrical breakthrough curve. When using extracted soil cores (or blocks) to run tracer experiments, initiation of the breakthrough curve on the y-axis usually indicates flow along the interface between the extracted core and the encasement material. This can bias results, so it must be corrected (see figure 10.5).

The basic approaches used to represent hydrodynamic dispersion have been described empirically by mathematics such as Hagen–Pouiselle's law (discussed in chapter 6) and by modeling, which is discussed in chapter 13. When using models, one usually conducts a field or laboratory experiment to determine the effects of hydrodynamic dispersion by analysis of breakthrough curves.

Mechanical Dispersion

Most dispersion is caused by the presence of solids through meandering soil pores of different sizes that vary spatially, and are dependent on the statistical distribution of velocity of the liquid. Dispersion takes place in two directions, longitudinal and transverse, with the greater amount occurring in the longitudinal direction (along the flow line). Transverse dispersion is smaller in scope, unless flow velocity slows enough for it to be equal-to, or greater-than longitudinal dispersion. At this point, however molecular diffusion normally dominates. The

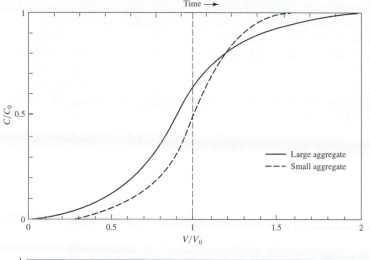

Figure 10.4   The shift of a breakthrough curve based on aggregate size

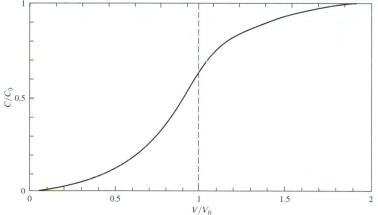

Figure 10.5   A typical breakthrough curve using a constant flux input

movement of the solute in each pore-channel depends on the deviation of path, length of path, and pore size. With larger pores, we obtain a more rapid and wider range in liquid velocities through the soil. The hydrodynamic-dispersion flux is written in the same form as that for molecular diffusion (Bear 1972), and can be expressed by

$$J_H = -D_H \frac{dC}{dz} \tag{10.3}$$

where D_H is the hydrodynamic dispersion coefficient ($L^2 T^{-1}$), C is the concentration (M L^{-3}), and z is the vertical distance (m). This coefficient is considered proportional to pore-water velocity, v, that is, q/θ (Biggar and Nielsen 1967; Bear 1972), and has been expressed as $D_H = \alpha v$, where α is the dispersivity (cm). The dispersivity depends on the length of path over which the water flux and solute diffusion is averaged. It is also an important calibration tool for dispersion in many models.

Molecular Diffusion

At low pore-water velocities, solute transport is dominated by diffusion. Molecular diffusion is best described by Fick's first law:

$$J_D = -D_0 \frac{dC}{dz} \tag{10.4}$$

where D_0 is the molecular-diffusion coefficient (L^2T^{-1}). Thermodynamically, the gradient in chemical potential is the driving force for the diffusion process. As water content decreases in a soil, molecular diffusion predominates over mechanical dispersion, since dispersion is dependent on flow velocity. In a soil with uniform pore-size distribution but low hydraulic conductivity, molecular diffusion is predominant as flow velocity approaches zero.

The larger the diffusion coefficient, the more completely the incoming solute mixes with stagnant water in immobile zones and as a consequence, its appearance in the effluent is delayed (Biggar and Nielsen 1962). When a breakthrough curve shifts to the left it is because of incomplete mixing and lack of fluid displacement. The further to the left the curve shifts, the smaller the volume of resident water that is displaced. This lack of displacement can be caused by: **(1)** large aggregate size causing an abundance of immobile (stagnant) water present in the medium (i.e., pore-water velocity is zero); **(2)** the exclusion of a solute due to solute–solid interactions; and **(3)** sometimes by increased solute concentration, which can cause incomplete mixing of the solute with the soil water and/or anion exclusion.

A breakthrough curve is shifted to the right due to: **(1)** the displacing fluid or its solutes are retained, either by precipitation or exchange; **(2)** chemical reaction of the solute or solid; and **(3)** exchange of the solute with the solid phase of the medium. Essentially, adsorption of the solute to the solid phase is often the driving force that affects curve shape. As a result, the tracer does not reach the end of the column until adsorption sites are filled. In the case of large soil particles or aggregates, (such as probably occurs under no-till agricultural conditions), the solute has to desorb back into solution in order for transport by mobile water to occur. Unless there is a significant recharge event, solutes diffused into these aggregates are transported solely by diffusion. The shape of the breakthrough curve is not determined by retention alone, but also by how chemical and physical processes are coupled with microscopic velocity distribution or other processes.

Relation of a Breakthrough Curve to the Solution of the ADE

To arrive at a representative solution to the advection–dispersion equation (ADE), we need to know initial and boundary conditions. For a typical core, the boundary conditions would be as follows: *Top Boundary* $C(0, t) = 0$ for $t < 0$; $C(0, t) = C_0$ for $t > 0$; *Initial Condition* $C(x, 0) = 0$; *Bottom Boundary* (semi-infinite media) $C(0, t) = 0$ for $t \leq 0$. With these boundary and initial conditions, a solution can be obtained for the advection–dispersion equation

$$D_z \frac{\partial^2 C}{\partial z^2} - v_z \frac{\partial C}{\partial z} = \frac{\partial C}{\partial t} \tag{10.5}$$

where D_z is the hydrodynamic dispersion coefficient in the longitudinal direction (assumed vertical, z, in this case) (L^2T^{-1}), C is solute concentration (ML^{-3}), v is the average linear ground water velocity LT^{-1}, and t is time (T). The solution for the ADE, given for the above-listed boundary and initial conditions is

$$\frac{C(t)}{C_0} = \frac{1}{2}\left[\mathrm{erfc}\left(\frac{z - vt}{2\sqrt{D_z t}}\right) + \exp\left(\frac{vz}{D_z}\right) \mathrm{erfc}\left(\frac{z + vt}{2\sqrt{D_z t}}\right)\right] \tag{10.6}$$

where erfc is the complimentary error function. When D_z, z, or t is large, the second term on the right-hand side of equation 10.6 is negligible. With this solution, we can compute the shape of a typical breakthrough curve for a medium for which the solution is appropriate.

A General Solution for Dispersion of a Displacing Solute Front

At this point we assume a column of saturated soil to which we will apply a solution with concentration, C_0, of bromide. Within the column (initially) there is no bromide; however, at

$x_1 < 0$ a bromide solution is applied to the end of the column, yielding a boundary-value problem such that

$$\frac{\partial C}{\partial t_1} = D\left(\frac{\partial^2 C}{\partial x_1^2}\right) \tag{10.7}$$

Using the previous discussion, the initial conditions for the column are $f(u) = C(x_1, t_1) = C_0$ for $x_1 < 0$ and $t_1 = 0$; also, $f(u) = C(x_1, t_1) = 0$ for $x_1 > 0$ where $t_1 = 0$. The limit as x_1 approaches infinity is expressed as $\lim C(x_1, t_1) = 0$ for $0 < x_1 < \infty$ and $0 < t_1 < \infty$. If one sets $C(x_1, 0) = C_0$ and institutes the initial conditions for $x_1 < 0$, then equation 10.7 can be expressed mathematically as

$$C(x_1, t_1) = \frac{1}{2\sqrt{\pi Dt_1}} \int_{-\infty}^{0} C_0 \exp\left[\frac{-(x_1 - u)^2}{4Dt_1}\right] du \tag{10.8}$$

To solve this equation we need to develop an error function, erf. To begin, we must change the variable of integration from u to $(x_1 - u)/[2(Dt_1)^{0.5}]$ (this is the square root of the negative exponent of e). The next step is to determine the limits of integration by letting $u = 0$ and $u = -\infty$, in order to determine the upper limit such that $x_1/[2(Dt_1)^{0.5}]$, with the lower limit as infinity. For equation 10.8 we must change du to $d(x_1 - u)[(Dt_1)^{0.5}]$ and from the relation

$$d\left[\frac{x_1 - u}{2\sqrt{(Dt_1)}}\right] = -\frac{du}{2\sqrt{Dt_1}} \tag{10.9}$$

we must multiply du in equation 10.8 by $-[2(D_t D_1)^{0.5}]^{-1}$ and must also multiply the complete integral by $-2(Dt_1)^{0.5}$, to avoid changing the value. Upon completing these steps we obtain

$$C(x_1, t_1) = -\frac{2\sqrt{Dt_1}}{2\sqrt{\pi Dt_1}} \int_{\infty}^{x} \frac{1}{2\sqrt{Dt_1}} C_0 \exp\left[\frac{-(x_1 - u)^2}{4Dt_1}\right] d\left[\frac{(x_1 - u)}{2\sqrt{Dt_1}}\right] \tag{10.10}$$

The last term on the right-hand side of the equation disappears upon integration and, as before, we need to assign a dummy variable to it, denoted by β, that, upon limit interchange, yields

$$C(x_1, t_1) = \frac{1}{\sqrt{\pi}} \int_{x_1/(2\sqrt{Dt_1})}^{\infty} C_0 \exp\left(-\beta^2\right) d\beta. \tag{10.11}$$

This will give an erf in the form

$$C(x_1, t_1) = \frac{1}{2} C_0 \left[\frac{2}{\sqrt{\pi}} \int_{0}^{\infty} \exp\left(-\beta^2\right) d\beta - \frac{2}{\sqrt{\pi}} \int_{0}^{x_1/(2\sqrt{Dt_1})} \exp\left(-\beta^2\right) d\beta\right] \tag{10.12}$$

where the error function, erf, is typically written as

$$\text{erf}(z) = \frac{2}{\sqrt{\pi}} \int_{0}^{z} \exp\left(-\beta^2\right) d\beta \tag{10.13}$$

The typical properties associated with the erf are $\text{erf}(-z) = -\text{erf}(z)$, $\text{erf}(0) = 0$, and $\text{erf}(\infty) = 1$. The values associated with $\text{erf}(z)$ range from 0 to $+1$ for $\text{erf}(z)$ and 0 to -1 for $-\text{erf}(z)$. Using the property $\text{erf}(\infty) = 1$ in equation 10.12 we obtain

$$C(x_1, t_1) = \frac{1}{2} C_0\left(1 - \text{erf}\left[\frac{x_1}{2\sqrt{Dt_1}}\right]\right) \tag{10.14}$$

This equation is for the moving-coordinate system; for the fixed-coordinate system we write

$$\frac{C(x, t)}{C_0} = \frac{1}{2}\left(1 - \text{erf}\left[\frac{x - vt}{2\sqrt{Dt_1}}\right]\right) \tag{10.15}$$

In addition to the standard error function, erf, there is also erfc, the complementary error function. It can be determined by $\text{erfc}(z) = 1 - \text{erf}(z)$. For values of $\text{erf}(z)$ and $\text{erfc}(z)$, see appendix 3.

Although this form is mathematically adequate in describing what is happening in a soil, it is not easily used to plot a breakthrough curve because it does not reflect the relation. The form is easily changed by introducing the Darcy velocity (discussed in chapter 7) and the pore-volume concept. For example, consider a soil column of fixed length, L. The Darcy velocity (also called the flux, q) is simply the quantity of flow, Q (L^3/T), divided by the cross-sectional area, A (L^2). However, in a soil, a particle of water moves faster than the bulk water standing over the soil. Consequently, the average linear velocity is written as $q = (QA^{-1}/\theta)$, where θ is the liquid-filled phase of the soil. The pore volume, PV, of the medium can be written as $PV = \theta V_t$ where V_t is the total volume of the column. For a given duration, a number of pore volumes, N_{pv}, will pass through the column, which is written as $Npv = Qt/PV$, and can be rewritten as $N_{pv} = Qt/\theta V_t$. We divide both sides by the area, A; by substituting LA for V_t and using $q = Q/A$, we have $N_{pv} = qt/L$, which lets us express $C(x,t)/C_0$ in terms of PV. Now, by letting $L = x$, we can rewrite equation 10.15 as

$$\frac{C(x,t)}{C_0} = \frac{1}{2}\left(1 - \text{erf}\left[\frac{(1 - N_{pv})}{2\sqrt{\frac{DN_{pv}}{qL}}}\right]\right) \tag{10.16}$$

Also, equation 10.16 can be written in terms of the erfc function since $\text{erfc} = 1 - \text{erf}$; simply substitute erfc into the equation where appropriate. If the reader wishes to find the solution for the dispersion of a "slug" (i.e., a volume of fluid per area of x_0/L) in a similar fashion to that described above, it yields

$$\frac{C(x,t)}{C_0} = \frac{1}{2}\left(\left[\text{erf}\frac{1 + \frac{x_0}{L} - N_{pv}}{2\sqrt{\frac{DN_{pv}}{qL}}}\right] - \left[\text{erf}\frac{1 - N_{pv}}{2\sqrt{\frac{DN_{pv}}{qL}}}\right]\right) \tag{10.17}$$

Determining the Error Function

Essentially, there are two ways to calculate the error function: (1) from tables of error functions and (2) from tables of related probability integrals, which are easier to find and are more common. We can write the normal probability integral as

$$N(z) = \frac{1}{\sqrt{2\pi}} \int_0^z \exp\left(-\frac{w^2}{2}\right) dw \tag{10.18}$$

Using equation 10.18, one can substitute $w/\sqrt{2}$ for β in equation 10.13 such that

$$\text{erf}(z) = \frac{2}{\sqrt{2\pi}} \int_0^{\sqrt{2}z} \exp\left(-\frac{w^2}{2}\right) dw \tag{10.19}$$

Consequently $\text{erf}(z) = 2N(\sqrt{2}z)$ or $2N(1.414z)$. As a result, $N(z)$ is simply the area under the curve which from standard rules of calculus can be written for this case as

$$f(w) = \frac{\exp\left(-\frac{w^2}{2}\right)}{\sqrt{2\pi}} \tag{10.20}$$

QUESTION 10.1

Using the relation $erf(z) = 2N\sqrt{2}z$, find erf 0.90.

QUESTION 10.2

What is z when $erf(z) = 0.797$?

Calculating the Displacing Front of a Breakthrough Curve

In order to calculate points on the breakthrough curve we need to know the number of pore volumes that have eluted through the column, N_{pv}, D, q, and L. We begin by using the definition of the error function (equation 10.13) and substituting $w^2/2$ for β^2 (as before) from the definition of the probability; then equation 10.16 can be rewritten as

$$\frac{C(x, t)}{C_0} = \frac{1}{2} - \frac{1}{\sqrt{2\pi}} \int_0^{(1-N_{pv})/\sqrt{2DN_{pv}/qL}} \exp\left(-\frac{w^2}{2}\right) dw \qquad (10.21)$$

With this equation, we plot a breakthrough curve; however, to determine D in the equation the derivative must first be obtained (we show later that there is a simpler way to obtain D). To save time and space, a few steps are skipped and the derivative dq/dp is expressed as

$$\frac{dq}{dp} = -\frac{\sqrt{\dfrac{2DN_{pv}}{qL}} - \left(\dfrac{1}{2}\right)\dfrac{(1 - N_{pv})\left(\dfrac{2D}{qL}\right)}{\sqrt{\dfrac{2DN_{pv}}{qL}}}}{\dfrac{(2DN_{pv})}{qL}} \qquad (10.22)$$

Letting the number of pore volumes equal one (1) and substituting, we may write

$$\frac{d\left[\dfrac{C(x, t)}{C_0}\right]}{dp}\Bigg|_{N_{pv}=1} = \left[2\sqrt{\frac{\pi D}{qL}}\right]^{-1} \qquad (10.23)$$

Now, we may set the left side equal to S and solve for D (the dispersion coefficient), which is expressed as

$$D = \frac{qL}{4\pi S^2} \qquad (10.24)$$

In summary, we know both L and q from experimental measurement where $q = (Q/A)/\theta$. Consequently, to determine the diffusion coefficient, D, we need to measure the slope, S, at $N_{pv} = 1$ of the breakthrough curve; measure $C(x, t)/C_0$ versus N_{pv}, which is set equal to the derivative of the left-hand side of the equation 10.23; and solve for D.

QUESTION 10.3

You are running a column experiment in the laboratory. The column measures 125-cm height by 50-cm diameter. Determine D given that $Q = 2810$ cm hr^{-1}, $\theta = 0.45$ (water-filled porosity), and S (slope of C/C_0) at $N_{pv} = 1$.

QUESTION 10.4

How would you find the derivative of $C(x, t)/C_0$ with respect to N_{pv}?

Calculating the Concentration in a Moving Slug of Fluid for a Breakthrough Curve

By rewriting equation 10.17 in terms of the normal probability integral we obtain

$$\frac{C(x, t)}{C_0} = \sqrt{2\pi}^{-1} \int_0^{[1+(x_0/L)-N_{pv}]/\sqrt{2DN_{pv}/qL}} \exp\left(-\frac{w^2}{2}\right) dw$$
$$- \sqrt{2\pi}^{-1} \int_0^{[(1-N_{pv})/\sqrt{2DN_{pv}/qL}]} \exp\left(-\frac{w^2}{2}\right) dw \tag{10.25}$$

This will allow one to find a moving front (or slug) of fluid. Without working out the derivation, the diffusion coefficient (D) for this equation can be determined, after Corey, Nielsen, and Kirkha (1967) such that

$$D = \frac{qL\left(\dfrac{x_0}{2L}\right)^2}{2\left(1 + \dfrac{x_0}{2L}\right)z^2} \tag{10.26}$$

where x_0 is the volume (mL) of the slug, V_s, added to the medium. Thus, $x_0 = V_s/A\theta$, where A is the cross-sectional area of interest. The value for z may be determined by

$$z = \frac{\left(\dfrac{x_0}{2L}\right)}{\sqrt{2D\dfrac{(1 + x_0)/2L}{qL}}} \tag{10.27}$$

or in evaluating the maximum $C(x, t)/C_0$ by changing the limit on the second integral from $-z$ to z, adding both integrals, and dividing by 2 so that

$$\frac{1}{2}\left(\frac{C(x, t)}{C_0}\right)_{\max} = \sqrt{2\pi}^{-1} \int_0^z \exp\left(-\frac{w^2}{2}\right) dw \tag{10.28}$$

Thus, to find D, measure $C(x, t)/C_0$ at its maximum value on the breakthrough curve, divide by two, and set it equal to the right-hand side of equation 10.28. As a result, when the maximum value of $[C(x, t)/C_0]$ is divided by two, we go to the table of normal probability functions to find z; this allows us to solve equation 10.26. Once D is obtained, equation 10.25 can be used to plot the breakthrough curve. We discuss procedures for determining D through experimental methods later in this chapter.

Thus far, we have considered only dispersion as the mixing process and have neglected diffusion. In discussing this parameter we assume that the viscosity and density of all fluids within the system are the same, and that the dispersion coefficient is independent of solute concentration. However, the discussion has served to introduce miscible-displacement mathematics. Later, in question 10.7, the student is asked to derive the ADE. Considering the answer to the question (in one dimension), the solution can be expressed as

$$\frac{C(x, t)}{C_0} = \frac{1}{2}\left[\mathrm{erfc}\left(\frac{1 - N_{pv}}{2\sqrt{\dfrac{DN_{pv}}{qL}}}\right) + \exp\left(\frac{qL}{D}\right)\mathrm{erfc}\left(\frac{1 + N_{pv}}{2\sqrt{\dfrac{DN_{pv}}{qL}}}\right) \right] \tag{10.29}$$

As before, D can be found by equating the slope of the breakthrough curve at $N_{pv} = 1$ to the derivative of equation 10.29, with respect to N_{pv} evaluated at $N_{pv} = 1$. This is done in the same method as before, by converting equation 10.29 to the normal probability integral to find that

$$D = \frac{qL}{4\pi S^2} \tag{10.24}$$

Similar to previous solutions, once D is obtained, known values of q and L can be substituted into equation 10.29 and one can plot $C(x, t)/C_0$ versus the number of pore volumes, N_{pv}. For a soil that has a low-flow velocity, a diffusion model can easily fit the experimental data. For a high-flow velocity, a dispersion model fits better. This is logical, since D depends on fluid velocity. In this case, D is described by the diffusion coefficient in the diffusion model, and the dispersion coefficient in the dispersion model; however, remember that the two are essentially inseparable. Some of the original research evaluating this topic (Nielsen and Biggar 1962) indicates that for soils with a wide range of microscopic-pore velocities (typical in the unsaturated zone), the use of an average-flow velocity in the model can cause deviations between the experimental curve of the data versus the theoretical curve of the model. This is primarily due to: ion adsorption and exchange; rate of diffusion; pore geometry; chemical reaction; precipitation; incomplete mixing of solute with solution; aggregate size, as well as other physical and chemical factors. All of these parameters affect the dispersion coefficient D and thus, can affect the appearance of the first detectable concentration, the shape of the breakthrough curve, whether the breakthrough curve is shifted left or right, and/or the outcome of the modeled solution.

The mathematical models presented here have made several simplifying assumptions. Despite assumptions, however, using mathematical or analytical models for comparison to experimental data can provide information and insight about a soil. If the models accurately predict the results of an experiment, one can reasonably assume that the hypothesis used for the model works well for both the medium of interest and the solute being investigated. However, solutions for a specific soil or medium type are not always readily transferable to another chemical solution or medium.

Figure 10.6 shows a comparison of a numerical solution using the United States Geological Survey program VS2DT (Healy 1990) versus an analytical solution. Using a soil column of 35-cm length, water flowing into the column is maintained at a concentration of C_0 for 160 s, after which the concentration is set to zero for an additional 320 s. Figure 10.6 shows that the numerical results of VS2DT produces a good match, with analytical results at a distance of 8 cm for the column inlet at all times for steady water flow, and both first-order decay and linear sorption for three different cases of decay and sorption.

QUESTION 10.5

What is the derivative to equation 10.29?

Calculation of the Dispersion Coefficient (D)

To simulate transport through soil practically, the effective dispersion coefficient D should be determined—usually in the longitudinal direction. The dispersion coefficient is often calculated for a homogeneous medium and normally, determination of D is accomplished with the use of packed laboratory columns and intact cores for the medium of interest by "trial and error." Once the solute concentration of the resident solution in the column is known, a feed solution of the same relative concentration, but containing a different solute, is leached

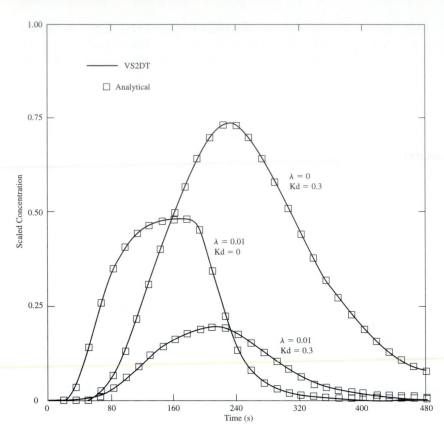

through the column. The dispersion coefficient is then determined as a function of time, distance, and concentration. When using packed columns, they should be uniformly packed so as to avoid layering, to ensure homogeneity. This helps prevent macroscopic variations in water content and pore-water velocity, which can result in additional spreading. Uniform packing also allows the use of a constant "effective" pore-water velocity for the experiment. Viscosity and density differences between resident and feed solutions during mixing are usually avoided in this experiment, because these differences can lead to fingering. As a result, low concentrations of a conservative tracer (i.e., bromide or chloride) do not affect the density and the viscosity of the solvent. Biggar and Nielsen (1964) show that fingering dominates for a density difference of 3.4×10^{-3} g cm^{-1} and a viscosity difference of 0.003 cP, obscuring the effect of molecular diffusion.

There are also other factors to consider in the determination of D. One is gravity segregation, discussed in detail by Rose and Passioura (1971b). However, this is more important in cases of saturated flow and saltwater intrusion, in which contaminated salt/water intrudes horizontally into a fresh-water aquifer. Since permeability k decreases rapidly during the desaturation process, gravity segregation is of little consequence in unsaturated-flow conditions. The reader is referred to Scheidegger (1974) for an in-depth discussion of gravity segregation. A major factor to consider is apparatus-induced dispersion. This is critically important when short columns are used under unsaturated-flow conditions. In the event of layering caused by insufficient mixing during column preparation, D can be as much as 40 percent less than obtained with a one-layer equation. This is because mixing is assumed to occur in the medium only, but in actuality it occurs in the "dead-volume" inside the column, but outside the medium. Thus, packing to reduce the "dead-volume" and obtain homogeneity is crucial. Also, by measuring concentration with depth as well as in the effluent, we obtain a more accurate D than by simply measuring concentration in the effluent alone. Measurements of concentration within the medium can be obtained from samples collected using suction lysimeters, or by measuring electrical sensitivity at a point in the medium with a conductivity probe if using salts such as sodium chloride.

The dispersion coefficient is often determined as a function of position at a certain time using an analytical inverse solution to the ADE. It can also be determined for varying time at a given depth. One method for determining D was given in equation 10.24; in this section we discuss other methods to calculate D. The one-dimensional ADE (equation 10.5) was used by Fried and Combarnous (1971). Assuming a longitudinal dispersion coefficient, an approximate solution to equation 10.5 can be written as equation 10.15. For a given time, the solution follows a normal distribution $1 - N[(x - \mu)/\sigma]$, with a mean displacement $\mu = vt$ and standard deviation $\sigma = (2Dt)^{0.5}$. The term $N[\ ]$ is the probability-density function (PDF) for a normal distribution with values $N[-1] = 0.16$ and $N[1] = 0.84$, as suggested by Lapidus and Amundson (1952). The width of the transition zone, 2σ, can be determined by plotting C/C_0 versus x (depth) such that

$$2\sigma = x_{0.16} - x_{0.84} = \sqrt{8D_L t} \tag{10.30}$$

where $x_{0.16}$ and $x_{0.84}$ are the positions for which C/C_0 is equal to 0.16 and 0.84 respectively. The longitudinal dispersion coefficient can be calculated by

$$D_L = \frac{(x_{0.84} - x_{0.16})^2}{8t} \tag{10.31}$$

For concentrations at a certain position as a function of time, we have

$$D_L = \frac{1}{8}\left[\frac{x - vt_{0.16}}{\sqrt{t_{0.16}}} - \frac{x - vt_{0.84}}{\sqrt{t_{0.84}}}\right]^2 \tag{10.32}$$

While Fried and Combarnous (1971) discuss how the transverse-dispersion coefficient can be determined, very little other research has been done to that end.

Rose and Passioura (1971a) used the Brenner number, B ($B = vL/D$; L is column length) and determined D by plotting $C_e = [(C - C_i)/(C_0 - C_i)]$ versus $\ln PV$ on probability paper where C_e is the exit concentration (dimensionless) and PV (pore volume) $= (vt/L)$. Nearly straight lines for particular values of B were obtained. Due to this linear relation we have

$$\text{erfc}^{-1}(2C_e - 1) = -m \ln PV - \beta \tag{10.33}$$

where m is the slope and β is the intercept. Rose and Passioura (1971a) also developed the following relation for $16 < B < 640$:

$$\log B = 0.1139(\log m)^3 - 0.3504(\log m)^2 + 2.3623(\log m) + 0.4732 \tag{10.34}$$

By plotting $-\ln PV$ versus $\text{erfc}^{-1}(2C_e - 1)$, the slope m and the intercept β can be determined. The Brenner number B can be obtained from equation 10.34; D can then be obtained using the relation $B = vL/D$. Van Genuchten and Wierenga (1986) use a slightly simpler version of equation 10.34.

Another method for obtaining D is by curve-fitting or least-squares analysis of the effluent curve. One of the more popular programs for determining D is called CXTFIT by Toride, Leij, and Van Genuchten (1995). These researchers use a nonlinear, least-squares inversion method to determine the dispersion coefficient, the retardation factor (R), and first-order degradation constants. This type of program is most useful on column-breakthrough experiments involving concentrations at fixed locations at different times, or for a specific time at different locations.

Additional methods for obtaining D involve the calculation of D from concentration versus distance curves, and the Boltzmann transformation. To obtain D in a concentration versus distance scenario, we simply section laboratory columns at specific distances or determine the solute concentration in situ in the field. The resulting equation for this method, given by Van Genuchten and Wierenga (1986), is $D = R/4\pi t\, S^2$, where R is the retardation factor (the complete form is shown later, in equation 10.93) and S is the slope of the experimental curve where the solute concentration is 0.50, or by calculation of the slope (β) $D = R/4\beta^2 t$. Note that in expression $R = vt/x_0$, x_0 is the value (distance) of x where the relative effluent concentration equals 0.5, and v is the average pore-water velocity. In using the Boltzmann transformation to transfer the partial-differential equation into an ordinary equation, we assume the dispersion coefficient to be dependent mainly on water content because of the relatively low velocity, the small value of the Peclet number, and the rapidly changing water content during infiltration.

The process of dispersion depends on several things: water content; molecular diffusion; velocity; viscosity; density; and hydraulic conductivity. To determine the influence of pore-water velocity and particle size, data are commonly analyzed by plotting D_L/D_0 versus $\mathcal{P}$ (the Peclet number) where $\mathcal{P} = vd/D_0$ on log–log paper. For such cases, d is a characteristic pore-size or particle-size dimension similar to that used in scaling (see chapter 16). Using the Peclet number, Bear (1979) designates the following dispersion regimes: (1) molecular diffusion is dominant when $\mathcal{P} < 0.4$; (2) molecular diffusion and mechanical dispersion are of the same order for $0.4 < \mathcal{P} < 5$ and thus, additive; (3) major mechanical dispersion occurs with some molecular diffusion in the range $5 < \mathcal{P} < 1000$—there is interference in these effects and they cannot be added; (4) mechanical dispersion is dominant for $1000 < \mathcal{P} < 1.5 \times 10^5$ with negligible molecular diffusion; and (5) mechanical dispersion occurs when the flow regime is out of the domain of Darcy's law ($\mathcal{P} > 1.5 \times 10^5$). As a note of caution, however, if we use experimental values for D, the solute front will exhibit additional spreading when solving the ADE by numerical methods. This computational artifact is called numerical dispersion and is most prone to occur directly around the front when using high values for P.

10.4   NONREACTIVE SOLUTES

Typical investigations of solute transport involve the movement of a conservative tracer through a medium that is both isotropic and homogeneous. A conservative tracer is one that does not react or interact with the medium, and which is completely miscible in the pore-water solution. The most commonly used conservative tracers are chloride and bromide (generally as KBr, and KCl or NaCl). If we assume no source or sink terms, the conservation of mass at a large scale (or what is termed a macroscopic scale) may be mathematically described as

$$\frac{\partial\, \theta C}{\partial t} = -\nabla \cdot [J_H + J_v C] = -\nabla \cdot J_s \tag{10.35}$$

where θ is the volumetric water content (unitless), C is solute concentration (ML^{-3}), t is time (T), J_H is flux due to hydrodynamic dispersion that includes both dispersive and diffusive fluxes ($ML^{-2}T^{-1}$), Jv is the volumetric-flux of the influent ($L^3 T^{-1}$), J_s is the total-flux of the solute ($ML^{-2}T^{-1}$), and $\nabla \cdot$ (pronounced nabla dot) is the divergence (L^{-1}).

Based on Fick's first law, the autonomous flux J_D is given in equation 10.3. However, hydrodynamic dispersion is written as

$$D_H(v, \theta) = \alpha v + D_0(\theta) \tag{10.36}$$

where α is the dispersivity (L), sometimes called the mass-transfer coefficient; v is the average advective pore-water velocity expressed as v, or q/θ where q is the Darcy flux (LT^{-1}); θ is the volumetric water content; and D_0 is as described in equation 10.3 written here as a function of water content (θ). The left term on the right-hand side of equation 10.36 is often termed the mechanical-dispersion coefficient (D_m). Rewriting equation 10.3 and substituting equation 10.36, we obtain

$$J_D = -\theta D_H \frac{\partial C}{\partial z} \tag{10.37}$$

where D_H is the hydrodynamic-dispersion coefficient ($L^2 T^{-1}$) and z is the vertical distance (L). D_H now represents the random effects of both molecular diffusion and mechanical dispersion.

By using the relation between Darcy's law and pore-water flux ($J_v = v\theta$), we can substitute equation 10.37 into equation 10.35 to obtain the one-dimensional transport equation for steady-state conditions; here we assume that v (average velocity) and θ are constant with depth, z:

$$\frac{\partial \theta C}{\partial t} = \frac{\partial}{\partial z}\left[\theta D_H(v, \theta)\frac{\partial C}{\partial z} - v\theta C\right] \tag{10.38}$$

We shall now drop the "H" on D_H and refer to it simply as D. Because θ is constant, the equation may be simplified to equation 10.5 by dividing through by θ:

$$\frac{\partial C}{\partial t} = D\frac{\partial^2 C}{\partial z^2} - v\frac{\partial C}{\partial z} \tag{10.5}$$

This is a linear, parabolic, second-order partial differential equation in which both v and D are independent of z and t (Risken 1984). For anisotropic, multi-dimensional flow, equation 10.5 can be rewritten

$$\frac{\partial C}{\partial t} = \frac{\partial}{\partial z_i}\left[D_{ij}\frac{\partial C}{\partial z_j}\right] - \frac{\partial v_i C}{\partial z_i} \tag{10.39}$$

where D_{ij} is the dispersion tensor and i, j are direction indices (Smith 1985).

Equation 10.5 is most useful for laboratory investigations in which packed columns are used to control: soil heterogeneity; structural differences; cracking; shrinking; swelling; and biological activity. For these conditions, equation 10.5 can be used to describe most of the phenomena represented in figure 10.1 (Bresler 1973a; Cho 1971; Cushman 1982; Robbin Jurinak, and Wagenet 1980 a, b; Van Genuchten and Cleary 1982).

10.5 SORPTION REACTIONS

When pesticides and other organic chemicals are applied to soils, the solutes do not act as a conservative tracer due to electrical-charge differences between the exchange sites in the soil matrix and the applied contaminant. For example, most soils found in the United States have a negative net charge; thus, the application of a pesticide or other organic compound would display reactive properties—that is, normally, the contaminant would be strongly bound to soil particles in the upper profile. However, many types of soil in various parts of the world contain variably charged soils (Sumner 1992) such as hydrous oxides and aluminum and iron sesquioxides, almost all of which exhibit ion-exclusion and adsorption phenomena. In the process of anion exclusion, negatively charged anions are repulsed from the negatively charged surfaces of soil particles, which results in more rapid flux of the anions. Consequently, breakthrough curves that exhibit anion exclusion are generally shifted left, that is, initial breakthrough occurs earlier than if no anions were present. While this phenomenon is commonly associated with the application of a conservative tracer like potassium bromide onto net-negative soils, it is also seen with the application of various reactive contaminants to variably charged soils, where the contaminant and soil can have a net positive charge.

Generally, a tracer (or chemical) that is applied intentionally to the soil is present in a single liquid, non-sorbed phase. Assuming the processes of adsorption and desorption apply, equation 10.35 can be written as

$$\frac{\partial}{\partial t}[\theta C + C_s] = -\nabla \cdot [J_D + J_v C] \tag{10.40}$$

where C_s is the concentration ($M\ L^{-3}$) of the adsorption–precipitation phase. Equation 10.40 can be rewritten as

$$\frac{\partial}{\partial t}[\theta C + C_s] = \frac{\partial}{\partial z}\left[D\theta \frac{\partial C}{\partial z}\right] - \frac{\partial v\, \theta C}{\partial z} \tag{10.41}$$

There are three common types of adsorption isotherms for which the ADE has been written: **(1)** linear; **(2)** Freundlich; and **(3)** Langmuir. Gupta and Greenkorn (1974) give the following expressions for the ADE with regard to each isotherm.

Linear:

$$\left[1 + \frac{K_1}{\theta}\right]\frac{\partial C}{\partial t} = D\frac{\partial^2 C}{\partial z^2} - v\frac{\partial C}{\partial z} \tag{10.42}$$

Freundlich:

$$\left[1 + \frac{nK_2}{\theta}C^{n-1}\right]\frac{\partial C}{\partial t} = D\frac{\partial^2 C}{\partial z^2} - v\frac{\partial C}{\partial z} \tag{10.43}$$

Langmuir:

$$\left[1 + \frac{a}{\theta(1 + bC)^2}\right]\frac{\partial C}{\partial t} = D\frac{\partial^2 C}{\partial z^2} - v\frac{\partial C}{\partial z} \tag{10.44}$$

where a, b, K, and n are empirical constants. The amount of chemical adsorbed in mass-per gram of soil in a solution with a known equilibrium concentration (C) in mass per volume, can be calculated by $S = KC^n$, where K is the adsorption coefficient and S can be expressed as the source/sink term. The exponent n may be taken as: **(1)** 1.0 if a linear isotherm is expected; **(2)** 0.87 if a nonlinear isotherm is likely (0.87 is an average of 26 pesticides as reported by Lyman, William, and Rosenblatt 1990); or **(3)** any value that has empirical basis. For case 3, we consider measured values for similar compounds to that being investigated. Scientists typically investigate linear isotherms as a first choice because of simplicity. However, for accuracy, nonlinear isotherms are preferred.

Due to the source/sink term, S, there is a retardation factor, R, involved in the ADE. The retardation factor is affected by soil-bulk density, volumetric water content, soil-adsorption characteristics, and other parameters; we describe it in the following paragraphs. Initially, we write ($S\rho_b/\theta$) which yields (ML^{-3}), however, when multiplying by the volume, $\nabla x \nabla y \nabla z$, we are left with total-mass adsorbed. The retardation factor for a volume basis is expressed by

$$\left(\frac{\rho_b}{\theta}\frac{\partial S}{\partial t}\Delta x\, \Delta y\, \Delta z\right) = MT^{-1} \tag{10.45}$$

Substitution into equation 10.5 gives

$$\frac{\rho_b}{\theta}\frac{\partial S}{\partial t} + \frac{\partial C}{\partial t} = D\frac{\partial^2 C}{\partial z^2} - \frac{v}{}\frac{\partial C}{\partial z} \tag{10.46}$$

Because $S = KC^n$, application of the chain rule (and assuming $n = 1$) yields

$$\frac{\partial S}{\partial t} = \frac{K}{}\frac{\partial C}{\partial t} \tag{10.47}$$

Substituting equation 10.47 into equation 10.46, we obtain

$$\frac{\rho_b}{\theta}\frac{K \,\partial C}{\partial t} + \frac{\partial C}{\partial t} = D\frac{\partial^2 C}{\partial z^2} - \frac{v \,\partial C}{\partial z} \tag{10.48}$$

This can now be rewritten in terms of one dependent variable, as

$$\frac{R \,\partial C}{\partial t} = D\frac{\partial^2 C}{\partial z^2} - \frac{v \,\partial C}{\partial z} \tag{10.49}$$

where

$$R = \frac{\rho_b}{\theta}K + 1 \tag{10.50}$$

R (retardation coefficient) is dimensionless. Thus, equation 10.49 is the ADE for one-dimensional flow at steady state with assumption of reaction between solute and solid phases. To obtain the slope of the exchange isotherm, we plot dS/dC. An exchange isotherm, in which all parameters are typically held constant (except concentration), is the amount of solute sorbed to the medium versus the amount of solute in concentration. This also includes the physical and chemical parameters of both the soil and solution. Note that in the consideration of a binary system, the amount of solute adsorbed S (usually written as μg adsorbed/g medium), and the solution concentration C (μg/mL or ML^{-3}), are fitted to the Freundlich equation, $S = KC^n$, to determine the adsorption coefficient, K, and the parameter, n.

A principal difficulty in using the ADE for either analytical determination or numerical modeling is that it is often difficult to obtain values for the parameters used in the equation. The simplest method for obtaining these various parameters is to utilize regression equations obtained from experimental data in log–log form. For a wide variety of pesticides and other chemicals, Kenaga and Goring (1980) give the following equation to calculate the adsorption coefficient:

$$\log K_{oc} = -0.55 \log S_w + 3.64 \tag{10.51}$$

where S_w (water solubility) is reported in mg L^{-1}. In this instance, -0.55 represents the a constant and 3.64 represents the b constant. Other equations reported by Kenaga and Goring (1980), mainly for pesticides, are

$$\log K_{oc} = 0.681 \log BCF(f) + 1.963 \tag{10.52}$$

and

$$\log K_{oc} = 0.681 \log BCF(t) + 1.886 \tag{10.53}$$

where $BCF(f)$ is the bioconcentration factor due to flowing-water tests and $BCF(T)$ is the bioconcentration factor using model ecosystems. Regression equations for other applications are listed in Table 10.1 and in chapter 15. The adsorption coefficient, K (from the Freundlich equation: $S = KC^n$), can be estimated from K_{oc} from the expression $K = K_{oc}(\text{percent oc})/100$, where $K_{oc} = (\mu$g adsorbed/g organic carbon)/(μg/mL solution). The adsorption coefficient K is not the same as K_{oc}. The adsorption coefficient K_{oc} is the extent to which an organic chemical partitions itself between the solution and solid phases in either unsaturated or saturated soil; K_{oc} is largely independent of soil properties and can be thought of as the ratio of the amount of chemical adsorbed per-unit weight of organic carbon (oc) in a soil at assumed equilibrium.

QUESTION 10.6

Estimate K_{oc}, K, and S for dicamba. Assume a water solubility of 0.04 mg L^{-1}, organic carbon content of 2 percent, and a solution concentration, C, of 10 mg L^{-1}.

TABLE 10.1   Regression Equations for the Estimation of K_{oc}

Eq. No.	Equation*	No.[†]	r^2	Chemical classes represented
1	$\log K_{oc} = -0.54 \log S + 0.44$ (S in mole fraction)	10	0.94	Mostly aromatic or polynuclear aromatics, two chlorinated
2	$\log K_{oc} = -0.557 \log S + 4.277$ (S in μ moles/L)	15	0.99	Chlorinated hydrocarbons
3	$\log K_{oc} = -0.544 \log K_{ow} + 1.377$	45	0.74	Wide variety, mostly pesticides
4	$\log K_{oc} = 0.937 \log K_{ow} - 0.006$	19	0.95	Aromatics, polynuclear aromatics, triazines and dinitroaniline herbicides
5	$\log K_{oc} = 0.94 \log K_{ow} + 0.02$	9	[††]	s-Triazines and dinitroaniline herbicides
6	$\log K_{oc} = 1.029 \log K_{ow} - 0.18$	13	0.91	Variety of insecticides, herbicides, and fungicides
7	$\log K_{oc} = 0.0067 (P - 45N) + 0.237$	29	0.69	Aromatic compounds: ureas, 1,3,5-triazines, carbamates, and uracils

Sources: 1. Karickhoff, Brown, and Scott 1979; 2. Chiou, Peters, and Freed 1979; 3. Kenaga and Goring 1978; 4. Brown and Flagg 1981; 5. Brown, D. S. (personal communication); 6. Rao and Davidson 1980; 7. Hance 1969

*K_{oc} = soil (or sediment) adsorption coefficients; S = water solubility; K_{ow} = octanol–water partition coefficient; P = parachor; N = number of sites in molecule which can participate in the formation of a hydrogen bond

[†] = number of chemicals used to obtain regression equation

[††] Not given

QUESTION 10.7

Derive Richard's equation and then, beginning with Fick's first law of diffusion, derive the general ADE (commonly referred as the solute-transport equation) for one-dimensional, steady-state flow.

10.6   EQUILIBRIUM CHROMATOGRAPHY

Most solute-transport studies involve solute interaction with the solid phase, and for a majority of these studies, the solution is obtained numerically by utilizing experimentally, exchange isotherms that are determined. For most studies, it is assumed that steady-flow conditions prevail and only two different ionic (cations) species are present during miscible displacement; equilibrium chromatography is in this category. Following Bolt (1982), it is convenient to express the adsorbed concentration S (which depends on the liquid concentration of a specific species of interest) in moles-of-charge per volume of soil ($mol_c\ m^{-3}$) while C is expressed in moles-of-charge per volume of solution ($mol_c\ m^{-3}$). To simplify the calculation, we assume concentration is constant during ion exchange. If we neglect effects due to dispersion and diffusion, a step-change in concentration occurs at the concentration front, mathematically expressed as $J_v t / \theta$; this is, actually the penetration depth. Because we assume the concentration is constant during ion exchange and thus, time t, depends only upon C for such, we rewrite equation 10.40 as

$$\left(\frac{dS}{dC} + \theta \right) - \frac{\partial C}{\partial t} = -J_v \frac{\partial C}{\partial x} \tag{10.54}$$

For this case, dS/dC is the differential capacity of the exchanger for the exchanging ion, J_v is the volumetric-water flux (LT^{-1}), and other parameters are as previously explained. The

$\partial C / \partial x$ term must be finite (assume negligible dispersive flux) to solve equation 10.54. Using the chain rule of calculus and rewriting equation 10.54, we obtain

$$\left(\frac{\partial x}{\partial t}\right)_C = \left(\frac{J_v}{\dfrac{dS}{dC} + \theta}\right) \tag{10.55}$$

Should the concentration profile exhibit a jump (i.e., the condition of the finite term is violated), the conservation of mass is expressed as

$$\left(\Delta \frac{dS}{dC} + \theta \, \Delta C\right) dx = \Delta C \, dV_i \tag{10.56}$$

where V_i is the input volume per unit area of the soil column. This also requires that equation 10.55 be rewritten such that

$$\left(\frac{dx}{dt}\right)_{\Delta C} = \frac{J_v}{\left(\dfrac{\Delta S}{\Delta C} + \theta\right)} \tag{10.57}$$

Following our previous assumptions, the position for a specific concentration (both dS/dC and θ are homogeneous with respect to location and $dV_i = J_v \, dt$) is found by integration of equation 10.55 so that

$$x_C = \int_0^t \left(\frac{J_v}{\dfrac{dS}{dC} + \theta}\right) dt = \frac{V_i - V_0(C)}{\left(\dfrac{dS}{dC} + \theta\right)} \tag{10.58}$$

where $V_0(C)$ is the volume of solution applied to the column, at the instant that the concentration at $x = 0$ reaches C. For a step-type displacement, $V_0 \equiv 0$. For the adsorbed and liquid phase, the average depth (position) of the concentration front is given by

$$x_p = \frac{\displaystyle\int_{C_i}^{C_0} x_c \left(\frac{dS}{dC} + \theta\right) dC}{\displaystyle\int_{C_i}^{C_0} \left(\frac{dS}{dC} + \theta\right) dC} \tag{10.59}$$

Thus, with the use of equilibrium chromatography we determine the propagation in the adsorbed phase. In order to investigate the shape of the solute front, equation 10.58 is differentiated ($V_0 = 0$) such that

$$\left(\frac{\partial x}{\partial C}\right)_{V_i} = -\frac{V_i \dfrac{dS'}{dC}}{\left(\dfrac{dS}{dC} + \theta\right)^2} \tag{10.60}$$

Based on the previous discussion, three types of ion exchange are of concern here. These include linear ($dS'/dC = 0$), favorable or convex isotherm ($d^2S/dC^2 < 0$), and unfavorable ($dS'/dC > 0$). In the case of linear exchange, $\partial x/\partial C = 0$ for any applied volume that indicates the initial profile is not altered during fluid flow through the soil. However, in the case of favorable exchange, there is a minor problem: according to equation 10.54, the slope of the solute front is negative ($\partial C/\partial x < 0$), but by using equation 10.60, we see that $\partial C/\partial x > 0$, implying that the solute front migrates faster at high-concentration than at low-concentration—physically impossible in the case of a step front. Thus, the rate of propagation should be calculated using equation 10.57. Also, if the particular case of interest is not a step change, a sharpening effect typically occurs until a step front is actually established. Equation 10.60 indicates that for an unfavorable exchange at a C of interest, the concentration front flattens with increasing V_i; this produces solute spreading, or in simple terms, a decrease in the concentration gradient.

10.7 MATHEMATICAL MODELING OF TRANSPORT PHENOMENA IN SOILS

In the previous sections we discussed the basic physical processes and movement of solutes, the types of fluid flow that occur in the unsaturated zone, breakthrough curves, and nonreactive and reactive solutes. We now attempt to simplify the basic equations given thus far, with separate discussions of the various equations used for longitudinal dispersion, dispersion of a displacing front and a pulse, and various solutions and numerical calculations of these simple models, prior to discussing general solutions to the ADE.

Assuming that a fluid with an input concentration of C_0 is displacing another fluid of equal density and viscosity, there are basically two coordinate systems. One system is the moving-coordinate system of the fluid (Lagrangian method); the other, a fixed-coordinate system of the medium that allows physical measurement at specific points (Eulerian method). Without the fixed-coordinate system, it is difficult to measure between points within the moving fluid. For simplification purposes we also assume a coordinate system with space-coordinate x_1 and a time-variable of t_1. The transformation to the moving-coordinate system would be $x = x_1 + vt_1$ and $t = t_1$ where v is the velocity of the moving fluid, which also happens to be the velocity of the fixed-coordinate system. This allows for a fixed plane in the moving (x_1, t_1) system. Since the plane moves at velocity v and initially was at the sharp boundary (or interface) between the displacing and displaced fluid, the plane must always be $x_1 = 0$ in the moving-coordinate system. We determine $C(x_1, t_1)$ by the inverse such that $x = x_1 - t_1$ and $t_1 = t$ yields $C(x, t)$. By assuming dispersion in a capillary tube (linear flow), we have

$$q = -D\left(\frac{\partial C}{\partial x_1}\right) \tag{10.61}$$

where all parameters are as previously discussed. Also, following the rationale and considering the area, as stated in the solution of question 10.7 (at the end of the chapter), longitudinal dispersion is given as

$$\frac{\partial C}{\partial t_1} = D\left(\frac{\partial^2 C}{\partial x_1^2}\right) \tag{10.62}$$

which has the initial condition $C(x_1, t_1) = f(x_1)$ for $t_1 = 0$. The general solution for the dispersion equation is expressed as

$$C(x_1, t_1) = \frac{1}{2\sqrt{\pi D t_1}} \int_{-\infty}^{\infty} f(u) \exp\left[\frac{-(x_1 - u)^2}{4 D t_1}\right] du \tag{10.63}$$

This general solution is valid for $-\infty < x_1 < \infty$. By treating u as a dummy variable of integration that disappears when the definite integral involving u is evaluated,

$$f(u) = [C(x_1, 0)]_{x=U} = C(u, 0) \tag{10.64}$$

This is the formulated initial condition. As a final-boundary condition, we have $C(\infty, t_1) = 0$ for t_1 (finite) and $C(x_1, \infty) = 0$ for finite x_1. Consequently, $C(x_1, 0) =$ the concentration in the capillary tube at time $t_1 = t = 0$ and at $x_1 = x$ for which $x_1 = x$ is the starting position in the capillary tube of the fluid pulse that moves with dispersion; $C(x_1, t_1)$ depends on $C(x_1, 0)$.

QUESTION 10.8

Beginning with the longitudinal-dispersion equation 10.62, derive equation 10.63, the general solution of the one-dimensional dispersion equation.

10.8   FURTHER SOLUTIONS TO THE ADE: INITIAL AND BOUNDARY CONDITIONS

To obtain a solution to the ADE, we must select the appropriate initial and boundary conditions of the given transport problem, necessary and sufficient to guarantee a unique solution to the transport equation. Three types of boundary conditions are normally used in solute transport. These are: **(1)** the Dirichlet condition (boundary condition of the first kind), in which the value of a dependent variable is specified at every point of a boundary—sometimes referred to as a constant head or constant pressure boundary; **(2)** the Neumann condition, boundary condition of the second kind; in which the gradient of the pressure and gradient of the piezometric head are imposed on the boundary—sometimes called a constant-flux boundary; and **(3)** the Cauchy condition (boundary condition of the third kind), a mixed boundary condition in which the state variable of piezometric head and its gradient are imposed on the boundary. Also, a solution to the ADE has to consider the necessary source functions and constitutive relations. In addition to the initial and boundary conditions we also have to consider the types of concentrations the posed problem deals with. The typical concentrations used in transport phenomena are: time-averaged; volume or spatial averaging; and flux-averaged concentration. Mathematically, the time-averaged concentration C is expressed as

$$\overline{C_t}(x, y, z, t_0) = \frac{1}{\Delta t} \int_{t_0-(\delta t/2)}^{t_0+(\delta t/2)} C(x, y, z, t)\, dt \tag{10.65}$$

The spatial-averaged concentration (microscopic in this instance) is expressed as

$$\overline{C_v}(x_0, y_0, z_0, t) = \lim_{\Delta\, \delta v} \frac{\displaystyle\int_{\Delta v_L} C\, dv}{\displaystyle\int_{\Delta v_L} dv} \tag{10.66}$$

where C_v is the volume-averaged concentration, Δv is the soil volume of interest, Δv_L is the volume of the liquid phase in the Δv, dv is the (microscopic) differential-volume element, and δv is the representative elementary volume; all measured in L^3. The coordinates x_0, y_0, and z_0 are positions that are fixed at the center of the medium of interest. A representative elementary volume (REV) was defined for soil by Lauren et al. (1988) as a rule-of-thumb requiring a measurement area to contain at least 30 peds in cross-section, in order to be representative of a soil. This represents an estimate based on a morphological soil-structure analysis; the sample volume is primarily a function of soil structure, not the type of extraction device used to obtain an in-situ sample. The flux-averaged concentration (at a position of interest), $\overline{C_f}$, is expressed as

$$\overline{C_f}(t) = \frac{J_s}{J_v} \tag{10.67}$$

where the flux-averaged concentration represents the mass of solute per-unit volume of fluid passing through a specific cross-section for a specific time interval. Often, solute-flux distribution is of greater interest than pore-fluid concentrations (Parker and Van Genuchten 1984b). The basic relation between flux-averaged and volume-averaged concentrations is expressed as

$$\overline{C_f} = \overline{C_v} - \frac{D}{v}\frac{\partial \overline{C_v}}{\partial x} \tag{10.68}$$

It is important to be able to distinguish between volume-and flux-averaged concentrations during an experiment as well as during data analysis. Most research typically expresses concentrations during transport as volume-averaged concentrations denoted by C, and not necessarily by $\overline{C_v}$.

We use a column of soil as an example for selecting initial and boundary conditions. To solve the ADE for transport through our column, we need to select the inlet and exit boundary conditions and initial condition; following Van Genuchten and Alves (1982), we write the initial condition as $C(x, t) = f(x)$ for $x > 0$ and $t = 0$ where $f(x)$ is an arbitrary function. The boundary condition at $x = 0$ for the Dirichlet problem is expressed as $C(x,t) = g(t)$ for $x > 0$ and $t > 0$. For the Cauchy problem (third type), the boundary condition is expressed as:

$$\left(-D\frac{\partial C}{\partial x} + vC\right)\bigg|_{x\downarrow 0} = vg(t) \quad t > 0 \tag{10.69}$$

For this expression, $g(t)$ is also an arbitrary function that describes the concentration of the influent.

Continuing to use our column as an example, the exit-boundary conditions can be described for a column of finite length or a column of semi-infinite length. For the finite-length column, the exit-boundary condition that must be satisfied is

$$\left(-D\frac{\partial C}{\partial x} + vC\right)\bigg|_{x\uparrow L} = vC_z \quad t > 0 \tag{10.70}$$

where C_z is the concentration at the exit, which we assume equals $C\big|_{x\uparrow L}$. As a result $[\partial C/\partial x](x, t) = 0$ for $x\uparrow L$ and $t > 0$. For a semi-infinite column the exit-boundary condition can be expressed as $[\partial C/\partial x](x, t) = 0$ for $x \to \infty$ and $t > 0$.

Analytical solutions for both conditions are similar. A mathematical solution for the Cauchy problem yields a conservation of mass-type equation and is therefore preferred for the inlet-boundary condition. A transition zone develops when the fluid moves through the soil column, that results in a fluid-concentration variance from the influent concentration. This occurs because the influent is not well mixed, which results in a boundary layer outside the medium. Thus, a certain amount of time is required for equilibrium to be achieved. Although the initial-boundary conditions given for the Dirichlet problem (first type) imply equal concentrations in both the medium and influent, this is not the case at first because the influent solution can only be injected at a specific rate. As a result, we have a displacement experiment which involves a step change in concentration such that $(g(t) < 0) = 0$ and $g(t > 0) = C_0$, and a Cauchy-type condition should be used as

$$\left(-D\frac{\partial C}{\partial x} + vC\right)\bigg|_{x\downarrow 0} = vC_0 \tag{10.71}$$

where C_0 is the influent concentration. We are also aware that a discontinuity in C across the inlet boundary increases with increasing D/v. Solutions that are subject to first-type boundary conditions generally lead to flux-averaged concentrations, while solutions subject to the third-type (mixed or Cauchy) lead to volume-averaged concentrations.

Using this basic information we determine several analytical solutions for various inlet- and exit-boundary conditions. If we assume an infinite system ($-\infty < x < \infty$), a solution for the general ADE (equation 10.5) can be obtained (by making a coordinates-transformation so that $\xi = x - vt$ and $\tau = t$) which will transform equation 10.5 into

$$\frac{\partial C}{\partial \tau} = D \frac{\partial^2 C}{\partial \xi^2} \tag{10.72}$$

which has the original boundary and initial conditions of $C = C_i$ for $x > 0$ and $t < 0$; $x \to \infty$ and $t > 0$; also, $C = C_0$ for $x < 0$ and $t < 0$; $x \to -\infty$ and $t > 0$. These must be transformed as well. If we use an alternative transformation, then

$$\xi = \frac{x - vt}{\sqrt{4Dt}} \tag{10.73}$$

which yields an ordinary differential equation, such that

$$\frac{d^2 C}{d\xi^2} + 2\xi \frac{dC}{d\xi} = 0 \tag{10.74}$$

This equation has transformed initial and boundary conditions where $C = C_i$ for $\xi \to \infty$ and $C = C_0$ for $\xi \to -\infty$ with the solution

$$\overline{C} = \frac{(C - C_i)}{(C_0 - C_i)} = \frac{1}{2} \operatorname{erfc} \xi \tag{10.75}$$

We need to remember that coordinate transformation is not always convenient, and may not work for other types of boundary conditions.

Van Genuchten and Alves (1982) and Carslaw and Jaeger (1959) have shown that Laplace transforms can be an efficient tool for solving equation 10.5. Table 10.2 gives some analytical solutions for several inlet-and exit-boundary conditions.

TABLE 10.2 Analytical Solutions for Various Inlet and Exit Boundary Conditions (R assumed constant)

Inlet boundary condition	Exit boundary conditions	Analytical solution for selected boundary conditon
$\left(-D\dfrac{\partial C}{\partial x} + vC\right)\Big\|_{x=0} = vC_0$ case 1	$\dfrac{\partial C}{\partial x}(L, t) = 0$	$c = 1 - \displaystyle\sum_{m=1}^{\infty} \dfrac{\dfrac{2vL}{D}\beta_m\left[\beta_m \cos\left(\dfrac{\beta_m x}{L}\right) + \dfrac{vL}{2D}\sin\left(\dfrac{\beta_m x}{L}\right)\right]\exp\left[\dfrac{vx}{2D} - \dfrac{v^2 t}{4DR} - \dfrac{\beta_m^2 Dt}{L^2 R}\right]}{\left[\beta_m^2 + \left(\dfrac{vL}{2D}\right)^2 + \dfrac{vL}{D}\right]\left[\beta_m^2 + \left(\dfrac{vL}{2D}\right)^2\right]}$ $\beta_m \cot(\beta_m) - \dfrac{\beta_m^2 D}{vL} + \dfrac{vL}{4D} = 0$
$C(0, t) = C_0$ case 2	$\dfrac{\partial C}{\partial x}(L, t) = 0$	$c = 1 - \displaystyle\sum_{m=1}^{\infty} \dfrac{2\beta_m \sin\left(\dfrac{\beta_m x}{L}\right)\exp\left[\dfrac{vx}{2D} - \dfrac{v^2 t}{4DR} - \dfrac{\beta_m^2 Dt}{L^2 R}\right]}{\left[\beta_m^2 + \left(\dfrac{vL}{2D}\right)^2 + \dfrac{vL}{2D}\right]}$ $\beta_m \cot(\beta_m) + \dfrac{vL}{2D} = 0$
$\left(-D\dfrac{\partial C}{\partial x} + vC\right)\Big\|_{x=0} = vC_0$ case 3	$\dfrac{\partial C}{\partial x}(\infty, t) = 0$	$c = \dfrac{1}{2}\operatorname{erfc}\left[\dfrac{Rx - vt}{2\sqrt{DRt}}\right] + \sqrt{\dfrac{v^2 t}{\pi DR}}\exp\left[-\dfrac{(Rx - vt)^2}{4DRt}\right]$ $- \dfrac{1}{2}\left(1 + \dfrac{vx}{D} + \dfrac{v^2 t}{DR}\right)\exp\left(\dfrac{vx}{D}\right)\operatorname{erfc}\left[\dfrac{Rx + vt}{2\sqrt{DRt}}\right]$
$C(0, t) = C_0$ case 4	$\dfrac{\partial C}{\partial x}(\infty, t) = 0$	$c = \dfrac{1}{2}\operatorname{erfc}\left[\dfrac{Rx - vt}{2\sqrt{DRt}}\right] + \dfrac{1}{2}\exp\left(\dfrac{vx}{D}\right)\operatorname{erfc}\left[\dfrac{Rx + vt}{2\sqrt{DRt}}\right]$

Sources: Data from Brenner (1962); Cleary and Adrian (1973); Lindstrom (1976); and Lapidus and Amundson (1952) for cases 1 through 4, respectively.

TABLE 10.3   Solutions for C_x in Terms of the Peclet Number ($\mathcal{P}$) and Pore Volume, N_{pv} for the Analytical Solutions in Table 10.2

Case	Relative concentration of effluent upon exit from column

1
$$C_x(N_{pv}) = 1 - \sum_{m=1}^{\infty} \frac{2\beta_m \sin(\beta_m) \exp\left[\dfrac{\mathcal{P}}{2} - \dfrac{\mathcal{P}N_{pv}}{4R} - \dfrac{\beta_m^2 N_{pv}}{\mathcal{P}R}\right]}{\beta_m^2 + \dfrac{\mathcal{P}^2}{4} + \mathcal{P}}$$

$$\mathcal{P}\beta_m \cot(\beta_m) - \beta_m^2 + \frac{P^2}{4} = 0$$

2
$$C_x(N_{pv}) = 1 - \sum_{m=1}^{\infty} \frac{2\beta_m \sin(\beta_m) \exp\left[\dfrac{\mathcal{P}}{2} - \dfrac{\mathcal{P}N_{pv}}{4R} - \dfrac{\beta_m^2 N_{pv}}{\mathcal{P}R}\right]}{\beta_m^2 + \dfrac{\mathcal{P}^2}{4} + \dfrac{\mathcal{P}}{2}}$$

$$\beta_m \cot(\beta_m) + \frac{\mathcal{P}}{2} = 0$$

3
$$C_x(N_{pv}) = \frac{1}{2}\operatorname{erfc}\left[\sqrt{\frac{\mathcal{P}}{4RN_{pv}}}(R - N_{pv})\right] + \sqrt{\frac{\mathcal{P}N_{pv}}{\pi R}}\exp\left[-\frac{\mathcal{P}}{4RN_{pv}}(R - N_{pv})^2\right] - \frac{1}{2}\left(1 + \mathcal{P} + \frac{\mathcal{P}N_{pv}}{R}\right)\operatorname{erfc}\left[\sqrt{\frac{\mathcal{P}}{4RN_{pv}}}(R + N_{pv})\right]$$

4
$$C_x(N_{pv}) = \frac{1}{2}\operatorname{erfc}\left[\sqrt{\frac{\mathcal{P}}{4RN_{pv}}}(R - N_{pv})\right] + \frac{1}{2}\exp(\mathcal{P})\operatorname{erfc}\left[\sqrt{\frac{\mathcal{P}}{4RN_{pv}}}(R + N_{pv})\right]$$

Source: Data from Van Genuchten and Wierenga (1986).

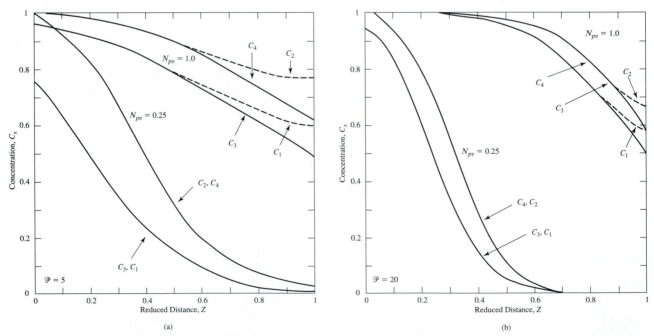

Figure 10.7   (a) A calculated concentration distribution for $R = 1$ and $\mathcal{P}$-value (Peclet number) of 5; (b) A calculated concentration distribution for $R = 1$ and $\mathcal{P}$-value for 20. N_{pv} refers to pore volume. These curves were obtained using the analytical solutions in table 10.3; C_1–C_4 refer to case 1 through case 4. Notice how solutions for C_1 and C_2 (dashed lines) deviate from remaining solutions.

We can rewrite equation 10.5 to reflect the Peclet number, $\mathcal{P}$, as one of its parameters. (The Peclet number and its effect on transport is discussed in detail in chapter 13.) The rewritten equation is expressed as

$$R \frac{\partial \left(\dfrac{C - C_i}{C_0 - C_i} \right)}{\partial \left(\dfrac{vt}{L} \right)} = \frac{1}{\mathcal{P}} \frac{\partial^2 \left(\dfrac{C - C_i}{C_0 - C_i} \right)}{\partial \left(\dfrac{x}{L} \right)^2} - \frac{\partial \left(\dfrac{C - C_i}{C_0 - C_i} \right)}{\partial \left(\dfrac{x}{L} \right)} \tag{10.76}$$

where $\mathcal{P} = vL/D$. Table 10.2 lists analytical expressions for $\overline{C}\big|_{(x/L)=1} = C_x$ where C_x is the exit concentration. The cases listed in table 10.3 correspond to the same initial and boundary conditions listed by case in table 10.2; the expressions listed in table 10.3 also involve the number of pore volumes (N_{pv}) along the column, which are expressed as vt/L.

Differences between the various boundary conditions on the concentration profiles are illustrated in figure 10.7a, b. Notice that at small t, the use of a first-type boundary condition (relative to other types) results in significant differences. Also, as solute concentration begins to increase at the column exit, a difference between the solutions for the finite and semi-infinite cases occurs. This is because the concentration at the inlet and outlet is not continuous nor is it likely to be. Generally, these solutions can predict solute concentration in a homogeneous media and can also be used to determine transport parameters such as D. As a general rule, the semi-infinite solution is favored.

10.9   NUMERICAL SOLUTIONS OF EQUILIBRIUM EXCHANGE

In the preceding sections, we discussed some fundamentals of solute transport, both with and without dispersion. In doing this, we obtained explicit expressions for the position of a solute front with approximate analytical solutions. This section focuses on numerical techniques that are normally required to solve the ADE because of the nonlinearity associated with exchange isotherms. An explicit finite-difference method was used by Lai and Jurinak (1971; 1972) to solve the ADE for nonlinear adsorption. The conditions assume use of a binary-exchange system for one-dimensional, steady-state flow and monovalent exchange. These authors introduce two dimensionless variables: $X_k = C_k/C_T$ and $Y_k = S_k/S_T$, which correspond to the solute-concentration in the liquid and adsorbed-phase, respectively. Lai and Jurinak (1972) obtain the following mathematical expression:

$$\frac{\partial X_k}{\partial t} = D(X_k) \frac{\partial^2 X_k}{\partial x^2} - v(X_k) \frac{\partial X_k}{\partial x} \tag{10.77}$$

where $D(X_k)$ represents the dispersion coefficient and $v(X_k)$ represents the velocity coefficient; both were adapted according to the following mathematical expressions

$$D(X_k) = \frac{D}{1 + \left(\dfrac{\rho_b S_T}{\theta C_T} \right) f'(X_k)}$$

$$v(X_k) = \frac{v}{1 + \left(\dfrac{\rho_b S_T}{\theta C_T} \right) f'(X_k)} \tag{10.78}$$

where $f'(X_k) = dY_k/dX_k$ and the denominator of equations 10.78 are simply the retardation factor R (as previously described), which depends on X_k because of nonlinearity in the exchange process.

For one-dimensional, steady-state conditions of multi-component equilibrium exchange in both homogeneous and layered media, Rubin and James (1973), showed that multiple fronts and plateau zones occur during multi-species transport problems. They used a varying total solute concentration in the liquid phase C_T. Using chromatography in a similar approach, Valocchi, Street, and Robert (1981) developed an analytical approach for multi-species transport of exchanging solutes. By including hydrodynamic dispersion and assuming electrolyte concentrations are not constant, the governing equation is

$$\theta \frac{\partial C_k}{\partial t} + \rho_b \frac{\partial S_k}{\partial t} = \theta D \frac{\partial^2 C_k}{\partial x^2} - J_v \frac{\partial C_k}{\partial x} \quad k = 1, 2, \ldots, n \tag{10.79}$$

and

$$C_T = \sum_{k=1}^{n} C_k$$
$$S_T = \sum_{k=1}^{n} S_k \tag{10.80}$$

where C_T is the total-solute concentration in the liquid phase (variable) and S_T is the total-solute concentration in the adsorbed phase (assumed constant since it usually equals CEC). The exchange coefficient assumes an associated valence with the adsorbed species v_k, with a valence for the exchanging species j_k. Thus, the exchange coefficient is expressed as

$$K_{jk} = \left(\frac{Y_j}{X_j} \right)^{v_k} \left(\frac{X_k}{Y_k} \right)^{v_j} \tag{10.81}$$

where X and Y are dimensionless concentrations. The exchange coefficient, K_{jk}, is important because it affects the shape of the solute front. Equation 10.81 is a generalized solution; the use of activity coefficients makes it more accurate. However, in order to obtain a solution to the ADE, the number of dependent variables must be reduced. This is achieved by expressing S_k in terms of C_k, and by using the $n - 1$ independent-equilibrium expressions in combination with the second equation of equation 10.80. This leads to a multi-component exchange isotherm for a species k such that

$$S_k = F_k(C_1, C_2, \ldots, C_n) \quad k = 1, 2, \ldots, n \tag{10.82}$$

Substitution of equation 10.82 into equation 10.79 yields a set of n transport equations such that

$$\theta \frac{\partial C_k}{\partial t} + \rho_b \sum_{k=1}^{n} f_{jk} \frac{\partial C_j}{\partial t} = D \theta \frac{\partial^2 C_k}{\partial x^2} - J_v \frac{\partial C_k}{\partial x}$$
$$f_{jk} = \frac{\partial F_k}{\partial C_j} = \frac{\partial S_k}{\partial C_j} \tag{10.83}$$

Equation 10.83 is easily solved by the Galerkin finite-element method. In many transport problems, the solute front is assumed to travel at a rate proportional to $t^{1/2}$; however, for a concave isotherm, $K_{jk} < 1$, the profile (front) of X_k generally travels at a rate proportional to t, which is similar to dispersion under Fick's law. In the case of a convex isotherm, $K_{jk} > 1$, the X_k front becomes steeper with distance traveled in the medium; both Helfferich (1962) and Brunauer (1945) discuss many different types of isotherms. Normally, ion exchange is influenced or induced by dispersion because of the mixing across the solute front as it passes through the medium. Because of nonlinear exchange, the retardation factor R is not constant, however. By assuming an effective, constant value for R, the effective velocity of the solute front is expressed as

$$v_e = \frac{v}{R} = \frac{dx_p}{dt} \tag{10.84}$$

where x_p is the equivalent depth of penetration of the solute front. This provides a way to calculate a constant value for R which may be used in cases where small or trace quantities of species are present in solution, for which linear exchange is almost always assumed. Generally, a constant R can be used with reasonable predictability if dispersion is negligible.

Various approaches are used for obtaining numerical solutions to the ADE. A popular approach used by Robbins, Jurinak, and Wagenet (1980a) is to use separate subroutines in the computer program to calculate cation exchange, complexation, and precipitation/dissolution. There are numerous computer packages on the market that facilitate the simulation of equilibrium chemistry, that involve various components and reactions (see chapter 13 for a list of such models). However, the authors and other researchers (Jennings, Kirkner, and Theis 1982) suggest direct insertion of these parameters (exchange, complexation, etc.) into the main transport-equation being used, which has been the methodology discussed thus far—that is, the inclusion of R. This is because convergence problems arise when the ADE and the model describing exchange, complexation, and so on (equilibrium chemistry), are combined. Jennings, Kirkner, and Theis (1982) express the dependency of solid-phase concentration to various quantities such that $S = f(C, S_T, t, x, \partial C/\partial t)$, which can easily be solved by the Galerkin finite-element method (Pinder and Gray, 1977; Kirkner, Jennings, and Theis 1985).

Abriola (1987) discusses various research which has been performed on contaminant-transport modeling. Generally, we carefully consider the type of chemical reaction when choosing the numerical-solution technique to be used for the problem. For example, Rubin (1983) described six broad classes of chemical reactions that occur during contaminant transport; each has its own mathematical expression unique to the reactions taking place.

10.10   NONEQUILIBRIUM CONDITIONS

In addition to instantaneous equilibrium, which is generally assumed between a solute in the liquid and adsorbed phase (both of which occur due to the exchange isotherm), two kinds of nonequilibrium exist in transport studies. These include physical and chemical nonequilibrium. As was discussed in chapter 8, liquid flow in the vadose zone is primarily by water films, which may be periodically discontinuous.

In contrast to physical nonequilibrium, chemical nonequilibrium is caused by kinetic-adsorption and exchange processes. Often, these processes are not instantaneous and the typical approach for a solution is to combine the ADE with a rate equation for chemical adsorption within the medium of interest. The physical constituents of a medium (clay, organic matter, sand, etc.) have a wide variety of exchange sites that are generally classified as type 1 (instantaneous) and type 2 (time-dependent). A rough approximation of first-order kinetics as applied by Selim, Davidson, and Mansell (1976a) expresses the general sorption rates for these phenomena as

$$\frac{\partial S_1}{\partial t} = \frac{\theta}{\rho_b} k_1 C - k_2 S_1 \qquad (10.85)$$

where S_1 is the concentration of the sorbed solute (ML^{-1}) and k_1 and k_2 are the forward-rate reaction coefficients. This equation can also be written for backward reactions. For equilibrium conditions, the concentration of the sorbed solute can be expressed as

$$S_1 = \frac{\theta}{\rho_b} \frac{k_1}{k_2} C = K_1 C \qquad (10.86)$$

and the sorbed concentration for all sites is $S = S_1 + S_2 = (K_1 + K_2)C = KC$. For time-dependent sites we write

$$\frac{\partial S_2}{\partial t} = \alpha(K_2 C - S_2) \qquad (10.87)$$

where α is the first order rate coefficient (T^{-1}). Substitution into equation 10.5 (the normal ADE) yields

$$\left(1 + \frac{Kf\rho_b}{\theta}\right)\frac{\partial C}{\partial t} + \frac{\rho_b}{\theta}\frac{\partial S_2}{\partial t} = D\frac{\partial^2 C}{\partial x^2} - v\frac{\partial C}{\partial x} \tag{10.88}$$

where f is $S_1/S = K_1/K$, and the equation for the sorbed solute is expressed as

$$\frac{\partial S_2}{\partial t} = \alpha[(1 - f)KC - S_2] \tag{10.89}$$

and for a one-site model, $f = 0$ and the first term on the left-hand side of equation 10.88 drops out; also, S_2 is expressed simply as S; this also follows for equation 10.89.

A positive distinction between the physical and chemical nonequilibrium is not usually possible. As a result, Cameron and Klute (1977) conceived a "black box"-approach for describing the sorption process. Basically, they described two types of sites that are the same as those discussed for each of the models presented here; instantaneous and time-dependent, in which sorption of the time-dependent reaction is described kinetically and takes into account both the physical and chemical nonequilibrium conditions. Cameron and Klute (1977) divided their sites into two types expressing equilibrium (S_1) and kinetics (S_2), where exchange between ions in sorbed and liquid phases is via Freundlich (S_1) and kinetic (S_2) processes. Their solutions to the ADE involve Laplace transforms, and they were able to model atrazine, phosphorus, and silver transport successfully.

Nkedi-Kizza et al. (1983) fitted both a first-order reversible kinetic model and a diffusion-controlled model utilizing breakthrough curves. How fast equilibrium is achieved through the ion-exchange process is determined by two mechanisms: solute supply of the influent to the liquid–solid interface, and the nature of the exchange reaction. According to Helfferich (1962), the ion diffusion to the exchange sites is the rate-limiting step, even in cases of chemical nonequilibrium and when no immobile water is present. The physical nonequilibrium model is characterized by instantaneous equilibrium at mobile sites and diffusion-controlled equilibrium at immobile exchange sites, which is described by equation 10.111, later. For time-dependent sites, the kinetic nature of the exchange process is described by equations 10.88–89. If we express both models in dimensionless terms, they are the same and have nearly identical breakthrough curves.

Dependent on the physical parameters and phenomena being measured, the ADE can be generalized to include most equilibrium phenomena of interest in unsaturated zone hydrology. This includes: adsorption; precipitation; dissolution; radioactive decay; and additional chemical reactions. The following expression of the ADE was presented by Parker and Van Genuchten (1984a)

$$\frac{\rho_b}{\theta}\frac{\partial S}{\partial t} + \frac{\partial C}{\partial t} = D\frac{\partial^2 C}{\partial z^2} - v\frac{\partial C}{\partial z} - \mu_w C - \frac{\mu_s \rho_b}{\theta}S + \gamma_w + \frac{\gamma_s \rho}{\theta} \tag{10.90}$$

where μ_w and μ_s are rate constants based on first-order decay in the liquid and solid phase (T^{-1}) and γ_w $(\mathrm{ML}^{-3}T^{-1})$ and γ_s (T^{-1}) are rate constants for zero-order production in the liquid and solid phases.

A category of reactions that are classified as chemical nonequilibrium are those of the radioactive-decay chain. Considering the transport of a single radionuclide species, the governing equation for two-dimensional transport in a variably saturated media is expressed as

$$\frac{\partial}{\partial x_i}\left(D_{ij}\frac{\partial C}{\partial x_j}\right) - \frac{\partial}{\partial x_i}(v_i C) = \frac{\partial}{\partial t}[\phi S_w C + \rho_s(1 - \phi)C_s] - qC^* + \lambda[\phi S_w C + \rho_s(1 - \phi)C_s] \tag{10.91}$$

where D_{ij} is the apparent hydrodynamic-dispersion tensor, C is solute concentration, v_i is the Darcy velocity, ρ_s is particle density, ϕ is effective porosity, C_s is the adsorbed concentration,

λ is the first-order decay coefficient, and $C*$ is the solute concentration in the injected fluid. By assuming that the relation between adsorbed and solution concentration is described by a linear-equilibrium isotherm equation 10.91 can be expressed as

$$\frac{\partial}{\partial x_i}\left(D_{ij}\frac{\partial C}{\partial x_j}\right) - \frac{\partial}{\partial x_i}(v_i C) = \frac{\partial}{\partial t}[\phi S_w RC] - qC* + \lambda\phi S_w RC \tag{10.92}$$

where S_w is the water-phase saturation expressed as a percentage (also referred to as θ in some publications) and R, the retardation factor, is expressed as

$$R = 1 + \frac{\rho_s(1 - \phi)k_d}{\phi S_w} = 1 + \frac{\rho_b k_d}{\phi S_w} \tag{10.93}$$

By expanding the convective and mass-accumulation terms of equation 10.92 and by using the continuity equation of fluid flow, and assuming that the time derivative of $(\rho_b k_d)$ is negligible, equation 10.92 reduces to

$$\frac{\partial}{\partial x_i}\left(D_{ij}\frac{\partial C}{\partial x_j}\right) - v_i \frac{\partial C}{\partial x_i} = \phi S_w R\left(\frac{\partial C}{\partial t} + \lambda C\right) + q(C - C*) \tag{10.94}$$

The term $q(C - C*)$ is zero in cases where q corresponds to the specific discharge of a pumped well because $C \equiv C*$. The hydrodynamic dispersion tensorial components can be computed by the method of Scheidegger (1961) such that

$$D_{11} = \frac{D_L(v_1)^2}{|v|} + \frac{D_T(v_2)^2}{|v|} + D^0 \tag{10.95}$$

$$D_{22} = \frac{D_L(v_2)^2}{|v|} + \frac{D_T(v_1)^2}{|v|} + D^0 \tag{10.96}$$

$$D_{12} = D_{21} = (D_L - D_T)\frac{v_1 v_2}{|v|} \tag{10.97}$$

where the subscripts L and T of D refer to longitudinal and transverse dispersion and D^0 (apparent molecular-diffusion coefficient) $= \phi\tau D_m$, where D_m is the free-water molecular-diffusion coefficient and τ is evaluated using the relation of Millington and Quirk (1961), such that $\tau = \phi^{4/3}S_w^{10/3}$.

For the transport of a chain of decaying-solute species, the following equation can be used

$$\frac{\partial}{\partial x_i}\left(D_{ij}\frac{\partial C_l}{\partial x_j}\right) - v_i\frac{\partial C_l}{\partial x_i} = \phi S_w R_l\left(\frac{\partial C_i}{\partial t} + \lambda_l C_l\right) + q(C_l - C_l^*)$$

$$- \sum_{m=1}^{M} \phi S_w \xi_{lm} R_m \lambda_m C_m, \quad l = 1, \ldots, n_c \tag{10.98}$$

where the subscript l denotes the chemical component, n_c is the number of components in the chain, ξ_{lm} is the mass fraction of the parent component m transforming to the daughter component l, and M is the number of parents transforming to l. For radioactive decay, λ_l is related to the half-life by $\lambda_l = ln\,2/(t_{1/2,l})$. Initial and boundary conditions associated with equation 10.98 can be expressed as

$$C_l(x_1, x_2, 0) = C_{l0} \tag{10.99}$$

$$C_l(x_1, x_2, t) = \overline{C_l} \quad \text{on} \quad B_1 \tag{10.100}$$

$$D_{ij}\frac{\partial C_l}{\partial x_j}\,n_i = q_{Cl}^D \quad \text{on} \quad B_2 \tag{10.101}$$

$$D_{ij}\frac{\partial C_l}{\partial x_j}\,n_i - v_i n_i C_l = q_{Cl}^T \quad \text{on} \quad B_3 \tag{10.102}$$

For a decaying source, the prescribed concentration, $\overline{C}$, or prescribed contaminant flux, q_{Cl}^{T}, is time-dependent and governed by Bateman's equation:

$$\frac{d\overline{C}_l}{dt} = -\lambda_l \overline{C}_l + \sum_{m=1}^{M} \xi_{lm} \lambda_m \overline{C}_m \quad \text{where} \quad \overline{C}_l(t=0) = \overline{C}_l^0 \tag{10.103}$$

For an n_c-member chain, the general solution of equation 10.103, subject to the prescribed concentration at $t = 0$, is expressed as

$$
\overline{C}_l = \overline{C}_l^0 \, e^{-\lambda_l t} + \xi_{l,l-1} \lambda_{l-1} \overline{C}_{l-1}^0 \sum_{m=l-1}^{l} \frac{e^{-\lambda_m t}}{l(\lambda_k - \lambda_m) \underset{\substack{k=l-1 \\ k \neq m}}{\Pi}}
$$

$$
+ \cdots + \xi_{l,l-1} \lambda_{l-1} \xi_{l-1,l-2} \ldots \xi_{21} \lambda_1 \overline{C}_l^0 \cdot \sum_{m=1}^{l} \frac{e^{-\lambda_m t}}{l(\lambda_k - \lambda_m) \underset{\substack{k=1 \\ k \neq m}}{\Pi}}
\tag{10.104}
$$

where l ranges from 1 to n_C, and $\overline{C}_l^0$ ($l = 1, \ldots, n_c$) are initial source concentrations of components 1 through n_c, and λ_1 ($l = 1, \ldots, n_C$) are source-decay coefficients. Equation 10.104 assumes a step release of dissolved contaminants at the source—that is, all waste material begins to dissolve at $t = 0$ and distribution is allowed to proceed continuously at a uniform rate. If we wish to perform a pulse release of duration T of the contaminant, equation 10.104 must be modified by multiplying the right-hand side by $f = [U(t) - U(t - T)]$ where $U(t - T)$ is the Heaviside unit function, defined by

$$U(t - T) = \begin{pmatrix} 1, t \geq T \\ 0, t < T \end{pmatrix} \tag{10.105}$$

The use of the equations for chained-decay transport in this section makes some major assumptions on their adequate use. These assumptions presuppose that the air phase is static—that is, water is the only flowing fluid phase; flow of the fluid phase is considered isothermal and governed by Darcy's law; the fluid is only slightly compressible and homogeneous; and the medium is also homogeneous. Also, solute transport is governed by Fick's law and adsorption and decay are described by a linear-equilibrium isotherm and a first-order decay constant. For the case of a straight three-member chain, an analytical solution can be found

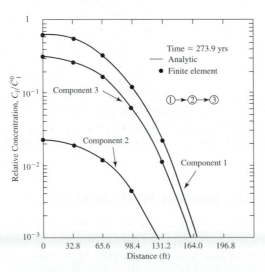

Figure 10.8   Simulated concentration profiles for the case of straight chain radionuclide decay, showing comparison of analytical and finite element solutions (data from Huyakorn et al. 1992)

TABLE 10.4   Parameter values for one-dimensional transport of three-member radioactive-decay chains

Component l	Radionuclide properties			
	$t_{1/2}$ yrs	λ_l yrs^{-1}	R_l	C_l^0
1	433	0.0016	9352	1
2	15	0.0462	9352	0
3	6540	0.0001	9352	0

in Coats and Smith (1964). Using both analytical and numerical solutions, the transport of a chain of three radionuclides released from a source located at $x = 0$ in a confined porous medium reservoir, is shown in figure 10.8. The properties used for transport calculation for the three radionuclides are listed in table 10.4.

10.11   COMBINED EFFECTS OF ION EXCHANGE AND DISPERSION

In chapter 7 we discussed how hydraulic conductivity could be calculated through a layered soil by summing the resistance associated with each layer. Using a similar idea and analytical techniques, the physical-dispersion effects for an unsaturated soil can be combined into one dispersion length (L_D). This was done by Bolt (1982), who attempted to include ion exchange as well. Frissel, Poelstra, and Reiniger (1970) showed that a column-averaged L_D value could be used. Assuming constant L_D, equation 10.49 can be used, and if one also assumes a linear exchange (dS/dC) is constant and can be divided through by R, the analytical solution for the resulting equation can be used, the basic ADE.

For favorable exchange, the solute front in the liquid phase generally lags behind the solute front in the adsorbed phase. In terms of effective retarded velocity (v_r), equal to the right-hand side of equation 10.57 at steady-state, solutes at low concentration will travel at velocities less than v_r, while those at high concentration will travel at velocities greater than v_r. With time, a steady front will be formed with respect to the moving coordinate, steady-state. For steady-state, the propagation of the solute front is expressed as

$$\left(\frac{\partial x}{\partial t}\right)_C = \frac{J_v}{(dS/dC) + \theta}\left[1 - L_D\frac{\partial(\partial C/\partial x)}{\partial C}\right] \tag{10.106}$$

for which Bolt (1982) determined (analytically) the adsorbed and liquid concentrations.

For unfavorable exchange, the ADE must be solved numerically to describe the solute front. This is due to the lack of development of a steady front with respect to the moving co-ordinate, because dispersion, diffusion, and ion exchange cause high concentrations to travel slower than low concentrations.

10.12   MOBILE AND IMMOBILE REGIONS IN SOILS

Due to the structure of soil, pockets of air are generally present within soil pores regardless of water content. It is well known, that after purging a medium with carbon dioxide and then saturating it from bottom to top, there is entrapped air that is virtually impossible to remove. The description of mobile and immobile regions within a medium has been widely used, in direct relation to this phenomena. The concept of mobile and stagnant areas of transport was first used by Coats and Smith (1964) later expanded on by Van Genuchten and Wierenga (1976), and has been widely used and explained by numerous other scientists since then. The concept is simple—wetted-pore space is represented by two different water contents: a mobile-water content θ_m, through which water flows; and an immobile water content $\theta_{im} = \theta_s - \theta_m$, in which water is stagnant and does not flow, but which can be pushed out at various times due to high-recharge events. The concentration of solute under these conditions is represented by the same subscripts. As we might suspect, a solute in the mobile phase is transported via advection and dispersion due to water flow, while the solute in the immobile (stagnant) area is transported via diffusion processes that are typically rate-limited.

Normally, as aggregate size increases (as in a well-structured clay) and velocity of flow decreases, the amount of immobile water will increase. The concept of mobile and immobile water not only applies to structured soils, but also to other media types according to De Smedt and Wierenga (1979a, b, 1984), who applied this argument to glass beads (an unstructured system); it was discovered that concentration of the effluent was accurately predicted only when accounting for both types of water content. Without use of the immobile phase, a very high dispersion coefficient is needed in order to fit the experimental data with the ADE. However, Philip (1968a, b) discovered that only during initial stages of the diffusion process was transient diffusion different for mobile and immobile regions. Philip (1968) also found that during initial infiltration, the dimensionless cumulative-diffusive flux obeyed a $t^{1/2}$ time law for the mobile region and the immobile region obeyed a $t^{3/2}$ time law, but shortly after the initiation of the diffusive process, the cumulative-diffusive flux of both regions again obeyed a $t^{1/2}$ time law. Consequently, he argued that dead-end porosity is inconsequential and that there is an unwarranted distinction between the two regions of flow; other researchers (Smiles et al. 1978b; Warrick, Biggar, and Nielsen 1971; Kirda, Nielsen, and Biggar 1974) agree with Philip, but only for non-aggregated media. This is due primarily to the presence of macropores and preferential pathways in various soils found to be significant in transporting herbicides and other chemicals (White 1985; Edwards et al. 1990; Delin and Landon 1993; Tindall and Vencill 1993, 1995).

Such disagreements are not uncommon due to spatial variability of the medium used by the researchers, as well as the physical, chemical, and biological characteristics of both the medium and the chemical being studied. The above argument is similar to the determination by various researchers, as to whether or not the dispersion coefficient D is either velocity-dependent or velocity-independent. Smiles and Philip (1978a) found no dependence of D on v during infiltration of KCl into a mixture of kaolinite and sand. Pfannkuch (1963) reported that D was independent of v but in order for this to be true, the Peclet number, $\mathcal{P}$, was less than one (i.e., $\mathcal{P} = vd/D_e$). Watson and Jones (1982) found D to be dependent on v, but they used high $\mathcal{P}$ numbers ($P > 35$), whereas the previous researchers were using $\mathcal{P} < 1$. This is plausible since dispersion is the more active process at higher pore-water velocities, and diffusion is more prevalent at low pore-water velocities.

Aggregated Soils

In the earlier section on breakthrough curves we discussed how aggregated media affects the shifting and shape of a typical breakthrough curve. Water flow through media with large aggregates is complex; it is directly related to both immobile and mobile regions within the medium, and the amount of preferential flowpaths within the medium (preferential flowpaths and macropores are discussed in the next section). As a result, there has been considerable recent attention to the transport of industrial and agricultural chemicals through aggregated media, because chemicals and water applied to them can be lost due to bypass, preferential flow, or macropore flow. Not only is flow different, but chemical and microbiological effects on the applied (or spilled) chemical to the medium can be quite different than in less-structured media; this applies particularly to inter- versus intra-aggregate pore space within the medium.

Because of these processes, the resulting source/sink terms are more complex, for which the ADE must be modified significantly to explain and describe exchange inside the solid (aggregate) particles. Within aggregated media, we normally need to distinguish the difference between micropores and macropores. Micropores are located inside the aggregates, through which flow is a diffusive process; macropores are located between the aggregates, through which flow is advective (sometimes called dispersive, viscous, or convective). When pore-water volume increases during rainfall or other recharge events, advective flow can completely dominate the system. It is within such systems that the definition of mobile versus immobile regions plays a major role in formulation of the ADE and its solution.

Passioura (1971) and Passioura and Rose (1971) obtained the following simplified form of the ADE to describe flow and transport in aggregated media

$$\frac{\theta_{im}}{\theta_{mo}}\frac{\partial C_{im}}{\partial t} + \frac{\partial C_{mo}}{\partial t} = D_{mo}\frac{\partial^2 C_{mo}}{\partial x^2} - v_{mo}\frac{\partial C_{mo}}{\partial x} \tag{10.107}$$

where the subscripts *mo* and *im* refer to mobile and immobile phases, respectively. The immobile phase inside the aggregate is treated as a distributed sink and D_{mo} is the diffusion coefficient of the mobile-liquid phase. The concentration of the chemical in both phases for equation 10.107 is not usually known, so we have to resort to a simpler form of the ADE (equation 10.5), that uses an effective dispersion coefficient D. This entails a combination of advective dispersion in the mobile phase of the aggregated media, and the stagnant phase within the aggregate. If one assumes steady-state, then the implication is that

$$\frac{\partial C_{mo}}{\partial x} = \frac{\partial C}{\partial x} \tag{10.108}$$

Following this assumption, Passioura (1971) developed the following expression for the dispersion coefficient within an aggregate

$$D_{im} = \frac{\theta_{im}}{\theta_T}\frac{v^2 a^2}{15 D_{eim}} \tag{10.109}$$

where θ_T is the total-volume fraction (air and water) and the subscript *e* denotes effective-dispersion coefficient. The diffusion coefficient for intra-aggregate (macropore) flow can be expressed by

$$D_{mo} = D_{emo} + v_{mo}a \tag{10.110}$$

where v is aggregate pore-water velocity and a is aggregate radius. Measurement of a strict θ in either phase is not easy to obtain; however, θ_{im} can be considered to be the water content at matric potentials less than 7–10 kPa (Rose and Passioura 1971a; Tindall and Vencill 1995). Thus, after obtaining θ_T and that for the immobile phase (based on matric potential), the water content for the mobile phase can be determined by difference.

For aggregated media, flow is also primarily in the void space between aggregates. For large-aggregate sizes, unsaturated flow is more likely to be discontinuous than for small-particle sizes. The flow in large aggregates can become discontinuous at higher-water contents due to entrapped air. Because of aggregate size, the influent solute will not reach all sorption sites immediately, and instantaneous exchange cannot occur. Mobile and immobile flow regions occur within an aggregated soil, where the immobile region is within the aggregates. This results in advective flow in the mobile regions (between aggregates) and diffusive flow within the aggregates, which causes the concentration within the immobile region to lag behind that in the mobile region; this is true for both the liquid and adsorbed phase. As one might suspect, the greatest number of adsorption sites is within the aggregate, and after an initial-recharge event, flow through the medium becomes a reversible, diffusion-controlled event between the mobile and immobile regions. For a mobile–immobile region medium, considering physical nonequilibrium, Van Genuchten and Wierenga (1976) expressed the ADE as

$$\theta_{mo}\frac{\partial C_{mo}}{\partial t} + \theta_{im}\frac{\partial C_{im}}{\partial t} + \rho_b f\frac{\partial S_{mo}}{\partial t} + \rho_b(1-f)\frac{\partial S_{im}}{\partial t} = \theta_{mo}D\frac{\partial^2 C_{mo}}{\partial x^2} - \theta_{mo}v_{mo}\frac{\partial C_{mo}}{\partial x} \quad (10.111)$$

This equation was modified after Coats and Smith (1964), where f is the fraction of sorption sites in direct contact with the mobile region of the liquid, and m and im subscripts refer to mobile and immobile regions. Equation 10.111 is exactly the same as equation 10.107, but with the sorption terms. For transfer in the liquid phase between the two regions we write

$$\theta_{im}\frac{\partial C_{im}}{\partial t} + \rho_b(1-f)\frac{\partial S_{im}}{\partial t} = \alpha(C_m - C_{im}) \quad (10.112)$$

where α is the mass-transfer coefficient with units T^{-1}. Other processes within the medium correspond to these same types of parameters. An excellent example is anion exclusion, in which the volume of exclusion is basically equivalent to the immobile region expressed above.

The presence of aggregates in soil causes limited accessibility of fluid to adsorption and exchange sites; this is generally due to physical nonequilibrium, where it is often appropriate to utilize an effective-dispersion coefficient for aggregates of particular shapes or sizes, since it allows omission of the source-sink term describing mass transfer in the stagnant-liquid phase. Adsorption in mobile and immobile regions within the vadose zone have been treated by Van Genuchten and Wierenga (1976), Selim, Schulin, and Fluhler (1987), Bolt (1982), and others. Without longitudinal diffusion or dispersion, as in the previous equation, the ADE is written as

$$\theta_{mo}\frac{\left(\dfrac{\Delta S}{\Delta C} + \theta_{im}\right)}{\theta_{mo}}\frac{\partial C_{im}}{\partial t} + \theta_m\frac{\partial C_{mo}}{\partial t} + J_v\frac{\partial C_{mo}}{\partial x} = 0 \quad (10.113)$$

which is easily solved by using a transformation to obtain scaled variables and choosing a position-dependent time. Van Genuchten (1981) calls this solution a Goldstein J-function.

While the approach above is useful, the stagnant-phase effect can be described after Crank (1975) by an equivalent length parameter (L_r). For spherical aggregates with radius r, Crank's expression in the event of ion exchange is

$$\frac{\partial\left(\dfrac{dS}{dC} + \theta_{im}C_{im}\right)}{\partial t} = \frac{15}{r^2}\theta_{im}D(C_{mo} - C_{im}) \quad (10.114)$$

and

$$k_a = \frac{15D}{r^2}\frac{\theta_{im}}{\theta_{mo}} \quad (10.115)$$

where k_a is a rate constant (T^{-1}) for diffusion inside an aggregate, for which instantaneous chemical equilibrium is assumed. This allows the expression of L_r in terms of k_a. The primary advantage of this method is that all effects are taken into account by one value for a dispersion length that allows the ADE to be solved analytically.

Using a nonreactive solute, Scotter (1978) found that theoretical-breakthrough curves, in the presence of large channels, indicated a considerable amount of solute appeared in the effluent before one pore volume had leached through. Using Scotter's (1978) pore-geometry approach for cylindrical channels, Van Genuchten, Tang, and Guennelon (1984) developed an analytical solution for transport through cylindrical macropores. In this instance, the cylindrical macropore represents the mobile phase and the surrounding medium represents the smaller pores or immobile-phase of liquid flow. Van Genuchten, Tang, and Guennelon (1984) also used fractions to represent adsorption sites in contact with the mobile and immobile phases, as well as separate retardation factors for each phase. During the initial or early-time, the medium surrounding a macropore is considered semi-infinite, and dispersion within the macropore can generally be ignored without loss in predictability. This approach is now widely accepted when describing solute transport in aggregated soils.

QUESTION 10.9

Assuming linear adsorption for both the mobile and immobile phase, write a general form of the ADE for reactive-solute transport.

Fractured Media

Interest in transport through fractured media has increased in recent years because of the desire to dispose of hazardous wastes in fractured bedrock. The formulation of the ADE for transport in fractured media is similar to that in aggregated media. For example, the macropore in aggregated soil becomes the fissure in fractured rock, which accounts for advective flow. Diffusive flow (or the immobile region) in fractured media is simply diffusion into the rock matrix rather than the aggregate, as in aggregated soils. Although some analytical solutions are available, most solutions are typically numerical.

The transport parameters and processes that need to be delineated within fractured media include: **(1)** advective flow and transport along the fracture length; **(2)** mechanical dispersion longitudinally along the fracture; **(3)** molecular diffusion from the fracture into the porous rock matrix; **(4)** molecular diffusion within the fracture itself (along the fracture axis); **(5)** adsorption onto the matrix face along the fracture; **(6)** adsorption within the porous rock matrix; and **(7)** radioactive (first-order) decay. As previously mentioned, diffusion of a chemical into an aggregate or rock matrix greatly reduces the transport time of the chemical, and allows for microbial decay, half-life degradation, and greater adsorption within the matrix. Tang, Frind, and Sudicky (1981) used the following equation for solute transport within a fractured media, using the general principles outlined so that

$$\frac{\partial C_{mo}}{\partial t} = \frac{D}{R_{mo}} \frac{\partial^2 C_{mo}}{\partial z^2} - \mu_b C_{mo} + \frac{\theta D_{eim}}{\partial x} \left. \frac{\partial C_{im}}{\partial x} \right|_{x=b} - \frac{v}{R_{mo}} \frac{\partial C_{mo}}{\partial z} \tag{10.116}$$

where μ_b is the decay constant (T^{-1}), $2b$ is the mean fracture width (L), $R_{mo} = 1 + (\rho_b/\theta)(K_{dmo})$ is the retardation factor, K_{dmo} is the coefficient of distribution within the fracture, and θ is the unsaturated water content. Transport inside the porous rock can be described by

$$\frac{\partial C_{im}}{\partial t} = \frac{D_{eim}}{R_{im}} \frac{\partial^2 C_{im}}{\partial x^2} - \mu_b C_{im} \tag{10.117}$$

where D_{eim} is the effective-diffusion coefficient within the rock matrix, $R_{im} = 1 + (\rho_b/\theta)(K_{dim})$ is the retardation factor, z is the vertical direction on the x, y, z plane, x is the horizontal direction, and K_{dim} is the coefficient of distribution within the rock matrix. For both equations 10.116 and 10.117, linear adsorption is assumed. By using Laplace transforms, Tang, Frind, and Sudicky (1981) solved these equations analytically for both specific initial and boundary conditions. Rasmuson and Neretnieks (1981) and others have also investigated the transport of decaying radionuclides through fractured rock. Generally, it has been discovered that advective dispersion within fractures results in longer travel distances for radionuclides, which has the effect of decreasing diffusion into the rock matrix; thus, movement is hinderered so that the radionuclide often decays before it reaches ground water. Consequently, matrix diffusion can prevent or reduce contamination of underlying aquifers with low-level, decaying radionuclides.

What has been termed a stratified approach was used by Neretnieks (1983) to determine transport of radionuclides through fractured rock; basically, this assumed flow through parallel channels of different size. This is analogous to Pouiselle's law, in which flow is assumed through a bundle of capillary tubes. For stratified flow, the width of the zone of dispersion, σ, is proportional to the distance traveled x instead of $x^{1/2}$, as in Fick's law of diffusion. Consequently, any increase in the observation distance yields a larger value for D if the ADE is used to describe transport in such media. The implication is that using Fick's law to extrapolate over large distances and times can have significant consequences, and error in some applications. Thus, for larger distance and residence times, the effects of stratified flow and matrix diffusion become major factors in comparison with the effects of hydrodynamic dispersion.

Layered Media

Transport of solutes through layered soil is a very important aspect of unsaturated zone hydrology because most natural systems (soil profiles) have a distinct stratification due to factors of profile development (see chapter 2). As a result, much research has been conducted on contaminant transport through layered soils and stratified aquifers. For example, many laboratory experiments are set up such that a loam-soil layer overlies a sand layer, or a sand or loam layer might overlay a clay layer so that there is a distinct discontinuity in characteristics. Typical experiments have investigated wetting fronts, water retention, and solute transport. By adding a short pulse of solute to the inlet end of layer one and measuring the effluent concentration at $z = L_1$, the time of solute travel can be determined for layer one. However, the travel time for the solute in underlying layer two cannot usually be measured directly because of the inability to apply a short pulse of solute to the inlet end of layer two. As a result, the concentration of the effluent is typically measured at $z = L_1 + L_2$ and time, $t = t_1 + t_2$.

To complicate matters further, each layer generally has different physical, chemical, and hydraulic properties; all of these affect sorption, microbial degradation, and subsequent transport of the chemical of interest. Thus, a different behavior in transport of the chemical should be expected in layered media. In cases where an overlying soil layer has significant advective transport compared to a minimal advective transport in the underlying layer, significant spreading and tailing of the solute front occurs when flow is parallel to stratification. In such cases, a large D is obtained if we use the generalized form of the ADE. It is naturally assumed that the media is heterogenous due to layering, and under such conditions Gillham et al. (1984) indicate that no Fickian dispersion would occur; but in the case of large transport times and nonuniform velocity fields, Fickian dispersion occurs (Gupta and Bhattacharya 1986). The latter case is generally expected, due to spatial variability of soils.

Sudicky, Gillham, and Frind (1985) investigated the transport of chloride in a thin sand layer between two silt layers that were water saturated. The solution was injected into the sand layer to investigate horizontal flow parallel to the silt layers; within the sand layer, advective flow occurred. From the sand to the silt layers, transverse diffusion occurred, and molecular diffusion occurred in the silt layers, which was transverse to the direction of flow. It is likely that complete mixing occurred in the sand, thus the general form of the ADE was

$$\frac{\partial C_{mo}}{\partial t} = D_T \frac{\partial^2 C_{mo}}{\partial y^2} - v \frac{\partial C_{mo}}{\partial x} \tag{10.118}$$

In this case, the diffusion from the sand to silt layers is transverse dispersion D_T, C_{mo} is the concentration of chloride in the sand layer, and the x direction is parallel to layering. The movement of the solute in the silt layers, for which there is no advective velocity, is represented by a diffusive form of the ADE such that

$$\frac{\partial C_{im}}{\partial t} = D_e \frac{\partial^2 C_{im}}{\partial y^2} \qquad (10.119)$$

where C_{im} is the concentration of solute in the silt layers and D_e is the effective-diffusion coefficient for the silt. This is a clear case of a nonuniform-velocity field, exhibiting strong dispersion due to interlayer transport of the chemical.

By assuming a linear-exchange constant K_d, Starr, Gillham, and Sudicky (1985) used the same system with a reactive solute, ^{85}Sr. The general form of the ADE could not be used for the reactive solute due to poor performance at high-flow velocities; thus, neglecting D_L, transport in the sand layer was described by

$$\frac{\partial C_{mo}}{\partial t} = \frac{\theta_{im}}{\theta_{mo}} \frac{D_e}{R_{mo}b} \frac{\partial C_{im}}{\partial y}\bigg|_{y=b} - \frac{v}{R_{mo}} \frac{\partial C_{mo}}{\partial x} - \mu C_{mo} \qquad (10.120)$$

where the width of the sand layer is $2b$ and other parameters are as previously discussed.

Marle et al. (1967) showed that depth-averaged moments can account for heterogeneity due to layering or stratification. Fischer et al. (1979) investigated transport in soil with depth h that was bounded by impermeable layers at $y = 0$ and $y = h$, and for which the magnitude of flow in the x-direction depends on y. For a moving-coordinate system, the transport equation is

$$\frac{\partial C}{\partial t} = D\left(\frac{\partial^2 C}{\partial \xi^2} + \frac{\partial^2 C}{\partial y^2}\right) - v' \frac{\partial C}{\partial \xi}$$

$$v' = v - \bar{v}; \qquad \xi = x - \bar{v}t \qquad (10.121)$$

where ξ is the cartesian coordinates in a moving-coordinate system, and the overbar notation indicates a depth-averaged value such that

$$\bar{v} = \frac{1}{d} \int_0^h v \, dy \qquad (10.122)$$

The spatial moments can be expressed as

$$m_p = \int_{-\infty}^{\infty} \xi^p C(\xi, y, t) \, d\xi \qquad (10.123)$$

The operator defined by equation 10.123 is applied to the individual terms of equation 10.34. If one assumes that $C = 0$ and $\partial C / \partial \xi = 0$ for $\xi \to \pm\infty$, then

$$\frac{\partial m_p}{\partial t} = D\left[p(p-1)m_{p-2} + \frac{\partial^2 m_p}{\partial y^2}\right] + v' p m_{p-1} \qquad (10.124)$$

Equation 10.124 is averaged over depth according to equation 10.122, resulting in

$$\frac{\partial \overline{m_p}}{\partial t} = Dp(p-1)\overline{m_{p-2}} + p\overline{v' m_{p-1}} \qquad (10.125)$$

The depth-averaged moment m_p is sequentially solved for $p = 0, 1, 2, \ldots, n$ and using the theory of statistical moments, we can characterize solute concentration and distribution by determination of the variance, skewness, mean, and other parameters. For an example, the reader can read Marle et al. (1967), which shows how the moment method can be applied to multi-layered soil. Other researchers (Fried and Combarnous 1971; Guven et al. 1984; Guven and Molz 1986) have used the moment method with good results.

Because soils often do not consist of homogeneous layers with distinct physical and hydraulic characteristics, this strongly influences the parameters within the ADE, leaving a complicated case of solute transport; this is especially true if velocity varies both longitudinally and transversely. As a result, many models use a black-box type approach, where each layer is assumed to have homogeneous properties. This approach has been used with fair success with the pesticide root-zone model, PRZM, developed by the U.S. Environmental Protection Agency (Carsel et al. 1984).

In undisturbed, naturally structured, and layered media with existing macropores and flow channels, only partial displacement of resident soil-water and solutes by incoming solutions occurs. Intact cores in the laboratory are a common tool to address questions involving the characterization and predictability of solute transport in such media. Also, core size has a significant effect on results; generally, the larger the core, the more comparable the results are to collected field data. Since heterogeneity of field soils normally is large, the use of laboratory-measured hydrodynamic-dispersion coefficients is sometimes not appropriate with the typical core sizes used. Bouma (1979) and Bouma et al. (1979) indicate that field variability is strongly reduced by using cores with a volume of 10 L or larger. Tindall, Hemmen, and Dowd (1992) and Tindall and Vencill (1995) have obtained good agreement between laboratory- and field-transport parameters for herbicides using intact cores 28 L in size—that is, 15 cm diameter by 40 cm height.

10.13   PREFERENTIAL FLOW PATHS, MACROPORES, AND FINGERING

In 1980, it was discovered that more than 1,000 wells on eastern Long Island, New York, had been contaminated by the herbicide aldicarb (Baier and Moran 1981). It has been generally perceived that herbicides and many other chemicals have a long residence time within soils, and are strongly adsorbed. Due to adsorption as well as biological and chemical processes, it was commonly assumed that most chemicals would be permanently degraded or "bound" within the soil profile, thereby preventing ground water contamination. Since the early 1980s, researchers have increasingly reported the unexpected rapid movement of chemicals through the vadose zone to ground water. This can be due in part to the greater technological advances in measurement techniques of various chemicals; however, the governing processes through which contamination has occurred have not changed.

Field and laboratory experiments on adsorption and leaching suggest that agricultural and industrial chemicals have relatively limited mobility, and most chemical residues are confined to the surface horizon of soils. However, when scientists observed chemical residues that were at greater depths in the profile than expected (based on classical assumptions of solute transport), doubts arose about sampling and analysis techniques, and unexplained point-source contamination. Eventually, scientists rediscovered the principle of preferential flow.

Lawes, Gilbert, and Warington (1882) described preferential flow over 100 years ago. These scientists described the basic difference between matrix flow and preferential flow, and illustrated that matrix flow will predominate in less-structured media, while in structured media such as clays, preferential flow will dominate; the preponderance of one type of flow compared to the other will have significant impacts on solute transport. Despite the fact that large pores were recognized as conduits for rapid flow, it was not until the 1970s that soil physicists started to quantify the impact of preferential flow on solute transport and related recharge of ground water. Also, the fact that preferential flow can occur does not mean that preferential transport of the chemical will also occur.

Preferential flow is simply the nonideal behavior of water flow in soil. To inform the student, nonideal behavior has a wide variety of terms associated with it: preferential flow;

macropore flow; fingering; wetting-front instability; bypass flow; channeling; short-circuiting; partial displacement; subsurface storm flow; and others. All of these depict essentially the same phenomena. This section attempts to describe nonideal flow by three commonly used terms. Preferential flow is the general term used to describe the movement of water and related solutes through preferred flow paths in soils.

We describe a preferential flow path as a pathway of preferred flow that has a lower bulk density than the surrounding soil matrix rather than a crack or fauna tunnel; thus, it offers the least resistance to flow under specific field conditions. This implies that part of the soil matrix is "bypassed" when flow occurs, especially during significant recharge events (Tindall and Vencill 1995). The preferential flow path is considered a mesopore by some, ranging in size from 10–1,000 μm (Luxmoore 1981). Macropores are defined as large cracks, worm holes, fauna tunnels, and channels created from decayed roots, with a size range of > 1,000 μm. However, for such large cracks to transport much fluid, the head (or pressure potential) within the medium has to exceed atmospheric potential. For example, when a large crack (larger than a micropore of 0–10 μm) within the profile is filled with air, water can enter the crack only if the potential within the matrix exceeds atmospheric potential. This is likely to happen only during significant runoff or recharge events, or in cases where a capillary fringe occurs directly above the crack. In this case, ponded water overcomes atmospheric potential, thereby forcing fluid into the crack where it can rapidly flow to deeper depths.

The pore sizes mentioned above for micro-, meso-, and macropores are based on the senior author's research, but are comparable to those of Luxmoore (1981). Various researchers have pore-classification scales describing these pore sizes (Luxmoore 1981; Jongcrius 1957; Johnson and McClelland 1960; Everett 1972; Greenland 1977; Landon 1984); thus, it is no surprise that the need for standard pore-size classification was raised by Luxmoore (1981), who suggested the sizes listed above. While strict agreement on the pore sizes representing the various pathways of preferential flow has yet to occur, the effects of these pathways on ground water contamination are in mutual agreement.

The terms fingering, wetting-front instability, and partial-volume flow are applied differently than macropore flow; they are associated with layered soils, air compression, hydrophobic soils, and water redistribution following ponding. Fingering is not usually visible in the field (as is the case with macropores) but has become visible when using dyes to stain such areas and then excavating these areas (Starr, Gillham, and Sudicky 1986). Numerous laboratory and field studies have been conducted investigating fingering, using applied and theoretical approaches and involving both analytical- and numerical-modeling procedures (Miller and Gardner 1962; Hill and Parlange 1972; Baker and Hillel 1990).

Many of these experiments have been performed using a layer of fine-textured soil over a layer that is much coarser. Raats (1973) and Philip (1975a, b) used the Green and Ampt (1911) infiltration model to approximate fingering, a reasonable approach when using a uniform, coarse-textured initially dry soil. However, the underlying assumptions are that the soil is subject to one-dimensional absorption and the lack of entrapped air; in this instance, an initially dry soil will exhibit a sharp wetting-front rather than a diffuse front. Also assumed is that the soil is initially dry, so this approach is limited. Small amounts of initial wetness are sufficient to reduce instabilities that occur if the media is at an initially dry state (Diment and Watson 1983, 1985; Philip 1975a, b). This is because the distribution of initial wetness affects the pathways that the fingers would take; also, Glass et al. (1989) showed that fingers can repeatedly form in the same location during successive infiltration cycles in initially wet sands.

Hillel and Baker (1988) suggested that when a wetting front is characterized by a high value of suction, the front will not penetrate into a coarse-textured layer from a fine-textured layer until the suction at the interlayer boundary falls to an effective water-entry suction ψ_e, which is characteristic of the sublayer. They concluded that if the sublayer conductivity, K_u, at its water-entry suction (expressed as $K_u(\psi_e)$), exceeds the flux through the top layer, Q_t, the pore-water velocity increases across the interlayer and only a small portion of the underlying layer conducts any water delivered to it, which results in fingering. Baker and Hillel (1990) proposed that fingering would not be initiated unless $K_u(\psi_e) > Q_t$, assuming steady-state flow, and that by using the quotient of these terms the wetted fraction, F, of the underlying layer could be predicted such that

$$F = \frac{Q_t}{K_u(\psi_e)} \tag{10.126}$$

Using a top layer of very fine sand overlying a layer of coarser texture, Baker and Hillel (1990) obtained good results with this model. Also, utilizing a linear regression of the relation between particle size and the effective suction of water entry ψ_e, then

$$\psi_{e(\text{pred})} = \frac{4370}{d_u} + 0.074 \tag{10.127}$$

where d_u is the median particle diameter (μm) of the underlying layer, and ψ_e is in cm. The conductivity (cm/s) is determined from the expression

$$K_u(\psi_e)_{\text{pred}} = 10^{[-(299/d_u) - 0.916]} \tag{10.128}$$

By utilizing Darcy's law (see chapter 7) for the top-layer flux (Hillel and Baker 1988),

$$Q_t = K_t \frac{(H_0 + z_i + \psi_e)}{z_i} \tag{10.129}$$

If one assumes $K_t \approx K_{ts}$ (the saturated hydraulic conductivity of the top layer), then by substitution

$$F_{\text{pred}} = \frac{Q_t}{K_u(\psi_e)} = \frac{K_{ts}\left(H_0 + z + \left[\left(\frac{4370}{d_u}\right) + 0.074\right]\right)}{z_i(10^{[-(299/d_u) - 0.916]})} \tag{10.130}$$

Generally, fingering is a nondeterministic process, unlike macropore flow. Upon wetting, the water or solute is (or is not) transported down the same finger as before, during subsequent recharge events. Hillel and Baker (1988) and Baker and Hillel (1990) suggest that particle size and θ are the common properties underlying the promotion and suppression of fingering. Despite this evidence, other properties such as particle-size distribution, initial-water content, hysteresis, and so on, need to be investigated since these parameters also appear to affect fingering.

The primary factors that cause nonideal transport of solutes are nonhomogenous physical and chemical media properties, and transport and sorption nonequilibrium. Preferential flow is usually a function of soil porosity that is separated into the portion of media occupied by gases and fluid, and more specifically, to the medium's ability to retain water at various matrix potentials (pressure or suction gradients). Pores can form naturally, especially in the upper profile, where microbiological processes are greatest. Other voids (such as cracks) can form during wetting and drying events and can further bifurcate, allowing possible connection with other cracks to form continuous channels and even peds. In agricultural and industrial settings in which no tillage or other type of surface-sealing activities take place, soil can have a high density of continuous macropores (Dick, Edwards, and Haghiri 1986). The mode of fluid flow is also dependent on initial water content of the media. Thus, edaphic, management, and meteorological factors ensure that both porosity and the potential for preferential flow are dynamic variables, both temporally and spatially.

QUESTION 10.10

Using a sublayer particle size, d_u, of 1000–2000 μm, determine the wetted fraction for this media. Assume the ponding depth H_0 is 1.0 cm, the layer thickness z_i is 5.8 cm, and the saturated conductivity K_{ts} is 2.0×10^{-3} cm/s. Also, ψ_e is 3.0 cm, $K_u (\psi_e)$ is 7.7×10^{-2} cm/s, and Q_t is 3.4×10^{-3}.

QUESTION 10.11

If the particle size d_u decreases, will the wetted fraction F increase or decrease?

Modeling Chemical Transport under Conditions of Preferential Flow

Chemical transport under nonideal conditions is dependent on pore-size distribution for which there are two general cases: **(1) Uniform distribution**—in the presence of a relatively narrow, uniform pore-size distribution (i.e., there is little variability); nonideality usually increases with a reduction in water content from saturation. With uniform saturated pores, dispersion is minimal. When such a medium desaturates, the pores empty slowly and fluid flow occurs mostly by thin water films attached to each particle. Thus, the pores become not only relatively "empty," but also the immobile domain. At a low-water content, much dispersion (diffusion in this case) is exhibited and the resulting breakthrough-curve shifts left. If the medium continues to desaturate, the emptiness of the pores results in a narrowly wetted pore-size distribution, and chemical transport becomes more ideal once again. As a result, nonideality might be compared to a bell-shaped curve: it increases as θ decreases from saturation, reaching a maximum value at some critical θ, then decreases as θ is further reduced. **(2) Nonuniform distribution**—generally, structured media have a wide range of pore sizes, many of which can be macropores. During saturated transport, nonideality is high because of pore-size distribution; however, as the macropores empty during desaturation, a narrowing of the effective porosity or pore-size distribution results. This reduces nonideality because the pores now conducting fluid are more uniform in size. It has been suggested that θ_{im} can be considered to be the water content at matric potentials less than about −7 to −10 kPa (Rose and Passioura 1971a, b; Tindall and Vencill 1995). Thus, the macropores transporting water empty at very low matric potentials near saturation; below this threshold, transport is via matrix flow and nonideality decreases as the soil desaturates. Despite this, there is a critical point at which further reduction in water content can create immobile regions and increased nonideality results, as in case 1.

In structured soils, chemical transport and fluid flow between the mobile and immobile regions are commonly determined by the use of Fick's law to describe physical, diffusive transfer (explicitly), by using empirical first-order mass-transfer expressions to obtain an average transfer; and implicitly, by using an effective-dispersion coefficient that includes a source/sink diffusion term, hydrodynamic dispersion, and axial diffusion. In this case, the lumped D replaces the usual hydrodynamic-dispersion coefficient.

Diffusion-based models that use Fick's law normally assume that aggregates have a spherical geometry. A similar type of model that utilizes a cylindrical geometry was developed by Van Genuchten, Tang, and Guennelon (1984) for predicting solute transport in media with macropores. A geometric description of the medium is required to develop the ADE for these model types. For mathematical simplicity, most of these models assume uniform-size aggregate structure, but for natural media these conditions do not usually exist. As a result, aggregate-size distribution and aggregate-shape variations have to be taken into account for these models to apply to a range and variety of cases. Rao, Jessup, and Addiscott (1981) showed that nonspherical aggregates can be represented by the use of equivalent-spherical aggregates whose radii are such that the volume of the sphere is equal to the volume of the nonspherical aggregate. Rasmuson (1982) showed that one aggregate shape could be approximated with another by employing the same ratio of external-aggregate surface area to total porosity. Van Genuchten (1985) developed a geometry-dependent shape factor (f) that can be used to transform an aggregate of specific shape and size into an equivalent sphere that has similar diffusion characteristics to the original aggregate.

Both distribution and shape variations of aggregates must be described to extend diffusion-based models to field situations. Rao et al. (1980a, b) developed a method that works well, utilizing a range of aggregate sizes to compute an equivalent radius from the volume-weighted radii of each size-class; this was further validated by Nkedi-Kizza et al. (1982). A combination of the shape transformations used by Van Genuchten (1985) with the aggregate-size distribution of Rao, Jessup, and Addiscott (1981), for a soil that contains varying sizes of nonspherical aggregate, allows the transformation of the medium to one of equivalent, spherical aggregates of uniform size. With such an approach, the diffusion model can be extended to a variety of field situations. Many models are currently available that attempt to predict preferential transport of chemicals in the vadose zone. Cooney, Adesanya, and Hines (1983) found that aggregate-size distribution ranges of less than ≈ 0.15 mm are narrow enough that the effects of nonuniform distribution can be ignored. Thus, the diffusion model used for predicting transport should be chosen carefully.

The physical mass-transfer model replaces the mechanistic description of diffusive transfer in the diffusion model, with a kinetic first-order mass-transfer expression. This no longer requires a description of the medium's structure because chemical transfer is now described by a mass-transfer coefficient, and is assumed to be a function of the difference in chemical concentration between the mobile and immobile regions. Coats and Smith (1964) performed early work on mass transfer in two-region flow, and subsequent expressions for the mass-transfer coefficient have been presented by Rao et al. (1980a, b), Raats (1984), and Park, Parker, and Valocchi (1986). Using these allows an independent determination of the mass-transfer coefficient and reduces its limitation on various soils.

The lumped (or effective-dispersion) model accounts for diffusive transfer between the mobile and immobile regions, implicitly. As the medium approaches equilibrium, the symmetry of the experimental breakthrough curve increases to a point at which asymmetry is very difficult to ascertain. At this point, nonequilibrium is evidenced by the increased dispersion. This model fits experimental data very well in such cases (Nkedi-Kizza et al. 1982; Lee et al. 1988). The diffusion and mass-transfer models have been compared by Rao et al. (1980b), Goltz (1986), and Miller and Weber (1986) and in all cases, the diffusion model gave similar results to the mass-transfer model. However, the increase in accuracy gained with the diffusion model does not justify the additional mathematical complexity or the larger number of parameters associated with it. As a result, the mass-transfer model is preferred by some researchers because of the major advantage it has in not requiring a description of soil structure.

Problems Encountered in Modeling Preferential Flow

The complex nature of soils and the formation of structured media—which can have large voids that may or may not be continuous—creates a lack of continuity. This lack of continuity causes problems in estimating chemical fluxes under preferential-flow conditions. At present, scientists have no way to describe the continuity of macropores or preferential flow-paths and thus, modeling is difficult. Generally, soil is treated with traditional physical concepts, including local equilibration of potentials and fluid flux that is proportional to the hydraulic gradient. In preferential flow, this is extremely important because, during wetting of soil that exhibits preferential flow characteristics, the wetting front penetrates the profile to significant depths, thus bypassing the intervening pore-space in the general matrix. This decreases the residence time and solid-phase interaction of solutes and also reduces the effects of retardation, adsorption, and degradation, thereby increasing the potential of ground water contamination.

The use of drainage rates and tracers indicates preferential-flow behavior and sometimes supports the validity of a given model under certain conditions, but it is often difficult to extrapolate information gained in one experiment, to other initial and boundary conditions for different sites and experiments. Much of our knowledge is based upon parameters obtained from destructive sampling, which have been limited to small samples or small column studies. A "representative elementary volume" (REV) is used to describe a volume of medium over which water content and potential are usefully defined. Bouma (1979) and Bouma, Dekker, and Haans (1979) recommended an REV of 10 L to escape the effects of spatial variability.

Analysis of the problems of previous research in modeling flow through structured media suggests that new models will have to consider **(1)** the variety of pore sizes in which preferential flow takes place, since it takes place not only in non-capillary-sized pores; **(2)** the continuity of preferential-flow paths except under saturated conditions; **(3)** since not all large pores are effective in conducting fluid, what part of the pore is interacting with the adjacent matrix; **(4)** investigation of laminar-film flow and its effectiveness in flow in large pores; **(5)** the climatological and other local conditions that are important in preferential flow; **(6)** colloidal facilitated transport through larger pores and the local geochemical reactions that may result in improved structure within large pore walls (i.e., cutans); **(7)** the effects of hysteresis in preferential flow; **(8)** the influence of bounding macropores on matrix flux in desaturated conditions; **(9)** the change, with time, of the structure of preferential-flow paths and macropores, that continually develop in undisturbed media and in soils that support agricultural crops, and trees, and others.

Various models have been developed in an attempt to deal with the predictability of transport under preferential-flow conditions. To summarize, these include: diffusion models; mass-transfer models; effective-dispersion models; multi-region flow models; mechanistic; statistical; statistical–mechanistic models; transfer-function models (described in chapter 13); and numerical models (Steenhuis, Parlange, and Andreini 1990). However, all of these are based on physical parameters that can be measured in the laboratory or the field, to characterize medium behavior. Theoretically, these parameters should be characteristic of a given medium at the scale of the measured REV but only for a profile that is homogeneous and unstructured. As a result, the values obtained for the measured parameters are not applicable in structured soils because such media are both heterogeneous and layered/structured, which requires a very large number of parameter values. These are nearly impossible to obtain without altering the nature of the system, so we generally ignore the problem and make an "assumption" (all encompassing) that the parameter of interest can be used at a profile or hill-slope scale. Interestingly enough, there is little guidance in estimating the "real" parameters, or on how to obtain effective values for them. At present, the best we can do is to use conservative tracers and various isotopes (see chapter 13) as well as intact cores of representative size.

Currently, both uncertainty and risk are associated with the predictions of various models; these need to be further evaluated in experimental studies. Models should be developed at the scale of interest for which they are intended to be used, or developed at a smaller scale and modified for use at larger scales, using accurately collected data. The current trend appears to be to develop more complex models because of the computational ability of advanced computer systems (Jarvis 1994). However, simply increasing the complexity of model structure at the expense of adding more parameters (extremely difficult to measure and identify), is not the best approach. For practical field- and regional-scale applications, models need to be developed for specific scales, and nondestructive measurement techniques must be developed that yield the necessary experimental information.

10.14   COLLOIDAL-FACILITATED TRANSPORT

The transport of contaminants to ground water via preferential pathways is of increasing concern as a major environmental problem in the U.S. Agricultural, industrial, and petroleum-derived chemicals may move to ground water as soluble constituents in soil water, or may be associated with soil-derived colloids that are capable of transporting such pollutants to the ground water. Pesticides, trace metals, and contaminant organics, that exhibit high adsorption and/or low water solubility, can move to ground water by this mechanism. The process is greatly exacerbated by the presence of preferential pathways.

Gaps exist in our current understanding of colloidal transport in soils and underlying geologic materials, a process likely ongoing for many years, that may explain the higher-than-expected concentrations of contaminants in ground water. Previous investigations have focused on homogeneous soil materials but have not fully accounted for the influence of aqueous and surface chemistry on colloid stability.

Previous investigations dealing with contaminant migration to subsurface environments have considered soil or ground water as a two-phase system, where contaminants could partition between the mobile-liquid phase and immobile-solid phase. Based on this approach, it was predicted that many contaminants would be relatively immobile because of low solubility or high-adsorption affinity for the solid phase. However, recent research suggests the existence of mobile–immobile conditions of the aqueous phase within soils, especially those containing large interconnecting systems of macropores. Thus, it is becoming increasingly evident that under certain (yet poorly defined) conditions, contaminant migration to and/or within ground water can be significantly enhanced by colloidal migration, where the colloidal phase itself is undergoing transport via preferential pathways (Rees 1987; Buddemeier and Hunt 1988; McCarthy and Zachara 1989). It has been suggested that radionuclides, organics (including PCBs, PAHs, and pesticides), and nonradioactive inorganic contaminants found in ground water may have migrated via colloidal-transport mechanisms, and currently, significant research efforts are focusing on this mechanism (Champ 1990; Nelson et al. 1985; Gschwend and Wu 1985; Jury, Elaboi, and Resketo 1986; Rees 1987; McDowell-Boyer, Hunt, and Sitar 1986; Buddemeier and Hunt 1988; Enfield, Bengtsson, and Lindquist 1989; Harvey et al. 1989; McCarthy and Zachara 1989; Ryan and Gschwend 1990; Penrose et al. 1990).

Colloidal-transport mechanisms have recently received considerable attention in popular scientific publications (Raloff 1990; McCarthy 1990; Jardine, Weber, and McCarthy 1990; Champ 1990; Looney, Newman, and Elzerman 1990; Gschwend 1990; Enfield and Kerr 1990). Fewer investigations have specifically addressed hydro-geochemical processes regulating the generation of stable colloidal suspensions (SCS) within soil, or the mechanisms operative in colloidal migration through the unsaturated zone. While a significant body of literature has addressed the theory and application of colloid stability and transport through well-defined, homogeneous media (Yao, Habibian, and O'Melia 1971; O'Melia 1980), the process governing the generation of SCS in intact soil and unconsolidated geologic materials has received less attention (McCarthy and Zachara 1989).

Colloid Migration

Qualitatively, the transport of colloids within soil is dependent on three fundamental processes: **(1)** generation of colloids; **(2)** stabilization of colloidal suspensions; and **(3)** unattenuated transport—that is, the transport without aggregation, sorption, or filtration within the soil as in preferential pathways (McCarthy and Zachara 1989). The generation and transport of three general classes of colloidal material in soils has been identified: inorganic colloids, microorganisms, and organic molecules. These materials can be important in contaminant mobility, both as facilitator of transport to ground water, and as a source of ground water colloids capable of remobilizing contaminants within the subsurface.

Eluvial and illuvial processes—whereby clay and other inorganic colloids are transported vertically from surface soils and redeposited in lower strata—have long been recognized as important pedogenic processes (Jenny and Smith 1935). Clay migration within soil profiles previously has been evaluated using a number of radio-labeled clay suspensions. Bertrand and Sor (1962) examined the influence of rainfall intensity on soil structure, and migration of colloidal materials via application of ^{86}Rb labeled clays. ^{32}P (Kazo and Gruber 1962), ^{90}Sr (Von Reichenbach and Von der Bussche 1963), ^{60}CO (Toth and Alderfer 1960), and ^{59}Fe (Woolridge 1965; Coutts et al. 1968a, b) have also been utilized to examine lateral and vertical transport of soil particles.

Generation of colloidal clay and iron, aluminum, or mixed iron–aluminum colloidal soils is achieved through dispersion of the particles or dissolution of cementing agents. These particles are also formed by dissolution and transport of dissolved constituents, with subsequent reprecipitation of clays or soils. The stabilization of the colloidal suspensions is highly dependent on soil-solution chemistry, with counterion type and concentration as two of the most important variables. Traditionally, colloidal-suspension stability has been explained using DLVO theory (Sposito 1984; see chapter 3). Whether colloids are transported through the soil matrix or deposited within that matrix is usually related to molecular sieving, with the intent of colloidal aggregation related to sieving efficiencies. Similar discussions of coagulation and filtration theory with regard to water and waste-water treatment (Yao, Habibian, and O'Melia 1971; O'Melia 1980) emphasize the importance of chemical factors in particle-removal efficiency.

Clay illuviation and all colloidal transport through soils implicate the role of preferential flow through macropores and other soil discontinuities. Akamigbo and Dalrymple (1985) generated illuvial-intrapedal cutans (e.g., clay coatings on soil peds) by applying dilute clay suspensions to undisturbed blocks of soil. That clay could be transported through intact B horizon material via natural channels (macropores) associated with ped faces was also demonstrated in this investigation, and was observed by Anderson and Bouma (1977a, b). Such preferential flow can occur within earthworm channels (Ehlers 1975), along ped faces (Anderson and Bouma 1977a, b); and other discontinuities within the unsaturated zone, including particle-size changes in unstructured material (Glass, Steenhusin and Pavlange 1988). The ability of soil macropores and other preferential pathways to conduct significant volumes of water, and the implications of this flow in terms of minimal interaction (e.g., sieving) of dissolved and suspended components by the bulk-soil matrix, has been discussed in detail by Thomas and Phillips (1979) and Beven and Germann (1982).

The transport of colloidal-sized macroorganisms in soils has also been observed. Wollum and Cassel (1978) note that *Streptomycete conidia* were easily transported through sand columns under saturated conditions. Transport of *Escherichia coli* through intact and disturbed soil cores (Smith et al. 1985) and in the field (Rahe et al. 1978), has also been observed. Indications are that soil structure (e.g., macropores) influences the extent of transport, with movement of *coliform* up to 830 m reported by Hagedorn et al. (1981). Additionally, recent studies have provided evidence of in situ transport of bacteria and microspheres through an aquifer, although the processes regulating transport could not be delineated (Harvey et al. 1989).

Soil-humic materials can also facilitate transport of contaminants via a number of mechanisms. Humic materials have been noted to stabilize clay-colloidal suspensions (Jenny and Smith 1935; Gibbs 1983). Natural organic matter is readily adsorbed by oxide and mineral surfaces, with nearly complete surface-coverage likely for hydrous iron oxides, alumina, and edge sites (faces) of aluminosilicates under typical conditions found in natural waters (Davis 1982). Tessens (1984) found that essentially all topsoil samples from the upland soils of Malaysia had significant amounts of water-dispersible clay, as did a set of Brazilian soils (Camargo and Beinroth 1978), with organic coatings being implicated as the primary controlling factor. Jekel (1986) found that soluble humic materials extracted from surface and ground water stabilized colloidal suspensions of kaolinite and silica, with the degree of stabilization dependent on pH and ionic strength. Based on electrophoretic mobility measurements and the observed high-adsorption affinity at low pH, it was concluded that stabilization was primarily caused by the sorption of higher molecular-weight neutral molecules (Jekel 1986). Additionally, Tipping and Higgins (1982) found adsorbed humic substances stabilized colloidal dispersions of hematite, and that dissolved humics were responsible for stabilizing aluminum-oxide colloidal suspensions, although in contrast to Jekel (1986), their electrophoretic-mobility measurements indicated that the suspended alumina particles were highly charged negatively, as a result of adsorbed organic matter. Beckett and Le (1990) stated that organic coatings were the controlling factors determining the surface-charge behavior of suspended inorganic particles in riverine and estuarine waters from the Tara River system in Australia.

In addition to enhancing the stability and mobility of inorganic colloids, organic coatings greatly facilitate the partitioning of nonionic contaminant organics to their surfaces, since the relative amount of organic matter has been shown to be the most important factor for predicting nonionic organic partitioning to soils and sediments (Chiou 1989).

It has been demonstrated recently that laboratory-modified inorgano–organo clays are efficient sorbents, in many instances as efficient as granulated-activated carbon, for removing a number of organic contaminants, including polychlorinated dibenzo dioxins (PCDDs), polychlorinated biphenyls (PCBs), tetrachloromethane, and polycyclic aromatic hydrocarbons (PAHs) from aqueous solutions and industrial wastewaters (Srinivasan et al. 1989; Lee et al. 1989; Nolan, Srinivasan, and Fogler 1989; Srinivasan and Fogler, 1989, 1990a, b; Smith, Jaff, and Chiou 1990). Many of the modified clays used in these studies are the laboratory cogeners of the hydroxy-interlayered 2:1 clay minerals (HIM), which are important components of the clay and mineral fraction in soils throughout the southeastern and midwest United States. HIM can be an important sink for contaminant organics in organic-poor soils, as well as a potentially mobile phase capable of facilitating colloidal transport.

Colloidal particles are also potentially transported through the solum to underlying strata. Enfield and Bengtsson (1988) found that blue dextran, a model macromolecule, was subject to size-exclusion as it percolated through soil columns, thus flowing through the larger pores and eluting prior to tritiated water. Similarly, the transport of water-soluble organic carbon (WSOC) through soil columns was found to be more rapid than tritiated water; this was also attributed to a size-exclusion mechanism (Bengtsson, Enfield, and Linduvist 1987; Enfield et al. 1989). Such observations have suggested that this mechanism could significantly enhance the transport of contaminant organics to subsequent environments (Enfield and Bengtsson 1988).

An earlier study on facilitated-DDT transport by humic substances (Ballard 1971) and corroborated by recent investigations, has provided evidence that dissolved organic macromolecules enhance the mobility of contaminant organics, that is, naphthalene, phenanthrene, and DDT (Kan and Tomason 1990). Few investigations have studied the sorption and transport of actual WSOC through the soil (Jardine, Weber, and McCarthy 1989). The vast majority of investigations in this area have employed either "model" organic macromolecules, or humic materials derived from base-extraction of solid-phase organic matter—both of which can be poor models for naturally-occurring soluble-humic substances. In this regard, Leenheer and Stuber (1981) found that the hydrophobic neutral-fraction of dissolved organic carbon (OC) from oil-shale process-water was preferentially adsorbed, compared to the more hydrophilic fraction. Transport of the various dissolved organic fractions through soil columns was found to parallel their sorption behavior in batch experiments of the same soil material (Leenheer and Stuber 1981). Likewise, Jardine, Weber, and McCarthy (1989) observed preferential sorption of the hydrophobic, relative to the hydrophillic fractions of WSOC in a number of soils, with the total sorption increasing with increased soil-profile depth.

The preceding discussion illustrates the general concerns regarding colloidal-transport processes in soil, and the potential for colloid-assisted transport of both inorganic and organic contaminants. Contaminants are viewed as being associated with two phases: either dissolved and hence mobile, or matrix-adsorbed and retained. Partition coefficients are often useful for predicting relative mobility of contaminants. Used in conjunction with advection–dispersion models (ADEs), contaminant transport can be simulated and the extent of migration with time estimated. Deviations from model predictions are frequently attributed to preferential pathways. However, such deviations can be partially explained by colloidal transport processes (Jury, Elaboi, and Resketo 1986).

Gschwend and Wu (1985) identified colloidal materials as important components in PCB-adsorption experiments that confounded the determination of partition coefficients. DDT was found to adsorb and concentrate up to 15,800 times its water solubility on colloidal materials, within the highly colored southeastern U.S. streams (Poirria, Bordelon, and Laseter 1972). Enhanced apparent solubility of hydrophobic compounds by humics was also noted by Chiou et al. (1986). Ninety-one percent of the DDT applied to a forested soil was associated with humic acids, while the remaining nine percent was contained within fulvic acid and dissolved fractions (Ballard 1971). Additionally, it was demonstrated that soil-humic material was readily dispersed by urea additions, with a 30-fold increase in DDT mobility noted. Similarly, polycyclic aromatic hydrocarbons also exhibited high affinity for dissolved-humic substances (McCarthy and Jimenez 1985) as did trace metals (Hoffman et al. 1981), and natural estuarine colloids exhibited high affinities for atrazine and linuron (Means and Wijayaratrne 1982). Mineral surfaces can also sorb large amounts of contaminants (e.g., Jenne 1968; Vinten and Nye 1985). Radionuclides were found to be associated principally with 2–3-nm-radius size-class (taken to be humic acids), when equilibrated with soils (Sheppard et al. 1980), while ^{239,240}Pu partitioning was dependent on a number of parameters and not clearly understood (Alberts et al. 1977). The deep (30 m) transport of plutonium and americium at a defense-program site at Los Alamos was also found to be colloid-facilitated, as the transported radiounuclides were shown by ultrafiltration to be present as colloids in the 0.025 to 0.45 m size range (Nylan et al. 1985; Nelson and Orlandini 1986). Similar results have recently been reported for radionuclide transport at the Nevada test site and at Los Alamos National Laboratory (Buddemeir and Hunt 1988; Penrose et al. 1990).

Despite the demonstrated potential for colloidal transport through soils and the viability of contaminant-colloid associations, the role of colloid-assisted, contaminant transport to ground water remains largely conjecture, although a limited number of systems studied provide indication of the significance of the process. Perhaps most notable is the observation of ground water turbidity resulting from application of low-salinity water to a recharge facility in Fresno, California (Nightingale and Bianchi, 1977). Application of water having a lower specific conductivity (from 147 to 100 dS/m) induced dispersion of native inorganic colloidal materials within the surface soils. The dispersed clays were then transported to the aquifer, resulting in high turbidity within monitoring wells. It was estimated that 148 metric tons (3.1 metric tons/hectare) of colloidal material moved out of the surface profiles into the ground water in 1975 (Nightingale and Bianchi 1977), illustrating the possible magnitude of colloidal transport. Gypsum applications were found to destabilize the colloidal suspensions effectively and clarify the waters (by flocculation). The influence of small-to-modest changes in chemistry upon soil-aggregate stability and the magnitude of mass of the material transported demonstrates the balance of forces operative within many soil/colloid systems.

Release of colloidal particles from sandstone formations during oil-field operations, and the properties and factors serving to generate and stabilize colloidal suspensions has also been reported (Kia, Fogler, and Reed 1987). Particle-release was attributed to an ion-exchange process on the surface of the clay particles, with slow release of fines observed under low pH conditions. More theoretical studies on diffusional-detachment processes (Kallay, Barouch, and Matijevic 1987), kinetics of particle detachment (Barouch, Wright, and Matijevic 1987; Adamczyk and Petlicki 1987), colloidal stability of variable-charge mineral suspensions (Bartoli and Philippy 1987), and precipitation-charge neutralization processes during coagulation (Deutel 1988) have also been provided recently. Gschwend and Reynolds (1987) reported on an aquifer system possessing monodispersed colloids resulting from in situ precipitation. They postulated that phosphate- and carbon-rich wastewaters initially percolated into the shallow aquifer, where oxygen concentrations were sufficiently reduced by respiration such that anoxic conditions were formed. Reduction of the indigenous iron in the presence of the high-phosphate loadings produced stable monodisperse vivianite particles approximately 100 nm in size. Ryan and Gschwend (1990) found elevated concentrations of predominantly organic-coated inorganic colloids in anoxic (compared to oxic) ground waters (up to 60 mg L^{-1} and < 1 mg L^{-1}, respectively) within an Atlantic Coastal Plain aquifer in the Pine Barrens of southern New Jersey. They suggested that the anoxic conditions resulted in the dissolution of Fe oxyhydroxides that coated the mineral phases, acting as a cementing agent.

Penrose et al. (1990) reported the presence of ^{238,239,240}Pu and ^{248}Am in ground water > 3,000 m down-gradient from the discharge, and these actinides were determined to be associated with colloidal material between 25 to 450 nm in size. Based on the low-water solubility, high-partitioning coefficients, and immobility demonstrated in laboratory experiments, it was predicted that these actinides would migrate only a few meters from the discharge, thus providing strong evidence for a colloid-facilitated transport mechanism within the subsurface environment.

However, many published reports concerning the presence of colloids in ground water have been received with skepticism, resulting from the inability to eliminate unequivocally the possibility that the observed colloids were formed as an artifact of sampling.

A Conceptual Model for Colloid Transport

A general model to conceptualize colloid transport describes two phases: **(1)** detachment and stabilization; and **(2)** transport processes with associated sub-processes (figure 10.9). The initial condition for colloid migration must include the mobilization of colloid-sized particles within the soil. The process is hypothesized to be both physically and chemically controlled. Detachment of colloid particles from the soil matrix (1a in figure 10.9) requires an energy input sufficient to overcome van der Waals forces, coulombic, or other forces binding the particle. These forces could arise from: shear forces of water flowing within individual pores; swelling or repulsive double-layer forces between colloids upon initial wetting, or due to a change in chemical environment; or physical disturbance (e.g., agricultural tillage operations or excavation). Emerson (1967) defined classes of soils that disperse spontaneously with wetting, those that disperse only with mechanical energy input; and those which are nondispersive; the majority of soils seem to fall into the mechanically dispersive category, requiring some form of kinetic energy to initiate the dispersion process (Rengasamy et al. 1984; Miller and Baharuddin 1986). Raindrops or irrigation hitting bare soils disperse large amounts of colloidal clays, some of which enter the soil-surface with infiltrating water; the movement of

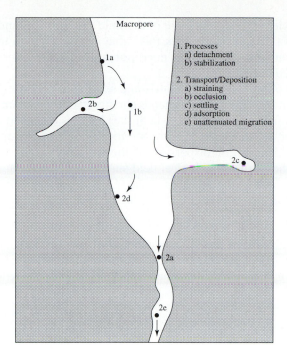

Figure 10.9   A conceptual model for colloidal-facilitated transport via a macropore

low ionic-strength rainwater into soils may also dilute soil solutions and enhance colloid mobility by increasing swelling and particle repulsion (McCarthy and Zachara 1989). Once detached from the matrix, the particle must be stabilized chemically (1b in figure 10.9) in order to avoid immediate redeposition via flocculation with other detached particles. Stabilization is accomplished largely through development of double-layer repulsive forces.

The second phase involved in colloid movement within soil is a transport phase, determined by both physical and chemical forces. Pore sizes must be sufficiently larger than the size of the particle to allow physical passage of the particle, or straining (2a in figure 10.9) occurs. Deposition of particles can also take place under other circumstances: occlusion of particles within micropores (immobile water) within the soil matrix (2b in figure 10.9) via Brownian motion; sedimentation of particles, if flow velocity is sufficiently low compared to settling velocity of the particle (2c in figure 10.9); and adsorption of the particle to the matrix, due to electrostatic or other forces (2d in figure 10.9). The latter process can considered to be largely chemical in nature, as charge characteristics of the soil matrix and particle control adsorption. However, adsorption may occur via van der Waals forces in narrow pores during unsaturated flow, where restricted water-film thickness increases the likelihood of particle approach to surfaces. Quantitative modeling of these processes has been attempted using filtration theory coupled with sub-models of specific processes described above (McDowell-Boyer, Hunt, and Sitar 1986). Straining is often empirically related to the ratio d_m/d_p, where d_m is the average matrix-particle diameter and d_p is the average colloid diameter; ratios < 10 indicate little colloid penetration into the matrix, ratios in the range 10–20 suggest significant straining and reductions in flow rate due to clogging, while ratios > 20 would result in limited straining. In soils with a distribution of particle and pore sizes, such a parameter would obviously need to be modified to be a useful predictor. Diffusion of suspended colloids within the flowing pore water can be described by the Einstein equation for Brownian motion

$$D_p = \frac{kT}{3\pi\eta d_p} \qquad (10.131)$$

where the particle diffusivity (D_p) is a function of the fluid viscosity (η) and particle diameter (d_p). At low-flow rates through small pores, this value is large enough to enable colloids to diffuse to the pore wall into connecting micropores, leading to occlusion (2b in figure 10.9), or adsorption (2d in figure 10.9).

Both adsorption of colloids to the soil matrix and settling of colloids (2c in figure 10.9) are highly dependent upon the surface and aqueous chemistry of the system. Particle settling can be predicted as a function of particle size via Stokes Law, as balanced against the flow-velocity of the pore water. Particle size, in turn, is critically dependent upon the chemical environment in terms of flocculation/dispersion of the (initially) colloidal material (Van Olphen 1977). The "thickness" of the electrical field surrounding the colloid (x^{-1}) is determined largely by ionic strength of the aqueous phase (I), as those ions partially balance the internal charge of the colloid

$$x^{-1} c \left(\sqrt{\bar{I}} \right) \tag{10.132}$$

where c is a constant incorporating the dielectric constant, temperature, and charge of an electron. Surface potential of the particle (ψ_p) and of the soil matrix (ψ_m), along with ion valence z, particle diameter d_p, determine the repulsive energy, σ_{dl}, forces acting between double layers

$$\sigma_{dl} = \left(\frac{\varepsilon d_p}{z} \right) \psi_p \, \psi_m \ln \left(1 - e^{-xh} \right) \tag{10.133}$$

where ε is the permittivity ($8.85 \times 10^{-12} \; C^2 \, N^{-1} m^{-2}$) and h is the distance of separation between particle and matrix, assuming surface potentials are roughly equal and < 50 mV. The attractive energy between a spherical colloid and a flat-plate (matrix) can be computed as follows:

$$\sigma_a = \left(\frac{A}{6} \right) \left[\ln \left(\frac{h + d_p}{h} \right) - \left(\frac{d_p}{h} \right) \left\{ \frac{\dfrac{h + d_p}{z}}{(h + d_p)} \right\} \right] \tag{10.134}$$

where A is Hamaker's constant and z is ionic valence (see chapter 3). The balance between these attractive and repulsive energies is controlled by variation in the repulsive component, which is highly sensitive to ionic strength of the aqueous phase.

In dealing with colloidal and subcolloidal organic molecules, adsorption to the soil matrix is a major factor retarding movement; adsorption is driven by both enthalpic and eutropic forces, although mechanistic interpretations of adsorption have not proved useful in quantifying sorption behavior. Instead, a number of empirical approaches have been used, particularly the simple approach of assuming a constant partitioning of the solute to the solid phase:

$$K_d = \frac{C_s}{C_f} \tag{10.135}$$

where the distribution coefficient (K_d) is the concentration ratio of adsorbed (C_s) to free (C_f) solute; the well-known Freundlich isotherm is a modification of this relation, adding a curve-fit exponential term to the equation

$$C_s = K_d (C_f)^n \tag{10.136}$$

thus allowing fitting of curvilinear adsorption data. Batch-adsorption experiment data are used to obtain K_d and n, whose values can then be used to account for attenuation during transport.

Approaches for quantifying attenuation and destabilization of migrating colloidal suspensions, such as those given above, can be coupled with basic one-dimensional advection–dispersion flow equations to yield breakthrough curves under a given set of boundary conditions. Parker and Van Genuchten (1984) offer a derivation of such an approach, and provide a useful integration of adsorption functions into the flow equations to account for partitioning of the solute flow through a one-dimensional soil volume, by the following equation:

$$\left(\frac{\rho_b}{\theta}\right)\left(\frac{\partial s}{\partial t}\right) + \left(\frac{\partial C}{\partial t}\right) = D\left(\frac{\partial^2 C}{\partial x^2}\right) - v\left(\frac{\partial C}{\partial z}\right) - \mu_w C - \left(\frac{\mu_s \rho_b}{\theta}\right)s + \gamma_w + \left(\frac{\gamma_s \rho_b}{\theta}\right) \quad (10.137)$$

where the major measured variables include solute concentration C, adsorbed solute in the solid phase s, vertical distance z, time t, dispersion coefficient D, pore-water velocity v, soil-bulk density ρ_b, and volumetric-water content θ. The μ and γ terms are (empirical) rate-coefficients for adsorption/desorption processes affecting solute concentrations in the liquid phase. The computer program CXTFIT (Toride, Leij, and Van Genuchten 1995) can provide computations for fitting this model to measured concentration-versus-time data. Adsorption parameters can be curve-fit, or varied independently to assess the fitness of the model assumptions with respect to flow conditions. Further refinement of such an approach can include additions of terms to the right-hand side of equation (10.137) to incorporate colloidal stability considerations, that allows a more complete modeling of colloidal transport in soil.

In summary, the most likely scenario for subsurface transport of pollutant-contaminated colloids is that in which colloids are detached and stabilized in surface soils or disturbed sub-soils, and are then transported through weathered-underlying-geologic materials by alternate saturated/unsaturated flow to ground water. The initial generation of the colloidal suspension will be highly dependent on the presence of detachment forces (physical shear forces, chemical repulsive forces), and on the chemical environment within that zone, which determines if the suspension can be stabilized and conducted into the underlying material. The variable nature of this subsurface material has a controlling influence on the proportion of colloid conducted to the aquifer below. Layers of weathered rock (saprolite) and/or unconsolidated sediments present varying pore sizes and geometries, as well as different chemical environments (pH, ionic strength, solution-ion composition) that may promote deposition by any of the mechanisms shown in Figure 10.9.

10.15   SOURCES OF CONTAMINATION

Municipal Sewage-Sludge Disposal and Wastewater Irrigation

This section is meant to introduce to the student the environmental impact and importance of various sources of contamination to land surface, surface water, below-surface disposal, and deep-well injections of contaminants. We lack space to treat each area in great detail; however, sufficient references are given that the reader may investigate each area of interest.

The use of manufactured nitrogen fertilizer has increased worldwide since the late 1940s. The application of nitrogen fertilizer in the United States increased from 500,000 metric tons (0.5 Tg) in 1945 to over 10 million metric tons (10 Tg) in 1976 (USDA 1977). Worldwide consumption in 1985 was 70.5 Tg (FAO 1985); Eichner (1990) predicted a consumption of about 94 Tg by 1997, based on World Bank estimates. However, during the past twenty years, the high cost of manufactured fertilizer has generated increased interest in municipal wastewater irrigation and land disposal of nitrogenous animal, municipal, and food-processing wastes.

Numerous studies of water quality in agricultural areas of the United States have documented nitrate concentrations in ground water and surface water greatly in excess of the regulated United States drinking water standard of 10 mg (N) L^{-1} NO_3^- (Mueller and Helsel

1996; Weil, Weismiller, and Turner 1990; Keeney 1982). Nitrate levels in household wells in some agricultural areas have been found to reach acutely toxic concentrations (Keeney 1982). Elevated nitrate concentrations in drinking water can cause methemoglobinemia in some livestock and in human infants under the age of six months. Nitrate consumption has also been circumstantially implicated in a greater incidence of stomach cancer in some localities (Lee and Dahab 1992). The same problems also plague other countries such as the Peoples' Republic of China, but are not as well-documented due to lack of adequate technology (in some cases), and research funds (Jiao Desheng, personal communication).

The application of municipal sewage-sludge to agricultural land can benefit the soil by improving the physicochemical condition and by recycling plant nutrients as well as, by providing a viable alternative to disposal in landfills or by incineration (Sopper and Kerr 1979; Khaleel, Reddy, and Overcash 1981). Investigation of the potential benefits of municipal sewage-sludge application to agricultural lands has been limited, but the practice has been studied extensively for reclamation purposes (Sopper and Kerr 1979). Municipal sewage sludge has been shown to provide substantial amounts of organic matter and plant essential nutrients when applied to soils (Elliott 1986; McCaslin and O'Connor 1982; Knudtsen and O'Connor 1987). However, depending on its source, this sludge may contain metals and other elements that can be harmful to plants and the food chain when applied in excessive amounts (U.S. Environmental Protection Agency 1978; Tindall, Lull, and Gaggiani 1994; Valdares et al. 1983). Sewage sludge can be applied in a variety of forms: dry, wet slurry, or through irrigation systems either to land-surface or tilled into the soil. Occasionally, it is stored in large piles that provide a significant point source for potential contamination. In rural settings, the presence of septic systems, with their associated drain fields, can contribute a significant amount of sewage, which may pose a hazard to ground water, depending on rural population density.

Millions of tons of sewage sludge are produced each year and while there are many benefits to the soil where it is applied, there are also disadvantages. First, sewage sludge contains a multitude of chemical constituents, including metals, organic compounds, and polynuclear aromatic hydrocarbons (PAHs) that are usually significantly higher than the normal range of concentrations found in agricultural soils without municipal-sludge amendments (Jones et al. 1989; Wild, McGrath, and Jones 1990). Second, there is concern that some chemicals have a propensity to transfer from municipal sewage sludge-amended soils into the human food chain, due to plant uptake by food crops grown in these soils (Scheunert and Klein 1985). Third, there is a major concern that, when any municipal sewage-sludge is applied to soil or a sensitive hydrogeologic setting, the ground water could be contaminated by nitrates, metals, organic compounds, or other constituents within the sludge (Berg et al. 1987). Nitrates are present in municipal-sewage sludge and when applied to the soil may reach concentrations greater than 60 mg L^{-1} (Berg, Morse, and Johnson 1987). As greater amounts of sludge are applied, nitrates increase in concentration, and because the nitrate ion is mobile in soil, it is readily leached from the upper soil-profile to deeper depths. In hydrogeologic settings such as fractured rock, karst, and porous soils (such as those with higher sand and rock contents), contamination of ground water and underlying, shallow aquifers is of grave concern (U.S. Environmental Protection Agency 1986).

Few guidelines have been developed for determining acceptable sludge-loading rates to soil in relation to nitrate loading, and the addition of harmful metals associated with sewage sludge. As a result, research has focused on what rates need to be applied with regard to the physical and chemical parameters of the soil (Tindall, Petrusak, and McMahon 1995). Haith (1983) developed a planning model for land-application of sewage sludge. This mathematical model estimated nitrate–nitrogen concentrations in percolation from sludge land-application sites. The model is based on an annual-mass balance of sludge and soil-inorganic nitrogen, and includes the processes of mineralization, ammonia volatilization, crop uptake, and leaching. The pollution risk of nitrate percolation from sludge land-application sites (Kaufman and Haith 1986) and development of models to evaluate application rates of wastewater sludge to various soils (Haith et al. 1992) are two important areas that are being investigated with regard to application of sludge loading. As the population increases, the need for research in these and other areas becomes a necessity.

Due to competition for valuable water resources, the application of municipal waste-water for irrigation purposes of agricultural crops has increased significantly in China during the past five years (Gao, Sun, and Qu 1991), in the U.S. (Tindall, Lull, and Gaggani 1994), and other countries as well. Irrigation of vegetable crops with wastewater has resulted in in-creased yields (Osburn and Burkhead 1992) however, Moreno (1981) reported that high lev-els of detergents in wastewater decreased calf-growth rate and that boron was toxic to both beans and lettuce at concentrations greater than 5 mg L^{-1}. In some instances, plant disease occurs as a result of irrigation with municipal wastewater, the most common diseases being Pythium and Phytophthora root-rots (Epstein and Safir 1982). Other studies have found no major limitations on crop growth by irrigating with wastewater (Neilsen et al. 1989; Lau 1981).

With regards to the application of fertilizers and wastewater, the movement of these chemicals plus associated herbicides and their relation to agricultural mulching practices, is relatively unknown. Lavy (1977) found that the fraction of atrazine free to move in a moist-soil environment is increased by increasing the soil-water content; this could prove to be a management problem where mulches are used for water-conservation purposes. Other research has found that, for advanced stages of redistribution, solute movement became neg-ligible, and that the maximum herbicide concentration was located at the same depth in the soil-profile regardless of irrigation intensity (Selim, Manzell, and Elzeftway 1976b). By applying wastewater as an irrigation source, we are essentially using "chemigation" method-ology for nutrient application. Chemigation can have a negative impact on ground water-quality depending on the occurrence of preferential flow (Jennings and Martin 1990). In the presence of preferential flow paths that can occur in all soil types, the ability and likelihood of increased chemical movement is greatly enhanced. For commercial fertilizers, urea added to localized irrigation systems (i.e., trickle, drip, and spray) can be hydrolyzed rapidly in the soil to ammonium, and then oxidized to nitrate (Clothier and Sauer 1988). Also, repeated applications of effluent as wastewater can reduce the assimilative capacity of the soil, leading to increased risk of ground water-contamination.

The effects of irrigation with municipal wastewater on soil components, and subse-quent impact on ground water-quality have been varied. It has been found that spray irriga-tion with municipal wastewater can reduce soil infiltration capacities under certain crops sig-nificantly (Sopper and Richenderfer 1978; Sopper and Richenderfer 1979). Kayser (1988) reported no increase in heavy metals in Wolfsburg, Germany, soils, but in Braunschweig, higher levels of heavy metals did accumulate in the upper soil horizon. However, Davis, Grieg, and Kirkham (1988) reported that extractable concentrations of heavy metals (cad-mium, copper, iron, lead, manganese, and zinc) were not higher in soils irrigated with waste-water than in soils treated with typical commercial (N-P-K) fertilizer and irrigated with tap water. Page and Chang (1981) reported little or no accumulation of trace metals in the root zone following long-term use of wastewater irrigation, and Schirado et al. (1986) showed that heavy metals may have migrated through the cultivated layer of soil and were distributed throughout the soil profile or leached below the depth of sampling. Lau (1981) showed that, through crop rotation and wastewater dilution, application of wastewater maintained sugar-cane yield without polluting the ground water. Additionally, soils irrigated with wastewater exhibited low-denitrification potentials due to a shortage of available organic carbon and rapid nitrification (Miller, Barr, and Logan 1977; Barr, Miller, and Logan 1978). Present research suggests that transport of chemicals from wastewater is dependent on soil type, application rate, application timing, and on constituent concentrations in the wastewater. The extent to which these waters can be used depends on the nature of the contaminant and the purpose for which the crop is to be used (Abeliuk, Riqueline, and Matthey 1993).

In addition to environmental problems, wastewater irrigation and sewage-sludge application also pose human health hazards. Current research indicates there is strong circumstantial evidence supporting the hypothesis of typhoid and cholera transmission by wastewater-irrigated crops, and demonstrate that wastewater irrigated vegetables are the main mode of transmission (Shuval 1993). In a long-term study of results from Middle East countries, Shuval, Yekutiel, and Fattal (1985) indicated that both *Ascaris* and *Trichuria* infections were actively and massively transmitted to the general public, who consumed vegetables irrigated with raw wastewater. Oron et al. (1991) showed that, in order to reduce the human-health risk associated with fecal coliform contamination of sweet corn, the harvesting schedule should be delayed as long as possible after the last irrigation. The development of a model using empirical evidence, (Shuval, Yekutiel, and Fattal 1986a) suggests that the highest risk of pathogen transmission, infection, and sickness is associated with the helminths, followed in order by bacterial infections, and last by viral infections. Although certain health risks are definitely associated with the use of raw wastewater in agriculture, the epidemiological evidence assembled by Shuval et al. (1986a, b) suggests that the very stringent wastewater-irrigation standards developed in many of the industrialized countries are overly restrictive, and that a guideline for unrestricted wastewater irrigation be based on an effluent with less than one nematode egg (*Ascaris* or *Trichuris*) per liter and a geometric mean fecal coliform concentration of 1,000 per 100 mL. In addition to the threat posed by direct application to land, some researchers have suggested (Shuval et al. 1989; Kowal, Pahren, and Akin 1981) an airborne threat from spray applications of wastewater irrigation; that public access to applied-to areas be limited to a distance of 200 m and humans should have minimal contact with treatment sites them.

Radioactive Waste Disposal

In certain ways, the transport of radionuclides is similar to what we have already discussed. However, the greatest difference is that radionuclides are both nonconservative, and extremely hazardous; because of this, research results are not readily available in this area. The transport of radionuclides is particularly important in the storage of radioactive wastes. Any new radioactive waste disposal site to be established must meet applicable federal and state regulatory criteria to become a licensed site, such as 10 CFR Part 61 under the authority of the Nuclear Regulatory Commission (NRC). This regulation (Part 61.5), contains criteria covering technical requirements for low-level waste (LLW) disposal. For a site to become licensed it must be able to be characterized, modeled, analyzed, and monitored. The disposal zone at such sites must also be well above the water-table fluctuation zone. One of the principal reasons that modeling is required, is to demonstrate that the migration of radionuclides away from the disposed waste, either in liquid or gaseous state, will meet regulatory concentration limits at the boundary of the site or at points of exposure. Current regulations require site-performance criteria to be satisfied for at least 500 years. Due to the hazardous nature of such wastes, it is likely that future regulations will become even more restrictive, especially as our technology for predicting transport improves.

For low-level waste sites, subsurface hydrology is a particularly important aspect of site characterization, because subsurface water and vapor transport are the primary natural pathways for the movement of contaminants from a disposal facility to the accessible environment. Such transport may move in any direction—vertically downward to the water table,

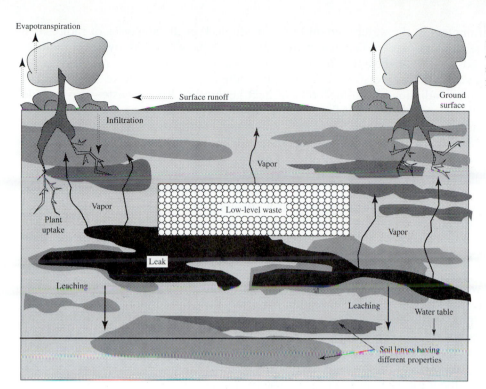

laterally into adjacent soils and root systems, or upward towards the atmosphere; this is illustrated in figure 10.10. The analysis of subsurface contaminant transport is complicated by a number of factors; including the difficulty of observing subsurface processes, the long time-scales required for experiments, and the need to account for soil-water interactions and spatial variability. The situation is particularly complex in the unsaturated zone, where soil properties change dramatically with moisture content; it is even more complicated in the presence of hazardous wastes. Because spatial variability and nonlinear behavior greatly complicate the equation that describes unsaturated flow and transport, performance assessments for unsaturated sites rely heavily on numerical modeling. Also, it is important to note that small-scale fluctuations in soil properties are important because they have large-scale consequences (e.g., anisotropy and macrodispersion). However, small-scale fluctuation in moisture content or concentration is not particularly important, at least from a licensing point-of-view.

Similar conditions are encountered at radioactive-waste disposal sites in comparison to other vadose-zone studies. These conditions include dry media with very low rates of fluid movement, especially for shallow land burial where evapotranspiration losses can have great effects on site hydrology. For arid conditions, evapotranspiration can represent a very large portion of the total water balance. Other conditions include strongly hysteretic and anisotropic conditions, nonlinear sorption, multi-species chain decay, highly variable transient-moisture conditions, dual porosity, and unusual boundary conditions such as wicking schemes, capillary-barrier trench caps, and deep-boring emplacement.

If the unsaturated zone is to be considered as host medium for the emplacement of radioactive wastes, the medium must be an effective barrier to radionuclide transport. The design of suitable repositories depends on the capability to predict the effectiveness of the medium as a barrier. The usefulness of predictive models is limited by the difficulty of analyzing and predicting solute transport as influenced by large reaction-networks. Additionally, there is a general lack of information relating solute-transport mixing parameters and reaction parameters to the water content of the medium. Typical categories of radioactive waste include: **(1)** high-level radioactive waste, **(2)** transuranic waste, **(3)** low-level radioactive and mixed low-level radioactive and hazardous wastes, and **(4)** uranium mill tailings.

High-level radioactive wastes include fission products that initially have a high level of beta and gamma radiation, and a high rate of heat generation. These wastes also include transuranic elements with long half-lives. Transuranic waste contains long-lived alpha emitters at concentrations greater than 100 nCi g^{-1}, and generate little or no heat. The term "low-level waste" has carried a changing and imprecise definition over the years; currently, it generally means waste that does not fit the definition of high-level waste, and in which the concentration of transuranic elements is less than 100 nCi g^{-1}. In part, these wastes consist of miscellaneous solid materials that have been irradiated and contaminated through use, as well as products of reactors and fuel-reprocessing plants. Some low-level wastes are mixed with materials that are on the list of hazardous materials maintained by the U.S. Environmental Protection Agency, and are thus subject to relevant environmental laws for such substances.

A variety of methods have been developed for storage of various radioactive wastes. Hazardous wastes with lower potential for contamination are often placed in individual containers made out of steel, concrete, and other materials, then stored directly on top of land surface. A similar method involves covering the specially engineered container with soil that generally extends the life-expectancy of the container. Another method is to use the same type of engineered containers and bury them below land surface, then backfill over them or surround the container with an engineered hydro-geological environment to provide enhanced containment for the system. For very deep storage, containers are sometimes placed in boreholes that can be in excess of 30-m deep; they are then either backfilled with the original material, or most often backfilled with bentonite clay. Because bentonite is so adsorptive, and due to its large surface area ($750 \text{ m}^2 \text{ g}^{-1}$) and physical properties, it is a very effective containment material as long as it is not in the presence of a rising and falling water table. The presence of such a water table can destroy the integrity of the bentonite, and cause significant swelling and the possible development of long-term cracks; such cracks can serve as flow paths for deeper movement of the waste material in the event of container leakage.

The preferred climate for burial of wastes is one with an arid environment and a very thick vadose zone. Yucca Mountain, Nevada, 160 km northwest of Las Vegas, meets these basic requirements. The U.S. Department of Energy (DOE) has been conducting a hydrogeological characterization of the soil at the site, as a potential repository for the disposal of commercially generated high-level radioactive wastes and high-level radioactive wastes from DOE defense facilities, as directed by the U.S. Congress. The process and schedule for this program were specified in the Nuclear Waste Policy Act of 1982. In May 1986, the DOE recommended (and the President approved) the Yucca Mountain, Nevada, site as one of three candidate-sites for detailed study. In December 1987, in the Nuclear Waste Policy Amendments Act, the Yucca Mountain site was designated by the Congress for characterization as the single candidate-site for a geological repository (10 CFR Part 60).

It should be made clear that the Yucca Mountain Site has not been selected as a repository; it is at this time the only site designated for study to assess its suitability for containment of high-level radioactive wastes (Nuclear Waste Policy Act, Section 113). Due to the very nature of radioactive wastes, the study of the Yucca Mountain site and its potential as a repository is surrounded by controversy, especially since a 7.5-magnitude earthquake occurred near Landers, California, on June 28, 1992 (O'Brien 1993). Since that time, the State of Nevada and many of its residents have openly expressed criticism of the site and often refer to 10 CFR Part 60 as the "Screw Nevada Bill." While this may sound somewhat humorous, finding a repository for high-level wastes is a necessity, since we continue to produce them.

For a site to be capable of storing radioactive, hazardous wastes, it must be isolated from subsurface-flow regimes and fractured bedrock because of the complexity of delineating flow paths. The flow velocity of the site and therefore of the radionuclides and other wastes, should be low; this indicates a greater residence-time of the chemical within the medium, allowing for physical, chemical, and biological changes such as retardation, degradation, and adsorption. The site should be structurally stable and should lack subsurface flowlines which can lead directly to ground water zones of potable water, or to the atmosphere.

Landfills

Landfills are used for the disposal of solid wastes that are produced by municipalities and local industries. Due to the mass of such wastes (about 3.5 kg of solid waste per capita per day), many cities are seeking disposal in other areas (and states) with more space for disposal, but also with smaller populations. The waste disposed of in landfills may contain many toxic substances, such as liquid industrial wastes, oil, solvents, and other chemicals.

The disposal of such wastes may have a detrimental effect on ground water. In humid regions in the northeastern and southeastern United States, infiltration of water from precipitation events can cause water-table mounding, within or directly below the landfill site; this causes leachates to flow downward and outward from the site. This in turn can induce leachate springs at the edges of the site where chemicals collect, and can then be transported via overland flow during significant-recharge events and potential seepage, into streams or nearby surface waters.

If sites could be ideally selected using the basic criteria described in the section on radioactive wastes, contamination could be minimized. However, due to spatial variability of soils, such sites are extremely difficult to find when considering economical disposal of solid wastes due to the costs of: collection; processing; transportation distances; and dumping. In addition to leachate migration, biochemical decomposition of organic matter (and byproducts within the waste) occurs, giving off common gases such as methane, carbon dioxide, and hydrogen sulfide. These common gases can cause extensive damage to vegetation surrounding the site, and can result in odor problems. The generation of methane, while odorless, can also cause hazards due to its explosive potential.

Landfill sites are becoming safer because of better design and technological capabilities; however, many sites are not well-monitored and thus, there is no realistic way to determine how much contamination they cause. Also, without long-term monitoring, conclusions about future contamination by landfills are simply conjecture.

Other Sources of Contamination

There are many other potential sources for contamination of ground- and surface-waters. These include agricultural activities: crop; poultry; swine and cattle production; greenhouses; golf courses; ball fields; and other areas that use high rates of irrigation, along with the significant application of fertilizers and pesticides. In many instances, agricultural activities result in what is termed "non-point source pollution", because they influence water quality some distance away from where actual operations are occurring.

Perhaps the most threatening problem from agricultural activities, other than nitrate leaching, is the potential hazard resulting from transport of herbicides (widely applied in the United States) through the soil profile. Currently, the most threatening of these chemicals may be atrazine. In 1990, about 29 million kilograms of atrazine were applied to soils in the United States to control broad-leaf weeds (Gianessi and Puffer 1991). Such chemicals are commonly sprayed on crops in early spring, when the soil is relatively moist, evapotranspiration rates are low, and spring rains are prevalent. Because the fate of atrazine is primarily controlled by chemical, biological, and hydraulic properties of soil—as well as by weather conditions—the timing of application of this and other herbicides is extremely important, not only for effective weed control, but also for minimal ground water contamination risks. Research has shown that, for many soils, atrazine and other herbicide transport is directly related to the amount and timing of rain that follows spring application and the occurrence of flooding (Eckhardt and Wagenet 1996; Tindall and Vencill 1995; Kolpin and Thurman 1993). Two other factors were also found to be critical in transport during wet years: **(1)** the variability of herbicide-application rate, and **(2)** degradation rate of the herbicide (Eckhardt and Wagenet 1996). Such research has shown that the coincidence of heavy rain soon after herbicide application can cause the herbicide to move below the rooting-zone depths at which biodegradation rates are assumed to be low, but are often unknown. Once an herbicide has been leached below the rooting zone, it can persist in the underlying soil and can subsequently be transported into ground water as the soil drains, usually after the growing season.

Industry also contributes a large amount of contaminants to the environment. In the U.S., there are about two million underground storage tanks, 90,000 confirmed releases of contaminants, 150 superfund sites, 15,000 annual oil spills, deep-well injections of hazardous wastes, and other contaminants from industrial and urban sources. These can include mine tailings, fly-ash deposits, salts applied to roads during winter months that run off in the spring, both killing vegetation and contaminating surface water, industrial-waste lagoons, and other activities.

Problems arising from these potential sources of contamination will not diminish, but worsen due to increasing population and its encroachment on aquifer recharge areas. It is imperative that scientists carry out well-planned, applied and theoretical research, and develop accurate, predictive models to provide answers that outline proper design and management techniques to deal with these problems.

10.16   CASE STUDY: RADIONUCLIDE DISTRIBUTION AND MIGRATION MECHANISMS AT MAXEY FLATS

The primary purpose of this case study is to help the reader develop an understanding of chemical processes that significantly influence the migration of radionuclides at commercial low-level waste burial sites. The chemical measurements of waste-trench leachate and identification of chemical changes in leachate during migration will provide a basis for geochemical waste-transport models. Two general kinds of radionuclides were determined in the research at Maxey Flats (near Morehead in northeastern Kentucky): endogenous (natural) radionuclides originating from soil parent material; and exogenous (human-induced) radionuclides originating from the nuclear fuel process. Cobalt-60 and ^{137}Cs are exogenous radionuclides and both are constituents of global fallout, as well as being components of low-level radioactive waste stored at Maxey Flats. Potassium-40 and ^{228}Th are endogenous radionuclides originating from the underlying rock strata and soil. In a typical gamma-ray spectrum, most of the peaks that were found at the site were due to endogenous radionuclides, chiefly natural uranium, thorium (and their daughter-products), and ^{40}K.

As with all multidisciplinary large-scale projects, the Maxey Flats study followed a systematic approach. The first part of the study involved an unsaturated-zone hydrology investigation to determine the pathway(s) of water entry into the low-level radioactive waste burial-trenches, and to investigate possible countermeasures for reducing water infiltration into the trenches. A transect across trench 19S (figure 10.11) and the adjacent area was instrumented with electrical soil-moisture sensors to allow measurements of relative porous media matric potential and with mini-porous cups (i.e., suction lysimeters) for extraction of

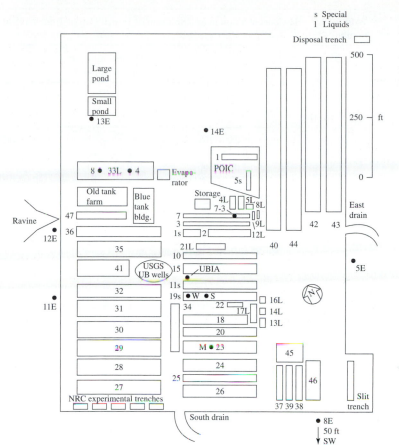

Figure 10.11 Map of Maxey Flats disposal site, showing the location and identification of disposal trenches (data from Kirby 1984)

soil water in the region of the soil-moisture sensors. This type of porous cup allows extraction of soil water under unsaturated conditions down to about -80 kPa and thus are good for sampling in moist unsaturated media. Tritium concentrations in the porous cups in the trench transect indicated that near the trenches, the principal source of tritium was the waste, and not the evaporator located on site (table 10.5). The investigation of transport of tritium to the atmosphere by plant-root uptake (Kentucky fescue grass) of trench water, and leaf transpiration was also part of this investigation. Additionally, a large lysimeter was installed to account for all precipitation at the site, and to investigate the management of countermeasures for reducing water infiltration into the trenches (figure 10.12). Tritium present in the waste was employed as a tracer of fluid movement. It was discovered that the principal mode of water entry into the trench was by percolation through the trench caps, and that lateral movement within the soil profile was very slow; the deep vertical movement of water through the undisturbed soil profile was very slow as well. The Kentucky fescue grass grown and sampled on-site extracted significant amounts of tritium from the trenches and, by transpiration, transported the tritium to the atmosphere (table 10.6). Also, tritium in the plant transpiration stream could have served as a tool for mapping trench boundaries (table 10.7). Data from the measurements conducted on the west side of the restricted area (figure 10.13) indicate that no major tritium sources originated from seepage at the lower elevations leading from the site.

Ethylenediaminetetraacetic acid (EDTA) was found to be the major organic compound in water from various waste trenches at the site and from inert atmosphere wells. Inert atmosphere wells were installed near the experimental trench to help maintain the

TABLE 10.5 ³H Concentration in Soil Solution of Trench Cap of Trench 19S at Station number 4*

	³H concentration (plastic applied)		
	2/80	9/81	8/82
Soil depth (m)		pCi/l	
0.15	0.28×10^6	—	0.59×10^6
0.30	0.65×10^6	0.57×10^6	0.43×10^6
1.50	1.07×10^6	0.97×10^6	0.43×10^6
2.10	2.07×10^6	1.85×10^6	0.86×10^6
2.40	7.93×10^6	5.23×10^6	3.70×10^6

Source: Data from Kirby 1984

* There is a marked increase in ³H with depth into the cap toward the waste, indicating that the waste is the principal source of the ³H, not the evaporator. Installation of the plastic membrane on the trench cap surface did not have a marked effect on the ³H distribution in the trench cap.

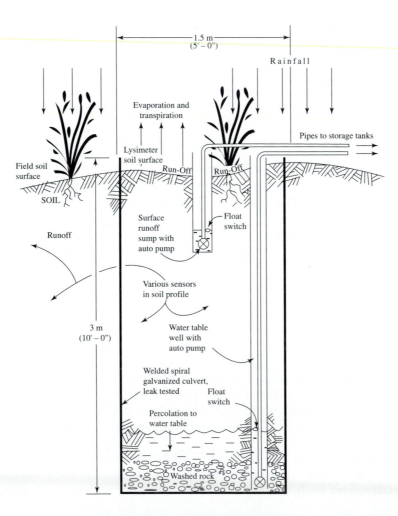

Figure 10.12 Cross-section of an installed lysimeter. The system gives complete accountability of all rain falling on soil surface (data from Kirby 1984).

TABLE 10.6 ^{3}H Concentration in Leaves of Kentucky Fescue Grass Grown on Trench Caps and Nearby Field (Controls) at the Southeast Corner of the Maxey Flats Site, Sampled February 1982*

Trench number	^{3}H concentration in plant sap[†] (^{3}H$_2$O)	^{3}H concentration in trench water (^{3}H$_2$O)	^{3}H concentration in leaf organic material[††] (C^{3}H$_2$O)
	pCi/l		
37	6.9×10^7	3.7×10^7	9.7×10^4
38	3.5×10^7	3.3×10^7	4.3×10^4
39	10.5×10^7	10.1×10^7	14.0×10^4
control	1.1×10^7	—	2.0×10^4
control	1.0×10^7	—	2.0×10^4
control	0.9×10^7	—	1.0×10^4

Source: Data from Kirby 1984

*This was a preliminary study with one grass sample taken from each trench cap. The ^{3}H in the transpiration stream was about seven times higher in the plants grown in the trench caps than in that of the controls, but it was only 1/1000 of the trench water. The organic (structural) ^{3}H was twice as high as it should have been at equilibrium, with the ^{3}H present in the transpiration stream indicating higher ^{3}H concentration probable in the soil solution and transpiration stream during the growing season of the past fall.
[†] Determined by vacuum distillation of fresh leaf material, then liquid scintillation of the ^{3}H$_2$O distilled off.
[††] Determined by reacting vacuum-dried leaf material with O$_2$ at 900 °C, then liquid scintillation of the resulting ^{3}H$_2$O.

TABLE 10.7   Titrium in Transpiration Stream of Fescue Grass*

Sample no.	Trench no.	^{3}H pCi/l	Sample no.	Trench no.	3H pCi/l
1N	38	0.12×10^6	7N	39	1.18×10^6
1E	38	0.14×10^6	7E	39	0.91×10^6
1S	38	0.43×10^6	7S	39	6.97×10^6
1W	38	[†]	7W	39	108.76×10^6
1C	38	0.44×10^6	7C	39	3.79×10^6
2N	38	0.13×10^6	8N	38	0.84×10^6
2E	38	0.65×10^6	8E	38	0.71×10^6
2S	38	[†]	8S	38	1.05×10^6
2W	38	[†]	8W	38	2.09×10^6
2C	38	0.40×10^6	8C	38	1.05×10^6
3N	38	0.12×10^6	9N	37	0.25×10^6
3E	38	[†]	9E	37	0.27×10^6
3S	38	[†]	9S	37	0.24×10^6
3W	38	[†]	9W	37	0.28×10^6
3C	38	0.36×10^6	9C	37	0.27×10^6
4	38	0.27×10^6	10N	37	0.67×10^6
4	38	0.50×10^6	10E	37	1.07×10^6
4	38	0.30×10^6	10S	37	1.31×10^6
4	38	0.26×10^6	10W	37	0.18×10^6
4	38	0.27×10^6	10C	37	0.63×10^6
5	38	1.63×10^6	11N	37	0.44×10^6
5	38	[†]	11E	37	0.67×10^6
5	38	[†]	11S	37	0.37×10^6
5	38	[†]	11W	37	0.07×10^6
5	38	0.29×10^6	11C	37	0.46×10^6
6	39	0.30×10^6	12	38	0.12×10^6
6	39	[†]	12	38	0.12×10^6
6	39	[†]	12	38	0.18×10^6
6	39	[†]	12	38	0.13×10^6
6	39	0.43×10^6	12	38	0.14×10^6

Source: Data from Kirby 1984

*Purpose of sampling was to locate a uniform area to use for repeated sampling to determine the annual transport of ^{3}H to the atmosphere by the transpiration stream of fescue. Location of trench boundaries are not well-known and, evidently, some samples were not taken from over trench. This observation suggests that ^{3}H concentration in the transpiration stream may serve as an accurate indicator of the location of trench boundaries .
[†]Samples collected but not yet analyzed.

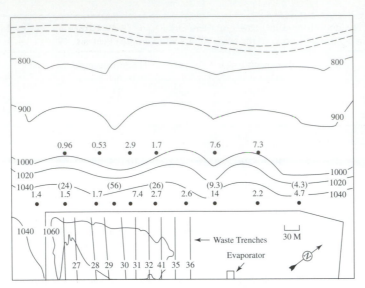

Figure 10.13 Tritium concentrations in systemic water from trees in the west drainage at Maxey Flats in July, 1981 (bracketed numbers) and July, 1982 (unbracketed numbers). Concentration units are 10^4 pCi/L and contour intervals are in feet. The absence of elevated tritium levels at the illustrated elevations demonstrates that there is probably very little subsurface flow from the site (data from Kirby 1984).

initial-oxidation state of the dissolved radionuclides in ground water samples; they were generally filled with inert gases such as nitrogen and argon. EDTA and EDTA-like species coelute with alpha-emitting radionuclides and ^{60}Co. Data collected at Maxey Flats showed that migration of plutonium was as an EDTA complex. Other phases of the study included geochemistry of trench leachates and the influence of soil chemistry on mobility of radionuclides. These studies are not within the scope of this text; the reader is referred to Kirby (1984) for additional information on the Maxey Flats study.

Another phase of the Maxey Flats study was ecological monitoring to develop efficient and statistically valid ecological field-sampling procedures and methods for post-closure monitoring at commercial shallow-land burial sites. The kinds and amounts of radionuclides and trace elements in environmental samples are normally measured with emphasis on biotic uptake, bio-accumulation, biotic transport, and ecological pathways in semi-wild ecosystems. To evaluate potential biotic pathways, forest-floor litter, newly fallen leaves, newly opened leaves and surface soil were sampled in the forest surrounding the restricted area, and then analyzed by high-resolution gamma-ray spectroscopy. Most of the radionuclides in the forest were associated with the soil; relatively small amounts of radionuclides were associated with the litter and leaves. The ^{60}Co content of newly fallen leaves (figure 10.14) was more variable than the other radionuclides, suggesting that the burial site has influenced the ^{60}Co concentrations at a few of the sampling locations and newly opened leaves from one hickory tree suggested that the source of elevated ^{60}Co concentrations was the rooting substrate. The other radionuclide concentrations exhibited little variation between tree species from location to location. Leaf water data indicated that tritium above ambient fallout levels migrated into the edge of the forest at Maxey Flats.

As a potential biomonitor of subsurface-water movement, maple trees (*Acer sacharum*), were also sampled as part of the ecological phase because of their unusual ability to move water through their trunk before new leaves emerge in the spring; if the roots tap tritiated water, tritium will appear in the sap stream. The maple trees were tapped at various locations around the site (figure 10.15) during the most vigorous sap flow of the season. Samples of sap from the maple trees showed concentrations ranging from < 430 pCi L^{-1} at a location about 12 miles south of the site, to a high value of 2.9×10^5 pCi L^{-1} at a location near the west side of the site.

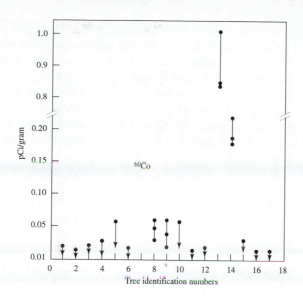

Figure 10.14   Cobalt-60 in newly fallen leaves in the oak–hickory forest at Maxey Flats, Kentucky (data from Kirby 1984)

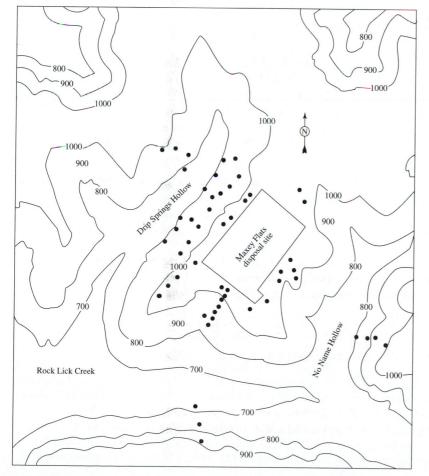

Figure 10.15   Locations of sap sampling trees (•) near the Maxey Flats site, 1983 (data from: Kirby 1984)

From the studies involving the fescue grass, biomass, and maple trees, we hope to illustrate a point to the reader. Generally, much research is accomplished in soil investigations that totally disregard the effects of the ecological system—that is, the presence of plants, microorganisms, and fauna. However, as we have discussed, soil is a complex and dynamic system, especially in the presence of flora and fauna. As a result, most real-world problems are not enhanced, nor is the transport of chemicals accurately predicted by ignoring the presence of living flora and fauna inherently existent in most field situations. Both the studies of fescue grass and ecological studies in this investigation clearly indicate the potential of plants to serve as biomonitors of subsurface movement of water from burial sites, and the ability of such studies to be both cost-effective and biologically nondestructive. In the Maxey Flats study, tritium uptake by deep-rooted trees illustrated the potential of certain plant species to serve as a monitor of subsurface water flow off-site and for species which root in the upper 10–15 m of soil, there is an economic advantage in wet climates to use trees as a tool. The primary reason: the cost of using a tree versus the greater expense of a monitoring well. While the presence of plants may make the system of interest more complex in some instances, they can also serve as valuable research tools and thus, should not necessarily be discounted in favor of purely physical studies of the media (a less-dynamic system).

SUMMARY

This chapter has focused on transport of chemicals in the unsaturated zone. We have discussed the physical processes that cause contaminant transport, types of fluid flow, breakthrough curves and hydrodynamic dispersion, as well as piston flow and mechanical dispersion. We also discussed: nonreactive and reactive solutes; mathematical modeling; general solutions for a displacing-solute front; determining the concentrations of these fronts; various boundary conditions; nonequilibrium conditions; equilibrium chromatography; the effects of ion exchange and dispersion; and numerical solutions of equilibrium exchange. The discussion also focused on aspects of mobile and immobile regions in soils, the effects that aggregated, fractured, and layered-media have on transport, as well as problems associated with preferential pathways and the difficulty of modeling them. We also showed the reader how to calculate dispersion coefficients, and discussed colloidal-facilitated transport and various sources of contamination that affect ground water quality. Finally, we discussed the transport of radionuclides and presented a case study involving low-level, radioactive-waste storage.

Contaminant transport is dependent on a wide range of physical and chemical parameters. Because our knowledge of transport is limited concerning such topics as preferential flow, colloidal-facilitated transport, waste storage, loading rates of municipal wastes and other activities, it is imperative that we increase our knowledge and understanding of these processes to combat environmental pollution successfully in the face of an increasing population and demand on our environmental resources.

ANSWERS TO QUESTIONS

10.1. Determine erf(0.90); erf(0.90) = $2N(1.414)(0.90)$. We find that by multiplying $1.414 * 0.90 = 1.273$. This value can be evaluated in the Normal Probability Function Table (from a standard text on statistics); looking up the area for 1.273 we find by extrapolation it is about 0.3985, which we now use in the equation above such that erf(0.90) = $2N(1.414)(0.90) = 2(0.3985) = 0.797$.

10.2. You will notice that this value (0.797) is the answer obtained for 10.1 thus, when erf $z = 0.797$, z can be found by the relation erf $z = 2N\sqrt{2z}$. Consequently, $N(\sqrt{2z}) = $ erf $z/2 = 0.797/2 = 0.3985$ which matches the area found for question 10.1, thus, $z = 0.90$. This answer could be found straight/forward by looking up $z = 0.90$ in the table of complimentary error functions (appendix 3) to find that $erf(0.90) = 0.796908$.

10.3. The first step is to examine equation 10.24 to see which parameters we need. From the diameter of the column we need to know the area A, to determine q. Thus, $A = 1963.5$ cm^2 and $q = (Q/A)\theta = (280/1963.5)/0.45 = 0.317$ cm hr^{-1}, and the slope $S = 5.0$. Now, using equation 10.24, $D = (0.317 * 125)/(4\pi5) = 0.631$ cm^2 hr^{-1}.

10.4. This requires taking the upper limit of the integral in equation 10.21, equal to q by using the chain rule to begin with, such that $d[C(x, t)/dp]$ is given by

$$\frac{d\left(\dfrac{C(x, t)}{C_o}\right)}{dq} = -\sqrt{2\pi}^{-1} \exp\left[-\frac{(1 - N_{pv})^2}{\left(\dfrac{4DN_{pv}}{qL}\right)}\right] \tag{10.138}$$

Because this derivative is a function of $-w^2/2$ raised to the exp and evaluated at $[(1 - N_{pv})/(2DN_{pv}/qL)^{0.5}] \times (2\pi)^{-0.5}$, the derivative of dq/dp is given by equation 10.22.

10.5. Considering equation 10.16 and its relation to the complimentary error function, equation 10.29 can be converted to the normal probability integral such that

$$\frac{C(x, t)}{C_o} = \frac{1}{2}\left(1 - \frac{2}{\sqrt{\pi}}\int_0^{(1-N_{pv})/2\sqrt{DN_{pv}/qL}} \exp\left(-\beta^2\right) d\beta\right. \tag{10.139}$$
$$\left. + \exp\left[\frac{qL}{D}\right]\left[1 - \frac{2}{\sqrt{\pi}}\int_0^{1+N_{pv}/2\sqrt{DN_{pv}/qL}} \exp\left(-\beta^2\right) d\beta\right]\right)$$

Now, by substitution of $w^2/2$ for β^2 as done earlier, we obtain

$$\frac{C(x, t)}{C_o} = \frac{1}{2} - \sqrt{2\pi}^{-1}\int_0^{(1-N_{pv})/\sqrt{2DN_{pv}/qL}} \exp\left[-\frac{w^2}{2}\right] dw + \frac{\exp\left[\dfrac{qL}{D}\right]}{2} \tag{10.140}$$
$$- \frac{\exp\left[\dfrac{qL}{D}\right]}{\sqrt{2\pi}}\int_0^{(1+N_{pv})/\sqrt{2DN_{pv}/qL}} \exp\left[-\frac{w^2}{2}\right] dw$$

This results (skipping several steps) in a derivative of the above equation with respect to N_{pv}, which is evaluated at $N_{pv} = 1$ of

$$\frac{d\left(\dfrac{C(x, t)}{C_o}\right)}{dp}\Bigg|_{N_{pv}=1} = \left(2\sqrt{\frac{\pi D}{qL}}\right)^{-1} \tag{10.141}$$

A slight manipulation will yield equation 10.24 as previously discussed.

10.6. We can obtain an estimate of K_{oc} by using equation 10.51. First substitute the value for the solubility of dicamba (0.04 mg/L) for S and then calculate to obtain $K_{oc} = 25,637$. For K we can use the expression $K = K_{oc}$ (percent oc)/100 to obtain $K = 25,637(2)/100 = 512.7$. To obtain S, we may now use equation $S = KC^n$ such that $S = 512.7(10)^{0.87} = 3801$ μg/g.

10.7. For the following derivations of Richard's equation and the ADE, the reader should refer to figure 10.16 to visualize conceptually the mass-change, storage, and flux processes.

Derivation of Richard's equation

$$q = \frac{v}{At} = \frac{cm^3}{cm^2/sec} = \frac{cm}{sec} \tag{10.142}$$

flow in   *flow out*

$$q \times \Delta y \, \Delta z \qquad \left(q_x + \frac{\partial qx}{\partial x}\Delta x\right)\Delta y \, \Delta z \tag{10.143}$$

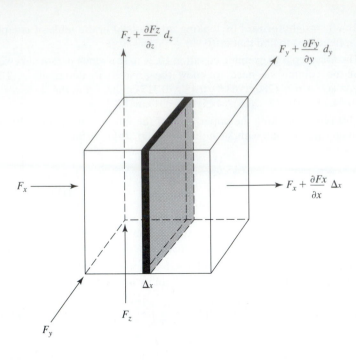

Figure 10.16   Flux across an interval (Δx) within a three-dimensional box

Mass change = flow in − flow out

$$z - \left[qx + \frac{\partial qx}{\partial x} \Delta x \right] \Delta y\, \Delta z = -\frac{\partial qx}{\partial x} \Delta x\, \Delta y\, \Delta z = -\left(\frac{\partial qx}{\partial x} + \frac{\partial qy}{\partial y} \right) \tag{10.144}$$

What is the volume change of H_2O?

$$\frac{\partial}{\partial t}\left(\theta\, \Delta x\, \Delta y\, \Delta z \right) = \frac{d\theta}{dt} \Delta x\, \Delta y\, \Delta z \tag{10.145}$$

$$\frac{d\theta}{dt} \Delta x\, \Delta y\, \Delta z = -\left(\frac{\partial qx}{\partial x} + \frac{\partial qy}{\partial y} + \frac{\partial qz}{\partial z} \right) \Delta x\, \Delta y\, \Delta z \tag{10.146}$$

$$\frac{d\theta}{dt} = -\frac{\partial qx}{\partial x} - \frac{\partial qy}{\partial y} - \frac{\partial qz}{\partial z} \tag{10.147}$$

Equation 4 is the three-dimensional continuity equation. In one dimension

$$\frac{d\theta}{dt} = -\frac{\partial qx}{\partial x} \tag{10.148}$$

By Darcy's law

$$q_z = -K\frac{\partial H}{\partial z} \qquad \text{where } H = h + z \tag{10.149}$$

Consequently,

$$\frac{\partial qz}{\partial z} = -\frac{\partial}{\partial z}\left[-K_z \frac{\partial H}{\partial z} \right] \tag{10.150}$$

Thus

$$\frac{\partial \theta}{\partial t} = \frac{\partial}{\partial z}\left[K_z \frac{\partial H}{\partial z} \right] \qquad \text{one dimension (vertical)} \tag{10.151}$$

However, for unsaturated flow $K(\theta)$ and $K(\psi_m)$, then

$$\frac{\partial \theta}{\partial t} = \frac{\partial}{\partial z}\left[K_z(\theta) \frac{\partial H}{\partial z} \right] \tag{10.152}$$

Since $H = h + z$, then

$$\frac{\partial \theta}{\partial t} = \frac{\partial}{\partial z}\left[K_z(\theta)\left(\frac{\partial h}{\partial z} = \frac{\partial z}{\partial z}\right)\right] \tag{10.153}$$

and

$$\frac{\partial \theta}{\partial t} = \frac{\partial}{\partial z}\left[K_z(\theta)\left(\frac{\partial h}{\partial z} + 1\right)\right] \tag{10.154}$$

$$\frac{\partial \theta}{\partial t} = \frac{\partial}{\partial z}\left[K_z(\theta)\frac{\partial h}{\partial z}\right] + \frac{\partial K_z(\theta)}{\partial z} \tag{10.155}$$

Equations 10.153–155 are all one-dimensional forms of Richard's equation to describe water flow under transient conditions. Now, we wish to enter the term D (diffusivity) to describe Darcian flow under unsaturated conditions.

$$D = K_z(\theta)\frac{\partial h}{\partial z} \qquad \text{where } D(\theta) \tag{10.156}$$

By substitution and assuming $\int \theta$ and $\int h(t)$ is continuous, and also by use of the chain rule

$$K(\theta)\frac{dh}{dz} = K(\theta)\frac{\partial h}{\partial z}\frac{\partial \theta}{\partial z} \tag{10.157}$$

So

$$\frac{\partial \theta}{\partial t} = \frac{\partial}{\partial z}\left[D(\theta)\frac{\partial \theta}{\partial z}\right] + \frac{\partial K_z(\theta)}{\partial z} \tag{10.158}$$

Equation 15 is the basic form of **Richard's equation** to describe water flow under transient conditions.

Derivation of Solute Transport Equation

Starting with Fick's first law of diffusion

$$J = -D_s\frac{dc}{dx} \tag{10.159}$$

Where hydrodynamic dispersion (D_H) is

$$D_H(V, \theta) = \alpha\bar{V} + D_s(\theta) \tag{10.160}$$

By substituting equation 10.159 into equation 10.160

$$J = -D_{S,H}\frac{dC}{dx} \tag{10.161}$$

For mass flow

$$J = qC \tag{10.162}$$

Thus

$$J = \theta D_{S,H}\frac{dC}{dx} - qC \tag{10.163}$$

where

$$q = -K\frac{dH}{dx} \tag{10.164}$$

Now, utilize figure 10.16 for conceptual visualization

$$(J\text{in} - J\text{out})\,\Delta y\,\Delta z = -\frac{\partial J}{\partial x}\,\Delta x\,\Delta y\,\Delta z \tag{10.165}$$

Written in terms of x, since we are interested in the incremental volume Δx in figure 10.16. Here, the mass change must account for θ and solute concentration (C) with time. Thus

$$\theta\,\Delta x\,\Delta y\,\Delta z \qquad \text{(mass water change)} \tag{10.166}$$

So

$$[C\theta]\, \Delta x\, \Delta y\, \Delta z \tag{10.167}$$

and

$$\frac{\partial(C\theta)}{\partial t}\, \Delta x\, \Delta y\, \Delta z = -\frac{\partial J}{\partial x}\, \Delta x\, \Delta y\, \Delta z \tag{10.168}$$

Remembering that

$$J = \theta D_{S,H}\frac{dC}{dx} - qC \tag{10.169}$$

Then

$$\frac{\partial(C\theta)}{\partial t} = \frac{\partial}{\partial x}\left[\theta D_{S,H}\frac{\partial C}{\partial x}\right] - \frac{\partial qC}{\partial x} \tag{10.170}$$

Since $v = q/\theta$, we divide by θ to obtain

$$\frac{\partial C}{\partial t} = \frac{\partial}{\partial x}\left[D_{S,H}\frac{\partial C}{\partial x}\right] - V\frac{\partial C}{\partial x} \tag{10.171}$$

This is the general form of the solute-transport equation (or ADE). However, this is not the final form since we must account for source/sink terms in the soil. Thus,

$$\frac{\partial S}{\partial t}\, \Delta x\, \Delta y\, \Delta z \tag{10.172}$$

where S is $+$ for source, $-$ for sink, and $S = KC$

$$S = \frac{\mu g}{g} \cdot \frac{g}{cm^3} = \frac{\mu g}{cm^3} \tag{10.173}$$

$$\frac{S\rho_b}{\theta}\, \Delta x\, \Delta y\, \Delta z = \mu g \tag{10.174}$$

$$\frac{\rho_b}{\theta}\frac{\partial S}{\partial t}\, \Delta x\, \Delta y\, \Delta z = \frac{\mu g}{sec} \tag{10.175}$$

By substituting equation 10.175 into equation 10.171, we obtain

$$\frac{\rho_b}{\theta}\frac{\partial S}{\partial t} + \frac{\partial C}{\partial t} = D_{S,H}\frac{\partial^2 C}{\partial x^2} - V\frac{qC}{\partial x} \tag{10.176}$$

Using the relation $S = KC$ then

$$\frac{\partial S}{\partial t} = K\frac{\partial C}{\partial t} \tag{10.177}$$

Substituting this into 33, we have

$$\frac{\rho_b}{\theta}K\frac{\partial C}{\partial t} + \frac{\partial C}{\partial t} = D_{S,H}\frac{\partial^2 C}{\partial x^2} - V\frac{qC}{\partial x} \tag{10.178}$$

$$\left(\frac{\rho_b}{\theta}K + 1\right)\frac{\partial C}{\partial t} = D_{S,H}\frac{\partial^2 C}{\partial x^2} - V\frac{qC}{\partial x} \tag{10.179}$$

If we let

$$R = \left(\frac{\rho_b}{\theta}K + 1\right) \tag{10.180}$$

Then

$$R = \frac{\partial C}{\partial t} = D_{S,H}\frac{\partial^2 C}{\partial x^2} - V\frac{qC}{\partial x} \tag{10.181}$$

This is the final form, in one dimension, of the solute-transport equation; other parameters dealing with radioactive decay, et cetera (usually expressed by λ) would be added to the equation depending on the specific nature of the transport problem.

10.8. This is an exercise in mathematics to sharpen your calculus skills. The initial condition for equation 10.62 is $C(x_1, t_1) = f(x_1)$ for $t_1 = 0$. With this in mind, we can find a solution (generalized form) using the separation-of-variable technique (one of the simpler methods). For simplification, we shall drop the subscript 1 for x and t prior to continuing. Using the separation-of-variable technique, we have a solution form of $C(x, t) = F(x)G(t)$ where $F(x)$ is a function of x and $G(t)$ is a function of t only. Substituting this into equation 10.63 yields $G'(t)F(x) = DF''(x)G(t)$. At this time we can divide through by $DF(x)G(t)$ to obtain $G'(t)/DG(t) = F''(x)/F(x)$. Since $F(x)$ and $G(t)$ are functions of x and t only, the left and right side have to be equal to the same constant, which we shall denote as α^2. This will allow us to write the previous equation as two ordinary differential equations such that $F''(x) + \alpha^2 F(x) = 0$ and $G'(t) + \alpha^2 DG(t) = 0$. To solve the first ordinary differential equation, we can assume a solution of the form $F(x) = \exp^{mx}$ which after substitution gives $m^2 \exp^{mx} + \alpha^2 \exp^{mx} = 0$. For validity, m must equal $i\alpha$ where α is an arbitrary constant thus, $F(x) = \exp^{i\alpha x}$. From Euler's principle we obtain $F(x) = \cos \alpha x + i \sin \alpha x$. The second ordinary differential equation has a solution form of $G(t) = \exp(-\alpha^2 Dt)$. Using a similar procedure then, the solution is given by $C(x, t) = \exp(-\alpha^2 Dt)(A \sin \alpha x + B \cos \alpha x)$ where the imaginary part is multiplied by constant A and the real part is multiplied by constant B. However, to obtain the best solution, a full-range of Fourier series must be used such that the solution becomes

$$C(x, t) = \sum_{n=1}^{\infty} \exp(-\alpha_n^2 Dt)(A_n \sin \alpha_n x + B_n \cos \alpha_n x) \qquad (10.182)$$

for the initial conditions (for which $t = 0$), we have

$$C(x, 0) = \sum_{n=1}^{\infty} (A_n \sin \alpha_n x + B_n \cos \alpha_n x) \qquad (10.183)$$

For the evaluation of A_n, B_n, and α_n: a valid interval for $C(x, 0)$ must be specified. This can be $-a < x < a$. At this point we shall define $\alpha_n = n\pi/a$, so that

$$A_n = \frac{1}{a} \int_{-a}^{a} C(x, 0) \sin\left(\frac{n\pi x}{a}\right) dx \qquad (10.184)$$

and

$$B_n = \frac{1}{a} \int_{-a}^{a} C(x, 0) \cos\left(\frac{n\pi x}{a}\right) dx \qquad (10.185)$$

In both equations 10.184 and 10.185, x will vanish during integration; thus to avoid confusion, x is replaced by a dummy variable u and $f(u)$ is defined by

$$f(u) = [C(x, 0)]_{x=u} \qquad (10.186)$$

The results will be substituted into equation 10.182 to find (after using equation 10.186) the following equation.

$$C(x, t) = \sum_{n=1}^{\infty} \exp\left(\left[-\left(\frac{n\pi}{a}\right)^2 Dt\right]\left(\left[\sin\frac{n\pi x}{a}\right]\left[\frac{1}{a}\int_{-a}^{a} f(u) \sin\frac{n\pi u}{a} du\right] + \qquad (10.187)$$

$$\cos\frac{n\pi x}{a}\left[\frac{1}{a}\int_{-a}^{a} f(u)\cos\frac{n\pi u}{a} du\right]\right)$$

by placing $\sin(n\pi x/a)$ and $\cos(n\pi x/a)$ inside the integral and combining the two integrals in equation 10.187, we obtain

$$C(x, t) = \sum_{n=1}^{\infty} \exp\left[-\left(\frac{n\pi}{a}\right)^2 Dt\right]\frac{1}{a}\int_{-a}^{a} f(u)\left[\sin\frac{n\pi x}{a}\sin\frac{n\pi u}{a} + \cos\frac{n\pi x}{a}\cos\frac{n\pi u}{a}\right] du \qquad (10.188)$$

By moving the exponential term inside the integral of equation 10.188 and applying the formula (401.04) of Dwight 1961 we obtain

$$C(x, t) = \frac{1}{a} \sum_{n=1}^{\infty} \int_{-a}^{a} f(u) \exp\left[-\left(\frac{n\pi}{a}\right)^2 Dt\right] \cos \frac{n\pi(x - u)}{a} \, du \tag{10.189}$$

Equation 10.189 gives us a solution for u between $-a$ and a—however, we need a solution between minus and plus infinity. Thus, for large values let us write $\Delta\lambda = \pi/a$ and we can express the right side of equation 10.189 as

$$\frac{\Delta\lambda}{\pi} \sum_{n=1}^{\infty} \int_{-a}^{a} f(u) \exp[-(n\,\Delta\lambda)^2 Dt] \cos n\,\Delta\lambda\,(x - u)\,du \tag{10.190}$$

By placing $\Delta\lambda$ after the integral in equation 10.190 we can rewrite it as

$$\frac{1}{\pi} \sum_{n=1}^{\infty} \left[\int_{-a}^{a} f(u) \exp[-(n\,\Delta\lambda)^2 Dt] \cos n\,\Delta\lambda\,(x - u)\,du\right] \Delta\lambda \tag{10.191}$$

For simplification we define the bracketed expression in equation 10.191 as a function like $g(n\,\Delta\lambda)$, and write the expression as

$$\frac{1}{\pi} \sum_{n=1}^{\infty} [g(n\,\Delta\lambda)]\,\Delta\lambda \tag{10.192}$$

The limit (if it exists) of equation 10.192 as $\Delta\lambda$ approaches zero is given by

$$\lim_{\Delta\lambda \to 0} \sum_{n=1}^{\infty} [g(n\,\Delta\lambda)]\,\Delta\lambda = \int_{0}^{\infty} g(\lambda)\,d\lambda \tag{10.193}$$

To determine when this limit exists we can look at our expression for $g(n\,\Delta\lambda)$ and examine it as $\Delta\lambda$ approaches zero. From equations 10.192 and 10.193, our $g(n\,\Delta\lambda)$ is the integral inside the brackets of equation 10.192. As $\Delta\lambda$ approaches zero it may be seen (from the relation $\Delta\lambda = \pi/a$) that the integral becomes

$$\int_{-\infty}^{\infty} f(u)\,du \tag{10.194}$$

where $f(u)$ is as defined before and we can see that equation 10.193 is valid when $f(u)$ is sectionally continuous, and that equation 10.194 does indeed exist. As a result, we can use equation 10.194 and rewrite equation 10.192 as

$$\frac{1}{\pi} \int_{0}^{\infty} \left[\int_{-\infty}^{\infty} f(u) \exp(-\lambda^2 Dt) \cos \lambda(x - u)\,du\right] d\lambda \tag{10.195}$$

Since equation 10.195 is the right side of equation 10.189, which has been expanded for $-\infty < x < \infty$, then $C(x, 0)$ can be expressed as

$$C(x, 0) = \frac{1}{\pi} \int_{0}^{\infty} \left[\int_{-\infty}^{\infty} f(u) \cos \lambda(x - u)\,du\right] d\lambda \tag{10.196}$$

Equation 10.196 is the Fourier integral formula of $f(x) = C(x, 0)$ and upon executing a reverse order of integration, equation 10.195 is written as

$$\frac{1}{\pi} \int_{-\infty}^{\infty} f(u) \left[\int_{0}^{\infty} \exp(-\lambda^2 Dt) \cos \lambda(x - u)\,d\lambda\right] du \tag{10.197}$$

where the inside integral is given by

$$\int_{0}^{\infty} \exp(-\lambda^2 Dt) \cos \lambda(x - u)\,d\lambda = \frac{\sqrt{\pi}}{2\sqrt{Dt}} \exp\left[-\frac{(x - u)^2}{4Dt}\right] \tag{10.198}$$

By placing the right side of equation 10.198 in equation 10.197 and also moving the constants outside of the integral, we obtain

$$\frac{1}{2\sqrt{\pi D t}} \int_{-\infty}^{\infty} f(u) \exp\left[\frac{-(x-u)^2}{4Dt}\right] du \tag{10.199}$$

As one will notice, the right-hand side of equation 10.199 is also the right-hand side of equation 10.189, which is expanded for a near ∞. Thus, for the interval $-\infty < x < \infty$, we rewrite the equation as

$$C(x_1, t_1) = \frac{1}{2\sqrt{\pi D t_1}} \int_{-\infty}^{\infty} f(u) \exp\left[\frac{-(x_1-u)^2}{4Dt_1}\right] du \tag{10.200}$$

Note that we have now reintroduced the subscript 1 for both x and t. This then is our general solution, which is valid for $-\infty < x_1 < \infty$ and for which the formal boundary condition for our solution is given as $C(x_1, \infty) = 0$ for finite x_1; $C(x_1, 0) =$ the concentration in the medium at time $t_1 = t = 0$ and at $x_1 = x$, where $x_1 = x$ is the starting position in the medium (usually a column or tube) of the slug of fluid that moves with dispersion. The value of $C(x_1, t_1)$ will always depend on the initial concentration, or rather its distribution within the column (i.e., on $C(x_1, 0)$). It is hoped that this problem will help sharpen the reader's calculus skills, help demonstrate the relationship between the ADE and a solution for it, and that one can see where certain parameters are derived from.

10.9. As explained previously, each phase would require its own retardation factor. Thus, the general form would be expressed as

$$\theta_{mo} R_{mo} \frac{\partial C_{mo}}{\partial t} + \theta_{im} R_{im} \frac{\partial C_{im}}{\partial t} = \theta_{mo} D \frac{\partial^2 C_{mo}}{\partial x^2} - \theta_{mo} v \frac{\partial C_{mo}}{\partial x} \tag{10.201}$$

10.10. Using equation 10.130, the wetted fraction is about 0.04.

10.11. A decrease in particle size, assuming the same relative parameters as given in question 10.10, would cause an increase in F. For example a particle size of 500–700 μm would yield an $F = 0.12$ while a particle size of 100–250 μm would yield an $F = 1.0$. What causes this? Is it a linear relationship with one of the other parameters?

ADDITIONAL QUESTIONS

10.12. A colleague has called to obtain your opinion about transport time at a local spill site. He is unsure of the chemical that has been spilled, but wishes to know the worst-case scenario for transport time from soil surface to ground water, 7 m below. The only values he can give are $\theta = 0.08$ and the drainage rate is 0.3 m per year. What is the amount of time it would take, worst-case, for a chemical to reach the water table?

10.13. Assuming the same parameters as above and $\rho_b = 1.35$, how long will be required for dicamba ($K_d = 2.2$) and trifluralin ($K_d = 13,700$) to reach ground water?

10.14. Your instructor has assigned you the task of demonstrating the effect of structural voids on contaminant transport. How can you illustrate—visually and mathematically—that structural voids have a greater effect on transport than relatively homogeneous soil profiles?

10.15. As an unsaturated-zone hydrologist, a person will not always work with mobile ions such as nitrate, but will often find it necessary to apply his/her skills to work involving other compounds such as aniline, ethyl acetate, isopropyl iodide, benzene, toluene, radioactive constituents, et cetera. What is the diffusion coefficient of aniline ($C_6H_5NH_2$) in water at 25 °C? *Hint:* use Hayduk and Laudie (1974) method. Assume $\eta = 0.8904$ and the LeBas molal volume is 6(C) $=$ 6(14.8) $= 88.8$; 7(H) $= 7(3.7) = 25.9$; 1(N-primary amine) $= 10.5$; 1 (6-membered ring) $= -15.0$. Total $= V'_B$ (total molal volume) $= 110.2$ cm^3 mol^{-1}.

10.16. What is the diffusion coefficient for ethyl acetate ($CH_3CO_2CH_2CH_3$) in water at 25 °C? Assume $\eta = 0.8904$ and the LeBas molal volume is 4(C) $= 59.2$; 8(H) $= 29.6$; 2(O) $= 19.8$. Total $= V'_B$ (total molal volume) $= 108.6$ cm^3 mol^{-1}.

11

Effects of Infiltration and Drainage on Soil-Water Redistribution

INTRODUCTION

Infiltration is the process by which water passes across the atmosphere–soil interface and enters a given soil column. The time-rate at which water (or another liquid) infiltrates the soil across the atmosphere–soil interface is known as the infiltration rate. The total volume of liquid crossing the interface is known as the cumulative infiltration. Quantitatively, infiltration rate is the volume of liquid entering the soil per unit area in a unit time.

This chapter defines terms and conditions essential to understanding the infiltration process during a water-input event, although infiltration of other liquids can be understood as an extension of the liquid characteristics presented in other chapters. This chapter also describes approaches to theoretical and practical calculation of infiltration and quantitatively discusses the variables that control the process. Some of the well-known and practical infiltration models are presented and discussed, along with examples of the infiltration process. Conditions affecting infiltration also change as a result of antecedent infiltration events; such processes are briefly discussed here. The reader is referred to chapter 13 for applied modeling of the redistribution process as a result of infiltration from a single event and antecedent events.

In engineering, infiltration capacity is often used and is defined as the maximum rate at which liquid can be moved into the soil in a given condition, and as such, signifies soil sorptivity. It is this characteristic that determines how much of the incident rainfall will run off and how much will enter the soil and either percolate downward or be evapotranspired.

Under ponded conditions, infiltration into an initially dry soil profile has a high rate early in time, decreasing rapidly and then more slowly until the rate reaches a nearly constant rate as shown on figure 11.1. As water redistributes through a soil profile, it displaces air and fills (or partially fills) the pores. The average hydraulic gradient, as well as the infiltration rate, continues to decrease during the infiltration process. The reason for the decrease in hydraulic gradient is that water is transmitted to the wetting front through an already wet-zone of soil that is continuously increasing in length as infiltration proceeds. This increases the resistance to flow and decreases the infiltration rate. It must be pointed out that this process is modified in soils that are subject to shrinking or swelling, and in materials or soils where the infiltrating liquid reacts chemically or physically with the media.

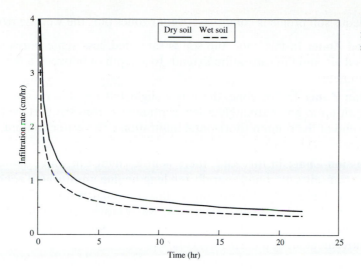

Figure 11.1   Examples of infiltration rate as a function of time and initial soil-water content

11.1   PROFILE-MOISTURE DISTRIBUTION

Consider the infiltration of water into a semi-infinite, uniform soil-profile having an initial volumetric-water content at residual saturation, θ_r. The infiltration process can be described—whether it is a horizontal or a vertical profile—by four distinct zones, as shown in figure 11.2. At the surface (or wet end of a horizontal column), the soil is nearly saturated for a few centimeters; there may be some pores where there is entrapped air. Below, or adjacent to, this saturated zone is a transmission zone where water is being transmitted to the wetting front as it moves downward (horizontally). The transmission zone is increasing in length as time goes on, with a uniform water content very near saturation. Within the wetting zone, the water content decreases until it merges into the wetting front, which forms a sharp boundary between the wet and the initially dry soil (figure 11.2). Both the wetting zone and the wetting front move continuously during the infiltration process. Thus, the wetted portion of

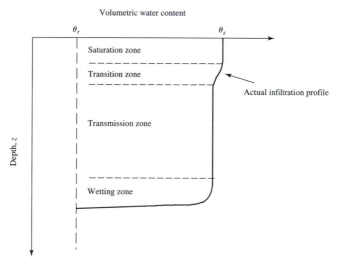

Figure 11.2   Idealized infiltration water profile distribution

the homogeneous soil profile is comprised of four zones plus the wetting front (figure 11.2):

1. **Saturated Zone:** In this zone, the soil is saturated, less some pores which may have entrapped air, and this saturation extends to a depth of between a few millimeters and one centimeter.

2. **Transition Zone:** In this zone, there is a slight but rapid decrease of water content with depth up to approximately a few centimeters from the surface (vertical infiltration) or end of the column (horizontal infiltration). The water content is still very near saturation.

3. **Transmission Zone:** In this zone, there is little change in water content from saturation. The transmission zone generally is a lengthening unsaturated zone with a uniform water content. For vertical infiltration in this zone, matric potential gradients are small compared to gravity gradients, which cause water movement; for horizontal infiltration, diffusive gradients are much larger than gravity (assumed to be negligible), causing water movement along a horizontal column.

4. **Wetting Zone:** In this zone, the water content decreases sharply with distance from the near-saturation values of the transmission zone to the initial residual-saturation value.

The wetting front is the zone of steep-water content and potential gradients, and forms a sharp boundary between the wet and dry (residual saturation) soil. Beyond the wetting front, the water content of the soil is at its initial value and there is no visible penetration of water. The moisture distribution presented in figure 11.2 is idealized, and actual moisture distributions can depart from those in figure 11.2, most likely in major qualitative ways.

Figure 11.3 shows the vertical cumulative infiltration under a thin film of water at the ground surface in three uniform soil materials consisting of a sand, a loam, and a clay. Figure 11.3 shows that at a time of 5 minutes (0.083 hours) the sand, loam, and clay profiles absorb approximately 1.3, 0.7, and 0.2 cm of water, respectively. The cumulative infiltration indicates one of the factors, soil texture, that is important in affecting infiltration rate. The factors that affect the infiltration rate can be divided into four general groups, representing **(1)** soil factors; **(2)** liquid-property factors; **(3)** rainfall or other liquid-arrival factors; and **(4)** other soil surface factors.

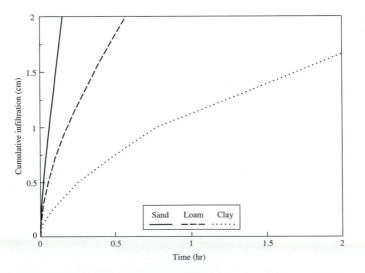

Figure 11.3   Cumulative infiltration into uniform profiles

Soil Characteristics

Soil characteristics affecting infiltration include the surface-entry characteristics and soil-transmission characteristics: texture; structure; organic matter; compaction; hydraulic conductivity; soil-water content; pore-size distribution; clay content; and behavior. Reduction of infiltration rate has been observed in soils due to surface sealing; this surface sealing can be caused by the impact of raindrops, animal activity, or vehicular traffic. The cause of the surface sealing is the movement of fine particles in between coarser particles, to form a relatively impermeable layer. Even soils exhibiting macropore flow could have the macropores clogged due to the movement of finer particles into the macropores.

When a soil material swells or increases in volume, this increase is almost always at the expense of pore volume. Macropores (non-capillary pores) (Beven and Germann 1982; Morrison 1993; Dingman 1994) can become capillary in size and smaller capillary pores are sealed. These factors affect the pore size and distribution in a soil, changing the rate of water movement and decreasing infiltration. Surface layers generally undergo larger changes in volume due to swelling and shrinking than deeper layers, even though the latter contains larger amounts of clay. Because water cannot enter a soil layer faster than it can be transmitted downward, surface conditions cannot increase the infiltration rate unless transmission characteristics of each layer of the profile are adequate. At saturation, the infiltration rate is limited to the lowest transmission rate encountered by the infiltrating water up to that time.

The total infiltration capacity of any layer depends upon its porosity, thickness, and quantity of water or other liquid present. Soil texture, structure, organic matter, root activity, and other physical properties determine the magnitude of the porosity of a given soil. Both the total porosity and the pore-size distribution determine the water-holding capacity of the soil. The initial infiltration will depend upon the volume, size, continuity, and relative stability of capillary and non-capillary pores, which provide paths for the percolating water or other liquid. Sandy materials have relatively stable pores. Well-graded silty and clayey materials, while having relatively larger pores in the dry state, on wetting, can have reduced pore sizes due to disintegration of the larger pores into smaller pores. Figure 11.3 shows the relation between total infiltration with time for a sand, loam, and clay soil representing high, moderate, and low infiltration rates, respectively. For the sand material to absorb 1 cm of water takes approximately 4 minutes; whereas for the loam material, it takes approximately 10 minutes; and for the clay material, approximately 45 minutes to absorb the same 1 cm of water.

During wetting, if air is trapped in the pores as liquid is applied from the surface to the wetting front, the entrapped air has the effect of causing a reduction on infiltration rate (Rawls et al. 1993). The reduction in rate can be large, because hydraulic conductivity is often reduced over one-half its saturated value (Bouwer 1966). The wetting front continues to advance slowly as a result of capillary forces, and the air pressure continues to increase until there is sufficient pressure to cause upward release of air through the pores holding water, and through the saturated surface layer.

The type of soil structure (e.g., fragmental, platy, cubical, blocky, prismatic, single grained, and massive), relative dimensions of structural aggregates, as well as the amount and direction of overlap of aggregates, determines the hydraulic conductivity classes. However, such secondary factors as compaction, direction of natural breakage, silt content, degree of mottling, and climatic factors are also important. Organic and inorganic fluids, along with residues, also influence the hydraulic conductivity. Dragun (1988) summarizes the effects of bulk hydrocarbons on the hydraulic conductivity and intrinsic permeability of selected soils. In general, bulk hydrocarbons result in substantial increases in hydraulic conductivity for the reasons given in the following section.

Microorganisms in the soil also are known to influence infiltration rates. Soils under prolonged saturation tend to have any entrapped air dissolved into the water, thus increasing the infiltration rate. However, microbial-activity byproducts such as slimes, gums, and gases can clog capillary pores, resulting in decreases in infiltration rates. On the other hand, microbial decomposition of plant and animal residues can markedly increase infiltration rates; earthworms are known to increase infiltration rates in compact, fine-textured soils.

Initial volumetric-water content has a great influence on the infiltration rate. As shown in figure 11.1, the infiltration rate decreases as initial water content increases, and as the time of water or other liquid application increases, the effect of antecedent moisture decreases. Additionally, the increase in initial water content decreases the infiltration rate during the early stages of infiltration, but increases the velocity of the wetting-front advance at all times. During the first few hours of infiltration, the antecedent moisture content will probably be the major factor determining the infiltration rate of a given soil.

The initial soil-water content affects the infiltration rate in three ways: **(1)** partially full pores reduce infiltration; **(2)** wetting of an initially dry soil causes increased capillary forces that increase infiltration; however, with the increase in the depth of the wetting front, infiltration is decreased; and **(3)** wetting of the soil may cause swelling of the soil materials, which decreases infiltration.

Liquid Properties

The physical characteristics of the infiltrating fluid (e.g., water or NAPL) also affect infiltration rate. When water enters a soil, fine clays, organics, salts, and other materials contaminate the water. These suspended and dissolved materials in the infiltrating fluid not only block pores, but also may affect the viscosity, density, and surface tension of the water. Some of these materials, such as salts, can affect the swelling potential of some soils by forming complexes with colloidal materials.

The temperature, viscosity, and surface tension of the fluid affect the rate of movement of the fluid into and within the soil. The relation between the infiltration rate and factors such as temperature, viscosity, surface tension, pore size, depth of wetting, head of water, wettability of solids by solution, and density of the wetting solution, indicates that: **(1)** when temperature of the fluid is increased, there is a corresponding increase in the infiltration rate (Moore 1941); **(2)** increases in viscosity decrease the infiltration rate hyperbolically; **(3)** decreases in surface tension increase the infiltration rate linearly; **(4)** as pore size increases, infiltration rate increases parabolically; and **(5)** as depth of wetting increases, the infiltration rate decreases hyperbolically (Ghildyal and Tripathi 1987).

Rainfall or Other Liquid-Arrival Factors

The characteristics of how the fluid is deposited on the surface can have both direct and indirect effects on infiltration rate. If the maximum infiltration rate exceeds the rate at which the fluid is applied, this fluid infiltrates and a direct relation between rate of infiltration and rate of fluid application occurs. However, when fluid application rate (or application intensity) exceeds the infiltration rate, an inverse relation between application rate and infiltration rate is obtained. For example, an increase in rainfall intensity causes increased compacting forces as the raindrops hit the soil surface, thus decreasing infiltration rate. Rainfall or other fluid characteristics can also affect infiltration rate indirectly by increasing the initial soil moisture content, and the activities of earthworms and other animals.

At time $t = 0$, liquid begins arriving at the surface at a rate, w, which continues for a specified time t_w. For the idealized condition of constant liquid arrival, two cases can be considered: **(1)** the liquid-input rate at the surface is less than the saturated hydraulic conductivity; and **(2)** the liquid-input rate at the surface is greater than the saturated hydraulic conductivity. The following discussion is valid for a homogeneous layer of soil or for several layers where the underlying layers have successively higher saturated hydraulic conductivities.

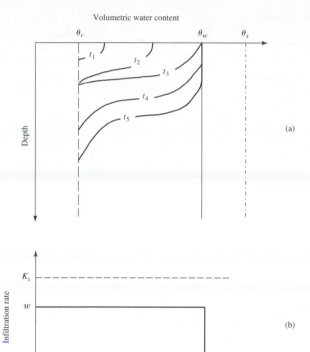

Volumetric water content

Figure 11.4   Water content profiles and infiltration rate in a soil layer where the liquid-input rate is less than the saturated hydraulic conductivity

Liquid-input rate at the surface is less than the saturated hydraulic conductivity   Let us consider a thin layer of homogeneous soil at the instant that liquid input at the surface begins. If it is assumed that $w < K_s$ (figure 11.4(b)), liquid will enter this layer faster than it is leaving, resulting in an increase in volumetric-water content, θ. The increase in volumetric-water content causes an increase in hydraulic conductivity consistent with the unsaturated hydraulic-conductivity versus water-content relation for the soil. Therefore, the flux out of the layer also increases as the water content increases. However, as long as the water content in the layer is less than that at which the hydraulic conductivity, K, equals the liquid-input rate, w, the water content will continue to increase. When the water content reaches θ_w (which is less than the saturated water content, θ_s), the hydraulic conductivity is K_w, which is equal to the liquid-input rate, w, so that the rate of outflow from the layer is equal to the rate of input and there is no further change in water content for the constant liquid-input rate.

This process occurs successively in each underlying soil layer as liquid input continues. Figure 11.4a shows schematically the resulting water-content profiles in a uniform vertical soil layer. Analysis of Figure 11.4b confirms the relation that, for a liquid-input rate at the surface less than the saturated hydraulic conductivity, the infiltration rate is constant with a value of w for all time between 0 and t_w, and the infiltration rate is zero for all other times.

Liquid-input rate at the surface is greater than the saturated hydraulic conductivity   When $w > K_s$, liquid still enters a thin, homogeneous surface layer faster than it is leaving, but only during the early stages of infiltration. Liquid generally arrives at each layer faster than it can be transmitted downward and initially goes into storage, increasing the water content and the hydraulic conductivity. However, because the water content cannot exceed its saturated value, θ_s, the hydraulic conductivity cannot increase beyond K_s. After the surface layer reaches saturation, liquid will accumulate on the surface and ponding will begin until surface-detention storage is satisfied. After this time, runoff will occur. Figure 11.5(a) shows the resulting water-content profiles for the case of the liquid-input rate greater than the

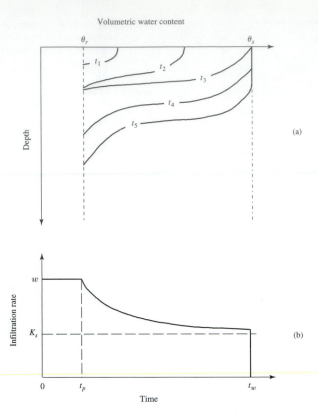

Figure 11.5 Water content profiles and infiltration rate in a soil layer where the liquid-input rate is greater than the saturated hydraulic conductivity

saturated hydraulic conductivity. The time at which the surface layer becomes saturated is called the time of ponding, t_p, which is shown schematically in Figure 11.5b. Up to the ponding time, the infiltration rate is equal to the liquid-input rate, w. As liquid input continues after the time of ponding, infiltration continues but at an asymptotically decreasing rate towards K_s (figure 11.5b), until cessation of liquid input.

Other Soil-Surface Factors

Surface slope, vegetative cover, and roughness all have an effect upon the infiltration rate (Ghildyal and Tripathi 1987). Steep, smooth slopes with little vegetative cover allow little time for infiltration; on flatter, rougher slopes with vegetation, infiltration is encouraged. Clearly, one of the factors governing infiltration is time available for the process to occur. Thus, surface conditions which slow runoff increase the infiltration volume and rate, other factors being equal. Vegetation can be a particularly important factor, because vegetative cover tends to retard surface-water runoff, allowing time for infiltration to occur. Additionally, vegetation reduces the compaction of the soil surface due to influence of falling raindrops. The plant roots may increase hydraulic conductivity of the surface layers, thereby increasing infiltration. Types of natural and cropping vegetative cover have a significant effect on infiltration (Brakensiek and Rawls 1988; Branson et al. 1981).

11.2 INFILTRATION THEORIES

As shown in chapter 8, combining Darcy's law with the one-dimensional continuity equation results in the one-dimensional Richards' equation for vertical flow in an unsaturated soil.

No known closed-form analytical solutions exist for Richards' equation in layered soil, unless it is assumed that the flux rate, q, is constant (see chapter 13). For nonlayered soil with a uniform initial-water content, some closed-form solutions for variable flux rates (infiltration rates) are available. These physically based solutions include those by Green and Ampt (1911), Philip (1957a, b, c, d), Morel-Seytoux and Khanji (1974), and Smith and Parlange (1978). Solutions to Richards' equation can also be obtained numerically under appropriate boundary and initial conditions.

In addition to physically based, closed-form solutions to Richards' equation for homogeneous soils, several empirical equations are often used in operational hydrology. These equations, such as the Horton equation (Horton 1939; 1940), Holtan equation (Holtan 1961), and Kostiakov equation (Kostiakov 1932), relate measured infiltration rate or cumulative-infiltration to elapsed time modified by various soil and vegetation parameters. These parameters have to be obtained by fitting them to measured data. These infiltration theories are reviewed in more detail in the following sections.

Green–Ampt Approach

One of the earliest physically based approaches to infiltration of water into soils was formulated by Green and Ampt (1911), for soils that exhibit a sharp wetting front. A sharp wetting front is most common in soils having uniform pore shapes and being generally coarse-textured. The following assumptions for the Green–Ampt model permit an analytical solution to the infiltration equation: **(1)** the soil under consideration is homogeneous with respect to water retention and transmission properties; **(2)** a distinct and precisely definable wetting front exists; **(3)** matric suction at the wetting front remains constant throughout infiltration; and **(4)** the soil is uniformly wet behind the wetting front.

For horizontal infiltration into a uniform soil column where gravitational forces can be neglected, a Darcy-type equation can be written as:

$$i = \frac{dI}{dt} = K_s \frac{\psi_0 - \psi_f}{L_f} \tag{11.1}$$

where i is the infiltration rate (flux rate) into the soil and through the transmission zone, I is the cumulative infiltration, K_s is the saturated hydraulic conductivity of the transmission zone, ψ_0 is the saturated suction (matric potential = 0), ψ_f is the effective suction at the wetting front, and L_f is the distance from the soil surface to the wetting front (length of the total infiltration zone). Because $\psi_0 - \psi_f$ is constant and is independent of time and space as long as the soil surface is maintained at a matric potential of zero and the ponding depth is negligible, equation 11.1 can be rewritten as

$$\frac{dI}{dt} = K_s \frac{\Delta\psi}{L_f} \tag{11.2}$$

where $\Delta\psi$ is the change in potential from the soil surface to the wetting front. Equation 11.2 also indicates that the infiltration rate i is a linear inverse function of the total infiltration zone length, L_f. Because the distance from the surface to the wetting front is uniform, cumulative infiltration I can be given by

$$I = L_f(\theta_t - \theta_i) \tag{11.3}$$

where θ_t is the transition-zone volumetric water content during infiltration, and θ_i is the initial soil volumetric water content. For the special case where θ_t is the volumetric water content at saturation (porosity) and θ_i is zero, then cumulative infiltration in equation 11.3 is given by

$$I = \phi L_f \tag{11.4}$$

where ϕ is the effective porosity. Thus,

$$\frac{dI}{dt} = (\theta_t - \theta_i) \frac{dL_f}{dt} \tag{11.5}$$

where dL_f/dt is the rate of advance of the wetting front during infiltration. It should be recognized that

$$K_s \frac{\Delta\psi}{L_f} = K_s \frac{\Delta\theta\,\Delta\psi}{I} \tag{11.6}$$

or the infiltration rate i is an inverse function of the cumulative infiltration I.

Substituting the value of L_f in equation 11.3 into equation 11.2 gives

$$\frac{dI}{dt} = K_s \frac{\Delta\psi\,(\theta_t - \theta_i)}{I} \tag{11.7}$$

Rearranging equation 11.7 and integrating

$$\int I\,dI = [K_s\,\Delta\psi\,(\theta_s - \theta_i)]\int dt \tag{11.8}$$

or

$$\frac{I^2}{2} = K_s\,\Delta\psi\,(\theta_t - \theta_i)t + C \tag{11.9}$$

where C is the constant of integration.

For typical initial conditions $t = 0$ and $I = 0$, and therefore $C = 0$. The cumulative infiltration I is then given by

$$I = [2K_s\,\Delta\psi\,(\theta_t - \theta_i)]^{1/2}\,t^{1/2} \tag{11.10}$$

Writing $(\theta_t - \theta_i)$ as $\Delta\theta$, equation 11.10 can be expressed as

$$I = \Delta\theta\,(2\widetilde{D}t)^{1/2} \tag{11.11}$$

where $\widetilde{D}$ is the effective diffusivity of the draining soil profile, given by

$$\widetilde{D} = K_s \frac{\Delta\psi}{\Delta\theta} \tag{11.12}$$

It should be noted that the quantities in the brackets on the right-hand side of equation 11.10 are constant for all times greater than zero. This implies that the relation between cumulative infiltration I and $t^{1/2}$ will be linear, with the straight line passing through the origin and having a constant slope, S, given by

$$S = [2K_s\,\Delta\psi\,(\theta_t - \theta_i)]^{1/2} = \Delta\psi\,(2\widetilde{D})^{1/2} \tag{11.13}$$

where S is called the sorptivity. Differentiating equation 11.13 gives an expression for infiltration rate i

$$i = \frac{dI}{dt} = \left[\frac{1}{2}K_s\,\Delta\psi\,(\theta_t - \theta_i)\right]^{1/2}\,t^{-1/2} \tag{11.14}$$

Equations 11.10 and 11.14 show that the cumulative infiltration I increases, whereas the infiltration rate i decreases linearly with the square root of time.

Gravitational forces (vertical infiltration) can be taken into account by adding the distance from the soil surface to the wetting front to equation 11.1, which is the Green–Ampt model for horizontal infiltration, or

$$i = \frac{dI}{dt} = K_s \frac{\psi_b + L_f - \psi_f}{L_f} \tag{11.15}$$

which can be rewritten as

$$i = \frac{dI}{dt} = K_s \left[1 + \frac{\psi_b - \psi_f}{L_f} \right] \tag{11.16}$$

In this case, ψ_b is defined as the bubbling pressure (potential), which is the potential at which the largest pore begins to drain.

From equation 11.16, it is apparent that the infiltration rate i is maximum early in the infiltration process, when the length of the total infiltration zone L_f is minimum, and decreases as L_f increases. Also, i approaches a constant value K_s as $t \to \infty$, for which L_f becomes very large, or

$$i(t = \infty) = K_s \tag{11.17}$$

If L_f in equation 11.3 is substituted into equation 11.16, we obtain the common form of the Green–Ampt equation

$$i = K_s \left(1 + \frac{(\theta_t - \theta_i)\,\Delta\psi}{I} \right) \tag{11.18}$$

If equation 11.5 is substituted into equation 11.16, we obtain

$$(\theta_t - \theta_i)\frac{dL_f}{dt} = K_s \left(1 + \frac{\psi_b - \psi_f}{L_f} \right) \tag{11.19}$$

Rearranging the terms and integrating equation 11.19 gives

$$\frac{K_s}{(\theta_t - \theta_i)} \int dt = \int \frac{L_f}{(\psi_b - \psi_f) + L_f}\, dL_f \tag{11.20}$$

or

$$\frac{K_s t}{(\theta_t - \theta_i)} = (\psi_b - \psi_f) + L_f - (\psi_b - \psi_f)\ln(\psi_b - \psi_f + L_f) + C \tag{11.21}$$

where C is the constant of integration that can be evaluated for the initial conditions of $t = 0$ and $L_f = 0$, so that

$$C = -(\psi_b - \psi_f) + (\psi_b - \psi_f)\ln(\psi_b - \psi_f) \tag{11.22}$$

and equation 11.21 finally becomes

$$\frac{K_s t}{(\theta_t - \theta_i)} = \left[L_f - (\psi_b - \psi_f)\ln\left(1 + \frac{L_f}{\psi_b - \psi_f}\right) \right] \tag{11.23}$$

Equation 11.23 relates the total infiltration depth to the wetting front at time t to the cumulative infiltration volume. In general, as time increases, L_f increases such that at very long times, equation 11.23 can be approximated by

$$K_s t = (\theta_t - \theta_i)L_f + A \tag{11.24}$$

where

$$A = -(\theta_t - \theta_i)\left[(\psi_b - \psi_f)\ln\left(1 + \frac{L_f}{\psi_b - \psi_f}\right) \right] \tag{11.25}$$

From equation 11.3, equation 11.24 can also be written as

$$I = K_s I - A \tag{11.26}$$

where A changes much more slowly as L_f becomes very large. The infiltration rate i at large L_f thus becomes

$$i = K_s \qquad (11.27)$$

The Green–Ampt model is satisfactory for both unponded and ponded surface conditions. For infiltration under ponded conditions, the soil in the total infiltration zone is nearly saturated. This near-saturation condition develops a viscous resistance to air flow, which reduces infiltration rate. To account for this effect, Morel-Seytoux and Khanji (1974) introduced a correction factor to the Green–Ampt equation for a homogeneous soil (discussed later in this section).

To apply the Green–Ampt model, the saturated hydraulic conductivity K_s, suction at the wetting front ψ_f, initial volumetric water content θ_i, and transition-zone volumetric water content θ_t all must be measured or estimated. Obviously these variables can be calculated by fitting them to experimental infiltration data; however, other data may also be readily available that can be used.

It can be assumed that θ_t is approximately equal to porosity ϕ, which can be obtained from other geotechnical tests or calculated from bulk density and particle density. K_s can be obtained from laboratory tests, as can θ_i. If moisture-characteristic curve data are available for a specific soil, ϕ should be available from these data. Regression equations available from Rawls and Brakensiek (1983) or Rawls et al. (1993) can also be used to estimate detailed soil properties.

The Green–Ampt effective suction at the wetting front can be estimated from Rawls et al. (1993) and the Brooks–Corey variables (Brooks and Corey 1964) as

$$\psi_f = \frac{2 + 3\lambda}{1 + 3\lambda} \frac{\psi_b}{2} \qquad (11.28)$$

where ψ_f is the Green–Ampt effective suction at the wetting front, λ is the Brooks–Corey pore-size distribution index, and ψ_b is the Brooks–Corey bubbling pressure (suction). Both λ and ψ_b are variables determined by fitting a Brooks–Corey relation to moisture-characteristic curve data. The methodology for estimating these two fit-variables is detailed in chapter 13. Rawls and Brakensiek (1983) and Rawls et al. (1993) present regression equations for estimating λ and ψ_b from a soil sample if the percent sand, percent clay, and porosity of the sample are known.

Figure 11.6 is an example of the Green–Ampt model to calculate infiltration rate (equation 11.18) versus time for various assumed cumulative infiltration values I for two different soils. A river sand, having $\Delta\psi = 16$ cm, $K_s = 26.4$ cm/hr, $\theta_t = 0.39$, and $\theta_i = 0.20$; and the other a Laveen Loam, having $\Delta\psi = 25$ cm, $K_s = 0.45$ cm/hr, $\theta_t = 0.38$ and $\theta_i = 0.30$. In this case, I was assumed and the time t was calculated as the ratio of the initially assumed cumulative infiltration to the infiltration rate calculated by equation 11.18. The coarser-grained river sand experiences larger values of cumulative infiltration and higher initial and final infiltration rates.

Horton and Kostiakov Equations

Horton (1939; 1940) proposed an empirical formula for infiltration, based upon the many natural processes that move spontaneously to their final value at a rate determined by the difference between their initial and final values. Horton's derivation of infiltration as a function of time began with

$$\frac{di}{dt} = -k(i - i_f) \qquad (11.29)$$

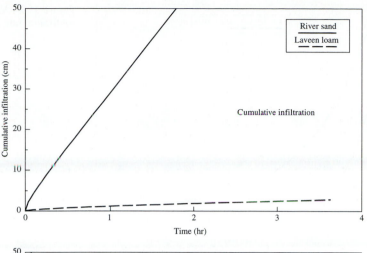

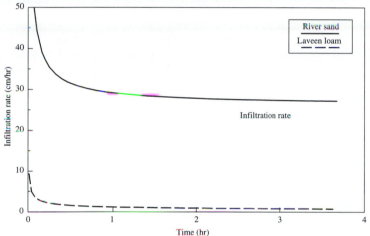

Figure 11.6   Green–Ampt cumulative infiltration, and infiltration rates for two different soils

where i is the infiltration rate, i_f is the final (constant) infiltration rate, k is a proportionality constant ($k > 0$) dependent on the soil type and vegetative cover, and t is the time since the liquid input began. The negative sign in equation 11.29 indicates that infiltration decreases toward its final constant value. If the initial infiltration rate is i_o at zero time, then equation 11.29 can be integrated to yield

$$\ln \frac{(i - i_f)}{(i_o - i_f)} = -kt \tag{11.30}$$

or, solving for i

$$i = i_f + (i_o - i_f) e^{-kt} \tag{11.31}$$

It is clear from equation 11.31 that as $t \to \infty$, i approaches a constant rate, i_f at $k > 0$. Cumulative infiltration can be found by integrating equation 11.31 and applying the initial conditions of $I = 0$ when $t = 0$. The resulting cumulative infiltration is

$$I = i_f t + \frac{i_o - i_f}{k} (1 - e^{-kt}). \tag{11.32}$$

Values of i_o, i_f, and k depend upon soil and vegetative cover complexes. According to Skaggs and Khaleel (1982), typical values for agricultural soils range from approximately 30

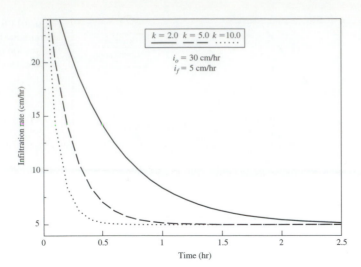

Figure 11.7 Variation of k on Horton infiltration rate

to 90 cm/hr for i_o; less than 1 to 30 cm/hr for i_f; and up to 50/hr for k. Values of i_o, i_f, and k should be evaluated using observed infiltration data. Figure 11.7 shows changes in infiltration rate for various assumed values of k. The curves in figure 11.7 were calculated using equation 11.30, with $i_o = 30$ cm/hr and $i_f = 5$ cm/hr.

Kostiakov (1932) proposed an empirical infiltration equation of the form

$$I = at^b \tag{11.33}$$

where I is cumulative infiltration, t is time since infiltration started, and a and b are constants that depend upon the soil and initial conditions. The values of a and b may only be estimated using observed infiltration data.

The Kostiakov equation can only be used if a set of observed infiltration data is available for variable estimation. Therefore, it cannot be applied to soil conditions which differ from those used to estimate a and b. The Kostiakov equation has been used primarily for analysis of irrigation applications (Rawls et al. 1993).

Holtan Model

Holtan (1961) presented an empirical infiltration equation, based on the concept that the infiltration rate is proportional to the unfilled pore volume of a soil. Holtan's equation was developed by assuming that the unfilled pores included surface-connected porosity and the effects of root paths, which are probably the most important features affecting infiltration rate. Holtan's (1961) original equation was of the form

$$i = aI_p^n + i_f \tag{11.34}$$

where i is the infiltration rate, i_f is the final (constant) infiltration rate, I_p is the unfilled capacity of the soil to store water, and a and n are constants. The Holtan model has the advantage over the Horton model in that it can describe infiltration rate and the recovery of infiltration capacity during periods of little or no rainfall.

Holtan et al. (1975) modified equation 11.34 to give

$$i = GAI_p^{1.4} + i_f \tag{11.35}$$

where G is the growth index of the vegetative cover in percent maturity, varying from 0.1 to 1.0 during the growing season; A is the infiltration capacity of available storage and is an index representing surface-connected porosity and the density of plant roots which affect infiltration; I_a is the available storage in the soil-surface layer (A horizon); and n is now held constant at 1.4. The value of I_a ranges from zero to a maximum of the available water-holding capacity of the soil. Values of water-holding capacity are given for many soils by Rawls et al. (1993). Values of the vegetative parameter A are given in table 11.1 (Frere, Onstad, and Holtan 1975). Musgrave (1955) provided values of i_f that were related to U.S. Soil Conservation Service (SCS) hydrologic soil groups. These values of i_f are presented in table 11.2, along with the SCS definitions for each hydrologic soil group (SCS 1980).

TABLE 11.1 Estimates of Vegetative Parameter A in the Holtan Infiltration Equation

	Basal area rating* (in.$^{1.4}$/hr)	
Land use or cover	Poor condition	Good condition
Fallow[†]	0.10	0.30
Row crops	0.10	0.20
Small grains	0.20	0.30
Hay (legumes)	0.20	0.40
Hay (sod)	0.40	0.60
Pasture (bunch grass)	0.20	0.40
Temporary pasture (sod)	0.20	0.60
Permanent pasture (sod)	0.80	1.00
Woods and forests	0.80	1.00

Source: Data from Frere, Onstad, and Holtan (1975).
*Adjustments needed for "weeds" and "grazing."
[†]For fallow land only, poor condition means "after row crop" and good condition means "after sod."

TABLE 11.2 Final Infiltration Rates by SCS Hydrologic Soil Group for the Holtan Infiltration Equation

SCS Hydrologic Soil Group[1]	i_f (cm/hr)
A	0.76
B	0.38–0.76
C	0.13–0.38
D	0.0–0.13

Sources: Data from Musgrave (1955) and SCS (1980).
[1]**Group A** soils have low runoff potential and high infiltration rates even when thoroughly wetted. They consist mainly of deep, well-to-excessively drained sand or gravel. The USDA soil textures normally included in this ground are sand, loamy sand, and sandy loam. These soils have a transmission rate greater than 0.76 cm/hr.
Group B soils have moderate infiltration rates when thoroughly wetted and consist mainly of moderately deep to deep, moderately well to well-drained soils with moderately fine to moderately coarse textures. The USDA soil textures normally included in this group are silt loam and loam. These soils have transmission rate between 0.38 and 0.76 cm/hr.
Group C soils have low infiltration rates when thoroughly wetted and consist mainly of soils with a layer that impedes downward movement of water and soils with moderately fine to fine texture. The USDA soil textures normally included in this group is sandy clay loam. These soils have a transmission rate between 0.13 and 0.38 cm/hr.
Group D soils have high runoff potential. They have very low infiltration rates when thoroughly wetted and consist mainly of clay soils with a high swelling potential, soils with a permanent high-water table, soils with a claypan or clay layer at or near the surface, and shallow soils over a nearly impervious material. The USDA soil textures normally included in this group are clay loam, silty clay loam, sandy clay, silty clay, and clay. These soils have very low rate of water transmission (0.0 to 0.13 cm/hr).

Philip Model

Using a series approximation, Philip (1957a, b, c, d) found the cumulative infiltration I, in a horizontal, semi-infinite column to be

$$I = St^{1/2} + (A_2 + K_o)\, t + A_3\, t^{3/2} + A_4\, t^2 + \cdots + A_m\, t^{m/2} + \cdots \qquad (11.36)$$

where $S, A_2 + K_o, A_3, A_4, \ldots$ are functions of volumetric water content that are the solutions of a series of ordinary equations. Philip found that the power series in $t^{1/2}$ given by equation 11.36 converged for all except very large t. Philip (1957d) called S the sorptivity, which he claimed was a property of the medium. For a uniform, semi-infinite horizontal-soil column having a sharp wetting front, the sorptivity S is approximately given by

$$S \approx \frac{(\theta_s - \theta_i)x}{t^{1/2}} \qquad (11.37)$$

where x is the horizontal distance of the wetting front from the end of the column, θ_i is the initial volumetric water content of the column, θ_s is the saturated volumetric water content behind the wetting front, and t is the time since water started entering the column.

The horizontal infiltration rate i can be determined by taking the derivative of equation 11.36, or

$$i = \frac{dI}{dt} = \frac{1}{2} St^{-1/2} \qquad (11.38)$$

where only the first term in equation 11.36 is significant. As with other physically based infiltration equations, water is transferred to the wetting front through a wet zone that is continuously increasing in length. This increases the resistance to flow and decreases the infiltration rate. The equation of cumulative infiltration I can be used to determine sorptivity S, provided the infiltration is known at some time from field or laboratory experiments.

Philip (1957d) provided an approximate solution to the vertical-infiltration equation for a homogeneous soil with water ponded on the surface as

$$I = St^{1/2} + At \qquad (11.39)$$

where I is the cumulative infiltration, S is sorptivity, t is time, and A is a soil parameter. The value of A is related to the hydraulic conductivity of the soil. For a saturated soil surface, A is identical to the saturated-hydraulic conductivity K_s. If the soil surface does not have a thin film of ponded water, then A is equal to the unsaturated hydraulic conductivity K of the transmission zone.

The equation for vertical infiltration rate can be obtained by differentiating equation 11.39, or

$$i = \frac{dI}{dt} = \frac{1}{2} St^{-1/2} + A \qquad (11.40)$$

where A is equal to K_s if water is ponded on the surface and equal to K, otherwise.

In the Philip model, the variables S and A can be evaluated from experimental infiltration data using best-fit regression techniques; however, these variables may also be estimated from laboratory data using approximations presented by Rawls et al. (1993). Sorptivity S in the Philip model can be approximated using an equation originally developed by Youngs (1964)

$$S = [2(\theta_t - \theta_i)K\psi_f]^{1/2} \qquad (11.41)$$

where ψ_f is the Green–Ampt effective suction at the wetting front given by equation 11.31, θ_t is the transition-zone volumetric water content which can be taken as the soil porosity if a thin film of water is ponded on the surface, and θ_i is the initial water content. For homogeneous soil conditions with no macropore flow, surface crust, or vegetative cover, the value of K in equation 11.41 can be assumed to be the saturated hydraulic conductivity K_s. Bouwer (1966) suggests that the value of K should be the hydraulic conductivity at residual air saturation, or a value of 0.5 K_s. Other conditions affecting K are discussed in following sections. The variable A in Philip's model was found by Youngs (1964) to range from 0.33 K_s to K_s, with K_s being the recommended value (Rawls et al. 1993).

Morel-Seytoux and Khanji Model

Morel-Seytoux (1973) was the first to show the important effect of air movement on water infiltration into an initially dry soil, under ponded conditions. Morel-Seytoux and Khanji (1974) recognized, as have others, that the Green–Ampt equation for infiltration could lead to errors of prediction by an excess of between 10 and 70 percent, especially when infiltration was occurring under ponded conditions. In the case of ponded conditions, the soil in the transmission zone (see figure 11.2) is nearly saturated, and develops a viscous resistance to air flow; this resistance reduces the infiltration rate. To correct for this, Morel-Seytoux and Khanji (1974) found it necessary to introduce a viscous correction factor, β. No particular pattern of β emerges as a function of soil type. Morel-Seytoux and Khanji (1974) suggest that the values of infiltration rate given by the Green–Ampt model (equation 11.18) be divided by β to give

$$i = \frac{K_s}{\beta}\left(1 + \frac{(\theta_t - \theta_i)(H_o + \Delta\psi)}{I}\right) \qquad (11.42)$$

where H_o is the depth of ponded water.

The correction factor, β, varies with the soil type and ponding depth, ranging from approximately 1.1 to 1.7, with an average of 1.4. This factor should be used to reduce the Green–Ampt infiltration rate by dividing that rate by the correction factor, when entrapped air in the soil is a significant factor.

Smith–Parlange Model

Smith and Parlange (1978) considered two extreme cases concerning the behavior of unsaturated hydraulic conductivity K, near saturation during infiltration. In one case they assume that K varies slowly near saturation and leads to an expression for ponding time and infiltration rate. For initially ponded conditions (a thin film of water on the soil surface), the ponding time was zero and Smith and Parlange's model resulted in the Green–Ampt model

$$i = K_s\left(1 + \frac{S}{K_s I}\right) \qquad (11.43)$$

where S is the sorptivity previously given by equation 11.13.

In the other second extreme, Smith and Parlange (1978) assumed that near saturation K varies exponentially. This model will hold for both the non-ponding and ponding cases, and for ponded conditions the expression for infiltration rate is given by

$$i = \frac{K_s \exp\left(\dfrac{IK_s}{S}\right)}{\exp\left(\dfrac{IK_s}{S - 1}\right)} \qquad (11.44)$$

Equations 11.43 and 11.44 both use only two parameters, the saturated hydraulic conductivity and the sorptivity. As seen above both of these parameters are related to measurable soil properties based upon either field or laboratory tests.

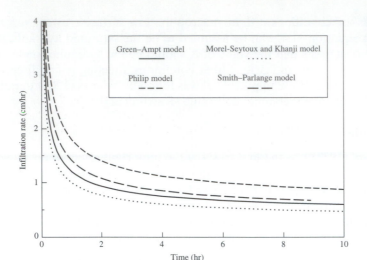

Figure 11.8   Comparison of physically based infiltration models for a Laveen loam

Comparison of the Various Physically Based Infiltration Models

Figure 11.8 shows the infiltration rates versus time for the Green–Ampt, Philip, Morel-Seytoux and Khanji, and Smith and Parlange models, for a Laveen loam having the properties presented above. The curves in figure 11.8 assume that ponding is immediate, but that the depth of ponding is infinitesimally small. A value of β equal to 1.35 was assumed for the Morel-Seytoux and Khanji model.

The Philip model appears to predict the highest infiltration rates, whereas the Morel-Seytoux and Khanji model predicts the lowest. The other two models predict between these two extremes. All models appear to provide acceptable predictions for the Laveen loam.

11.3   EFFECTS OF MACROPOROSITY

Non-capillary pores (macropores) have been categorized by Beven and Germann (1982) into four morphological groups: **(1)** pores formed by soil fauna; **(2)** pores formed by plant roots; **(3)** crack and fissures; and **(4)** natural soil pipes. A single dead-root channel or worm hole can govern both the drainage of water and air movement through a block of soil. Beven and Germann (1982) observed that there should be no doubt that, under saturated conditions, water will move through the large pores. Because water flowing through macropore channels encounters minor viscous resistance, macropores have large conductivities.

Kirkby (1988) defined macropores as being at least one order-of-magnitude (10 times) larger than indigenous pores. Even though the porosity contributed by macropores to the total soil porosity is usually small (Watson and Luxmoore 1986), they may transmit more than 70 percent of the total water-flux at saturation. Typically, macropores constitute 0.001 to 0.05 percent (1 to 50 per 10,000 volumes) of total soil volume (Morrison 1993). However, many questions remain as to how and when flow occurs in macropores, and how these pores interact with the surrounding soil matrix under unsaturated conditions.

Rawls et al. (1993) provide two approaches to modeling infiltration with macropores. One approach is to adjust the saturated hydraulic conductivity to account for the macropores. A second approach is to consider the soil as two domains comprised of the macropores and the soil matrix, with an interaction between these two domains.

Adjustment of the saturated hydraulic conductivity to account for macroporosity is a practical approach, providing an effective hydraulic conductivity with macropores as the product of the saturated-hydraulic conductivity K_s, without macropores times a macro-porosity factor (Rawls, Brakensiek, and Savabi 1989; Brakensiek and Rawls 1988)

$$K_{mp} = AK_s \qquad (11.45)$$

where K_{mp} is the effective hydraulic conductivity with macropores, K_s is the saturated hy-draulic conductivity without macropores, and A is the macroporosity factor. Rawls et al. (1993) present two macroporosity factors A for two different land uses. For areas that do not undergo mechanical disturbance on a regular basis—such as rangeland—the prediction equation for the undisturbed macroporosity factor is

$$A = \exp\,(2.82 - 0.099PS + 1.94\rho_b) \qquad (11.46)$$

where PS is the percent sand (2.0 to 0.05 mm) for the less-than-2.0-mm fraction of a given soil and ρ_b is the dry-bulk density of the less-than-2.0-mm fraction of a given soil. For areas that undergo mechanical disturbance on a regular basis—such as cultivated land—the pre-diction equation for the disturbed macroporosity factor is

$$A = \exp\,(0.96 - 0.032PS + 0.04PC - 0.032\rho_b) \qquad (11.47)$$

where PC is the percent clay (< 0.002 mm) fraction for the less-than-2.0-mm fraction of a given soil and the other variables are as defined above. The values of A given by equa-tions 11.49 and 11.47 are constrained to be not less than 1.0 (Rawls et al. 1993).

The more physically based approach to macropore effects on infiltration is that, under rainfall or spill conditions, the flow into macropores begins only after ponding on the soil surface. Infiltration into the soil matrix before and after ponding can be modeled by the physically based approaches given in section 11.2. After ponding, the free fluid available at the surface is allowed to flow into the macropores. The flow in macropores can be adsorbed by the drier soil matrix via diffusion below the transient wetting front, or laterally. The lateral infiltration begins at the uppermost point below the soil matrix wetting front, and proceeds downward until the available fluid is absorbed or the lowest point of interest is reached.

11.4 LAYERED SOILS

Water cannot enter the soil at a faster rate than it is transmitted downward. Therefore, soil-surface conditions cannot increase infiltration rate unless the transmission-zone characteris-tics of each layer of the system are adequate. The infiltration rate of the surface layer can be low due to compaction induced by traffic or heavy machinery, raindrop impact, or the nature of the soil texture and structure, this compaction is often called "crusting" and will be dis-cussed in the following section.

At saturation, the rate of infiltration is limited to the lowest transmission rate encoun-tered by infiltrating water (or other fluid), up to that time. In a soil-layered profile, consider three layers: A, B, and C. If the B-layer has a rate of transmission lower than that of the A- and C-layers, the initial infiltration will be governed by the transmission rate of the A-layer until it is saturated, and later by the B-layer. Because the B-layer has a lower transmis-sion rate than the C-layer, the infiltration into C will be limited by the B-layer, so C may not reach saturation.

It is interesting to note that the infiltration rate into a layered soil (exclusive of crust-ing) generally decreases when a boundary between layers is encountered, whether the un-derlying layer is sand or clay. This phenomenon is shown graphically in figure 11.9 (Miller and Gardner 1962). When the wetting front encounters a smaller-pored material—such as a clay layer—than that in which it has been moving, the smaller pores begin to fill rapidly be-cause of their greater attraction for water (larger negative matric potential). As the wetting

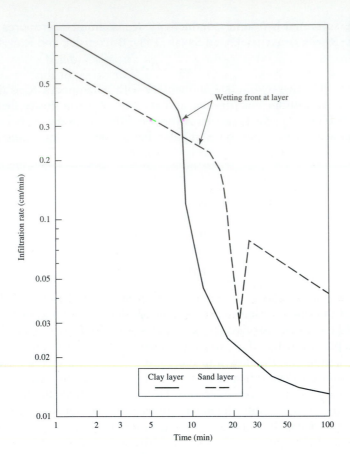

Figure 11.9 Infiltration rate into a palouse silt loam underlain by a clay and sand layer

front advances into the clay layer, water is transmitted through the smaller pores that have filled. For clayey soils, resistance to flow is often large due to the small pores, and can be so great that flow rates are markedly reduced. Figure 11.9 shows over an order-of-magnitude decrease in infiltration rate when the wetting front encounters the clay layer, with a continued decrease (but not as rapid) as time goes on.

When the wetting front (after moving through a relatively small-pored material) encounters a larger-pored material, the pore sizes capable of holding water at the matric potentials existing in the small-pored material are few in number. Before the wetting front can advance, the matric potentials in the small-pored material must be equal to those in a majority of the large-pored material, so that these larger pores may begin filling with water. However, some water continues to enter the larger-pored material, in response to a small potential gradient. Figure 11.9 shows that when the wetting front encounters the sand layer, the infiltration rate immediately decreases, but not as much as for the clay layer. As the overlying smaller-pored material increases in saturation (and decreases in negative matric potential), the larger pores in the sand begin to fill and the infiltration rate rebounds, but not to the original pre-encounter rate. The infiltration rate then resumes at a rate determined by the overlying smaller-pored—and presumably smaller-hydraulic conductivity—material. The relative size of the pores in the coarse material, when compared to the size of the pores in the overlying layer, determines the degree to which infiltration is affected (Miller and Gardner 1962).

The effect of layers of larger-pored materials on infiltration should be somewhat similar, regardless of the direction of flow, and should vary only in magnitude. This is because of the effects of gravity on downward-versus-downward or -horizontal infiltration. Therefore, data obtained for downward wetting can be used to infer similar infiltration patterns, if upward or horizontal layering were present instead.

The Green–Ampt equations 11.18 and 11.23 can be modified to describe infiltration into layered materials if the saturated hydraulic conductivity of the successive layers decreases with depth (Childs and Bybordi 1969; Hachum and Alfaro 1980). If the wetting front is in the top layer, no changes in equations 11.18 or 11.23 are needed. Rawls et al. (1993) indicate that after the wetting front enters the second layer, the saturated-hydraulic conductivity can be set to the harmonic-mean hydraulic conductivity, or

$$K_{s_h} = \sqrt{K_{s_1} K_{s_2}} \tag{11.48}$$

for wetted depths of the first and second layers, respectively. Additionally, the change in matric potential $\Delta\psi$ is set to that of the second layer. This is valid both with and without ponding, if modifications in $\Delta\psi$ are made for the ponding depth. This same principle can be used for three or more layers as well.

For a layered soil in which the saturated hydraulic conductivity of a lower layer is greater than that of an overlying layer (typical of the crusted condition), no simple variable-substitution method for the Green–Ampt equations is available after the wetting front enters the second (higher-K_s) layer. An approximation for such cases is available by assuming that infiltration through the higher-K_s layer continues to be governed by the harmonic mean of the upper layers (Rawls et al. 1993). As we see under a special case in the next section, the Green–Ampt equations can be modified to account for a crust at the surface.

11.5   CRUSTED SOILS

Crusted soils are a special case of layered soils. Often the crust is very thin, between 1.5 and 3.0 mm (Rawls et al. 1993). Assuming that the depth of ponded water on top of the crust is negligible, the horizontal-infiltration rate into soils with a surface crust can be written as

$$i = \frac{dI}{dt} = K_s \left(\frac{\psi_f - \psi_i}{L_f} \right) \tag{11.49}$$

where K_s is the saturated hydraulic conductivity of the transmission zone beneath the crust; ψ_f is the effective suction at the wetting front; and ψ_i is the effective suction at the interface of the crust and the soil. Note that $\psi_f - \psi_i$ is the reduced driving force for infiltration into the soil beneath the crust. L_f is still defined as the distance from the soil surface to the wetting front if the crust thickness is small.

Alternatively, equation 11.49 can be written as

$$\frac{dI}{dt} = \frac{\psi_f}{R_e + R_s} \tag{11.50}$$

where R_e is the hydraulic resistance of the crust, and R_s is the hydraulic resistance of the subcrust transmission zone. R_s can be given, for a thin crust, by

$$R_s = \frac{L_f}{K_s} \tag{11.51}$$

Combining equations 11.50 and 11.51 gives

$$i = \frac{dI}{dt} = \frac{K_s \psi_f}{K_s R_e + L_f} \tag{11.52}$$

Substituting L_f from equation 11.3 into equation 11.52 and rearranging, we obtain

$$(I + K_s R_e \, \Delta\theta) \, dI = K_s \psi_f \, \Delta\theta \, dt \tag{11.53}$$

where $\Delta\theta$ is the difference in volumetric water content in the transition zone during infiltration and the initial volumetric water content, as defined above.

For constant R_e, which is true if the crust is thin and unchanging, equation 11.53 can be integrated using the initial conditions of $I = 0$ at $t = 0$, to give

$$I^2 + 2K_s R_e \,\Delta\theta\, I = 2K_s \psi_f \Delta\theta\, t \tag{11.54}$$

Solving equation 11.54 using the quadratic formula gives a relation for cumulative-horizontal infiltration as

$$I = \Delta\theta \left[\left(K_u^2 R_e^2 + 2K_s \frac{\psi_f}{\Delta\theta} t \right)^{1/2} - K_s R_e \right] \tag{11.55}$$

It should be noted that, in the absence of a crust, equation 11.55 reduces to equation 11.10, and is valid for horizontal infiltration as well as for early periods of vertical infiltration. Because gravitational forces become important as the infiltration process continues in time, for vertical infiltration, equation 11.52 can be modified by adding the distance from the soil surface to the wetting front, or

$$\frac{dI}{dt} = \frac{K_s(\psi_f + L_f)}{K_s R_e + L_f} \tag{11.56}$$

which, from equation 11.5, can also be written as

$$\Delta\theta \frac{dL_f}{dt} = \frac{K_s(\psi_f + L_f)}{K_s R_e + L_f} \tag{11.57}$$

Integrating equation 11.57 with $L_f = 0$ at $t = 0$ gives

$$L_f - (\psi_f - K_s R_e) \ln\left(\frac{\psi_f + L_f}{\psi_f} \right) = \frac{K_s t}{\Delta\theta} \tag{11.58}$$

It should be noted that for large times, the second term on the left-hand side of equation 11.58 is very small compared to L_f. Therefore, L_f for the crusted case at large times can be written as

$$L_f = \frac{K_s t}{\Delta\theta} + \delta(t) \tag{11.59}$$

where $\delta(t)$ is a small-error term.

Substituting equation 11.59 into equation 11.58 yields L_f as a function of time

$$L_f = \frac{K_s t}{\Delta\theta} + (\psi_f - K_s R_e) \ln\left(\frac{\psi_f + \dfrac{K_u t}{\Delta\theta} + \delta(t)}{\psi_f} \right) \tag{11.60}$$

Neglecting $\delta(t)$ for very large times, the cumulative infiltration for vertical infiltration through a crusted soil surface can be expressed as

$$L_f = K_s t + \Delta\theta(\psi_f - K_s R_e) \ln\left(1 + \frac{K_u t}{\Delta\theta\, \psi_f} \right) \tag{11.61}$$

Figure 11.10 compares cumulative infiltration and infiltration rate for an uncrusted and crusted soil. For this example, the soil was assumed to have a saturated-hydraulic conductivity of 0.0004242 cm/min, a wetting-front effective suction of 47.99 cm, a value of $\Delta\theta$ of 0.4367, and a thin film of water (but no ponding) on the soil surface. For the crusted condition, a crust

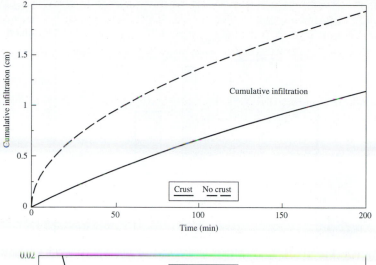

Figure 11.10   Green–Ampt cumulative infiltration and infiltration rates for crusted and uncrusted conditions

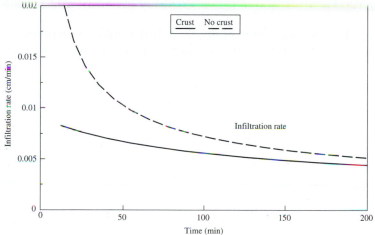

hydraulic resistance of 4,318 minutes and a thin crust was assumed. Equation 11.23 was solved for the uncrusted case by substituting $I/\Delta\theta$ for L_f. Using the same substitution, equation 11.58 was solved for the crust condition. Figure 11.10 indicates that the steady-state infiltration rate, given by equation 11.27, is approached sooner in crusted soil.

11.6   RUNOFF

The effects of infiltration rate on runoff quantity, duration, and distribution are well-known in hydrologic engineering. Often, runoff is estimated from strictly infiltration-based models; however, many runoff models take into account lumping of all losses—including infiltration, surface-depression storage, and interception of rainfall by vegetation growing on the surface. When rainfall (or other liquid inputs) begins on soil surfaces, a portion of the liquid is retained on vegetation or other surfaces, and is thus intercepted before it reaches the soil surface. Additionally, a portion of the liquid input may be retained at the soil surface in depressions, ranging in size from soil-grain-size cavities to ponds and lakes. A portion of the liquid is passed across the soil–atmosphere interface as infiltration, which is the most important liquid-input loss process under normal circumstances. Infiltration is a direct loss that governs the volume and rate of runoff, and highly influences the shape of the runoff hydrograph.

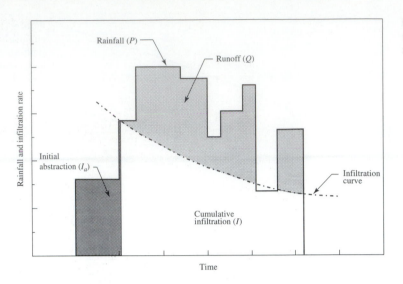

Figure 11.11   Schematic of the role of infiltration on runoff

Figure 11.11 shows schematically, the role that infiltration plays in runoff under storm conditions. The SCS (1980) combines infiltration with initial abstractions (interception and detention storage) to estimate the rainfall excess, which would appear as runoff. The runoff volume Q is estimated by the relation

$$Q = \frac{(P - I_a)^2}{P - I_a + S} \tag{11.62}$$

where Q is the total runoff expressed as a depth over the drainage area; P is the rainfall depth; I_a is the depth of the initial abstractions; and S is the potential maximum retention after runoff begins. As indicated above, initial abstractions include losses before runoff begins, and include water retained in surface depressions, water intercepted by vegetation and other surfaces, evaporation, and infiltration. Values of I_a are highly variable for a given drainage basin; however, SCS data from many small basins indicate that I_a can be approximated by

$$I_a = 0.2S. \tag{11.63}$$

If I_a from equation 11.63 is substituted into equation 11.62, then the SCS runoff relation becomes

$$Q = \frac{(P - 0.2S)^2}{P + 0.8S}. \tag{11.64}$$

Because S is related to the soil- and vegetative-cover properties of the drainage basin, it can be derived through the use of SCS runoff curve numbers CN (SCS 1980). Curve numbers have a range of between 30 and 100, and S is related to CN by

$$S = \frac{1000}{CN} - 10. \tag{11.65}$$

In addition to the above relation, the cumulative infiltration I also can be calculated from

$$I = P - I_a - Q. \tag{11.66}$$

It also can be noted that runoff does not occur until initial abstractions are satisfied. This leads to the two following relations which must be satisfied before runoff can occur

$$P \geq I_a \tag{11.67}$$

$$S \geq I_a + I \tag{11.68}$$

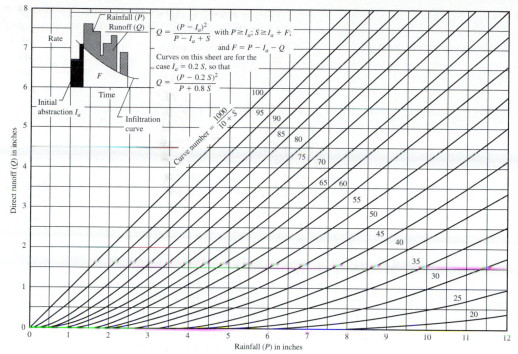

Figure 11.12   Graphical solution of equation 11.64

Figure 11.12 (SCS 1980) shows the relation between runoff and precipitation for various *CN*s, and is a solution to equation 11.64 for the particular case assumed by equation 11.63. Other definitions of the relation between I_a and S would result in slightly different solutions.

11.7   REDISTRIBUTION AND INTERNAL DRAINAGE

Moisture redistribution, as well as internal flux of infiltrated water, is highly dependent upon several factors described previously. These factors include the quantity of liquid reaching the soil surface and its distribution over time; the physical and chemical characteristics of the soil and the liquid; and other factors such as initial-moisture content, surface slope, vegetative cover, and surface roughness.

Water Redistribution

After an infiltration event, liquids infiltrating the soil surface are subject to redistribution downward by matric and gravity potentials, and removal across the soil–atmosphere interface by evapotranspiration. The quantity and rate of redistribution of infiltrated water in relation to the cumulative infiltration over time is discussed below, for a simple, rectangular water-redistribution profile (Charbeneau 1989), as discussed in chapter 13. Water redistribution is altered if the infiltration rate is less than the effective saturated-hydraulic conductivity, as opposed to an infiltration rate that is greater than the effective saturated-hydraulic conductivity (Bouwer 1966). These different infiltration rates also affect the water flux rates (internal drainage) within the soil profile.

Figure 11.13 shows the effects of infiltration on moisture redistribution in a sandy loam, whose properties have been defined by Rawls, Brakensiek, and Savabi (1982). Two cumulative infiltration values are shown, 4 cm in 4 hours (average infiltration rate i of 1.0 cm/hr) and

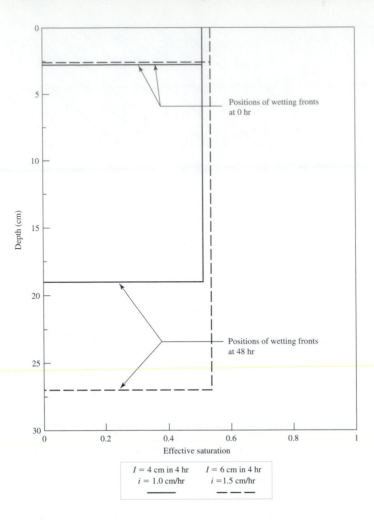

Figure 11.13   Water redistribution with changing infiltration rate

6 cm in 4 hours (i = 1.5 cm/hr). The simplified, rectangular water-redistribution profile, shown in figure 11.13, indicates that an average infiltration rate of 1 cm/hr (less than the effective saturated-hydraulic conductivity of 1.3 cm/hr) gives a wetting-front position of 19 cm below ground surface, with an effective saturation of 0.511 after 48 hours. However, an average infiltration rate of 1.5 cm/hr (greater than the effective saturated hydraulic conductivity) gives a wetting-front position of 27 cm below ground surface with an effective saturation of 0.537 after 48 hours. Both of these calculations assume that the initial soil-moisture content was at residual saturation. The difference in wetting-front movement is attributable to the increased moisture content and hydraulic conductivity from more infiltrated water. In general, a higher-average infiltration rate and/or a higher cumulative-infiltration results in wetting fronts that proceed deeper and faster into a soil column, with uniform properties, resulting in a higher degree of saturation.

Internal Drainage

Figure 11.14 shows the time-rate of change of Darcian flux, q, in the same sandy loam soil at a depth of 30 cm below ground surface, for the two cumulative infiltration values of 4 cm in 4 hours and 6 cm in 4 hours. For the average infiltration rate of 1.0 cm/hr, the wetting front arrives at a depth of 30 cm at a time of approximately 89.5 days, with an effective saturation

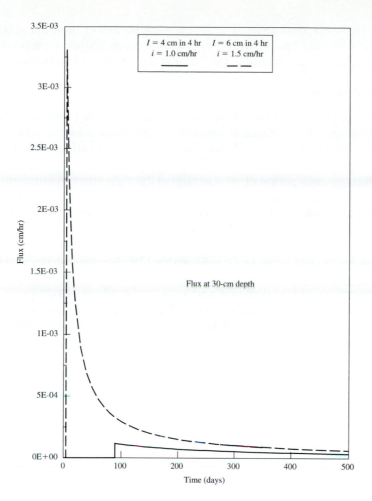

Figure 11.14   Change in water flux with changing infiltration rate

of 0.324. The wetting front for the higher infiltration rate of 1.5 cm/hr arrives at a depth of 30 cm at a time of approximately 4.6 days, with an effective saturation of 0.485. The peak flux rate for $i = 1.0$ cm/hr was calculated to be 1.2×10^{-4} cm/hr and the peak flux-rate for $i = 1.5$ cm/hr, was 3.3×10^{-3} cm/hr. The difference in these two wetting-front arrival times and peak flux rates is attributable to the additional increase in liquid from the higher-average infiltration rate. In general, a higher average-infiltration rate and/or a higher cumulative-infiltration results in substantially higher Darcian flux rates at a given depth in a soil column with uniform properties. The peak flux rates appear to change by orders of magnitude with less than a doubling of the cumulative infiltration.

11.8   FIELD MEASUREMENTS

Because infiltration is a complex process that varies both spatially and temporally, the selection of a measurement or data-analysis technique should consider the complexities involved. Infiltration-measurement techniques can be broadly categorized into two methods: **(1)** areal measurements (or data analysis); and **(2)** point measurements. Areal measurements and data analysis usually involve drainage basin or large-plot scales. Point measurements involve measurement of infiltration at a point in a small plot, using various techniques. It is generally felt that areal data-analysis techniques have the advantage over point measurements of infiltration because they relate more directly to prevailing conditions of precipitation and field conditions. However, data-analysis techniques are no better than the precision with which both precipitation and runoff hydrographs are measured. Some typical measurement techniques for both areal and point-field infiltration are given in the following sections.

Areal Measurements or Data Analysis

Infiltration rate curves can be estimated for drainage basins and field plots by analyzing runoff hydrographs. In hydrograph methods, the infiltration rate is estimated by records of rainfall and runoff from a given drainage basin, with evaporation neglected. Theoretically, a detailed study of rainfall and runoff with time should give a reasonable estimate of infiltration rate. Because each period of intense rainfall during a given storm event produces a peak on the discharge hydrograph, the infiltration rate can be determined by subtracting the amount of runoff under the hydrograph peak from the amount of the rainfall. An example of this generalized technique is as follows: if the first rainfall peak with x cm of rainfall produced y cm of runoff, then $(x - y)$ cm is the total amount of water infiltrated during that part of the storm event. The value of $(x - y)$, divided by the time during which the rainfall occurred, gives the average infiltration rate for that time period. Similarly, infiltration rates for other time periods can be calculated, and an overall infiltration-rate curve can be obtained.

Hydrograph methods apply for relatively small drainage basins where there is a clear relation between rainfall and runoff. In drainage basins where the hydrograph is comprised of the large ground water flow component, it is difficult to separate the surface-water and ground water contributions to the runoff hydrograph, making the methods difficult to use. Other sources of difficulty and of possible errors in estimating infiltration rates, include the effects of evaporation/evapotranspiration, surface storage, and interception effects. In general, the validity of hydrograph methods for the estimation of infiltration rates is inversely related to the size of the drainage basin—because of the lag-times associated with the rainfall-runoff process, antecedent moisture conditions, changing meteorological conditions over the basin, and spatial and temporal variability of other basin hydrologic conditions.

There are three common hydrograph methods that use one of the following techniques: **(1)** detention-flow relations to derive rainfall excess; **(2)** time-condensation technique to eliminate periods of inadequate rainfall; and **(3)** block methods of dividing the rainfall into a series of blocks to obtain a succession of average infiltration rates over time. Each of these methods has been treated in detail by Musgrave and Holtan (1964), and are summarized below.

Detention-flow relation technique   The essential features of this infiltration-rate estimation technique assume: **(1)** that the rain storm is of sufficient duration and intensity such that the infiltration rate will have become relatively constant; and **(2)** that the rainfall is intermittent so that the hydrograph rises and falls over the time-periods of interest; a single-peaked hydrograph would not provide enough information to determine an infiltration-rate curve. A step-by-step procedure for this method is available in Musgrave and Holtan (1964).

Time-condensation technique   In this technique, variations in rainfall are eliminated by condensing time in order to obtain a constant rainfall rate, and then applying the same techniques used in analyzing rainfall simulator data. Time is condensed in a such a way as to cause the mass-rainfall curve to be a straight line over all time used in the analysis. Infiltration rate is estimated based on the assumptions that: **(1)** initial abstractions have been satisfied at the time runoff starts; **(2)** the curve of cumulative infiltration is parallel to the retention curve during periods of runoff; and **(3)** during periods of no runoff, the cumulative infiltration curve intersects the retention curve. Infiltration rate is then calculated as the slope of the cumulative infiltration-curve segments for various time periods. Musgrave and Holtan (1964) present more details on the use of this method.

Block technique   The block technique is probably best-suited to estimating infiltration rates on larger drainage basins, because rainfall on larger basins is seldom known closely enough to warrant the use of either the detention-flow or time-condensation techniques. The block, or average infiltration, technique involves preparing an isohyetal map for each storm. Each of the resulting storm hydrographs can be separated from the succeeding ones by transposing recession curves (Musgrave and Holtan 1964). The average infiltration rate is calculated for each storm event. Nachabe and others (1997) present methods for estimating infiltration over heterogeneous watersheds under the influence of spatially varying parameters, such as sorptivity and saturated-hydraulic conductivity.

Point Measurements

Point measurements of infiltration rate include specific types of infiltrometers whose results are applicable only at the locations and for the particular soil, vegetative, and other hydrologic conditions where the measurements are made. Point measurements have the advantage that rainfall/runoff data are not needed and that infiltration-rate tests can be repeated quickly, compared to areal analyses. Point-infiltration measurements fall into four categories: **(1)** rainfall simulators; **(2)** ring or cylinder infiltrometers; **(3)** tension infiltrometers; and **(4)** furrow methods (Rawls et al. 1993). Each of these infiltration-measurement techniques was specifically designed to mimic a given infiltration-process cause. For example, if the effect of raindrops is important at a given site, then rainfall simulators that use sprinkling are appropriate. If infiltration rates for flooded conditions are desired, then the ring or cylinder infiltrometer is used. Tension infiltrometers are used to measure infiltration rates on soil surfaces where macropores may be present. Furrow techniques are used where flowing water may affect infiltration rates.

Rainfall Simulators   Simulation of rainfall using spray nozzles, drip screens, or drip towers has been used for infiltration and erosion measurements. The simulators attempt to reproduce the characteristics of natural rainfall, including raindrop-size distribution, raindrop velocity, as well as rainfall intensity and duration.

Spray nozzles, spaced at different intervals and at different heights above the soil surface, have been developed to produce raindrops of different sizes. The nozzles produce artificial rain when water is sprayed upward into the air and the drops are allowed to fall downward onto the surface of interest. Nozzles can be obtained commercially in a variety of types and sizes, for various water-pressure ranges.

Drip screens and drop towers have also been widely used to study the effects of rainfall on soil-surface erosion and to evaluate infiltration rates. The artificial raindrop size is controlled by the screen-mesh size and/or pieces of yarn of varying sizes, to obtain uniform raindrops. Raindrop velocity is controlled by varying the height of the apparatus, and intensity by varying the head of water used to produce the raindrops, or by reducing the size of the holes supplying water to the drip screen. Glass capillary tubes and hypodermic needles have been used to form simulated raindrops instead of pieces of yarn. Peterson and Bubenzer (1986) have inventoried and categorized rainfall simulators, and provide the details for constructing a rainfall simulator.

Ring or cylinder infiltrometers   These infiltrometers are sometimes referred to as "flooding-type" infiltrometers. They are made of metal rings with diameters ranging from approximately 30 to 100 cm, and are 30 to 60 cm high. The metal rings are driven (jacked) into the soil to a depth of between 5 and 15 cm. The most common type of ring infiltrometer is the double-ring infiltrometer, which is a ring infiltrometer with a second, larger ring around it. Both rings have water applied to them, but measurements are only taken inside the inner ring; the outer ring serves as a buffer to help minimize lateral spreading and maintain vertical flow beneath the rings. The double-ring infiltrometer has been adopted by the American Society for Testing and Materials (ASTM 1994) as a standard test.

One of the major limitations of using ring infiltrometers is disturbance of the soil while the rings are being driven into place. The interface between the soil and the sides of the rings can be a preferred pathway for water and could result in abnormally high infiltration rates.

However, if the rings are left in place for a long period of time (say, several hours), the impact of this disturbance is reduced. A second problem associated with ring infiltrometers is entrapped air, which rises when a constant head of water is applied to the surface. This confined air in saturated soil impedes the downward movement of water. Ring infiltrometers usually produce higher steady-state infiltration rates compared to sprinkler infiltrometers (Rawls et al. 1993).

Generally, the larger the area used for infiltration, the more accurate the estimation of infiltration rate; to this end, infiltration basins have been proposed. These basins range in size from a few square meters to 1,000 m^2, and usually have border arrangements for impounding water. A given amount of water is applied to the basin and the rate of infiltration is recorded. It should be noted that small basins have limitations similar to those of ring infiltrometers, with entrapped air being the largest. In larger basins, the volume of water needed for infiltration is quite large, and the heads across the basin may be non-uniform due to varying topography.

Tension infiltrometers   The tension, or disk, infiltrometer consists of a tension-control tube, a tube to measure change in water level, and a large disk with a porous membrane approximately the same diameter as the inner ring of a double-ring infiltrometer. These infiltrometers are commercially available and have been well tested (Watson and Luxmoore 1986; Ankeny, Kaspar, and Horton 1988; Perroux and White 1988). The advantage of a tension infiltrometer is that, by varying the tension on the device, certain pores in the media of interest can be eliminated from the flow process and the effects of pore size distribution can be determined. This is especially important where macropores are present.

Furrow techniques   Infiltration-rate measurements for flowing water—and in particular furrow irrigation systems—can be made with either a blocked-furrow infiltrometer, recirculating-furrow infiltrometer, or an inflow/outflow measurement. The blocked-furrow technique consists of blocking off a section of up to 5 m of furrow, and then ponding water in the furrow. The water level in the furrow is kept at a constant depth by continually adding water until a constant infiltration rate (constant water addition) is achieved. The infiltration rate is calculated as the time-rate of water additions to the furrow.

The recirculating furrow technique involves a modification of the blocked furrow technique by continuously recycling water over the 5-m furrow segment. It is felt that this technique better simulates conditions of actual flowing water (Rawls et al. 1993).

In the inflow/outflow technique, infiltration rates along the furrow are calculated by taking the difference in measured flow rates, at both the upstream and downstream ends of the furrow segment, over time. In this case, the length of the furrow segment is more important, because this length determines the early-time infiltration rate. Flow rates are usually measured by flumes. The impact of water lost to storage in the soil needs to be considered during calculation of infiltration rates, as well.

Additional information on field-infiltration measurement techniques can be found in Klute (1986); this also contains many references and specific step-by-step procedures.

QUESTION 11.1

Using Darcy's equation for vertical flow:

$$q = -K \frac{dH}{dz} = -K \frac{d}{dz}(H_p - z)$$

where q is flux, H is total hydraulic head, H_p is pressure head, z is vertical distance from the soil surface downward, and K the hydraulic conductivity, and the Green–Ampt approximations develop the relation of the form of $i = i_c + b/I$ that expresses the infiltration rate as a function of the parameters given in the following figure.

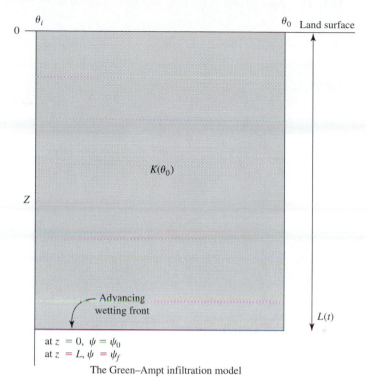

The Green–Ampt infiltration model

QUESTION 11.2

Considering question 11.1, what happens to the infiltration rate when t is small? What happens to the infiltration rate when t is large?

QUESTION 11.3

Given:

$$K(\theta_o) = 0.001 \text{ cm/s} = 3.6 \text{ cm/hr}$$

$$\Delta h = h_o - h_f = 40 \text{ cm}$$

$$\Delta\theta = \theta_o - \theta_i = 0.35$$

Calculate and plot cumulative infiltration I versus time and plot infiltration rate $(\Delta I/\Delta t)$, I versus time for the first three hours of infiltration, for both the horizontal and vertical case.

SUMMARY

In this chapter, we presented the topics of infiltration; profile-water distribution; soil characteristics that affect infiltration; several of the most common infiltration models; and a comparison of various physically based infiltration models. We discussed the effects macroporosity, layered soils, and crusted soils each have on infiltration. Additionally, we examined the effects of infiltration rate on runoff quantity, duration, and distribution, redistribution and internal drainage, and various methods used for field measurement of infiltration.

ANSWERS TO QUESTIONS*

11.1. The general form of the Green–Ampt equation is

$$i = i_c + \frac{b}{I}$$

We want to derive an expression for I in terms of i_c and b. From

$$q = K\frac{d\psi}{dz} + K$$

we can state that

$$i = q = K\frac{d\psi}{dz} + K$$

Substituting

$$i = K(\theta_o)\left(\frac{\psi_f - \psi_o}{L}(t) - 0\right) + K(\theta_o) \tag{11.69}$$

The cumulative infiltration I is simply the total volume of water that has entered the profile at a given time t:

$$I = (\theta_o - \theta_i) L(t)$$

Rearranging:

$$L(t) = \frac{I}{\theta_o - \theta_i}$$

Substituting into equation 11.69, we have

$$i = K(\theta_o)\frac{\Delta\psi\,\Delta\theta}{I} + K(\theta_o) \tag{11.70}$$

where $\Delta\psi = \psi_f - \psi_o$ and $\Delta\theta = \theta_o - \theta_i$. Therefore, $i_c = K(\theta_o)$ and $b = K(\theta_o)\,\Delta\psi\,\Delta\theta$.

11.2. From equation 11.70, we can see mathematically that when t is small, the infiltration rate is very large. However, as t increases, the infiltration rate will asymptotically approach $K(\theta_o)$.

11.3. Using the Green–Ampt model, the following relations can be derived:

For one-dimensional horizontal infiltration,

$$I = \Delta\theta\,(2D_o t)^{1/2}$$

where

$$D_o = K(\theta_o)\frac{\Delta h}{\Delta\theta}$$

For one-dimensional vertical infiltration,

$$t = (1 - \xi)\ln\left(1 + \frac{1}{\xi}\right)\frac{1}{K(\theta_o)}$$

where

$$\xi = \Delta h\,\Delta\theta$$

---

*Questions 11.1–3 and the figure for question 11.1 were extracted from the unsaturated zone hydrology course taught at the United States Geological Survey's National Training Center in Denver, Colorado.

and

$$\Delta h = h_o - h_f > 0, \quad \Delta\theta = \theta_o - \theta_i > 0$$

Note that the equation for the vertical case is implicit.

Using these relations, we can tabulate the data, shown below. We leave it to the student to perform the actual plot of the data obtained.

One-Dimensional Horizontal			One-Dimensional Vertical		
1 cm^3	time (hours)	$i \text{ cm}^3/\text{hr}$	1 cm^3	time (hours)	$i \text{ cm}^3/\text{hr}$
0.000000	0.00	44.90	0.000000	0.00	55.15
2.244994	0.05	14.97	2.000000	0.04	23.80
4.489989	0.20	10.09	3.000000	0.08	18.02
5.499091	0.30	8.51	4.000000	0.13	13.72
6.349803	0.40	7.14	6.000000	0.28	10.82
7.776889	0.60	6.02	8.000000	0.46	9.21
8.979978	0.80	5.30	10.000000	0.68	8.19
10.039920	1.00	4.79	12.000000	0.93	7.48
10.998182	1.20	4.41	14.000000	1.19	6.96
11.879394	1.40	4.10	16.000000	1.48	6.57
12.699606	1.60	3.85	18.000000	1.79	6.25
13.469967	1.80	3.64	20.000000	2.10	6.00
14.198591	2.00	3.47	22.000000	2.44	5.79
14.891608	2.20	3.31	24.000000	2.78	5.62
15.553778	2.40	3.18	26.000000	3.14	5.47
16.188885	2.60	3.06	28.000000	3.51	5.34
16.800000	2.80	2.95	30.000000	3.88	5.05
17.389652	3.00		40.000000	5.86	

From the plots of I and i versus time, we see that at early times (< 2 hours), the amount of infiltrated water and the infiltration rate are very similar for horizontal and vertical infiltration. This is because the dominant driving force is the suction gradient and the driving force due to gravity is relatively small. However, as time increases, the suction gradient decreases, hence the relative magnitude of the effect of gravity increases. Therefore, vertical infiltration occurs at a rate that is greater than horizontal infiltration. At very large times, the suction gradient diminishes. In the case of horizontal infiltration, this results in a continually decreasing infiltration. In the case of vertical infiltration, it results in an infiltration rate that asymptotically approaches the saturated-hydraulic conductivity. This means that for early times, the shape of the wetting front for three-dimensional infiltration will be spherical. As time progresses, the horizontal movement of the wetting front will slow down relative to the vertical movement, and eventually most flow will be in the vertical direction.

ADDITIONAL QUESTIONS

11.4. A Laveen loam has the following infiltration hydraulic properties: $\Delta\psi = 25$ cm, $K_s = 0.45$ cm/hr, $\theta_t = 0.38$, and $\theta_i = 0.30$. Find the vertical infiltration rates and plot them versus time, for the Green–Ampt, Philip, Morel-Seytoux and Khanji, and Smith and Parlange models. Assume a value of $\beta = 1.35$ for the Morel-Seytoux and Khanji model, and that there is a thin film of water on the soil surface.

11.5. A water district disposes of reclaimed wastewater by sprinkler irrigating a large area of fallow land. The application rate is spatially uniform and constant at -3.0×10^4 cm^3 sec^{-1} cm^{-2} and has been continuous for some time, such that steady-state conditions exist in the subsurface. The negative sign indicates that the flux is downward. The soil beneath the site is layered, as sketched in

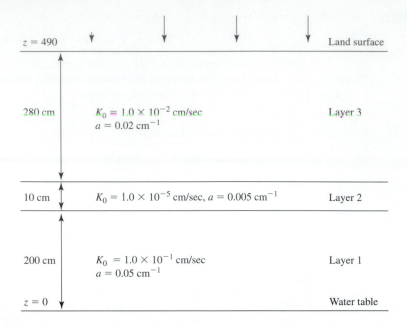

the accompanying figure. The $K(\psi)$ relation for each layer is $K(\psi) = K_o e^{-a\psi}$ and the parameters for each layer are given in the sketch below. The water table is located 4.9 m below the surface. Determine and plot the profiles of matric suction head $\psi(z)$ and the hydraulic head $H(z)$, between the water table $z = 0$ cm, and the ground surface $z = 490$ cm. Indicate saturated and unsaturated regions as appropriate. Explain the features of the profiles such as sign changes, magnitude of gradients, and so on.

12

Field Water in Soils

INTRODUCTION

Water and energy transport and retention processes are very complex in field soils. The soil is continually exposed to changes in water, heat, and chemical fluxes, both at and near the surface (Jury, Gardner, and Gardner 1991). Some of the material presented here was originally discussed in chapter 9; this chapter focuses on field water, radiation and energy balances, and provides practical methods to solve the water- and energy-balance equations from an applications or empirical point of view, as commonly used by many consulting and engineering firms in the United States and abroad.

Many of the techniques used to solve water-balance problems practically rely on empirical formulas, many of which use mixed units of English engineering, centimeter-gram-second (cgs) and Système International (SI). The units used in the source material for this chapter are preserved; this means that some formulas have English engineering units, whereas others have cgs or SI units. Conversions are given where appropriate, so that the results of the formulas can be presented in common units for a final water balance. For additional conversion factors, see the appendices.

12.1 FIELD WATER BALANCE

The general hydrologic equation describing the water balance of the unsaturated zone, from the soil surface to below the root zone, is given by (Thornthwaite and Mather 1955; 1957)

$$P - Q \pm \Delta S_w - E \pm \Delta S_s - D = 0 \tag{12.1}$$

where P is precipitation, Q is runoff, ΔS_w is change in storage of water ponded on the surface, E is evapotranspiration (evaporation plus transpiration), ΔS_s is change in soil-moisture storage, and D is deep percolation (unrecoverable by vegetation). Figure 12.1 shows each of these water-balance components in a schematic of the unsaturated zone. Water balances are used to evaluate and design water supply and water storage facilities for public and industrial uses as well as to evaluate environmental issues (Fenn, Hanley, and DeGeare 1975; van Zyl, Hutchinson, and Kiel 1988). When working with forests, interception (adherence of precipitation to the trunk, leaves, and stems of trees) of precipitation will occur. In such instances, interception, I, must become a variable of equation 12.1. For forests that have frequent, light precipitation, interception can account for as much as 50 percent of the precipitation. This intercepted precipitation does not fall to the ground and is not quickly evaporated into the atmosphere. Thus, equation 12.1 can be rewritten as $(P - I) - Q \pm \Delta S_w - E \pm \Delta S_s - D = 0$.

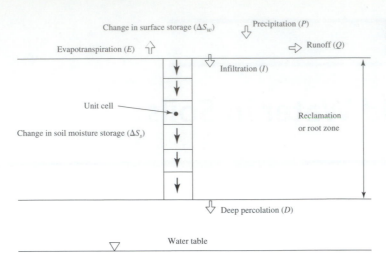

Figure 12.1   Schematic of water balance components (data from Kunkel and Murphy 1983)

Each of the terms in equation 12.1 represent inflows, outflows, or changes in storage over an arbitrary time period. All the water-balance terms in equation 12.1 are considered to be positive except the change in storage terms, which can be either positive or negative. Not all of the terms in equation 12.1 are computed directly. For example, infiltration is calculated as the difference between precipitation and runoff. Additionally, climatological variables, such as air temperature and solar radiation, are needed to calculate evapotranspiration. In general, deep percolation, D, is the unknown in equation 12.1 and can be determined by measuring or estimating the remaining terms.

The issue of what time increment to use in evaluating the individual terms in equation 12.1 is important. Obviously, all terms must be estimated using the same time increment: hour, day, month, or year. The question of whether a daily, monthly, or other time increment is used depends in part on the problem being solved, and the available data and estimation techniques. Most climatological data (precipitation, air temperature, evaporation, relative humidity, and solar radiation) are available on at least a daily basis throughout the world. However, sometimes only monthly summaries are published and easily obtainable, especially in developing countries. A general rule is to use the time increment that has the best available data. Often, an extreme event (such as the daily precipitation depth that falls, on the average, every 100 years) can be embedded into a more general time-series of daily precipitation values, and used for design purposes.

The following sections discuss each term in equation 12.1. Some common methods of estimation are presented that allow for practical use of the water-balance equation for design purposes. Other estimation methods also are often used throughout the world, some of which can be found in the references.

Precipitation, Air Temperature, and Solar Radiation

Precipitation and air temperature are the most common climatological data collected. Values of precipitation collected over at least a 20-year period at the location where the water balance is being calculated are the best data. If a short precipitation record (several years of daily values) is available on-site, and a longer record is available at a nearby site, interstation correlation of the on-site precipitation record with the longer-term record can be made. The result is a long-term synthetic precipitation record which is usually acceptable for design purposes. Figure 12.2 shows a typical interstation correlation for annual precipitation data.

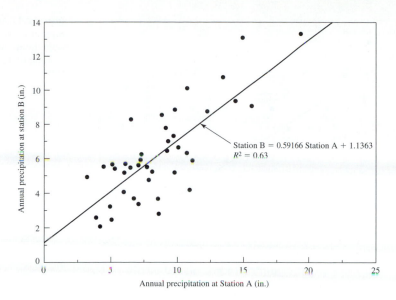

Figure 12.2 Interstation correlation of annual precipitation

Station B = 0.59166 Station A + 1.1363
$R^2 = 0.63$

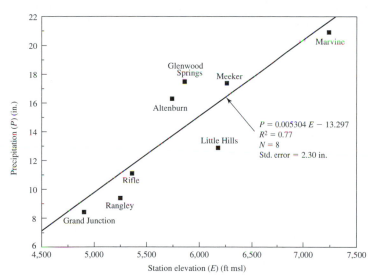

Figure 12.3 Average annual precipitation versus elevation for a mountainous area in Western Colorado (data from Wymore 1979)

$P = 0.005304 E - 13.297$
$R^2 = 0.77$
$N = 8$
Std. error = 2.30 in.

If on-site precipitation data are unavailable, precipitation data (at a daily or longer time scale) may be "imported" from nearby stations. Transfer of precipitation (and air temperature) is possible if the data from the nearby station are adjusted for elevation and/or aspect prior to use at the site. Figures 12.3 and 12.4 show elevation adjustment of average annual precipitation and air temperature, respectively, for a mountainous area of western Colorado (Wymore 1979). Similar analyses can and should be made on a monthly basis. After obtaining the elevation adjustment from a long-term precipitation station to the site of interest, daily data can be imported using the adjustment variables applied to records for each day. Figure 12.5 shows relations between average annual precipitation and elevation in several regions of the world. It is important to note that a linear precipitation relation with elevation is not universal, and the relation may even be adverse—that is, precipitation may decrease with increasing elevation. Care is advised if large distances or large changes in elevation between stations are used; particular attention should be given to mountain ranges and large water bodies in the area of interest.

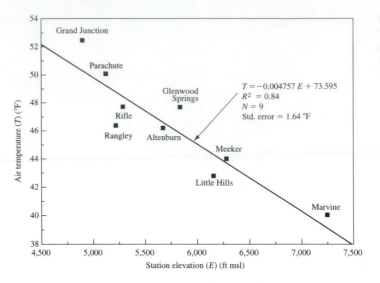

Figure 12.4 Average annual air temperature versus elevation for a mountainous area in Western Colorado (data from Wymore 1979)

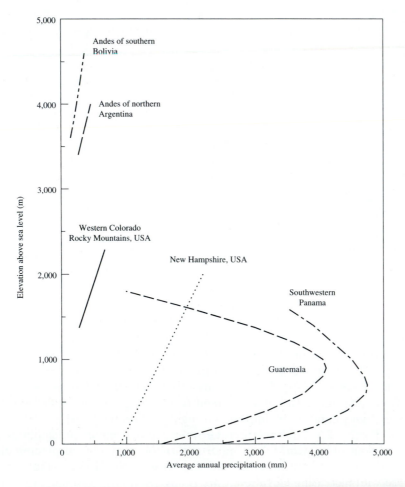

Figure 12.5 Relations between average annual precipitation and elevation in several regions of the world

If daily data are not available, they may be generated using a synthetic weather generator developed by the U.S. Agricultural Research Service (Richardson and Wright 1984). The most common source of this synthetic weather generator is the Hydrologic Evaluation of Landfill Performance (HELP) model (Schroeder et al. 1994a, b). Because daily precipitation and air temperature are usually available for many sites, the synthetic weather generator is most useful for calculating solar radiation, which often is not measured, is of poor quality, or is of short period of record.

Runoff

Runoff estimates for a water balance can be obtained from several possible sources. The best source is gaged-streamflow data in the drainage basin of interest. On-site or nearby gaged-streamflow data having a period-of-record of at least 20 years is ideal, but shorter records may be used. These gaged records can be extended using stochastic techniques (Fiering and Jackson 1971; Lane 1991; Lane and Frevert 1990; Grygier and Stedinger 1990; McLeod and Hipel 1978; Salas, Smith, and Markus 1992; Salas and Markus 1992). Both single-site and multisite runoff time series may be extended.

If sufficient runoff data are available in the basin, one method of transporting these data upstream or downstream is simply to adjust the runoff data for drainage-basin size or stream-channel length. Reliable estimates of runoff percentiles should include consideration that the flow rates can vary along a stream channel. In extreme cases, such as where a perennial stream discharges onto an alluvial fan, the losses can be sufficiently large to cause the stream to become ephemeral. Thus, care should be exercised in transporting data from one part of a drainage basin to another part of the same basin, using drainage area or channel length.

In hydrologically similar areas, a long-term gaged record can be correlated with another gaged record of shorter duration to obtain a predictive relation between the two records. Alternatively, multiple-regression methods can be used to develop a longer runoff record if some runoff records are available, along with those for longer-term climatological variables such as precipitation, air temperature, and pan evaporation. Multiple linear regression of runoff with the known variables provides a predictor for runoff. A typical regression-equation predictor is

$$Q = aB^eC^fD^g \tag{12.2}$$

where a, e, f, and g are regression constants, Q is runoff, and B, C, and D are hydrological and climatological variables of interest. Equation 12.2 is best-used to estimate annual values, but can be used with care to make seasonal estimates. A typical monthly rainfall-runoff correlation applying 32 months of data for a station in the Peruvian Andes—using only runoff and monthly precipitation—is shown in figure 12.6. The correlation in this case is fair, but might be improved by using air temperature and pan evaporation, in addition to precipitation, in the regression analysis.

Equation 12.2 can also be used to estimate runoff within a basin (or series of basins) where the hydrology is considered to be similar, but the variables B, C, and D are drainage-basin characteristics (e.g., basin area, channel slope, mean basin elevation, basin relief, basin perimeter), or other measurable basin variables that can be correlated with runoff. The multiple-correlation variables can also be a mixture of basin characteristics and climatological parameters. In the United States, the U.S. Geological Survey (USGS) and others have made studies of many drainage basins or regions using this technique. Examples can be found in Lowham (1976), Parker (1977), Craig and Rankl (1978), Masch (1984), and Christensen, Johnson, and Plantz 1986.

Where no runoff records are available for a stream, discharges for the time-period of interest are generated using climatological and drainage-basin data and information, such as

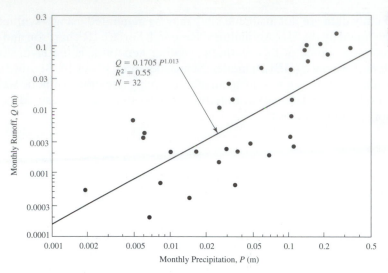

Figure 12.6   Typical rainfall/runoff relation, Peruvian Andes

$Q = 0.1705\ P^{1.013}$
$R^2 = 0.55$
$N = 32$

precipitation, drainage-basin area and slope, basin soil and vegetation types, and general use. Many rainfall/runoff models, both simple and sophisticated, have been proposed to estimate runoff from climatological and drainage-basin data.

One of the easiest models to apply for estimating runoff from rainfall is the National Resources Conservation Service (NRCS) (formerly the SCS) model that was discussed in chapter 11, section 11.6. This model uses a curve number (CN) approach and is applicable to both specific storm events and to daily precipitation values (SCS 1985). The model uses precipitation, soil characteristics, and vegetative cover to develop runoff curve numbers.

Soil characteristics   The SCS has classified over 5,000 soils into four hydrologic soil groups (Rawls, Brakensiek, and Saxton 1982). The definition of each soil group, along with limiting-infiltration rates by soil group, are given in chapter 11, table 11.2. The four hydrologic soil groups are designated as A, B, C, and D, in order of decreasing infiltration rate and increasing runoff potential. In general, soils in Group A are well-drained sands, whereas, soils in Group D are clay soils with high runoff potential (SCS 1985).

Vegetative-cover characteristics   Some of the abstractions (losses) from precipitation falling on a basin are the result of vegetative cover. Additionally, vegetative cover or land use is an important factor not only in the volume, but in the rate of runoff as well. Generally, higher vegetative-cover densities result in lower runoff rates. Thus, curve numbers tend to be smaller for higher vegetative densities or for more pervious surfaces (SCS 1985).

Table 12.1 shows storm-event curve numbers based upon soil characteristics and vegetative cover for an average antecedent soil-moisture condition (SCS 1986). As shown in table 12.1, SCS runoff curve numbers have a wide range, depending upon the soil and vegetation characteristics; impervious surfaces and water surfaces are usually assigned curve numbers between 98 and 100. These curve numbers have not been well defined for areas of the world outside the United States; therefore, engineering judgment is necessary in order to select the correct curve number under different soil and vegetative characteristics overseas.

The curve numbers given in table 12.1 are for an average antecedent soil-moisture condition (AMC); the average condition is AMC II (SCS 1985). Recognizing that abstractions from precipitation depend upon the antecedent moisture condition existing at the time of the precipitation event, the SCS has defined three AMCs: AMC I is for conditions when little precipitation precedes an event, and the root-zone soils are relatively dry; AMC II is used when average antecedent precipitation has occurred prior to an event; and AMC III is used when considerable precipitation has occurred in the five days immediately prior to the precipitation

TABLE 12.1   Runoff Curve Number for Selected Land Uses (Antecedent Moisture Condition II)

Land use or cover	Treatment or practice	Hydrologic condition	A	B	C	D
Fallow	Bare soil	—	77	86	91	94
	Crop residue[1]	Poor	76	85	90	93
		Good	74	83	88	90
Row Crops	Straight row	Poor	72	81	88	91
	Straight row	Good	67	78	85	89
	Contoured	Poor	70	79	84	88
	Contoured	Good	65	75	82	86
	Contoured and terraced	Poor	66	74	80	82
	Contoured and terraced	Good	62	71	78	81
Small grain	Straight row	Poor	65	76	84	88
	Straight row	Good	63	75	83	87
	Contoured	Poor	63	74	82	85
	Contoured	Good	61	73	81	84
	Contoured and terraced	Poor	61	72	79	82
	Contoured and terraced	Good	59	70	78	81
Close-seeded legumes or rotation meadow	Straight row	Poor	66	77	85	89
	Straight row	Good	58	72	81	85
	Contoured	Poor	64	75	83	85
	Contoured	Good	55	69	78	83
	Contoured and terraced	Poor	63	73	80	83
	Contoured and terraced	Good	51	67	76	80
Pasture or range[2]		Poor	68	79	86	89
		Fair	49	69	79	84
		Good	39	61	74	80
	Contoured	Poor	47	67	81	88
	Contoured	Fair	25	59	75	83
	Contoured	Good	6	35	70	79
Meadow		Good	30	58	71	78
Woods[3]		Poor	45	66	77	83
		Fair	36	60	73	79
		Good	30	55	70	77
Farmsteads		—	59	74	82	86
Roads	Dirt	—	72	82	87	89
	Hard surface	—	74	84	90	92
	Paved with curb and gutter	—	98	98	98	98
Commercial business areas	(85% impervious)	—	89	92	94	95
Industrial districts	(72% impervious)	—	81	88	91	93
Open spaces, lawns, parks, golf courses, cemeteries[4]		Poor	68	79	86	89
		Fair	49	69	79	84
		Fair	49	69	79	84
Residential lots	1/8 acre (65% impervious)		77	85	90	92
	1/4 acre (38% impervious)		61	75	83	87
	1/3 acre (30% impervious)		57	72	81	86
	1/2 acre (25% impervious)		54	70	80	85
	1 acre (20% impervious)		51	68	79	84
	2 acres (12% impervious)		46	65	77	82
Herbaceous[5]	Mixture of grass, weeds, and low growing brush, with brush the minor part	Poor		80	87	93
		Fair		71	81	89
		Good		62	74	85
Oak–Aspen[5]	Mountain brush mixture of oakbrush, aspen, mountain mahogany, bitter brush, maple, and other brush	Poor		66	74	79
		Fair		48	57	63
		Good		30	41	48

(continued)

TABLE 12.1   (*Continued*)

Land use or cover	Treatment or practice	Hydrologic condition	Hydrologic soil group			
			A	B	C	D
Piñon–Juniper[5]	Piñon, juniper or both with grass understory	Poor		75	85	89
		Fair		58	73	80
		Good		41	61	71
Sagebrush[5]	with grass understory	Poor		67	80	85
		Fair		51	63	70
		Good		35	47	55
Desert Shrub[5]	major plants include salt-bush, greasewood, creasotebush, blackbush, bursage, paloverde, mesquite and cactus	Poor	63	77	85	88
		Fair	55	72	81	86
		Good	49	68	79	84

Source: Data from SCS (1986)
[1] Applies only if residue is on at least 5% of surface throughout the year
[2] Poor: < 50% ground cover or heavily grazed with no mulch
  Fair: 50% to 75% ground cover and not heavily grazed
  Good: > 75% ground cover and lightly or only occasionally grazed
[3] Poor: Forest litter, small trees, and brush are destroyed by heavy grazing or regular burning
  Fair: Woods are grazed but not burned, and some forest litter covers the soil
  Good: Woods are protected from grazing, and litter and brush adequately cover the soil
[4] Poor: Grass cover < 50%
  Fair: Grass cover 50% to 75%
  Good: Grass cover > 75%
[5] Poor: < 30% ground cover (litter, grass, and brush overstory)
  Fair: 30 to 70% ground cover
  Good: > 70% ground cover

of interest. For most hydrologic designs, AMC II is used. Table 12.2 shows the AMC I and III curve-number values equivalent to those for AMC II.

The curve numbers are used to estimate runoff for a given depth of precipitation. Because the SCS curve numbers were originally developed for short-term storm events with rather high-intensity precipitation, they need to be adjusted if daily precipitation values are used. Schroeder et al. (1994a, b) use a correlation between minimum infiltration and soil type to calculate curve numbers for use with daily precipitation values. These curve numbers are shown for various soil textures and vegetation covers in figure 12.7. Both the Unified Soil Classification System (USCS) and the U.S. Department of Agriculture (USDA) soil-classification systems are shown in figure 12.7. The curve numbers shown in figure 12.7 are similar to those shown in the HELP model (Schroeder et al. 1994a, b), but can vary slightly for fair, good, and excellent grass covers. The definitions of both USDA and USCS soil classifications are given in table 12.3; the USDA soil textures shown in table 12.3 can be converted to USCS classifications using a soil-classification triangle provided by McAneny et al. 1985.

QUESTION 12.1

Extension of climatological records using interstation correlation is a common method used to augment data at a site having a short period of record. In the answer to this question at the end of the chapter, pan evaporation data for four climatological stations are tabulated. Stations A and B have 24 and 27 years, respectively, of pan evaporation data. Station C has 13 years of data. The objective is to extend the pan evaporation record of Station Z using interstation correlations with the other three climatological stations. Try several methods, including multiple linear and nonlinear regression.

TABLE 12.2   Equivalent Curve Numbers for AMC I and AMC III Conditions Given AMC II Curve Number

CN for AMC II	CN for AMC Conditions		CN for AMC II	CN for AMC Conditions	
	I	III		I	III
100	100	100	60	40	78
99	97	100	59	39	77
98	98	99	58	38	76
97	91	99	57	37	75
96	89	99	56	36	75
95	87	98	55	35	74
94	85	98	54	34	73
93	83	98	53	33	72
92	81	97	52	32	71
91	80	97	51	31	70
90	78	96	50	31	70
89	76	96	49	30	69
88	75	95	48	29	68
87	73	95	47	28	67
86	72	94	46	27	66
85	70	94	45	26	65
84	68	93	44	25	64
83	67	93	43	25	63
82	66	92	42	24	62
81	64	92	41	23	61
80	63	91	40	22	60
79	62	91	39	21	59
78	60	90	38	21	58
77	59	89	37	20	57
76	58	89	36	19	56
75	57	88	35	18	55
74	55	88	34	18	54
73	54	87	33	17	53
72	53	86	32	16	52
71	52	86	31	16	51
70	51	85	30	15	50
69	50	84	25	12	43
68	48	84			
67	47	83	20	9	37
66	46	82	15	6	30
65	45	82			
64	44	81	10	4	22
63	43	80	5	2	13
62	42	79	0	0	0
61	41	78			

Source: Data from SCS, 1985.

Snowmelt   Prediction of snowmelt runoff generally uses one of two approaches: the energy-balance approach and the temperature-index approach. The physically based energy-balance approach uses a multiple-phase technique (solid and liquid water) of bringing the snowpack from a temperature less than the freezing point of water to 0 °C, ripening the snowpack and finally, computing the melt on a time-dependent basis. The time period used in many energy-balance snowmelt models is hourly. Clearly, the energy-balance snowmelt approach requires measurement of many climatological and snowpack variables; these include: air temperature, relative humidity, wind speed, cloud cover or solar radiation, precipitation,

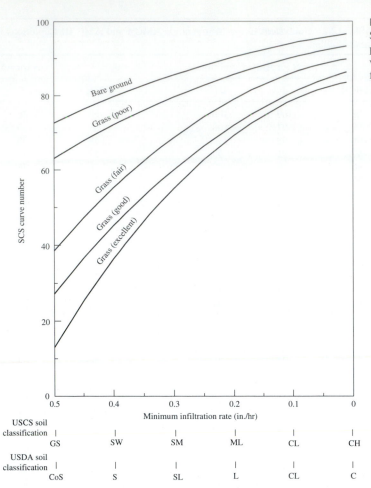

Figure 12.7 Relation between SCS curve number for daily precipitation and soil texture for various vegetative covers (data from Schroeder et al. 1994b)

TABLE 12.3 Definition of USCS and USDA Soil-Texture Classifications

USCS	Definition	USDA	Definition
G	Gravel	G	Gravel
S	Sand	S	Sand
M	Silt	Si	Silt
C	Clay	C	Clay
P	Poorly graded	L	Loam (sand, silt, clay, and humus mixture)
W	Well-graded	Co	Coarse
H	High plasticity or compressibility	F	Fine
L	Low plasticity or compressibility		

Source: Data from Schroeder et al. 1994b.

snow temperature, snow density, and initial snow water content. These data need to be available hourly, or at a minimum, as daily averages. Therefore, the energy-balance approach is often data-limited, especially in remote areas. More details on this approach can be found in U.S. Army Corps of Engineers (USACOE, 1956), Dingman (1994), Rantz (1964), Federer (1968), Anderson (1968; 1976), and are not included in detail here.

The temperature-index approach is the easiest and most common method used to esti-mate snowmelt, especially in areas where detailed climatological data are limited or not

TABLE 12.4   Selected Equations for Snowmelt Coefficient[1]

Equation	Reference
$A = 0.4f_S(1 - a)e^{-4F}$	Male and Gray (1981)
$A = f_F(0.7 + 0.0099J)f_S, J < 183$	Federer and Lash (1978)
$A = 0.07(1 + F) + [0.029 + (1 - 0.7F)(0.0084w) + 0.007Pr]$	Rantz (1964)
$A = 0.00629(z_a z_b)^{-1/6}(P/P_o)w$	Corps of Engineers (1956)

[1]Values for A in the equation $Q = A(T_a - T_m)$.

Notes:

A = melt coefficient in cm °C^{-1} day^{-1} for Male and Gray (1981); in inches °F^{-1} day^{-1} for Federer and Lash; Rantz; and Corps of Engineers

a = albedo

F = fraction of forest cover in Male and Gray (1981); forest density as a decimal in Rantz

f_S = ratio of solar radiation received at the site of interest to that on a horizontal surface

f_F = vegetative factor equal to 3.0 for open areas, 1.75 for hardwood forests, and 1.0 for conifer forests

J = Julian day

Pr = daily rainfall in inches

P = vapor pressure in mb

P_o = saturation vapor pressure at T_a in mb

T_a = average air temperature in °C in Male and Gray (1981), °F in the other equations

T_m = average snow surface temperature in °C in Male and Gray (1981), °F in the other equations

w = mean wind velocity in mph at 50 feet above ground in Rantz; mean wind velocity near surface of ground in mph for Corps of Engineers

z_a = height above ground, where T_a and P are measured in feet

z_b = height above ground, where w is measured in feet

available. The approach was popularized by the USACOE (1956), later detailed by Federer and Lash (1978) and Gray and Prowse (1993). Snowmelt for a typical daily time period, using the temperature-index approach, is a linear function of average air temperature given as

$$Q = A(T_a - T_m) \tag{12.3}$$

where Q is snowmelt in equivalent depth of water, T_a is air temperature, and T_m is snow melting-point temperature (usually taken as 0 °C), and A is a melt coefficient. The melt coefficient, A, is not a simple constant but includes such variables as latitude, elevation, basin slope and aspect, and type of forest cover. Therefore, A should be developed from data empirically; many empirical equations have been proposed for the snowmelt coefficient. Table 12.4 presents some of these snowmelt-coefficient equations and their units.

Evaporation/Transpiration

Free-water surface and wet-soil surface evaporation   Pan-evaporation data are a good estimate of free-water surface and wet-soil surface evaporation. Because year-to-year variations in evaporation are usually small, a few years of data (typically three or more) provide a satisfactory estimate of annual and even monthly values. A good source of evaporation data in the United States is the Climatic Atlas (Environmental Science Service Administration (ESSA) 1983). In other countries, local climatological data often include pan- or lake-evaporation data.

Free-water surface evaporation can be calculated, in the absence of measured data, using a form of Dalton's (1802) law

$$E = C(e_w - e_a) \tag{12.4}$$

where E is evaporation rate, C is an empirical coefficient, e_w is maximum (saturation) vapor pressure, and e_a is actual vapor pressure. The vapor-pressure gradient, $e_w - e_a$, is the driving force for free-water surface and wet-soil surface evaporation. Several factors can affect either the vapor-pressure gradient or the empirical coefficient in equation 12.4; these factors include: air and water temperature, wind over the wet surface, atmospheric pressure, soluble solids in the water evaporating, and the nature and shape of the water surface.

Temperature generally increases the vapor-pressure gradient and therefore, evaporation rates. Increased evaporation is caused either by increasing the temperature of the water body, or by increasing the temperature of the air so that more water can be stored in the air without causing vapor saturation. Wind velocities over the wet surface also tend to increase the evaporation rates by keeping the air at less-than saturation.

Decreased atmospheric pressures, usually caused by increases in elevation, can cause increased vapor-pressure gradients that result in increased free-water surface evaporation. Increases in dissolved solids in the water decrease the vapor-pressure gradient, thus decreasing evaporation. The shape of the water surface also influences the quantity of free-water surface evaporation; a flat-water surface has a greater vapor-pressure gradient than a concave-upward surface. Also, evaporation through small surface openings such as tanks, is proportional to the diameter or perimeter of the opening, rather than its surface area.

Table 12.5 summarizes some of the commonly used free-water surface evaporation equations, physically based upon Dalton's law. These equations have the basic form of equa-

TABLE 12.5 Free Water-Surface Evaporation Equations Based upon Dalton's Law

Equation	Reference
$E = C(e_w - e_a)$	Dalton (1802)
$E = \psi(e_w - e_a)$, where $\psi = 0.4 + 0.199w$	FitzGerald (1886)
$E = C(e_w - e_a)\psi$, where $\psi = 1 + 0.1w$	Meyer (1915)
$E = 0.4(\psi e_w - e_a)$, where $\psi = 2 - e^{-0.2w}$	
For large areas, E is multiplied by $(I - P) + P\dfrac{\psi - 1}{\psi - h}$	Horton (1917)
$E = 0.771(1.465 - 0.186B)\,\psi\,(e_w - e_a)$, where $\psi = 0.44 + 0.188w$	Rohwer (1931)
$E = 0.00177w(e_w - e_a)$	Harbeck and others (1954)
$E = 0.001813w(e_w - e_a)t[1 - 0.03(T_a - T_w)]$	Harbeck and others (1958)
$E = 0.7[E_{\text{pan}} \pm 0.00064P_a\alpha(0.37 + 0.00255w)|T_{\text{wpan}} - T_a|^{0.36}]$ when $T_{\text{wpan}} > T_a$, sign is +; when $T_{\text{wpan}} < T_a$, sign is −	Kohler and others (1955)
$E = [0.0211 + (9.24 \times 10^{-4})(T_w - T_a) + (1.19 \times 10^{-4})w](e_w - e_a)$	Dingman and others (1968)

Notes:

B = mean barometric reading in inches of Hg at 32 °F

C = coefficient depending upon various uncounted factors affecting evaporation; or $C = 15$ for small, shallow water and $C = 11$ for large, deep water in monthly Meyer equation, $C = 0.5$ and 0.37 for daily Meyer equation

e_a = actual vapor pressure in air based upon monthly mean air temperature and relative humidity at 30 ft above water surface in Meyer equation; mean pressure at water-surface temperature in mb in Harbeck and Dingman equations; inches of Hg in other equations

e_w = maximum (saturation) vapor pressure in inches of Hg based upon monthly mean air temperature corresponding to water temperature in Meyer equation; mean vapor pressure at water-surface temperature in mb in Harbeck and Dingman equations; inches of Hg in other equations

E = evaporation rate in inches per 30-day month or daily in Meyer equation; in inches per t days in Harbeck equation; cm per day in Dingman and Kohler equations; inches per day in other equations

E_{pan} = measured pan evaporation in cm per day

h = relative humidity

P = percentage of time during which wind is turbulent

P_a = atmospheric pressure in mb

t = number of days in period for evaporation

T_a = average air temperature, °C + 1.9 °C in Harbeck equation; °C in other equations

T_w = average water surface temperature in °C

T_{wpan} = evaporation pan water-surface temperature in °C

w = monthly wind velocity in mph at 30 ft above ground in Meyer equation; mean wind velocity near surface of ground in knots for Harbeck (1958) equation; average wind velocity at 15 cm above pan for Kohler equation; average wind velocity near surface of ground in kilometers per day for Dingman equation; mph in other equations

α = proportion of energy lost from the evaporation pan in the Kohler equation, $\alpha = 0.34 + 0.0117T_w - (3.5 \times 10^{-7})(T_w + 17.8)^3 + 0.0135w^{0.36}$

ψ = wind factor

tion 12.4, with terms for increasing or decreasing the evaporation based upon the factors discussed above.

Monthly evaporation: Meyer equation   Data from Viehmeyer and Brooks (1954) indicate that evaporation from a wet-soil surface (where the soil is near saturation) is approximately the same as that from a free-water surface. Thus, the Meyer equation—or some similar equation based upon Dalton's law—is a good approximation of evaporation from both a free-water surface and a wet-soil surface. Other free-water surface-evaporation equations shown in table 12.5 can also be used if sufficient data are available. The Meyer equation is described as

$$E = C(e_w - e_a)\psi, \qquad \psi = 1 + 0.1w \qquad (12.5)$$

In the Meyer equation applied monthly, $C = 15$ for small, shallow-water bodies and $C = 11$ for large, deep-water bodies. The maximum and actual vapor pressures e_w and e_a, respectively, are in inches of mercury (Hg), corresponding to the monthly mean-air temperature and relative humidity. The wind factor, ψ, is calculated from the monthly mean-wind velocity, w, in miles per hour (mph) at a height of 30 feet (10 meters) above the water or wet-soil surface. The maximum vapor pressure (saturation-vapor pressure) over water is given by

$$e_w = (0.0041T + 0.676)^8 - 0.000019|T + 16| + 0.001316 \qquad (12.6)$$

where $-60\,°F \le T \le 130\,°F$ and e_w is in inches of Hg. The actual vapor pressure is given by the definition of relative humidity rH and

$$rH = \frac{e_a}{e_w} \qquad (12.7)$$

or

$$e_a = e_w rH \qquad (12.8)$$

The Meyer equation can also be used daily if the coefficient C is divided by 30 days and the temperature, wind velocity, and relative humidity are all daily mean values. In this case, the values of $C = 0.50$ for small, shallow-water bodies and $C = 0.37$ for large, deep-water bodies.

Evapotranspiration   Evapotranspiration is composed of four components: **(1)** evaporation from the wet soil near the vegetation; **(2)** transpiration of water by the vegetation; **(3)** evaporation from the moist membrane surfaces of the vegetation; and **(4)** use of water by the vegetation to build new plant tissue (Blaney and Hanson 1965). Evapotranspiration is the same as consumptive use, a term often used by water-rights engineers.

Several factors affect evapotranspiration, among them air temperature, wind speed, solar radiation, and available soil moisture. As indicated by Dalton's law (equation 12.4), the driving force for evaporation as well as evapotranspiration is vapor-pressure gradient, itself highly dependent upon air temperature; the vapor-pressure gradient and resulting evapotranspiration increase with increasing air temperature. Wind speed also affects evapotranspiration by removing water vapor immediately above wet soil and plant membranes; therefore, increased wind speed results in increased evapotranspiration. The primary source of evapotranspiration energy is solar radiation; increased solar radiation results in increased evapotranspiration.

Even though the above climatological factors can act to increase evapotranspiration, vegetation only transpires if sufficient soil moisture is available. If soil-water contents are limiting, evapotranspiration is limited by the ability of the vegetation to extract water from the soil. Additionally, plant factors are important in transpiration. Research on revegetation of retorted-oil shale in Colorado (Berg et al. 1979; Herron, Berg, and Harbert 1980; Kilkelly 1981; Jump and Sabey 1983) has shown that during times of water stress, natural vegetation transpires less water, reduces growth, sheds leaves, and closes their stomata to conserve water. Thus, plant factors can be important in controlling evapotranspiration in arid and semiarid areas. In general, increased soil water results in increased—or at least does not limit—evapotranspiration. Soil moisture has to be above wilting point for evapotranspiration to occur at the rate predicted by the theoretical equations.

There are three common methods for estimating evapotranspiration: **(1)** temperature methods; **(2)** radiation methods; and **(3)** combination of temperature and radiation methods (Jensen, Burman, and Allen 1990). It has been shown that combination methods provide the best models for estimating evapotranspiration. All estimation methods predict potential evapotranspiration (PET) for a reference crop (such as short grass or alfalfa). Actual evapotranspiration for a given type of vegetative cover or crop is estimated by multiplying PET by an appropriate crop coefficient. Vegetative or crop coefficients are both annual and seasonal, and can be obtained from various sources in the literature. Often these coefficients are site-dependent, so care needs to be taken to assure that the vegetative coefficients in use were determined for conditions similar to those occurring at the site of interest; coefficients for selected vegetation are given below.

Temperature method: Blaney–Criddle The generally recommended temperature method for estimating monthly evapotranspiration is the Blaney–Criddle method (Blaney and Criddle 1950; 1957; 1962). The original method was developed for estimation of evapotranspiration seasonally (growing season). The assumption is that evapotranspiration varies directly with the sum of the products of mean-monthly air temperature and monthly percentage of daytime hours, for an actively growing crop with adequate soil moisture. This relation is given mathematically by

$$U = KF = \sum_{i=1}^{m} u_m = \sum_{i=1}^{m} k_c k_t f \qquad (12.9)$$

where U is the seasonal evapotranspiration (consumptive use) in inches for the growing season of interest; K is an empirical consumptive-use coefficient (growing season/growing period); F is the sum of monthly consumptive-use factors; f, for the season or period; u_m is the monthly consumptive use in inches; k_c is the monthly empirical-crop coefficient; and k_t is the monthly temperature coefficient. The monthly temperature coefficient is given by

$$k_t = 0.0173t - 0.314 \qquad (12.10)$$

where t is the monthly mean-air temperature in °F. The monthly consumptive-use factor is given by

$$f = \frac{tp}{100} \qquad (12.11)$$

where p is monthly percentage of daytime hours of the year. Individual monthly evapotranspiration is given by $u_m = k_c k_t f$ for the month of interest. Table 12.6 presents monthly percentages of daytime hours by latitude (SCS 1970). Monthly and annual empirical-crop coefficients for various crops and natural vegetation are shown on table 12.7. SCS (1970) presents curves of crop coefficients for use with the Blaney–Criddle method; Pochop, Borrell, and Burman (1984) recommend an elevation-correction for the Blaney–Criddle equation that increases the evapotranspiration by approximately 10 percent for each 1,000-meter increase in elevation. This correction adjusts for lower temperatures that occur at higher elevations, at a given level of solar radiation.

Doorenbos and Pruitt (1977) modified the Blaney–Criddle method to compute evapotranspiration for a grass-related reference crop. The FAO-24 Blaney–Criddle method requires the intermediate step of estimating a grass-related reference crop evapotranspiration, E_{to}, prior to applying grass-related crop coefficients.

Radiation method: Jensen–Haise The Jensen–Haise method (Jensen and Haise 1963) is recommended for estimating evapotranspiration every five days, but can also be used to

TABLE 12.6   Monthly Percentage of Daytime Hours, p, of the Year

Latitude (deg.)	Jan.	Feb.	March	April	May	June	July	Aug.	Sept.	Oct.	Nov.	Dec.
						North						
60	4.70	5.67	8.11	9.69	11.78	12.41	12.31	10.68	8.54	6.95	5.02	4.14
55	5.44	6.04	8.18	9.44	11.15	11.53	11.54	10.29	8.51	7.23	5.63	5.02
50	5.99	6.32	8.24	9.24	10.68	10.92	10.99	9.99	8.46	7.44	6.08	5.65
45	6.40	6.54	8.29	9.08	10.31	10.46	10.57	9.75	8.42	7.61	6.43	6.14
40	6.75	6.72	8.32	8.93	10.01	10.09	10.22	9.55	8.39	7.75	6.73	6.54
35	7.04	6.88	8.35	8.82	9.76	9.76	9.93	9.37	8.36	7.88	6.98	6.87
30	7.31	7.02	8.37	8.71	9.54	9.49	9.67	9.21	8.33	7.99	7.20	7.16
25	7.54	7.14	8.39	8.62	9.33	9.24	9.45	9.08	8.31	8.08	7.40	7.42
20	7.75	7.26	8.41	8.53	9.15	9.02	9.24	8.95	8.29	8.17	7.58	7.65
15	7.94	7.37	8.43	8.45	8.98	8.81	9.04	8.83	8.27	8.25	7.75	7.89
10	8.14	7.47	8.45	8.37	8.81	8.61	8.85	8.71	8.25	8.34	7.91	8.09
5	8.32	7.67	8.47	8.29	8.85	8.41	8.67	8.60	8.24	8.41	8.07	8.30
0	8.50	7.67	8.49	8.22	8.49	8.22	8.50	8.49	8.21	8.49	8.22	8.50
						South						
10	8.86	7.87	8.53	8.09	8.18	7.86	8.14	8.27	8.17	8.62	8.53	8.88
20	9.24	8.09	8.57	7.94	7.85	7.43	7.76	8.03	8.13	8.76	8.87	9.33
30	9.70	8.33	8.62	7.73	7.45	6.96	7.31	7.76	8.07	8.97	9.24	9.85
40	10.27	8.63	8.67	7.49	6.97	6.37	6.76	7.41	8.02	9.21	9.71	10.49

Source: Data from SCS, 1970

TABLE 12.7   Typical Seasonal Blaney–Criddle Growth-Stage Coefficients (k_c)

Crop	Jan.	Feb.	March	April	May	June	July	Aug.	Sept.	Oct.	Nov.	Dec.	Annual[1]
Alfalfa	0.63	0.73	0.86	0.99	1.08	1.13	1.11	1.06	0.99	0.91	0.78	0.64	0.80–0.90
Pasture Grasse	0.49	0.57	0.73	0.86	0.90	0.92	0.92	0.91	0.87	0.79	0.67	0.55	0.75–0.85
Greasewood[2]	0.50	0.58	0.63	0.85	0.99	1.56	2.00	2.40	2.50	2.10	1.49	0.85	—
Greasewood[3]	0.35	0.39	0.40	0.45	0.52	0.84	1.10	1.29	1.35	1.12	0.80	0.45	—
Greasewood[4]	0.15	0.16	0.18	0.19	0.20	0.34	0.44	0.52	0.55	0.45	0.32	0.19	—
Saltcedar	—	—	—	—	—	—	—	—	—	—	—	—	1.40–1.50
Cottonwood	—	—	—	—	—	—	—	—	—	—	—	—	1.15–1.25
Phreatophytic Bushes	—	—	—	—	—	—	—	—	—	—	—	—	0.90–1.00
Bermuda grass (high values)	1.17	1.07	1.27	1.28	1.22	0.81	0.68	0.73	0.74	1.03	0.95	0.72	—
Bermuda grass (low values)	—	0.44	0.42	0.83	0.81	0.76	0.58	0.53	0.58	0.77	1.02	0.44	—

Sources: Data from Blaney and Hanson, 1965; SCS, 1970; Pochop and others, 1984
[1] Values of K. Lower values are for humid areas and higher values are for arid climates.
[2] 0–12 in. of water
[3] 18–36 in. of water
[4] 36–60 in. of water

estimate evapotranspiration daily, if care is used. The Jensen–Haise method uses the equation

$$E_{tp} = C_t(T - T_x)R_s \tag{12.12}$$

where E_{tp} is daily evapotranspiration in inches, C_t is the temperature coefficient, T is mean-air temperature in °F, T_x is the temperature-scale intercept when $ET/R_s = 0$, and R_s is solar radiation in inches/day of evaporation.
        Also,

$$C_t = \frac{1}{C_1 + C_2 C_h} \tag{12.13}$$

and

$$C_h = \frac{50 \ mb}{e_2 - e_1} \tag{12.14}$$

where e_1 and e_2 are saturated vapor pressures at the mean-maximum and -minimum air temperatures for the warmest months of the year; and

$$C_2 = 13 \ °F \tag{12.15}$$

Also,

$$C_1 = 68 - \frac{3.6E}{1000} \tag{12.16}$$

where E = site elevation in feet, and

$$T_x = 27.5 - 0.25(e_2 - e_1) - \frac{E}{1000} \tag{12.17}$$

If daily solar radiation data are not available, they can be calculated using either stochastic techniques (Richardson and Wright 1984; Schroeder et al. 1994a, b), or from cloudless-sky radiation (see table 12.8), using

$$R_s = R_{so}\left[1 - (1 - k)\frac{n}{10}\right] \tag{12.18}$$

TABLE 12.8 Mean Solar Radiation for Cloudless Skies

Latitude	Mean solar radiation per month (cal/cm²/day)*											
	Jan.	Feb.	March	April	May	June	July	Aug.	Sept.	Oct.	Nov.	Dec.
60 N	58	152	319	533	671	763	690	539	377	197	87	35
55 N	100	219	377	558	690	780	706	577	430	252	133	74
50 N	155	290	429	617	716	790	729	616	480	313	193	126
45 N	216	365	477	650	729	797	748	648	527	371	260	190
40 N	284	432	529	677	742	800	755	674	567	426	323	248
35 N	345	496	568	700	742	800	761	697	603	474	380	313
30 N	403	549	600	713	742	793	755	703	637	519	437	371
25 N	455	595	629	720	742	780	745	703	660	561	486	423
20 N	500	634	652	720	726	760	729	697	680	597	537	474
15 N	545	673	671	713	706	733	706	684	697	623	580	519
10 N	584	701	681	707	684	700	681	665	707	648	617	565
5 N	623	722	690	700	652	663	645	645	710	665	650	606
0 N	652	740	694	680	623	627	616	623	707	684	680	619
5 S	648	758	690	663	590	587	577	590	693	690	727	677
10 S	710	772	681	640	571	543	526	558	680	690	727	710
15 S	729	779	665	610	516	497	497	519	657	687	747	739
20 S	748	779	645	573	474	447	445	481	630	677	753	761
25 S	761	779	626	533	419	400	406	439	600	665	767	777
30 S	771	772	600	497	384	353	358	390	567	648	767	793
35 S	774	754	568	453	335	300	310	342	530	629	767	806
40 S	774	729	529	407	281	243	261	290	477	603	760	813
45 S	774	704	490	357	229	183	203	235	447	571	747	813
50 S	761	669	445	307	174	127	148	177	400	535	727	806
55 S	748	630	397	250	123	77	97	123	343	497	707	794
60 S	729	588	348	187	77	33	52	74	283	455	700	787

Source: Data from Budyko 1963
*1 cal/cm²/day = 1004.515 kilowatts/m²

where R_s is solar radiation at the ground surface in langleys per day (cal cm^{-2} d^{-1}), R_{so} is cloudless-sky radiation in langleys per day, k is a mean-annual coefficient (see table 12.9), and n is cloudcover in tenths ($0 \leq n \leq 10$). Note that, to convert radiation from langleys per day (cal cm^{-2} d^{-1}) to inches per day, multiply by 0.000673; and to convert langleys per day to kilowatts per square meter (kW/m^2), multiply by 0.000484.

Crop coefficients for the Jensen–Haise method have been published by Jensen et al. (1971; 1990). Generalized Jensen–Haise crop coefficient for various crops, including alfalfa and pasture grass, are presented in figures 12.8 through 12.11. These generalized coefficients can also be used with the FAO–Penman method (presented following), or other PET methods.

Combination method: FAO–Penman   This method (Doorenbos and Pruitt 1977) is recommended for estimating daily evapotranspiration. Details on the use of this method—as well as comparison of the results of this method to other evapotranspiration methods—are presented in Jensen, Burman, and Allen (1990), and were discussed in chapter 9. The basic equation for the FAO–Penman method is

$$E_o = c\left[\frac{\Delta}{\Delta + \gamma}(R_n - G) + \frac{\gamma}{\Delta + \gamma}2.7W_f(e_z^w - e_z)\right] \tag{12.19}$$

where E_o is the evapotranspiration for a grass-reference crop, c is an adjustment factor based on local climate, and W_f is wind factor given by $(1 + 0.864\,u_2)$; the other variables are defined below. In equation 12.19, E_o and R_n are in mm/day, u_2 is in m/s, and e is in kPa; G is assumed to be zero for daily periods in the FAO–Penman method. The vapor-pressure deficit is

TABLE 12.9   Mean Annual Values of k for Use in Converting Cloudless Sky Radiation to Actual Solar Radiation

Latitude (degrees North or South)												
0	5	10	15	20	25	30	35	40	45	50	55	60
0.35	0.34	0.34	0.33	0.33	0.32	0.32	0.32	0.33	0.34	0.36	0.38	0.40

Source: Data from Jensen, Burman, and Allen 1990

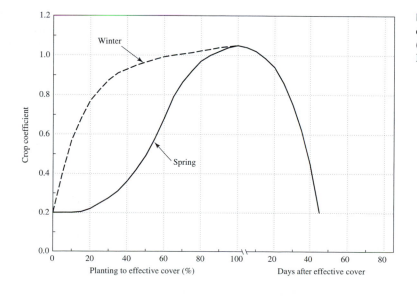

Figure 12.8 Jensen–Haise crop curves for small grains (data from Jensen, Wright, and Pratt 1971)

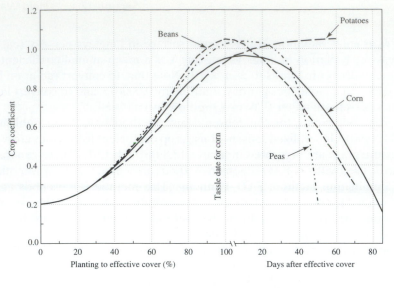

Figure 12.9 Jensen–Haise crop curves for beans, potatoes, peas, and corn (data from Jensen, Wright, and Pratt 1971)

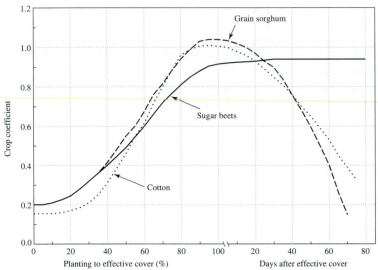

Figure 12.10 Jensen–Haise crop curves for sugar beets, grain sorghum, and cotton (data from Jensen, Wright, and Pratt 1971)

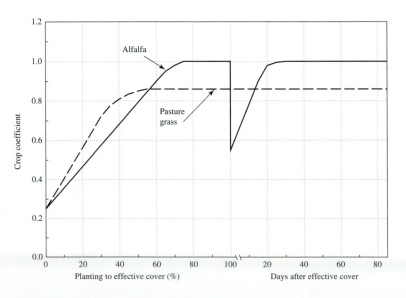

Figure 12.11 Jensen–Haise crop curves for alfalfa and pasture grass (data from Jensen, Wright, and Pratt 1971)

calculated by Method 1 (equation 12.20) when dewpoint-temperature data are available, otherwise Method 2 (equation 12.21) is used:

Method 1

$$(e_z^w - e_z) = e_w(T_{\text{mean}}) - e_w(T_{\text{dew}}) \tag{12.20}$$

Method 2

$$(e_z^w - e_z) = e_w(T_{\text{mean}})(1 - \text{rH}/100) \tag{12.21}$$

The adjustment factor c is given by

$$c = 0.68 + 0.0028\,\text{rH}_{\text{max}} + 0.018\,R_s - 0.068 U_d + 0.013 U_d/U_n$$
$$+\ 0.0097 U_d(U_d/U_n) + 0.430 \times 10^{-4}\,\text{rH}_{\text{max}}\,R_s U_d \tag{12.22}$$

where rH_{max} is maximum daily relative humidity in percentage; R_s is in mm/day; U_d/U_n is the ratio of the daytime to nighttime wind speed; U_d is the daytime (0700–1900 hrs) windspeed in m s^{-1}; R_n is net-solar radiation; R_s is incident-solar radiation; G is soil-heat flux (assumed to be zero for daily time periods); and Δ is the slope of the saturation vapor pressure versus air temperature curve, and is given by

$$\Delta = \frac{de_w}{dT} = 0.200(0.00738\,T + 0.8072)^7 - 0.000116 \tag{12.23}$$

for $T \geq -23\,°\text{C}$ and $\Delta = \text{kPa}/°\text{C}$. The psychrometric constant γ is given by

$$\gamma = \frac{c_p P}{0.622 L_v} \tag{12.24}$$

where the units are in kPa/°C, and P is atmospheric pressure in kPa, c_p is the specific heat of moist air at a constant pressure of 1.013 kJ/kg $-$ °C, and L_v is the latent heat of vaporization of water, in kJ/kg.

Pan-evaporation method   Often, pan-evaporation data are available. Estimates of evapotranspiration can be made from pan-evaporation data using the following method from Doorenbos and Pruitt (1977):

$$E_g - C_{et} E_{\text{PAN}} \tag{12.25}$$

where E_g is the evapotranspiration calculated from pan evaporation, C_{et} is a coefficient relating pan evaporation to evapotranspiration, and E_{PAN} is pan evaporation. Values of C_{et} are presented in table 12.10. The coefficients shown in table 12.10 are given for two cases: **(1)** sites in cropped fields (Case A); and **(2)** sites in noncropped, dry-surface fields (Case B), as shown in figure 12.12. Jensen, Burman, and Allen (1990) state that adjustments are needed to relate to C_{et} for taller vegetation (such as alfalfa or tall grass), especially in hot, drier climates where crop height and aerodynamic roughness have a greater effect on E_g than in humid climates. For taller vegetation and aerodynamically rougher crops, the values of C_{et} are greater and vary less with differences in wind and relative humidity than do values for shorter, smoother grass surfaces.

In addition to the variation in coefficients with wind and relative humidity, there also is an interaction with radiation intensity (Jensen, Burman, and Allen 1990). The coefficients in table 12.10 apply to Class A pans, annually painted inside and out with aluminum paint, then mounted on a standard wooden platform so that the top of the pan is 40 cm above ground level. If pans are not well maintained or have been screened to protect them from birds or other animals, the coefficients need to be modified. For example, Jensen, Burman, and Allen (1990) recommend that pans with 12.5-mm mesh screen have coefficients increased by 5 to

TABLE 12.10 Suggested values of C_{et} for relating class-A pan evaporation to evapotranspiration from 8–15 cm tall, well-watered grass crop[1]

| Wind speed (km/day) | Case A: Pan surrounded by short green crop | | | | Case B: Pan surrounded by dry-surface gound[2] | | | |
| | Upwind fetch of green crop (m) | Relative humidity (percent)[3] | | | Upwind fetch of dry fallow (m) | Relative humidity (percent)[3] | | |
		Low 20–40	Medium 40–70	High >70		Low 20–40	Medium 40–70	High >70
Light <170 km/day	0	0.55	0.65	0.75	0	0.70	0.80	0.85
	10	0.65	0.75	0.85	10	0.60	0.70	0.80
	100	0.70	0.80	0.85	100	0.55	0.65	0.75
	1000	0.75	0.85	0.85	1000	0.50	0.60	0.70
Moderate 170–425 km/day	0	0.50	0.60	0.65	0	0.65	0.75	0.80
	10	0.60	0.70	0.75	10	0.55	0.65	0.70
	100	0.65	0.75	0.80	100	0.50	0.60	0.65
	1000	0.70	0.80	0.80	1000	0.45	0.55	0.60
Strong 425–700 km/day	0	0.45	0.50	0.60	0	0.60	0.65	0.70
	10	0.55	0.60	0.65	10	0.50	0.55	0.65
	100	0.60	0.65	0.70	100	0.45	0.50	0.60
	1000	0.65	0.70	0.75	1000	0.40	0.45	0.55
Very Strong >700 km/day	0	0.40	0.45	0.50	0	0.50	0.60	0.65
	10	0.45	0.55	0.60	10	0.45	0.50	0.55
	100	0.50	0.60	0.65	100	0.40	0.45	0.50
	1000	0.55	0.60	0.65	1000	0.35	0.40	0.45

[1] $E_a = C_{et} * E_{pan}$—Doorenbos and Pruitt 1974

[2] These coefficients apply only to conditions when the soil surface is indeed dry. Following rains for a day or two in midsummer and longer periods in the fall, winter, and spring, such pans are essentially equivalent to Case A type pans with large fetches. Rains also affect Case A type pans by essentially increasing the 0- and 10-meter fetch to an effective moist surface fetch of 1000 m or more. One large advantage of Case-B type pans is the minimal upkeep involved in siting of such a pan; that is, only an effective weed control program is needed for the site.

[3] Mean of maximum and minimum relative humidity

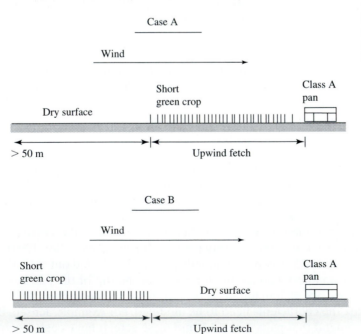

Figure 12.12 Schematic relating class-A pan evaporation to evapotranspiration (data from Jensen, Burman, and Allen 1990)

10 percent. If the pan is made of Monel metal, or is an old, unpainted, galvanized pan, coefficients need to decrease by 5 percent.

Change in Surface Storage

In water-balance calculations, surface storage can occur as ponds, reservoirs, or storage tanks that hold water temporarily or permanently, with and without seepage and/or evaporation losses. In a given time frame, water can enter or leave surface storage. The primary components of surface storage that are taken into account in the water-balance calculations are its water-surface elevation, surface area, and storage capacity. For a given storage reservoir, these components are usually lumped into a set of curves called elevation–area–capacity curves, as shown in figure 12.13. Such curves are easily constructed from the geometry of the storage reservoir, taken either from topographic maps or from prismatic shapes such as frustrums of regular pyramids or circular cones.

The factors that cause the reservoir storage to change include: precipitation falling directly on the reservoir; evaporation from the reservoir-water surface; inflow or outflow of ponded water; and seepage/infiltration from/to the reservoir. The elevation–area–capacity curve can also change over time if sediment is added to the reservoir through inflow, or is removed through natural or man-induced mechanisms.

Change in Soil-Moisture Storage

Change in soil-moisture storage is caused by precipitation falling on the soil surface, evaporation occurring directly from a wet-soil surface, evapotranspiration, and seepage/infiltration. The quantitative estimates of soil-moisture storage can be made using a running balance of soil-moisture storage as a "reservoir," subject to operating rules that describe when deep percolation (see figure 12.1) occurs, as well as the distribution of actual evapotranspiration with depth and currently available soil moisture. Typical operating rules related to retention

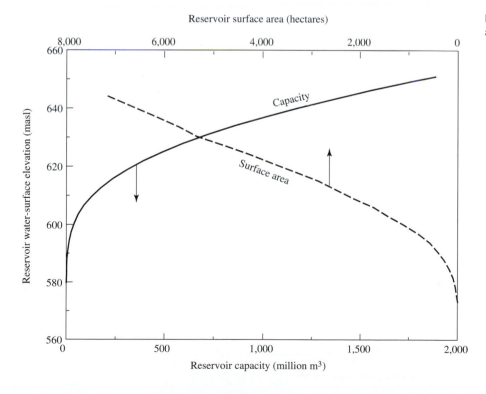

Figure 12.13   Typical elevation–area–capacity curves

(or movement) of soil moisture use the concepts of field capacity and permanent-wilting point. These two variables are generally defined as the volumetric-water content at matric potentials of approximately −340 and −15,300 cm of water, respectively, and are highly dependent upon soil texture. Figure 12.14 shows a typical relation between water content and matric potential for a sandy soil and a clay soil. Holtan et al. (1968), Rawls, Brakensiek, and Saxton (1982), and Rawls et al. (1993) provide soil-moisture versus matric-potential data for over 5,000 soils in the United States.

Operating rules in various soil-moisture balance calculations are proposed by Schroeder et al. (1994a, b). In simple water-balance models, soil moisture is allowed to move only if the water content exceeds field capacity. As shown on figure 12.1, the root-zone soil is often divided into several isotropic and homogeneous cells, whose soil properties are assumed to be uniform. Soil moisture is then moved from cell to cell at each time-step, until the water is evapotranspired or discharges from the lowermost cell as deep percolation. If the water content of the soil falls to permanent wilting point, then it is assumed that plants cannot extract moisture. This part of the soil-moisture rule is a lower bound on minimum soil moisture. If all pores are filled with water (saturation), then all additional water trying to enter the soil is assumed to run off.

Root-zone extraction of water by plants is not uniform with depth, because root densities usually decrease from near-ground surface to maximum rooting depth. Several models assume that water extraction from the soil by plant roots takes place in a linearly decreasing mode (Schroeder et al. 1994a, b; Davis and Neuman 1983). Kunkel and Murphy (1983) proposed that removal of water by plant roots is a parabolically decreasing function with depth, based upon test plot data.

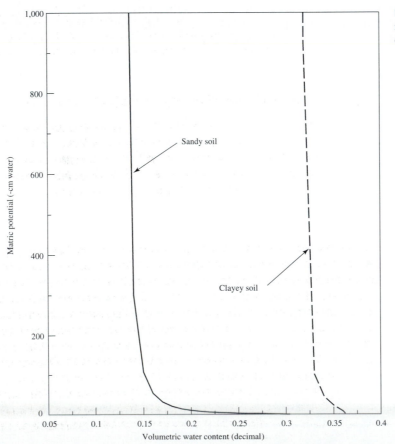

Figure 12.14 Typical soil-moisture characteristic curves

12.2   FIELD RADIATION AND ENERGY BALANCE

The radiation- and energy-balance equations can be derived exactly in the same manner as the water-balance equation. Radiation and energy are important because solar radiation is the primary external energy source that drives heat and water movement in the soil. Radiation includes various components of solar radiation such as incoming short-wave radiation, reflected short-wave radiation, and long-wave radiation. Energy components include—in addition to the radiation components—sensible and latent-heat fluxes from water and soil bodies (Rasmusson et al. 1993).

Radiation Balance

Averaged over the globe, the earth's surface absorbs approximately 124 kilolangleys (kly) (124,000 cal cm^{-2}) of solar radiation each year. Of this amount, approximately 52 kly yr^{-1} is radiated to the atmosphere from long-wave energy, and the remaining 72 kly yr^{-1} is the net radiation at the earth's surface (Sellers 1965). The earth's atmosphere absorbs only 45 kly yr^{-1} of solar radiation and radiates 117 kly yr^{-1} of long-wave energy; the net atmospheric radiation is -72 kly yr^{-1}. Thus, the atmospheric radiation losses (-72 kly yr^{-1}) exactly balance the earth's surface radiation gains (72 kly yr^{-1}), on the average (Sellers 1965). The surface net-radiation balance (or the net radiation) is given by

$$R_n = C_a + A_a - C_r - A_r + R_s(1 - a) - aR_s - R_b \tag{12.26}$$

This expression says that the radiation incident on a horizontal surface at the top of the atmosphere can be reflected and scattered back into space by clouds (C_r); by dry-air molecules, dust, and water vapor (A_r); or by the surface of the earth aR_s, where R_s is the sum of the direct (Q) and diffuse (q) solar radiation, respectively, incident on the earth's surface, and a is the surface albedo; and R_b is the effective outgoing radiation from the surface. Alternatively, this solar radiation can be absorbed by clouds (C_a); by dry-air molecules, dust, and water vapor (A_a); or by the earth's surface [$R_s(1 - a)$]. This radiation balance is shown schematically in figure 12.15.

As indicated in equation 12.18, direct and diffuse solar radiation can be estimated from cloudless-sky radiation (see table 12.7) and cloud cover. The effective outgoing radiation R_b can be estimated from the net long-wave radiation at the earth's surface. The net long-wave radiation is the sum of radiation being transmitted from the atmosphere to the earth, and from the earth to the atmosphere. Long-wave radiation is a function of the temperature of the radiating body. For both the atmosphere and the earth's surface—each with temperatures above absolute zero—the long-wave radiation R_b can be calculated using the radiation law:

$$R_b = \varepsilon \sigma T^4 \tag{12.27}$$

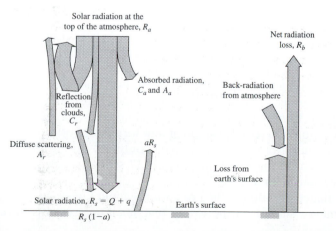

Figure 12.15 Schematic of the radiation balance at the Earth's surface (data from Sellers 1965 and Shuttleworth 1993)

where ε is emissivity (Stefan–Boltzmann constant), and T is the absolute temperature of the radiating body $(273.2 + °C)$ (Kelvin). Values of the Stefan–Boltzmann constant are: $\sigma = 1.17 \times 10^{-7}$ cal cm^{-2} K^{-4} d^{-1} or $\sigma = 4.903 \times 10^{-9}$ MJ m^{-2} K^{-4} d^{-1}. Values of emissivity typical of surfaces on earth (water, snow, soil, vegetation) range from approximately 0.90 to 0.99. Typical values of emissivity can be found in Sellers (1965), or calculated by equations found in Jensen, Burman, and Allen (1990).

A portion of the solar radiation incident at the top of the atmosphere (R_a) reaches the ground (R_s). Typically, R_s is between 25 and 50 percent of R_a, whereas A_r is between 15 and 100 percent of R_s. Both of these values are influenced by albedo, which is typically 0.23 for land surfaces and 0.08 for water surfaces (Shuttleworth 1993).

Energy Balance

If one considers a soil or water column extending from the earth's surface to a depth where vertical heat exchange is negligible, the net rate, G, at which heat energy in this column is changing is equal to the sum of the rates at which heat energy is being added or lost by the various heat processes. The energy-balance equation at the earth's surface is given by

$$G = R_s(1 - a) + I_\downarrow - I_\uparrow - H - E + F_i - F_o \tag{12.28}$$

where $R_s(1 - a)$ is the heat added by absorption of solar radiation; $I_\downarrow$ is heat added by absorption of longwave counterradiation from the atmosphere; $-H$ is the heat added by downward transfer of sensible heat from the air, when the air is warmer than the soil or water surface; F_i is the horizontal transfer of heat into the soil or water column from the surroundings; $I_\downarrow$ is heat being lost by longwave radiation to the atmosphere; H is the transfer of sensible heat to the air if the air is cooler than the soil or water surface; E is the heat lost by evaporation (given by the product of mass of water evaporated times the latent heat of vaporization (590 cal g^{-1}); and F_o is the horizontal transfer of heat out of the column. These components are shown schematically on figure 12.16.

As indicated by equation 12.26, the first three terms of equation 12.28 are from the radiation balance R_n. The energy-balance equation is often written as

$$R_n = H + E + G + \Delta F \tag{12.29}$$

where $\Delta F = F_o - F_i$ is the net-subsurface flux of sensible heat out of a soil or water column. This flux is usually only important for large water bodies, where currents can transport considerable heat energy from one region to another. Over land, ΔF is negligible and

$$R_n = H + E + G \tag{12.30}$$

which implies that the net-available radiative energy is used to warm air, evaporate water, and warm the soil.

At night, when there is no solar radiation, the heat energy equation is

$$R_n = -I = H + E + G. \tag{12.31}$$

Because the effective outgoing radiation I is nearly always positive, the surface loses heat by radiation at night. In order to preserve the heat balance, sensible heat is usually transferred to the cooler surface from the warmer air ($-H$), and from the warmer deeper soil layers ($-G$). Negative evaporation (condensation or dew formation) often occurs also. However, in arid climates, positive evaporation customarily continues during the night, although at a much lower rate than during the day. Figure 12.17 schematically shows the daytime and nighttime energy balance for a soil column.

The heat-balance equation in the proper form applies over any time period. However, it is an approximate equation to the extent that small components have been neglected.

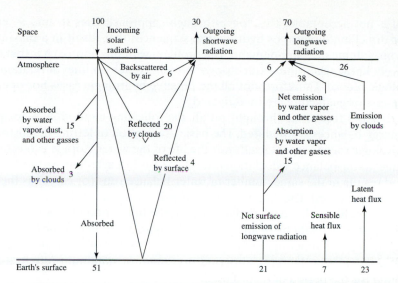

Figure 12.16 Schematic of the energy balance at the top of the atmosphere at the earth's surface; numbers indicate percentage of total incoming solar radiation (data from Rasmusson et al. 1993)

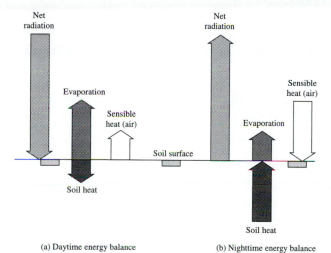

Figure 12.17 Schematic of daytime and nighttime energy balance (data from Rasmusson et al. 1993)

These might be important locally at a particular time, and include

- snowmelt, which may require approximately 10 ly d^{-1};
- dissipation of mechanical energy of wind, which can be between 1 and 10 ly d^{-1};
- heat transfer by precipitation, whose magnitude can reach 40 ly hr^{-1};
- expenditure of heat by photosynthesis, estimated at less than 5 percent of the incoming solar radiation, R_s;
- heat gain by biological oxidation processes such as forest fires, which can give off 850 ly d^{-1} of heat;
- other sources including combustion, volcanic eruptions, earthquakes, street lighting, and flux from the earth's interior.

12.3 WATER- AND ENERGY-BALANCE METHODOLOGY

The use of water balance for the sizing of water-storage ponds and reservoirs is often applied incorrectly. In order to have confidence that the water-storage facility is correctly sized according to the available hydrologic data, a traditional "once-through" use of the available record is not acceptable. This "once-through" approach tries to find a "critical period" of wet or dry climatological or hydrological sequences that result in a maximum or minimum reservoir-volume or water-demand. The problem with this approach is that the designer does not likely know when the water-storage project comes on line, in relation to the historical hydrologic record. Therefore, not all the possible hydrologic cases nor operational cases for the water-storage facility are investigated.

To avoid this "once-through" pitfall, the following methodology for performing water and energy balances is stipulated. The basic assumption of letting the historical climatologic or hydrologic record repeat itself over the life of the water-storage project is still used. However, the water-balance calculations should begin in each year of the historic record. This method results in the same number of water-balance runs (or cases) as the number of years of hydrologic record. The data for the years prior to the start-date are transposed to the end of the project life cycle, so all analyses are for the full period of record. In this way, the initial project start-date could be in any year, resulting in a number of maximum or minimum values equal to the number of operational cases. The results of the sizing of water-storage facilities guarantees that the critical-period maximum or minimum required-reservoir volume are found for the period of record used.

SUMMARY

In this chapter we discussed field water, radiation and energy balances, and provided practical approaches to solve water- and energy-balance equations. The evapotranspiration methods in this chapter are similar to those discussed in chapter 9 and are frequently the ones used by the consulting industry. The evapotranspiration methods herein may be compared to the methods detailed in chapter 9.

ANSWERS TO QUESTIONS

12.1. Among the many possible relationships which could be developed the following were selected as potential predictors. The coefficient of determination R^2 is shown as a measure of the goodness-of-fit of the predictor equation. You may find other equations as well.

$$E_z = 0.7732072\, E_A + 34.09690 \qquad\qquad R^2 = 0.718 \qquad (12.32)$$

$$E_z = 1.0139667\, E_B - 50.01599 \qquad\qquad R^2 = 0.717 \qquad (12.33)$$

$$E_z = 0.4333176\, E_A + 0.561208\, E_B - 34.66132 \qquad R^2 = 0.799 \qquad (12.34)$$

$$E_z = 0.8920611\, E_A - 2.32 \times 10^{-4}\, (E_A)^2 - 20.269656 \qquad R^2 = 0.718 \qquad (12.35)$$

The nonlinear regression (equation 12.35) is not an improvement upon the linear regression shown in equation 12.32. Therefore, equations 12.32–34 constitute good predictors, but not the only ones, of the pan evaporation at Station Z. Additionally, because only three coincidental values were available at Station C, this station was not used in the regression analyses.

	Pan evaporation (mm)					Pan evaporation (mm)			
Date	Station A	Station B	Station C	Station Z	Date	Station A	Station B	Station C	Station Z
Jan-74	452.3				Jan-77	348.7	351.1		312.0
Feb-74	418.3				Feb-77	264.0	297.3		240.5
Mar-74	476.3				Mar-77		335.9		276.1
Apr-74	259.5				Apr-77	200.9	254.3		198.5
May-74	176.5				May-77	227.5	219.8		184.9
Jun-74	138.6				Jun-77	146.3	194.4	136.1	178.9

(continued)

Pan evaporation (mm)					Pan evaporation (mm)			
Date	Station A	Station B	Station C	Station Z	Date	Station A	Station B	Station C	Station Z
Jul-74	164.5				Jul-77	172.5	238.1	199.5	135.9
Aug-74	197.0				Aug-77	195.2	260.1	200.1	166.0
Sep-74	260.5				Sep-77	241.6	335.0		256.2
Oct-74	334.5				Oct-77	317.4	303.5		308.7
Nov-74	371.0			128.1	Nov-77	357.5	363.8		326.4
Dec-74	373.6	340.3		255.7	Dec-77	358.2	399.6		359.5
Jan-75	342.4	303.1		188.4	Jan-78	349.9	390.7		319.5
Feb 75	247.5	263.9		102.9	Fcb-78	263.8	305.5		289.1
Mar-75	290.7	267.8		199.0	Mar-78	217.2	381.1		318.9
Apr-75	239.4	215.8		226.0	Apr-78	193.6	272.8		236.6
May-75	162.8	166.4		140.0	May-78	181.5	255.0		184.4
Jun-75	129.8	148.3			Jun-78	175.1	216.5		121.6
Jul-75	143.4	157.2			Jul-78	138.1	199.8		135.3
Aug-75	197.3	201.2			Aug-78	188.8	230.4		190.3
Sep-75	245.7	240.8		228.0	Sep-78	235.1	242.0		217.3
Oct-75	341.7	317.3		239.3	Oct-78	319.6	266.1		314.4
Nov-75	344.8	328.7		316.6	Nov-78	346.1	282.6		279.6
Dec-75	361.0	339.4		367.1	Dec-78	391.0	336.7		334.2
Jan-76	301.0	297.7		227.0	Jan-79	377.4	348.6		265.9
Feb-76	293.0	268.1		230.4	Feb-79	317.6	332.0		276.1
Mar-76	295.0	287.2		337.8	Mar-79	320.8	364.1		287.0
Apr-76	232.4	248.8		128.1	Apr-79	222.0	312.9		196.0
May-76	180.7	174.1		132.3	May-79	148.7	244.4		185.5
Jun-76	121.3	169.8		127.7	Jun-79	124.8	221.2		133.1
Jul-76	158.9	177.4		120.4	Jul-79	153.1	267.1		168.1
Aug-76	176.2	191.1		136.7	Aug-79	184.5	268.5		205.0
Sep-76	258.5			206.2	Sep-79	263.8			219.5
Oct-76	322.8	294.4		244.5	Oct-79	299.0			284.2
Nov-76	352.2	324.1		324.4	Nov-79	308.5	288.8		
Dec-76	365.0	345.0		351.37	Dec-79	318.6	126.0		

ADDITIONAL QUESTIONS

12.2. The following table contains average annual values of precipitation and air temperature for various climatological stations at different elevations. Find a relation between average-annual precipitation and elevation, and annual-air temperature and elevation. Find the average annual precipitation and air temperature at elevation 3,000 meters above sea level (masl).

Station number	Elevation (meters asl)	Average annual precipitation (mm)	Average annual air temperature (°C)
1	3,750	—	4.1
2	5,150	—	−6.5
3	1,450	83.3	—
4	3,850	179.2	3.1
5	1,200	80.4	—
6	674	49.1	—
7	1,385	68.9	—
8	1,100	60.8	17.3
9	560	45.6	19.3
10	3,500	164.1	4.1
11	373	38.3	—

12.3. Use the various methods presented for calculation of evapotranspiration (Blaney–Criddle, Jensen–Haise, and pan-evaporation methods) to determine evapotranspiration in mm/day. Compare the results. Assume that the site is located in the southern hemisphere at latitude 40 °S and elevation 1,200 meters above sea level (366 feet) during the month of December. Further assume that the crop of interest is alfalfa. Mean climatological data for the site are given as follows.

Estimated atmospheric pressure = 887 mb

Mean maximum air temperature = 86 °F = 30 °C

Mean minimum air temperature = 53 °F = 11.7 °C

Mean air temperature = 69.5 °F = 20.8 °C

Mean dewpoint temperature = 49 °F = 9.4 °C

Mean vapor pressure = 11.8 mb

Mean wind movement at a height of 366 cm = 128 miles/day = 206 km/day

Mean relative humidity = 48%

Mean extraterrestrial solar radiation = 961 ly/day

Mean cloudless-day solar radiation = 740 ly/day

Mean observed solar radiation = 64 ly/day

Mean net radiation = 382 ly/day

Mean annual cloudless-sky radiation coefficient = 0.33

Mean cloud cover in tenths = 4

Estimated mean soil-heat flux = −6 ly/day

Mean Class A pan evaporation = 0.35 inches/day = 8.9 mm/day

13

Applied Soil Physics: Modeling Water, Solute, and Vapor Movement

INTRODUCTION: MODELING APPROACHES

Variably-saturated-zone mathematical models are useful tools for predicting the extent of subsurface contamination and conducting pre-monitoring studies for the placement of detection devices, in a format understood by field response personnel as well as regulatory entities. Modeling approaches range from simple analytical and semianalytical solutions, to complex numerical codes. The assumptions inherent in each type of model are a key element in understanding the uncertainties associated with each type of approach. The primary emphasis here is the use of simple formulas and comprehensive tables to create a practically oriented and readily usable guide to some of the basic unsaturated-flow problems commonly encountered. In general, three different levels of sophistication are presented in this chapter: **(1)** simple analytical solutions (based upon applicable differential equations) formed by simplifying idealizations of the soil and boundary conditions, and resulting in estimates of flow and contaminant transport; **(2)** semi-analytical methods based upon the concept of soil potential, which provides for both steady-state fluid flow and approximate transient-fluid flow of a contaminant, corresponding to an arbitrary number of contaminant sources; and **(3)** sophisticated numerical models that can account for nonhomogeneous soil, dispersion, diffusion, and chemical processes (e.g., sorption, precipitation, decay, ion exchange, and degradation).

As indicated above, the emphasis in this chapter is on analytical solutions that are readily used to solve problems in unsaturated-zone flow; these solutions are easily used on spreadsheets. It is felt that analytical solutions—in addition to their use as check solutions for numerical models—can provide solutions even when large quantities of unsaturated-zone data are not available. Analytical solutions also provide guidance on the sensitivity of a given solution to various unsaturated-zone properties, thus permitting guidance on where to collect more data or to concentrate data-collection efforts.

Analytical Models

Analytical and semianalytical models are comprised of formulas that represent simplified physical and/or chemical conditions of the real world, based on the complete partial differential equations used in the more sophisticated numerical models. Analytical and semianalytical model solutions are usually linearized versions of the nonlinear partial differential equations of flow in soil (e.g., Richards' equation). Generally, simplified boundary conditions are required (i.e., infinite or semi-infinite areal extent of the soil, or linear-vertical or horizontal boundaries) in order to obtain analytical solutions. Also, the nonlinear partial differential equations are linearized, usually accomplished through the use of mathematical transforms.

Analytical solutions give exact answers for the geometries and soil physical limitations required for their use. They are also used as confirmation for the solutions of more sophisticated numerical models, under the same geometries and physical parameters. Because analytical solutions are less expensive to set up and run—and often require less data than numerical models—they are valuable in assessing simple problems or for use as pre-screening models. Their use under well-defined uncertainty constraints (or in the face of small data sets) is important because often, decisions related to regulatory issues and contaminant-cleanup timing are based on use of analytical models. Regulators, as well as the general public, often find analytical solutions easier to understand than complex numerical solutions.

Nevertheless, very few analytical solutions are available for flow and transport in unsaturated soil. Solutions are available for selected flow and transport problems, such as: determination of water content, matric potential, and unsaturated-hydraulic conductivity in layered soils (López-Bakovic and Nieber 1989); one-dimensional redistribution of moisture (Charbeneau 1989); three-dimensional steady-state and time-dependent moisture distributions from point sources (Bumb et al. 1988; McKee and Bumb 1988); three-dimensional steady-state and time-dependent distribution of nonwetting fluids from point sources (Murphy, Bumb, and McKee 1987); and three-dimensional time-dependent vapor diffusion from an initial contaminant distribution (Lawrence Livermore National Laboratory (LLNL) 1990). The derivation and use of these analytical-model solutions are discussed in detail in this chapter.

These analytical models are chosen from available analytical solutions because they represent basic flow and soil properties that are commonly encountered in typical experiences related to unsaturated flow. It should be noted that both one-dimensional and three-dimensional solutions for vertically homogeneous and layered-flow systems are presented.

Numerical Models

One of the decisions that needs to be made is whether to use an analytical model or a numerical model, to solve a particular problem. Because of the many simplifying assumptions inherent in analytical solutions, there exists some doubt as to the tractability of the solution, and its defensibility in light of those assumptions and simplifications (Javendel, Doughty, and Tsang 1984).

Numerical models are much less burdened by simplifications and assumptions, and are therefore inherently capable of addressing more complicated problems; they require significantly more data, however, and their solutions are still only numerical approximations. Assumptions regarding homogeneity and isotropy are unnecessary in a numerical model, due to its ability to assign nodal or elemental values for many variables of interest. In addition, the capacity to incorporate complex boundary conditions does not require the infinite-areal-extent assumption often needed in analytical models. Other choices, such as the time-step, distance-step, and numerical-solution scheme are chosen by the model user; improper choices of these variables can render the results of the numerical model incorrect and useless.

Several types of numerical models (methods) are generally available for solving unsaturated-flow and transport problems; the two principal ones are the finite-difference method and the finite-element method (Istok 1989). An overview of the current numerical models for solving variably saturated-soil problems is presented in the following sections. Listings of selected numerical models also are presented, but not detailed descriptions. The reader is referred to the references section for more details on individual numerical models.

13.1   ONE-DIMENSIONAL DETERMINISTIC LIQUID-FLOW MODELS

Analytical Models

Analytical solutions to the unsaturated water flow equation (Richards' equation) are based upon homogeneous soil layers, power and/or exponential relation for unsaturated hydraulic conductivity and soil-moisture characteristic curves. Both one-dimensional solutions by López-Bakovic and Nieber (1989) and Charbeneau (1989) and a three-dimensional solution by McKee and Bumb (1988), are derived by assuming Darcian water flow in homogeneous, unsaturated soil by representing Darcy's equation as

$$q = -K \frac{\partial \psi_h}{\partial z} \tag{13.1}$$

where q is the Darcian flux rate, K is the hydraulic conductivity as a function of water content, ψ_h is the total hydraulic potential with respect to a datum, and z is the vertical distance component. We note that ψ_h is represented as the sum of the gravitational potential ψ_z, the matric potential ψ_m, and the pressure potential ψ_p. For unsaturated conditions, ψ_p is zero, while ψ_z is represented by the distance z from an arbitrary datum. Thus, equation 13.1 becomes

$$q = -K \frac{\partial \psi_m}{\partial z} - K \tag{13.2}$$

or

$$q = -K \left(1 + \frac{\partial \psi_m}{\partial z} \right) \tag{13.3}$$

A general analytical solution to equation 13.3 is not available, because both K and ψ_m are a function of volumetric water content θ, and often z; therefore, the equation is nonlinear. If K and ψ are linear functions of θ, then equation 13.3 is linear. Two possibilities exist for linearizing equation 13.3; these include both integral and nonintegral linearization. An analytical solution to equation 13.3 is obtained by simplification, using a Kirchhoff integral transformation (McKee and Bumb 1988):

$$K \frac{\partial \psi_m}{\partial z} = \frac{\partial H}{\partial z} \tag{13.4}$$

where H is defined as the matric-flux potential by Gardner (1958) and Warrick (1974), and is given by

$$H = \int_{-\infty}^{p} K \, dp \tag{13.5}$$

where p is a variable of integration.

López-Bakovic and Nieber (1989) assume an exponential law relating hydraulic conductivity and matric potential of the form

$$K = K_s \exp\left(\alpha \psi_m\right) \tag{13.6}$$

where K_s is saturated hydraulic conductivity, and α is a constant (fitting parameter) related to the moisture-characteristic curve for a given unsaturated soil. Using the exponential relation of equation 13.6, the matric flux potential becomes

$$H = \int_{-\infty}^{\psi} K_s \exp(\alpha p) \, dp = \frac{K_s}{\alpha} [\exp(\alpha p)]_{-\infty}^{\psi} \tag{13.7}$$

Evaluating equation 13.7 gives

$$H = \frac{1}{\alpha} K_s \exp(\alpha \psi_m) = \frac{K}{\alpha} \tag{13.8}$$

McKee and Bumb (1988) assume a power law relating hydraulic conductivity and matric potential of the form

$$K = K_s(S_e)^n \tag{13.9}$$

where S_e is the effective saturation and n is an exponent. S_e is given by Corey (1977) and Bumb (1987) as

$$S_e = \frac{\theta - \theta_r}{\theta_m - \theta_r} = \exp\left(-\frac{\psi_m - \psi_1}{\beta}\right) \tag{13.10}$$

where θ is the volumetric water content, θ_m is the maximum volumetric water content, θ_r is the residual (or irreducible) volumetric water content, and β and ψ_1 are fitting parameters to the moisture-characteristic curve. Using the power-law relation of equation 13.9, the matric flux potential becomes

$$H = \int_{\psi}^{\infty} K_s(S_e)^n \, dp = -\frac{\beta K_s}{n} [(S_e)^n]_{\psi}^{\infty} \tag{13.11}$$

and evaluating equation (13.11) gives

$$H = \frac{\beta}{n} K_s(S_e)^n = \frac{\beta}{n} K \tag{13.12}$$

Comparison of equation 13.8 with equation 13.12 indicates that $\alpha = n/\beta$. Therefore, using either the exponential law or the power law to relate hydraulic conductivity and water content results in essentially the same relation for matric flux potential H, used to linearize the differential equation of unsaturated flow. Substituting the result of H (equation 13.8) into equation 13.3 results in the following linear differential equation

$$q = -\frac{\partial H}{\partial z} - \alpha H \tag{13.13}$$

Solving the transformed equation 13.13 and using the exponential law proposed by López-Bakovic and Nieber (1989), we obtain the following relation:

$$-q \exp(\alpha z) = \frac{\partial[H \exp(\alpha z)]}{\partial z} = \exp(\alpha z)\frac{\partial H}{\partial z} + \exp(\alpha z)\alpha H \tag{13.14}$$

Integrating both sides of equation 13.14 gives

$$H \exp(\alpha z) = -\int^z q \exp(\alpha t) \, dt + c \tag{13.15}$$

and solving for matric flux potential H gives

$$H = -\exp(-\alpha z)\int^z q \exp(\alpha t) \, dt + c \exp(-\alpha z) \tag{13.16}$$

Because we assume the flow to be steady-state, the water-flux rate q is constant throughout the vertical profile; q can be moved outside the integral sign and equation 13.16 is written as

$$H = -q \exp\left(-\alpha z\right) \int^z \exp\left(\alpha t\right) dt + c \exp\left(-\alpha z\right) \tag{13.17}$$

Performing the integration in equation 13.17 gives

$$H = -q \exp\left(-\alpha z\right) \frac{\exp\left(\alpha z\right)}{\alpha} + c \exp\left(-\alpha z\right) \tag{13.18}$$

Simplifying equation 13.18 gives the final expression for matric-flux potential H, as a function of depth z, in the unsaturated-soil profile as

$$H(z) = -\frac{q}{\alpha} + c \exp\left(-\alpha z\right) \tag{13.19}$$

where c is the constant of integration, defined by the boundary conditions for the particular problem being solved.

Analytical solution for a homogeneous soil   A typical unsaturated-soil problem involves computation of matric potential ψ_m, volumetric water content θ, and unsaturated hydraulic conductivity K—all as a function of depth, z. Using figure 13.1 for definition, we assume a homogeneous and isothermal soil having a water table; with z increasing upward from the water table, and the flux rate q upward from the water table, the constant of integration in equation 13.19 can be evaluated. At the water table, $\psi_m = 0$, and $z = 0$, so equation 13.8 becomes

$$H(z = 0) = H_0 = \frac{1}{\alpha} K_s \exp\left(\alpha 0\right) = \frac{K_s}{\alpha} \tag{13.20}$$

Substituting equation 13.20 into equation 13.19 at $z = 0$ gives

$$H_0 = \frac{K_s}{\alpha} = -\frac{q}{\alpha} + c \exp\left(-\alpha 0\right) \tag{13.21}$$

Solving equation 13.21 for c gives

$$c = \frac{q + K_s}{\alpha} \tag{13.22}$$

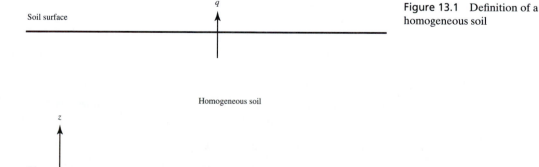

Soil surface

q

Homogeneous soil

z

Water table

Figure 13.1   Definition of a homogeneous soil

Substitution of c into equation 13.19 gives the complete solution for matric-flux potential H for a homogeneous unsaturated soil, as defined by figure 13.1:

$$H(z) = -\frac{q}{\alpha} + \frac{q + K_s}{\alpha} \exp(-\alpha z) \tag{13.23}$$

López-Bakovic and Nieber (1989) derive analytical solutions (based on the above equations) for matric potential, volumetric water content, and unsaturated-hydraulic conductivity as a function of depth z, for the homogeneous soil described in figure 13.1. Their solution for matric potential ψ_m is derived by rearranging equation 13.8 and taking the natural logarithm of both sides:

$$\ln\left[\frac{\alpha H}{K_s}\right] = \ln[\exp(\alpha \psi_m)] = \alpha \psi_m \tag{13.24}$$

Solving for ψ gives the matric potential as a function of H

$$\psi_m = \frac{1}{\alpha} \ln\left[\frac{\alpha H}{K_s}\right] \tag{13.25}$$

Substituting equation 13.23 into equation 13.25 gives the matric potential as a function of depth z above the water table in a homogeneous soil

$$\psi_m(z) = \frac{1}{\alpha} \ln\left\{\frac{\alpha}{K_s}\left[-\frac{q}{\alpha} + \frac{q + K_s}{\alpha} \exp(-\alpha z)\right]\right\} \tag{13.26}$$

López-Bakovic and Nieber (1989) derive an expression for water content similar to the one for matric potential, using the following expression:

$$\frac{K}{K_s} = \left(\frac{\theta}{\theta_s}\right)^n \tag{13.27}$$

where θ is the unsaturated volumetric-water content, θ_s is the saturated volumetric-water content, and n is a dimensionless coefficient related to the pore-size distribution index, as defined by Brooks and Corey (1964). This n is the same as the exponent in equation 13.9. From equation 13.6, equation 13.27 is rewritten as

$$\frac{K}{K_s} = \exp(\alpha \psi_m) = \left(\frac{\theta}{\theta_s}\right)^n \tag{13.28}$$

and by taking the nth root of both sides of equation 13.28, gives

$$\theta = \theta_s \exp\left(\frac{\alpha \psi_m}{n}\right) \tag{13.29}$$

Substituting ψ_m from equation 13.26 into equation 13.29 gives

$$\theta = \theta_s \exp\left(\frac{\alpha}{n}\left\{\frac{1}{\alpha} \ln\left[\frac{\alpha}{K_s}\left(-\frac{q}{\alpha} + \frac{q + K_s}{\alpha} \exp[-\alpha z]\right)\right]\right\}\right) \tag{13.30}$$

which, upon simplification, becomes the volumetric water content with distance z above the water table for a homogeneous soil,

$$\theta(z) = \theta_s \left\{\frac{\alpha}{K_s}\left[-\frac{q}{\alpha} + \frac{q + K_s}{\alpha} \exp(-\alpha z)\right]\right\}^{1/n} \tag{13.31}$$

López-Bakovic and Nieber (1989) also present a solution for hydraulic conductivity by substituting the expression for ψ_m (equation 13.26) into equation 13.6, or

$$K = K_s \exp\left(\alpha\left\{\frac{1}{\alpha} \ln\left[\frac{\alpha}{K_s}\left(-\frac{q}{\alpha} + \frac{q + K_s}{\alpha} \exp\left[-\alpha z\right]\right)\right]\right\}\right) \quad (13.32)$$

which, upon simplification, becomes the hydraulic conductivity with distance z above the water table for a homogeneous soil,

$$K(z) = \alpha\left[-\frac{q}{\alpha} + \frac{q + K_s}{\alpha} \exp\left(-\alpha z\right)\right] \quad (13.33)$$

The above derivation presents equations for a homogeneous soil. These equations are now used to solve for matric potential, volumetric water content, and unsaturated hydraulic conductivity for a layered-soil profile.

Analytical solution for a layered soil   Consider a soil profile comprised of discrete layers, where equation 13.3 applies within each layer [López-Bakovic and Nieber (1989)]. The solution, given by equation 13.19, also applies to each layer; however, the coefficient c in equation 13.19 is unique for each layer, and is determined by the conditions at each interlayer boundary. In addition, the matric potential must be continuous across each boundary.

Referring to figure 13.2, define i as the number of a layer, and $i + 1$ as the number of the layers above layer i. Then the matric-potential continuity across layer boundaries requires that ψ_i be equal to ψ_{i+1} at the boundary between layer i and layer $i + 1$. Therefore, using equations 13.18 and 13.25, it follows [López-Bakovic and Nieber (1989)] that

$$H_{i+1} = \frac{K_{si+1}}{\alpha_{i+1}} \exp\left(\alpha_{i+1} \psi_{i+1}\right) = \frac{K_{si+1}}{\alpha_{i+1}} \exp\left(\alpha_{i+1}\psi_i\right) \quad (13.34)$$

$$H_{i+1} = \frac{K_{si+1}}{\alpha_{i+1}} \exp\left\{\alpha_{i+1}\left[\frac{1}{\alpha_i} \ln\left(\frac{\alpha_i H_i}{K_{si}}\right)\right]\right\} \quad (13.35)$$

$$H_{i+1} = \frac{K_{si+1}}{\alpha_{i+1}} \left[\frac{\alpha_i H_i}{K_{si}}\right]^{\alpha_{i+1}/\alpha_i} \quad (13.36)$$

Again referring to figure 13.2, define L_i as the length from the water table to the boundary between the i and $i + 1$ layer. Using equation 13.35 and equation 13.19, the following

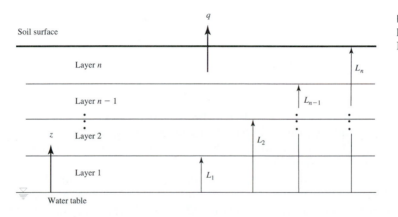

Figure 13.2   Definition of a layered soil (data from López-Bakovic and Nieber 1989)

equation results [López-Bakovic and Nieber (1989)]:

$$-\frac{q}{\alpha_{i+1}} + c_{i+1} \exp\left(-\alpha_{i+1}L_i\right) = \frac{K_{si+1}}{\alpha_{i+1}}\left\{\frac{\alpha_i}{K_{si}}\left[-\frac{q}{\alpha_i} + c_i \exp\left(-\alpha_i L_i\right)\right]\right\} \tag{13.37}$$

The relation between c_{i+1} and c_i is given by

$$\exp\left(\alpha_{i+1}L_i\right)\left\{\frac{q}{\alpha_{i+1}} + \frac{K_{si+1}}{\alpha_{i+1}}\left[\frac{\alpha_i}{K_{si}}\left(-\frac{q}{\alpha_i} + c_i \exp\left\{-\alpha_i L_i\right\}\right)\right]\right\}^{\alpha_{i+1/\alpha_i}} \tag{13.38}$$

knowing that

$$c_i = \frac{q + K_{si}}{\alpha_i} \tag{13.39}$$

In summary, as defined by figure 13.2, the resulting equations for a layered system are:

$$H_i(z) = -\frac{q}{\alpha_i} + c_i \exp\left(-\alpha_i z\right) \tag{13.40}$$

$$\psi_i(z) = \frac{1}{\alpha_i} \ln\left\{\frac{\alpha_i}{K_{si}}\left[-\frac{q}{\alpha_i} + c_i \exp\left(-\alpha_i z\right)\right]\right\} \tag{13.41}$$

$$\theta_i(z) = \theta_{si}\left\{\frac{\alpha_i}{K_{si}}\left[-\frac{q}{\alpha_i} + c_i \exp\left(-\alpha_i z\right)\right]\right\}^{1/n_i} \tag{13.42}$$

$$K_i(z) = \alpha_i\left[-\frac{q}{\alpha_i} + c_i \exp\left(-\alpha_i z\right)\right] \tag{13.43}$$

These equations apply to values of z between L_i and L_{i+1}. The coefficient c_i can be calculated for the appropriate layer from equations 13.38 and 13.39.

As an example of the application of the above equations, consider a soil with two layers, each 60-cm thick; a water table is located 1.2 meters below the soil surface. The properties of the layers are given as

Property	Top layer	Bottom layer
K_s	1×10^{-7} m/s	5×10^{-8} m/s
α	15/m	10/m
n	3	3
θ_s	0.50	0.60

Find the matric potential, volumetric water content, and hydraulic conductivity as a function of depth above the water table, if the following flux rates q are occurring:

$$q = 3.00 \times 10^{-14} \text{ m/s}$$
$$q = 0.0 \text{ m/s}$$
$$q = -3.00 \times 10^{-11} \text{ m/s}$$
$$q = -3.00 \times 10^{-8} \text{ m/s}$$

Note that negative flux rates mean infiltration is occurring; positive flux rates mean evaporation is occurring from the soil surface. The above flux rates were intentionally chosen to be less than the saturated hydraulic conductivities to avoid positive matric potentials, for which the solutions given are not valid.

Results of the example two-layer soil are presented in figure 13.3 for matric potential, figure 13.4 for volumetric water content, and figure 13.5 for hydraulic conductivity. The effect of the layering is clearly visible in figures 13.4 and 13.5, but only slightly visible in figure 13.3.

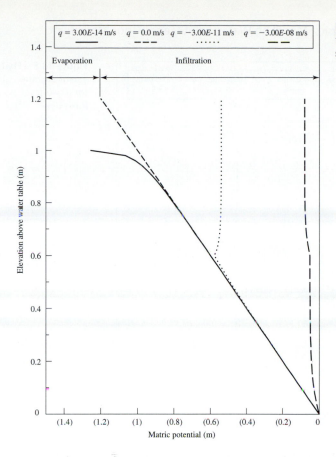

Figure 13.3 Matric potential versus elevation above the water table for a two-layer system with variable flux rates

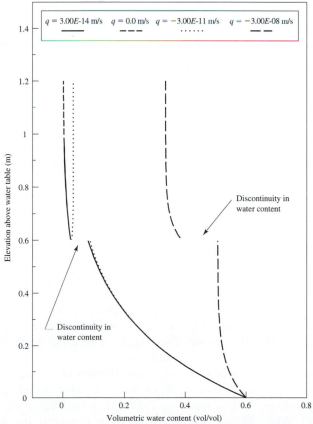

Figure 13.4 Volumetric water content versus elevation above the water table for a two-layer system with variable flux rates

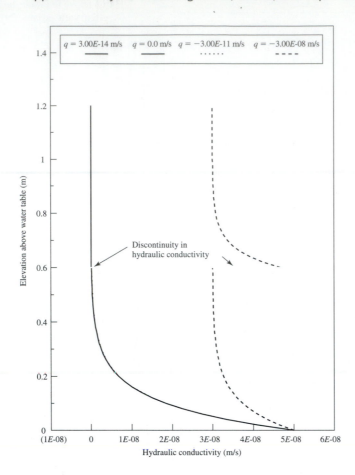

Figure 13.5   Hydraulic conductivity versus elevation above the water table for a two-layer system with variable flux rates

This is because matric potential is continuous across soil interfaces, whereas water content and hydraulic conductivity are both highly discontinuous.

The flux rate of zero in figure 13.3 gives a straight line with a slope of 1 between matric potential and elevation. Curves above this line indicate infiltration (negative or downward flux rates), and curves below this line indicate evaporation (exfiltration) associated with positive or upward flux rates. It should be noted that there is a limit to the positive flux rate that can be used. Mathematically, this limit to exfiltration is obtained when the matric flux potential H is set to zero in equation 13.40. Higher exfiltration rates require that H is less than zero, which is not physically possible. If water-vapor flow were included in the model in addition to liquid-water flow, then it would be possible to exceed this mathematical limit slightly.

In summary, the above analytical solutions for soil-water content, matric potential, and unsaturated hydraulic conductivity are possible for a layered soil between the water table and the ground surface, if appropriate boundary conditions for each layer are used. As with most analytical solutions, the above equations apply to steady-state flux rates with neither water extractions nor water additions to individual layers within the soil profile. Therefore, these equations are not appropriate for transient conditions and are most likely to apply to "average" conditions within the profile. The effects of layering using these equations can, however, be investigated for a variety of problems.

Analytical solution for liquid-moisture redistribution in a homogeneous soil   Redistribution refers to the continued movement of water or other liquid through a soil profile after infiltration, irrigation, or other input has ceased at the surface of the soil (Jury, Gardner, and Gardner 1991). When surface storage of liquid is depleted, the movement of liquid added to the soil does not immediately cease, but can continue for a long time as it redistributes within the profile. The major difference between redistribution and infiltration is that the wetting front continues to move as a result of water coming from the transmission zone, rather than from flux across the surface of the soil. At the same time, a significant amount of liquid is lost from the profile due to evaporation. According to Charbeneau (1989), a number of questions are directly related to redistribution processes. These include: **(1)** the length of time for the soil profile to drain; **(2)** the water content and matric potential near the surface of the soil; **(3)** the recharge (or flux) at depth; **(4)** the influence of evaporation on each wetting event; **(5)** the influence of multiple-wetting events; and **(6)** the effects of spatial variability in the physical parameters.

There have been relatively few analytical or semi-analytical solutions for liquid redistribution, compared to those for infiltration (Charbeneau 1989). Gardner, Hillel, and Benyamini (1970a) consider approximate solutions to the unsaturated-flow equation to describe the soil-water content above the initial wetting front, as a function of time. They also calculate the cumulative drainable water for a rectangular-redistribution profile. Gardner, Hillel, and Benyamini (1970b) show that redistribution reduces evaporation, and they develop expressions for obtaining an estimate of the amount of reduction in evaporation due to redistribution when the redistribution rate is known. Dagan and Bresler (1983) derive a rectangular-profile model for vertical redistribution of liquid in a homogeneous soil column. This approximate solution includes both the time distribution of soil-water content, and flux. Morel-Seytoux (1987) formulates a redistribution model using a kinematic approach, by assuming that the hydraulic conductivity function can be represented by a power function. Charbeneau (1989) summarizes two liquid-redistribution models—a rectangular and kinematic approach—with and without evaporation. These models are discussed in detail next.

For purposes of developing the analytical solution to one-dimensional (vertical) liquid-moisture redistribution, assume that a homogeneous soil lies in a horizontal plane x and y, with a vertical coordinate z, directed downward as shown in figure 13.6. Let the flux rate q be positive upward; let θ be the volumetric water content; ψ_m be the matric potential; and K the unsaturated hydraulic conductivity. Both ψ_m and K are functions of volumetric water content θ. The two basic equations are Darcy's equation (equation 13.3, above) and the continuity equation given by

$$\frac{\partial \theta}{\partial t} = -\frac{\partial q}{\partial z}$$

(13.44)

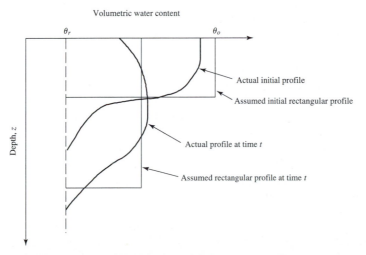

Figure 13.6   Actual and approximate rectangular moisture profile (data from Charbeneau 1989)

Substituting equation 13.3 into equation 13.44 gives

$$\frac{\partial \theta}{\partial t} = \frac{\partial}{\partial z}\left(K\frac{\partial \psi_m}{\partial z}\right) + \frac{\partial K}{\partial z} \tag{13.45}$$

Assuming smooth, single-valued functions of θ and ψ_m versus z, the chain rule gives

$$\frac{\partial \theta}{\partial t} = \frac{\partial}{\partial z}\left(K\frac{d\psi_m}{d\theta}\frac{\partial \theta}{\partial z}\right) + \frac{\partial K}{\partial z} \tag{13.46}$$

which is Richards' equation. We further recognize that the diffusivity D, also a function of volumetric water content, is given by

$$D = K\frac{d\psi_m}{d\theta} \tag{13.47}$$

Thus, Richards' equation can also be written in terms of diffusivity as

$$\frac{\partial \theta}{\partial t} + \frac{\partial}{\partial z}\left(D\frac{\partial \theta}{\partial z}\right) + \frac{\partial K}{\partial z} \tag{13.48}$$

For purposes of this analytical solution, a power law relating hydraulic conductivity and matric potential of the form used in equations 13.9 and 13.10 is used to give physically based parameters for the redistribution solution. These power equations can be written as

$$K = K_s(S_e)^n \tag{13.49}$$

and

$$S_e = \frac{\theta - \theta_r}{\theta_m - \theta_r} = \left(\frac{\psi_b}{\psi_m}\right)^\lambda \tag{13.50}$$

where: K is hydraulic conductivity as a function of volumetric-water content θ; K_s is saturated-hydraulic conductivity; n is an exponent; S_e is effective saturation; θ_r is residual or irreducible water content that occurs as K approaches zero; θ_m is the maximum volumetric water content that is close to the value of porosity, unless entrapped air reduces its value; ψ_m is matric potential; ψ_b is the air-entry (or bubbling) pressure, assumed to be the matric potential at which the largest pore in the soil begins to drain; and λ is an exponent. Brooks and Corey (1964) show that the exponent n in equation 13.49 is related to the pore-size distribution index λ, through $3 + 2/\lambda$. Small values of λ correspond to a wide range of pore sizes, while large values correspond to a narrow distribution. Bumb et al. (1988) present typical values of λ, along with typical values of porosity θ_r, and air-entry pressure ψ_b, for soil-texture data provided by Rawls, Brakensiek, and Sarabi (1982). These soil variables are treated in greater detail in later sections of this chapter.

During infiltration, the volumetric water content does not reach full saturation due to entrapped air. Brooks and Corey (1964) present values of θ_m (the maximum water content) as a result of entrapped air in various soils. These values are presented in a subsection of section 13.5, following. Based on empirical evidence, Bouwer (1966) suggests that the maximum effective-hydraulic conductivity is about one-half the saturated hydraulic conductivity for values of θ_m. This concept can be presented mathematically as

$$K_e = \frac{K_s}{2} \tag{13.51}$$

where K_e is the effective hydraulic conductivity. According to equation 13.49—assuming that K is equal to K_e—the corresponding effective saturation is given by

$$S_e = \left[\frac{1}{2}\right]^{1/n} \tag{13.52}$$

The analytical solutions presented here require that the antecedent volumetric-water content is known. For most unsaturated soil, the antecedent volumetric-water content is closely related to the average annual flux rate. The analytical solutions that follow assume that the antecedent water content of the soil is equal to the annual flux rate i_a, assuming uniform and steady infiltration. Rearranging equation 13.49 and assuming K is uniform, a steady infiltration rate of i_a gives

$$S_{ea} = \left[\frac{i_a}{K_s}\right]^{1/n} \tag{13.53}$$

where S_{ea} is the effective antecedent-water content. Because n usually has a value of 3 or greater (Brooks and Corey 1964; McKee and Bumb 1988), the value of S_{ea} is relatively insensitive to estimated values of i_a.

Charbeneau (1989) presents two approximate-profile models to simulate liquid-moisture redistribution in an unsaturated soil after infiltration has ceased. These simple profile solutions are the rectangular profile, taken from Gardner, Hillel, and Benyamini (1970a) and Dagan and Bresler (1983); and kinematic profiles (or Buckley–Leverett models), taken from Charbeneau (1984) and Morel-Seytoux (1987). These two approximate moisture-redistribution analytical solutions are discussed in detail next.

The rectangular profile   The simplest shape of a soil-moisture redistribution profile is the rectangular profile, shown in figure 13.6. Using this profile, it is assumed that the initial soil-water content within the profile—prior to infiltration from the wetting event—corresponds to either residual saturation θ_r or effective antecedent saturation S_{ei}. The case of an initially dry soil θ_r is examined first.

The initial effective water content S_{ei} for the rectangular profile immediately behind the wetting front (after the wetting event) is found by comparing the average infiltration rate i during the wetting event, to the natural effective hydraulic conductivity K_e. If the average infiltration rate exceeds K_e, then the natural effective-saturation conditions occur, and S_{ei} is equal to S_e. Otherwise, for the average infiltration rate i equal to K, and using equation 13.49, gives

$$S_{ei} = \left[\frac{1}{2}\right]^{1/n}; \quad i > K_e \tag{13.54}$$

or

$$S_{ei} = \left[\frac{i}{K_s}\right]^{1/n}; \quad i < K_e \tag{13.55}$$

With S_{ei} known from either equation 13.54 or equation 13.55, the initial depth of the wetting front z_{fi} as a result of the wetting event, is given for the rectangular profile by

$$z_{fi} = \frac{I}{S_{ei}(\theta_m - \theta_r)} \tag{13.56}$$

where I is the cumulative infiltration depth associated with the wetting event.

As the wetting front moves deeper into the soil, the water content of the rectangular profile decreases with time. Neglecting evaporation and other losses, and letting S_{ea} equal zero, I is equal to $(\theta_m - \theta_r)S_e z_f$, with S_e and z_f both functions of time. Letting i be the rate of change in cumulative infiltration with respect to time, the derivative of I with respect to time is given by

$$i = \frac{dI}{dt} = S_e(\theta_m - \theta_r)\frac{dz_f}{dt} + z_f(\theta_m - \theta_r)\frac{dS_e}{dt} \tag{13.57}$$

Additionally, if the initial water content of the soil prior to wetting is assumed to be at residual, then

$$\frac{dz_f}{dt} = \frac{K(S_e)}{S_e(\theta_m - \theta_r)} \tag{13.58}$$

Using $z_f = I/S_e(\theta_m - \theta_r)$ the decrease in water content as the front moves deeper must satisfy

$$\frac{I}{S_e}\frac{dS_e}{dt} + K_s(S_e)^n = 0 \tag{13.59}$$

Integration of equation 13.59 and application of the initial condition $S_e = S_{ei}$ from either equation 13.54 or equation 13.55, gives the deterministic model to predict S_e as a function of time as

$$S_e = \frac{1}{\left[(S_{ei})^{-n} + \dfrac{nK_s t}{I}\right]^{1/n}} \tag{13.60}$$

A related relation includes the flux rate q (Darcian velocity) within the moisture-wetting front, which is numerically equal to equation 13.49, or $q = K_s(S_e)^n$. The Darcian flux, then, is given by

$$q = \frac{K_s}{(S_{ei})^{-n} + \dfrac{nK_s t}{I}} \tag{13.61}$$

Using $z_f = I/S_e(\theta_m - \theta_r)$, the deterministic model of the depth of the wetting front as a function of time is

$$z_f = \frac{I}{(\theta_m - \theta_r)}\left[(S_{ei})^{-n} + \frac{nK_s t}{I}\right]^{1/n} \tag{13.62}$$

Kinematic profiles Charbeneau (1984; 1989) has proposed using Buckley–Leverett models, or kinematic-wave models, to represent a second type of soil-moisture redistribution model. Figure 13.7 shows the progression of a soil-moisture wave moving downward. Figure 13.7(a) is the shape of the assumed wave immediately after the end of the wetting event. This wave consists of a rectangular wave similar to that shown in figure 13.6; as with the rectangular profile, a sharp wetting front is evident. After a short period of time, drainage within the profile causes increasing soil-water content with depth, resulting in a curved portion of the wave as shown in figure 13.7(b). Figure 13.7(b) also indicates that a constant soil-water content (plateau in the wave) still remains. This is because the draining part of the wave has not yet reached the wetting front, and is separated from the front by the plateau part of the wave. Figure 13.7(c) shows the wave at a later time after drainage has caught up with the wetting front. In this case, the plateau is no longer present.

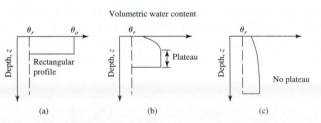

Figure 13.7 Kinematic wave moisture profiles (data from Charbeneau 1989)

The basic assumption in the application of kinematic models is that pressure gradients are negligibly small, and therefore, equation 13.3 (Darcy's equation) and equation 13.44 (continuity equation) become

$$(\theta_m - \theta_r) \frac{\partial S_e}{\partial t} + \frac{dK}{dS_e} \frac{\partial S_e}{\partial z} = 0 \tag{13.63}$$

Equation 13.63 can be solved using the method of characteristics (Sisson, Ferguson, and Van Genuchten 1980). According to the method of characteristics, S_e is constant along paths that satisfy

$$\frac{dz}{dt} = \frac{1}{(\theta_m - \theta_r)} \frac{dK}{dS_e} \tag{13.64}$$

Because dK/dS_e is only a function of S_e, and S_e is constant along each characteristic path (see figure 13.8), the images of these paths are straight lines in the z-t plane. Sisson, Ferguson, and Van Genuchten (1980) show that if the soil-water content changes abruptly at the wetting front (from $S_e = S_{ei}$ to $S_e = 0$) during redistribution, equation 13.64 is integrated and rearranged to determine the soil-water content as a function of z and t. Assuming the power law (equation 13.49), the integration of equation 13.64 gives

$$\frac{z}{t} = \int \frac{1}{\theta_m - \theta_r} \frac{dK}{dS_e} = \frac{nK_s(S_e)^{n-1}}{\theta_m - \theta_r} \tag{13.65}$$

Solving equation 13.65 for S_e (Sisson, Ferguson, and Van Genuchten 1980; Charbeneau 1984; 1989), gives the deterministic model to predict S_e versus time and depth for the kinematic model, as long as the profile remains continuous and no wetting front is encountered. This solution is given by

$$S_e = \left[\frac{(\theta_m - \theta_r)z}{nK_s t} \right]^{1/n-1} \tag{13.66}$$

where t is measured from the time at which the wetting event ends. As with the rectangular profile, the kinematic profile of Darcian flux is given by $q = K_s(S_e)^n$, or as a function of z and t by

$$q = \left[\frac{1}{K_s} \left(\frac{(\theta_m - \theta_r)z}{nt} \right)^n \right]^{1/n-1} \tag{13.67}$$

For the kinematic profile, the drainable water W in storage above any depth z is given by

$$W = (\theta_m - \theta_r) \int_0^z S_e \, dz = \frac{(n-1)z(\theta_m - \theta_r)}{n} S_{ez} \tag{13.68}$$

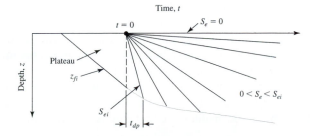

Time, t

$t = 0$

$S_e = 0$

Depth, z

Plateau

z_{fi}

$0 < S_e < S_{ei}$

S_{ei}

t_{dp}

Figure 13.8   Characteristic plane of the kinematic model (data from Charbeneau 1989)

where S_{ez} is the effective water content above depth z in the soil. Equations 13.63 through 13.68 are appropriate as long as the kinematic profile remains continuous and no wetting fronts are encountered (Charbeneau 1989). At the wetting front, the water-content gradient becomes large (infinitely large in the kinematic profile), and therefore, equation 13.63 is no longer valid and the rate of change of the wetting front with respect to time becomes

$$(\theta_m - \theta_r)S_{ei}\frac{dz_f}{dt} = K_s(S_{ei})^n \tag{13.69}$$

Referring to figure 13.7, two separate cases describing the kinematic profile are identified. Figure 13.7(b) shows a plateau region between the draining upper part of the profile and the wetting front. This implies that S_{ei} is constant in equation 13.69, and the rate of movement of the wetting front also is constant. Therefore, the image of the wetting front in the z-t plane in figure 13.8 is a straight line; this is shown in figure 13.8 as the lower boundary of the plateau. At some time (designated t_{dp} in figure 13.8), the plateau has disappeared from the profile and only the wave behind the wetting front remains as shown in figure 13.7. In this case, the moisture arriving at the wetting front decreases with time, and the rate of movement of the wetting front also decreases (see figure 13.8). Combining equations 13.65 and 13.69 gives

$$\frac{dz_f}{dt} = \frac{K_s}{(\theta_m - \theta_r)}(S_e)^{n-1} = \frac{z_f}{nt} \tag{13.70}$$

Equation 13.70 can be integrated to give the following at the initial point (z_{fdp}, t_{dp}), which is the point where the plateau disappears

$$\frac{t}{t_{dp}} = \left(\frac{z_f}{z_{fdp}}\right)^{n-1} \tag{13.71}$$

The location of the initial point (z_{fdp}, t_{dp}) needs to be found in order to provide a method of calculating the time and location where the plateau disappears. Charbeneau (1989) shows that the initial point lies at the intersection of the wetting-front path originating from $z = z_{fi}$ at $t = 0$, and the characteristic with water content S_{ei} originating from the ground surface at $t = 0$ (see figure 13.8). These initial points are

$$z_{fdp} = \frac{n}{n-1}z_{fi} \tag{13.72}$$

$$t_{dp} = \frac{(\theta_m - \theta_t)z_{fi}}{(n-1)K_s(S_{ei})^{n-1}} \tag{13.73}$$

where z_{fi} remains as given by equation 13.56.

To demonstrate the use of the above equations, find the rectangular and kinematic profiles at times of 8, 48, and 240 hours, for a sandy loam as defined by Rawls, Brakensiek, and Saxton (1982). The sandy loam is assumed to have the following characteristics: $\theta_m = 0.453$; $\theta_r = 0.041$; $K_s = 2.59$ cm/hr; $\lambda = 0.378$; and $n = 3 + 2/\lambda = 8.29$ (Brooks and Corey 1964). Also assume that a wetting event occurs, consisting of water that infiltrates at a rate of 1.0 cm/hr for 4 hours, giving $I = 4$ cm. We need to find the moisture flux rate q for each of the rectangular and kinematic profiles at a depth of 30 cm below ground surface.

The moisture profiles for both the rectangular and kinematic cases must know S_{ei} and z_f for the wetting event. Because i (1.0 cm/hr) is less than K_e (1.3 cm/hr), S_{ei} is given by equation 13.55 as 0.891. The initial depth of the wetting front z_{fi} is calculated from equation 13.56 as 11.1 cm. Values of S_e and z for the rectangular profile are given by equations 13.60 and 13.62, respectively. For the kinematic profiles, the first step is to determine the time and depth where the plateau disappears, to check to see which equations apply. Using equation 13.71, t_{dp} is calculated as 0.56 hrs, and using equation 13.72, z_{fdp} is calculated as 12.62 cm; therefore, for times of 8, 48, and 240 hours, the plateau has disappeared. Thus, the depth of the wetting front z_f at various times is calculated using equation 13.71, and S_e for each of the z_f values is calculated using equation 13.66. After establishing the depth and effective saturation of the kinematic profile at the wetting front for the three times, values of S_e behind the wetting front are calculated using equation 13.66. Figure 13.9 presents the results for a sandy loam soil for both the rectangular and kinematic profiles.

The moisture flux rate (Darcian velocity) for the sandy loam at a depth of 30 cm below the ground surface is calculated for the rectangular profile using equation 13.61, and for the kinematic profile using equation 13.67. First, the time to reach the 30-cm depth (as well as the effective saturation at that depth) is calculated for both profiles. For the rectangular profile, the time to reach 30 cm is calculated using equation 13.62 and S_e at 30 cm is calculated from $z_f = I/[S_e(\theta_m - \theta_r)]$. For the kinematic profile, the time to reach the 30-cm depth is calculated using equation 13.71 and S_e is calculated using equation 13.66. Figure 13.10 shows the flux rates at the 30-cm depth for the sandy loam soil.

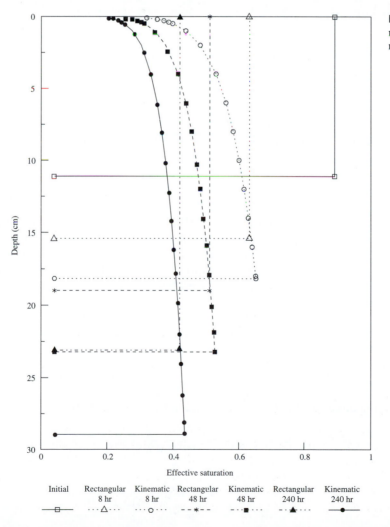

Figure 13.9   Example of rectangular and kinematic moisture profiles

Initial	Rectangular 8 hr	Kinematic 8 hr	Rectangular 48 hr	Kinematic 48 hr	Rectangular 240 hr	Kinematic 240 hr
—□—	···△···	···○···	– – ✳ – –	– ·■· –	– ·▲· ··	—●—

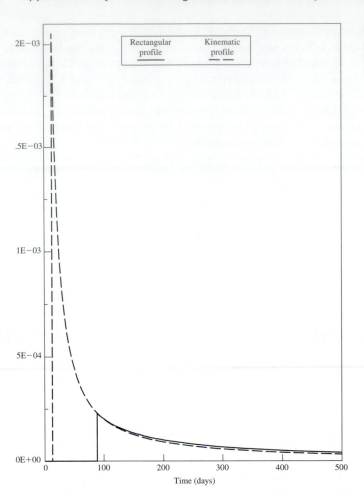

Figure 13.10 Flux rates at 30-cm depth for rectangular and kinematic profiles with an initial water content at residual saturation

Up to this point, we have assumed that the wetting event and the subsequent moisture redistribution occurred in soil whose initial water content was at residual saturation. A more realistic approach is to estimate an initial soil-water content within the soil profile that is higher than residual saturation. Equation 13.53 provides an estimate of antecedent soil-water content based on average annual recharge. For the rectangular model, the water balance across the wetting front is calculated, where the soil-water content behind the wetting front is S_e, and water content ahead of the wetting front is the antecedent moisture given by S_{ea}. In a control volume, the one-dimensional water balance is given by

$$(\theta_m - \theta_r)z_f \frac{dS_e}{dt} + (\theta_m - \theta_r)(S_e - S_{ea}) \frac{dz_f}{dt} + K(S_{ea}) = 0 \qquad (13.74)$$

Combining equation 13.74 with the speed of the wetting front and noting that

$$z_f = \frac{I}{(\theta_m - \theta_r)(S_e - S_{ea})} \qquad (13.75)$$

gives

$$\frac{I}{S_e - S_{ea}} \frac{dS_e}{dt} = K(S_e) = 0 \qquad (13.76)$$

While equation 13.76 looks similar to equation 13.59, it is no longer possible to determine a general analytical solution in terms of simple functions. Using separation of variables, S_e is written as a function of t, such that

$$\frac{K_s}{I} t = \int_{S_e}^{S_{ei}} \frac{dx}{(x - S_{ea})x^n} \qquad (13.77)$$

Equation 13.77 is easily evaluated numerically, by assuming a reasonably small value of dx (say 0.001 or 0.0001) and letting x take on values slightly larger than S_{ea} to S_{ei}, and presenting the results in a table. The value of the initial time at S_e is obtained by looking up the value of the integral in the table and then solving for t. After S_e is known, the flux rate q is calculated from equation 13.49 $[q = K_s(S_e)^n]$.

For the kinematic profile, the wetting front initially follows a straight path at the base-characteristic plane, with drainage characteristics given by equation 13.65. One approach to solving the kinematic problem with antecedent moisture is to write the water-balance equation in the vicinity of the wetting front as

$$\frac{dW}{dt} + q(z_f, t) = 0 \qquad (13.78)$$

where W is the water-storage depth within the profile given by equation 13.68. Recognizing that the water content remains constant at depth z_f up to time t, equation 13.78 is integrated at its antecedent water content to give

$$W(z_f, t) - W(z_f, 0) + Q_a = 0 \qquad (13.79)$$

where Q_a is the cumulative drainage depth from the profile at time t, with the soil-water content constant at its antecedent value:

$$Q_a = K_s(S_{ea})^n t \qquad (13.80)$$

The water content (within the kinematic profile to depth z_f at the beginning of the re-distribution process) consists of the moisture present at its antecedent value, plus the moisture added through infiltration. Thus, equations 13.79 and 13.80 become

$$(\theta_m - \theta_r)S_{ea} z_f + I = W(z_f, t) + K_s(S_{ea})^n t \qquad (13.81)$$

Equation 13.81 states that the sum of the drainable water at the antecedent water content (down to depth z_f) plus the infiltration added to the profile, is equal to the drainable water depth above the wetting front at time t, plus the amount of soil-water that has drained (from the profile above z_f) during the time period since the end of the infiltration event. Combining equations 13.66, 13.68, and 13.81 gives

$$(\theta_m - \theta_r)S_{ea} z_f + I = \frac{(n-1)z_f(\theta_m - \theta_r)}{n} \left(\frac{z_f(\theta_m - \theta_r)}{nK_s t} \right)^{1/n-1} + K_s(S_{ea})^n t. \qquad (13.82)$$

Equation 13.82 provides the kinematic wetting-front depth as a function of time for the antecedent water content case, and is valid once the plateau has disappeared from the profile.

As an application of the above equations for antecedent-moisture conditions, consider the same sandy loam as above, with a cumulative infiltration of 4 cm over a period of 4 hours. We want to find the time-history of water flux at a depth of 150 cm if the average annual recharge is 40 cm, using both the rectangular- and kinematic-profile assumptions. From equation 13.53, S_{ea} is calculated to be 0.465. For the rectangular profile, equation 13.75 shows that

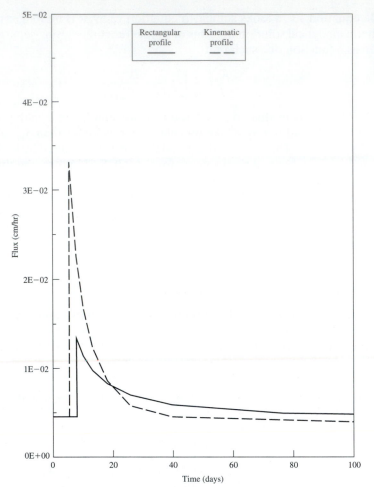

Figure 13.11 Flux rates at 150-cm depth for rectangular and kinematic profiles with an antecedent water content caused by average annual recharge

the wetting front reaches $z_{zt} = 150$ cm with a water content $S_e = S_{ea}I/(\theta_m - \theta_r)z_{zt} = 0.530$. With $S_e = 0.530$ in equation 13.77, the corresponding time for the wetting front to reach 150 cm is 188 hours or 7.8 days. The flux rate is found from $q = K_s(S_e)^n$, and the maximum flux is 0.0134 cm/hr. Figure 13.11 shows the flux rate as a function of time for the rectangular profile.

The corresponding results for the kinematic profile are also shown in figure 13.11. According to equation 13.82, the wetting front reaches the 150-cm depth at 133 hours or 5.5 days, and the maximum flux rate is, according to equation 13.67, 0.0331 cm/hr. Figure 13.11 indicates that most of the drainage occurs earlier for the kinematic model than for the rectangular model.

Other analytical models Other analytical and semi-analytical models are also available for use in the unsaturated zone. These models are summarized in table 13.1, along with the model source and reference. The analytical solutions presented in the models given in table 13.1 solve problems generally similar to the one-dimensional solutions presented above. As with most analytical solutions, the models assume that the input-flux term q is constant over the time of interest in the solution. A description of many of the one-dimensional analytical models presented in table 13.1 can be found in van der Heijde (1994), who compiled a description of over 90 unsaturated-flow models. The reader is referred to the individual reference for details on a particular analytical model.

TABLE 13.1   Other Analytical and Semianalytical Liquid-Flow Models

Model name	Source
FLO	National Hydrology Research Institute, Inland Waters Directorate, Ottawa, Ontario, Canada—Vandenberg (1985)
SOILMOP	Colorado State University, Department of Civil Engineering—Ross and Morel-Seytoux (1982), Morel-Seytoux (1979)
HSSWDS (Hydraulic Simulation of Solid Waste Disposal Sites)	EPA—Perrier and Gibson (1982)
HELP (Hydrologic Evaluation of Landfill Performance)	EPA—Schroeder and others (1994a, b)
B&W analytical solution	Broadbridge and White (1988)
McWhorter analytical solution	McWhorter and Sunada (1990), McWhorter and Kueper (undated)
Warrick analytical solution	Warrick et al. (1971; 1990; 1991)
Fokas analytical solution	Fokas and Yortsos (1982)

Numerical Models

In numerical models, a discrete solution is obtained in both space and time by using numerical approximations of the governing partial differential equations. Because the numerical solutions are only approximations, the conservation of mass and accuracy in prediction are not always assured. Therefore, numerical models need to be verified by either field measurements or comparison to analytical solutions. Because the unsaturated-zone solutions involve nonlinear equations, sophisticated solution techniques are often required.

The primary solution techniques used for approximating the spatial components of the governing flow equations in the unsaturated zone are: **(1)** finite-difference methods (FDM); **(2)** integral finite-difference methods (IFDM); and **(3)** finite-element methods (FEM). In most models, time is approximated by finite-difference techniques resulting in explicit, implicit, or fully implicit solution schemes. In the FDM, the solution is obtained by approximating the derivatives of the differential equations. In the FEM approach, integral equations are formulated first, followed by numerical evaluation of the integrals over the flow domain.

There are many considerations in selecting a numerical simulation model for the unsaturated zone. Simulating flow in nearly saturated-soil systems requires expression of Richards' equation in terms of hydraulic head, matric-potential head, or suction head, especially when parts of the modeled system become fully saturated. This application of Richards' equation causes significant convergence problems when simulating an infiltration front in soil where the initial soil-water content is at residual saturation. Also, significant mass balance problems can occur when site-specific conditions result in highly nonlinear model relations. Other issues that should be addressed in selecting an unsaturated-zone model for flow simulation are: possible needs for double-precision versus single-precision variables; the time-stepping approach; the definition of intercell conductance; and the method in which steady-state simulation is achieved.

The International Ground Water Modeling Center (IGWMC) has identified, compiled, and published a description of over 90 unsaturated-zone models (van der Heijde 1994). The compilation includes models for: flow only; flow and solute transport; solute transport requiring a given head distribution; flow and heat transport; and flow, solute and heat transport in the unsaturated zone. Table 13.2 summarizes the numerical flow-only models documented by the IGWMC (van der Heijde 1994), along with the model source and reference.

TABLE 13.2 Numerical Liquid-Flow Models

Model name	Source
UNSAT2	Department of Hydrology and Water Resources, University of Arizona, Tucson, Arizona—Davis and Neuman (1983)
TRUST	Lawrence Berkeley Laboratory, Earth Sciences Division, University of California, Berkeley, California—Reisenauer et al. (1982)
FLUMP	Lawrence Berkeley Laboratory, Earth Sciences Division, University of California, Berkeley, California—Narasimhan, Neuman, and Witherspoon (1978)
MUST (Model for Unsaturated flow above a Shallow water Table)	International Institute for Hydraulic and Environmental Engineering, Delft, The Netherlands de Laat (1985)
UNSAT1D	Battelle Pacific Northwest Laboratory, Richland, Washington—Bond, Cole, and Gutknech (1984)
SWACROP (Soil WAter and CROP production model)	Winand Staring Centre, Department of Agrohydrology, Wageningen, The Netherlands—Wesseling et al. (1989)
SEEPV	Water, Waste and Land, Inc., Fort Collins, Colorado—Davis (1980)
The One-Dimensional Princeton Unsaturated Code	Princeton University, Princeton, New Jersey—Celia, Bouloutas, and Zarba (1990)
FEMWATER/FECWATER	Pennsylvania State University, University Park, Pennsylvania—Yeh and Ward (1980), Yeh and Strand (1982)
UNSAT-1*	USDA Salinity Laboratory, University of California at Riverside, Riverside, California—van Genuchten (1978)
INFIL*	Institute de Mecanique de Grenoble, St. Martin D'Heres, France—El-Kadi (1983)
GRWATER	Department of Civil Engineering, Colorado State University, Fort Collins, Colorado—Kashkuli (1981)
UNSAT-H	Battelle Pacific Northwest Laboratory, Richland, Washington—Fayer and Gee (1985)
INFGR	Oak Ridge National Laboratory, Environmental Sciences Division, Oak Ridge, Tennessee—Craig and Davis (1985)
FLOWVEC	Simons, Li and Associates, Newport Beach, California—Li, Eggert, and Zachman (1983)
LANDFIL	Department of Civil and Environmental Engineering, Rutgers University, New Brunswick, New Jersey—Korfiatis (1984)
WATERFLO	Soil Science Department, University of Florida, Gainesville, Florida—Nofziger (1985)
SEEP/W (PC-SEEP)	Geo-Slope Programming, Ltd, Calgary Alberta, Canada—Krahn, et al. (1989)
SIMGRO (SIMulation of GROundwater flow and surface water levels)	Institute for Land and Water Management Research, Wageningen, The Netherlands—Querner (1986)
UNSAT	New Mexico Institute of Mining and Technology, Socorro, New Mexico—Khaleel and Yeh (1985)

Source: Summarized from van der Heijde (1994).
*Code and documentation available from the IGWMC, Colorado School of Mines, Golden, Colorado 80401

13.2   THREE-DIMENSIONAL DETERMINISTIC LIQUID- AND VAPOR-FLOW MODELS

Analytical Models

Three-dimensional analytical solution for liquid-water content from a point source leak: Linearization technique   As we have seen in section 13.2, the governing equation for liquid-moisture flow in the vadose zone derives from Richards' (1931) equation. Among other parameters, Richards' equation includes matric potential, volumetric water content, and unsaturated hydraulic conductivity as a function of water content. Matric potential (or capillary pressure) is related to volumetric water content through the moisture-characteristic curve. Due to interrelation between matric potential and volumetric water content, as well as hydraulic conductivity as a function of water content, obtaining solutions to Richards' equation is not easy, even with the help of numerical methods. Because of these difficulties, flow in the unsaturated zone is not well understood by practicing engineers and soil and environmental scientists. McKee and Bumb (1988) and Bumb et al. (1988) describe a computer model that uses an analytical solution to Richards' equation, applying an exponential function to describe the moisture-characteristic curve as shown in equation 13.10. This type of exponential function, along with other empirical forms for relating matric potential and volumetric water content are described here. Bumb, Murphy, and Everett (1992) compare three of the functional forms for representing moisture-characteristic curves and recommend use of a particular functional form.

A typical plot relating matric potential and volumetric water content is presented in figure 13.12. The two variables θ_m and θ_r are approximately established, as shown in the

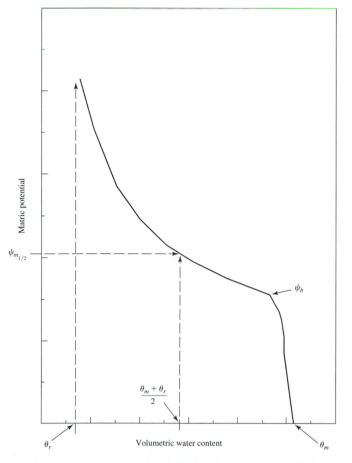

Figure 13.12   Typical soil moisture-characteristic curve

figure. The residual water content θ_r specifies the maximum amount of water in the soil that will *not* contribute to liquid flow, due to blockage from the flow paths or strong adsorption to the solid phase. Mathematically, θ_r is defined as the soil-water content at which both $d\theta/d\psi_m$ and K go to zero when ψ_m becomes large. Therefore, the residual-water content is an extrapolated (or fit) value, and does not necessarily represent the smallest possible water content in a soil. This is especially true for arid regions, where vapor-phase transport can dry the soil to water contents well below θ_r. The maximum-water content θ_m denotes the maximum volumetric water content of a soil. This maximum water content is not equated to the porosity of the soil θ_s for field conditions, because maximum field-water contents are generally 5 to 10 percent smaller than the porosity, due to entrapped air. Therefore, according to Van Genuchten, Leij, and Yates (1991), θ_r and θ_m are viewed as essentially empirical constants in soil moisture-characteristic curves, and hence, without much physical meaning.

Several expressions have been proposed to relate matric potential ψ_m to volumetric-water content θ or more commonly, effective saturation S_e, as given by

$$S_e = \frac{\theta - \theta_r}{\theta_m - \theta_r} \tag{13.83}$$

where θ is the volumetric water content, θ_m is the maximum volumetric water content, and θ_r is the residual volumetric water content. The maximum water content θ_m is generally less than 1.0 due to entrapped air, but can be nearly 1.0 at high-overburden pressures (Way et al. 1985). Fourteen closed-form expressions for water-retention data are presented by Leij, Rossell, and Lesch (1997).

Brooks and Corey (1964) plot S_e versus ψ_m on log–log paper and suggest a relation as given by equation 13.50, or

$$S_e = \frac{\theta - \theta_r}{\theta_m - \theta_r} = \left(\frac{\psi_b}{\psi_m}\right)^\lambda, \quad \text{for} \quad \psi_m > \psi_b \tag{13.84}$$

where ψ_m is matric potential, ψ_b is the air-entry pressure—assumed to be the matric potential at which the largest pore in the soil begins to drain—and λ is an exponent. Brooks and Corey (1964) show that the exponent λ (the pore-size distribution index) often has a typical value in soils of approximately 2. Equation 13.84 is best for use only for the drainage cycle; for the wetting cycle, Su and Brooks (1975) suggest an expression given by

$$\psi_m = \psi_b \left(\frac{\theta - \theta_r}{c}\right)^{-m} \left(\frac{1 - \theta}{d}\right)^{dm/c} \tag{13.85}$$

where c, d, and m are constants. Equation 13.85 is designed to fit data for $\psi_m > 0$. When $c = d$, equation 13.85 simplifies to

$$\frac{\theta - \theta_r}{1 - \theta_r} = \frac{1}{1 + \left(\dfrac{\psi_m}{\psi_b}\right)^{1/m}} \tag{13.86}$$

Van Genuchten (1980) suggested an empirical relation similar to equation 13.86, given by

$$S_e = \frac{1}{[1 + (\alpha\psi_m)^n]^m} \tag{13.87}$$

where α, n, and m are constants. Van Genuchten's variable n is related to Brooks and Corey's variable λ by $n = \lambda + 1$. The variable m is related to λ by $m = \lambda/(1 + \lambda)$. (Rawls et al. 1993).

Laliberte (1969) points out that equation 13.84 overestimates S_e for ψ_m near ψ_b, and suggests another expression of the form

$$S_e = 0.5(1 + \text{erf } \xi) \tag{13.88}$$

where erf is the error function (for which a series expansion is given in appendix 3) and ξ is given by

$$\xi = \frac{a}{(\psi_m + b)} - c \tag{13.89}$$

where a, b, and c are properties of the soil. Equation 13.89 is valid for $\psi_m > -b$, where $|b| < \psi_b$. For Laliberte's expression (equation 13.88), we need only one equation to fit data over the entire range of S_e. This is also true for Van Genuchten's relation (equation 13.87), and for the expression of Su and Brooks (equation 13.85).

McKee, Bumb, and Deshler (1983) and McKee and Bumb (1984) suggest an exponential form for the relation between effective saturation and matric potential, referred to as the "Boltzmann distribution" (see equation 13.10), given by

$$S_e = \exp\left(-\frac{\psi_m - \psi_1}{\beta}\right) \tag{13.90}$$

where β and ψ_1 are adjustable variables. This expression is very successful in matching data for $\psi_m > \psi_b$, but like the Brooks–Corey relation (equation 13.84), it is not meant to be used for $\psi_m < \psi_b$. For data that closely follow the trend of the curve in figure 13.9, ψ_1 is roughly equivalent to the air-entry (bubbling) pressure. As ψ_1 changes, the whole curve shifts up or down. The β fitting parameter generally gets smaller as the pore-size distribution becomes more uniform and the moisture-characteristic curve becomes flatter in the middle.

The Brooks–Corey relation and Boltzmann distribution are valid for matric potentials greater than the air-entry pressure, but are not valid near maximum moisture contents or under fully saturated conditions that occur in the capillary-fringe region. McKee and Bumb (1987) and Bumb (1987) suggest a new distribution, called the "Fermi distribution"— so-named due to its functional similarity to the energy relation among electrons. The Fermi distribution relates water content to matric potential over the complete range of saturations, given by

$$S_e = \frac{1}{1 + \exp\left(\dfrac{\psi_m - \psi_{m_{1/2}}}{\beta}\right)} \tag{13.91}$$

The Fermi distribution achieves a fit using only two curve-fitting parameters, $\psi_{m_{1/2}}$ and β. As shown in figure 13.12, the first parameter, $\psi_{m_{1/2}}$, is the matric potential when the effective saturation is halfway between the maximum and residual saturations. As with ψ_1 in the Boltzmann distribution, changes in $\psi_{m_{1/2}}$ shift the Fermi distribution up or down. The second parameter β is related to the pore-size distribution. The smaller β is, the more uniform the pore sizes, and the moisture-characteristic curve becomes flatter in the middle section.

Most of the equations above do not result in linear differential equations when substituted into Richards' equation. These include the power equation of Brooks and Corey, as well as the equations of Su and Brooks, Van Genuchten, and Laliberte. The Boltzmann distribution results in an analytical solution to Richards' equation using the Kirchhoff transform of equations 13.4 and 13.5. The Fermi distribution can also be integrated using the Kirchhoff transform under many conditions that cover most practical cases (Way et al. 1985).

Bumb, Murphy, and Everett (1992) compare the Brooks–Corey, Boltzmann, and Fermi relations for moisture-characteristic data assembled by Rawls, Brakensiek, and Saxton 1982. Way et al. (1985) fit the Fermi distribution to data found in Brooks and Corey (1964) as well as Laliberte, Corey, and Brooks (1966). McKee and Bumb (1988) fit the Brooks and Corey relation, and Bumb et al. (1988) fit the Boltzmann distribution to data from Brooks and Corey (1964). Bumb, Murphy, and Everett (1992) used least-squares, relative least-squares, minimization of absolute error, and minimization of relative error curve-fitting techniques to match the Brooks–Corey, Boltzmann, and Fermi relations to data. For implementing the least-squares method (the most commonly used fitting technique), the Brooks–Corey, Boltzmann, and Fermi equations were rearranged to provide linear equations:

Brooks–Corey:

$$\frac{1}{\lambda} \ln S_e = \ln \psi_b - \ln \psi_m \tag{13.92}$$

Boltzmann distribution:

$$\beta \ln S_e = -\psi_m + \psi_1 \tag{13.93}$$

Fermi distribution:

$$\beta \ln \left[\frac{(1 - S_e)}{S_e} \right] = \psi_m - \psi_{m_{1/2}} \tag{13.94}$$

These equations are suitable for standard least-squares analysis and can be plotted on log–log or semi-log graph paper for visual analyses as well. Generally, a plot is produced for one or more of these methods and a final selection is made manually, to determine which of the methods results in the best visual fit to the data (Bumb, Murphy, and Everett 1992). Because the Brooks–Corey power relation results in no analytical solution, it is used only for comparison purposes. Bumb, Murphy, and Everett (1992) conclude that, in general, the Boltzmann distribution and the Brooks–Corey relation produced equally good fits to the data. The Fermi distribution, while providing excellent matches to data in several cases, did not achieve the high rate of success that the Brooks and Corey equation or the Boltzmann distribution attained (Bumb, Murphy, and Everett 1992).

Because equations 13.92 through 13.94 are linear in λ or ψ and S_e, the optimum values of the fit variables are obtained. Before least-squares curve-fitting techniques are utilized, saturation data has to be converted to effective saturation using equation 13.83. This can be done using volumetric water content θ or degree of saturation S, related to volumetric-water content by $S = \theta/\phi$, where ϕ is the porosity. Values of S_r and S_m are selected using, the extreme points of the capillary pressure versus saturation data, in most cases. In some cases, data for high values of ψ_m—where S is near S_r—were ignored because these data seemed to be unduly controlling the shape of the curve. It is believed that this does not represent a serious error, because most of the flow takes place at volumetric water contents somewhat removed from S_r. Figure 13.13 shows the linearized least-squares best-fit for equations 13.92 through 13.94 to data for a Touchet silt loam from Brooks and Corey (1964). Note that the fits are done using degree of saturation S rather than volumetric water content. In some cases, the values of S_r were increased (or decreased) slightly to force the curves to turn upward sooner than would have been the case otherwise; this is especially true for the Fermi-distribution fit. This is necessary for the distributions that are quite flat, to begin to show a gradual increase of ψ_m with decreasing saturation. By adjusting S_r, the curve is made to bisect this region of the data. Brooks and Corey (1964) and Corey (1977) also use an interpolative procedure to adjust the value of S_r so that the functional form fits the data well. For the Fermi

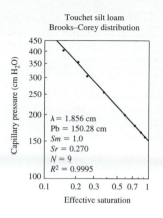

Touchet silt loam
Brooks–Corey distribution

$\lambda = 1.856$ cm
Pb = 150.28 cm
Sm = 1.0
Sr = 0.270
N = 9
$R^2 = 0.9995$

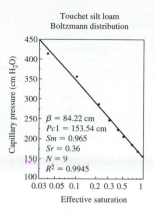

Touchet silt loam
Boltzmann distribution

$\beta = 84.22$ cm
Pc1 = 153.54 cm
Sm = 0.965
Sr = 0.36
N = 9
$R^2 = 0.9945$

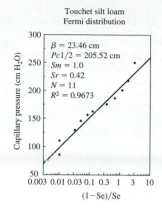

Touchet silt loam
Fermi distribution

$\beta = 23.46$ cm
Pc1/2 = 205.52 cm
Sm = 1.0
Sr = 0.42
N = 11
$R^2 = 0.9673$

Figure 13.13 Linearized least-squares fit to various distributions to data from Brooks and Corey (1964). (Data converted to an equivalent water–air system using Brooks and Corey's equation 17)

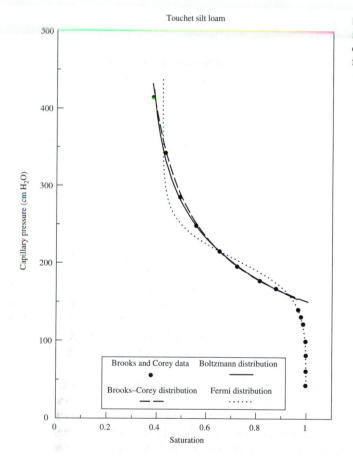

Figure 13.14 Brooks–Corey, Boltzmann, and Fermi distribution matches to capillary pressure versus saturation data for touchet silt loam

distribution, only the interior points ($S_r < S < S_m$) are used to obtain values of the variables $\psi_{m_{1/2}}$ and ψ, because the $\ln(1 - S_e)/S_e$ is undefined at $S_e = 0$ and $S_e = 1$. Figure 13.14 shows the moisture-characteristic curve matches. Table 13.3 summarizes the variables of various soil types and textures, for fits by the Brooks–Corey relation, Boltzmann distribution, and Fermi distribution. Bumb, Murphy, and Everett (1992) present fitted variables for the Brooks–Corey relation as well as the Boltzmann and Fermi distributions to 12 soil-texture

TABLE 13.3 Fitted Variables[1] for Selected Media

Brooks–Corey Relation

Sample	S_r (%)	S_m (%)	ψ_b (cm)	λ (cm)
Touchet-silt loam	27.0	100.0	150.3	1.86
Fine sand	16.7	100.0	82.0	3.70
Hygiene sandstone	57.7	100.0	108.0	4.17
Berea sandstone	29.9	100.0	86.0	3.69
Volcanic sand	15.7	100.0	32.0	2.29
Fragmented sandstone	30.0	100.0	20.6	1.92
Fragmented mixture	27.6	100.0	34.4	2.89
Glass beads	8.5	100.0	58.0	7.30

Source: Data from Brooks and Corey (1964)

Boltzmann Distribution

Sample	S_r (%)	S_m (%)	ψ_1 (cm)	β (cm)
Touchet-silt loam	36.0	96.5	153.5	84.2
Fine sand	17.4	94.5	75.0	33.7
Hygiene sandstone	58.0	97.5	107.2	33.2
Berea sandstone	31.0	96.0	86.1	27.9
Volcanic sand	15.5	98.0	31.0	22.2
Fragmented sandstone	33.0	97.0	16.9	17.3
Fragmented mixture	30.0	96.0	33.2	14.9
Glass beads	9.5	97.0	57.1	9.2

Source: Data from McKee and Bumb (1988)

Fermi Distribution

Sample	S_r (%)	S_m (%)	$\psi_{m_{1/2}}$ (cm)	β (cm)
Touchet-silt loam	42.0	100.0	205.5	23.5
Fine sand	19.0	99.5	104.4	14.8
Hygiene sandstone	62.0	100.0	129.5	10.2
Berea sandstone	32.0	100.0	107.5	10.4
Volcanic sand	16.0	100.0	47.0	5.19
Fragmented sandstone	33.0	100.0	34.2	6.39
Fragmented mixture	32.0	99.0	48.7	6.06
Glass beads	9.7	100.0	65.2	6.85

Source: Data from Way et al. (1985)

[1]Converted to an equivalent water–air system using Equation 17 of Brooks and Corey (1964)

types obtained from Rawls, Brakensiek, and Saxton (1982) for 1,323 soils from about 5,380 horizons in 32 states of the United States.

Three-dimensional analytical solution Bumb et al. (1988) and McKee and Bumb (1988) propose a three-dimensional analytical solution to Richards' equation. McKee and Bumb (1988) detail both an exact steady-state solution as well as an approximate time-dependent solution. Steady-state solutions for constant-rate infiltration from a point source is presented by Philip (1969), Raats (1972), and Lomen and Warrick (1978). All of these solutions assume that the unsaturated hydraulic conductivity K decreases as an exponential function of the matric potential ψ, or as a power function of the effective saturation S_e. The resulting solution to Richard's equation becomes linear in the matric-flux potential H. This was shown in section 13.2 by equation 13.8, for the exponential case, and equation 13.12 for the power-equation case. Additionally, McKee and Bumb (1988) show that if a Boltzmann distribution is fit to the moisture-characteristic curve data, a linear relation for matrix flux potential results, of the form of equation 13.12.

For the three-dimensional LaPlace equation with z taken as positive downward, McKee and Bumb (1988) write the equation for matric flux potential for an isotropic medium:

$$\frac{\partial^2 H}{\partial x^2} + \frac{\partial^2 H}{\partial y^2} + \frac{\partial^2 H}{\partial z^2} - \alpha \frac{\partial H}{\partial z} = \frac{1}{D} \frac{\partial H}{\partial t} \tag{13.95}$$

where H is the matric flux potential defined by equation 13.12 for a Boltzmann-distribution fit to the moisture-characteristic curve data as $(\beta/n)K$, with K being the unsaturated hydraulic conductivity in the x, y, and z directions. The constant α is defined as n/β, and is the same α defined by López-Bakovic and Nieber (1989) in Section 13.2. D is defined as the diffusivity and is (McKee and Bumb, 1988)

$$D = \frac{\beta K_s}{(\theta_m - \theta_r)} S_e^{n-1} \tag{13.96}$$

Equation 13.96 is still nonlinear in D for all $n \neq 1$. However, for $n = 1$, D simplifies to

$$D = \frac{\beta K_s}{(\theta_m - \theta_r)} \tag{13.97}$$

If we assume that the anisotropy is in the x and y directions, and that each z-layer is isotropic over its thickness, then transformations are needed for changing the results from an isotropic to an anisotropic system. The change involves scaling the x, y, and z axes using the following transformations

$$x^* = x \left[\frac{K_y}{K_x} \right]^{1/4} \tag{13.98}$$

$$y^* = y \left[\frac{K_x}{K_y} \right]^{1/4} \tag{13.99}$$

$$z^* = z \frac{[K_x K_y]^{1/4}}{[K_z]^{1/2}} \tag{13.100}$$

and

$$\alpha^* = \frac{n}{\beta} \frac{[K_z]^{1/2}}{[K_x K_y]^{1/4}} \tag{13.101}$$

where x, y, z are the untransformed coordinates, and x^*, y^*, z^* are the actual coordinates of the anisotropic media, and coincide with the principal axes of the hydraulic-conductivity tensor.

The *time-dependent solution* to equation 13.95 is available only for the value of $n = 1$, and is analogous to the problem of heat flow from a point source of constant strength moving through a uniform, infinite medium (McKee and Bumb 1988). The solution for a source at the origin of a Cartesian-coordinate system is (Carslaw and Jaeger 1959)

$$\Delta H = H - H_0 = \frac{Q}{8\pi\sqrt{\pi D}} \int_0^t \frac{\exp\left\{ -\frac{[z/\alpha D(t - t')]^2 + x^2 + y^2}{4D(t - t')} \right\}}{(t - t')^{3/2}} dt' \tag{13.102}$$

where H_0 is the value of H at the initial saturation, and ΔH is the change in H. If the initial saturation is near residual saturation, then H_0 is small compared to H, and ΔH can be approximated by H. Evaluating the integral for a constant-strength point-source infiltration rate Q located at coordinates x', y', and z' in an infinite soil gives (McKee and Bumb 1988)

$$H - H_0 = \frac{Q e^{\alpha(z-z')/2}}{8\pi R} \left(e^{\alpha R/2} \operatorname{erfc}\left[\frac{R}{2\sqrt{Dt}} + \frac{\alpha\sqrt{Dt}}{2} \right] + e^{-\alpha R/2} \operatorname{erfc}\left[\frac{R}{2\sqrt{Dt}} - \frac{\alpha\sqrt{Dt}}{2} \right] \right)$$

(13.103)

where $R = \sqrt{(x - x')^2 + (y - y')^2 + (z - z')^2}$ and x, y, and z are coordinates of any point in space.

The above solution is for a point source only. An areal source—such as leakage from a pond or land treatment facility—is obtained by superposition of a large number of points. This superposition is used to sum any number of solutions of the form of equation 13.103, because the solution is linear when $n = 1$ or in the steady-state case presented following. The effective saturation as a function of space and time is obtained after calculation of the matric flux potential H, by using the inverse transform from equations 13.9, 13.10, and 13.12 to obtain

$$S_e = \left(\frac{nH}{\beta K_s} \right)^{1/n}$$

(13.104)

The nonlinearity in equation 13.95 occurs only in the diffusivity D because of the S_e term. McKee and Bumb (1988) suggest that a conservative estimate of the time-dependent solution is possible for $n \neq 1$ by substituting a value of 1 for S_e in equation 13.96, but by doing so, both D and spreading of the moisture are overestimated. McKee and Bumb (1988) point out that for large times, the time-dependent solution is the same as the steady-state solution, indicating some confidence in this approximation.

The *steady-state solution* does not depend on diffusivity, because the right-hand side of equation 13.95 is zero. Therefore, equation 13.95 is linear for all values of n at steady-state. The boundary and initial conditions for the solution of equation 13.95 are obtained by recognizing that: **(1)** the soil is initially assumed to be at a constant volumetric-water content; **(2)** at a large distance from the point source, the soil is unaffected; and **(3)** the point source is of constant strength. Carslaw and Jaeger (1959) and Philip (1969) presented a steady-state solution in an infinite soil, with a point source Q located at x', y', and z' as

$$\Delta H_\infty = H_\infty - H_0 = \Delta H_\infty[r, (z - z')]$$

(13.105)

$$= \frac{Q}{4\pi\sqrt{r^2 + (z - z')^2}} \exp\left\{ \frac{\alpha}{2}\left[(z - z') - \sqrt{r^2 + (z - z')^2} \right] \right\}$$

where $r = \sqrt{(x - x')^2 + (y - y')^2}$. The subscript ∞ indicates that the solution is for a column of soil of infinite x, y, and z.

The previous time-dependent and steady-state solutions were for an infinite depth and areal extent of unsaturated soil. Raats (1972) gives a solution for the steady-state case when no flow is allowed through the soil surface. This upper boundary is common under a lined pond, where the lining forms an impermeable boundary. In general, solutions for impermeable boundaries can be developed using superposition techniques, by imposing image sources across the impermeable boundary and then superimposing the solutions to obtain the solution with the boundary. In this analysis, only one image is used for impermeable boundaries, and therefore, the solutions are approximate but adequate for most practical problems.

Raats' (1972) solution for an impermeable boundary at a distance d above the source (see figure 13.15) is given by

$$\Delta H = \Delta H_\infty[r, z - d] + e^{-\alpha d}\Delta H_\infty[r, z + d]$$

$$-\frac{Q\alpha}{4\pi}e^{\alpha d}E_i\left[\frac{\alpha}{2}\left(z + d + \sqrt{r^2 + (z + d)^2}\right)\right]$$

(13.106)

where E_i is the exponential integral function; the exponential integral function is closely approximated by the series expansion given in appendix 3. Equation 13.106 is the superposition of an image source a distance d above the impermeable boundary, where the notation $\Delta H_\infty[r, z - d]$ and $\Delta H_\infty[r, z + d]$ are for the infinite cases defined by equation 13.105. In this case the value of z' is replaced by d. Equation 13.106 is applicable to any impermeable boundary a distance d above the point source, as shown in figure 13.15 (McKee and Bumb 1988).

For a horizontal impermeable boundary a distance $z = a$ beneath the surface and also the real source (see figure 13.16), McKee and Bumb (1988) show that the no-flux condition is represented by

$$q = -K\frac{\partial}{\partial z}\left(\psi_m + \frac{g \cdot \hat{e}_z z}{|g|}\right) = 0, \quad \text{at } z = a$$

(13.107)

where q is the flux, g is the acceleration due to gravity and $\hat{e}_z$ is the unit vector in the positive z direction (downward). This condition, represented in terms of matric flux potential H is

$$q = \left.\frac{\partial H}{\partial z}\right|_{z=a} - \frac{g \cdot \hat{e}_z}{|g|}K\Big|_{z=a} = 0$$

(13.108)

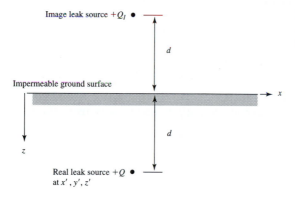

Figure 13.15   Schematic of real and image sources used to model an impermeable surface above the real source

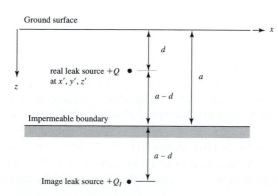

Figure 13.16   Schematic of real and image sources used to model a horizontal impermeable boundary beneath the real source

The theory of images, shown schematically in figures 13.15 and 13.16, has been used extensively in hydrology to model impermeable and constant-head boundaries, and has also been used in heat flow in solids by Carslaw and Jaeger (1959). For these applications, H represents the potential, and the no-flow boundary condition is given by $\nabla \cdot H = 0$ at the boundary. In that case, the exact method of images results in $H = H_R + H_I$, where H_R is the solution for the real source and H_I is the solution for the image source. In the saturated case ($\alpha = 0$), the real and image solutions have the same mathematical form but for unsaturated flow, the presence of gravity causes an asymmetry between them. McKee and Bumb (1988) present a modified method of images with the same form as the classical method, or

$$\Delta H = \Delta H_R + \Delta H_I^h \tag{13.109}$$

In equation 13.109, ΔH_R is the solution for a real point source at x', y', z' in an infinite porous medium, which may either time-dependent (equation 13.103) or steady-state (equation 13.105), and ΔH_I^h is the solution for an image point source for a horizontal impermeable boundary at $x', y', 2a - z'$ in an infinite porous medium with the gravitational force acting upward, or for the time-dependent solution

$$H_I^h - H_0 =$$

$$\frac{Q e^{\alpha(z - 2a + z')/2}}{8\pi R_I} \left(e^{\alpha R_I/2} \operatorname{erfc}\left[\frac{R_I}{2\sqrt{Dt}} + \frac{\alpha\sqrt{Dt}}{2} \right] + e^{-\alpha R_I/2} \operatorname{erfc}\left[\frac{R_I}{2\sqrt{Dt}} + \frac{\alpha\sqrt{Dt}}{2} \right] \right) \tag{13.110}$$

where $R_I = \sqrt{(x - x')^2 + (y - y')^2 + (z - 2a + z')^2}$.

The steady-state image solution is obtained by changing the sign of $z - z'$ in equation 13.105 and replacing R with R_I. In this case, the real solution H_R is downward (gravity pulling water downward), whereas the image solution H_I^h is upward (the direction of gravity is reversed). However, the total net flux at the impermeable boundary is approximately zero. The error caused by reversing the sign of gravity is evident if equation 13.109 is substituted into equation 13.95, because all the terms do not cancel. The remaining terms are shown to vanish exponentially with distance from the real source, as long as R and z are positive (McKee and Bumb 1988), and therefore are negligible. Thus, the image solution for a horizontal boundary beneath the source, while not exact, is still a preferred approximation to an otherwise very complex problem.

For a vertical impermeable boundary at $x = b$ away from the source, as shown in figure 13.17, the flux is given by

$$q = -K \frac{\partial}{\partial x} (\psi_m + z) = 0 \tag{13.111}$$

or equivalently by $\nabla \cdot H = 0$. This is easily satisfied (with no change in the sign of gravity) by

$$\Delta H = \Delta H_R + \Delta H_I^v \tag{13.112}$$

where ΔH_R is the solution for the real point source and ΔH_I^v is the solution for the image-point source located at $2b - x', y', z'$. The solution for ΔH_I^v is obtained by substituting $2b - x'$ for x' in the equations for the time-dependent or steady-state cases.

To demonstrate the use of the above equations, a steady-state leak of 500 liters of water per day into a homogeneous and isotropic Touchet silt loam is modeled using equations 13.104 and 13.105. In this example, the leak is located 5 m below ground surface. The Touchet silt loam soil has the properties shown in table 13.1 for the Boltzmann distribution. In addition, the soil has a saturated hydraulic conductivity of 0.35 m/day, a porosity of 50.1 percent, a maximum saturation of 96.5 percent, and an irreducible saturation of 36.0 percent.

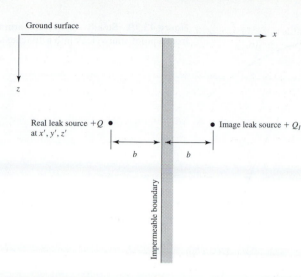

Figure 13.17   Schematic of real and image sources used to model a vertical impermeable boundary

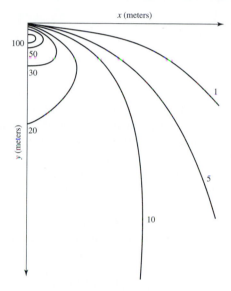

Figure 13.18   Steady-state effective water saturations (S_e = %) from a point-source leak into a Touchet silt loam

The solution is to find the steady-state water-content distribution as a result of the point-source leak.

The steady-state effective saturations S_e and actual saturations S are shown in figures 13.18 and 13.19, respectively. The highest saturations are directly below the leak, with measurable saturations spreading away from the leak laterally, a distance of over 20 m. Such an analysis is used to assess the locations of monitoring devices or to assess the relative location of conservative contaminants.

Three-dimensional semianalytical solution for vapor diffusion from an initial contaminant distribution in the vadose zone   During the early stages of liquid-phase releases to the subsurface, liquid advection occurs. The above section presented an analytical solution for the advective transport of this liquid downward under gravity for a point source. This solution can also be extended to a plane source by using superposition. If the liquid phase can volatilize—such as is the case with most non-aqueous phase liquids (NAPLs)—downward migration of the gaseous-phase continues by gravity-driven density gradients, if the vapor

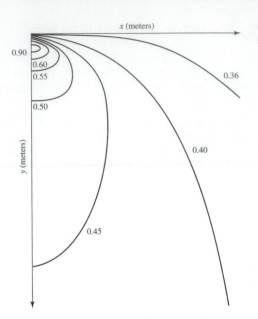

Figure 13.19 Steady-state actual saturations (S = decimal) from a point-source leak into a Touchet silt loam

density is heavier than air. If the vapor is lighter than air, it will migrate upward. This process is highly dependent on the concentration of the vapor as well as its density.

Gaseous diffusion occurs by moving vapor from areas of high concentration to areas of low concentration. This process involves the movement of vapor by intermolecular collisions. Gaseous diffusion and gaseous advection are significantly reduced by high soil-water content. Aqueous diffusion is similar to gaseous diffusion, but is very slow and is assumed to be negligible, compared to gaseous diffusion.

Aqueous advection is the passive transport of a volatile NAPL dissolved in the aqueous phase. Aqueous fluxes typically occur from infiltration of rainfall into the unsaturated zone; this transport mechanism dominates if the fluxes are high. Volatile NAPLs migrate through the unsaturated zone by one or more of the transport mechanisms described above, until the capillary fringe is encountered. Volatile NAPLs are mobilized to the ground water from the capillary fringe by advection, aqueous- and gas-phase diffusion, or flucuations in the water table.

The governing equation for transport of a single NAPL in the unsaturated zone is (LLNL 1990)

$$\frac{\partial}{\partial t}(\phi S_l C_l + \phi S_g C_g + C_s) = -\nabla \cdot |C_g q_g| - \nabla \cdot |C_l q_l|$$
$$+\nabla \cdot |\rho_g \phi S_g \tau_g D_g \nabla(C_g/\rho_g)|$$
$$+\nabla \cdot |\rho_l \phi S_l \tau_l D_l \nabla(C_l/\rho_l)| \qquad (13.113)$$
$$-\lambda_l \phi S_l C_l - \lambda_g \phi S_g C_g$$
$$-\lambda_g \phi \rho_B K_d C_l + q_{src}$$

where t is time; C is volatile NAPL concentration (mass/volume); ϕ is effective porosity; S is saturation; l is the subscript for the aqueous phase; g is the subscript for the gaseous phase; s is the subscript for sorption onto the solid phase; src is the subscript for a source term; τ is a tortuosity factor; ρ_b is bulk density; D is the diffusion coefficient of the NAPL; λ is the decay constant for NAPL degradation; q is mass flux; ρ is density; $\nabla\cdot$ is the divergence operator, and ∇ is the gradient operator.

The three terms in parentheses on the left-hand side of equation 13.113 are the changes in accumulated mass of volatile NAPL in the aqueous, gaseous, and liquid phases, respectively. The first two terms in brackets on the right-hand side are the advective fluxes in the gaseous and aqueous phases; the next two terms in brackets are the diffusive fluxes in the gaseous and aqueous phases; the next three terms are the chemical or biological degradation in the three phases; and the last term is the source term.

In the absence of free product, liquid advection is nonexistent and the total concentration of contaminant is given by the total-concentration equation, which represents partitioning among the solid, aqueous (pore water), and gaseous phases in the soil. This total contaminant concentration C_T is $C_s + C_l + C_g$, where C_s is the concentration of the solids, C_l is the concentration of the liquid, and C_g is the concentration of the gas for a given contaminant. Equilibrium partitioning between the solid and aqueous phases is described by a linear isotherm given by

$$C_s = \rho_b K_d C_l \qquad (13.114)$$

where ρ_b is the soil bulk density, and K_d is the partitioning coefficient between the aqueous phase and solid phase. Henry's law is used to partition between the gaseous phase and the aqueous phase for a contaminant, or

$$C_g = K_H C_l \qquad (13.115)$$

where K_H is Henry's law constant for an individual contaminant.

Gaseous diffusion occurs by the movement of vapors from areas of high concentration to areas of low concentration in the vadose zone, due to intermolecular collisions. Both gaseous advection and gaseous diffusion are significantly reduced by increases in water content in the vadose zone. If the relative humidity of the vadose zone is high, then vapor-to-solid sorption is assumed to be negligible. Laboratory studies (LLNL 1990) indicate that soils with less than 90 percent relative humidity have a very high sorptive capacity due to the direct sorption of the vapor into the soil phase. Under these conditions, the upper layers of the soil retard vapor flux diffusing to the surface.

The boundary of the ground surface often acts as a sink to gaseous diffusion, because the concentration at the surface is often approximately zero due to dispersion into the atmosphere by wind currents. An exception occurs when sufficient vegetative cover creates a stagnant boundary layer, resulting in a buffer zone between the air in motion above the surface and the upper boundary of the vadose zone (Jury 1986); transport across this zone is primarily by diffusion. If the surface is paved, an additional buffer zone exists that has a diffusion coefficient specific to the paving material. If the paving material is completely impervious, a no-flow boundary applies instead of the diffusion sink.

When gaseous- or aqueous-flow velocities are low (such as in the vadose zone), diffusion is the important contributor to dispersion and transport, and the dispersion equals the diffusion coefficient ($D_L = D^*$). The decision, as to the relative significance of aqueous advection or gaseous diffusion in the vadose zone, can be made using the Peclet number which provides a quantitative estimate of the dominant mode of transport. The Peclet number $\mathcal{P}$ is given as (Freeze and Cherry 1979)

$$\mathcal{P} = \frac{vL}{D^*} \qquad (13.116)$$

where v is the linear velocity of water percolating through the vadose zone, L is the characteristic length (or distance traveled) of the aqueous front, and D^* is the apparent diffusion coefficient for the gaseous contaminant. The linear velocity v, is determined by dividing the recharge rate by the product of effective porosity and the degree of saturation S_l of the soil of interest. For a recharge rate q of water percolating through the vadose zone, the advective mass flux, q_a, is given by

$$q_a = \frac{C_g q}{K_H R} \tag{13.117}$$

where R is defined as the unsaturated retardation coefficient, given by

$$R = S_l + S_g K_H + \frac{\rho_b K_d}{\phi} \tag{13.118}$$

where S_l and S_g, respectively, are the degree of saturation of water and air in the soil ($S_l + S_g = 1$); ρ_b is the bulk density of the soil; K_H is Henry's law constant; K_d is the partitioning coefficient between the aqueous and solid phases; and ϕ is the porosity of the soil. The apparent gaseous-diffusion coefficient is given by

$$D^* = \frac{S_g \phi \tau D_g}{R} \tag{13.119}$$

where D_g is the actual gaseous-diffusion coefficient for the contaminant of interest, and τ is the tortuosity factor to account for reduced cross-sectional areas and increased path length in the soil for the gaseous transport around the water-filled pores (see chapter 7). The tortuosity is determined empirically by Millington and Quirk (1961) as

$$\tau = S_g^{7/3} \phi^{1/3} \tag{13.120}$$

The Peclet number provides a quantitative estimate of the relative importance of aqueous advection (compared to gaseous diffusion) as the dominant mode to transport, and is given by

$$\mathcal{P} = \frac{qL}{S_g \phi \tau D_g K_H} \tag{13.121}$$

Peclet numbers must be greater than 10 for advection to become important (Dragun 1988) and for Peclet numbers less than 1, chemical exchange is completely independent of flow velocities. Peclet numbers between 1 and 10 indicate that chemical exchange is essentially diffusion controlled. For most unsaturated soil, chemical exchange should be diffusion controlled (Dragun 1988). Additionally, if the diffusion rate of one contaminant is known, the diffusion rate D of a structurally similar chemical is estimated using Graham's law of diffusion (Dragun 1988)

$$\frac{D_1}{D_2} = \left(\frac{\rho_2}{\rho_1}\right)^{0.5} = \left(\frac{M_2}{M_1}\right)^{0.5} \tag{13.122}$$

where ρ is the density and M is the molecular weight of the contaminants of interest.

The characteristic length L is the distance the aqueous front moves through the vadose zone during a recharge event, and is equal to

$$L = \frac{q \Delta T}{\phi R} \tag{13.123}$$

It is likely that temporal variations in volumetric and mass flux in the vadose zone are highly attenuated by the time infiltrating water reaches a depth where contaminants are located, and an almost constant flux exists down to the water table. In this case, the characteristic length becomes the distance from the center of the contaminant plume down to the water table. Therefore, the Peclet number is calculated based on the average annual recharge rate and the distance to the water table. While this Peclet number may be larger than that calculated using equations 13.121 and 13.123, it still results in a Peclet number less than 10.

Another way to compare advective versus diffusive transport modes is to note the movement of aqueous fronts based on average annual water balances, including both precipitation and evapotranspiration. Often in arid and semiarid environments, this movement is only a few centimeters per year. This should be compared to the order-of-magnitude movement of the diffusive-contaminant front computed by

$$L = \sqrt{(S_g \tau D_g K_H)} \times 1 \ \text{year}/R \qquad (13.124)$$

which gives a distance greater than the aqueous-front movement.

The Peclet number can also be used to assess the relative importance of gaseous advection through density gradients, compared to gaseous diffusion. If a vapor is denser than air, it sinks toward the ground water in the gaseous phase; otherwise it rises to the ground surface. For example, consider a denser-than-air vapor. The density gradients in the gaseous phase that drive the vapor toward the ground water are significant during the early and intermediate stages of the release, because of elevated concentrations. However, in time the concentrations in the vadose zone are diluted by gaseous diffusion, and the transport of the contaminant by gravity-driven density gradients is unimportant compared to gaseous diffusion (Falta et al. 1989).

Falta et al. (1989) use an approximate expression for the maximum, nonretarded-Darcian flux q_g under vapor-density gradients as

$$q_g = K_g \frac{M_g - M_a}{M_g} \left(\frac{C_g}{\rho_g} \right) \qquad (13.125)$$

where K_g is the vapor conductivity, M_g is the molecular weight of the vapor, M_a is the molecular weight of air, and C_g is the vapor concentration (mass/volume). This approximation is accurate to within ± 20 percent (LLNL 1990). The Peclet number for the gaseous case is now given by

$$\mathcal{P} = \frac{q_g L}{S_g \phi \tau D_g} \qquad (13.126)$$

The maximum Peclet number occurs at the maximum concentration of the vapor. Again, if the Peclet number is less than 10, then gaseous diffusion is assumed to be the dominant mode of transport.

To solve for transport in the vadose zone under gaseous diffusion, LLNL (1990) proposes a semi-analytical solution of the diffusion equation with the initial mass distribution approximated by a three-dimensional, radially symmetric Gaussian function. If there is no degradation, the soil concentrations C_s satisfy the diffusion equation given by

$$D^* \left(\frac{\partial^2 C_s}{\partial x^2} + \frac{\partial^2 C_s}{\partial y^2} + \frac{\partial^2 C_s}{\partial z^2} \right) = \frac{\partial C_s}{\partial t} \qquad (13.127)$$

where D^* is the apparent gaseous-diffusion coefficient given by equation 13.119. C_s is used because the primary process is gaseous diffusion from the soil. LLNL (1990) assumed radial symmetry so that the concentration distribution was a function in cylindrical coordinates, r and z, as well as a function of time. The initial soil-contaminant distribution was approximately equal to the product of Gaussian functions in the vertical z, and horizontal r, directions as shown schematically in figure 13.20. The equation for the initial distribution is

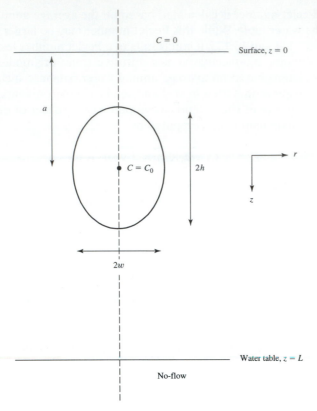

Figure 13.20 Initial gaseous-diffusion contaminant distribution variables

given as

$$C_s(r, z, t = 0) = C_0 \exp\left(-\frac{(z - a)^2}{2h^2}\right) \exp\left(-\frac{r^2}{2w^2}\right) \tag{13.128}$$

where h is the standard deviation half-width of the initial release in the vertical direction, w is the half width in the horizontal direction, and a is the distance from the center of the contaminant mass from the ground surface (figure 13.20). The constant C_0 represents the peak soil concentration at the center of the contaminant distribution.

The boundary condition at the ground surface is zero concentration while the top of the ground water can be approximated as a no-flow boundary. Therefore,

$$C_s(r, z = 0, t) = 0 \tag{13.129}$$

and

$$\frac{\partial C_s(r, z = L, t)}{\partial z} = 0 \tag{13.130}$$

are applied to the top ($z = 0$) and bottom ($z = L$) of the vadose zone. By making the variable substitution $z = z/L$, $r = r/L$, and $t = t/\sqrt{L^2/D^*}$, the diffusion equation 13.127, becomes

$$\frac{\partial^2 C_s}{\partial x^2} + \frac{\partial^2 C_s}{\partial y^2} + \frac{\partial^2 C_s}{\partial z^2} = \frac{\partial C_s}{\partial t} \tag{13.131}$$

Substituting $C_s(r, z, t) = R(r, t)Z(z, t)$ gives

$$\frac{1}{r}\frac{\partial R}{\partial r} + \frac{\partial^2 R}{\partial r^2} = \frac{\partial R}{\partial t} \tag{13.132}$$

and

$$\frac{\partial^2 Z}{\partial z^2} = \frac{\partial Z}{\partial t} \tag{13.133}$$

Solutions that are Gaussian at time zero are given by

$$R(r, t) = \frac{1}{t + \dfrac{w^2}{2}}\exp\left\{-\frac{r^2}{4\left(t + \dfrac{w^2}{2}\right)}\right\} \tag{13.134}$$

and

$$Z(r, t) = \frac{1}{\sqrt{t + \dfrac{h^2}{2}}}\exp\left\{-\frac{z^2}{4\left(t + \dfrac{h^2}{2}\right)}\right\} \tag{13.135}$$

To satisfy the boundary conditions, periodic image solutions are superimposed to obtain the final solution of

$$C_s(r, z, t) = R(r, t)\sum_{n=-\infty}^{\infty}\{(-1)^n Z(z - [2n + a], t) + (-1)^{n+1} Z(z - [2n - a], t\} \tag{13.136}$$

This solution is adjusted to give the correct peak concentration at the center of the release, by using a multiplication factor. Because of the superposition, the variables a, h, and w do not correspond exactly to their definitions (as above), and need to be readjusted to match the shape of the initial distribution. For all practical purposes, the solution using equation 13.136 is approximated by using only one image for the soil surface and one image for the no-flow boundary at the water table.

The solution given by equation 13.136 is applicable when there is no degradation. Therefore, the solution obtained using equation 13.136 must be multiplied by the appropriate exponential time-decay factor, to provide the solution for first-order contaminant decay. The first-order decay factor λ is given by

$$\lambda = 2^{(-t/T)} \tag{13.137}$$

where t is time and T is the decay half-life of the contaminant of interest. Equation 13.136 is multiplied by the factor λ to obtain a solution when there is degradation. Table 13.4 presents values for relative vapor density, Henry's constant K_H, molecular weight M, degradation half-life T, and vapor-diffusion coefficient for selected constituents.

An example of the use of the above vapor-diffusion equations is given for trichloroethylene (TCE) contamination at a site. The original concentrations of TCE are observed by laboratory testing of soil-vapor samples. Both the age and amount of the TCE releases are unknown; however, the maximum measured soil TCE concentration is far below the free-phase concentration. Free-phase advection is terminated well above the water table, and the highest DNAPL concentrations are at approximately 6.1 m below ground surface. Equation 13.128 is used by trial and error to estimate the values of w and h, by assuming that the initial TCE concentration that occurred at a depth of 6.10 m below ground surface was 6,000 ppb.

TABLE 13.4   Properties for Selected Constituents

Constituent	Relative vapor density* (g/m³)	Henry's law constant (atm-m³/mole)	Molecular weight (g/mole)	Degradation half-life (yrs)	Vapor diffusion coefficient (cm²/sec)
Acetone	1.30	3.7E–05	58.08	0.06	—
Ammonia	0.79	—	17.03	—	0.28[†]
Benzene	1.21	0.0054	78.12	0.19–0.30	0.088[†]
Carbon Tetrachloride	1.51	0.023	153.8	0.5–1.0	0.080[†]
Chlorobenzene	1.05	0.0036	112.6	0.10	0.075[‡]
Chloroform	1.62	0.0028	119.4	0.27	0.099[†]
Chlorine	—	—	70.91	—	—
Gasoline	1.10	—	114.2	—	0.060[†]
n-Hexane	1.31	1.85	86.18	—	0.084[†]
Isopropyl Alcohol	1.06	0.0000081	60.09	0.01	0.137[‡]
Mercury	—	—	200.6	—	—
Nitrobenzene	1.00	0.000024	123.1	0.03–0.53	0.072[‡]
Methylene Chloride	2.10	0.0027	84.94	0.02–0.08	0.102[‡]
Phenol	1.00	0.00000040	94.11	0.25	0.085[†]
Tetrachloroethylene (PCE)	1.09	0.0149	165.8	0.82	0.078[†]
Toluene	1.08	0.0059	92.15	0.10	0.049[†]
Trichloroethylene (TCE)	1.27	0.0103	131.5	0.82	0.081[†]
Xylene	1.02	0.0051	106.2	0.088	0.071[†]
Air	1.00	—	29.0	—	—
Water	0.99	0.000001	18.01	—	0.256[†]

Note: Additional selected data on DNAPLs can be found in Cohen, Mercer, and Matthews 1993

* Relative to dry air at 20°C and 760 mm Hg

[†] At 25°C

[‡] At 30°C

The final initial-Gaussian TCE distribution is shown in figure 13.21. The half-width of the Gaussian distribution in the horizontal direction is $w = 6.31$, and in the vertical direction is $h = 4.11$ (see figure 13.21). These values are obtained from the field data as follows: **(1)** the variable a is chosen so that the peak of the Gaussian distribution (actually the sum of Gaussian distributions chosen to satisfy the boundary conditions) matched the peak concentration of the measured field distribution; **(2)** the variable h is adjusted until the depth of the 1-ppm concentration closest to the water table matched the field data; and **(3)** the variable w is adjusted to match the volume of the region estimated to exceed 1 ppm in concentration.

Equation 13.136 was solved for TCE migration into the future at 50 and 100 years as shown in figures 13.22 and 13.23, respectively, and the peak concentration is adjusted at the center of the release by using a multiplication factor determined by solving equation 13.136 for $t = 0$ years. First-order decay is used (equation 13.137) with the half-life of TCE assumed to be 50 years for this particular case, even though table 13.4 shows a half-life of 0.82 years. Other parameter values used: bulk dry density $\rho_b = 1.8$ g/cm³; effective porosity $\phi = 0.30$; solid sorption, $\rho_b K_d / \phi = 5.0$; Henry's law constant $K_H = 0.39$; and liquid saturation of the soil $S_l = 0.50$.

Other analytical models   Other fate and transport analytical solutions are often used to assess cleanup levels in the unsaturated zone, in order to minimize exposure to the underlying ground water or overlying air. Because these models rely on the quantification of relations between specific variables to simulate the effects of natural processes, a close match between the natural processes and those of the selected models has to exist if the modeling

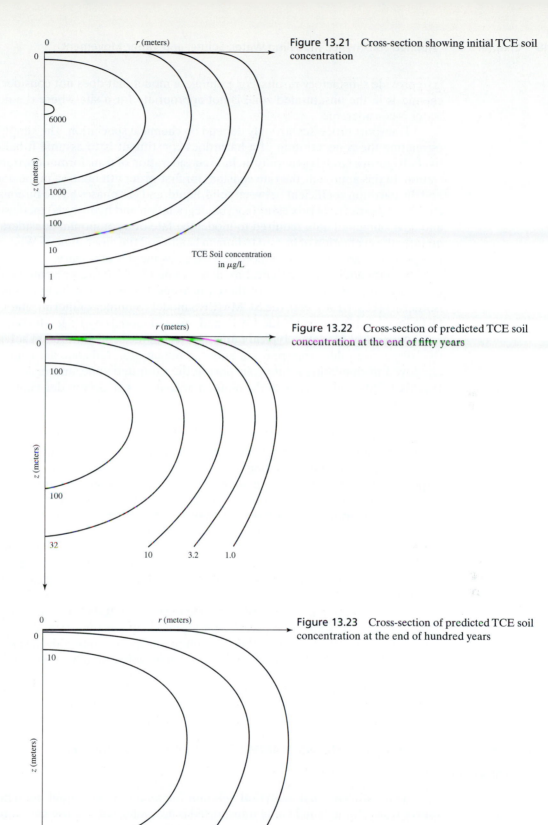

Figure 13.21 Cross-section showing initial TCE soil concentration

TCE Soil concentration in μg/L

Figure 13.22 Cross-section of predicted TCE soil concentration at the end of fifty years

Figure 13.23 Cross-section of predicted TCE soil concentration at the end of hundred years

is to provide satisfactory results. For example, a model that does not consider attenuation of chemicals in the unsaturated zone is not appropriate for a site where the depth to ground water is considerable.

Transport processes strongly depend on chemical speciation. The simplest approach to estimating the concentration of a hazardous constituent is to assume it behaves conservatively. Rigorous models generally include consideration of transformation, transport, and speciation. In this approach, the rate constant for first-order attenuation in the unsaturated zone and the partition coefficient between solid, liquid, and gas phases has to be considered. The inclusion of degradative processes (e.g., biodegradation and hydrolysis) increases the chemical and environmental data required to model the fate of a compound considerably and consequently, the evaluation of hazard to human health and the environment. Where such degradative processes are suspected, a more-refined assessment is necessary.

Several additional analytical models—some of which are presented elsewhere in this text—account for one or more of these processes. These models include: simple Fruendlich isotherms (U.S. EPA 1989); the SUMMERS model (Summers, Gherini, and Chen 1980); the CHAIN model (van der Heijde 1994); and selected models by ES&T-IGWMC (1992).

Numerical Models   Typical numerical techniques encountered in solving the convective–dispersive solute-transport equations in the unsaturated zone are comparable to those employed in simulating solute transport in the saturated zone, and include various FDMs, IFDMs, FEMs, and variants of the method of characteristics (van der Heijde 1994). As with flow, time is generally approximated by finite-difference techniques resulting in explicit, implicit and fully implicit solution schemes.

Typical problems associated with applying traditional finite-difference and finite-element techniques to simulate contaminant transport in the unsaturated zone include numerical dispersion and oscillations. Numerical dispersion occurs when the actual physical-dispersion mechanism of the contaminant transport cannot be distinguished from the front-smearing effects of the computational scheme. For the FDM, this problem is reduced by using the central difference approximation. Spatial-concentration oscillations can occur near a sharp concentration front in an advection-dominated transport system. Remedies for these problems are found (to some extent) in the reduction of grid increments (or time-step size), or by using upstream weighting for spatial derivatives. The use of weighted differences or the selection of other methods significantly reduces the occurrence of these numerical problems.

The International Ground Water Modeling Center (IGWMC) has identified, compiled, and published a description of over 90 unsaturated-zone models (van der Heijde 1994). The compilation includes models for flow and solute transport; solute transport requiring a given-head distribution; flow and heat transport; and flow, solute, and heat transport in the unsaturated zone. Table 13.5 summarizes selected flow and transport models documented by the IGWMC, along with the model source and reference.

13.3   THREE-DIMENSIONAL DETERMINISTIC IMMISCIBLE LIQUID-FLOW MODEL

Analytical Model

Three-dimensional analytical solution for non-aqueous liquid content from a point-source leak   If the liquid being transported is not water but a dense non-aqueous phase liquid (DNAPL) or a light non-aqueous phase liquid (LNAPL), a modification to the analytical solutions in sections 13.2 and 13.3 has to be made. Previously, it was assumed that only two phases were present: a liquid phase consisting of water and constituents dissolved in the water; and air that can contain water vapor. If a DNAPL or LNAPL is introduced to the two-

TABLE 13.5 Selected Numerical Unsaturated-Zone Flow and Transport Models

Model name source	Fate and transport processes
VIP (Vadose Zone Interactive Processes)/U.S. Environmental Protection Agency, Stevens, Grenney, and Yan (1991)	Conservative transport (advection, dispersion, isotropic, diffusion); first-order chemical/microbial decay
NMODEL/University of Florida	Conservative transport (advection, dispersion, isotropic, anisotropic, diffusion); solid/liquid-phase transfers; first-order radioactive decay; single mother/daughter decay; first-order chemical/microbial decay
PMODEL/Louisiana State University	Conservative transport (advection, dispersion, isotropic, anisotropic, diffusion); solid/liquid-phase transfers; ion exchange; reduction/oxidation reactions; first-order chemical/microbial decay; plant uptake
SATURN (Saturated-Unsaturated Flow and Radionuclide Transport)/Geotrans, Inc.	Conservative transport (advection, dispersion, isotropic, anisotropic, diffusion); solid/liquid-phase transfers; first-order radioactive decay; single mother/daughter decay; first-order chemical/microbial decay
MMT/DPRW/Pacific Northwest Laboratory, Ahlstrom and Baca (1974)	Conservative transport (advection, dispersion, isotropic, anisotropic, diffusion); solid/liquid-phase transfers; first-order radioactive decay; single mother/daughter decay; first-order chemical/microbial decay
GS3/U.S. Nuclear Regulatory Commission, Davis and Segol (1985)	Conservative transport (advection, dispersion, isotropic, anisotropic, diffusion)
GS2/U.S. Nuclear Regulatory Commission, Davis and Segol (1985)	Conservative transport (advection, dispersion, isotropic, anisotropic, diffusion); solid/liquid-phase transfers; reduction/oxidation reactions; first-order radioactive decay; single mother/daughter decay; first-order chemical/microbial decay
VADOSE/Analytic & Computational Research, Inc.	Conservative transport (advection, dispersion, isotropic, anisotropic, diffusion); reduction/oxidation reactions; first-order radioactive decay; single mother/daughter decay; first-order chemical/microbial decay
CADIL (Chemical Adsorption and Degradation In Land)/AGTEHM/Oak Ridge National Laboratory, Emerson, Thomas, and Luxmoore (1984)	Conservative transport (advection, dispersion, diffusion); reduction/oxidation reactions; first-order radioactive decay; single mother/daughter decay; first-order chemical/microbial decay
FLOTRA/Analytic & Computational Research, Inc.	Conservative transport (advection, dispersion, isotropic, diffusion); solid/liquid phase transfers; first-order radioactive decay; single mother/daughter decay; first-order chemical/microbial decay, zero-order production
CXTFIT/U.S. Salinity Lab, ARS, Riverside, California (1995)	Conservative transport (advection, dispersion, isotropic, diffusion); solid/liquid–liquid/solid-phase transfers; ion exchange; reduction/oxidation reactions; first-order radioactive decay; single mother/daughter decay; first-order chemical decay
DISPEQ/DISPER/PISTON, Fluhler and Jury (1983)	Conservative transport (advection); solid/liquid/gas-phase transfers; reduction/oxidation reactions; first-order chemical/microbial decay; plant uptake
CREAMS/U.S. Department of Agriculture (1984)	Conservative transport (advection, dispersion, isotropic, anisotropic, diffusion); solid/liquid-phase transfers; first-order radioactive decay; single mother/daughter decay; first-order chemical/microbial decay; plant uptake
SUTRA*/U.S. Geological Survey, Voss (1984)	Conservative transport (advection, dispersion, isotropic, anisotropic, diffusion); solid/liquid-phase transfers; reduction/oxidation reactions; first-order radioactive decay; single mother/daughter decay; first-order chemical/microbial decay
TRACER3D/Los Alamos National Laboratory, Travis (1984)	Conservative transport (advection, diffusion); solid/liquid-phase transfers; first-order chemical/microbial decay
FEMTRAN/Sandia National Laboratory, Martinez (1985)	Conservative transport (advection, dispersion, isotropic, anisotropic, diffusion); solid/liquid-phase transfers; first-order radioactive decay; single mother/daughter decay
SBIR/Simons, Li and Associates, U.S. Bureau of Reclamation (1984)	Conservative transport (advection, dispersion, isotropic, anisotropic, diffusion)
GASOLINE/U.S. Geological Survey	Conservative transport (advection, dispersion, isotropic, diffusion); solid/liquid/gas-phase transfers; reduction/oxidation reactions; first-order chemical/microbial decay

(continued)

TABLE 13.5   Selected Numerical Unsaturated-Zone Flow and Transport Models (*continued*)

Model name source	Fate and transport processes
MOTIF (Model of Transport In Fractured/Porous Media)/Atomic Energy of Canada, Ltd.	Conservative transport (advection, dispersion, isotropic, anisotropic, diffusion); solid/liquid phase-transfers; ion exchange; reduction/oxidation reactions; first-order radioactive decay; single mother/daughter decay; first-order chemical/microbial decay
VS2D/VS2DT*/U.S. Geological Survey	Conservative transport (advection, dispersion, isotropic, anisotropic, diffusion); solid/liquid-phase transfers; ion exchange; first-order radioactive decay; single mother/daughter decay; first-order chemical/microbial decay; plant uptake
NITROSIM/University of Florida	Conservative transport (advection, dispersion, isotropic, diffusion); solid/liquid phase-transfers; ion exchange; reduction/oxidation reactions; first-order chemical/microbial decay; plant uptake
FEMWASTE/FECWASTE/Oak Ridge National Laboratory	Conservative transport (advection, dispersion, isotropic, anisotropic, diffusion); solid/liquid-phase transfers; ion exchange; first-order radioactive decay; single mother/daughter decay; first-order chemical/microbial decay
FLAMINCO/Geotrans, Inc., Huyakorn and Wadsworth (1985)	Conservative transport (advection, dispersion, isotropic, anisotropic, diffusion); solid/liquid-phase transfers; ion exchange; first-order radioactive decay; single mother/daughter decay; first-order chemical/microbial decay
SESOIL* (Seasonal Soil Compartment Model)/U.S. Environmental Protection Agency	Conservative transport (advection, dispersion, isotropic, diffusion); solid/liquid-phase transfers; ion exchange; substitution/hydrolysis; reduction/oxidation reactions; acid/base reactions; complexation; first-order chemical/microbial decay; plant uptake
CTSPAC/University of Oregon, Lindstrom, Garfield, and Boersma (1988)	Conservative transport (advection, dispersion, isotropic, diffusion); solid/liquid-phase transfers; plant uptake
DOSTOMAN (Dose to Man)/Savannah River Laboratory	Conservative transport (advection); solid/liquid phase transfers; first-order radioactive decay; single mother/daughter decay; first-order chemical/microbial decay
VAM2D (Variably Saturated Analysis Model in 2 Dimensions)/HydroGeologic, Inc.	Conservative transport (advection, dispersion, isotropic, anisotropic, diffusion); solid/liquid phase transfers; first-order radioactive decay; single mother/daughter decay; first-order chemical/microbial decay
PRZM (Pesticide Root Zone Model)/U.S. Environmental Protection Agency	Conservative transport (advection); solid/liquid/gas-phase transfers; first-order chemical/microbial decay; plant uptake
RITZ (Regulatory and Investigative Treatment Zone Model)/U.S. Environmental Protection Agency	Conservative transport (advection); solid/liquid/gas-phase transfers; first-order chemical decay
CHEMRANK/University of Florida	Conservative transport (advection); solid/liquid/gas-phase transfers; first-order chemical decay.
ICE-1/International Ground Water Modeling Center	Conservative transport (advection)
PATHRAE/Clemson University	Conservative transport (advection, dispersion, isotropic, diffusion); first-order radioactive decay; single mother/daughter decay; first-order chemical/microbial decay
BIOSOIL/Occidental Chemical Corporation	Conservative transport (advection, dispersion, isotropic); solid/liquid-phase transfers; first-order radioactive decay; single mother/daughter decay; first-order chemical/microbial decay; biotransformation; aerobic/anaerobic
CMIS (Chemical Movement in Soils)/University of Florida	Conservative transport (advection); solid/liquid-phase transfers; ion exchange; first-order chemical/microbial decay
CMLS (Chemical Movement in Layered Soils)/University of Florida	Conservative transport (advection); solid/liquid phase-transfers; ion exchange; first-order chemical decay
GLEAMS (Groundwater Loading Effects on Agricultural Management Systems)/U.S. Department of Agriculture (ARS)	Conservative transport (advection, diffusion); solid/liquid-phase transfers; first-order chemical/microbial decay
CHEMFLO/U.S. Environmental Protection Agency	Conservative transport (advection, dispersion, isotropic, diffusion); solid/liquid-phase transfers; first-order chemical/microbial decay

(*continued*)

TABLE 13.5   Selected Numerical Unsaturated-Zone Flow and Transport Models (*concluded*)

Model name source	Fate and transport processes
MOUSE (Method of Underground Solute Evaluation)/Cornell University	Conservative transport (advection, dispersion, isotropic, diffusion); solid/liquid-phase transfers; first-order chemical/microbial decay
PESTAN* (Pesticide Analytical Model)/U.S. Environmental Protection Agency	Conservative transport (advection, dispersion, isotropic, diffusion); solid/liquid-phase transfers; first-order chemical/microbial decay
LEACHMP (Leaching Estimation and Chemistry Model-Pesticides)	Conservative transport (advection, dispersion, isotropic, diffusion); solid/liquid-phase transfers; first-order chemical/microbial decay; plant uptake
MLSOIL/DFSOIL (Multi-Layer Soil Model)/Oak Ridge National Laboratory	Conservative transport (advection); first-order radioactive decay; single mother/daughter decay; chain decay.
MOFAT/Environmental Systems & Technologies, Inc.	Conservative transport (advection, dispersion, isotropic, anisotropic, diffusion); solid/liquid/gas phase transfers; first-order chemical/microbial decay.
PORFLOW-3D/Analytic & Computational Research, Inc.	Conservative transport (advection, dispersion, isotropic, anisotropic, diffusion); solid/liquid-phase transfers; first-order radioactive decay; single mother/daughter decay; first-order chemical/microbial decay
VENTING/Environmental Systems & Technologies, Inc.	Conservative transport (advection); solid/liquid/gas-phase transfers; first-order chemical/microbial decay
VADOFT/HydroGeologic, Inc.	Conservative transport (advection, dispersion, isotropic, diffusion); solid/liquid phase transfers; first-order radioactive decay; single mother/daughter decay; first-order chemical/microbial decay.
MOTRANS/Environmental Systems & Technologies, Inc.	Conservative transport (advection, dispersion, isotropic, anisotropic, diffusion); solid/liquid/gas phase transfers; first-order chemical/microbial decay.
NITRO/Environmental Systems & Technologies, Inc.	Conservative transport (advection, dispersion, isotropic, anisotropic, diffusion); solid/liquid/gas-phase transfers; first-order chemical/microbial decay
TDFD1O/Slotta Engineering Associates, Inc.	Conservative transport (advection, dispersion, isotropic, anisotropic, diffusion); solid/liquid-phase transfers; first-order radioactive decay; single mother/daughter decay; first-order chemical/microbial decay
VSAFT2 (Variable Saturated Flow and Transport in 2 Dimensions)/University of Arizona	Conservative transport (advection, dispersion, isotropic, anisotropic, diffusion)
RUSTIC/U.S. Environmental Protection Agency	Coupled-root-zone (PRZM), unsaturated-zone (VADOFT) and saturated-zone (SAFTMOD) modeling package

Source: Summarized from van der Heijde (1994).
*Code and documentation available from the IGWMC, Colorado School of Mines, Golden, Colorado 80401

phase air–water system, a three-phase system is created that can consist of water, the NAPL, and a vapor phase containing both water and NAPL vapors. Parker, Lenhard, and Kuppusamy (1987) proposed a method to scale the air–water, moisture-characteristic curve to obtain a two-phase air–NAPL or water–NAPL system, if one of the three phases is assumed to be at residual saturation. This scaling technique is valid for monotonic wetting-phase drainage from near saturation (Parker, Lenhard, and Kuppusamy 1987). Before continuing with the presentation of a scaling methodology, an understanding of the concepts of surface tension and wetting are needed.

Whether or not a fluid is wetting or nonwetting depends on its ability to adsorb to solid particles, as well as its ability to adsorb to other fluids. Liquid fluids (e.g., water) are often wetting fluids, and gaseous fluids (e.g., air) are often nonwetting fluids. NAPLS can be either wetting or nonwetting, and generally fall into an intermediate category between air and water. Figure 13.24 shows a schematic of a soil with two fluid phases, where one phase is a wetting (w) phase and the other a nonwetting (a) phase. Figure 13.25 shows a schematic of a soil with three fluid phases, where one is a wetting (w) phase, another is a nonwetting (a) phase, and the third is an intermediate (o) phase.

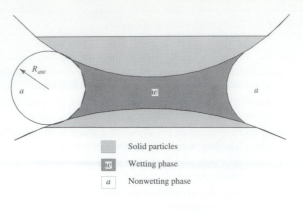

Figure 13.24   Schematic of soil with two fluid phases

Solid particles

w Wetting phase

a Nonwetting phase

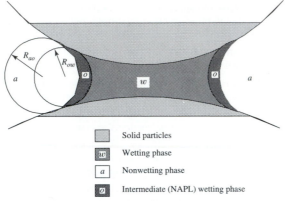

Figure 13.25   Schematic of soil with three fluid phases

Solid particles

w Wetting phase

a Nonwetting phase

o Intermediate (NAPL) wetting phase

Scaling an air–water system to a two-phase Air–NAPL system   Parker, Lenhard, and Kuppusamy (1987) propose that for a given soil, the saturation/matric potential (capillary pressure) function or moisture-characteristic curve, is written in a generalized form. This generalized form is presented as the effective saturation/matric potential $(\overline{S}/\psi_m)$ function, where S is the effective saturation as defined by equation 13.83. The generalized function of the saturation/matric potential curve is written by Murphy, Bumb, and McKee (1987) as

$$\overline{S}^*(\psi_m^*) = \overline{S}^*(\eta_{ij}\psi_{mij}) = \overline{S}_j^{ij}(\psi_{mij}) \tag{13.138}$$

where $\overline{S}^*(\psi_m^*)$ is the generalized function and $\overline{S}_j^{ij}(\psi_{mij})$ is the specific function for a two-phase system in which i is the nonwetting phase and j is the wetting phase. In this case, η_{ij} is a scaling factor, and a function of the interfacial tension between the two phases, which relates the specific to the generalized function for one of the phases. If we assume that the nonwetting phase is air, then we can let $i = a$. If the wetting phase is water $j = w$, or if it is NAPL, $j = o$. Generally, it is convenient to let $\eta_{aw} = 1$ Then,

$$\overline{S}^*(\psi_m^*) = \overline{S}^*(\psi_{maw}) = \overline{S}_w^{aw}(\psi_{maw}) \tag{13.139}$$

Thus, if the air–water characteristic curve for a given soil is known, then that same curve is scaled to define the saturation/matric potential curve for any other two-phase system. For example, an air–NAPL system is described as

$$\overline{S}_o^{ao}(\psi_{mao}) = \overline{S}^*(\eta_{ao}\psi_{mao}) = \overline{S}_w^{aw}(\eta_{ao}\psi_{mao}) \tag{13.140}$$

and a NAPL–water system as

$$\overline{S}_w^{ow}(\psi_{mow}) = \overline{S}^*(\eta_{ow}\psi_{mow}) = \overline{S}_w^{aw}(\eta_{ow}\psi_{mow}) \tag{13.141}$$

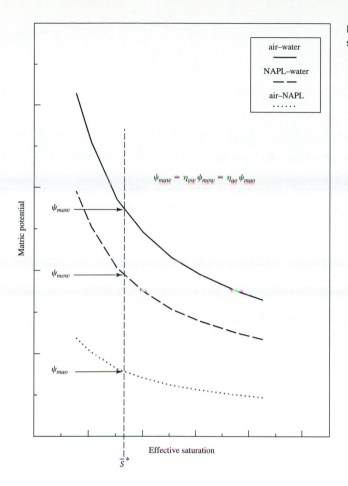

Figure 13.26   Schematic of scaling of saturation/matric potential functions

Figure 13.26 illustrates of the effects of scaling. At a given effective saturation, the corresponding matric potential on the NAPL–air characteristic curve is ψ_{mao}, while the corresponding potential on the water–air characteristic curve is ψ_{maw}, and they are related by

$$\psi_{maw} = \eta_{ao}\,\psi_{mao} \tag{13.142}$$

If η_{ao} is constant for any arbitrary values of $\overline{S}(\psi_{mao})$, then the NAPL–air characteristic curve is scaled to the water–air characteristic curve through the scaling factor η_{ao}.

Lenhard and Parker (1986) show that the values of η depend on the interfacial tensions σ_{ij}. They also point out that at a given effective saturation, it is evident that $\psi_{maw} = \eta_{ao}\,\psi_{mao} = \eta_{ow}\,\psi_{mow}$, and the capillary-rise equation can be applied to give

$$\frac{\sigma_{aw}}{R_{aw}} = \frac{\eta_{ao}\,\sigma_{ao}}{R_{ao}} = \frac{\eta_{ow}\,\sigma_{ow}}{R_{ow}} \tag{13.143}$$

where R is the radius of curvature of the fluid interfaces, as defined in figures 13.24 and 13.25. Lenhard and Parker (1986) also note that for idealized monotonic drainage paths in a soil, the fluid interfaces have the same geometry for any two-phase system if the solid surfaces are completely wet by the wetting fluid, such that the contact angle with the surface is zero. Thus, assuming that $R_{aw} = R_{ao} = R_{ow}$ at a fixed-wetting phase, saturation yields

$$\sigma_{aw} = \eta_{ao}\sigma_{ao} = \eta_{ow}\sigma_{ow} \tag{13.144}$$

Thus,

$$\eta_{ao} = \frac{\sigma_{aw}}{\sigma_{ao}} \qquad (13.145)$$

and

$$\eta_{ow} = \frac{\sigma_{aw}}{\sigma_{ow}} \qquad (13.146)$$

Table 13.6 gives physical constants for selected fluids; this table can be used to obtain selected values of density, dynamic viscosity, and interfacial tension for two-phase systems.

For the Boltzmann distribution saturation- matric-potential function, scaling is presented by Murphy, Bumb, and McKee (1987) for a NAPL–air system as

$$\bar{S}_o^{ao}(\psi_{mao}) = \exp\left(-\frac{\psi_{mao} - \psi_{1ao}}{\beta_{ao}}\right) \qquad (13.147)$$

and for an air–water system as

$$\bar{S}_w^{aw}(\psi_{maw}) = \exp\left(-\frac{\psi_{caw} - \psi_{1aw}}{\beta_{aw}}\right) \qquad (13.148)$$

where ψ_{1ao} and β_{ao} are the curve-fitting variables for the NAPL–air characteristic curve, and ψ_{1aw} and β_{aw} are the curve-fitting variables for the air–water characteristic curve. Combining

TABLE 13.6 Physical Constants for Selected Liquids

Liquid*	Interfacial tension (dynes/cm)	Viscosity (cp)	Density (g/mL)
Acetone	23.7[†‡]	0.33	0.79
Ammonia	20[†]	0.2	—
Benzene	29[†‡]	0.65	0.88
Carbon Tetrachloride	27[†‡]	0.97	1.59
Chlorobenzene	34[‡]	0.80	0.96
Chloroform	27[†]	0.58	1.49
Chlorine	19[†]	0.3	—
Gasoline	21[†]	0.48	0.73
n-Hexane	18.4[†]	0.33	0.66
Isopropyl Alcohol	21.7[†‡]	2.5	0.8
Mercury	471[†]	1.87	13.6
Nitrobenzene	43.9[†‡]	2.03	1.2
Methylene Chloride	26.5[†]	0.43	1.33
Phenol	40.9[†‡]	10	1.06
Tetrachloroethylene (PCE)	31.7[‡]	0.9	1.62
Toluene	27.7[‡]	0.59	0.87
Trichloroethylene (TCE)	29.3[‡]	0.57	1.46
Xylene	30.1[†]	—	0.86
	28.9[‡]		
Water	73[†]	1.01	1.00

Note: Additional selected data on DNAPLs can be found in Cohen, Mercer, and Matthews (1993).

*At 20°C.

[†]In contact with air.

[‡]In contact with vapor.

equations 13.142 and 13.148 gives

$$\bar{S}_w^{aw}(\psi_{maw}) = \bar{S}_w^{aw}(\eta_{ao}\psi_{mao}) = \exp\left(-\frac{\eta_{ao}\psi_{mao} - \psi_{1aw}}{\beta_{aw}}\right) \tag{13.149}$$

or

$$\bar{S}_w^{aw}(\eta_{ao}\psi_{mao}) = \exp\left(\frac{\psi_{mao} - \dfrac{\psi_{1aw}}{\eta_{ao}}}{\dfrac{\beta_{aw}}{\eta_{ao}}}\right) \tag{13.150}$$

Equating equation 13.147 and equation 13.150, gives

$$\exp\left(-\frac{\psi_{mao} - \psi_{1ao}}{\beta_{ao}}\right) = \exp\left(-\frac{\psi_{mao} - \dfrac{\psi_{1aw}}{\eta_{ao}}}{\dfrac{\beta_{aw}}{\eta_{ao}}}\right) \tag{13.151}$$

or

$$\psi_{1ao} = \frac{\psi_{1aw}}{\eta_{ao}} \tag{13.152}$$

and

$$\beta_{ao} = \frac{\beta_{aw}}{\eta_{ao}} \tag{13.153}$$

where η_{ao} is given by equation 13.145. Equations 13.152 and 13.153 permit the calculation of a characteristic curve for an NAPL–air system, given the curve-fitting variables for the water–air characteristic curve for the same soil. Figures 13.27 and 13.28 show the characteristic curves for a water–air and TCE–air, and a water–air and gasoline–air system, respectively, for a Touchet–silt loam soil. These curves are calculated using the water–air characteristic curve and equations 13.152 and 13.153.

Relating a two-phase system to a three-phase system   After scaling NAPL–air and NAPL–water systems to the reference water–air system, Parker, Lenhard, and Kuppusamy (1987) and Murphy, Bumb, and McKee (1987) applied commonly accepted (Leverett 1941) correspondences of three-phase interfaces to two-phase systems, to reduce the three-phase problem to sets of two-phase problems. Figure 13.29 shows a soil with a three-phase fluid system, and provides an illustration for the correspondence theorem as well. The S/ψ_m function characterizing the behavior of water in the three-phase system (see figure 13.25) depends only on the radius of curvature at the water–NAPL interface R_{ow}. The corresponding water saturation is S_w^{III}, and is a function of the matric potential across the water–NAPL interface ψ_{mow}. According to the correspondence theorem, $S_w^{III}(\psi_{mow})$ the functional relation is identical to that for water saturation in the two-phase system, or

$$S_w^{III}(\psi_{mow}) = S_w^{ow}(\psi_{mow}) \tag{13.154}$$

since this function also depends solely on the radius-of-curvature of the water–NAPL interface.

The S/ψ_m function characterizing the behavior of air in the three-phase system (see figure 13.25) depends only on the radius of curvature at the air–NAPL interface R_{ao}. The corresponding air saturation is S_a^{III}, and is a function of the matric potential across the air–NAPL interface ψ_{mao}. According to the correspondence theorem, the $S_a^{III}(\psi_{mao})$ functional relation is identical to that for the air saturation in the two-phase system, or

$$S_a^{III}(\psi_{mao}) = S_a^{ao}(\psi_{mao}) \tag{13.155}$$

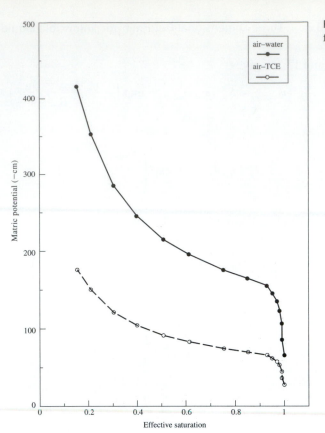

Figure 13.27   Moisture-characteristic curves for water and TCE

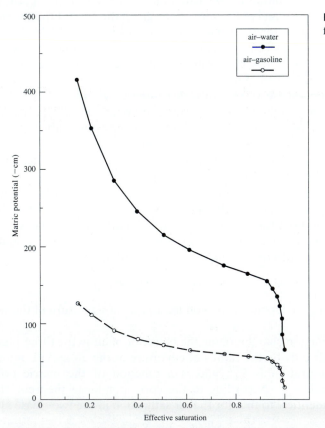

Figure 13.28   Moisture-characteristic curves for water and gasoline

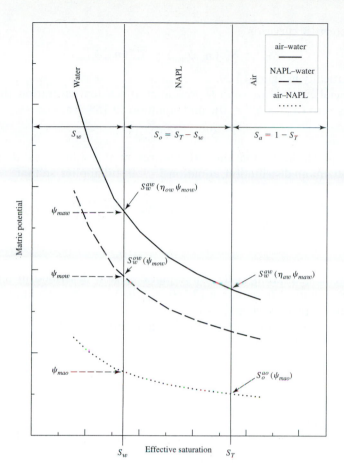

Figure 13.29   Correspondence of three-phase and scaled two-phase system

because this function also depends solely on the radius of curvature of the air–NAPL interface. However, in the two-phase system $S_a = 1 - S_o$, where S is the effective saturation of air or NAPL, while in the three-phase system, $S_a^{III} = 1 - S_T^{III}$, and $S_T^{III} = S_w^{III} + S_o^{III}$, where S_T is the total liquid saturation, the sum of the water saturation and NAPL saturation.

Both air and water (the nonwetting and wetting fluid phases of the three-phase system) behave as they would in a two-phase system. The NAPL, however, is of intermediate wettability and has two fluid interfaces affecting its S/ψ_m characteristic curve in the three-phase system. One of these two interfaces in the water–NAPL interface and the other is the air–NAPL interface (see figure 13.25). The small radius is established by the water saturation and the water–NAPL interface. The larger radius is established by the total liquid saturation (water plus NAPL) and the air–NAPL interface. Thus, the NAPL S/ψ_m characteristic curve in the three-phase system corresponds to both the water–NAPL and the air–NAPL systems as expressed by (Murphy, Bumb, and McKee 1987)

$$S_o^{III} = S_T^{III}(\psi_{mao}) - S_w^{III}(\psi_{mow}) \tag{13.156}$$

For effective saturations, this corresponds (functionally) to two-phase systems (Murphy, Bumb, and McKee 1987) as

$$\overline{S}_o^{III} = \overline{S}_o^{ao}(\psi_{mao}) - \overline{S}_w^{ow}(\psi_{mow}) \tag{13.157}$$

Scaling equation 13.157 in accordance with equations 13.140 and 13.141, we see that the difference is analytically equivalent to the difference between two points on the air–water reference characteristic curve, or

$$\overline{S}_o^{III} = \overline{S}_w^{aw}(\eta_{ao}\psi_{mao}) - \overline{S}_w^{aw}(\eta_{ow}\psi_{mow}) \tag{13.158}$$

as shown schematically in figure 13.29.

For the special case where water is present at its residual saturation, the effective saturation of water is zero ($\overline{S}_w^{ow} = \overline{S}_w^{III} = 0$), and equation 13.158 becomes

$$\overline{S}_{or}^{III} = \overline{S}_o^{ao}(\psi_{mao}) \tag{13.159}$$

where the subscript r indicates that the water is present at residual saturation.

For the Boltzmann distribution, equation 13.150 still applies, so that

$$\overline{S}_{or}^{ao}(\psi_{mao}) = \exp\left(-\dfrac{\psi_{mao} - \dfrac{\psi_{1aw}}{\eta_{ao}}}{\dfrac{\beta_{aw}}{\eta_{ao}}}\right) \tag{13.160}$$

which shows NAPL saturation in the three-phase system described in terms of a single, scaled-two-phase equation. In equation 13.160, ψ_{mao} is the matric potential at the air–NAPL interface in the three-phase system.

If water is present at other than residual saturation, the effective water saturation is greater than zero, thus the NAPL saturation is not so easily defined. In such a case, equation 13.158 still holds and the corresponding Boltzmann distribution equation is (Murphy, Bumb, and McKee 1987)

$$\overline{S}_o^{III} = \exp\left(-\dfrac{\psi_{mao} - \dfrac{\psi_{1aw}}{\eta_{ao}}}{\dfrac{\beta_{aw}}{\eta_{ao}}}\right) - \exp\left(-\dfrac{\psi_{mow} - \dfrac{\psi_{1aw}}{\eta_{ow}}}{\dfrac{\beta_{aw}}{\eta_{ow}}}\right) \tag{13.161}$$

In equations 13.160 and 13.161, ψ_{mao} refers to the matric potential at the interface between the air and NAPL phases for the three-phase system, and ψ_{mow} refers to the matric potential at the interface between water and NAPL for the three-phase system.

Murphy, Bumb, and McKee (1987) have shown that the unsaturated hydraulic conductivity K for a three-phase system—where the water is at residual saturation—also follows a power expression of the same form as equation 13.9. Because the saturated hydraulic conductivity K_s used in the power expression varies directly with fluid density and inversely with fluid viscosity, its value must be revised for the NAPL of interest. Thus,

$$K_{so} = K_{sw}\left(\dfrac{\rho_o}{\rho_w}\right)\left(\dfrac{\mu_w}{\mu_o}\right) \tag{13.162}$$

where ρ is density and μ is viscosity (throughout text we use η for viscosity, but use μ here to avoid confusion), and o and w are subscripts referring to the NAPL and water phases, respectively. Values of density and viscosity for selected liquids also are presented in table 13.6.

To demonstrate the use of the above equations, a steady-state leak of 500 liters of gasoline per day into a homogeneous and isotropic Touchet silt loam is modeled using equations 13.104 and 13.105. In this example, the leak is located 5 m below ground surface. The Touchet silt loam soil has the properties shown in table 13.3 for the Boltzmann distribution, but these properties have been scaled to gasoline by equation 13.153, using the ratio of the interfacial tension of water to gasoline. The saturated hydraulic conductivity for the loam soil is scaled to gasoline using equation 13.162. The soil has a saturated hydraulic conductivity to gasoline of 0.54 m/day (scaled using equation 13.162, a porosity of 50.1 percent, a maximum saturation of 96.5 percent, and an irreducible saturation of 36.0 percent. The solution consists of finding the steady-state gasoline-content distribution as a result of the point-source leak.

Ther steady-state effective saturations and actual saturations are shown in figures 13.30 and 13.31, respectively. The highest saturations are directly below the leak, with measurable saturations spreading away from the leak laterally, a distance of approximately 12 m. The gasoline spreading is much less than for water (see figures 13.18 and 13.19) and extends deeper, because of the interfacial tension and higher saturated hydraulic conductivity of the gasoline. Such an analysis can be used to assess the locations of monitoring devices, or to assess the relative location of conservative contaminants.

Numerical Models

Selected additional numerical models for transport of NAPLs are presented in table 13.5 and in van der Heijde (1994).

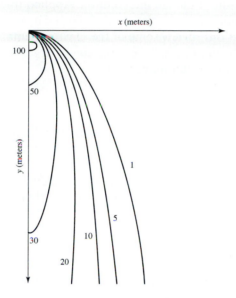

Figure 13.30   Steady-state effective gasoline saturations (S_e = %) from a point-source leak into a Touchet silt loam

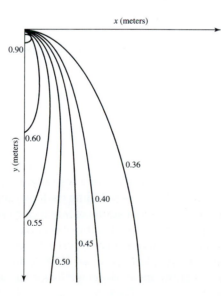

Figure 13.31   Steady-state actual gasoline saturations (S = decimal) from a point-source leak into a Touchet silt loam

13.4 USE OF TRACERS IN UNSATURATED SOIL STUDIES

Environmental and applied tracers have been used to measure water and contaminant movement in the unsaturated zone (Phillips 1994). This is especially true in arid and semiarid areas, where the diffuse downward flux (recharge) is typically very small compared to either annual precipitation or annual evapotranspiration. Because at low volumetric water content both matric potential and unsaturated hydraulic conductivity vary by several orders of magnitude with changes in water content, tracers are often used in place of traditional water-balance techniques to calculate the small fluxes associated with annual recharge. Environmental and applied tracers allow direct measurement of the displacement of water and/or solutes, thus eliminating the need for indirect calculations using uncertain variables, as is required in the water-balance technique. Tracers can also be the method of choice for assessing the impacts of preferential flow paths in the unsaturated zone.

Environmental tracers include both naturally occurring tracers as well as contaminants added to the environment as a result of human intervention. Applied tracers, in contrast to environmental tracers, are those input on a one-time basis, followed by sampling of the tracer pulse with time. The following sections review some of the environmental and applied tracers used for various purposes in the unsaturated zone; the single largest use of tracers in the unsaturated zone has been to measure water movement.

Theory of Unsaturated Liquid Movement Using Tracers

Many studies have reported attempts to use unsaturated-zone hydraulic characteristics K and ψ, to solve either Darcy's law or Richards' equation in the unsaturated zone, and to estimate soil-flux rates for time periods ranging from months to years (Sophocleous and Perry 1985; Stephens and Knowlton 1986). If the liquid (water) flux is calculated at a point below the root zone—where no further extraction by plant roots occurs—then the flux rate is equal to the ground water recharge, or

$$q = K(\theta)\, \Delta\psi_h \tag{13.163}$$

where q is the flux rate, K is the unsaturated hydraulic conductivity, and $\Delta\psi_h$ is the hydraulic potential gradient. If the concentration of salts in the unsaturated-soil water is sufficiently dilute so that osmotic potential is considered negligible, then

$$\Delta\psi_h = \Delta\psi_g + \Delta\psi_m \tag{13.164}$$

where ψ_g is the gravity potential and ψ_m is the matric potential. Because values of K and ψ are often difficult to measure with decreasing water content, equation 13.163 often leads to substantial errors in flux and recharge estimates. If a tracer-mass balance can be undertaken in the unsaturated zone—again assuming the flux rate is beneath the root zone—then the flux rate is given by

$$q = \frac{c_P P}{c_r} \tag{13.165}$$

where c_r is the tracer concentration of the unsaturated-zone water beneath the root zone, c_P is the average tracer concentration in precipitation—including both wetfall (rain) and dryfall (dust)—and P is the precipitation. If artificial applications of water are made, then P and c_P are defined for the artificial water applications. For tracers, obtaining reliable estimates of wetfall and dryfall over long time scales is a large source of error (Cook et al. 1994). Some studies (Peck, Johnston, and Williamson 1981; Phillips et al. 1988; Walker et al. 1992) have used a mass-balance approach to estimate the mean-drainage flux $\overline{q}$ to some depth z within the plant-root zone, or conversely, to estimate the age of water at a given depth within the root zone. In this case, the mean flux rate becomes

$$\bar{q}(z) = \frac{c_P P}{\bar{c}(z)} \tag{13.166}$$

where $\bar{c}(z)$ is the mean tracer concentration to depth z, given by the total tracer to depth z divided by the total water content. Cook et al. (1994) point out that such an approach is valid only if steady-state piston flow occurs in the unsaturated zone. If infiltration flows along preferential flow pathways, then equation 13.166 is not valid. If the flux rate is transient, then the tracer concentration is higher than that given by equation 13.165. Under these circumstances, the displacement of a tracer can be used to estimate the flux rate, defined by the tracer front z_{cf} (Walker et al. 1992) as

$$\int_0^{z_{cf}} \theta(z)\, dz = \int_0^{z_b} \theta(z) \frac{c_b - c(z)}{c_b - c_n}\, dz \tag{13.167}$$

where $\theta(z)$ is the volumetric water content; c_b is the equilibrium tracer concentration at a previous time; c_n is the new equilibrium tracer concentration (given by equation 13.165, above); and z_b is the depth below which $c = c_b$. The amount of water that has drained below the plant-root zone z_r since some previous time, is given by

$$Q_d = \int_{z_{cf}^o}^{z_{cf}^n} \theta(z)\, dz + \int_{z^r}^{z_{cf}^o} \delta\theta(z)\, dz + \left[\int_0^{z_r} \delta\theta(z)\, dz \right] \frac{c_n}{c_b} \tag{13.168}$$

where z_{cf}^n and z_{cf}^o are the depths to the tracer front under the new and previous times, respectively, and $\delta\theta$ is the change in water content (from the previous to the new time). This formulation is for changes in land use where infiltration and recharge increase as a result of this change, and tracer stored in the unsaturated zone is moved downward. The drainage rate can be approximated by

$$q = \frac{Q_d}{t} \tag{13.169}$$

where t is the time since the change in land use. Note that the tracer from z_{cf}^n is analogous to the center of mass c_c of an applied tracer introduced at depth z_{cf}^o at the time of a change in land use. The water flux is then given by the velocity of the tracer front, with a correction for changes in water content within the root zone.

The position of the tracer in the soil profile is described by using the center of mass z_c (median depth), or by the position of the peak concentration. The velocity of the soil water is inferred from the movement of either the center of mass or the peak concentration. The center of mass z_c is defined (Cook et al. 1994) by

$$\int_0^{z_c} \theta(z)c(z)\, dz = \frac{1}{2} \int_0^{\infty} \theta(z)c(z)\, dz \tag{13.170}$$

where $\theta(z)$ and $c(z)$ are the volumetric water content and the solute-concentration profiles, respectively. In practice, the volume of water in the soil profile above the center of mass (or peak concentration) is usually assumed equal to the total-water flux over the relevant time period. This flux is divided by the number of years that have elapsed since the center of mass of the wetfall and dryfall, or the year of highest fallout. If water movement in the unsaturated zone is by piston flow only, then using either the center of mass or peak concentration accurately measures the movement of the water (Cook et al. 1994). If water movement is via preferential flow, then the center-of-mass technique is preferred.

For the special case of tritium (^{3}H), which can be evapotranspired and decays, the mass balance is written as

$$q = \frac{\int_0^\infty \theta(z)c(z)\,dz}{\sum\limits_{i=1}^\infty \omega_i c_i e^{-t\lambda}} \qquad (13.171)$$

where q is the mean water flux below the soil surface, $c(z)$ is the tritium concentration of the soil water at depth z, $c_i e^{-t\lambda}$ is the tritium concentration in precipitation i years before the present (corrected for decay), and ω_i is a weighting function that takes into account year-to-year variations in drainage ($\omega_i = q_i/\bar{q}$). Different researchers have used different functions to assign the relative contribution of each year's precipitation to the total tritium in the soil profile. Most authors weight according to the mean annual precipitation, or $\omega_i = P_i/\bar{P}$, whereas others have used weighted fluctuations in ground water levels.

Environmental Tracers

Because of atmospheric nuclear testing in the late 1950s and early 1960s, significant increases in the concentrations of tritium (^{3}H) and chlorine-36 (^{36}Cl) occurred, and then decreased to lower levels. The ^{3}H and ^{36}Cl fallout is different for different locations on the earth's surface. Figure 13.32 shows tritium concentrations and chlorine-36 fallout distributions from the early 1950s until the late 1980s (Cook et al. 1994) at Adelaide, South Australia. Both tritium and chorine-36 have been used extensively for unsaturated-zone tracers and recharge studies (Zimmerman, Ehhalt, and Munnich 1967; Gvirtzman and Margaritz 1986; Phillips et al. 1988; Cook et al. 1994). Tritium has a half-life of approximately 12.3 years, whereas ^{36}Cl has a half-life of approximately 301,000 years. Another-often-used environmental tracer is ^{14}C, which has a half-life of approximately 5,700 years. The concentrations of these three radioactive tracers have changed dramatically over the last 30 years. Other naturally-occurring, nonradioactive tracers commonly used in unsaturated-zone studies include ^{15}N, ^{18}O, ^{2}H (deuterium), ^{13}C, and Cl. Input concentrations of these isotopes have also changed over time but

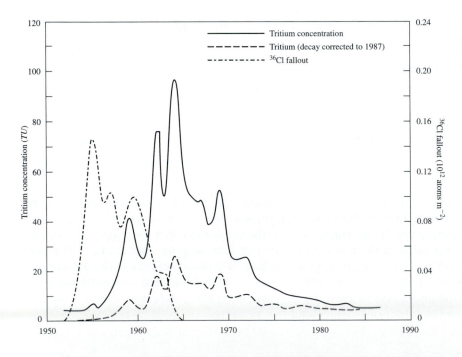

Figure 13.32 Fallout distributions for ^{3}H and ^{36}Cl at Adelaide, South Australia. Reproduced from Cook et al. 1994, copyright by the American Geophysical Union

on a much longer time scale, primarily due to changes in global climatic patterns (Allison, Gee, and Tyler 1994). Recently, Cl has been used extensively as a natural tracer, even though little is known of the temporal changes in the fallout of Cl.

Of the tracers mentioned above ^{3}H, ^{2}H, and ^{18}O probably simulate the movement of water in unsaturated soil most accurately, because they form part of the water molecule. In most unsaturated soils, ^{36}Cl and Cl also move with the water; however, in some clayey soils, anion exclusion can cause these tracers to move more rapidly than the water being traced (Allison, Gee, and Tyler 1994). While steady-state, piston flow is often capable of explaining the behavior of tracers in the field, there is mounting evidence that water movement along preferential pathways is the rule, rather than the exception. Thus, nonpiston flow has to be assessed in any unsaturated-zone tracer analysis. Unsaturated-zone flow in preferential pathways has been found to occur in both humid and arid sites (Gish and Shirmohammadi 1991).

Tritium   Studies using tritium have made use of the fact that the peak of ^{3}H in precipitation has been preserved in the unsaturated zone. As indicated in section 13.5, the position of ^{3}H in the unsaturated zone is best described using the center-of-mass technique. Many studies on the estimation of recharge using natural ^{3}H in the unsaturated zone are given in the literature (Allison, Gee, and Tyler 1994; Cook et al. 1994). To estimate recharge from ^{3}H profiles, the ^{3}H concentration of the effective input to the unsaturated-zone water system has to be known. It is noted that, in the northern hemisphere, mean annual concentrations of ^{3}H reached several hundred times the natural levels. In the southern hemisphere, mean annual ^{3}H concentrations exceeded 5 to 10 times the natural levels in the early 1960s, but have since decayed to near-natural fallout levels (see figure 13.32) (Cook et al. 1994). The techniques for measuring ^{3}H are not valid in areas where the root zone is deep, and interpretation of data can become difficult if flow regimes are not reasonably uniform. It has also been suggested that in arid areas, where soils are sandy and have a low water content, it is possible for water vapor with high ^{3}H concentrations to diffuse into unsaturated-zone water or ground-water systems. Estimates of recharge made using ^{3}H in arid areas, therefore, are high. This is especially true for recharge rates less than 1 mm/yr. Careful consideration should be given to this possibility when interpreting data in arid areas.

Chlorine-36   Bomb-pulse ^{36}Cl has been used in a number of studies (Allison, Gee, and Tyler 1994; Cook et al. 1994). Concentrations of ^{36}Cl were over 1,000 times greater than natural fallout levels between 1952 and 1965 (Bentley, Phillips, and Davis 1986). In studies in arid and semiarid areas, the ^{36}Cl profile appeared to match that of the input signal; however, the pulse was still very near the soil surface, often within the root zone. Thus, the bomb peak of ^{36}Cl, like ^{3}H, is not an ideal tracer in areas of low recharge or in areas of changing land use, because its root-zone movement can be affected by water uptake. Chlorine-36 is probably best-suited for regions where the local recharge is expected to be higher than 30 to 50 mm/yr, but this depends on the water-holding capacity of the soil and the root-zone depth. Other difficulties in using ^{36}Cl is analyzing near-background concentrations, and/or the cost of analyses. Per-sample costs for ^{36}Cl range from a few hundred to a few thousand dollars, depending on the concentration.

Chloride   Input of Cl occurs at the soil surface both as precipitation and as dryfall. The Cl can be either atmospheric or terrestrial in origin. Several researchers have found that Cl of oceanic origin can exist several hundred miles inland. Because most plant species do not take up significant quantities of Cl from unsaturated-zone water, Cl is concentrated in the root zone by evapotranspiration. If we assume that flow of water in the unsaturated zone is by piston flow, the Cl concentration in the soil increases through the root zone, obtaining a constant value beneath the root zone. If the water table is deep or has the same Cl concentration as the unsaturated-zone water, a profile such as that shown in figure 13.33 results. Under steady-state conditions, the flux of Cl is given by equation 13.163. The Cl mass-balance technique has been used in the unsaturated zone to evaluate recharge in a range of

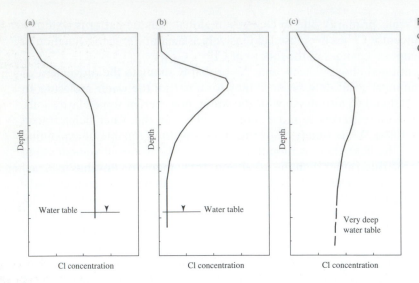

Figure 13.33   Schematic of Cl concentration depth profiles of soil water (data from Allison, Gee, and Tyler 1994)

environments successfully (Allison, Gee, and Tyler 1994). Many of the depth profiles of Cl concentration in soil water show a more complex shape than that of figure 13.33a. Some idealized examples of these more-complex shapes are given for comparison, in figures 13.33b and 13.33c. While figure 13.33a is an example of steady-state piston flow with extraction of water by plant roots, figure 13.33b is an indication of extraction of water by roots, but also either by preferred-pathway flow of water through and below the root zone, or diffusive loss of Cl to the water table. Figure 13.33c is a possible profile reflecting paleo-recharge conditions (Allison, Gee, and Tyler 1994).

Oxygen-18 and deuterium   The stable isotopes ^{18}O and ^{2}H have been used successfully in determining the origin of ground water. Variations of isotopic composition with rainfall intensity, as well as the changes that occur following evaporation, have led to a determination of the possible sources of ground water in arid areas. However, relatively few studies using these stable isotopes have been done in the unsaturated zone. The studies using ^{18}O and ^{2}H in unsaturated zone have been carried out in temperate areas where recharge was on the order of 200 mm/yr or more. In more arid areas, strongly positive values of the displacement of either ^{18}O or ^{2}H concentration from the concentration found in precipitation can occur near the surface due to evaporation through the soil surface, leading to the possibility of identifying an annual marker in unsaturated-zone water. Barnes and Allison (1988) developed detailed models for the movement of stable isotopes in both the liquid and vapor phases in unsaturated soil.

Nitrate   Nitrate (NO_3) is an involuntary tracer that can be used to give information on the rate of water movement in the unsaturated zone. Because NO_3 has come into increasing agricultural use since the 1950s and some of it has leached below the root zone, the change from higher to lower concentrations of NO_3 in unsaturated-zone water at depth, is an indication of the position in the profile of recharge, originating at the time of increased-use of NO_3. A knowledge of the amount of water stored in the profile enables an estimate of recharge then. In some situations the reverse effect occurs where, once vegetation is cleared, NO_3 associated with the native vegetation can possibly be a marker associated with the time of this clearing.

Applied Tracers

In contrast to the above tracers (input to the unsaturated zone each year in precipitation), applied tracers are a one-time application followed by sampling of the pulse of the tracer over time and space. Ideally, the applied tracer is applied beneath the root zone to reduce uptake by plants, and a sufficient time allowed to elapse between injection and sampling to allow the depth interval traversed by the tracer peak to be measured accurately (Allison, Gee, and Tyler 1994). In temperate areas, where the root zone is relatively shallow and the annual recharge is high, this method is ideal; in semiarid and arid areas, where root zones are usually deeper and recharge fluxes lower, this technique is less useful if natural precipitation is the only mechanism used to move the tracer downward.

Tracer Flux through the Root Zone

Because root-zone effects of natural environmental and applied conservative tracers are important, the processes that affect the tracer flux are presented here, along with the possible equations to account for tracer flux and extraction within the root zone. The processes that affect environmental and applied conservative tracers in recharge studies are: precipitation; evaporation; transpiration; overland flow; and vapor transport. While many arid systems can have a net upward flux from the water table, this section concerns those systems with a net downward flux. Precipitation falling on the ground surface either runs off or infiltrates the unsaturated-soil profile. Infiltrated water is partitioned within the root zone into evaporation from the wet surface and, just below the wet surface, transpiration by plants and downward flux. If the overland-flow component is assumed to be negligible, the recharge q at the bottom of the root zone is given by equation 13.163. Across the root zone, the net downward flux is also given by equation 13.163, and varies from a maximum at the ground surface to a minimum at the bottom of the root zone. Because of the large ranges of root-zone fluxes encountered in arid and semiarid areas (two to four orders of magnitude between the ground surface and the bottom of the root zone), the velocity of a tracer within the root zone is not constant for most soil profiles. Also of importance is root uptake of water, which strongly affects water movement, mixing, and dispersive processes acting on the tracer. Because recharge velocities are often very small in arid areas, diffusive transport can likely dominate deep in the root zone, while dispersive fluxes can dominate in the upper portion of the root zone, where velocities are usually higher (Tyler and Walker 1994). Therefore, a simple Fickian model probably does not describe the mixing processes in the root zone.

To further complicate the process, plant uptake of water is likely not uniformly distributed with depth in the unsaturated zone. The resulting flow field is strongly controlled by root density and distribution in both horizontal and vertical directions. Therefore, a simple one-dimensional steady-flow approach (or model) can be invalid under these conditions. In contrast, below the root zone, plant uptake is negligible and the flow field is more likely to be one-dimensional and steady-state. While the equations in this section do not account for the complexities of root-zone processes, it is not clear what magnitudes of error are introduced by assuming one-dimensional, steady-state, plug flow. The following equations for a simple model for downward root-zone water flux are taken from Raats (1974), as modified by Tyler and Walker (1994).

Consider a homogeneous soil profile subject to a steady volumetric flux of water P at the soil surface, and a steady-state downward flux q at the bottom of the root zone z_r. Assume that the flux is Darcian, and solely through the soil matrix without any macropore flow. Water is extracted by plant roots throughout the root zone at a rate of $q_{ex}(z)\ dz$. Between the ground surface $z = 0$ and the bottom of the root zone $z = z_r$, the cumulative extraction E is given (Tyler and Walker 1994) by

$$E = \int_0^{z_r} q_{ex}(z)\ dz \equiv P - q \tag{13.172}$$

where $q_{ex}(z)$ is the root extraction of water as a function of depth per unit time. This root-extraction function can be assumed to combine both evaporation and transpiration into a single term, and z represents the depth at which active roots terminate and plant uptake of water ceases. The average downward velocity of water $v(z)$, and of a conservative tracer at any point in the soil profile, is given as

$$v(z) = \frac{P}{\theta(z)} - \frac{1}{\theta(z)} \int_0^z q_{ex}(z)\, dz \qquad 0 < z \le z_r$$

$$v(z) = \frac{q}{\theta(z)} \qquad z > z_r$$

(13.173)

where $\theta(z)$ is the water content at depth z. Equation 13.173 can be integrated to determine the travel time of a tracer pulse to reach a depth z, provided $q_{ex}(z)$ is known. For a uniform root extraction with depth, the extraction function is given by

$$q_{ex}(z) = \text{constant} = \frac{P - q}{z_r} \qquad 0 < z \le z_r$$

$$q_{ex}(z) = 0 \qquad z > z_r$$

(13.174)

Raats (1974) proposes an exponential root-extraction function, which is probably more realistic compared to a uniform extraction function. This exponential form of the root-extraction function is given by

$$q_{ex}(z) = q_m \exp\left(-\lambda z/z_r\right) \qquad 0 < z \le z_r$$

$$q_{ex}(z) = 0 \qquad z > z_r$$

(13.175)

where q_m is the value of the root-extraction function at the ground surface and λ is a constant. From equation 13.172 q_m is given by

$$q_m = \frac{\lambda(P - q)}{z_r(1 - e^{-\lambda})}$$

(13.176)

The uniform and exponential root-extraction function models are shown in figure 13.34. The larger the value of λ, the more water extraction that takes place near the ground surface, whereas the smaller values of λ produce an extraction rate approaching the uniform root-extraction rate. Tyler and Walker (1994) suggest that for most arid areas λ is large, accounting for both high density of shallow roots and bare-ground evaporation.

If the water content θ is constant with depth, or $\theta(z) = \theta$, then equation 13.173 can be integrated to obtain the travel time $t(z)$ of a tracer pulse injected at the ground surface (Tyler and Walker 1994)

$$t(z) = \int_0^z \frac{dz}{\left(\dfrac{P}{\theta}\right) - \left(\dfrac{1}{\theta}\right) \displaystyle\int_0^z q_{ex}(z)\, dz}$$

(13.177)

Equation 13.177 accounts for the variation in water content with depth. Raats (1974) indicates that for most arid conditions, the variation of water content with depth is usually small, except for the upper 5–10 cm that are subject to variations due to evapotranspiration. If the two equations for the uniform root-extraction function (equation 13.174) and the two equations for the exponential root-extraction function (equation 13.175) are inserted into equation 13.177 and integrated, simple expressions for travel time as a function of P and q are obtained. The uniform root-extraction function travel time is

$$t(z) = \frac{z_r \theta}{q - P}\left[\ln\left(1 - \frac{z(P - q)}{Pz_r}\right)\right]$$

(13.178)

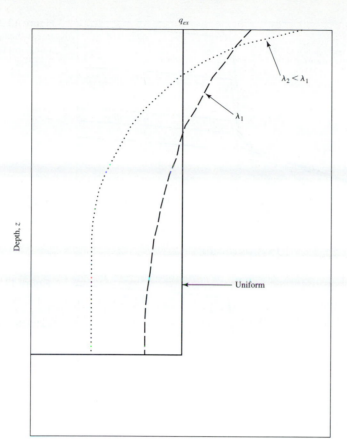

Figure 13.34   Definition of a uniform and exponential root-extraction model

For the exponential root-extraction function, the travel time is

$$
t(z) = \cfrac{z\theta}{P - \cfrac{P-q}{1-e^{-\lambda}}} + \cfrac{z_r\theta}{\lambda\left(P - \cfrac{P-q}{1-e^{-\lambda}}\right)}
$$

$$
\times \left[\ln\left(P - \frac{P-q}{1-e^{-\lambda}} + \frac{P-q}{1-e^{-\lambda}} \exp\left(-\frac{\lambda z}{z_r}\right) \right) \right] \tag{13.179}
$$

$$
- \cfrac{z_r\theta}{\lambda\left(P - \cfrac{P-q}{1-e^{-\lambda}}\right)} \ln(P)
$$

Equations 13.178 and 13.179 are applicable for the root zone from ground surface to z_r. For depths below the root zone, the simple constant-velocity model is written in terms of travel time, given by

$$
t(z) = \frac{z\theta}{q} \tag{13.180}
$$

Figures 13.35 and 13.36 show the travel times for the uniform root-extraction function and a strongly exponential ($\lambda = 5$) root-extraction function, respectively, for selected ratios of recharge flux q to precipitation P, and with $z_r = 100$ cm, $\theta = 0.10$, and average annual P of 25 cm/yr, typical of many arid to semiarid areas. The ratio of recharge to precipitation has little effect on travel times from either root-extraction model in the upper portion of the root

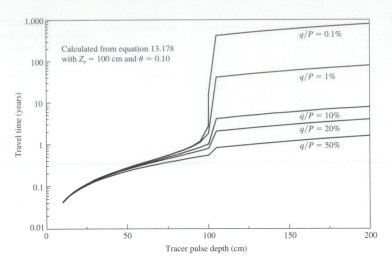

Figure 13.35   Travel time of a tracer through the root zone for a uniform root-extraction function

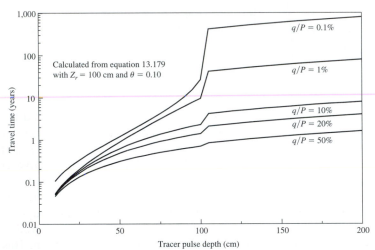

Figure 13.36   Travel time of a tracer through the root zone for an exponential root-extraction function

zone; the difference is apparent as the tracer pulse approaches the bottom of the root zone ($z = 100$ cm). The choice of root-extraction model is obvious within the root zone, with the uniform model showing the most rapid transport through the root zone. Figure 13.37 shows the comparison of tracer travel times for various root-zone extraction functions for a uniform ratio of recharge to precipitation of 1 percent. For fixed values of q, P, θ and z_r, the travel-times calculation (taking into account plant root uptake of water) is much less than that calculated by the simple steady-state, piston-flow model. If a tracer is injected at the ground surface at $t = 0$, its velocity through the upper portion of the root zone is much faster than the final recharge velocity. As the tracer moves to the bottom of the root zone, its instantaneous velocity asymptotically approaches the recharge velocity as seen in figures 13.35 and 13.36. Therefore, a constant-velocity estimate of the recharge flux made while the tracer is still in the root zone is erroneous. Only after the tracer has moved below the root zone can the constant-velocity model estimate of recharge begin to approach that of the actual recharge.

Tyler and Walker (1994) investigate the errors associated with the constant-velocity model by calculating the relative errors in travel time between assuming a constant-velocity in the root zone (equation 13.180) versus a root-zone extraction model (equations 13.178 or

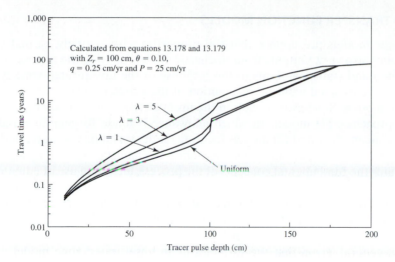

Figure 13.37   Travel time of a Tracer through the root zone for various root-extraction functions, for a 1 percent Recharge-to-precipitation ratio

13.179). The relative error between actual recharge and that calculated by the position of the tracer is given by

$$RE = \frac{z(t)\theta}{t_a q} = \frac{\hat{q}}{q} \qquad (13.181)$$

where RE is the relative error between the actual recharge rate and the constant-velocity model recharge rate; $z(t)$ is the position of the tracer pulse at any time t; and depth z; t_a is the actual tracer travel time; q is the actual recharge rate; and $\hat{q}$ is the estimated recharge rate from equation 13.180. Tyler and Walker (1994) show that the relative errors are small for large values of recharge; that is, ratios of q/P are near 50 percent. For recharge ratios approaching those typical of arid areas (between 0.1 and 1 percent), the estimated recharge is two to three orders of magnitude higher than actual recharge through the root zone. Additionally, Tyler and Walker (1994) point out that these errors persist well below the root zone and can still be within a factor of two at twice the root-zone depth. This level of accuracy, however, is well within the estimation errors of other recharge-estimation techniques.

The time for a tracer to reach twice the root zone depth is strongly related to the recharge rate as shown in figures 13.34 and 13.35. Figure 13.34 shows the travel time for a uniform root-extraction function, and indicates that the travel time to twice the root-zone depth (200 cm) ranges from approximately 2 years (when 50 percent of the precipitation is recharged) to over 700 years (when only 0.1 percent of the annual precipitation is recharged). The strongly exponential root-zone functions do not significantly change this conclusion, as shown in figure 13.35. Therefore, it is important to rethink the use of environmental radioactive-tracer pulses, such as [36]Cl and ^{3}H, in studies of recharge in arid climates. Because these environmental tracers were introduced through precipitation up to 50 years ago, their applicability to estimate recharge accurately is strongly related to the actual-recharge flux. For typical arid area-recharge variables ($P = 25$ cm/yr, $z_r = 100$ cm, and $\theta = 0.10$), the time to reach twice the root-zone depth is 10 years or less when q/P is 10 percent, but approaches 80 years for q/P equal to 1 percent. Tyler and Walker (1994) conclude that radioactive tracers are therefore limited to use in those areas where actual recharge is perhaps 10 percent or more of the actual average-annual precipitation. Local climate and vegetation can, however, combine to produce high-recharge rates that make radioactive (and other) environmental tracers appropriate. Therefore, local conditions have to be investigated prior to the selection of an appropriate tracer.

13.5   STOCHASTIC AND TRANSFER FUNCTION MODELS

Stochastic models presuppose that soil properties vary spatially, so that water and solute movement also vary. Outputs from stochastic models provide the moments (mean, variance, skew, etc.) and statistical limits of the response of the unsaturated-zone system. Stochastic models have evolved with the recognition of the problems caused by variability for deterministic models. Stochastic models in the unsaturated zone can be broadly categorized into two approaches: **(1)** models in which allowance is made for spatial variability in existing mechanistic models; and **(2)** models focusing on the variability of water and solute transport that do not take the mechanism into account. "Mechanistic models" mean that the model incorporates the fundamental concepts of the process, including use of equations derived from Darcy's law for water movement and mechanisms of advection and dispersion for solute transport. Nonmechanistic models have their approach in transfer functions, as we will discuss in the next section. Stochastic models are often difficult to use because of the shortage of suitable field studies against which to validate them.

In general, given that the variables in an unsaturated-zone model vary spatially, the usual approach is to model the resulting flow and transport in terms of: a stochastic representation of those variables; an appropriate simulation of the process; and application of Monte Carlo-techniques (Charbeneau 1989; Addiscott and Wagenet 1985). In this type of modeling, the formulation of the simulation is critical in terms of computational effort. Addiscott and Wagenet (1985), Hill (1986), and Adamson (1976) indicate that the particular model to be randomized can be quite approximate, yet still lead to reasonable results.

Analytical Model

Charbeneau (1989) presents relatively simple stochastic models of soil-water content that can be used in addition to Monte Carlo methods; this simple model is presented below. As a particular example, Charbeneau (1989) considers the water content at a depth z in a soil that was initially at a uniform natural saturation $\widetilde{S}_e$. The simple model formulation assumes that the greatest variability in the field occurs in saturated hydraulic conductivity K_s. Therefore, it was assumed that K_s was the only random variable; additionally, it was assumed that K_s had a log-normal distribution. This is not unreasonable, because many hydrologic variables are shown to be log-normally distributed. Saturated hydraulic conductivity has repeatedly been shown to be log-normally distributed (Paschis, Kunkel, and Keonig 1988). If $y = \ln(K_s)$, then the random variable y has a mean μ_y and a variance σ_y^2. The probability density function of K_s is given by (Benjamin and Cornell 1970)

$$f_{K_s}(y) = \frac{1}{K_s \sigma_y \sqrt{2\pi}} \exp\left\{ -\frac{1}{2}\left[\frac{1}{\sigma_y} \ln\left(\frac{K_s}{m_y} \right) \right]^2 \right\} \tag{13.182}$$

where $f_{K_s}(y)$ is the log-normal probability density function for K_s and $\overline{m}_y$ is the median of K_s. Remembering the kinematic-profile model of Section 13.2, the effective water content S_e at depth z is equal to that given by equation 13.52 if the lowermost-drainage characteristic has not reached a depth L, and is given by equation 13.66 otherwise. The lowermost-drainage characteristic moves downward with a velocity given by equation 13.65. At depths greater than this lowermost-drainage characteristic, the water content is equal to the natural-saturation value, whereas at depths above this characteristic, the water content is equal to its corresponding value on the drainage curve. In terms of the random saturated hydraulic conductivity value, S_e is equal to $\widetilde{S}_e$ (Charbeneau 1989) so long as

$$K_s < \frac{(\theta_m - \theta_r)z}{nt\left(\dfrac{1}{2}\right)^{n-1/n}} \tag{13.183}$$

otherwise, S_e is given by equation 13.66. Because K_s is considered a random variable, so is S_e. Thus, of interest here is the development of the probability-density function for S_e, or the first two moments of S_e (mean and variance). For this simple problem, the probability density functions for K_s and S_e are related through the general rule for transformation of random variables (Benjamin and Cornell 1970). If two random variables x and y are related such that the function $y = g(x)$ relating the two random variables, is always increasing or decreasing and there is only one value of y for each value of x, then it can be shown (Benjamin and Cornell 1970) that the probability density functions for x and y [$f_X(x)$ and $f_Y(y)$] are related by

$$f_Y(y) = \left| \frac{dx}{dy} \right| f_X(x) = \left| \frac{dg^{-1}(y)}{dy} \right| f_X(g^{-1}(y)). \tag{13.184}$$

For this simple example, $x = K_s$ and $y = S_e$. Because S_e is defined differently over separate ranges of K_s, the resulting probability density function is a mixture of a single-density function defined by $\tilde{S}_e$, and a continuous probability density defined by the drainage wave. The condition defined by equation 13.183 specifies the probability mass associated with the soil moisture when $S_e = \tilde{S}_e$. The continuous part of the probability density function is found from the transformation rule of equation 13.184, applied to equation 13.66 as

$$\left| \frac{dK_s}{dS_e} \right| = \frac{(n-1)(\theta_m - \theta_r)z}{nt(S_e)^n} \tag{13.185}$$

Combining equations 13.182, 13.184 and 13.185 and using equation 13.66 to eliminate K_s gives the continuous part of the probability density function as

$$f_{S_e}(\theta) = \frac{n-1}{\sigma_y \theta \sqrt{2\pi}} \exp \left\{ \frac{\left[\ln \left(\frac{(\theta_m - \theta_r)z}{nt\overline{m}_y \theta^{n-1}} \right) \right]^2}{2\sigma_y^2} \right\} \tag{13.186}$$

for K_s not satisfied by equation 13.183.

Figure 13.38 shows the continuous probability-density function given by equation 13.186 at times of 100, 1,000, and 10,000 hours for the sandy loam soil example given in section 13.2 at a depth of 30 cm; with $\theta_m = 0.453$; $\theta_r = 0.041$; $\overline{m}_y = 2.59$ cm/hr $n = 8.29$; and $\sigma_y = 1.2$. As indicated in figure 13.38, for small times the probability density function is truncated and most of the mass is associated with the undrained natural-water content. After about 1,000 hours, the singular density no longer contributes to $f_{S_e}(\theta)$.

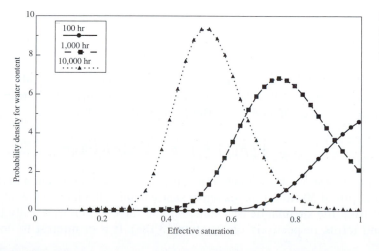

Figure 13.38   Water-content probability density function at times of 100; 1,000; and 10,000 hours for a sandy loam soil

The greatest interest, however, lies in the first two statistical moments, specifically the mean and variance of S_e. The mean (or expected) value of S_e is found from (Benjamin and Cornell 1970)

$$E[S_e] = \int_0^1 \theta f_{S_e}(\theta)\, d\theta \qquad (13.187)$$

Charbeneau (1989) shows that with a change in variables ($\alpha = (n - 1) \ln \theta$), the integral of equation 13.185 can be simplified to give the expected value of S_e in terms of the standard normal probability distribution, with the final expected value given as

$$E[S_e] = \tilde{S}_e F_K \left\{ K_s < \frac{(\theta_m - \theta_r)z}{nt\left(\frac{1}{2}\right)^{(n-1)/n}} \right\} \exp + \left\{ \frac{\ln\left[\dfrac{(\theta_m - \theta_t)z}{nt\overline{m}_y}\right] + \dfrac{\sigma_y^2}{2(n - 1)}}{n - 1} \right\}$$

$$\times N\left\{ \ln\left[\frac{nt\overline{m}_y(\tilde{S}_e)^{n-1}}{(\theta_m - \theta_r)z}\right]^{1/\sigma_y} - \frac{\sigma_Y}{n - 1} \right\} \qquad (13.188)$$

where $F_K\{\ \}$ is the cumulative normal frequency distribution for K_s and $N\{\ \}$ is the cumulative normal frequency distribution for the terms in braces.

There is no simple expression for the above cumulative normal distribution, but it has been evaluated numerically and tabulated for the standardized random variable. For the general case of the cumulative normal distribution for K_s

$$F_K(x) = P[X \le x] = P\left[U \le \frac{x - m_X}{\sigma_X} \right]$$

$$= F_U\!\left(\frac{x - m_X}{\sigma_X}\right) = F_U(u) \qquad (13.189)$$

$$= \frac{1}{\sqrt{2\pi}} \int_{-\infty}^{u} e^{-1/2 v^2}\, dv \qquad -\infty \le u \le \infty$$

in which $u = (x - m_X)/\sigma_X$; tables yield values of $F_U(u)$. Because of the symmetry of the probability density function, tables give only half the range of u, usually $u \ge 0$. Because of the case presented here, $y = \ln K_s$, $F_U(u)$ is evaluated from

$$F_U(u) = F_U\!\left(\frac{\ln (y/\overline{m}_y)}{\sigma_{\ln Y}}\right) \qquad (13.190)$$

where

$$u = \frac{1}{\sigma_{\ln Y}} \ln \frac{y}{\overline{m}_Y} \qquad (13.191)$$

The cumulative normal distribution can still be evaluated using the table of the cumulative normal distribution [$F_U(u)$], but now u is in terms of the logarithmic-transformed variable. The variance of S_e is computed from (Benjamin and Cornell 1970)

$$VAR[S_e] = \int_0^1 (\theta - E[S_e])^2 f_{S_e}(\theta)\, d\theta \qquad (13.192)$$

Figure 13.39 shows the expected (mean) effective water content and the standard deviation at a depth of 30 cm for the example sandy loam soil. Figure 13.39 also shows the contributions from the singular-density function and the continuous-density function (the first and second terms, respectively, of equation 13.188). It is of interest to note that while the

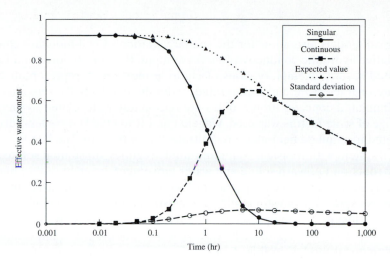

Figure 13.39 Components of stochastic water content at 30-cm depth for a sandy loam soil

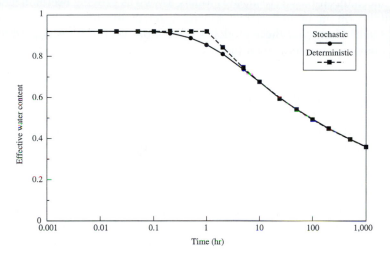

Figure 13.40 Comparison of stochastic and deterministic models at 30-cm depth for a sandy loam soil

standard deviation of S_e is small relative to the mean, this variability is important because both volumetric flux and recharge are related by $(S_e)^n$.

Figure 13.40 presents the water content versus time at the 30-cm depth as predicted by the stochastic model (equation 13.188) and the deterministic model of equation 13.66, with K_s values equal to the mean and median of the distribution. The median value predicts the stochastic mean quite well, as would be expected, because the median value of K_s corresponds to the mean value of $y = \ln K$.

Another stochastic analytical model is VADSAT (ES&T, Inc. and HydroGaia, Inc. 1994). VADSAT is based on coupled analytical solutions of the unsaturated- and saturated-zone flow and transport equations. With appropriate initial and boundary conditions, the model can estimate peak concentrations of contaminant, as well as the time to reach the peak concentration for downgradient receptors in the saturated zone. Uncertainty analyses can be conducted via Monte Carlo simulations to assess effects of soil and waste-property uncertainty on the risk of ground water contamination at land-disposal sites. The model is available as an interactive computer program to simulate the movement of conservative inorganic or reactive organic species. The model considers evaporation of volatile organic contaminants, leaching of soluble contaminants, advective transport, dispersive transport, adsorption, and microbial decay.

Numerical Models

Numerical models also allow for the spatial variation of soil-moisture characteristic curves and relative hydraulic-conductivity curves as input. Bresler, Bielorai, and Laufer (1979) have generated several exemplary cases of chloride movement, representative of the range of moisture-characteristic and unsaturated hydraulic conductivity relations measured in the field. Simulated concentration profiles are reasonably well-related to measured ones when the range of scale factors was used. Wagenet and Rao (1983) were less successful when they attempted to simulate field-measured nitrate concentrations. The fact that they worked on a cropped system with transient water and solute regimes, substantial solute and water extraction by roots, and upward solute and water movement, brought out an important point: the variability in soil hydraulic properties did not produce as much variability in solute concentrations as expected from stochastic-mechanistic simulations of uncropped systems. The apparent conclusion that the presence of roots diminishes the impact of the variability in hydraulic properties—if supported by further work—should prove important.

Numerical models in this class give a conceptual framework for the development of further mechanism-based stochastic models, as well as proving useful in assessing the impact of variability on the input variables. Further refinements in technique can be expected; however, it seems likely that these models will be more useful in research rather than environmental management in the near future.

An interesting alternative approach to numerical mechanistic models (i.e., entirely nonmechanistic models) lies in the transfer functions applied to industrial processes by Dankwaerts (1953) and hydrological processes by Ericksson (1971) and suggested by Raats (1978) for unsaturated porous media. As used by Jury (1982) and Jury and Roth (1990), this stochastic approach measures the distribution of solute travel times from the soil surface to some reference depth. A distribution function of the form

$$P_L(I) = \int_0^I f_L(I)\, dI \tag{13.193}$$

is constructed in which $f_L(I)$ represents the probability density function summarizing the probability (P_L) that a solute added at the soil surface will arrive at a given depth L, as the quantity of water applied at the surface increases from I to $(I + dI)$. The model considers the unsaturated soil composed of twisted capillaries of different lengths, within which water moves by piston flow. An estimate of the travel-time probability density function, $f_L(I)$, is obtained using solution samples of the soil located at depth L at various field locations. In general $f_L(I)$ is log-normally distributed (Biggar and Nielsen 1976; van der Pol, Weirenga, and Neilsen 1977). Calibration of this model using $C(z, I)$ values from one depth, z, provided all the information necessary to predict concentration, C, at deeper depths as well as larger values of I, or values of $C(z)$ at one I. Field comparisons show good agreement between measured and predicted bromide concentrations (Jury, Stolzy, and Shouse 1982), and indicate that these types of models are useful as a stochastically based environmental management model for solute movement. The most important characteristic of transfer-function models is that they attempt to simulate spatially variable field processes with only the minimum of input data. It is not yet known whether such an approach is satisfactory in vertically nonhomogeneous unsaturated soil, or whether such a model can give accurate estimates of flux rate as well as solute concentration.

SUMMARY

In this chapter, we have discussed how variably-saturated-zone mathematical models are useful tools for predicting the extent of subsurface contamination, and conducting premonitoring studies for the placement of detection devices in a format understood by both field-response personnel and regulatory entities. The emphasis of this chapter was on those analytical solutions readily used to solve problems in unsaturated-zone flow. Thus, we discussed analytical models and numerical models in several frameworks: one-dimensional deterministic liquid-flow models; three-dimensional deterministic liquid- and vapor-flow models; and three-dimensional deterministic immiscible-liquid-flow models. Additionally, we described the use of tracers in unsaturated-soil studies and stochastic and transfer-function models. Finally, we have provided several tables listing various models, their developers, and descriptions of the approach the model uses for predictability. For the reader who has little programming language experience, the best approach to understanding modeling may be to understand fully the parameters incorporated within various models, and how changes in these parameters affect predictability; having accomplished this, the reader will understand the physics involved in modeling. Then, once he or she has learned a programming language such as Fortran or C, the principles involved in modeling will be more fully realized.

ADDITIONAL QUESTIONS

13.1. Average characteristics for a sandy loam are $n = 8.29$, $K_s = 18.90$ cm/hr, $\varphi = 0.423$, and residual water content $= 0.048$; $i = 1.5$ cm/hr for 4 hours. Find the soil moisture profiles using **(1)** rectangular profile, and **(2)** kinematic profile at times of 6, 48, and 240 hours. *Hint:* for rectangular profile use equations 13.56–13.60 and 13.62; for kinematic profile you may need to use equations 13.66 and 13.71–13.72.

13.2. Use the same soil in question 13.1 and the same value for i; assume average annual infiltration of 40 cm. Find the time history of water flux (drainage) at 1.5 m using rectangular and kinematic profiles. How does the drainage history of the soil differ with the type of profile assumed?

13.3. Brooks and Corey found the following moisture characteristics for a Touchet silt loam soil.

Degree of saturation (decimal)	Matric potential (cm)
1.00	−15.6
1.00	−25.6
1.00	−35.6
1.00	−45.6
0.998	−65.6
0.995	−85.6
0.992	−105.6
0.984	−125.6
0.978	−135.6
0.967	−145.0
0.946	−155.6
0.892	−164.6
0.821	−175.4
0.719	−195.6
0.641	−215.2
0.562	−246.0
0.492	−285.2
0.424	−354.0
0.383	−414.4

For the above data, Brooks and Corey found the porosity to be 0.485. Find fit parameters for the Brooks–Corey relation (λ, S_m, S_r, and ψ_b); the Boltzmann distribution (β, S_m, S_r, and P_{cl}), and the Fermi distribution (β, S_m, S_r, and $P_{cl/2}$). Plot the original Brooks–Corey data along with your fit curves for each of the above three relations.

14

Drainage in Soil Water and Ground Water

INTRODUCTION

In many areas throughout the world, there is either an abundance or a shortage of water. For agriculture, an ideal condition is one where the water table is deep, and there is unrestricted flow of excess water or salt from the root zone through the soil. In an industrial setting such as that at a landfill, one would prefer conditions with restricted movement of moisture through the soil away from the site or a very deep (even nonexistent) water table is desired. Also, in the latter case, it would be an advantage to have minimal amounts of rainfall since this means less potential for contamination from chemicals moving away from the site to ground water via infiltration, runoff, and drainage. Ideal conditions are rare, and drainage of excess water can pose a problem. Anytime excess water from drainage operations enters the soil, the potential for environmental hazards and a threat to water quality exists. The water-quality aspects due to runoff, infiltration, and other parameters was previously discussed (chapters 11 and 12); the hydraulic aspects of drainage will be discussed in this chapter. Also, the purpose here is to give a brief introduction to the principles involved in drainage. For a more detailed discussion, the reader is referred to Luthin (1957), Luthin (1966), and Van Schilfgaarde (1974).

14.1 PROBLEMS ASSOCIATED WITH DRAINAGE

Drainage is the removal of excess surface and subsurface water by means of various water-conveying devices (e.g., artificial drains, pipes, and ditches). Poor drainage can cause a variety of pollution problems; prime examples are the rising of a water table beneath a landfill (or other waste-storage site), or developed wetlands, as well as the possibility of leaching and surface runoff (see chapter 11). Leaching and runoff can affect the potential contamination by accidental spills. When the water table rises beneath a spill, the ground water comes in direct contact with the contaminant, which can then be transported through the soil at (typically) much faster velocities than normally occur through the vadose zone. As the ground water rises and falls, it causes a washing (or leaching effect) on the contaminant that increases its concentration in ground water. In agriculture, poor drainage usually enhances the development of saline and sodic soils, and can cause severe plant growth problems. Regardless of the industry, anaerobic conditions produced by poor drainage can cause the reduction of various oxidized forms of both organic and inorganic compounds; the resulting toxic substances accumulate and can cause serious harm to the environment. Also, such reducing conditions favor the denitrification process, and can produce large amounts of nitrous oxide. Nitrous oxide emitted as an end-product of denitrification contributes to the destruction of stratospheric ozone; it is transported slowly to the stratosphere where it either photolyses to $N_2 + O_2$, or reacts with singlet oxygen to produce N_2 plus O_2 or NO. The NO thus produced reacts with stratospheric ozone to produce NO_2 and O_2 (Finlayson-Pitts and Pitts 1986).

Although drainage can reduce contamination of the environment in many instances—and provide a greater land base for construction of homes, businesses, and industry—drainage is not appropriate for all situations. For instance, in some areas there will always be a "trade-off" or "balancing-act" between what is perceived as good for man and what is best for the environment; examples of this are wetland areas. Wetlands generally are characterized by shallow fresh water; some may even be dry part of each year. They can be called marshes, bogs, or swamps, but can also include coastal beaches and estuaries, lakes, rivers, and poorly drained farmlands. The main types of wetlands are defined by the dominant vegetation and—because the water within a wetland does not flow like that in a stream or movement in a lake—a wetland can have extreme spatial heterogeneity. For example, a small pool lined with cattails or reeds can be directly adjacent to a large patch of saw grass inhabited by a wide variety of birds and other wildlife, which can be interspersed with small stands of trees or pockmarked by open pools of water that are filled with submerged water weeds or have inundated bottoms.

In recent ecological and regulatory literature, the term "wetland" refers to any site whose soil development, biotic community, or hydrologic behavior is dominated by a periodic saturation of water. Since saturation can be a nuisance, much time and effort has been spent in attempts to drain wetlands or in some instances, to fill them. Because wetlands have become rare due to residential, industrial, and agricultural development, their importance has finally been realized and more effort is now spent to understand and protect them. The bulk of legal and technical work associated with wetlands has been in their identification and delineation. The federal government is committed to a "no net loss" policy because of the benefits associated with wetlands; laws at the local, state, and federal level have been instituted with regard to the use and development of these areas. Wetlands show an increasing value for improving water quality by: acting as riparian zones around various industrial sites to effectively treat polluted water, including municipal wastewater (Ward and Elliot 1995); promoting deposition of soil erosion; acting as sites for ground water recharge; becoming flood control buffers; acting as detention areas to slow the flow of runoff. Because wetlands are shallow, pollution hazards associated with them are intensified when a drainage network is poorly designed, or when the drainage system deteriorates due to inadequate maintenance. Consequently, it is the development of these wetland areas that result in pollution, not just the fact that they may be poorly drained. As a result, the benefits of wetlands and whether or not they should be drained, need to be carefully balanced against their development for commercial purposes, as well as the need to minimize effects of such development—by proper drainage versus the benefits of leaving the wetlands undeveloped for recreational, esthetic, or wildlife habitat resources.

Several factors influence drainage. These include: **(1) Recharge rate**—the rate at which water is added to ground water. If supply is greater than the discharge or drainage rate, the water table will rise; when supply equals drainage rate, a steady-flow condition will exist, and when recharge is less than the drainage rate, the water table will fall; **(2) Hydraulic conductivity**—the greatest effect observed with this parameter is in the event of soil-profile layering, in which one layer greatly retards water flow, thus causing a difference in the flow pattern; **(3) Hydraulic pressure** of the ground water and subsequent water-table configuration that can affect the horizontal level of the water table and—in instances where a confined aquifer is present—can exhibit artesian pressure; and **(4) Physical parameters of the drain/drainage**

device—these include drain diameter/ditch size; inlet openings in drainage devices; depth to drain from land surface; horizontal drain spacing; embedding materials (typically, coarse gravel); tendency of drains to clog; and of course, the type of medium in which the drains/ditches are installed.

14.2   TYPICAL DRAINAGE SITUATIONS UNDER FIELD CONDITIONS

Flow in an Unconfined Aquifer

Aquifers that are close to the soil surface, and with continuous layers of various materials of high permeability that extend to the base of the aquifer, are called unconfined aquifers. These aquifers have a ground water table that is at atmospheric pressure. From the principles discussed in chapter 6, we see that water flow in this case is primarily caused by gravity. The upper boundary of the unconfined aquifer is typically taken as the water table, where pressure is atmospheric. A drainage theory to investigate flow in an unconfined aquifer was developed by Dupuit (1863) and later extensively used and popularized by Forchheimer (1930); consequently, the theory is aptly named the Dupuit–Forchheimer (D–F) theory. This theory assumes that: **(1)** streamlines in a gravity-flow system are horizontal and uniform throughout the aquifer's depth; and **(2)** the flow velocity associated with each streamline is proportional to the slope of the water table, but independent of the depth of the water-saturated medium. In this case, the proportionality factor is the hydraulic conductivity used in Darcy's law.

Strict adherence to the theory, although offering useful results, can be meaningless in the physical sense. For example, observe the four streamlines above the water table in figure 14.1 at points A, B, C, and D. To take them purely as horizontal (as suggested in the first assumption) is absurd. Despite this, the exact discharge from a canal to a ditch—as in the Florida Everglades case study (discussed at the end of this chapter)—or to a well tapping an

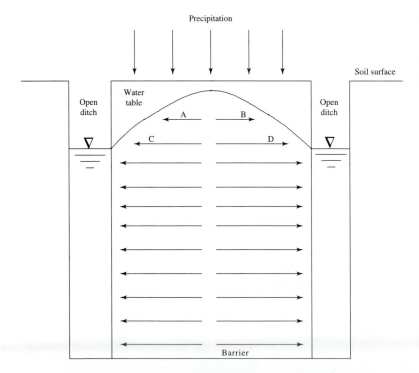

Figure 14.1   Horizontal flow lines of the Dupuit–Forchheimer theory for open-ditch drainage (after Kirkham 1971). Note that the barrier is arbitrarily drawn at the bottom of the ditches. In field situations, the bottom of the ditch could be any distance above the barrier.

unconfined aquifer, can be computed using D–F theory. Also, at distances larger than about twice the aquifer thickness from the drain or well, D–F theory provides excellent predictions of the water-table height. The D–F theory is widely used in drainage design for land reclamation in many parts of the world. An excellent treatise on the Dupuit–Forchheimer theory is given by Kirkham (1971), who shows its applicability to drainage between two parallel ditches in a homogeneous soil.

Flow Toward a Well

Assuming a homogeneous and isotropic soil, the idealized solution for steady, radial ground water flow to a well in an unconfined aquifer can be found. Using the representation in figure 14.2, the rate of flow q_r of water into the well is determined by using Darcy's law in the following form:

$$q_r = K \frac{A \, \Delta H}{\Delta r} \tag{14.1}$$

where q_r is the flow rate ($L^3 \, T^{-1}$); A is the area in the medium through which water flows toward the well ($2\pi r h$); H is the height of the water table above the impermeable stratum (L) and is found as previously discussed ($h_2 - h_1$); r is the radius of the cylindrical area (L); and $\Delta H/\Delta r$ is the hydraulic gradient.

Substituting the value for A ($2\pi r h$), separating the variables, and integrating between the limits $r = r_0$ where $h = h_0$, as well as $r = r_1$ where $h = h_1$, then

$$\int_{h_0}^{h_1} h \, dh = \frac{q_r}{2\pi K} \int_{r_0}^{r_1} \frac{1}{r} \, dr \tag{14.2}$$

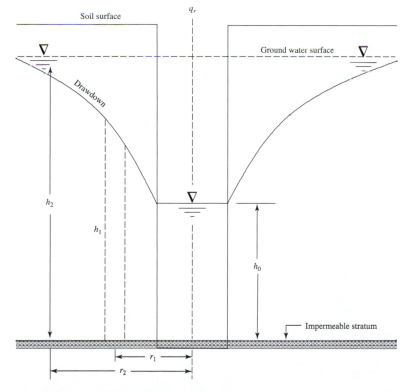

Figure 14.2   Steady radial flow (seepage) toward a fully penetrating well in an unconfined aquifer

or

$$\frac{h_1^2 - h_0^2}{2} = \frac{q_r}{2\pi K} \ln \frac{r_1}{r_0}$$

(14.3)

Solving for q_r we have

$$q_r = \pi K \frac{h_1^2 - h_0^2}{\ln \left(\dfrac{r_1}{r_0} \right)}$$

(14.4)

which yields the discharge into a fully penetrating well from steady, radial ground water flow.

Flow between Parallel Ditches

For a condition in which two parallel ditches have a different water-table elevation—as in points A and B in figure 14.3—the flow Q_r can be determined using Darcy's law. The difference in elevation between points A and B establishes a steady flow between the two ditches, for which the discharge through a unit width of the medium (perpendicular to the drawing plane) is

$$Q_r = -Kh \frac{dh}{dx}$$

(14.5)

A decrease in pressure head with flow distance is indicated by a minus sign in the above equation. Where Q_r is the volume of flow per unit time through the aquifer (unit width basis), h is the height of the free water surface (often called the depression line) above the impermeable stratum (L), and dh/dx is the hydraulic gradient. Performing the same mathematical

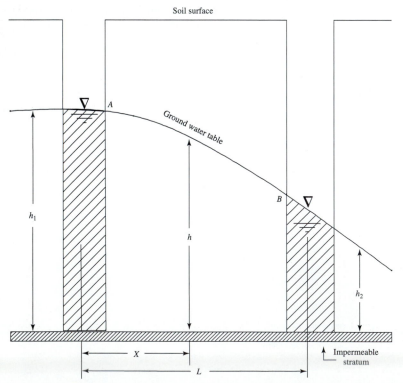

Figure 14.3   Steady flow of ground water between two parallel open ditches. Note that the impermeable stratum does not have to be uniform across the study area as depicted in the drawing. For example, the stratum could be much deeper under point B than point A.

operations as for equation 14.4 and integrating between the limits for h from h_1 to h_2, and for x from 0 to L, then

$$\frac{Q_r}{K} \int_0^L dx = -\int_{h_1}^{h_2} h \, dh = \int_{h_2}^{h_1} h \, dh \tag{14.6}$$

and

$$Q_r = K \frac{(h_1^2 - h_2^2)}{2L} \tag{14.7}$$

Equation 14.7 is commonly called the Dupuit equation. A full derivation can be found in Fetter (1994). Equation 14.7 gives the discharge per unit length along the ditch. To obtain the equation for the height of the water table, we apply the continuity principle so that

$$Q_r = K \frac{(h_1^2 - h^2)}{2x} \tag{14.8}$$

which results in

$$h_x = \sqrt{h_1^2 - \frac{x}{L} (h_1^2 - h_2^2)} \tag{14.9}$$

Equation 14.9 yields the height of the water table above an impermeable aquifer base.

Flow with Uniform Recharge

By assuming steady, uniformly distributed precipitation or irrigation, flow between two parallel ditches can be illustrated (see figure 14.4). For steady conditions, a maximum height h_m is maintained midway between the ditches, no flow occurs through the vertical plane h_m, and water on either side of the plane flows toward the nearer ditch. Horizontal flow toward the ditch (through vertical plane x) is equal to surface recharge between vertical plane x and the line denoting h_m. As a result, the flow rate Q_x, moving horizontally through the vertical plane (of unit width), is expressed as

$$Q_x = -R\left(\frac{S}{2} - x\right) \tag{14.10}$$

where R is the rainfall or recharge rate (units of $L \, T^{-1}$), S is the distance between ditches (m), and the term in parentheses is the surface area per unit width over which recharge takes

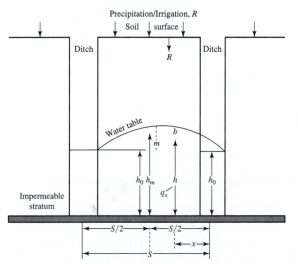

Figure 14.4   Steady water flow toward parallel open ditches with uniform recharge, where $q_x = R(S/2 - x)$

place. If the Dupuit (1863) assumption is used, the flow rate Q_x, moving horizontally through the vertical plane (of unit width) is expressed as

$$Q_x = -Kh \frac{dh}{dx} \tag{14.11}$$

where h is the height of the water table above the impermeable stratum (L), and x is the distance from the center of the right ditch to the point on the impermeable stratum that is intersected by the line h (m). If we equate the last two equations, separate the variables, integrate and substitute $h = h_m$ and $x = S/2$, we obtain

$$(h^2 - h_0^2) = \frac{R}{K} (Sx - x^2) \tag{14.12}$$

This is Hooghoudt's equation (Ghildyal and Tripathi 1987; Hooghoudt 1937) for the elliptical shape of the water table between drains.

Evaluation of Falling Water Tables

Most of the equations discussed above are based on the assumptions that fixed volumes of soil will be drained when drains or ditches are installed, and that the soil is homogeneous. This tends to imply to the reader that there is also a fixed (or drainable) porosity, and that the specific volume of soil being considered will drain instantaneously. However, based on the physical and chemical factors discussed in chapters 2, 4, 8, 9, and 10, the volume of water drained gradually increases, due to increased matric potential. Consequently, the drainable porosity is a function of the matric potential and can be written (Luthin 1966) as

$$V_{dw} = \int_{\psi_{m_2}}^{\psi_{m_1}} \phi_d(\psi_m) \, d\psi_m \tag{14.13}$$

where V_{dw} is the volume of water drained per unit volume of soil (L^3), ψ_m is the matric potential or soil pressure/soil suction (kPa), and ϕ_d is the drainable porosity (unitless). By considering a soil that has a falling water table, an approximate expression of drainable porosity can be written such that

$$V_{dw} = \frac{a}{2} (\psi_{m_1}^2 - \psi_{m_2}^2) \tag{14.14}$$

where a is the drained porosity at a specified matric potential ($\phi_d = \psi_m$). Equation 14.14 gives a reasonable prediction of the volume of water drained from the profile.

Drainage problems that deal with falling water tables normally consider the transient process as a series of steady-state drops, but can also include time dependency for vertical drainage. Many such cases have been investigated by researchers (Gardner, 1962; Jackson and Whisler, 1970; Jensen and Hanks, 1967; Youngs, 1960). As with any modeling scheme, the validity of assumptions made determines the predictability of the total drainage.

QUESTION 14.1

Calculate the discharge to a fully penetrating well from steady, radial ground water flow. Assume $K = 2.4 \times 10^{-3}$ m/s, $h_1 = 14.2$ m, $h_0 = 12$ m, $r_1 = 40$ m, and $r_0 = 80$ m.

QUESTION 14.2

Calculate the height of the water table above base level h, between a ditch at $x = 0$ and a ditch at $x = L$, assuming $h_1 = 4$ m, $h_2 = 3.2$ m, $x = 25$ m and $L = 100$ m.

14.3   GROUND-WATER DRAINAGE

Water Tables and the Capillary Fringe

The water table is the upper surface of the saturated zone of free ground water. Free ground water is defined as water neither confined by artesian conditions nor subject to the forces of surface tension. At the water table, the total water potential is zero (atmospheric pressure). Thus, the water table is the imaginary surface separating the unsaturated zone from the saturated zone. The water table in soil is not an observable, physical surface, because capillary water (capillary fringe) is just above the water table and decreases in amount gradually upward. Auger holes, piezometers, wells, and drains that are open to the atmosphere fill to the true water-table level when bored or driven into the water table.

When an auger hole is drilled to locate the water table in a fine- or medium-textured soil, it is often difficult to recognize the top of the saturated zone due to the gradual change from unsaturated to saturated soil at the capillary fringe. It may take hours—or even days—for an auger hole to register the water table in low-conductivity soils. Small wells or piezometers react more quickly than large ones, because less water needs to flow through the soil to fill the smaller openings. Water in the capillary fringe can be a significant proportion of ground water moving toward subsurface drains; as much as 20 percent or more under some conditions (SCS, 1971).

An auger hole or piezometer should penetrate the saturated zone only a short distance (< 1 m) if the water-table location is to be measured accurately. This is particularly important where upward flow (or water under confined conditions) is tapped by a deeper hole. An auger hole that penetrates two or more aquifers in a stratified soil containing confined water, registers the highest hydraulic head modified by water movement through the well from the aquifers of higher hydraulic head to those of lower hydraulic head. These characteristics of the water table have a significant impact on the kinds of field measurements to be made, on the types of measurement devices to be used, and on the data interpretations for a ground water drainage system design.

At the water table, the component of potential energy is zero relative to atmospheric pressure. Therefore, the hydraulic head h (of a point at the water table) is the distance of that point above the datum or above an impermeable boundary (see figure 14.5). The water-table slope represents the hydraulic gradient of flow only under certain conditions. Hydraulic gradient differs greatly from the water-table slope if there is a significant upward or downward component of flow such as that occurring: in the vicinity of a pumping well or subsurface drain; in flow from artesian aquifers; and in unsaturated seepage from canals. Figure 14.5 shows this difference between the slope S_L of the water table and hydraulic gradient. The hydraulic gradient of the water table is the difference in hydraulic head h at two points, divided by the distance between the points measured along the flow path L_P.

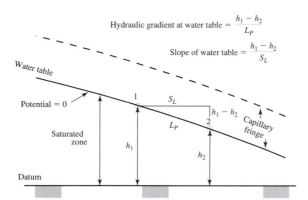

$$\text{Hydraulic gradient at water table} = \frac{h_1 - h_2}{L_P}$$

$$\text{Slope of water table} = \frac{h_1 - h_2}{S_L}$$

Figure 14.5   Difference between slope of the water table and hydraulic gradient

On flat surfaces and with parallel flow, the water-table slope is essentially the hydraulic gradient, because S_L is nearly equal to L_P. It needs to be noted that the water table is not invariably a flowpath; water can be flowing upward into (or downward from) the unsaturated zone, thus crossing the water table.

Equipotential Lines, Streamlines, and Potential Flow

In a given flow system, each "particle" of water in the system has its corresponding hydraulic head. All particles (or points) with the same hydraulic head lie in an equipotential surface. The force that tends to produce flow acts in the direction of greatest hydraulic gradient (i.e., normal to the equipotential surface). Whether or not the flow actually moves in the same direction as the line of force depends on whether the soil has the same hydraulic conductivity in all directions. If the soil is isotropic—that is, if its hydraulic conductivity is the same in all directions—the path of flow is along the lines of force, perpendicular to the equipotential surfaces.

If the soil has a higher hydraulic conductivity in one direction than in another direction, the flowpath is not perpendicular to the equipotential surface; such a soil is said to be anisotropic. Soil and rock often have bedding planes and fractures, causing them to be anisotropic. The paths of flow in anisotropic soils are perpendicular to the equipotential surfaces at points where the lines of force are exactly parallel to (or normal) to the bedding planes or fractures.

Figure 14.6 shows a two-dimensional flow system in the X–Z plane, with an isotropic soil. Figure 14.7 shows another soil, with a line of force normal to the equipotential line, at an angle b to the vertical axis Z. In this soil—which is anisotropic—the horizontal hydraulic conductivity K_h is larger than the vertical hydraulic conductivity K_z. The direction of flow is not along the line of force but rather along a line closer to the horizontal axis. It can be shown (Viessman et al. 1977) that the angle the flowpath makes with the horizontal is (see figure 14.7)

$$a = \tan^{-1} \frac{K_z}{K_h \tan b} \tag{14.15}$$

Thus, the flow pattern can be computed and drawn for an anisotropic system if the equipotential lines are known, and if the relative hydraulic conductivities K_z and K_h are known. In analyzing the flow direction of ground water, the investigator needs to be aware of the effects of anisotropy on the flow pattern.

Flow in the saturated zone is often studied by using graphic representations of the hydraulic head and the flowpaths. Cross-sections are taken through the flow problem area, usually in vertical planes. Lines connecting points of equal hydraulic head plotted on such planes are called equipotentials. Lines indicating flowpaths plotted on the planes are called

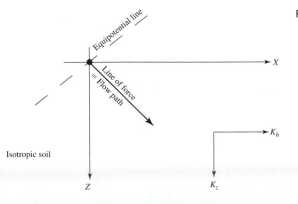

Figure 14.6   Flow direction in an isotropic soil

Figure 14.7   Flow direction in an anisotropic soil

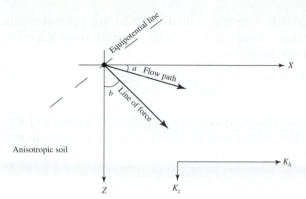

streamlines, which are normal to the equipotentials. The two-dimensional graph showing equipotentials and streamlines for a flow system (or part of a flow system) is called a flow net.

There are actually an infinite number of equipotentials and streamlines, but it is convenient to draw only a limited number such that the rate of flow between each pair of streamlines is equal, and potential drop between successive equipotentials is the same. For isotropic systems, the distance between the streamlines is made equal to the distance between the equipotentials, thus forming a series of squares. Where the streamlines are curved, the squares are distorted, but they become more nearly "perfect" as the number of lines is increased, and approaches true squares as the number of equipotentials and streamlines increases.

Construction of Flow Nets

For two-dimensional flow, the manner in which a flow net is used in problem solving is explained by considering figure 14.8. This schematic shows a portion of a flow net constructed so that each square is formed by a pair of equipotentials and streamlines. After a flow net is constructed, it can be used to analyze the flow rate using geometry and Darcy's law.

Remembering that the potential h is given by $(p/\gamma + z)$ and referring to figure 14.8, the hydraulic gradient G_h between two equipotentials is given by

$$G_h = \frac{\Delta h}{\Delta s} \tag{14.16}$$

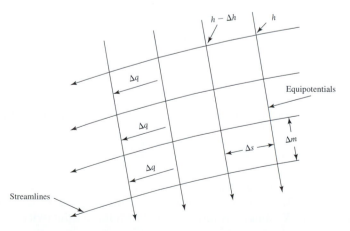

Figure 14.8   Schematic of an orthogonal flow net in an isotropic soil

where Δh is the change in hydraulic potential between two equipotentials, and Δs is the orthogonal distance between two equipotentials. By applying Darcy's law, the incremental flow between two adjacent streamlines is

$$\Delta q = K \, \Delta m \left(\frac{\Delta h}{\Delta s} \right) \tag{14.17}$$

where Δm represents the cross-sectional area for a flow net, of unit width normal to the plane of the diagram. If the flow net is constructed in an orthogonal manner and composed of approximately square elements, then $\Delta m \approx \Delta s$ and

$$\Delta q = K \, \Delta h \tag{14.18}$$

If there are n equipotential drops between the equipotential lines, then

$$\Delta h = \frac{h}{n} \tag{14.19}$$

where h is the total potential change over the n spaces. If the flow is divided into m sections by the streamlines, then the discharge per unit width of the medium is

$$Q = \sum_{i=1}^{m} \Delta q = \frac{Kmh}{n} \tag{14.20}$$

When saturated hydraulic conductivity is known, the discharge can be computed using equation 14.20, with knowledge of the flow-net geometry. Where the flow net has a free surface (or line of seepage), the entrance and exit conditions are useful. A comprehensive discussion of entrance and exit conditions is given in U.S. Department of Interior (1978).

Construction of flow nets is difficult where the hypothetical-flow velocity becomes either infinite or zero. These conditions can occur when: a boundary coincides with a streamline; there is a discontinuity along the boundary that abruptly changes the slope of the streamline; or where a source (or sink) exists in the flow net (e.g., a drain or a well). A description of flow-net construction for these types of flow conditions can be found in Davis and DeWiest (1966) and Viessman et al. 1977.

Flow nets can be constructed in anisotropic media, for which the saturated-hydraulic conductivity in one direction is greater than in another direction. This is the case for most stratified sediments, in which the flow proceeds more easily along the planes of deposition than across them. Figure 14.9 illustrates two flow nets in saturated soils, each taken in a vertical plane at right angles to a drain with the soil saturated to the surface and an impermeable layer at twice the drain depth d ($2d$). Figure 14.9a is for an isotropic soil; whereas figure 14.9b is for the same boundary conditions, except that the soil has a horizontal hydraulic conductivity 16 times its vertical conductivity (anisotropic). Numbers on each streamline indicate the percent of the total flow that occurs to the left of that streamline. Note that 50 percent of the flow that reaches the drain through the isotropic soil originates in a strip over the drain, and covers approximately one-fourth of the source area. For the anisotropic soil, approximately one-half of the flow originates in a much wider strip, covering nearly one-half of the source area.

To construct the flow net for the anisotropic case, a change of variable is made

$$x_t = x \sqrt{\frac{K_v}{K_h}} \tag{14.21}$$

and an isotropic flow net is constructed in the transformed section, then replotted to true scale. In equation 14.21, x_t is the transformed horizontal-space variable for the horizontal hydraulic conductivity K_h, and K_v is the vertical hydraulic conductivity.

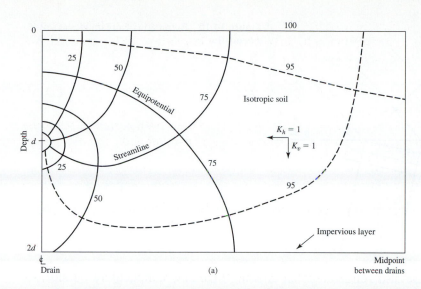

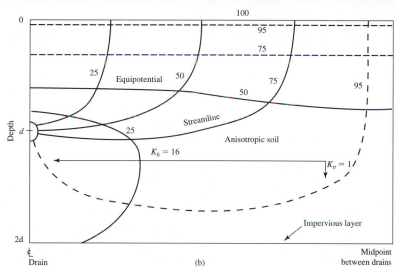

Figure 14.9   Flow nets in (a) isotropic and (b) anisotropic soils

When streamlines from a soil with a given hydraulic conductivity K_1 cross the boundary of a soil with a different hydraulic conductivity K_2, the streamlines are refracted similar to the optical refraction of light rays as shown in figure 14.10. This refraction adheres to the following law:

$$\frac{K_1}{\tan \theta_1} = \frac{K_2}{\tan \theta_2} \tag{14.22}$$

This relation shows that flow proceeds from a coarse-grained medium (I) to a fine-grained medium (II). The flow elements of figure 14.10 are squares in medium I, but rectangles in medium II. If a flow net with squares on both sides of the boundary and like-flow quantities is required, the potential drops Δh_1 and Δh_2 in the media must satisfy

$$\frac{\Delta h_2}{\Delta h_1} = \frac{K_1}{K_2} \tag{14.23}$$

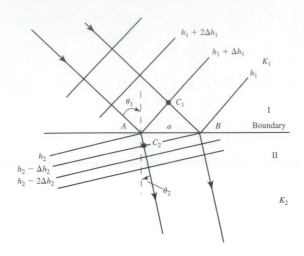

Figure 14.10   Refraction of streamlines at soil boundaries

Equipotentials are also refracted when crossing hydraulic conductivity boundaries, such as two different, horizontal strata. The relation for this is

$$\frac{K_1}{K_2} = \frac{\tan \alpha_2}{\tan \alpha_1} \tag{14.24}$$

where α is the angle between the equipotential line and a normal to the hydraulic conductivity boundary.

14.4   DRAINAGE DESIGN

Elements of Drainage Design

Installation of a drainage system, much like any similar application of science, includes a desired goal; a survey of existing conditions; previous experience with similar conditions; and preparation of designs, plans, and specifications for construction. At several stages during the design procedure, it can be necessary to choose between alternative locations, methods, or materials. The choice depends on the management and economic aspects of the application, as well as on the physical requirements of the site. The designer may have to present alternative methods or intensities of drainage to the owner, so that he or she can make a final choice.

The technical design elements are generally the same for small or large drainage systems, but public and institutional factors are often involved, also. These factors include: the drainage organization (drainage enterprise); legal requirements for rights-of-way and water disposal (or use); financial arrangements; and cost allocation. Drainage projects require complete and detailed documentation of the surveys, plans, and construction.

Development of Drainage-Design Criteria

Design criteria are developed in two general ways: **(1)** from empirical data collected through evaluation of existing drainage systems; and **(2)** from a theoretical analysis of the problem, applying physical laws and testing the theory through evaluation of existing drainage systems. An example of empirical criteria are the drainage coefficients used in design of drains in humid areas (SCS 1971). Such coefficients are the removal rates for excess water—which provide a certain degree of protection—found by experience with many installed drainage systems. Such protection has been carefully assessed against observed responses; measurements of flow from drainage systems providing good drainage; and measured water-table heights. Because empirical criteria are based substantially on experience and assessments of numerous interrelating factors, care has to be taken in transposing their use from one location to another.

Drainage Coefficients

The SCS (1971) provides step-by-step methods for applying drainage coefficients for the design of drains. The drainage coefficient is that rate of water removal necessary to obtain the desired protection (or lowering) of the water table from excess surface and subsurface water. In humid areas, it is common practice to express the drainage coefficient for ground water drainage in units (of depth) of water removed in 24 hours. This coefficient is closely related to the climate and infiltration characteristics of the soils; therefore, there is similarity in drainage coefficients within areas of similar climatic and soil characteristics. The reader is referred to SCS (1971) for a detailed treatment of the drainage-coefficient method.

Theoretical analysis applies proven principles (or laws) to problems that have known limiting conditions. The resulting mathematical expression explains the observed action of existing drainage systems, thereby permitting the rational design of new systems. Usually several variable-site factors enter the expression. An example of the theoretical analysis is the Hooghoudt equation (equation 14.12) for spacing subsurface drains (Hooghoudt 1937). In the form of Hooghoudt's equation (given alone) the variables are: hydraulic conductivity of the soil, depth to an impermeable layer; depth to the water table at midpoint between the drains; and the rate water is removed. By substituting known—or estimated—values for these variables, the equation can be applied to a variety of sites, as long as the site conditions are within the limits for which the equation was derived. This last requirement is important when using this kind of theoretical approach. Another method to approximate drain spacing is to use equation $S^2 = 4K/R(h^2 - h_0^2)$. Thus, to determine the required drain spacing to maintain the water table at a desired level: the depth to the impermeable stratum; the desired (or target) height of the water table; average infiltration flux; and K all must be known. Average drain spacing for various soil types is given in table 14.1 Whichever method is used to establish drainage-design criteria, it is evident that its value depends not only on sound analysis of the drainage situation, but also on an evaluation of installed drains to check their performance.

Theoretical Analysis

Equation 14.12 presented Hooghoudt's equation for spacing of vertical-walled ditches. This equation can be extended to a layered soil and to pipe drains. Rearranging equation 14.12

$$R = \frac{4K(h_m^2 - h_0^2)}{S^2}$$

(14.25)

TABLE 14.1   Average Depth and Spacing of Drain Tubes versus Soil*

Medium	Spacing (m)	Depth (m)	Hydraulic conductivity (cm/day)
Peat	70	2	18.5
Sandy loam	50	2	9.5
Fine, sandy loam	35	1.5	4.3
Loam	25	1.5	1.3
Clay loam	20	1	0.30
Clay	15	1	0.15

Source: Data from Ghildyal and Tripathi (1987)
*Columns 1–3 represent average values (midrange) based on soil type.

and replacing $h_m^2 - h_0^2$ by $(h_m + h_0)(h_m - h_0)$ and noting (see figure 14.4) that $(h_m + h_0) \cdot (h_m - h_0) = (2h_0 + m)m$. Then,

$$R = \frac{8Kh_0m + 4Km^2}{S^2} \tag{14.26}$$

When $h_0 = 0$,

$$R = \frac{4Km^2}{S^2} \tag{14.27}$$

The R term in equation 14.27 represents the contribution to the drains per-unit area from the soil above the water level in the drain. It follows that the other term in equation 14.26 $(8Kh_0m/S^2)$ is the contribution from the soil between the plane of the drain-water level and that of the impermeable boundary. If the hydraulic conductivity above the plane of the drain-water level is K_1 and that below is K_2, then the solution for such a two-layered soil system is

$$R = \frac{8K_2h_0m + 4K_1m^2}{S^2} \tag{14.28}$$

If the vertical-walled ditches are replaced by circular drain pipes, so that the water level in the drain pipe is the same as in the ditches, the streamlines to the drain are no longer horizontal—as they were for the vertical-walled ditch—but radial. As the streamlines converge on the pipe drain, flow velocity and therefore—by Darcy's law—head loss also, increases. To account for the convergence losses of the pipe drain, Hooghoudt developed a second equation in the same form as equation 14.28, but replaced h_0 by an equivalent depth d, which is smaller than h_0 and is a function of h, S, and pipe radius r_0. If d is substituted for the real depth to the impermeable boundary instead of h, the equation for spacing of a pipe drain is

$$R = \frac{8K_2dm + 4K_1m^2}{S^2} \tag{14.29}$$

The diameter of the pipe drain does not have a large effect on drain spacing (Withers and Vipond, 1980). In practice, the drain backfill generally is more permeable than the surrounding soil, and flow is greater than that predicted by the above equations.

The steady-state drainage assumption is invalid when inflow occurs over short, widely separated time periods. This type of situation often occurs when irrigation of an agricultural area or "dosing" of a soil-treatment system causes a rise in the water table, which falls during the remainder of a given time period. The Bureau of Reclamation (U.S. Department of the Interior, 1978) has presented a drainage equation for the case of a fluctuating water table. The basic assumptions for this equation are similar to those for steady flow. These assumptions are: **(1)** the soil is homogeneous and isotropic; **(2)** the problem is two-dimensional; **(3)** streamlines are horizontal, a correction necessary for the effect of convergence to a pipe; **(4)** the velocity along the streamlines is proportional to the slope of the water table; and **(5)** an impermeable boundary exists below the drains.

The solution to such a problem is found from the "heat-flow" equation, and takes the form of a parabola. To arrive at an exact solution, an initial shape of the water-table surface has to be assumed. The Bureau of Reclamation's (U.S. Department of the Interior, 1978) solution assumes that this initial shape is described by a fourth-degree parabola of the form

$$h = \frac{8m}{S^4}(S^3x - 3S^2x^2 + 4Sx - 2x^4) \tag{14.30}$$

where m is the initial height of the water table in meters above the plane of the drains, S is the drain spacing in meters, and h is the height of the water table at a distance x from the drain. A simplified version of the drain spacing equation is

$$S^2 = \frac{\pi^2 KDt}{\phi_d \log_e \left(1.16 \frac{h_0}{h_t} \right)} \tag{14.31}$$

where ϕ_d is the drainable porosity in percent by volume, D is the average thickness (in meters) of the aquifer from the impermeable boundary to the water table, h_0 is the initial height of the water table, and h_t is the height of the water table after a time interval t. Allowance for convergence can be made by substituting the equivalent depth to the impermeable boundary for the actual depth, as in Hooghoudt's method for steady-state flow.

14.5   CASE STUDY: THE FLORIDA EVERGLADES

Drainage in the form of open ditches and canals has been practiced extensively in the Florida Everglades for the past several decades, primarily because the areas now under production were previously under water, and would have remained so if not for man's efforts to tame the land. This is due primarily to the heavy rainfall received in the region (75–175 cm annually) and the desire of sugar cane, sod, citrus, vegetable crops, and cattle industries to produce these products on the very rich, typically nearly saturated soils that predominate in the area. In addition, the increase in population in the region has required greater levels of water control and management to allow for residential development.

For years, many inhabitants of Florida looked at the Everglades as a swamp to be drained, while many others considered it a wonderful ecosystem to be preserved and protected. Before development, the Everglades began in the center of the Florida peninsula (near Orlando) and expanded south through the marshy Kissimmee River area to Lake Okeechobee (meaning "land of big water" in the Seminole Indian language) and on down the peninsula to Miami and Homestead, to the Florida Bay and Gulf of Mexico. The Everglades is really a "river of grass," and was one of the most perfectly balanced ecosystems in the world. Since drainage projects first began about 110 years ago—and in earnest about 60 years ago—the vast areal expanse of the Everglades has steadily decreased. Because of the devastating threat of complete destruction from the drainage projects and onslaught of increasing population, the Everglades National Park was established in 1947 and signed into law by President Harry Truman. However, long before this—beginning in 1837, during the three Seminole Wars—the U.S. Army drove the Seminole Indians deep into the swamp, then could not force them out. Because of the difficult terrain and vicious mosquitos as well as other insects, the Army eventually left the Indians alone.

Slowly, very affluent people began migrating to places like West Palm Beach and other areas further south, toward Miami; many of these looked at Florida as a developer's dream. As a result, politicians began pushing for the transformation of the Everglades and instructed the U.S. Army Corps of Engineers to dredge, dike, divert, and provide flood control and drainage—all for the purpose of creating dry lands for establishing new homes, businesses, industry, and farms.

By the early 1950s, even the Seminole Indians had become involved. Although many did not like what was happening to their habitat, they needed to find employment and wanted to make their homes once again away from the deepest parts of the swamp. As more tourists and entrepreneurs moved into the Dade and Broward County areas (which include the cities of Miami and Fort Lauderdale), the Seminoles set up shop and began selling Indian handicrafts to tourists, serving as guides in the swamp, and staging very popular "alligator wrestling" shows. With a steady population growth, increased pressure was put on the Everglades. More and more water was pumped out, resulting in a loss of over 50 percent of the Everglades' original surface area, as well as the destruction of much wildlife habitat. The drainage and flood-control projects initiated by the U.S. Army Corp of Engineers have made it possible for the inhabitants of Florida to live there. Without these projects and the accompanying mosquito control, as well as current technologies such as air conditioning, very few individuals would want to live in the Everglades region of Florida. Today, approximately 1,500 miles of canals, ditches, and levees scar the landscape and constrict the flow of water, generally from the north to the south.

Prior to the beginning of the drainage projects, rain that fell in the northern part of the Everglades near Orlando produced a gradual rise in the overall water level. The slow flow took about a year to reach the area within and surrounding Everglades National Park, and the flow was very consistent well into the dry season, providing habitat for over 600 species of wildlife and 900 species of plants. Currently—even with the existing level of technology—the U.S. Corp of Engineers and South Florida Water Management District cannot duplicate the natural conditions of flow that existed prior to drainage initiation. This has had a devastating effect on both the Everglades and Everglades National Park, truly about all that remains of the Florida Everglades; the Park is comprised of about 1.5 million acres. In addition to the ditches and canals, it was decided that a plant that requires an abundant amount of water could help dry the "Glades," thus melaleuca trees were imported and now grow as noxious trees all over south Florida. Many counties have declared war on them by physically cutting them down and spraying with herbicides, but they are very well adapted and persistent, making control difficult.

The drainage and flood-control projects have created man-made droughts in the region, killing off large areas of cypress and saw grass swamps. Additionally, flow of freshwater from the main peninsula into Florida Bay had been greatly curtailed, resulting in hypersalinity of the Bay. This resultant salinity has killed sea grass and many other marine organisms, and denuded the bay bottom. Compounding the problem, Lake Okeechobee has become a storage basin for providing water for the inhabitants of South Florida, but agriculture, cattle, and related industries around the Lake have contributed large amounts of phosphorus and nitrates from fertilizers and animal waste to the Lake itself. The phosphorus has caused eutrophication in many areas around the Lake. Activity from sod, citrus, vegetable crop, and sugarcane industries have contaminated the water that drains into the Everglades region with nitrates, phosphates, herbicides, and pesticides, further compounding the devastating effects of both low water levels and contamination.

During the 1980s, the U.S. Congress finally decided they had to act to save the Everglades and passed the Everglades National Park Protection and Expansion Act, which was signed into law by President George Bush. More recently, an "Everglades Initiative" has begun with funding from Congress. The goal of the initiative is to attempt to solve the problems that are causing the devastation of the Everglades. Many scientists are cooperating in the project from agencies such as the U.S. Geological Survey, U.S. Army Corps of Engineers, South Florida Water Management District, University of Florida, Florida State University, University of Miami, and many other state, federal, and local representatives. However, because so many agencies are involved, lack of bureaucratic coordination makes cleanup still more difficult. This has resulted in a federal task-force being established for coordination purposes.

The Everglades Initiative has obtained solutions to some of the water-management problems faced by South Florida, in terms of amounts of storage before pumping from one area to another. In the beginning of the drainage projects, the U.S. Army Corps of Engineers straightened the Kissimmee River for more effective flow. Prior to this, it had been a long, meandering river. Because of the initiative, it has been discovered that the river (in its original course and form) acted as a filtering system for contaminants, especially fertilizers used in agriculture. As a result, the river is being returned to its natural course and the contaminant level downstream is gradually being reduced. Although positive actions are occurring, it should be remembered that it took decades to cause the problem, and so will likely take several decades to correct it.

The main predicament in this instance is that—although drainage of wet areas is ideal for construction of businesses, homes, farms, and other purposes—a plan should be developed that will allow harmony between man and the existing environment. The Everglades is a very large and complex ecosystem. As with other projects or ecosystems, it should have been thoroughly studied before any drainage projects were developed or initiated; due to impatience and greed for development, this was not done. This is not unique to the Florida Everglades, but is occurring worldwide. In each instance, however, there is a price to pay. Whenever man competes with the environment, the environment is usually devastated and eventually, the populace within the region also suffers. Economically, it would be much cheaper to develop a sound plan before starting such a project than to pay for the results afterward. However, the criteria by which such a plan might be judged can be highly subjective. For example: had the canals and ditches been sized, constructed, and located in different areas—then perhaps the need to purchase additional areas from private citizens for wetlands control; to pay hundreds of agencies; contractors, and scientists to investigate the problem; and to spend countless hours trying to find solutions—the problem could have been avoided. Although drainage projects can serve a very useful purpose, simply knowing how to perform drainage is not enough. We need to consider both the purpose of the drainage and the overall effects of the drainage on the environment. If we learn from past mistakes, we can minimize costly future mistakes.

SUMMARY

This chapter has briefly discussed the basic principles involved in drainage and some of the common problems associated with it. While drainage can be thought of as a water-quality issue, this is not entirely true. Essentially, the effects of drainage can be harmful to the environment, but the causes of drainage are hydraulic in principle. The reader is urged to consult the references for texts and other articles related to drainage principles.

ANSWERS TO QUESTIONS

14.1. Using figure 14.2 as an example and equation 14.4, we obtain:

$$q_r = \pi(2.4 \times 10^{-3}\,\text{m/s}) \frac{[(14.2\,\text{m})^2 - (12\,\text{m})^2]}{\ln\left(\dfrac{40\,\text{m}}{80\,\text{m}}\right)} = -0.63\,\text{m}^3\,\text{s}^{-1}$$

What does the negative sign indicate?

14.2. Using equation 14.9, we obtain:

$$h_x = \sqrt{(4\,\text{m})^2 - \frac{25\,\text{m}}{100\,\text{m}}\,[(4\,\text{m})^2 - (3.2\,\text{m})^2]} = 3.82\,\text{m}$$

ADDITIONAL QUESTION

14.3. Approximate the drain spacing to be used for a landfill; assume $K = 2.3 \times 10^{-5}$ m/s, that $R = 0.003$ m/d, $h = 4$ m, and $h_0 = 3.2$ m.

15

Soil Remediation Techniques

INTRODUCTION

Corrective-action objectives for remediating soil contamination can be achieved by a large number of technologies. Numerous remedial-action alternatives that have proved to be effective in contaminated soil cleanups are reviewed here, and discussed as having potential implementation. Alternative remediation techniques discussed do not include free-product recovery. This chapter presents techniques for determining corrective-action criteria, and provides brief descriptions and analyses of potential alternative remediation techniques that can be applicable to contaminated soil. Site-specific differences need to be taken into account when selecting a preferred alternative.

15.1 SOIL CORRECTIVE-ACTION CRITERIA

Background

While many different kinds of contaminants can be remediated in soils, the most common are hydrocarbons and other organics. The question then, is "how clean is clean?" when considering practical conditions for making a given contaminated site available for public or other uses. The U.S. Environmental Protection Agency (EPA 1989a) has studied soil-response action levels based on potential contaminant migration to the underlying ground water. In general, EPA considers that a remediation level in soil does not have to be as clean as the ground water, because ground water is a drinking source. A general rule-of-thumb is that soil remediation levels may have 100 times more contaminants than ground water. Thus, if there were 100 milligrams (mg) of total petroleum hydrocarbons (TPH) in one kilogram (kg) of soil, and that contaminant level were acceptable to a regulator, the ground water would have to be 100 times cleaner, or about 1 mg per liter (mg/L) to be acceptable to the same regulator. In the United States, some individual states have the authority to set remediation levels for both soil and water. For soil, there is a very large range of remediation levels in considering TPH. This range varies from background (or ambient) TPH concentrations to over 1,000 parts per million (ppm) TPH. Figure 15.1 shows TPH remediation levels for all 50 of the United States. Several orders of magnitude separate the recommended TPH soil remediation levels. For example, New Hampshire has a 1-ppm remediation level for TPH, whereas California has a 1,000-ppm TPH soil remediation level. In light of this large TPH remediation level difference from state to state, and because certain remediation technologies work well for cleaning soil to certain levels, it is probably best to know what the remediation

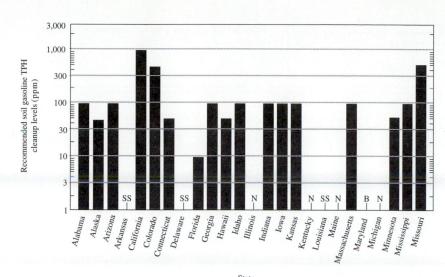

Figure 15.1 Recommended soil TPH remediation levels (data from Marencik 1991)

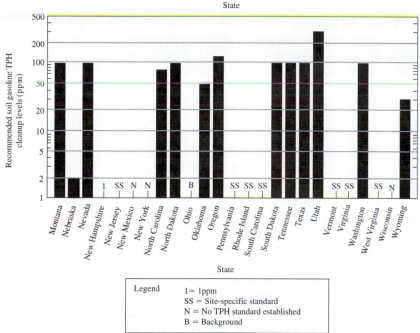

Legend	1 = 1ppm
	SS = Site-specific standard
	N = No TPH standard established
	B = Background

level for a soil will be before selecting a remediation technology. Additionally, it is best to negotiate a remediation level with the regulator prior to attempting remediation; it might even be advantageous to negotiate a remediation level with the option of changing the level if the technology cannot meet the negotiated soil remediation level. States in figure 15.1 that do not have established TPH remediation levels must negotiate the soil-remediation level prior to acceptance of the remediation plan.

General Principles of Application

The level of remediation required depends on many site-specific factors, as well as the hazardous substances involved, and the degree of public exposure at a given site. The methods and models used as tools in the analysis of contaminant migration to ground water vary from

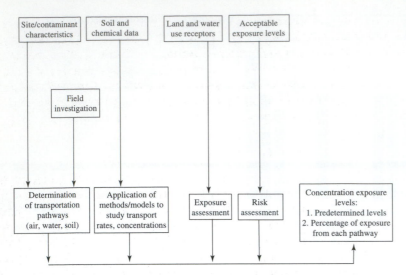

Figure 15.2   Process for determining soil cleanup levels

one site to another, depending on the unique site characteristics and the chemicals involved. In general, the milestones in the determination of cleanup levels are depicted as shown in figure 15.2 (EPA 1989a). The aspects of figure 15.2 discussed here deal only with fate and transport of contaminants from soil to the ground water. Other aspects in figure 15.2 are treated in other documents (EPA 1989b).

Fate and Transport Models for Evaluating Migration to Ground Water

The fate and transport factors affecting subsurface contaminant-migration processes within the environment are broadly classified as physical, chemical, and microbial. These processes and the factors affecting their relative significance at a given site are listed in table 15.1. The variety and quantity of such factors make the exposure-route determination more difficult for ground water than for other exposure pathways. Consequently, arriving at an acceptable cleanup level based on potential contaminant migration to the ground water can warrant a detailed characterization of the site, and careful selection of analytical tools.

Transport and speciation models rely on the quantification of relations between specific parameters and variables to simulate the effect of natural processes. A close match between the natural processes at the site and those of the selected model has to exist if the modeling exercise is to provide a satisfactory result. Transport processes strongly depend upon chemical speciation. The simplest approach to estimating the concentration of a hazardous constituent is to assume it behaves as a conservative substance (i.e., it does not react with its surroundings). The inclusion of degradative processes required (e.g., biodegradation and hydrolysis) to model the fate of a given compound considerably increases the chemical and environmental data needed. Where such degradative processes are suspected, a more sophisticated assessment is necessary.

The prediction of contaminant transport and transformation involves the following six steps (EPA 1989a):

Step 1 Determination of fate-influencing processes such as transport parameters and partition coefficients

Step 2 Delineation of environmental compartments

Step 3 Representation of soil/hydrogeologic processes

Step 4 Mathematical representation of speciation processes such as acid–base and sorption

TABLE 15.1   Fate and Transport Processes Affecting Subsurface
Contaminant Migration

Category	Process	Factors affecting process
Physical	Advection	Topography
	Dispersion	Climate
	Flow in fractures	Precipitation
	Diffusion	Soil type
	Precipitation	Vegetative cover
	Dissolution	Depth to ground water
		Soil-hydraulic conductivity
		Soil-void ratio
		Soil-moisture characteristics
		Geology
		Hydrology
		Morphology
Chemical	Partitioning	Physical, chemical prop-
	• sorption/desorption	erties of contaminants
	• ion exchange	Geology
	• volatilization	
	Equilibrium speciation	
	• acid/base equilibration	
	• organic complexation	
	• inorganic complexation	
	Abiotic transformation	
	• hydrolysis	
	• oxidation/reduction	
Microbial	Oxidation/reduction and	Geology
	hydrolysis	Contaminants
		Microbial environment

Source: Data from EPA (1989a)

Step 5 Mathematical representation of transport and transformation processes such as precipitation, degradation, dissolution, advection, and solubility

Step 6 Determination of contaminant load and mode of entry into the environmental media.

The following fate and transport methods can be considered for use in estimating soil levels for contaminated sites. These models are analytical, and include techniques appropriate for both organics and inorganics. These methods include: the Freundlich equation; allowable concentration model; Summers model; modified Summers model; Single Pathway Preliminary Pollutant Limit Values (SPPPLV) model; contaminant profile model; decision-tree process; and the Regulatory and Investigative Treatment Zone (RITZ) model. Additional models are also available and are presented in more detail in EPA (1989a). Depending on the complexity of the site and the desires of the regulator, other models can also be appropriate.

Freundlich equation   The Freundlich method is designed for use with organic compounds. It is used to determine a dry-soil contaminant concentration that would elevate the ground-water contaminant levels above a given ground water quality goal, such as a maximum contaminant level (MCL), set by a regulator. A dry-soil contaminant level is calculated for each individual or group of organic contaminants of concern. The Freundlich equation is:

$$Q_e = (K_d)(C_e)^{(1/n)} \tag{15.1}$$

where Q_e is the dry-weight concentration of the nonionic organic compound in the soil (mg/kg), C_e is the equilibrium pore-space aqueous concentration (mg/L), n is an experimentally derived exponential adjustment factor to the adsorption isotherm (unitless), and K_d is the soil:water partition coefficient [(mg/kg)/(mg/L)].

The soil:water partition coefficient K_d is usually normalized to the organic-carbon content of the mixture (K_{oc}), which can be given as:

$$K_d = (K_{oc})(f_{oc}) \tag{15.2}$$

where K_{oc} is the amount of chemical adsorbed per unit weight of organic carbon (oc) in the soil to the concentration of the chemical in solution at equilibrium and f_{oc} is the fraction of organic-carbon content of the soil. Equation 15.2 is valid for carbon contents above about 0.1 percent. K_{oc} has been related to the aqueous solubility (S) and the octanol:water partition coefficient (K_{ow}) of a chemical. The relations usually take the form of

$$\log K_{oc} = (c) \log (S \quad \text{or} \quad K_{ow}) + d \tag{15.3}$$

where c and d are empirical regression coefficients. Table 15.2 lists the selected equations which use either S or K_{ow} to estimate K_{oc}. The valid use of the equations in table 15.2 requires that the medium of interest possess the organic contents that fall within the range of those used to develop the equations. It is important to recognize that these equations are not universally applicable to all organic chemicals in all soil systems and should, therefore, be used with caution (Dragun 1988). Table 15.3 is a summary of selected octanol:water partition coefficients K_{ow}, and water solubility S for organic and inorganic chemicals (Dragun 1988; Cohen, Mercer, and Matthews 1993). These values can be used with the equations in table 15.2 to estimate values of K_{oc} in order to compute values of K_d for use in the Freundlich equation.

In order to use the Freundlich equation, a value of C_e has to be derived; this requires several calculations. The first involves using the following equation (CH$_2$M-Hill 1985):

$$\frac{C}{C_o} = \text{erf} \left[\frac{Z}{2(D_t X)^{1/2}} \right] \cdot \text{erf} \left[\frac{Y}{4(D_t X)^{1/2}} \right] \tag{15.4}$$

where C_o is the original ground water concentration at the source area (μg/L), C is the concentration at the receptor well or compliance area (soil) (μg/L), Z is the saturated-zone thickness (m), X is the distance of the compliance point from the source (m), Y is the width of the lateral extent of the source (m), D_t is the transverse dispersivity (m), and erf is the error function.

By using a health-based level or ground water quality value (e.g., an MCL for C), the desired concentration in the ground water directly beneath the contaminant source C_o that correlates with this source can be calculated. After the original source concentration C_o is obtained, the percolation rate through the unsaturated zone is calculated.

$$PR = (p)(A) \tag{15.5}$$

where PR is the percolation rate (m^3/yr), p is the percolation (amount of rainfall per year at the site) (m/yr), and A is the total area of the site contaminated with a specific contaminant (m^2).

The next calculation required to estimate migration is the lateral ground water flow (LGWF):

$$LGWF = (Z)(v)(L) \tag{15.6}$$

where LGWF is the lateral ground water flow (m^3/yr), Z is the aquifer saturated thickness (m), V is the ground water velocity (m/yr), and L is the lateral source length (m).

TABLE 15.2   Predictive Equations for K_{oc}

Equation:	$\log K_{oc} = -0.54 \log S + 0.44$
	S = water solubility (mole fraction)

$$\log K_{oc} = \log K_{ow} - 0.21$$

Reference: Karickhoff, Brown, and Scott (1979)

Soil percent-organic-carbon range: 0.09 to 3.29

Organic chemicals utilized to develop K_{oc}:

anthracine	2-methylnaphthalene
benzene	naphthalene
hexachlorobiphenyl	phananthrene
methoxychlor	pyrene
9-methylanthracene	tetracene

Equation: $\log K_{oc} = -0.557 \log S + 4.277$
S = water solubility (micromoles/liter)

Reference: Chiou, Peters, and Freed (1979)

Soil percent-organic-matter content (one soil):   1.6

Organic chemicals utilized to develop K_{oc}:

DDT	1,2-dichloroethane
1,2-dibromo-3-chloropropane	1,2-dichloropropane
1,2-dibromomethane	2,2',4,4',5,5'-hexachlorobiphenyl parathion 1,1,2,
	2-tetrachloroethane
1,2-dichlorobenzene	lindane 2,2',5,5'-tetrachlorobiphenyl tetrachloroethene
2,4'-dichlorobiphenyl	1,1,1-trichloroethane

Equation: $\log K_{oc} = 0.937 \log K_{ow} - 0.006$

Reference: Brown and Flagg (1981)

Sediment organic carbon content (one sediment):   0.033

Organic chemicals utilized to develop K_{oc}:

atrazine	simazine
cyanazine	trietazine
ipazine	trifluralin
propazine	trifluralin photodegradation products

Equation: $\log K_{oc} = 1.029 \log K_{ow} - 0.18$

Reference: Rao and Davidson (1980)

Soil organic carbon range: not specified

Organic chemicals utilized to develop K_{oc}:

atrazine	DDT	diuron	methylparathion
bromacil	dicamba	lindane	simazine
carbofuran	dichlobenil	malathion	terbacil
2,4-D			

Equation: $\log K_{oc} = 0.524 \log K_{ow} + 0.855$

Reference: Briggs (1973)

Soil percent-organic-matter range: 1.0–4.0

(continued)

TABLE 15.2 Predictive Equations for K_{oc} (*continued*)

Organic chemical utilized to develop K_{oc}:

3-(3-bromophenyl)urea	1-methyl-3-(3-chlorophenyl)urea
3-(4-bromophenyl)urea	1-methyl-3-(3-chloro-4-methoxyphenyl)urea
3-(3-chlorophenyl)urea	1-methyl-3-(3,4-dichlorophenyl)
3-(3-chloro-4-methoxyphenyl)urea	1-methyl-1-methoxy-3-(4-bromo-3-
3-(3,4-dichlorophenyl)urea	chlorophenyl)urea
1,1-dimethyl-3-(4-chlorophenyl)urea	1-methyl-1-methoxy-3-(4-bromophenyl)urea
1,1-dimethyl-3-(3-chloro-4-methoxyphenyl)urea	1-methyl-1-methoxy-3(4-chlorophenyl)urea
1,1-dimethyl-3-(3,4-dichlorophenyl)urea	1-methyl-1-methoxy-3-(3,4-dichlorophenyl)urea
1,1-dimethyl-3-(3-trifluoromethylphenyl)urea	phenylurea
3-(3-fluorophenyl)urea	3-(4-sulfophenyl)urea
3-(4-fluorophenyl)urea	3-(3-trifluoromethylphenyl)urea
3-(3-hydroxyphenyl)urea	

Equation: $\log K_{oc} = -0.82 \log S + 4.07$
                 S = water solubility (ppm)

Reference: Means et al. (1980)

Soil percent-organic-carbon range: 0.11 to 2.38

Organic chemicals utilized to develop K_{oc}:

debenzanthracene	3-methylcholanthrene
7,12-dimethylbenz[a]anthracene	pyrene

Equation: $\log K_{oc} = 0.72 \log K_{ow} + 0.49$

Reference: Schwartzenback and Westall (1981)

Soil percent-organic-matter range: < 0.01–33.0

Organic chemicals utilized to develop K_{oc}:

n-butylbenzene	1,2,4,5-tetramethylbenzene
chlorobenzene	toluene
1,4-dichlorobenzene	1,2,3-trichlorobenzene
1,4-dimethylbenzene	1,2,4-trichlorobenzene
1,2,3,4-tetrachlorobenzene	1,2,3-trimethylbenzene
1,2,4,5-tetrachlorobenzene	1,3,5-trimethylbenzene
tetrachloroethylene	

Equation: $\log K_{oc} = -0.594 \log S - 0.197$
                 S = water solubility (mole fraction)
                 $\log K_{oc} = 0.989 \log K_{ow} - 0.346$

Reference: Karickhoff (1981)

Soil percent-organic-carbon range: 0.66–2.38

Organic chemicals utilized to develop K_{oc}:

anthracene	phenanthrene
benzene	pyrene
naphalene	

Equation: $\log K_{oc} = -0.55 \log S + 3.64$
                 S = water solubility (mg/L)
                 $\log K_{oc} = 0.544 \log K_{ow} + 1.377$

Reference: Kenega and Goring (1980)

Soil percent-organic-matter range: not specified

(continued)

TABLE 15.2   Predictive Equations for K_{oc} (concluded)

Organic chemicals utilized to develop K_{oc}:

aldrin	dinitramine	norfluorazon
ametryn	dinoseb	oxadiazon
anthracene	dipropetryn	parathion
asulam	disulfoton	pebulate
atrazine	diuron	2,2',4,5,5'-pentachlorobiphenyl
benefin	EPTC	pentachlorophenol
benzene	ethion	phenanthrene
bromacil	ethylene bromide	phenol
sec-bumeton	fenuron	picloram
butralin	fluchloralin	phorate
carbaryl	fluometuron	profluralin
carbophenothion	2,2',4,4',5,5'-hexachlorobiphenyl	prometon
chloramben	hexachlorobenzene	prometryn
chloramben methyl ester	ipazine	pronamide
chlobromuron	isocil	propachlor
chloroneb	isopropalin	propazine
6-chloropicolinic acid	leptophos	propham
chloroxuron	lindane	pyrazon
chlorpropham	linuron	pyrene
chlorpyrifos	methaxole	pyroxychlor
chlorpyrifos-methyl	metobromuron	silvex
chlorthiamid	methomyl	simazine
crotoxyphos	methoxychlor	2,4,5-T
cyanazine	2-methoxy-3,5,6-trichloropyridine	tebuthiuron
cycloate	9-methylanthracene	terbacil
2,4-D acid	methyl isothiocyanate	terbutryn
DDT	2-methylnaphthalene	tetracene
diallate	methyl parathion	thiabendazole
diamidaphos	metribuzin	triallate
ibromochloropropane	monolinuron	3,5,6-trichloro-2-pyridinol
dicamba	monuron	triclopyr
dichlobenil	naphthalene	trietazine
3,6-dichloropicolinic acid	napropamide	trifluralin
cis-1,3-dichloropropene	neburon	urea
trans-1,3-dichloropropene	nitralin	
diflubenzuron	nitrapyrin	

The LGWF is then added to the percolation rate to obtain the total flow in the saturated zone underlying the contaminated area.

$$Q_t = \text{LGWF} + PR \tag{15.7}$$

where Q_t is the total flow (m³/yr), LGWF is the lateral ground water flow (m³/yr), and PR is the percolation rate (m³/yr). Next, to determine the annual mass of contaminant (X) leaching from the unsaturated zone, the following relation is used:

$$\frac{X}{Q_t} = C_o \tag{15.8}$$

where X is the annual mass of contaminant leaching from the unsaturated zone (mg/yr) and Q_t is the total flow in the saturated zone underlying the site (m³/yr). Thus, to cause a contaminant level of C_o in the saturated zone underlying the contaminated area requires:

$$X = (C_o)(Q_t) \tag{15.9}$$

mass of contaminant leaching from the unsaturated zone annually.

TABLE 15.3 Solubility and Octanol–Water Partition Coefficients for Selected Organic and Inorganic Constituents

Compound	Solubility in water at 25 °C (mg/L)	Log_{10} Octanol–water partition coefficient	Compound	Solubility in water at 25 °C (mg/L)	Log_{10} Octanol–water partition coefficient
Ethers			*Polynuclear aromatic*		
Bis(chloromethyl)	22,000	−0.38	*hydrocarbons*		
Bis(2-chloroethyl)	10,200	1.58	2-Chloronaphthalene	6.74	4.12
Bis(2-chloroisopropyl)	1,700	2.58	Benzo(a)anthracene	0.014	5.16
2-chloroethyl vinyl	15,000	1.28	Benzo(b)fluoranthene	6.75	
4-chlorophenyl phenyl	3.3	4.08	Benzo(k)fluoranthene	0.00055	6.84
4-Bromophenyl phenyl		5.15	Benzo(a)pyrene	0.0038	6.04
Bis(2-chloroethoxy)methane	81,000	1.26	Indeno(1,2,3-cd)pyrene	0.62	7.66
			Dibenzo(a,h)anthracene	0.0005	5.97
Phthalates			Benzo(ghi)perylene	0.00026	7.23
Dimethyl	4,300	2.12	Acenaphthene	3.42	4.33
Diethlyl	1,000	3.22	Acenaphthylene	3.93	4.07
Di-*m*-butyl	13	5.2	Anthracene	0.073	4.45
Di-*n*-octyl	3	9.2	Chrysene	0.002	5.61
Bis(2-ethylhexl)	0.4	8.73	Fluoranthene	0.26	5.33
Butyl benyl	2.9	5.8	Fluorene	1.98	4.18
			Naphthalene	34.4	3.37
Nitrogen compounds			Phenanthrene	1.29	4.46
N-nitrosodimethylamine	miscible	0.06	Pyrene	0.14	5.32
N-nitrosodiphenylamine		2.57			
N-nitroso-di-n-propylamine		1.31	*PCBs and related compounds*		
Benzidine	400	1.81	Aroclor 1016	0.42	4.38
3,3′-Dichlorobenzicine		3.02	Aroclor 1221	15	2.8
1,2-Diphenylhydrazine	221	3.03	Aroclor 1232	1.45	3.2
Acrylonitrile	73,500	−0.14	Aroclor 1242	0.24	4.11
			Aroclor 1248	0.054	5.75
Phenols			Aroclor 1254	0.012	6.03
Phenol	67,000	1.46	Aroclor 1260	0.0027	7.14
2-Chlorophenol	28,500	2.17			
2,4-Dichlorophenol	4,500	2.75	*Halogenated hydrocarbons*		
2,4,6-Trichlorophenol	800	3.38	Methyl chloride	6,450–7,250	0.91
Pentachlorophenol	14	5.01	Methylene chloride	16,700	1.25
2-Nitrophenol	2,100	1.76	Chloroform	9,600	1.97
4-Nitrophenol	16,000	1.91	Carbon tetrachloride	800	2.64
2,4-Dinitrophenol	5,600	4.09	Chloroethane	5,740	1.54
2,4-Dimethylphenol	17,000	2.50	1,1-Dichloroethane	5,500	1.79
p-Chloro-*m*-cresol	3,850	2.95	1,2-Dichloroethane	8,300	1.48
4,6-Dinitro-*o*-cresol	250	2.85	1,1,1-Trichloroethane	950	2.17
			1,1,2-Trichloroethane	4,500	2.17
Aromatics			1,1,2,2-Tetrachloroethane	2,900	2.56
Benzene	1,800	2.13	Hexachloroethane	50	3.34
Chlorobenzene	472	2.84	Vinyl chloride	1.1	0.60
1,2-Dichlorobenzene	145	3.38	1,2-Dichloropropane	2,700	2.28
1,3-Dichlorobenzene	123	3.38	1,3-Dichloropropene	2,700	1.98
1,4-Dichlorobenzene	79	3.39	Hexachlorobutadiene	2	3.74
1,2,4-Trichlorobenzene	30	4.26	Hexachlorocyclopentadiene	0.805	3.99
Hexachlorobenzene	6	6.18	Methyl bromide	900	1.1
Ethylbenzene	206	3.15	Dichlorobromomethane		1.88
Nitrobenzene	1,900	1.85	Chlorodibromomethane		2.09
Toluene	535	2.69	Bromoform	3,190	2.30
2,4-Dinitrotoluene	270	2.01	Dichlorodifluoromethane	280	2.16
2,6-Dinitrotoluene		2.05			

(continued)

TABLE 15.3   Solubility and Octanol-Water Partition Coefficients for Selected Organic and Inorganic Constituents (*continued*)

Compound	Solubility in water at 25 °C (mg/L)	Log$_{10}$ Octanol–water partition coefficient †	Compound	Solubility in water at 25 °C (mg/L)	Log$_{10}$ Octanol–water partition coefficient†
Trichlorofuoromethane	1,100	2.53	*Selected metal compounds*		
Trichloroethylene	1,100	2.29	Antimonic acid and oxides	Very slightly soluble	
1,1-Dichloroethylene	5,000	1.48	Arsenic oxide @ 16 °C	1.5×10^6	
1,2-Trans-dichloroethylene	6,300	1.48	Arsenic oxide @ 20 °C	3.7×10^6	
Tetrachloroethylene	150	2.88	Beryllium oxide @ 30 °C	0.2	
			Cadmium chloride @ 20 °C	1.4×10^6	
Pesticides			Cadmium sulfide @ 18 °C	1.3	
α-Endosulfan	0.53	3.55	Cadmium hydroxide @ 25 °C	2.6	
β-Endosulfan	0.28	3.62	Chromium oxide		
α-BHC	2.0	3.81	(as H$_2$CrO$_4$) @ 0 °C	6.17×10^5	
β-BHC	0.24	3.80	Copper chloride @ 0 °C	7.06×10^5	
δ-BHC	31.4	4.14	Lead oxide @ 20 °C	17	
gamma-BHC	7.5	3.72	Lead chloride @ 20 °C	9.9×10^3	
Aldrin	0.011	5.17	Mercuric oxide @ 25 °C	53	
Dieldrin	0.20		Mercuric sulfide (α) @ 18 °C	0.01	
4,4'-DDE	0.12	5.69	Mercuric sulfide	Insoluble	
4,4'-DDT	0.006	3.98	Mercuric chloride @ 20 °C	6.9×10^4	
4,4'-DDD	0.02	5.98	Nickel sulfide @ 18 °C	3.6	
Endrin	0.26	5.6	Nickel chloride @ 20 °C	6.42×10^5	
Heptachlor	0.056	4.40	Selenium dioxide @ 14 °C	3.84×10^5	
Heptachlor epoxide	0.35	3.65	Selenium trioxide	Very soluble	
Chlordane	1.85	2.78	Silver oxide @ 20 °C	0–13	
Toxaphene	1.75	3.3	Silver chloride @ 10 °C	0.89	
			Thallium sulfide @ 20 °C	2.0×10^2	
			Thallium chloride @ 16 °C	2.9×10^3	
Oxygenated compounds			Zinc oxide @ 29 °C	1.6	
Acrolein	400,000	−0.090	Zinc chloride @ 25 °C	4.32×10^6	

†Not available for metals.

The next step in the procedure is to determine the average unsaturated pore-space aqueous concentration C_e that causes the C_o in the saturated zone directly beneath the site to exceed the calculated C_o from equation 15.4. This is estimated by dividing the annual mass of the contaminant X escaping from the unsaturated zone by the percolation rate PR, or:

$$C_e = \frac{X}{PR} \tag{15.10}$$

Using a literature K_d value or K_d calculated from equation 15.2 and an estimated value for $1/n$ gives an estimate of Q_e. The resultant dry-soil concentration is the suggested level of cleanup that derives a contaminant concentration in compliance with a ground water quality goal or a health-based value at a receptor well.

The basic limitations of the Freundlich isotherm method include the assumptions that adsorption is completely reversible, and that the rates of adsorption and desorption result in instantaneous equilibrium. These assumptions are probably not correct, thus may never be achieved. Also, it is often assumed that the (*n*) value used to obtain the adjustment factor is unity, making the isotherm linear. In reality, values of *n* are experimentally derived and are different for different ranges of the Freundlich isotherm. The time investment required to derive *n* is often a constraint on using this method.

QUESTION 15.1

Find the original groundwater concentration C_o at a source area for a Polynuclear Aromatic (PNA), whose health-based level C is 0.029 $\mu g/L$ (10^{-6} unit cancer-risk factor direct-ingestion level) if the saturated-zone thickness Z is 16 feet; the distance of the compliance point from the source X is 1,000 feet; the width of the lateral extent of the source Y is 2,400 feet; and the transverse dispersivity D_t is 13 feet.

QUESTION 15.2

Find the equilibrium pore-space aqueous concentration for the PNA value of C_o in Question 15.1 if the annual precipitation at the site is 11.15 in/yr, the site area is 28 acres, and the saturated ground water velocity of the underlying soil is 60 ft/yr.

Allowable concentration model   This method estimates the allowable cleanup concentration of a soil contaminant that corresponds to the maximum allowable contaminant level in the underlying ground water at some down-gradient receptor. The maximum allowable contaminant level in ground water is usually available from local regulators.

To use this method, the initial source concentration C_o is calculated from equation 15.4. After calculating the initial source concentration, the next step is to convert the maximum allowable ground water contamination concentration (assumed to be C in equation 15.4) to an allowable soil-contaminant concentration. This is done by using the partition coefficient K_d, defined by the following, where C_o is assumed to be C_{water}:

$$K_d = \frac{C_{\text{soil}}}{C_{\text{water}}} \qquad (15.11)$$

Methods for determining K_d are available from EPA (1989a) or from sources in the literature. The required soil-cleanup concentration level can be determined by multiplying the allowable concentration in the ground water directly beneath the site by the partition coefficient:

$$C_{\text{soil}} = (C_{\text{water}})(K_d) \qquad (15.12)$$

Summers model   The Summers model (Summers, Gherini, and Chen 1980) was developed to estimate the point at which contaminant concentrations in soil produce ground water contaminant concentrations at acceptable levels. The resultant soil-contaminant concentration can then be used to specify a cleanup goal. The model assumes that a percentage of the rainfall at a given site infiltrates the surface and desorbs contaminants from the soil, based on equilibrium K_d values. The model also assumes that this contaminated infiltration mixes completely with the ground water beneath a given contaminant site, resulting in an equilibrium ground water concentration.

The model begins by estimating the concentration of the contaminant infiltration that results in ground water concentrations at or below target level. The mixing rates of uncontaminated ground water with contaminated infiltration, and the resultant concentrations in ground water can be calculated from:

$$C_{gw} = \frac{(Q_p C_p)(Q_A C_A)}{Q_p + Q_A} \qquad (15.13)$$

where: C_{gw} is the contaminant concentration in the ground water ($\mu g/L$); $Q_p = (VD_z)(A_p)$ is the volumetric flow rate of infiltration (pore water) into the aquifer (m^3/day); $VD_z = (V_s)(e)$ is the Darcian velocity in the downward direction (m/day); V_s is the ground water seepage velocity (m/day); e is the void ratio (ground water volume:volume of solid); A_p is the horizontal area of the spill (m^2); C_p is the concentration of pollutant in the infiltration at the unsaturated-, saturated-zone interface (μg/L); $Q_A = (V_D)(h)(w)$ is the volumetric flow rate of ground water (m³/day); V_D is the Darcian velocity of the aquifer (m/day); h is the thickness of the aquifer (m); w is the spill-width perpendicular to the flow direction of the aquifer (m); and C_A is the initial (or background concentration of) contaminant in the aquifer (μg/L). The maximum allowable contaminant concentration in the infiltration that does not result in a ground-water concentration exceeding a water-quality goal (i.e., an MCL) can be calculated by using this water-quality goal for C_{gw} in equation 15.13, and solving for the infiltration contaminant concentration C_p, or:

$$C_p = \frac{C_{gw}(Q_p + Q_A) - Q_A C_A}{Q_p} \tag{15.14}$$

Once the maximum allowable contaminant concentration in the leachate is calculated, the contaminant concentration in the soil can be calculated. This is the soil-cleanup level that needs to be attained in order to protect the ground water, and can be obtained from the soil:water partitioning equation

$$C_s = (K_d)(C_p) \tag{15.15}$$

where C_s is the soil concentration (μg/kg), C_p is the concentration in the infiltration, from equation 15.14 (μg/L), and K_d is the equilibrium partition coefficient [(μg/kg)/(μ/L)].

Modified Summers model   The modified Summers model (EPA 1989a) is a variation of the Summers model, developed to derive soil-cleanup criteria using established ground water regulatory and health-based levels, coupled with an equilibrium partitioning approach. Soil cleanup levels are calculated for saturated and unsaturated media, assuming equilibrium between dissolved and adsorbed phases for each contaminant. The following relation is used

$$S_{sat} = (K_d)(C_{sat}) \tag{15.16}$$

where S_{sat} is the concentration of contaminant adsorbed to the soil in the saturated zone (μg/kg), K_d is the partition coefficient [(μg/kg)/(μg/L)], and C_{sat} is the concentration of contaminant in the ground water in the saturated zone (μg/L). The desired contaminant concentration for ground water is determined using established health-based criteria (e.g., MCLs or cancer risk values). The cleanup criteria is calculated using equation 15.16.

Calculations to derive unsaturated soil-cleanup criteria include the assumption that dissolved contamination in the ground water recharge reaches equilibrium with the adsorbed phase on unsaturated soils, and that such recharge is fully diluted into the entire water column upon reaching the water table. Cleanup criteria for unsaturated soils are established using equation 15.16 as well as a dilution equation for calculating the contaminant concentration in the ground water in the saturated zone C_{sat}:

$$C_{sat} = \frac{(C_{unsat})(q)}{(q + Q)} \tag{15.17}$$

where C_{unsat} is the contaminant concentration of the ground water recharge (μg/L), q is the volumetric flow rate of recharge flowing downward through a unit area (m³/day), and Q is the volumetric flow rate of ground water in the saturated zone throughout the unit (m³/day). The equilibrium assumption

$$S_{unsat} = (K_d)(C_{unsat}) \tag{15.18}$$

combined with equation 15.17 yields the equation used to calculate the cleanup criteria for soil in the unsaturated zone:

$$S_{unsat} = \frac{(S_{sat})(e + Q)}{e} \tag{15.19}$$

where S_{unsat} is the concentration of contaminant adsorbed to the soil in the unsaturated zone (μg/kg) and Q is the ground water volumetric flow rate (m^3/day) in the saturated zone from Darcy's law

$$Q = (K)(i)(A) \tag{15.20}$$

where K is the saturated hydraulic conductivity (m/day), i is the hydraulic gradient (m/m), and A is the cross-sectional area of flow (unit width $\times$ the saturated thickness of the aquifer) (m^2).

SPPPLV model   The Single Pathway Preliminary Pollutant Limit Values (SPPPLV) and Preliminary Pollutant Limit Values (PPLV) methods were developed by the U.S. Army (1987) to determine site-specific cleanup levels. They require the identification and measurement of contaminants that are present; pathways of exposure; and the determination (or estimation) of an acceptable daily dose (D_t) for each contaminant to a receptor.

SPPPLVs for all pathways and contaminants are calculated from measured levels of contaminants at a particular site. The acceptable daily intake for each contaminant; distance to the receptor; rate of off-site migration; and rates of dilution and degradation are used in the model. Assuming that contaminants are in equilibrium along all exposure pathways from source to receptor, partition coefficients can be used to determine levels of contaminants in different media along the exposure pathways. Critical pathways are selected for each contaminant, and a preliminary pollutant limit value (PPLV) is then derived for each medium by normalization of the SPPPLV using the following equation (U.S. Army 1987):

$$\text{PPLV} = \cfrac{1}{\left(\cfrac{1}{\displaystyle\sum_{i=0}^{n} (\text{SPPPLV})_i}\right)} \tag{15.21}$$

In order to establish PPLVs, the best available toxicological information is used to estimate an acceptable daily dose D_t for human exposure to each compound. A PPLV is derived from consideration of the D_t, along with the probable exposure level. For example: for soil, two exposure pathways (ingestion and skin absorption) are considered for a given site. The equations for these two pathways are written as:

$$\text{Soil Ingest (SPPPLV)} = \frac{(D_t)\,(\text{body weight})}{\text{daily amount of soil ingested}} \tag{15.22}$$

$$\text{Skin Absorption (SPPPLV)} = \frac{(D_t)\,(\text{body weight})}{\text{weight of soil per day}} \tag{15.23}$$

The PPLV for soil when considering both ingestion and skin absorption is then calculated by:

$$\text{Soil PPLV} = \cfrac{1}{\cfrac{1}{\text{Soil Ingestion (SPPLV)}} + \cfrac{1}{\text{Skin Absorption (SPPPLV)}}} \tag{15.24}$$

Contaminant profile model   The contaminant profile model method (Williams et al. 1988) was developed as a tool for estimating the transport and fate of chemicals from sites where initial concentrations of contaminants are known as a function of depth. One of the objectives of the model is to provide an estimate of the amount of contaminant that leaves the unsaturated zone and enters the ground water. Transport of contaminants through the following five phases is considered: water; stationary inorganic; immobile organic; mobile organic; and vapor. In certain disposal or spill situations, the mobile organic phase—in addition to the water phase—flows through soil, thereby enhancing the mobility of potentially hazardous chemicals which are adsorbed to the mobile organic phase. The contribution from all five phases is described by the following relation:

$$C_T = (1 - n)(\rho_s)(C_s) + (\phi_i)(\rho_i)(C_i) + (\theta_a)(\rho_a)(C_a)$$
$$+ (\phi_m)(\rho_m)(C_m) + (\eta_v)(\rho_v)(C_v) \tag{15.25}$$

where C_T is the initial total concentration of the contaminants in the soil sample (kg/kg); n is the total pore fraction (porosity) of the system (m³/m³); C are the concentrations of the contaminant (kg/kg) in the solids s, water a, immobile organic i, mobile organic m, and vapor v phases; ρ are the densities of the solids s, water a, immobile organic i, mobile organic m, and vapor v phases (kg/m³); ϕ are the volume fractions of the immobile i and mobile m organic phases in the total sample (m³/m³); θ_a is the volume fraction of the water phase in the total sample (m³/m³); and η is the volume fraction of the vapor phase (m³/m³).

Independent relations (R) must be obtained between each of the phases and the total concentrations, allowing independent calculation of the concentration of each contaminant in each phase. This can be done by defining five new terms:

$$C_T = (R_s)(C_s) = (R_i)(C_i) = (R_a)(C_a) = (R_m)(C_m) = (R_v)(C_v) \tag{15.26}$$

Assuming linear partitioning and local equilibrium, the R terms can be defined in terms of partition coefficients. Another assumption is that the interface between each of the phases is water and the other phases do not contact each other. Therefore, contaminant transfer from one phase to another must include transfer through the water phase. The partition coefficients can be defined as follows:

$$C_s = (K_s)(C_a) \qquad C_m = (K_m)(C_a)$$
$$C_i = (K_i)(C_a) \qquad C_v = (K_v)(C_a)$$

where K_s is the solid:water partition coefficient, K_i is the immobile organic:water partition coefficient, K_m is the mobile, immiscible organic:water partition coefficient, and K_v is the vapor:water partition coefficient.

Now the equation can be rewritten in terms of the partition coefficients and the respective phase concentrations. For example, the total soil concentration and the R terms can be expressed as:

$$C_T = C_s \left[(1 - n)\rho_s + \frac{(\phi_i \rho_i K_i + \theta \rho_a + \phi_m \rho_m K_m + \eta_v \rho_v K_v)}{K_s} \right] \tag{15.27}$$

$$R_s = (1 - n)\rho_s + \frac{(\phi_i \rho_i K_i + \theta \rho_a + \phi_m \rho_m K_m + \eta_v \rho_v K_v)}{K_s} \tag{15.28}$$

Decision tree process   This method (State of California 1986) evaluates the movement of chemicals through the unsaturated zone, and is used to estimate the concentrations of organic chemicals in the saturated zone as water percolates through the unsaturated soil column. The concentration in the ground water depends on the residual concentration in the soil prior to percolation. The decision tree process incorporates a retardation factor based on the carbon and clay content in the soil. It is assumed that the unsaturated zone usually has a higher retardation factor than the saturated zone.

The equations following were developed to analyze a "batchwise" extraction of chemicals by percolating water from a soil column divided into several cells of equal size. Cell size is determined by factors such as the location of the cell (saturated or unsaturated zone) and the limitations of the computer used to run the model. The equations assume that mobile water is replaced by clean water and the system reaches equilibrium with each successive percolating cycle. The concentration of chemicals leaving the first cell can be expressed as:

$$C_w = \left(\frac{C_s}{K_d}\right)\left(\frac{(K_d)(M_s) + (M_w)(\alpha)}{(K_d)(M_s) + (M_w)}\right) \tag{15.29}$$

where C_w is the concentration of chemicals in the water after wetting (mg/L); C_s is the chemical concentration in the soil (mg/kg); K_d is the partition coefficient [(mg/L)/(mg/kg)]; M_s is the mass of solids per unit volume of soil (kg), M_w is the mass of water per unit volume of soil, assuming 50-percent water content (kg); and α is the fraction of immobile water. The chemical concentration in the water leaving the second cell can be expressed as:

$$C_{w2} = \frac{C_{w2*}(M_s K_d + M_w) - C_{w1}M_w(1 - \alpha)}{M_s K_d + M_w \alpha} \tag{15.30}$$

where C_{w2} is the concentration of chemicals in water leaving cell 2 (mg/L); C_{w1} is the concentration of chemicals in water entering cell 2 from cell 1 (mg/L); and C_{w2*} is the concentration of chemicals in water in cell 2 after one pore-volume flush (mg/L).

Water leaving cell 2 enters cell 3; this methodology uses a batchwise extraction of chemicals from the soil column. The resulting chemical concentration leaving the last cell is the concentration at the unsaturated–saturated boundary. Upon entering the saturated zone, the chemical concentrations are attenuated by the higher flow rates. The method also assumes that total mixing of chemicals occurs as water is leached out of the unsaturated zone. The amount of attenuation is calculated by using relative flow rates and chemical concentrations entering and leaving a control volume:

$$C_{H!} = \frac{(Qin)_H C_H + (Qin)_v C_v}{(Qin)_H + (Qin)_v} \tag{15.31}$$

where $C_{H!}$ is the resulting attenuated soil chemical concentration in the saturated zone (mg/kg); C_H is the initial soil chemical concentration in the saturated zone (mg/kg); $(Qin)_H$ is the flow rate entering the control volume in the horizontal direction from the saturated zone (L^3/T); $(Qin)_v$ is the flow rate entering the control volume in the vertical direction from the unsaturated zone (L^3/T); and C_v is the soil chemical concentration entering the control volume vertically from the unsaturated zone (mg/kg).

Two additional calculations are needed—the percolation rate and the dilution factor—to complete the evaluation of the chemical concentration in the unsaturated zone, compared to the estimated chemical concentration in the aquifer. The percolation rate is a fraction of the precipitation and is the principle contributor to chemical leaching from the unsaturated zone. The equation to calculate the monthly water balance (or mean percolation) is:

$$\text{PERC} = P - RO - \Delta S - \Delta ET \tag{15.32}$$

where PERC is the monthly percolation rate (m/mo); P is the mean monthly precipitation (m); RO is the mean monthly runoff (m); ΔS is the monthly change in soil-moisture storage (m); and ΔET is the monthly actual evapotranspiration (m). The dilution factor is defined as:

$$DF = \frac{(Qin)_v}{(Qin)_H + (Qin)_v} \tag{15.33}$$

where $(Qin)_v = (A_v)(\text{PERC})$ and A_v is the horizontal cross-sectional area of the control volume in the saturated zone (m²). The flow rate entering the control volume in the horizontal direction from the saturated zone, $(Qin)_H$, is given by Darcy's Law as:

$$(Qin)_H = (A_H)(K_H)(i) \tag{15.34}$$

where A_H is the cross-section area of the aquifer (m²), K_H is the aquifer hydraulic conductivity (m/T), and i is the aquifer hydraulic gradient (m/m). After obtaining the dilution factor and the chemical concentration at the unsaturated–saturated boundary, the predicted concentration of the chemical in the aquifer can now be calculated as:

$$\text{Aquif Conc} = (\text{Conc at the Unsaturated} - \text{Saturated Boundary})(DF) \qquad (15.35)$$

Equation 15.35 is an estimate of the resultant chemical concentration at the point of exposure, given the initial chemical concentration in the soil in the unsaturated zone. This method has limitations that include: extensive (and sometimes difficult to obtain) field measurements; assumed vertical movement of water through the unsaturated zone into the ground water; assumed total mixing of chemicals leaving the unsaturated zone; and the assumption that the soil column is flushed with clean water.

RITZ model   The Regulatory and Investigative Treatment Zone Model (RITZ) was originally developed and published by Short (1985). Nofziger and Williams (1988) subsequently published a user guide that incorporated microcomputer hardware and software, input guidance, and graphical and tabular output. The model incorporates the influence of oil in sludge applied to land areas, water movement, volatilization, and degradation upon the transport and fate of a hazardous chemical. This model is one of many models that can be used to assess the fate and transport of contaminants in the unsaturated zone, in order to calculate soil remediation levels. Chapter 13 gives a more complete listing of the available models for fate and transport of contaminants in the unsaturated zone.

Short (1985) makes several assumptions in developing the RITZ model. These include:

- Waste material is uniformly mixed in the plow zone;
- The oil in the waste material is immobile and never leaves the plow zone; only the contaminant moves with the soil water;
- The soil properties are uniform from the ground surface to the bottom of the treatment zone;
- The flux of water is uniform throughout the treatment site and throughout time;
- Hydrodynamic dispersion is insignificant and can be neglected;
- Linear isotherms describe the partitioning of the contaminant between the liquid, soil, vapor, and oil phases, and local equilibrium between phases is assumed;
- First-order degradation of the contaminant and oil is assumed and degradation constants do not change with soil depth or time;
- The contaminant partitions between the soil, oil, water, and soil vapor and does not partition to the remaining fractions of the sludge;
- The sludge does not measurably change the properties of the soil water or the soil, so the pore liquid behaves as water; and
- The water content of the soil is related to the hydraulic conductivity as described by Clapp and Hornberger (1978):

$$\frac{K}{K_s} = \left(\frac{\theta}{\theta_s}\right)^{2b+3} \qquad (15.36)$$

where K is the hydraulic conductivity at a volumetric water content of θ, K_s is the saturated hydraulic conductivity (or the conductivity of the soil) at the saturated water content, (θ_s), and b is the Clapp and Hornberger constant for the soil.

The User's Guide and RITZ program are available from EPA (Nofziger and Williams 1988). The user of the model is cautioned to consider the assumptions in the model, and to apply the model only where appropriate. The model presents results for the specific parameters entered without any measure of uncertainty in the calculated values. The user is encouraged to compare results for a series of simulations using parameters in the expected ranges for the site, in order to obtain an estimate of the uncertainty. Layered soils cannot be handled by the model, although if the site contains two layers, the user can run the simulation twice, one for the soil properties of each layer.

15.2 ALTERNATIVE TECHNOLOGIES FOR SOIL REMEDIATION

Corrective-action objectives for remediating soil can be achieved by a large number of technologies. Numerous remedial-action alternatives that have proven to be effective in contaminated-soil cleanups are available, and selected potential implementation technologies are discussed below; the alternatives discussed here do not include free-product recovery. Table 15.4 lists some of the soil (and ground water) treatment technologies for site remediation. Selected technologies are described in more detail below.

No Action

Under this alternative, no remedial action is undertaken. Two possible no-action sub-alternatives include passive bioremediation and monitoring. Effectiveness of both of these no-action sub-alternatives assumes that steps are taken to ensure that any further handling of products causing the contamination at a given site is conducted so as to prevent any further impact on the environment. If the contamination was caused by a leaking tank or other source, this source must be removed in order to implement these two sub-alternatives effectively.

Passive bioremediation Many regulators permit use of passive bioremediation to reduce contaminant plumes naturally and to complete eventual cleanup. In order to use such a technique, several conditions usually need to be met: **(1)** the contaminant can be metabolized and degraded by naturally occurring microorganisms; **(2)** the plume must have stabilized—that is, it is no longer moving; and **(3)** the source of the plume has been removed. Hydrocarbons and similar organic substances are most readily amenable to passive bioremediation. Generally, metals are not as effectively treated using passive bioremediation; however, bacteria are known to "take up" metals as part of their metabolic processes, as well as removed in soils by the processes of adsorption, precipitation, ion exchange, and complexation (EPA 1981).

TABLE 15.4 Soil and Ground Water Treatment Technologies for Site Remediation

Physical treatment	Biological treatment	Chemical treatment	Thermal treatment
Coagulation/flocculation	Activated sludge	Neutralization	Incineration
Oil–water separation	Activated sludge with PAC	Precipitation	Thermal desorption
Air and steam stripping	Aeration tank	Ion exchange	Vitrification
Carbon adsorption	Aerobic and anaerobic fixed film	Chemical oxidation	
Filtration	Anaerobic digester/tank	Chemical reduction	
Reverse osmosis	Fluidized bed	Photolysis (ultraviolet light)	
Sedimentation	Rock-reed filter	Wet-air oxidation	
Evaporation and distillation	Sequencing batch reactor	Stabilization	
Solvent extraction	Trickling filter	Dechlorination	
Freeze crystallization	Composting	Soil washing	
Centrifugation	In-situ biodegradation		
Sonic treatment	Land treatment		
Soil venting	Liquid/solid systems		
Air sparging	Wetlands treatment		

If the contaminant source is removed and the contaminant is resident in the soil at residual saturation, both aerobic and anaerobic microorganisms act to reduce the plume mass and complete cleanup (LLNL 1995). A contaminant spreads in unsaturated soil due primarily to the influence of gravity, until the point is reached at which the fluid no longer holds together as a single, continuous phase, but rather occurs in isolated, residual globules. At this point, the contaminant has largely become immobile under the usual soil-potential conditions and can migrate further only: **(1)** in water according to its solubility; or **(2)** in the gaseous phase within the unsaturated zone (Schwille 1988). Evaluation of over 1,500 leaking underground fuel tanks in California indicates that, in general, plume lengths change slowly and tend to stabilize in relatively short distances from the release site (LLNL 1995). The evaluation also indicates that once contaminant sources are removed and the plume is stabilized, hydrocarbons (at least in ground water) appear to degrade naturally at rates as high as 50 to 60 percent per year. Therefore, passible bioremediation is a potentially effective and low-cost method for final cleanup of selected organic contaminants in both the unsaturated and saturated zones.

Monitoring   A long-term ground water monitoring program needs to be established to identify and assess any potential contaminant migration. Existing boreholes and monitoring wells used in the site investigation should be used as a component of the ground water monitoring network. It may be necessary to install additional ground water monitoring wells both up- and down-gradient of the observed contamination to assess contaminant migration trends, and to document contamination levels at the site boundaries.

By sampling existing and new wells, this alternative establishes a mechanism for identifying any changes in ground water flow patterns, contaminant levels, and contaminant-plume migration. A monitoring program also provides an early warning of increases in contaminant levels. For the short term, this alternative is not protective of human health and the environment. Contaminated soil stays in place and there are not any mitigative measures taken to decrease the concentration of the contaminants in soil. The existing environmental impacts remain. For long-term effectiveness, the risks that exist without cleanup continue.

The residual risks—as a result of the no-action alternative—may prove to be acceptable to regulators, and this alternative is not to be overlooked as a possible solution if there are no potential receptors within a given site. Although there are no capital costs associated with the no-action alternatives, there are annual (or more frequent) monitoring costs. Relative to the costs for other alternatives presented below, this cost is low.

Capping

Capping is designed to minimize contact and infiltration of precipitation, thereby reducing the potential for leachate generation from contaminants in the soil. Capping also helps to eliminate erosion and storm-water transport of contaminants into local surface-water collection areas. The capping process requires a relatively impermeable barrier overlying the contaminated area. A variety of cap designs and capping materials are available, ranging from a single-layered synthetic type, to a multilayered, that supports a soil cover to protect the barrier and allow the growth of vegetation. The capping materials considered in practice include clay, synthetic membranes, asphalt or concrete, and multilayer combined media.

Historically, clay has been the most extensively used capping material in landfills or surface impoundments. However, exposed clay can shrink and crack when it dries. Additionally, clay caps have a low tolerance for heavy surface loads that can be expected at some sites. Consequently, consideration of clay as a capping material needs to take these parameters into account. Both bituminous asphalt and concrete have been demonstrated to perform as effective barriers for precipitation and surface-water infiltration.

Synthetic membranes can also be used as a capping construction material. The factors influencing the use of synthetic membranes include chemical compatibility, prevention of tears and punctures, and proper overlapping of seams and seam sealing; without proper protection from surface loads, synthetic membranes are subject to punctures and tearing.

Current environmental regulations require that the design of most caps conform to the performance standards in 40 CFR Part 264.310, which addresses Resource Conservation and Recovery Act (RCRA) landfill-closure requirements. Most cap designs are multilayered to conform to these standards. A typical multilayered cap design (see figure 15.3) includes an upper vegetative layer, underlain by a drainage layer and two low-permeability layers. The vegetation is supported by the loam layer; the drainage layer is composed of sand or a geosynthetic drainage net; and the low-permeability layer is formed by the combined synthetic liner and clay system. An additional layer of stone and gravel is often placed immediately above the contaminated soil, as structural backfill to support the cap. Prior to capping the fill must be graded and compacted, in order to enhance the stability of the cap and reduce settlement. The use of capping technology also necessitates compliance with local regulations. Closure considerations deal primarily with cover materials and designs, while post-closure concerns involve site maintenance and monitoring.

The effectiveness of a cap in reducing the potential for leachate generation is dependent on ground water levels. Ground water coming into contact with contaminated soil—more likely when the local water table is elevated—can produce leachate despite the existence of a well-designed cap. Capping can be combined with ground water extraction or other containment technologies, to prevent or reduce the movement of a contained plume. Therefore, capping with asphalt/concrete or a multimedia layer in conjunction with other technologies, is often an effective remedial-action method.

Capping is effective for the short-term time frame, since the contaminated soil remains in place and direct exposure to it is minimal. Occupational risks to workers are no greater

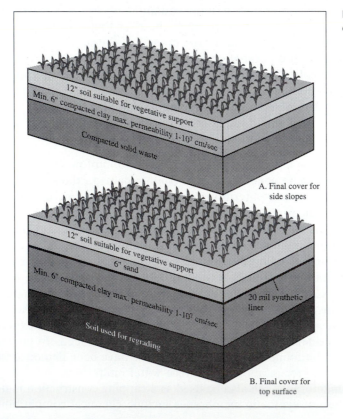

Figure 15.3 Typical multilayer cap design (data from McAneny et al. 1985)

12" soil suitable for vegetative support
Min. 6" compacted clay max. permeability 1-10^7 cm/sec
Compacted solid waste

A. Final cover for side slopes

12" soil suitable for vegetative support
6" sand
Min. 6" compacted clay max. permeability 1-10^7 cm/sec
20 mil synthetic liner
Soil used for regrading

B. Final cover for top surface

than those normally experienced in construction of this type. The risk to nearby, off-site people is negligible, because exposure to the contaminated soil does not occur if these soils are covered by a cap. The capping alternative is not effective for the long term, although the cap does prevent direct contact with the contaminated soil. If ground water intersects the contaminated soil, causing contaminates to be continuously washed (flushed) from it to ground water, then the cap would not prevent migration of contaminants and would not protect human health and the environment. The costs associated with this alternative would be in the low cost range compared to other alternative technologies.

Soil Venting

In situ soil venting implements a vacuum extraction system (VES) that removes volatile organic compounds (VOCs) from the soil matrix via vapor-phase pathways. In soil venting, VOCs are removed and can be further treated, either to completely degrade the compound or transfer it to another medium. The transfer to another medium does not destroy the contaminant, but does remove it from the soil. In addition to extracting contaminants directly, the pumping action draws a continuous supply of air through the soil, enhancing in situ biodegradation of the pollutant by naturally occurring aerobic bacteria. A soil venting system consists of the installation of wells and/or trenches in the contaminated area, with a pump to draw air out of the wells (see figure 15.4). In turn, air moves through the soil and into the wells, thus providing a continuously renewed oxygen source for the aerobic bacteria. A wide variety of environmental and soil factors determine whether this alternative is applicable to a given site; these factors include water content; porosity and permeability; clay content; as well as other factors. However, a VES pilot study can determine the feasibility of this type of alternative, and is encouraged at any site for which this alternative is considered.

The Superfund Innovative Technology Evaluation (SITE) Program (EPA 1991a) indicates that VES is the most widely used innovative technology for National Priorities List (NPL) sites. VES has proved successful in remediating petroleum-contaminated soils at many sites. Both horizontal and vertical VESs have proved successful in removing benzene, toluene, ethylene, and xylene (BTEX) contaminants from soils (EPA 1991b). If the water table is near the surface, the high vacuum required for vertical wells at some sites can cause the water table to be lifted into the extraction system, thus diminishing the effectiveness of the treatment system; horizontal VES requires less vacuum to achieve a greater radius of influence. Using either the vertical or horizontal design, the volatiles extracted are processed

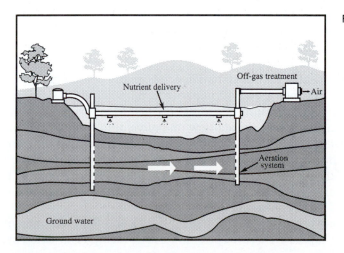

Figure 15.4   Typical soil venting system

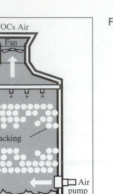

Figure 15.5 Air stripping system

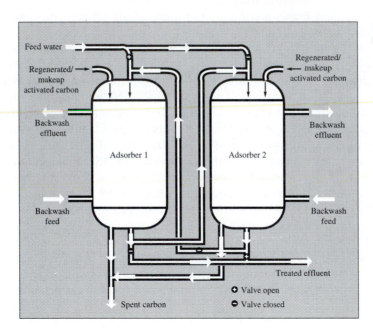

Figure 15.6 Carbon adsorption system

through a liquid-vapor separator or air stripping system (see figure 15.5), and the vapor is treated by activated-carbon adsorption or by catalytic converter, before being released to the atmosphere (see figure 15.6).

The effectiveness of a VES system depends on the contaminant vapor pressure—that is, the pressure of its vapor in equilibrium with its pure liquid (or solid) phase. The temperature at which the vapor pressure of a liquid is equal to atmospheric pressure, is the boiling point of the contaminant. For a fixed-flow rate of venting air, the maximum rate at which a contaminant is extracted from the soil is derived by assuming that the partial pressure of the contaminant in the vented vapor is equal to the vapor pressure of the contaminant. The molar density of the contaminant in the vapor phase is equal to its partial pressure, assuming an ideal vapor behavior. The vapor pressure at 40 °F and the maximum extraction rates of some common VOCs are shown in table 15.5.

This alternative is protective of human health and the environment during the short term, because the vapors are treated prior to release. Vacuum extraction provides permanent

TABLE 15.5   Maximum Vapor Extractability[1]

Compound	Vapor pressure at 40 °F (mm Hg)	lb/100 ft³	lb/day at 100 SCFM
Benzene	28.0	7.9	1134
Chlorobenzene	3.8	1.5	221
Chloroform	77.0	33.2	4782
1,1 DCA	89.0	31.7	4564
Methylene chloride	198.9	59.9	8622
Naphthalene	0.1	0.05	7
PERC	7.5	4.49	646
1,1,1 TCA	4.6	21.9	3154
TCE	28.0	13.1	1891
Toluene	9.0	3.0	430
Xylenes	3.0	1.1	165

[1]Assumes continuous vapor saturation

removal of the contaminants from the soil, for long-term effectiveness; therefore, the mobility and volume of contaminants is greatly decreased. Vacuum extraction has proved effective in controlling fugitive emissions from contaminated sites, and it is the most commonly used in situ remedial technology. Proper implementation of this technology does require a pilot test to determine the trench size; pump capacities; flow rates; well spacing; and operating pressures. Excavation of trenches and subsequent laying of pipe is also required to implement the alternative. Once flow paths are established, the vapor extraction is usually rapid. The more-volatile contaminants are removed the fastest, with the less-volatile contaminants being removed at a slower rate. However, old spills—where many of the volatiles have escaped—may not respond as well to the VES technology.

The cost of this alternative is a fraction of the cost for other innovative remedial-action technologies, such as on-site thermal treatment. The cost of VES is in the medium-cost range compared to the no-action or capping alternatives.

Air Sparging

Essentially, air sparging (see figure 15.7) creates a crude air stripper in the subsurface, with the saturated-soil column acting as the packing (Angell 1992). Injected air flows through the water column over the packing, and air bubbles contacting dissolved/adsorbed-phase contaminants cause the VOCs to volatilize. The entrained organics are then carried by the air bubbles into the unsaturated zone, where they are captured by a vapor extraction system—or if permissible—allowed to escape through the ground surface into the air. As a bonus, the sparged air maintains high dissolved oxygen, which enhances natural biodegradation.

The key to successful air sparging is good contact between the injected air and contaminated soil and ground water. Beneath the water table, the air bubbles have to travel vertically through the aquifer in order to strip the VOCs. Additionally, a permeability differential above the air-injection zone reduces the effectiveness of air sparging. As a consequence, low permeability or heterogeneities push the dissolved contamination concentrically from the injection point, thereby resulting in spread of contaminants, rather than cleanup. Thus, there are two primary concerns with air sparging: **(1)** the spread of dissolved contaminants; and **(2)** the acceleration of vapor-phase transport and subsequent accumulation of vapors in buildings (or other closed spaces). Because air sparging increases the pressure in the unsaturated zone, exhausted vapors from the ground water are drawn into the basement of buildings. In areas with potential vapor receptors, air sparging needs to be evaluated with a concurrent-vent system.

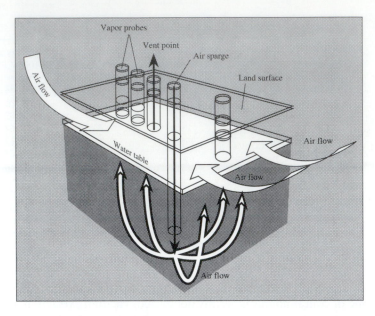

Figure 15.7   Air sparging system

Two factors are critical to the effective design and operation of an air sparging system: **(1)** the extraction system itself; and **(2)** the vapor-abatement system. The extraction system includes the number, spacing, and location of extraction wells; manifold layout; and the size (and type) of blowers. A properly designed extraction system operates with minimal adjustment; a poorly designed system requires the repeated installation of additional wells, piping, and blowers. Vapor-abatement systems—often required by regulatory agencies—consist of carbon or thermal treatment. Carbon is generally easy and inexpensive to install and permit, but is likely expensive to use for high VOC concentrations. Thermal systems, on the other hand, require higher capital costs and take more time to permit, but are relatively inexpensive to operate. The maximum venting efficiency occurs in a soil venting system when the following factors occur (Angell 1992):

- The induced-air flow directly contacts the contaminated soil;
- The radius of influence of the vent wells matches the area of contamination; and
- The correct size vacuum blower is chosen, based on site-specific soil-permeability conditions such as water content, texture, and mineralogy.

The information needed for an effective air sparging system design is as follows (Angell 1992):

- The location of potential ground water and vapor receptors;
- The geological conditions of the site (e.g., permeability, lithology, heterogeneity);
- The contaminant-mass distribution within the area, to be treated in both soil and ground water; and
- The radius of influence of the sparge wells at various flow rates/pressures.

A typical air sparging system design requires a field test that includes monitoring the following (Angell 1992):

- Pressure versus distance (indicator of radius of influence);
- VOC concentrations in ground water (indicator of what is being removed and should be done before, during, and after the test);

- CO_2 and O_2 concentrations in soil vapor (indicators of biological activity and should be done before, during, and after the test, and under both static and pumping conditions);
- Dissolved oxygen levels in the water (indicator of the effectiveness of the sparging, with changes being slower than for air flow); and
- Water levels before and during the test (air flow causes some rise in the water table and thus needs to be measured).

The cost of an air sparging system is similar to that for a VES system, and generally falls into the medium-cost range when compared to the no-action and capping alternatives. This alternative also is protective of human health and the environment in both the long- and short-term, as long as appropriate precautions are undertaken relative to venting and treatment of vapors from the sparge system.

Soil Flushing

In situ chemical treatment consists of flushing contaminants from soil through injection of a flushing agent (see figure 15.8). Organic and inorganic contaminants can be washed from contaminated soil by an extraction process called "soil flushing," "ground leaching," or "solution mining." Water (or an aqueous solution) is injected or sprayed into the area of contamination; the contaminated elutriate is collected and pumped to the surface for removal, recirculation, or on-site treatment and reinjection. During elutriation, the flushing solution mobilizes the sorted contaminants by dissolution or emulsification.

Soil flushing has been specified in the Records of Decision (RODs) for as many as 10 NPS sites. Studies have been conducted to determine the appropriate solvents for mobilizing various classes and types of chemical constituents. The difficulty in implementing this technology depends on the ability to flood the soil with the flushing solution, and to install collection wells or subsurface drains to recover all of the applied liquids; provision also must be made for disposal of the elutriate. The achievable level of treatment varies, dependent on the contact of the flushing solution with the contaminants, the appropriateness of solutions for the contaminants, and the hydraulic conductivity of the soil.

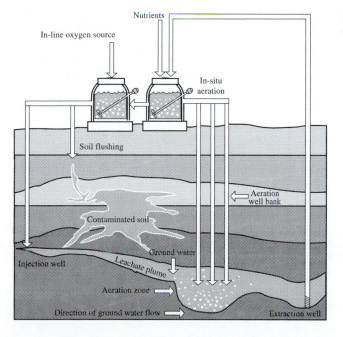

Figure 15.8   Flushing and bioreclamation of soil and ground water

A slight variation to soil flushing—an air stripping technology—has been developed. This method involves injecting superheated, compressed air into and below the zone of contamination, via vertical or horizontal perforated piping. Most volatiles go out with the soil water by evaporation, opening up void paths, thus increasing the permeability of the soil to air flow. As the water evaporates, it also removes volatile hydrocarbons by steam distillation. Air-injection rates increase as the soil permeability increases.

The advantages of using soil flushing for site remediation are that removal of contaminants is permanent and no additional treatments are necessary if the process is successful. Also, the technology can be easily applied to permeable areas. However, implementation of the process may be moderately expensive depending upon the flushing solution chosen. Soil flushing also introduces potential toxins (the flushing solution) into the soil system. An effective collection system is required to prevent contaminant migration.

Biological Treatment

In situ biological treatment (enhanced bioremediation) is the provision of nutrients and oxygen to enhance the rate of microbial biodegradation of volatile organic contamination in the soil (see figure 15.8). Biological remediation consists of the use of microbiological organisms to degrade the organic contaminant compounds to carbon dioxide and water. The system supplies nutrients and oxygen to the indigenous bacteria and fungi to stimulate their growth and reproduction, enabling them to use the contaminants as a substrate. As the contaminants are broken down, they are incorporated into the microorganisms' biochemical cycles, and transformed from toxic to nontoxic substances. Care should be taken to investigate the possibility that biological treatment does not produce intermediate products that are more toxic than the original contaminant.

Bioremediation has proved effective in the degradation of certain types of contaminated soil; however, certain conditions must exist for the microorganisms to achieve this effectiveness. These conditions pertain primarily to providing a suitable environment based on temperature, pH, moisture and nutrient contents, oxygen availability, and biodegradability of the contaminated soil constituents. Minimal permeability requirements must be met; sandy materials are far more amenable to in situ treatment than clayey materials. It should be noted that very high concentrations—as well as very low concentrations—of organic contaminants are difficult to treat with the biological process. Treatability studies are required prior to any bioremediation process design, to show that the proposed bioremediation will not accelerate contaminant migration, cause the release of additional contaminants without adequate capture mechanisms in place, or otherwise cause a contaminated site to become a greater hazard.

Bioremediation, if properly implemented, provides permanent removal of the contaminants from the soil media for long-term effectiveness. Therefore, the mobility and volume of contaminants is greatly decreased and this alternative is protective of human health and the environment, after any corrective action has been implemented. The cost of this alternative is in the medium to low range in comparison to all other alternatives except the no-action alternative.

Soil Excavation

Excavation involves the physical removal of the contaminated portion of the soil, using standard excavation practices and technology. Excavation is performed extensively in contaminated-soil site remediation, and satisfies the objective of preventing any future release of contaminants to the ground water. Typical equipment used in this alternative include backhoes, drag lines, and front-end loaders.

Materials handling is a major concern that affects the implementation of excavation; staging areas are necessary. These areas are used to prepare contaminated soil for disposal or treatment; graded to prevent ponding; lined with clay (or other liners) to prevent ground-water contamination; and bermed to prevent runoff. Backfilling, grading, and revegetation after excavation are necessary to prevent large, open areas that could trap precipitation and cause runon. Sampling of remaining soil confirms the removal of all contaminants. The excavated area is backfilled with clean soil obtained from off-site. The subsequent transportation of contaminated soil resulting from excavation must meet federal, state, and local shipping and manifesting regulations.

Excavation and removal of contaminated soil eliminates the environmental and health concerns associated with leachate production and subsequent ground water contamination. However, consideration must be given to the health and safety of remedial workers. Contaminated soil at the site can contain VOCs, and workers could be subject to exposure to these vapors. On-site air monitoring and dust-, vapor-, and odor-control provisions are necessary during all excavation operations. Excavation activities can also result in the release of fugitive dusts and runoff from disturbed areas. Dust controls include water sprays or application of chemical dust suppressants; surface-water controls might also be required. Excavation, in conjunction with other soil treatment technologies (e.g., incineration, soil washing, etc.), is an effective remedial alternative for many sites.

Soil Washing

Soil washing (see figure 15.9) is the extraction of contaminants from excavated soil by mixing them with water, solvents, surfactants, and chelating agents; the contaminated mixture is then treated for removal of contaminants. Heavily contaminated soil is commonly treated several times in a multistage, countercurrent treatment system. Soil washing technology has been developed independently by the U.S. EPA and by ECOVA Corporation. Soil washing is most effective on materials with less than 35 percent fine-grained materials. Soil washing is not particularly effective for remediation of contaminated fill, because of the heterogeneous mixture of materials and particle sizes.

Soil washing has demonstrated its effectiveness in removing various gasoline components, solvents, pesticides, phenols, and trace-metals to regulatory levels. The EPA's demonstration unit has concentrated on remediation of gasoline-contaminated soil; ECOVA's unit has demonstrated successful treatment of pesticides, phenols, and trace metals.

Treatability testing is required in advance to determine the appropriate solvents, surfactants, or acid washes that are required for a prototype soil washing operation. Additionally, permitting and regulatory acceptance is needed prior to implementation of remedial activities. Costs associated with this alternative are very high compared to other in situ alternatives.

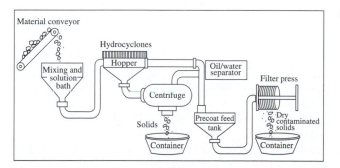

Figure 15.9   Soil washing system

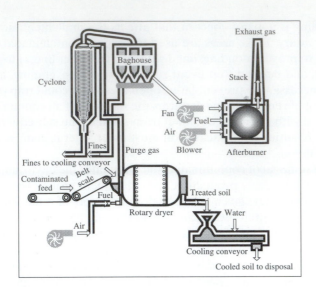

Figure 15.10   Thermal desorption system

On-Site Thermal Treatment

On-site (but not in situ) thermal treatment is carried out in conjunction with excavation, to address soil contamination (see figure 15.10). On-site thermal processes include rotary kiln, fluidized bed, and circulating-bed incineration. All of these processes are potentially acceptable to regulators, and are available from numerous vendors; different vendors provide different treatment processes. A treatability study/test burn is needed in order to select the appropriate treatment process for contaminated soil at a given site.

Incineration involves the thermal destruction of organic compounds, to a nonhazardous product. This process is completed using either an incinerator erected on-site, or a commercial facility located off-site. In general, the major technical considerations for effective incineration are: the ability to handle the physical properties of the contaminated soil involved; adequate size for a reasonable contaminated-soil throughput rate; and an ability to effectively destroy the soil contaminants, based on the chemical properties of the contaminated soil.

Cleanup of off-gases generated is required in order to meet pertinent air-quality standards. Gas cleanup technologies currently in use include: bag houses for particulate removal; various designs of scrubbers for particulate sulfur and chloride removal; and precipitators for particulate removal. Gas treatment, in turn, requires treatment of the resulting aqueous- and solid-waste streams by filtration, physical/chemical treatment, or land disposal.

Construction of a new on-site incineration facility requires compliance with all federal, state and local regulations, including RCRA and the Clean Air Act; RCRA design and permitting standards are listed under Subpart C, 40 CFR Part 264. The preferred alternative is an on-site mobile incinerator, since control of such an incinerator is possible.

New incinerator construction typically meets with strong public opposition, creating time-consuming siting and permitting problems. The entire process of siting, permitting, design, and construction can take several years. The capital cost for such a project is relatively high compared with other alternatives. For these reasons, construction of an on-site incinerator is not the best alternative for many sites.

Incineration technologies are capable of thermally destroying organic hazardous materials. Ash and sludge remaining from the incineration process have to be disposed of using an appropriate secondary technology; disposal of these treatment by-products needs to conform to all requirements of RCRA, most notably the Land Disposal Restrictions.

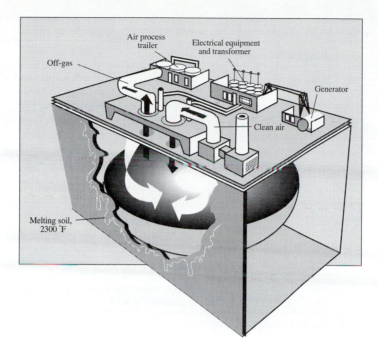

Figure 15.11   In-situ vitrification

Incineration effectively destroys the soil volatile contaminants that are contributing to ground water contamination. Because the contaminated soil is excavated, the risk of exposure of workers to contaminated material is higher than for in situ alternatives during the short term; however, these risks are considered to be minimal and temporary. During the burning of contaminated soil, the public and environment are protected, because equipment is installed for particulates and gas treatment. In the long term, this alternative provides protection to the public and the environment, because the soil is removed and then incinerated. Removing the soil reduces the mobility—and volume—of contaminants into the ground water. Incineration is a proven technology for use in hazardous-waste applications, and is commercially available from various vendors. For on-site treatment, fixed and mobile rotary kiln incineration are available. The capital and operating costs of on-site incineration are much higher than the costs of other alternatives we have considered, so far.

In-situ Vitrification

A much more complicated and expensive system is in-situ vitrification (figure 15.11) or other solidification or stabilization techniques. Generally, in-situ vitrification is used for inorganic wastes and metals, although it has been tested in some organic compounds. The principle is to physically hold the contaminant or solidify the contaminant into a soil matrix. It has little effect on the chemistry of the contaminant when used for inorganic metals, but does increase the solubility and, hence, the toxicity of the contaminant. In-situ vitrification may cause future leaching potential as a result of its use. The cost of this alternative treatment technology is the highest of those considered.

SUMMARY

Table 15.6 is a summary of the types of wastes that can be treated with a given remediation technology; additional details on the applications of each technology can be found in EPA (1991c). Most of the technologies presented primarily treat organic wastes, although some—

TABLE 15.6   Summary of Remediation Technology Applications

Remediation technology	Application
No action	Soils containing residual concentrations of soluble organics, sludges, oily wastes, inorganics and radioactive wastes
Capping	Soils containing soluble organics, inorganics, and radioactive wastes
Venting	Soils containing volatile organic compounds
Air Sparging	Soils containing volatile organic compounds
Flushing	Organic and inorganic compounds
Biological treatment	Soils or soil water containing soluble organics
Excavation	All types of contaminants
Washing	Soils containing organic and/or inorganic compounds
Thermal treatment	Soils containing organics
In-situ vitrification	Soils containing organics, inorganics, and radioactive wastes

Sources: Data from EPA (1991c)

TABLE 15.7   Qualitative Soil Remediation Technology: Costs and Effectiveness

Remediation technology	Qualitative cost	Effectiveness
No action	Low (monitoring)	Some long-term contamination risks
Capping	Low	Some long-term contamination risks
Venting	Medium	Long-term effectiveness
Air sparging	Medium	Long-term effectiveness
Flushing	Medium to high	Long-term effectiveness
Biological treatment	Low to medium	Long-term effectiveness
Excavation	Low to medium	Long-term effectiveness
Washing	Very high	Long-term effectiveness
Thermal treatment	Very high	Long-term effectiveness
In-situ vitrification	Very high	Long-term effectiveness

such as no action, capping, flushing, soil washing, carbon adsorption, and in-situ vitrification—can be used to treat inorganics, including metals.

The costs and effectiveness of alternative soil remediation technologies are highly site-specific. Table 15.7 presents a qualitative summary of the remediation costs for the technologies described above.

ANSWERS TO QUESTIONS

15.1. First find the arguments to be used in the error function (erf) terms. The two arguments are calculated by substituting the values into the variables of equation 15.4. $A = Z/[2(D_t X)^{1/2}] = 16$ ft/$[2(13$ ft $\times 1{,}000$ ft$)^{1/2}] = 0.0702$. $B = Y/[4(D_t X)^{1/2}] = 2{,}400$ ft/$[4(13$ ft $\times 1{,}000$ ft$)^{1/2}] = 5.262$. Evaluating the value of C_o: $C_o = C/[\text{erf}(A) \times \text{erf}(B)] = 0.029$ μg/L/$[\text{erf}(0.0702) \times \text{erf}(5.262)] = 0.367$ μg/L.

15.2. Using equations 15.5 through 15.10 we calculate the following: $PR = p \times A = 11.15$ in/yr $\times$ 28-ac $\times$ 1 ft/12 in $\times$ 43,560 ft^3/ac-ft $\times$ 28.32 L/ft^3 = 32,094,660 L/yr; LGWF $= Z \times v \times Y = 16$ ft $\times$ 60 ft/yr $\times$ 2,400 ft $= 2,304,000$ ft^3/yr $= 65,249,280$ L/yr; Q_t = LGWF + PR = (32,094,660 + 65,249,280) L/yr $= 97,343,940$ L/yr; $X = Q_t \times C_o = 97,343,940$ L/yr $\times$ 0.367 μg/L = 35.7 g/yr; and $C_e = X/PR = (35.7$ g/yr$)/(32,094,660$ L/yr$) = 1.11$ μg/L.

ADDITIONAL QUESTIONS

15.3. Find the dry-weight concentration of the PNA Benzo(a)anthracene in the soil for the site defined in Questions 15.1 and 15.2, using the Freundlich method and assuming that the percentage of organic carbon in the soil is 0.10 percent and $n = 1$.

15.4. Find the soil cleanup criteria for the site described in Questions 15.1 and 15.2 and in question one, above, using the Allowable Concentration Model.

16

Spatial Variability, Scaling, and Fractals

INTRODUCTION

In the previous chapters, traditional methods for investigating the unsaturated zone have been discussed: the importance of physical properties; microscopic parameters such as the double-layer theory; behavior of clays; water flow; gaseous diffusion; contaminant transport, as well as other parameters. The concepts discussed in this chapter deal with treating and analyzing the data gathered from unsaturated-zone studies. This discussion is meant to introduce the reader to the basic concepts involved with some of the tools available to us for treating data, while incorporating the various methods used. Because of the importance of spatial variability and understanding its importance in soils research, the reader is urged to study other texts that discuss such material in detail; the same is true for geostatistics and fractals.

Geostatistics has been used for several decades, and precedes the advent of both scaling and fractals. Indeed, it can be said that scaling begins where geostatistics ends; fractals are somewhat similar to scaling, but are much more complicated. The advantage of using geostatistics is the intuitive nature of the process, coupled with the fact that they provide a better tool for analysis than simple statistics. Scaling and fractals are not as intuitive and therefore, are more difficult to comprehend. However, they are additional tools in our arsenal for problem solving and as such, can be quite useful in various investigations of soils research. The concepts discussed here aim at whetting the reader's appetite to seek further knowledge concerning the use of geostatistics, scaling, and fractals.

16.1 FREQUENCY DISTRIBUTIONS OF SOILS

As soils developed throughout eons of time, they became the product of the very factors that helped to form them. These factors included such parameters as climate; parent material; topography; microbial organisms; and time. Soils are heterogenous rather than homogeneous, simply because of the variability in their formation processes due to freezing, thawing, shrinking, swelling, and so on (see chapter 2). The optimization of environmental resources and resource allocation make it necessary to quantify soil spatial variability, and to determine the scale of its occurrences. The need to generate explanations of observable variability and predictability through modeling efforts is addressed here. There are two categories of variability for most landforms: systematic, and random. The classification of a given medium into one of these categories is determined by the number of observations made. As more observations are made, the existing variability naturally decreases in importance, and confidence limits are smaller. In a strict statistical sense, the more the data distribution differs from a normal distribution, the larger the sample size has to be, for an adequate approximation. However, the cost of collecting large amounts of data is prohibitive and the number of samples needed is greatly influenced by spatial variability. Thus, analytical expressions that require simple solutions with a minimum number of samples are desired.

Because soils are variable, it is important to obtain an estimate of error associated with the specific parameter we wish to measure. This is done with statistics, in which—for simple estimates—the mean (average value of the parameter) and standard deviation (range of the parameter) can be obtained easily. Mathematically, an estimate of the mean is determined by

$$m = \frac{\sum x_i}{n} \tag{16.1}$$

where m is the mean, x_i is measured values for the parameter, and n is the sample number. An estimate for the standard deviation is given by

$$\sigma = \sqrt{\frac{\sum (x_i - m)^2}{(n - 1)}} \tag{16.2}$$

where σ is the standard deviation. For a homogeneous soil, σ values between samples are likely small, indicating that the properties of each of the collected samples are similar. However, if σ values are large, the samples are likely dissimilar. This refers more to variance than the mean; the mean can vary little between media types. An example of variance change is the comparison of bulk-density values for sandy-textured soil versus a clay-textured soil. While σ is likely to be similar within each soil, it is just as likely dissimilar between soil types. It has to be remembered that—statistically—the goal of analysis is an unbiased estimate of a specific parameter.

Parameters routinely measured in soil include: bulk density; particle-size distribution; soil-moisture characteristic curves; water flux; infiltration; water storage; hydraulic conductivity; and soil-water diffusivity. Here, we use bulk density as an example. If bulk-density samples are extracted from various depths within a research area, we might suspect—from a knowledge of statistical analysis and changes in soil texture with depth and space—that there is a distribution curve to describe the frequency of a given density to the number of observations or samples extracted. Figure 16.1 shows the frequency of distribution for bulk-density samples extracted from a clay soil. The normal frequency distribution is mathematically

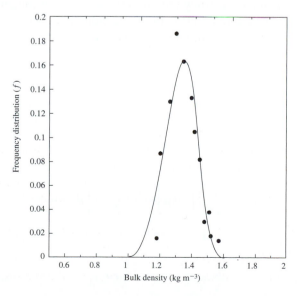

Figure 16.1   Measured (•••) and calculated (——) frequency distributions for soil-bulk density (data from Nielsen et al. 1973)

described by

$$f = \frac{1}{\sigma\sqrt{2\pi}} \exp\left[-\frac{(x_i - m)^2}{2\sigma^2}\right] \tag{16.3}$$

where f is the frequency. For a normal curve, the center position of the curve is determined by the mean of all samples extracted. Also, the less σ varies, the sharper and narrower the curve will be. Assuming a normal distribution, if a sample X were extracted at random from the population, the chance (probability) that it would fall between two other points x_1 and x_2 is expressed by

$$P\{x_1 < X \le x_2\} = \int_{x_1}^{x_2} f\, dx \tag{16.4}$$

Typically, some soil properties are also log-normally distributed. The frequency function of a log-normal distribution is

$$f = \frac{1}{\sigma(x_i - \beta)\sqrt{2\pi}} \exp\left\{-\frac{[\ln (x_i - \beta) - m]^2}{2\sigma^2}\right\} \tag{16.5}$$

for $x_i > \beta$ and $f = 0$ for $x_i < \beta$, where β is any constant to allow $\ln (x_i - \beta)$ to be normally distributed.

Several statistical methods are used to represent the variability of a given parameter on a relative basis such as the coefficient of variation $\{cv = (\sigma/m) * 100\}$. For example, both saturated-water content (θ_s) and bulk density have a low coefficient of variation when compared to textural analysis, and θ at varied soil pressures that have a medium coefficient of variation. Parameters such as saturated- and unsaturated hydraulic conductivity; pore-water velocity; apparent diffusion; and electrical conductivity all have a high coefficient of variation. As variability increases, more samples are needed for adequate analysis. Also, those parameters with high variability are likely log-normally distributed (owing to soil heterogeneity) while those of low variability have a normal distribution. As a result, most of the spatial variability within soil is described statistically. To prove this, we use hydraulic conductivity as an example, and because K is generally more variable than ρ_b.

Hydraulic conductivity is represented as a function of percentage of saturation, where the coefficient of variation generally increases with decreasing percentage saturation. Hydraulic conductivity is treated by regarding the sampling area as a homogeneous soil mass—in which case, both soil-water content and hydraulic conductivity are treated as two separate experimental variables, measured at given intervals. In the latter case, we suppose that 30 samples are taken from six separate depths, giving a total of 180 samples from the area. Using this approach, we consider the spatial variability of the sample area and wish to know if—for an average value of θ and its standard deviation—what is the corresponding hydraulic conductivity (mean value) and its standard deviation? Remember that θ most likely has a normal frequency distribution, and K has a log-normal frequency distribution. We can determine for any particular θ, the corresponding value of K (both at steady state) as well as its variability, by considering two independent variables, x_i and y_i; each is normal and their joint distribution function is expressed as

$$F(x_i, y_i) = \frac{1}{2\pi} \iint \exp\left[-\frac{(x_i + y_i)}{2}\right] dx_1\, dy_1 \tag{16.6}$$

By introducing two new variables x and y—both normally distributed and correlated—x and y can be related to x_1 and y_1 by

$$x = m_1 + \sigma_1 x_i \qquad \text{also} \qquad y = m_2 + r\sigma_2 x_1 + \sqrt{1 - r^2}\, \sigma_2 y_i \tag{16.7}$$

These two equations are arranged such that the new variables x and y have means m_1 and m_2, standard deviations σ_1 and σ_2, and the correlation coefficient r. Rearranging, we have

$$x_i = \frac{(x - m_1)}{\sigma_1} \tag{16.8}$$

and

$$y_i = \frac{1}{\sqrt{1 - r^2}} \left[-r\left(\frac{x - m_1}{\sigma_1}\right) + \frac{y - m_2}{\sigma_2} \right] \tag{16.9}$$

Thus, the joint distribution function, as a function of the new variables, becomes

$$f(x, y) = \frac{1}{2\pi} \iint D \exp\left[-\frac{Q(x, y)}{2} \right] dx\, dy \tag{16.10}$$

where

$$Q(x, y) = \frac{1}{1 - r^2} \left[\frac{(x - m_1)^2}{\sigma_1^2} - \frac{2r(x - m_1)(y - m_2)}{\sigma_1 \sigma_2} + \frac{(y - m_2)^2}{\sigma_2^2} \right] \tag{16.11}$$

and D is the Jacobian

$$D = \begin{vmatrix} \dfrac{\partial x_1}{\partial x} & \dfrac{\partial x_1}{\partial y} \\[2mm] \dfrac{\partial y_1}{\partial x} & \dfrac{\partial y_1}{\partial y} \end{vmatrix} \tag{16.12}$$

The joint frequency function of x and y (Cramer 1955) is

$$f(x, y) = \frac{1}{2\pi\sigma_1\sigma_2\sqrt{1 - r^2}} \exp\left[-\frac{Q(x, y)}{2} \right] \tag{16.13}$$

This is the general form of the two-dimensional, normal frequency distribution function. By letting $x = \ln z$, we can rewrite the equation as

$$f(z, y) = \frac{1}{2\pi\sigma_1\sigma_2 z\sqrt{1 - r^2}} \exp\left[-\frac{Q(x, y)}{2} \right] \tag{16.14}$$

where z represents depth of sample and

$$Q(z, y) = \frac{1}{1 - r^2} \left[\frac{(\ln z - m_1)^2}{\sigma_1^2} - \frac{2r(\ln z - m_1)(y - m_2)}{\sigma_1 \sigma_2} + \frac{(y - m_2)^2}{\sigma_2^2} \right] \tag{16.15}$$

Earlier we mentioned that K is log-normally distributed; by rewriting the log-normal distribution equation in terms of K at steady state (K_{ss}) and setting $\beta = 0$ in equation 16.5, we have

$$f_1(K_{ss}) = \frac{1}{\sigma_1 K_{ss}\sqrt{2\pi}} \exp\left[-\frac{(\ln K_{ss} - m_1)^2}{2\sigma_1^2} \right] \tag{16.16}$$

Also, because θ at steady state (θ_{ss}) is normally distributed, we can rewrite the normal frequency distribution (equation 16.3) as

$$f_2(\theta_{ss}) = \frac{1}{\sigma_2\sqrt{2\pi}} \exp\left[-\frac{(\theta_{ss} - m_2)^2}{2\sigma_2^2} \right] \tag{16.17}$$

The joint distribution function after equation 16.13 is

$$f(K_{ss}, \theta_{ss}) = \frac{1}{2\pi\sigma_1\sigma_2 K_{ss}\sqrt{1 - r^2}} \exp\left[-\frac{Q(K_{ss}, \theta_{ss})}{2} \right] \tag{16.18}$$

where

$$Q(K_{ss}, \theta_{ss}) = \frac{1}{1-r^2}\left[\frac{(\ln K_{ss} - m_1)^2}{\sigma_1^2} - \frac{2r(\ln K_{ss} - m_1)(\theta_{ss} - m_2)}{\sigma_1\sigma_2} - \frac{(\theta_{ss} - m_2)^2}{\sigma_2^2}\right] \quad (16.19)$$

Combining equations 16.17–16.19, the conditional frequency function of K_{ss} relative to another steady-state soil-water content (θ_{ss1}) is described statistically by

$$\frac{(K_{ss}, \theta_{ss1})}{f_2(\theta_{ss1})} = \frac{1}{\sigma_1 K_{ss}\sqrt{2\pi}\sqrt{1-r^2}} \exp\left[-\frac{\left(\ln K_{ss} - m_1 - \dfrac{r\sigma_1}{\sigma_2}(\theta_{ss1} - m_2)^2\right)}{2\sigma_{1ss1}(1 - r_{ss1})}\right] \quad (16.20)$$

Equation 16.20 is a normal frequency distribution function for K_{ss} with the mean value $m_3 = m_1 + r\sigma_1(\theta_{ss1} - m_2)\sigma_2^{-1}$ and the standard deviation $\sigma_3 = \sigma_1(1 - r^2)^{1/2}$. By letting $\varepsilon =$ arithmetic mean of K_{ss} and $\tau =$ corresponding standard deviation, the following relations are valid such that $\varepsilon = \exp[m_3 + \sigma_3^2/2]$ and $\tau = \exp[m_3 + \sigma_3^2/2][\exp(\sigma_3^2 - 1)]^{1/2}$. As a result, we have an estimation of the mean and standard deviation of K_{ss} relative to any given soil-water content, based on frequency distributions. This statistical approach is an appropriate method for a field study. Consequently, we see that the spatial variability of soils and the samples extracted from them can be statistically treated, so long as an adequate number of samples is taken. The number of samples are determined, assuming a normal distribution, from the following equation

$$n = \frac{Z_\alpha^2 v_s}{d^2 m^2} \quad (16.21)$$

where n is the required sample size, Z_α is the value of the standardized normal variate corresponding to the level of significance α (the value of Z_α can be obtained from a cumulative normal frequency distribution table), v_s is the sampling variance, and d is the margin of error expressed as a fraction of the plot mean. The information of primary interest to the scientist is usually the treatment mean (the average over all plots receiving the same treatment), rather than an average of a single plot or treatment area. Hence, the desired degree of precision is usually specified in terms of the margin of error of the treatment mean, rather than of the plot mean. In this particular case, sample size is computed by

$$n = \frac{Z_\alpha^2 v_s}{r(D^2 m^2) - (Z_\alpha^2 v_p)} \quad (16.22)$$

where n is the required sample size, Z_α and v_s are as defined previously, v_p is the variance between plots of the same treatment (i.e., experimental error), and D is the prescribed margin of error expressed as a fraction of the treatment mean. As an example, suppose a researcher wishes to measure the hydraulic conductivity within a research plot using extracted, intact soil cores with a larger coring sampler. The researcher wishes to determine the number of samples necessary to achieve an estimate of the treatment mean within 5% of the true value. This researcher knows from previous studies that the following values can be used: $Z_\alpha = 1.95$, $v_s = 5.043$ (i.e., a cv of 28.4 percent), $D = 0.05$, $m = 25$ (the average number of samples the researcher has taken in the past with a smaller sampler), and $v_p = 0.1832$—estimated from previous research. A sample size—number of cores required per plot—that can satisfy the researchers requirement at the 5% level of significance is computed as:

$$n = \frac{(1.95)^2(5.043)}{4(0.05)^2(25)^2 - (1.95)^2(0.1832)} = 3.45$$

Or, about four cores per research plot. Thus, by extracting an adequate number of samples, a great deal of spatial variability is nullified, and while calculated frequency distributions do not exactly fit measured values, they are a good approximation to work with when dealing with the effects of spatial variability on collected data. It is beyond the scope of this text to give a detailed treatise on statistics; the reader is therefore urged to consult a standard college text on statistics for a more in-depth discussion of the subject.

QUESTION 16.1

You have been given the following ten properties for an in-situ media: bulk density; water content at saturation; saturated and unsaturated conductivity; pore-water velocity; diffusion coefficient; particle-size analysis; water content at -100 kPa; and the scaling coefficient. For which properties would you expect a normal distribution?

Geostatistics

The frequency distribution just discussed typically deals with statistics as associated with univariate data and hypothesis testing. In some cases, it can extend to a correlation between two variables and linear regression. Generally, in the discussion of correlation between variables it is usually two different attributes that are considered. However, in the earth sciences we often wish to make a treatment of the correlation between values of a single variable, measured at different points in space. The ability to analyze many such measurements against a spatial framework is a necessity, and is where standard statistics are inefficient for obtaining solutions to more complex problems.

For example, consider typical histograms of two data sets. Visually, we see little difference between the two, and a Kolmogrov–Smirnoff test for comparisons—from which the sample data comes—would not reject (with a 5 percent significance level) a null hypothesis of no difference between the two populations. However, suppose we have contour maps of the same data. The first set of data shows a normal contour map, while the second set of data produces a contour map that is "busier"—so how is it possible that the histograms look the same? They look the same because the data were collected at the nodes of a regular two-dimensional grid, but it is within the grids where the contour map shows one data set busier than the other. Thus, the difference between the two contour maps is a reflection of the more random spatial arrangement of data in the busier map; in other words, there is less correlation between adjoining data pairs in the busier contour map. This is where geostatistics shows a difference and hence, is more valuable as an analysis tool than statistics alone.

One of the first scientists to recognize the necessity of accounting for spatial correlation between data was D. G. Krige (Journel and Huijbregts 1978). Krige derived empirically based "regression weights" that could be applied to the grades of channel-ore samples used in the estimation of slopes. Based on this work, others undertook a formal development of his theories, and the field of geostatistics was born. As a simple definition, geostatistics is the statistics of spatial- (or temporally) correlated data. It enables the scientist to measure spatial autocorrelation and evaluate the nature and quality of raw data. Included in this discussion is the term spatial variability; this is a common term used to indicate that geologic media and soils—most of which are heterogeneous—change with space and time. Such properties include bulk density, porosity, soil texture, water content, pH, and hydraulic conductivity. Because soils are spatially variable, it is important to have tools with which to measure that variability. We have already discussed some of these tools, but there are others that we now describe in our continuing discussion of geostatistics. (For an excellent treatise on spatial variability, the reader is referred to chapter 13 of Hillel 1980.)

Semivariogram

Geostatistics incorporates data taken from a grid (usually square), with a unit spacing. Upon analyzing the data, a histogram and contour map are usually generated. To describe the spatial correlation between samples in near proximity, a semivariogram is used. This is a basic geostatistical tool that allows us to visualize, model, and exploit the spatial autocorrelation of a regionalized variable. The function of the semivariogram is half the average squared difference between data pairs of points that are separated by displacement, $\vec{h}$. This can be calculated by

$$\gamma^*(\vec{h}) = \frac{1}{2N(\vec{h})} \sum_{i=1}^{N(\vec{h})} [Y(\vec{x}_i + \vec{h}) - Y(\vec{x}_i)]^2 \qquad (16.23)$$

where N is the number of data pairs in the region, and Y represents the value of the data at location $\vec{x}_i$.

A semivariogram needs to be calculated for a variety of directions, to allow recognition of anisotropic variability; for example, east to west and north to south. Commonly, neighboring sample pairs are closer in value than more separated pairs. Once an experimental semivariogram is obtained, it is modeled to obtain block estimates; usually, a spherical model is used. However, there is no one "correct" model for a particular situation, so the semivariogram of choice determines the amount of smoothing necessary in later steps. The accuracy of the modeling process depends on the number of data pairs used in the calculations; on the experimental semivariogram; and on the lag distance at which it is evaluated. The modeling process is complicated and can involve: polygonal estimation; inverse-distance weighting; inverse-distance-squared weighting; estimation variance; confidence limits; as well as other factors. Since this chapter is an introduction to various tools used in the earth sciences, discussing these parameters is beyond its scope; the reader is referred to Hohn (1988), Wackernagel (1995), or other texts on geostatistics.

Kriging

We now understand that it is possible to rank any estimation regime to its efficiency by calculating the estimation of variance. However, for this we need to determine the best set of weights for a particular block-sample configuration—that is, the set of weighting coefficients that minimize the estimation of variance. The process of calculating this optimal estimation of variance is called "kriging," named after D. G. Krige, who pioneered the work of geostatistics. Kriged estimates are used for drawing a contour map, such as for water-table depth. Since geostatistics are generally done on computers, the basics steps to computer contour mapping using the kriging technique are: (1) collect data to estimate the variance; (2) superimpose data on a regular grid; (3) interpolate values at each node on the grid; (4) construct contours; (5) smooth the contour lines by splining or other techniques, if necessary; and (6) draw the contour map.

Beginning, we assume the regionalized variable under study has the value $z_i = z(x_i)$, each representing the value at the point x_i. We also assume that this regionalized variable is a second-order stationary, with expectation E

$$E\{Z(x)\} = m \qquad (16.24)$$

where $Z(x)$ is the random variable observed at point x and m is the first-order moment. Three second-order moments are also useful in geostatistics; these include the variance of the random variable, the covariance, and the semivariogram function (discussed earlier). This is generally estimated by a centered covariance C

$$E\{Z(x + h)Z(x)\} - m^2 = C(h) \qquad (16.25)$$

and a variogram

$$E\{[Z(x + h) - Z(x)]^2\} = 2\gamma(h) \tag{16.26}$$

where h is the distance from point x. The Kriged estimator—that is, a linear combination of n values of the regionalized variable—is

$$z_{k^*} = \sum_{i=1}^{n} \lambda_i Z_i \tag{16.27}$$

where λ is the calculated weight. This ensures that the estimate is unbiased, and the estimation variance minimized. When the kriging theory is met, we quickly interpolate values of the sampling variable between measured points; then contour lines are drawn. Kriging allows a determination of the estimation of variance, useful for determining uncertain values of the function, as well as in subsequent sampling of the same area within the field. However, the calculation of λ and other parameters are beyond the scope of this text, so the reader is once again referred to Hohn (1988) and Wackernagel (1995) for an in-depth treatise on Kriging and geostatistics.

Because even the simplest calculations in geostatistics are performed on the computer (they are tedious and unwieldy for large data sets otherwise), the practitioner needs to be well-versed in some computer programs, as well as a variety of techniques in problem solving; these include: univariate statistics; multivariate statistics; means; histograms; scattergrams; semivariograms; variograms; interactive curve fitting; plotting; grid searching; equation solving; contouring; and map drawing. We have mentioned only the main parameters for a basic understanding of geostatistics here; others might include variogram cloud; variogram and covariance function; extension and dispersion variance; measures and plots of dispersion; linear model of regionalization; Kriging spatial components; and many more.

A concluding note on geostatistics: while it is widely used in the earth sciences, it is not the only tool for solving complex problems with large data sets. However, it is easier to understand intuitively than other tools that we now discuss: scaling and fractals.

Power-Law Distributions

Random variables   The concept of random variables is basic to modern statistics. A function of a variable is a rule (mathematically speaking) whereby one or more numerical values are associated with different values of a variable. For example, consider the function $f(x) = 2x + 3$. If $x = 0$, then $f(x) = 3$, if $x = 2$ then $f(x) = 7$, and so forth. Assuming x is a random variate, the probability density function of x, $f(x)$, gives the probability that the variate assumes the value x, that is, $f(x) = Pr(x)$. If we distinguish between the name of the variate (x) and the values the variate assumes, then $f(x = x') = Pr(x = x')$. Usually, $f(x)$ represents a model for the relative frequency (in the series of experiments), with which the variate x assumes specified values. These values normally vary both spatially and temporally, and each quantity measured is termed a random variable. Since these values fluctuate, we need to associate the random variable with a probability distribution, $F(x)$. The probability that the random variable x assumes a specific value x_j is given by $Pr(x = x_j) = f(x_j)$. The distribution function, as a probability law, is given as

$$Pr(x \leq x_j) = F(x_j) = \int_{-\infty}^{x_j} f(x)\, dx \tag{16.28}$$

Thus, the distribution function of a random variable x represents a cumulative probability. Any nonnegative function, whose integral over the entire range of the variate in the function is unity (1), defines a probability density. Consequently, a random variable x is said to have a density function $f(x)$ if

$$\int_{-\infty}^{x_j} f(x)\, dx = Pr(x \le x_j) \tag{16.29}$$

In science, the variation we measure is often defined as "measurement error." It then becomes important to overcome variability due to sampling and environmental problems. As a result, we need to use distribution functions to help interpret the data gathered. There are a number of distribution functions for random variables that we discuss in reference to power-law distribution, but for exact, technical detail on the evaluation of data fit to a particular distribution function, we refer the student to a standard statistics text. An example of a cumulative probability distribution is shown in figure 16.2.

For any given random variable, there is a probability distribution associated with it. If we let the symbol x denote such a random variable, the symbol $f(x)$ denotes the probability density for x. Suppose the probability of the random variable x assumes the specific x_i, this can be expressed as

$$Pr(x = x_i) = f(x_i) \tag{16.30}$$

It is important to remember that a probability measure on the random variable x is defined by a function that has the following properties:

$$0 \le Pr(x \le x_i) \le 1$$

$$Pr(-\infty \le x \le \infty) = 1 \tag{16.31}$$

$$\text{for } x_j > x_i \quad Pr(x_i \le x \le x_j) = Pr(x \le x_j) - Pr(x \le x_i)$$

The distribution for the random variable x is denoted by $F(x)$. As a probability law, the distribution is interpreted as that given in equation 16.28. In this instance, $f(x)\, dx$ is the product and defines the area of a rectangle with height $f(x)$ and width dx, and is called the probability element. It is normally convenient to assign the range of a random variable as $\pm\infty$. Thus, the density function is zero for all values of $x \le 0$ and the probability density is defined

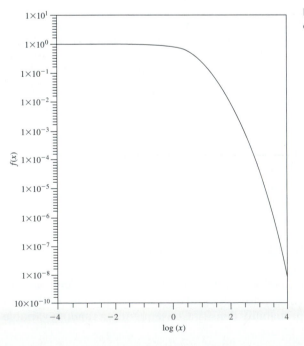

Figure 16.2   A cumulative probability distribution

as $Pr(x) = f(x)$, hence $Pr(x) = 0$. The random variable is either continuous or discrete. The distribution of a random variable is generally associated with parameters that are constants, determining certain characteristics of the distribution. An example of the density function and its relation to the distribution function is given in figure 16.3a. In figure 16.3b, the density at the point $x = x_j$ is given by the height $f(x_j)$.

Log-normal distribution   For log-normal distributions, the variable is a result of multiplication rather than a sum, as is the case for normal distributions (e.g., bulk density). Many natural phenomena tend to follow log-normal distributions since the log-normal distribution is not symmetric, and is defined only for positive x. For example (using hydraulic conductivity), suppose a researcher is extracting soil cores to obtain an "average" value of K over a specific area. Variability in each sample can include soil type, bulk density, occluded pore space, presence of cracks, or particle size. To obtain an average hydraulic conductivity, the value of K for each core is summed for the whole. However, each of the factors of variability influences the individual K value. Each value is therefore subjected to variability based on the physical factors listed as well as the K value obtained for each core, and for all cores is proportional to the product of all factors involved. As a result, the process is multiplicative and the value obtained for K for each core is likely distributed log-normally.

For a probability density function (pdf), the log-normal pdf can be expressed as

$$P(x) = \frac{\exp\left[-\frac{(\ln x - \alpha)^2}{2\beta^2}\right]}{x\beta\sqrt{2\pi}} \tag{16.32}$$

Because x can be expressed only positively, $P(x)$ approaches zero as x approaches zero. Since the expected value of x, $E[x] = \exp(\mu + \sigma^2/2)$, and the sampling variance, $Var[x] = \exp(2\mu + \sigma^2)[\exp(\mu^2) - 1]$ then, the coefficient of variation, CV is given by the expression

$$CV = [\exp(\sigma^2) - 1]^{0.5} \tag{16.33}$$

We write the cumulative log-normal distribution as

$$P[x] = \int_0^x e - [\ln(x) - \mu]^2/2\beta^2 \frac{dx}{\sqrt{2\pi}\beta x} \tag{16.34}$$

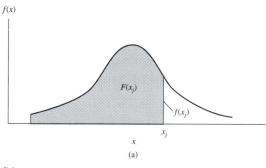

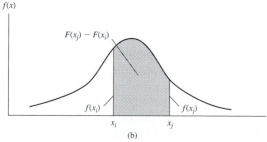

Figure 16.3   (a) An example of the density function, $f(x)$, and its relation to the distribution function, $F(x_j)$. (b) The area under the curve corresponds to the probability, $F(x_j) - F(x_i)$.

For this case, the normal curve of error is mathematically expressed by

$$P[x] = \frac{1}{2} + N\left(\frac{\ln(x) - \mu}{\beta}\right) \tag{16.35}$$

QUESTION 16.2

For the properties given in question 16.1, how would you represent the remaining soil properties? What makes a log-normal distribution more appropriate than a normal distribution for certain physical properties?

Exponential distribution If $f(x) = \gamma e^{-\gamma x} = 0$ (where both x and γ are greater than zero), the random variable is x, and the parameter is γ. We make the explicit distinction between the variable and the parameter by writing $f(x|\gamma) = \gamma e^{-\gamma x}$. For $\gamma = 1$, the probability density is written as $f(x|\gamma = 1) = e^{-x}$ (where $x > 0$). Consequently, the random variable defined by this density is called an "exponential variable." Also, the exponential distribution is really a *family* of distributions; an individual member can be specified by assigning a numerical value to the parameter γ. The distribution function for an exponential variable is mathematically expressed as

$$Pr(x \le x_i) = \int_0^{x_i} f(x)\,dx = \int_0^{x_i} \gamma e^{-\gamma x}\,dx \tag{16.36}$$

An example of an exponential distribution is the Poisson distribution. Suppose we want to study the emission of fast-moving neutrons from a neutron probe, for measuring soil volumetric-water content. We can investigate either a single event or no event, with related probabilities of $\lambda\Delta x \ll 1$ and $1 - \lambda\Delta x$. In this study, Δx is so short that it does not contain more than one event. Thus, if we consider a set interval Δx, the number of events within this interval $N(\Delta x)$ will follow the Poisson distribution, which can be expressed mathematically as

$$Pr[N(\Delta x) = k] = \frac{(\lambda\Delta x)^k e^{-\lambda\delta x}}{k!} \tag{16.37}$$

where $k = 1, 2, 3, \ldots$, $Pr[N(\Delta x) = k]$ is the probability of finding k events within Δx (also written as δx), and λ is the number of events per unit time, such as the neutrons being emitted from a probe (assumed constant).

QUESTION 16.3

(a) Explain the difference between sampling error and measurement error. **(b)** What is the difference between statistical true value and scientific true value? **(c)** between scientific bias, measurement bias, and sampling bias?

QUESTION 16.4

Why would someone purposely choose a sampling method that has a lower precision over a method having a higher precision?

16.2 SCALING AS A TOOL FOR DATA ANALYSIS OF PHYSICAL PROPERTIES

Scaling is a physical or geometrical difference between soil types or parameters. It is also defined as a statistical difference between soils and related parameters; a mathematical definition can also be applied. Miller and Miller (1956) introduced the "similar media" concept. In

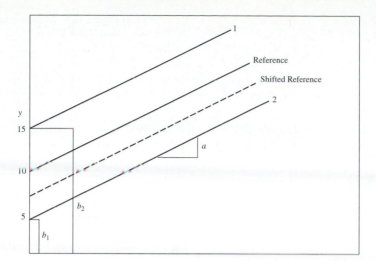

Figure 16.4   Example of shifted reference curve

principle, this concept allows description of soil-water behavior in one soil from either experimental or computed data; in another by employing reduced variables defined in terms of appropriate microscopic characteristic lengths. Basically, similar media differ only in the scale of their internal microscopic geometries, and thus have equal porosities—that is, scaling.

Scaling is used to simplify the description of the statistical variation of soil properties. By this simplification, the distribution of spatial variation is described by a set of scale factors α_r, relating the soil hydraulic properties at each location r to a representative mean.

Tillotson and Nielsen (1984) refer to functional normalization as an empirical method to determine scale factors. Its objective is to coalesce all relations in the set into a single reference curve that describes the set as a whole; the resulting scale factors have no physical significance. For example, consider two curves, 1 and 2 (see figure 16.4):

$$y_1 = ax + b_1$$
$$y_2 = ax + b_2 \tag{16.38}$$

where x and y are variables and a and b are constants. The reference curve is defined such that

$$y_{\text{ref}} = ax + b_{\text{ref}} : b_{\text{ref}} = y_{\text{ref}} - ax \tag{16.39}$$

and

$$\alpha_1 = \frac{b_{\text{ref}}}{b_1}; \quad \alpha_2 = \frac{b_{\text{ref}}}{b_2} \tag{16.40}$$

then

$$y_1 = ax + \frac{b_{\text{ref}}}{\alpha_1} \quad \therefore \quad y_1 = ax + y_{\text{ref}} - ax$$

$$y_2 = ax + \frac{b_{\text{ref}}}{\alpha_2} \quad \therefore \quad y_1 = y_{\text{ref}} \tag{16.41}$$

By performing this operation, both curves are coalesced to the reference curve (see figure 16.4), or a distribution set μ_α. For example:

$$\left.\begin{array}{l} b_1 = 15 \\ b_2 = 5 \\ b_{\text{ref}} = 10 \end{array}\right\} \quad \left.\begin{array}{l} \alpha_1 = 10/15 = 2/3 \\ \alpha_2 = 10/5 = 2 \end{array}\right\} \mu_\alpha = 4/3 \tag{16.42}$$

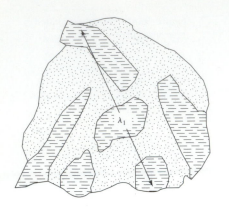

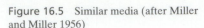

Figure 16.5 Similar media (after Miller and Miller 1956)

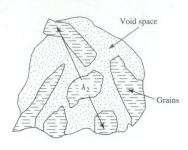

Now, by normalizing the α-distribution set $\mu_\alpha = 1.0$ by

$$\alpha_i = \frac{N\alpha_i}{\sum\limits_{i=1}^{N} \alpha_i} = 8/3 \tag{16.43}$$

thus, evaluating equation 16.27 at a distribution set, $\mu_\alpha = 1.0$ and b_{ref},

$$\left.\begin{array}{l} \alpha_1 = \dfrac{2*2/3}{8/3} = \dfrac{1}{2} \\[2ex] \alpha_2 = \dfrac{2*2}{8/3} = \dfrac{3}{2} \end{array}\right\} \begin{array}{l} \mu_\alpha = 1.0 \\[1ex] b_{\text{ref}} = \alpha_1 b_1 = 7.5 \end{array} \tag{16.44}$$

In doing so, we have shifted the reference curve (see figure 16.4)!

Similitude analysis (similar media) gives physical significance to the scale factors. Miller and Miller (1956) introduced microscopic length, and according to their concept, soils are similar if they are geometric scales of each other—that is, all the microscopic geometric details of the one medium could be multiplied by a constant to obtain the microscopic details of the other medium. The microscopic length characteristic is denoted by λ and could be described as: **(1)** average grain size; **(2)** average pore diameter; **(3)** maximum grain size; **(4)** maximum pore size; and **(5)** also, by combinations of 1 + 2 and 3 + 4. An example of similar media is depicted in figure 16.5. For further examples, we look at both capillary rise (described by the Laplace equation) and Poiseuille's law.

From the Laplace equation

$$\Delta P = \rho g h = \frac{2\sigma}{r} \tag{16.45}$$

where P is pressure, ρ is fluid density (m^3 kg^{-1}), g is gravitational force constant, h is height of rise (L), σ is surface tension (mN m^{-1}), and r is the maximum radius (L) of the water-filled pore. Equating r with λ, the scale factor between the two soils is given by $\alpha = \lambda_2/\lambda_1$. If $\alpha = 2$—the maximum water-filled pore radius of medium 2 is $2x$ as large as that of medium 1—then ΔP is $2x$ as small, or $|h|$ is $2x$ as small. Corresponding retention curves for similar media 1 and 2 using this concept gives equation 16.46, with a graphical representation shown in figure 16.6.

$$h_1 \lambda_1 = h_2 \lambda_2; \quad h_r = \frac{h_m}{\alpha_r} \tag{16.46}$$

where h_r is reference pressure head, h_m is the average pressure head, and α_r is the scaling coefficient. From Poiseuille's law we have:

$$K_r(\theta) = r^2 * f(n, l, \Delta P, g) \tag{16.47}$$

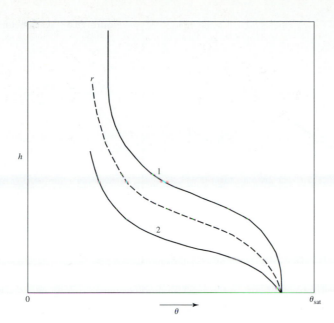

where K_r is the hydraulic conductivity at point r (i.e., the reference), n is the number of pore classes through which flow takes place, and l is the length of the material through which the fluid flows. In this case, if $\alpha = 2$, then K of medium 2 will be $4x$ as large as that of medium 1, and

$$\frac{K_1}{\lambda_1^2} = \frac{K_2}{\lambda_2^2}; \quad K_2 = \left(\frac{\lambda_2}{\lambda_1}\right)^2 * K_1 \tag{16.48}$$

or

$$K_i = \alpha_i^2 K_r \tag{16.49}$$

where the scale factor α_i is the characteristic length of medium i and (see figure 16.7).

From equation 16.46 and 16.49, we derive for sorptivity (S) such that

$$S_i = \sqrt{K_i h_i(\Delta\theta)}$$

$$S_r = \sqrt{K_r h_r(\Delta\theta)} \tag{16.50}$$

$$S_i = \sqrt{\alpha_i^2 K_r \left(\frac{1}{\alpha}\right) h_r(\Delta\theta)}; \quad S_i = S_r \alpha_i^{1/2}$$

In other words, soil-water transport characteristics of a set of soils are connected through scale factors. Thus, by collecting retention data of a soil set, we automatically infer conductivity, diffusivity, and sorptivity relations. An example of scaling of $\theta(h)$ and $K(\theta)$ is shown in figure 16.8 (Warrick, Mullen, and Nielsen 1977). This is physically described by setting $r = 1, \ldots, R$ locations and $i = 1, \ldots, I$ pressure increments for which $\theta(h)$ is determined, such that $s = \theta/\theta_{sat}$ where $0 \leq s \leq 1$ (similar media:equal porosity) and $h_m = \alpha_r h_r$; thus

$$\frac{\alpha_1 + \alpha_2 + \cdots + \alpha_R}{R} = 1.0 \tag{16.51a}$$

$$h_r = \log h_m = a_0 + a_1 S + a_2 S^2 + a_3 S^3 \tag{16.51b}$$

$$SS = \sum_{r,i} [h_m(S_i) - \alpha_r h_r(S_i)]^2 \tag{16.51c}$$

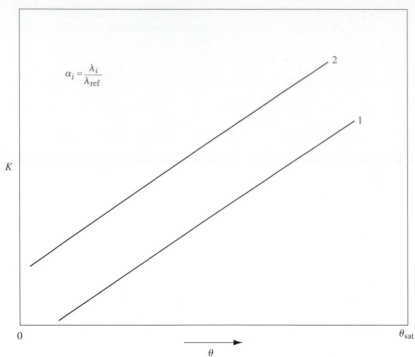

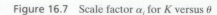

Figure 16.7   Scale factor α_i for K versus θ

$$\alpha_i = \frac{\lambda_i}{\lambda_{\text{ref}}}$$

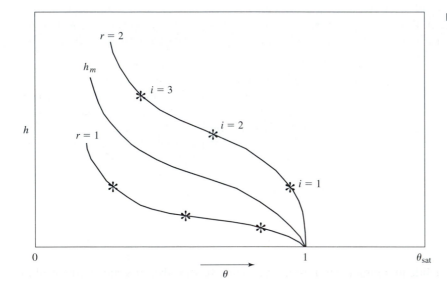

Figure 16.8   Example of scaling $\theta(h)$

where equation 16.51b is the reference curve. Further, we want to minimize *SS* as close to zero as possible; if it's zero, then *SS* is exact so scaling is exact. The basic steps are: **(1)** estimate $h_m(s)$ by multiple linear regression technique, that is, SAS, LMSL, and so on; **(2)** find α_r by minimizing equation 16.51c subject to 16.51a; **(3)** find new $h_m(s)$ such that

$$\log [h_m(S_i) = \log [\alpha_r h_r(S_i)] = a_0 + a_1 S(i) + a_2 S(i)^2 + a_3 S(i)^3 \tag{16.52}$$

and **(4)** repeat steps 2 and 3 until *SS* no longer changes.

Following this brief introduction to scaling, a generalized definition is helpful in understanding the concept. To that end, we use that of Shouse et al. (1989): *"Scaling is a systematic method for specifying a change of variable that transforms one system to another one with more desirable traits."* Scaling works very well for many problems where the soil is homogenous on a large scale. Additionally, scaling can be applied to systems where there is a separation of scales. For example, a soil can be heterogeneous at a small scale, but if viewed at a large scale, it appears homogeneous. The reason for this is that the averaging is large compared to the scale of heterogeneity.

In addition to using scaling as a tool for transforming systems, we now develop new conceptual models by using the concepts provided by fractal mathematics, especially in the way we investigate heterogeneity. For example, Feder (1989) uses the coastline of Norway to illustrate length in the fractal dimension. Every time the resolution is increased on the coastline, there is a subsequent increase in its measured length—that is, the length continues to increase on at least several smaller "recursive" levels. Because the outline of the coast is heterogeneous, this model is referred to as a "self-similar" process. A primary characteristic of the coastline of Norway (or of fractals) is that the image we observe generally looks the same at almost any scale in the real world. This property is referred to as "fractal scaling." Thus, scaling and fractals are closely related. The main advantage of fractal scaling is that traditional scaling techniques cannot be applied to materials that exhibit fractal scaling—such as physical description of topographic relief—to the development of consistent theories on both the formation and nature of fluid turbulence. Since a great number of shapes and processes directly affect soil and geologic formation, it is logical to apply the tools of fractal mathematics to the variability of hydraulic properties of soils and aquifers.

16.3   THE FRACTAL DIMENSION

This section is written to provide a brief introduction and overview of fractals, and some insight on how fractals facilitate the description of fluid flow and contaminant transport through soil, as well as other physical processes in soil. As a tool, it is much more suitable for these parameters than geostatistics. A complete description of fractals is not the objective of this text, so the reader is referred to the References section (at the back of the book) to gain more knowledge on the subject. Benoit Mandelbrot proposed the concept of fractal geometry in 1975 and since that time, fractal geometry has been used to render drawings of the surface of the planet Mars, trees, fractal shapes, paintings, movie scenes using computer graphics, and fluid flow and displacement. Indeed, computer programs using fractal generation have even been written to portray entire cities accurately. By running these programs, an individual can visit a given city via computer, and when he or she travels to that city in person, is already familiar with major streets, parks, restaurant locations, and other areas. The accuracy is phenomenal.

Prior to the development of fractal geometry, traditional Euclidean geometry was used to describe objects. What is the difference between the two forms of geometry? A comparison reveals that Euclidean geometry is: based on a characteristic size or scale; works well in describing man-made objects; and is described by formulas—while fractal geometry is independent of scale; appropriately describes natural shapes; and is described by recursive algorithms. What is a fractal? By definition, "a fractal is a set for which the Hausdorff–Besicovitch dimension strictly exceeds the topological dimension" (Mandelbrot 1982). However, since this definition is a bit rigorous for the uninitiated, a more aptly put, simple definition is that fractals are repeating patterns—that is, by looking at the whole, many small parts, similar in appearance to each other and to the whole, make it up. Some of the best examples of fractals in nature are clouds, trees, and mountainous landscapes. With the advent of fractal geometry, during the last decade we have incorporated fractal mathematics with both computer and natural sciences, and it has quickly become a necessary tool in physics, soil science, hydrology, and other natural sciences. Applications of fractals in soil science have recently been reviewed by Perfect and Kay (1995).

Triadic Von Koch Curves

The triadic von Koch curve (often called the snowflake curve) is a good example of the fractal dimension (D), where $D > 1$. The construction of this curve is either recursive or iterative. A line segment (termed the initiator) is divided into thirds, while the middle segment is replaced with two equal segments, forming an equilateral triangle. This process yields a line segment composed of 4/3, called the generator. This generator segment is iteratively applied to itself to generate the von Koch curve (see figure 16.9). At the beginning, the initial straight-line segment has $n = 0$, while the generator has $n = 2$ (once applied); before application, $n = 1$ for the generator. Thus, at any iterative portion of the generated curve at any stage n, we have a prefractal; by applying a reduced generator to all segments of a generation of the curve, a new generation is obtained—such a curve is called a prefractal and, each small portion, when magnified, reproduces a larger portion, exactly. As a result, the curve is invariant under changes of scale, much like a coastline or outline of a rough rock. Generation of the curve on a computer can squeeze an infinite length into a finite area without the curve intersecting itself. This denotes the concept of self-similarity and is one of the fundamental, central properties of fractal geometry.

Self-Similarity and Scaling

Because the original line segment is invariant with respect to both translation and scaling, the expression for D is easily obtained. For the von Koch curve, the length of the prefractal (the nth generation) is expressed mathematically as

$$L(\delta) = \left(\frac{4}{3}\right)^n \tag{16.53}$$

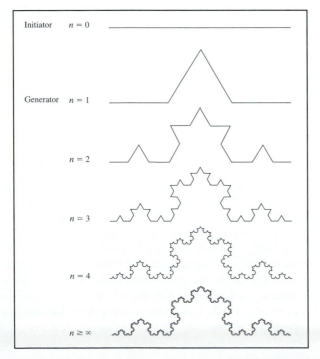

Figure 16.9   Example of the triadic von Koch curve showing the initiator and generator (data from Peitgen and Saupe 1988)

where L is the line segment of length (δ) and n is the iterative step of the curve generation. The length of each small segment of the line can be expressed as $\delta = 3^{-n}$, thus n (the generation number) is written as

$$n = -\frac{\ln \delta}{\ln 3} \tag{16.54}$$

Consequently, the length is expressed as

$$L(\delta) = \left(\frac{4}{3}\right) = \exp\left(-\frac{\ln \delta\,[\ln 4 - \ln 3]}{\ln 3}\right) = \delta^{1-D} \tag{16.55}$$

where $D = \ln 4/\ln 3 \approx 1.26$. We state that the number of segments in the line is $N(\delta) = 4^n = 4^{-\ln\delta/\ln 3}$ and is written in the form $N(\delta) = \delta^{-D}$, where D is the fractal dimension, 1.2628 for the triadic von Koch curve. As we see, unlike the more familiar Euclidean dimension, the fractal dimension need not be an integer.

If points are specified in some space, using the Cartesian coordinate system, the location of a line drawn through a certain point can be determined. Likewise, if the length scale is changed in the positive direction by the same factor (λ) for all components of x, then a new system of points can be mapped. Also, if the second set of points is adjusted by a factor $(1 - r)x$, the original set of points can be retrieved. With this reasoning, a plane is invariant under translation in that plane, as well as uncertain change of scale or length of scale. There is also statistical self-similarity in which, upon magnification, the segments of a line look alike, but are never exactly alike at different scales. For example, if we consider a coastline, as did Feder(1989), the more closely we follow the smaller indentations or curves, the longer the coastline becomes. In this case, each smaller section of coastline has the same appearance of the whole coastline, but not exactly. Thus, the total length of the coastline is the yardstick δ multiplied the number of measurements of size λ, $N(\delta)$, in measuring the coastline—that is, coastline length $= \lambda \cdot N(\delta)$, where $N(\delta)$ varies on the average of δ^{-D} and length: $\alpha\lambda \cdot \delta^{-D} = \delta^{D-1}$. Here, $D > 1$ and as the length of the yardstick used to measure the coastline length decreases, coastline length increases. For real coastlines D is about 1.15 to 1.25. Consequently, the similarity dimension D_s is

$$D_s = -\frac{\ln N}{\ln \lambda(N)} \tag{16.56}$$

This similarity dimension is relatively easy to determine for self-similar fractals such as the von Koch curve and its variants. The basic values to know for fractal dimensions are those for the set of points that make up a line in ordinary Euclidean space $D = 1$, for the set of points that form a surface in space $D = 2$, and for a ball or sphere $D = 3$.

Considering the von Koch curve to be the graph of a function $f(t)$, a scaling ratio of $\lambda = (1/3)^n$, where $n = 0, 1, 2, \ldots$, the property of the Koch curve is $f(\lambda) = \lambda^\alpha f(t)$ where the scaling component $\alpha = 1$. Because $f(t)$ is not a single value, the scaling relation is true for any point within the set. The power-law function $f(t) = bt^\alpha$, the same type of construction used on functions defined over all real positive numbers, satisfies the homogeneity relation $f(\lambda t) = \lambda^\alpha f(t)$ for all positive values of the scale factor λ. This function, and functions that satisfy this relation, are termed scaling.

Box Dimension

The fractal dimension of the coastline example is determined by covering it with a set of squares of edge length δ—the unit of length equal to the edge of the box width. The number of squares required to cover the coastline yields $N(\delta)$, from which the fractal dimension is

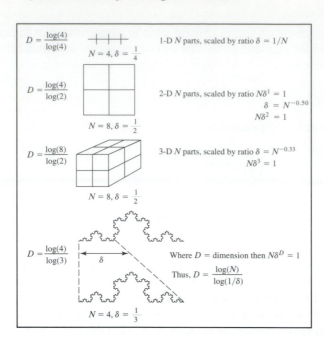

Figure 16.10 The box dimension and its various components (data from Peitgen and Saupe 1988)

determined by finding the slope of $\ln N(\delta)$, plotted as a function of $\ln \delta$ (see figure 16.10). This is an alternative definition of the fractal dimension, and is useful because the similarity dimension is of no value for either statistically self-similar or scale-invariant objects; this is because the "box" dimension measures how much space is filled by a geometrical object. The number of boxes required to cover a set S, if the entire S is contained within one box of size $\delta_{\max}$, is given by

$$N_{\text{box}}(\delta) = \left(\frac{\delta_{\max}}{\delta}\right)^D \tag{16.57}$$

or $N_{\text{box}}(\delta)$ is proportional to δ^{-D} (figure 16.10). This definition of box dimension is one of the most useful methods for estimating fractal dimension. For small sets, δ can be greater than the spacing between specific points, but less than the range of the set. If δ is less than the range of the set or greater than the spacing between the points, $N(\delta) = \delta^{-0}$ (single point), δ^{-1} (points on a line), and δ^{-2} (an even distribution on the plane). The box dimension is conveniently estimated by dividing the E-dimensional Euclidean space containing the set into a grid of boxes of size $\delta_{\max}^E$, and counting the number of boxes $N_{\text{box}}(\delta)$ that are not empty.

The length spanned by a line is also measured and since the line is straight, the dimension thus measured is termed the ruler dimension D_R (to be discussed later). If the measured line is fractal, $L(\delta)$ depends on the length characteristic δ, such that $L(\delta) \approx \delta^{(1-D_R)}$. From the previous definition, $D_R = 1$ if the line is smooth, suggesting $L(\delta)$ is constant for all values of δ, and the length of the line is simply $L(\delta) = \delta N(\delta)$. Based on the discussion on dimension, the fractal dimension is unique, and is simply the dimension explained by the most advantageous way to measure it. This, of course, depends on whether or not we measure a one- or two-dimensional plane. It should be noted here that the box dimension is widely accepted, easily used, and is close to the Hausdorff–Besicovitch fractal dimension, mathematically expressed as

$$M_d = \sum \gamma(d)\delta^d = \gamma(d)N(\delta)\delta^d \xrightarrow[\delta \to o]{} \begin{cases} 0 & \text{for} \quad d > D \\ & \text{or} \\ \infty & \text{for} \quad d < D \end{cases} \tag{16.58}$$

where M_d is the d-measure of the set and its value for $d = D$ is often finite, but can also be zero or infinite; $\gamma(d)\delta^d$ is a test function $h(\delta)$, and represents a disk, line, cube, square, or ball. If we have a line, cube, or square, the geometrical factor $\gamma(d) = 1$; $\gamma = \pi/6$ for spheres; and $\pi/4$ for disks. The position of the jump in M_d as a function of d is most important. Also, D in the Hausdorff–Besicovitch dimension is a local property because it measures properties of sets of points, in the limit of a decreasing size δ of the test function used to cover the set; D can therefore, depend on position. However, the Hausdorff–Besicovitch dimension is very difficult to compute for a real-life data set and therefore is of little functional use.

16.4 FRACTAL CONSTRUCTION

Fractal Dimension $0 < D < 1$

Both triadic and quadric cantor dust are examples of fractals of dimension $0 < D < 1$, and are also classified as "exactly self-similar" fractals. Exactly self-similar fractals are sets of points that cluster on a line segment, but do not fill that line segment. To construct the triadic cantor dust, we begin with a line segment of unit length ($n = 0$) and divide it into three equal parts. Now, eliminate the middle (central) third ($n = 1$) and repeat this procedure on the two remaining line segments ($n = 2$); repeat the procedure as many times as desired. It is readily apparent that as $n \to \infty$, the points cluster on the line segment of unit length in which their dimension is greater than a single point, but less than the dimension of the line segment of unit length. The dimension of the triadic cantor dust is

$$D_s = \frac{\log\,[N(\delta)]}{\log\,(1/\delta)} = \frac{\log\,(2)}{\log\,(3)} = 0.631 \tag{16.59}$$

Here (as discussed earlier), the initiator is a line segment of unit length, the generator consists of two segments ($\delta = 1/3$) that contain $N(\delta) = 2$ copies of the original. The construction of the quadric cantor dust is similar except that the two segments of the line are $\delta = 1/4$, which yields a $D_s = \log\,(2)/\log\,(4) = 0.500$.

In the above example, the central portion of the line segment was removed to create exactly self-similar fractals. By removing one-third of the line segment at random, the recursive construction is also randomized, and scale invariant—or statistically self-similar—fractals are constructed. By use of power-law distribution, a sequence of power-law-distributed gaps is generated. If a speck of dust is placed between each of these gaps, a cantor dust of any fractal dimension can be constructed. For cantor dusts, it is important to remember that the fractal dimension is really a measure of the degree of clustering on the line segment. Thus, if the dust is highly spread out, the fractal dimension is near zero, and as the dust tends to cluster along the line segment, the fractal dimension is near 1.

Fractal Dimension $1 < D < 2$

Earlier, we discussed the von Koch curves and how finer and finer detail is recursively added, an example of fractal construction beginning with an object of a lower dimension than the fractal dimension sought. Another way to achieve a desired fractal dimension is to begin with an object (volume, surface, etc.) that has a larger dimension than the fractal dimension desired, and cut holes in it until the desired dimension is obtained. Examples of this type of fractal dimension are the "Sierpinski gasket" and "Sierpinski carpet." The Sierpinski gasket is recursively generated using an equilateral triangle as a generator, and an initiator that has the central portion of this equilateral triangle removed. Sierpinski curves arrive from an

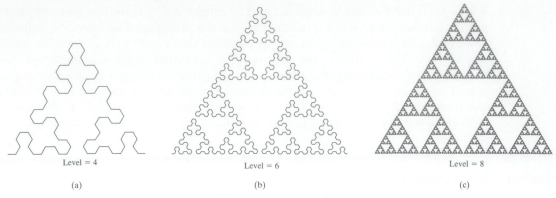

Level = 4   Level = 6   Level = 8

(a)   (b)   (c)

Figure 16.11   Various levels of a Sierpinski gasket (data from Barnsley et al. 1988)

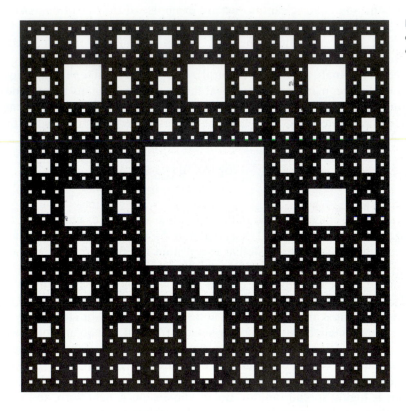

Figure 16.12   A Sierpinski carpet (data from Barnsley et al. 1988)

infinite number of generations of prefractals to leave a fractal curve; this fractal curve was obtained from the Sierpinski gasket (figure 16.11).

Sierpinski carpets have been used as models for the study of soil-water retention, Stokes' flow, pore geometry, pore-volume distribution, and other physical phenomena in soil. The relations to flow in soil is discussed more fully later. A simple way to obtain a Sierpinski carpet is to take a whole surface and randomly cut holes in this surface, where the holes cut have a power-law size distribution. For this case, the initiator contains $N(\delta) = 3$ duplicates of the original line segment of size $\delta = 1/2$ thus, the similarity dimension is $D_s = \log(3)/\log(2) = 1.59$. Unlike the Sierpinski gasket, the Sierpinski carpet (see figure 16.12) is recursively constructed beginning with a unit square, in which the generator is a square with the central square removed. The initiator contains $N(\delta) = 8$ duplicates of the original line

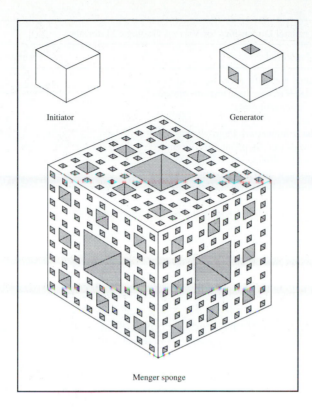

Figure 16.13   A Menger sponge shown with its initiator and generator (data from Barnsley et al. 1988)

Initiator

Generator

Menger sponge

segment of size $\delta = 1/3$. As a result, the similarity dimension $D_s = \log (8)/\log (3) = 1.893$. Both of these are examples of exactly self-similar fractals, and the pattern of distribution for holes within each type follows a power-law distribution.

Fractal Dimension 2 < D < 3

An example of this fractal type is the Menger sponge (see figure 16.13), which is a volumetric analog of the Sierpiński carpet. As we might suspect, the initiator is a unit cube divided into 27 smaller cubes to form the generator, and then 6 of the small cubes are removed from the center of each face, as well as the small cube at the center. Each of these smaller cubes is of size $\delta = 1/3$. The generator possesses $N(\delta) = 20$ duplicates of the original segment. The similarity dimension is $D_s = \log (20)/\log (3) = 2.727$; fractal dimensions can—and will—vary considerably for different materials.

16.5   SELF-SIMILAR FRACTALS: ESTIMATING THE FRACTAL DIMENSION

Now we discuss measuring the fractal dimension of self-similar fractals. As with self-affine fractals, we measure a characteristic of a data set that is related through some sort of power law to a length scale and as before, the results are plotted on a log-log scale. It is important to remember that more than 200–300 discrete measurements are needed at a variety of length scales, in order to make certain the data set has scale-invariant characteristics. From a physical viewpoint such as water flow or contaminant transport, it is important to obtain the fractal dimension because it can correspond to the effects of various physical processes that occur in soil. A partial list of fractal dimensions for various materials is given in table 16.1.

TABLE 16.1 Fractal Dimensions for Various Geologic Materials

Description of material	Fractal dimension
Upper Columbus dolomitic rock (Bellevue, Ohio)	2.91
Granitic rock (SHOAL nuclear test site, Nevada)	2.88
Soil (kaolinite, trace halloysite)	2.92
Porous silicic acid	2.94
Coal-mine dust from western Pennsylvania	2.52
Mosheim high calcium (Stephens City, Virginia)	2.63
Niagara (Guelph) dolomite (Woodville, Ohio)	2.58
Soil (mainly feldspar quartz and limonite)	2.29
Aerosil—nonporous fumed silica (Degussa)	2.02
Madagascar quartz from thermal syndicate	2.14
Graphite—Vulcan 3G (2700) (National Physical Laboratory, Teddington, United Kingdom	2.07
Iceland spar, massive (Chihuahua, Mexico)	2.16

Source: Data from Avnir, Farin, and Pfeifer (1984).

Grid Dimension

This measuring technique is so termed because boxes of linear size δ are used to measure a set; these boxes usually form a grid. The number of boxes required is termed $N(\delta)$ and the dimension has the relation $N(\delta) = \delta^{-D}$ (defined previously). The general technique is to count the boxes required of size δ, to cover the set of a range of values of δ. Then, $-D$ (the slope) is obtained by plotting $\log_{10} N(\delta)$ versus $\log_{10} \delta^{-1}$. As with the previous measurement techniques, if the set is fractal, the plot is a straight line with a negative slope equal to $-D$. For evenly spaced points on a log-log plot, the size of the box chosen follows a geometric progression.

Point Dimension

The point dimension—also referred to as the pointwise, cluster, or mass dimension—allows us to count the number of points within a set that has been encompassed by a circle of radius r. In this case, the circle does not include the entire set, but only that part of the set that fits within the defined radius. To imagine this, we draw a set of points in a straight or curving line, and select a circle of radius r to encompass a portion of the points, as in figure 16.14. This can

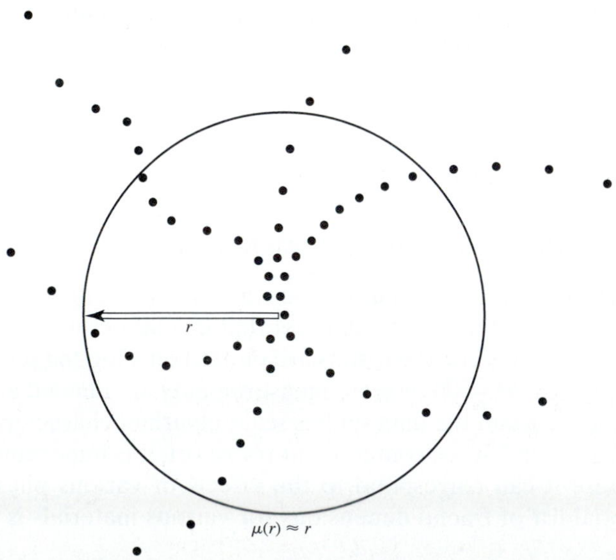

Figure 16.14 Exploded view of a point dimension

$\mu(r) \approx r$

also be done on a plane. The mathematical relation is

$$m(r) = \frac{P(r)}{P} \tag{16.60}$$

Since P is the number of points in the entire set, the mass is $m(r)$ within the circle of defined radius r, and the number of points within the circle is $P(r)$. The mass within the circle is proportional to r in one dimension, and r^2 in two dimensions. In real-life soil applications, a beginning measurement point (preferably at the center) is selected and circles of increasing radius are used to measure the mass $m(r)$. The next step is to plot $\log_{10} m(r)$ versus $\log_{10} r$ that follows a straight line if the set is fractal, as described in previous measurement methods. The relation for the mass in this case is $m(r) \approx r^D$, where D is usually written D_p to denote point dimension. This method is particularly applicable to sets that have a radial symmetry—that is, diffusion-limited aggregation (DLA).

Ruler (Divider) Dimension

An excellent example of the "ruler" dimension is the measurement of a coastline, given by Feder (1989) and other authors that describe fractal measurement. If we observe a world map and investigate the coastline of any country, at first glance it appears relatively smooth. However, upon closer inspection the mouths of rivers, bays, and various tributaries and features become more prominent. By the same example, if we use a ruler of length δ, the length of the coastline can be conveniently measured. If the ruler length were large—except for the basic outline of the coast itself—the detail of the various topological features (discussed above) is lost. By decreasing the ruler length δ, more detail of the coastline is apparent. The total length of the coastline is also increased because of this detail. The ruler dimension is normally expressed as D_R and is expressed mathematically as

$$L(\delta) \approx \delta^{1-D_R} \tag{16.61}$$

where $L(\delta)$ is the length of the coastline (or measured) object by the ruler (or divider) of length δ.

If the line is Euclidean, $D_R = 1$, the overall length is independent of δ if δ is sufficiently small compared to the measured object. However, if the ruler fills the space completely, the length of the line is linearly related to the ruler length. Once $L(\delta)$ has been determined by a ruler of length δ, we can plot the $\log_{10} L(\delta)$ versus the $\log_{10} \delta$ if the line is fractal, the plot will follow a straight line as before. The plot has a negative slope equal to $1 - D_R$. We note that the log-log plot of the measured length shows no sign of reaching a fixed value as δ is reduced.

Perimeter–Area Dimension

If we imagine a set of natural, yet geometrical, objects within a two-dimensional plane (e.g., tree leaves floating on the surface of a pond), we can parameterize the leaves as a group of Euclidean objects with area A and perimeter P. We can prove that A and P of each leaf within the group is related by

$$A = \pi r^2 = \frac{P^2}{4\pi} \approx P^2$$
$$\tag{16.62}$$
$$P = 2\pi r = r\sqrt{\pi A} \approx \sqrt{A}$$

where r is the radius of each leaf. Also, because r is independent, the proportionality between the area and perimeter is independent of r. If the perimeter of the leaves is fractal, then the relation of A to P is

$$A \approx P^{2/D_{P-A}} \approx P^2$$
$$\tag{16.63}$$
$$P \approx [\sqrt{A}]^{D_{P-A}} = A^{D_{P-A}/2}$$

where D_{P-A} is the perimeter–area dimension. Plot $\log_{10} A$ versus $\log_{10} P$ on the vertical versus horizontal axis; if the set is fractal, the slope is equal to $2/D_{P-A}$ in the positive direction. This is accomplished by using δ at a fixed, small finite scale or resolution.

16.6 SELF-AFFINE FRACTALS

The coastline example discussed earlier is statistically self-similar for any given value of the scaling ratio r, as well as for all scaling ratios between some minimum and maximum cutoff values. For these types of fractals, the box-counting method is used to estimate the fractal dimension. In nature, however, various cases of interest are not self-similar. For example, the motion of a Brownian particle has different physical quantities for both position X and time t. As a result, these two quantities do not scale to the same ratio, and the related fractal is termed "self-affine." The parts of these fractals need to be rescaled, by different ratios in different coordinates, to resemble the original. Mathematically, a bounded set S is self-affine with respect to a ratio vector such that $r = (r_1, \ldots, r_E)$ if S is the union of N nonoverlapping subsets $S_1, \ldots, S_N$. Each set has to be congruent (i.e., the set of points S_i is identical to the set of points $r(S)$ after possible translations of the set) to the set $r(S)$, obtained from S by the affine transform defined by r. The affine transform converts a point $x = (x_1, \ldots, x_E)$ into new points where the scaling ratios $r_1, \ldots, r_E$ are not all equal such that $x' = (r_1 x_1, \ldots, r_E x_E)$. In these equations, the subscript E refers to the Euclidean dimension.

Hurst's Empirical Law (Exponent H)

From a lifetime study of the Nile River, Hurst (1951) invented a new statistical method for analysis termed the "rescaled range analysis" (R/S analysis). After investigating river discharges, mud sediments, and other natural phenomena surrounding the Nile, Hurst used the dimensionless ratio R/S for data analysis, where R is the range and S is the standard deviation. During years of data collection and analysis, Hurst developed the following mathematical relation:

$$\frac{R}{S} = \left(\frac{\tau}{2}\right)^H \tag{16.64}$$

where H is the Hurst exponent and τ is the number of years (number of tree rings, number of times a coin is tossed, and so on). Actually, τ can be thought of as the number of observations. If we consider a self-affine fractal such as the Devil's staircase (Feder 1989; see figure 16.15), we obtain an exact copy of the original only by scaling portions of the curve by different factors, such as $r_x = 3$ in the x direction and $r_y = 2$ in the y direction. These are the

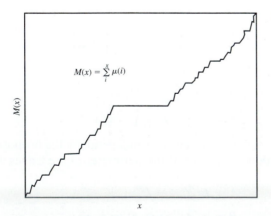

$$M(x) = \sum_{i}^{x} \mu(i)$$

Figure 16.15   Illustration of the "Devil's Staircase" (data from Feder 1989)

rescaling factors that can be rewritten to obtain the Hurst exponent H, such that $r_x = r$ and $r_y = r^H$. Solving for H yields

$$H = \frac{\log{(r_y)}}{\log{(r_x)}} \tag{16.65}$$

which is the general form of the equation and for which H in the Devil's staircase example is 0.631—that is, $\log{(2)}/\log{(3)}$. Also, because r_y is written as a power of r_x, the relation is scale invariant (i.e., independent of length). The standard deviation S is given by

$$S = \left(\frac{1}{\tau} \sum_{t=1}^{\tau} [\xi(t) - (\xi)_\tau]^2 \right)^{1/2} \tag{16.66}$$

where τ is the window (period) of time over which the observations are taken, $\xi(t)$ is the specific time and value of an observation, and ξ_τ is the average (or mean) value of observations for the time over which all observations are taken. From data gathered by Hurst, S is about 0.09 and is (as a rule) symmetrically distributed about a mean of 0.73. Hurst also shows that for numerous natural phenomena, $H > 0.5$. For statistically independent processes with finite variances in the absence of long-term statistical dependence, $R/S = (\pi\tau)^{1/2}$, it is also asymptotically proportional to $\tau^{1/2}$.

Brownian Motion and Random Walks in One Dimension

Consider the tossing of a coin. Each time the coin is tossed, there is a probability $p = 0.5$ of that coin being a head or a tail. Thus, if we assign a value of $+1$ for heads and -1 for tails, we can obtain a record of the trace of the compilation of tosses made (see figure 16.16). This is analogous to a particle moving on a line (assume the line is the x-axis), on which the particle jumps a step-length of $+\xi$ or $-\xi$ for a given time interval t (time interval in second, minute, hour; sometimes referred to as "collision time" in random walks). Such a stepping motion is referred to as a "random walk" (or Brownian motion) just as can be seen observing floating dust particles against a dark background. In one dimension, the random walk takes place only in the vertical coordinate. Also, the displacement of the individual particle in the given time interval is independent of the displacement of the same particle during another time interval. By consideration of itself, Brownian motion is self-similar; however, when considering particle position as a function of t, Brownian motion is self-affine. In the latter case, the Hurst exponent $H = 0.50$.

Considering a Gaussian distribution of zero mean, Brownian motion is a random walk in which the length of each step has nothing to do with the length of a neighboring step.

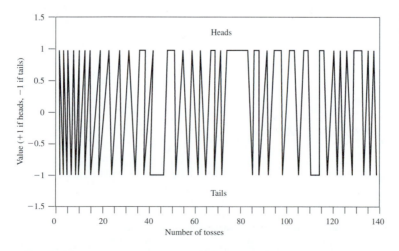

Figure 16.16   A compile trace of coins tosses (data from Feder 1989)

Consequently, the normal probability distribution (npd) is given by

$$p(\ell, t) = \frac{1}{\sqrt{4\pi Dt}} \exp\left(-\frac{\ell^2}{4Dt}\right) \tag{16.67}$$

where D is the diffusion coefficient given by the Einstein relation

$$D = \frac{1}{2t}\langle \ell^2 \rangle \tag{16.68}$$

The parameter $\langle \ell^2 \rangle$ is the mean-square jump distance. The Einstein relation is only valid under general conditions—that is, jumps do not occur at regular intervals, and when the npd for ξ is continuous, discrete, or has some arbitrary shape. The step-length ξ has to be chosen at random for any time interval t, such that the probability of finding ξ from ξ to $\xi + d\xi$ is $p(\xi, t)\, d\xi$. This yields a set of independent, random variables (Gaussian), for which the variance is

$$\langle \ell^2 \rangle = \int_{-\infty}^{\infty} \ell^2 p(\ell, t)\, d\ell = 2Dt \tag{16.69}$$

By letting ξ become $\xi/(2Dt)^{1/2}$, a normalized Gaussian random process is obtained. In the normalized process, ξ has a zero average with a variance $\langle \xi^2 \rangle = 1$. An example of a Gaussian random walk is shown in figure 16.17.

Scaling Properties of Random Walks in One Dimension

Consider sitting on a chair in a sunny room, observing floating dust particles. As sunlight reflects off each particle, the particle is clearly seen against shadows within the room or against a dark background. By focusing on only one of the dust particles, it appears to float effortlessly in a constant motion. As a result, we cannot clearly and finitely resolve the position of the particle; we can see it only at intervals bt, where b represents the first, third, or some other time step. Suppose we see the second time step of a particle's position; any increment ξ of the particle position is now a sum of the two steps (i.e., ξ_1 and ξ_2). Regardless of the number of b collision (or microscopic) time steps, the change in particle position is an independent Gaussian process with $\langle \xi \rangle = 0$ and a variance of $\langle \xi^2 \rangle = 2Dt$, where $t = b\tau$.

Because Brownian motion is self-affine, transformation of the normal probability distribution is accomplished by replacing $\tau = b\tau$ and $\xi = b^{1/2}\xi$; thus, the length scale is changed by a factor of $b^{1/2}$ and the time scale by a factor of b, which yields a scaling relation of

$$p\{\xi = \sqrt{b}\xi, \tau = b\tau\} = -\sqrt{b}\, p(\xi, \tau) \tag{16.70}$$

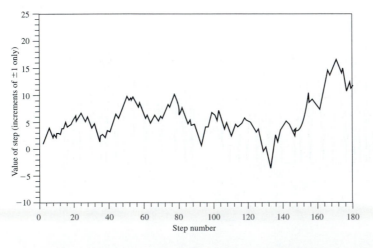

Figure 16.17   A Gaussian random walk (data from Feder 1989)

Hence, the scaled property appears the same, as if not scaled. Consequently, we effectively scaled and changed the resolution, and the Brownian record looks the same (i.e., it is scale invariant). Also, $b^{-(1/2)}$ ensures that the normal probability distribution is appropriately normalized.

Fractional Brownian Motion: Mathematical Models

Fractional Brownian motion, fBm, has become one of the most useful mathematical models for generating random fractals like those found in nature. This includes self-similar fractal landscapes, clouds, and mountainous terrain. Figure 16.18 shows fBm traces, usually denoted $V_H t$, plotted as a single-valued function consisting of only one variable t (time). The traces shown in figure 16.18 represent the point of location or differences of fBm between successive intervals. As we suspect, the scaling behavior of each trace is characterized by the Hurst exponent H, with a range $0 < H < 1$. The closer H approaches to 1, the smoother the trace, and as H gets closer to 0, the rougher the trace. H relates any change in quantity V ($\Delta V = V(t_2) - V(t_1)$), to a difference in time ($\Delta t = t_2 - t_1$) by the scaling law $\Delta V \propto \Delta t^H$. The sum of independent steps for random walks leads to a variation that scales as the square root of the sum of steps; in this instance $H = 1/2$ corresponds to a trace of fBm. The Gaussian distribution of fBm has a variance

$$\langle |V_H(t_2) - V_H(t_1)|^2 \rangle \propto |t_2 - t_1|^{2H} \tag{16.71}$$

where $\langle$ and $\rangle$ are grouping averages consisting of many samples of $V_H(t)$ and H has the value $0 < H < 1$. The mean-square steps depend only on Δt; each t is statistically equivalent. It is important to note that fBm is not differentiable, even though $V_H(t)$ is continuous. Constructs (based on averages $V_H(t)$) are developed to give meaning to a derivative of fBm. The derivative of normal Brownian motion is $H = 1/2$ and for $H > 1/2$ there is a positive correlation for increments of $V_H(t)$ and its derivative fractional-Gaussian noise, while for $H < 1/2$, there is a negative correlation.

A statistical scaling behavior for $V_H(t)$ is shown when the time scale t is altered by a factor τ, such that the steps ΔV_H are altered by a factor τ^H. This scaling behavior is given by

$$\langle \Delta V_H(\tau t)^2 \rangle \propto \tau^{2H} \langle \Delta V_H(t)^2 \rangle \tag{16.72}$$

Here, the t coordinate has a special status. While each t corresponds to only one value of V_H, any specific value for V_H can occur at multiple t's. This nonuniform scaling is termed self-affine.

Global Dimension of Self-Affine Fractals

The global dimension of self-affine fractals is $D = 1$, and is equal to the fractals' topological dimension; the local dimension is fractal and $D = 1/H$. The local dimension is often termed

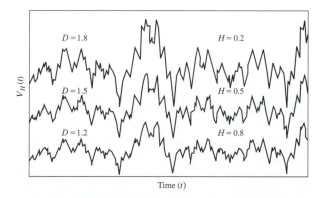

$D = 1.8$　　　　　$H = 0.2$

$D = 1.5$　　　　　$H = 0.5$

$D = 1.2$　　　　　$H = 0.8$

$V_H(t)$

Time (t)

Figure 16.18　A trace of fractional Brownian motion (fBm); the point of location or differences of fBm between successive intervals (data from Peitgen and Saupe 1988)

the "latent fractal" dimension because it is related to the fractal dimension associated with the trail of a specific Brownian particle. When $D = 1$, the fractal trace resembles a straight line; this infers that there is no vertical amplitude. The amplitude of the trace is also known as the "cross-over" scale, and it characterizes the overall roughness of the fractal surface. Thus, we can obtain two profiles with a similar fractal dimension, but totally different surface roughness. An example of this is the similar fractal dimensions of a concrete runway and mountainous terrain (Sayles and Thomas 1978), however, the cross-over scales for each of these is completely different. As a result of these parameters, two observations are made. First, there is a ratio of the vertical range (R) over the horizontal range L. The ratio can be written as $R(L)/L = w^H - 1$, where w is a window length (size) in which measurements are taken and H is the Hurst exponent. As the length L of the trace becomes much larger than the cross-over scale, any irregularity within the trace is minimal or if $H < 1$, the ratio approaches zero as L approaches infinity. Second, the self-affine record has a global value of the fractal dimension $D = 1$—that is, globally, a self-affine record is not fractal. As a result, it is important to measure the local dimension.

Measurement of Self-Affine Fractals

In discussing the global dimension of self-affine fractals, we have indicated that the ratio between the vertical and horizontal ranges is scale dependent (i.e., each varies with changes in scale). Also of note, the techniques used to measure self-similar fractals are inappropriate for self-affine fractals. For example, if the length of the ruler used to measure a self-affine fractal exceeds the cross-over scale discussed earlier, that measured length normally remains constant and the resulting dimension is always $D = 1$. Thus, since we want to measure the local dimension and not the global dimension, this is inadequate. Additionally, if we have a large scale in comparison to the cross-over scale, D is also one. In contrast to using boxes, grids, or rulers, as with self-similar fractals, we have to measure the change in vertical range (roughness) over different horizontal scales that yield an estimate of H (Hurst exponent). If the Hurst exponent is obtained, it can be converted to a fractal dimension, as proven earlier. To obtain this measurement, we need a variety of length scales that vary on an order of magnitude and a few hundred samples.

Roughness/length   This method of measurement is used when there is no constant sampling interval. However, we assume that each sampling interval (window) has about 5–10 samples. Instead of using the vertical range, we calculate the standard deviation of roughness data for windows of size w; this is related to the Hurst exponent by $\sigma(w) \approx w^H$. The entire length is measured by a series of windows and for each window, the roughness is determined after subtracting any local linear trends. As a final estimate, the average roughness for each window $\langle\sigma(w)\rangle$ is determined. For a number (series) of window lengths, plot $\log_{10} \langle\sigma(w)\rangle$ versus $\log_{10} w$. For a self-affine trace this plot is a straight line. The slope of this line is the Hurst exponent and is determined by dividing the horizontal range by the vertical range—that is, $H = \log(x)/\log(y)$. Once H is calculated, the fractal dimension D for a self-affine trace is given by $D = 2 - H$, and for a surface is given by $D = 3 - H$.

Rescaled range (R/S)   To use this method, the data measured must have a constant sampling interval. This is because the expected difference between successive values of $y_1 \cdots y_n$ is a function of the distance between each y value. We already have the equation for R/S from the beginning of this section, so it is not repeated here. Suffice it to say that R represents the range of values extracted over the y interval that is measured as a line segment, connecting the first point to the last point within a window (i.e., with respect to a trend within the window); S is simply the standard deviation. In this method, R is the average of a number of values. As with the roughness/length method, if the trace is self-affine, the plot follows a straight line. The slope of this line is the Hurst exponent H. Also, the fractal dimension D for both a trace and a surface are the same as for the roughness/length method, except that the R/S method has to be employed to obtain D. Thus, we write $D_{R/S}$.

Variograms   This measuring method can also be applied to a series that does not have a constant sampling interval, provided that points are selected such that $\Delta\tau$ is a small percentage of τ. By squaring the difference between two y values in a trace, separated by the interval τ, we have a variogram, which is related to the autocovariance function (Journel and Huijbregts 1978). This is written as

$$\gamma(h) = \frac{1}{N} \sum_{\alpha=1}^{N} (z(x_\alpha + h_{\alpha\beta}) - z(x_\alpha))^2 \tag{16.73}$$

where $\gamma(h)$ is the dissimilarity in the two values (squared difference), $z(x)$ is the regionalized variable at points γ and β, and N is number of points sampled. Dividing equation 16.73 by 2 (commonly seen in the literature) yields a semi-variogram. Previously, we listed the standard deviation in relation to fractional Brownian motion as $\sigma(\tau) = \tau^H$; if we now consider a normal distribution with a squared difference, we can write $V(\tau) \approx \tau^{2H}$. To obtain H, the $\log_{10} V(\tau)$ is plotted versus the $\log_{10} \tau$. Again, the slope has to be a straight line if the fractal is self-affine and represents $2H$ because of squaring. Thus, for true H, divide the slope by two. The fractal dimension D_v for this method for a trace and a surface is the same as the previous. Through measurement, an average trend appears that has to be subtracted before estimating range or roughness. The variogram accurately measures the average squared value of the trend. It should be noted that the average trend needs to be subtracted before range or roughness is estimated, to prevent error from creeping into the calculations.

Power-spectral density   As a function decomposes into a sine or cosine (harmonic) function, it is termed the "power-spectral density," PSD: the PSD is the squared amplitude of a harmonic function. To obtain the fractal dimension, the power spectrum $P(k)$ must be determined; $\log_{10} P(k)$ is then plotted versus $\log_{10} k$. The harmonic function is expressed as wavelength λ, frequency f, or wavenumber k. Both the frequency and wavenumber is $1/x$, where x is either selected as time or length. The relation between all three is $k = 2\pi f = 1/2\pi\lambda$. As a consequence, the power spectrum for self-affine fractals follows a power law (scale-invariant); the exponent is equal to $-\beta$. As with the other methods already discussed, if the trace is self-affine, the plot follows a straight line with a negative $-\beta$. The reader should recall that for a self-affine trace, $D = 2 - H$. Also, the method can be extended to two-dimensional surfaces, by a measure of $V(w)$ as the average squared distance between points. The points are separated by some distance w in all directions. Thus, the appropriate fractal dimension of the surface could be expressed as $D = 3 - H$.

The scale-invariant form of power spectrum of a self-affine trace and surface can be shown with Parseval's theorem, relating the spectral density $P(k)$ of a function $y(x)$ to its variance σ^2:

$$\int_{-\infty}^{\infty} y(x)^2 \, dx = \sigma^2 = \int_{-\infty}^{\infty} P(k) \, dk = 2 \int_{-\infty}^{\infty} P(k) \, dk \tag{16.74}$$

where $y(x)$ has zero mean and $P(k)$ is an always-positive function symmetric about zero. Parseval's theorem illustrates that the power spectral density $P(k)$ quantifies how the total variance in a function is partitioned in components of varying wave number, k. Because the variance of a self-affine function depends on the length of the window w, and the length scale is proportional to wavelength λ, the inverse of equation 16.74 is expressed as

$$(k) = \frac{d}{dk} \sigma^2(w) \approx \frac{d}{dk} w^{2H} \approx \frac{d}{dk} \lambda^{2H} \approx \frac{d}{dk} k^{-2H} \approx k^{-2H} \approx k^{-(2H+1)} = k \tag{16.75}$$

The power-spectral density for a self-affine trace, as seen in equation 16.75, takes the form of a power law with an exponent equal to $-\beta$. Using equation 16.75, β is related to the fractal dimension D to obtain

$$D = \frac{5 - \beta}{2} \qquad (16.76)$$

Since we are discussing a self-affine surface, $P(k)$ is a function of wave numbers in k_x and k_y, respectively. For two-dimensional surfaces, $P(k)$ is a function of wave numbers in both the x and y direction; however, by defining k as independent of direction, $P(k)$ is a function of only one variable. This is defined by

$$k = \sqrt{k_x^2 + k_y^2} \qquad (16.77)$$

In this instance, the fractal dimension $D = (8 - \beta)/2$. This method is the most difficult of all that we have discussed for obtaining the fractal dimension, although there are some who believe it the most cited in the literature. One reason for this: using the expression for k—taken as a squared-Fourier transform of a finite series—is a poor estimate of $P(k)$. This squared transform is termed a "periodogram," whose estimate of power at various frequencies is very noisy, the amplitude of the noise proportional to the function's spectral power. If the amplitude is filtered to make it more like the amplitude of "white noise" (for typical Brownian increments $H = 1/2$; the noise in this case is an independent Gaussian process normally called white noise), a more reliable estimate of the fractal dimension is obtained. In summary, we have presented four methods for measuring self-affine fractals: **(1)** roughness/length, for use when there is no constant sampling interval; **(2)** rescaled range, to be used when data measured have a constant sampling interval; **(3)** variograms, a method to be applied to a series that does not have a constant sampling interval; and **(4)** power-spectral density, a statistical tool commonly used to study the frequency content of signals. The first two methods are the easiest to use; however, before attempting the latter two, the reader should consult current literature on the subject. Because there is so much literature, we have conducted only a brief discussion here, and the reader is referred to the References section at the back of the book for a more complete list of articles on the subject.

   Hyperbolic distribution   The pdf for a random variable written in terms of a hyperbolic distribution (also called a power law) is expressed as $p(x) = bCx^{-(b+1)}$, where b is some power and C is a constant. This expression is only true for $x > 0$ and $b > 0$. The modeling of random variables using a hyperbolic distribution is only useful for positive random variables, and is most effective where the number of smaller members (within the group being sampled) is greater than the number of larger members. Here, b characterizes the hyperbolic distribution since C is only a constant and not very important, except that it depends on the minimum value of x chosen. Essentially, the hyperbolic distribution has all the characteristics of scale invariance. In such cases, the exponent b is invariant under multiplication. Unlike the previously discussed distributions, there is no scale length associated with hyperbolic distributions. Thus, hyperbolic distributions are scale invariant and any process that creates a variable in this manner must also be scale invariant.

16.7   VISCOUS FINGERING AND DIFFUSION-LIMITED AGGREGATION

Multifractals (Feder 1988) are shapes and measures that require more than one dimension, and have received broad attention in the study of such dynamical systems as Poincare maps and electric-field strength in aggregation problems. Consequently, the measurement of multifractals is related to the study and distribution of physical and other quantities on the surface of a sphere or an ordinary plane. Additionally, the support itself can be a fractal. Fractals provide a general language for the classification of various pathological or indescribable shapes ("animals") encountered in the natural sciences. The various ways in which matter condenses on the microscopic scale seem to generate fractals; this is an example of percolation, often confused with diffusion (Feder 1988; Peitgen and Saupe 1988). An example of diffusion is the random motion of dye particles mixed in a solution. In contrast, percolation is displayed when the randomness is attached to the medium itself. For example, consider water beading on the hood of an automobile. Initially it is several atomic layers thick but rather than a uniform coverage, the water beads, due to surface tension. Analysis of the connected clusters (droplets) shows irregular branching structures of finite size. As the amount of water increases, the clusters increase in size, and eventually connect across a certain sample length. The unique characteristic associated with percolation is that of a percolation threshold, p_t. If the rate of buildup is below p_t, then the spreading of fluid is confined to a usually small, finite area. Assuming that a point source is constant and that water is then applied, is the resulting contaminant going to be contained locally within a droplet, or is it going to spread across the hood and connect to other beading droplets, causing a large puddle that can connect to other large puddles? The transport of immiscible fluids (e.g., oil and water) is also a good example that can create fingering, invasion percolation, or preferential flow.

Diffusion-Limited Aggregation (DLA)

DLA is a simple model that reproduces many natural shapes, such as electrochemical deposition, electrostatic discharge and—for purposes considered here—fluid–fluid displacement. It is easily implemented on a computer, with resulting structures resembling those of root distributions within a soil profile. The DLA process is one in which a particle or monomer diffuse in the "random walk" process (see section 16.6). Imagine a solution of randomly mobile particles; DLA begins with a single fixed, sticky particle at an origin point within this solution. Each of the mobile particles moves on its own random path, one random unit at each time step. At some point, a random particle finds itself next to the origin point and "sticks" to it. This point (beginning from one site) slowly grows into a cluster with the addition of other random particles, to complete the random-walk process. A structure much like that of the distribution of roots within a medium forms with many branches (see figure 16.19). The open branches do not fill, because a particle trying to reach the innermost point of the branch invariably contacts one of the sticky sides first. The cluster thus formed is an example of a

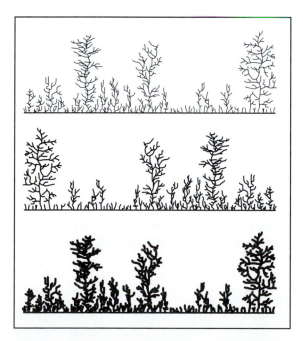

Figure 16.19 A schematic of diffusion-limited aggregation (DLA). The formed clusters are termed "percolation transition," and are normally fractal in nature (data from Feder 1989)

"percolation transition," and is usually fractal in nature. The cluster is both porous and random but also, the cluster has to exhibit decreasing density upon increase in size described by

$$\rho(r) \approx R_o^{-D} r^{D-E} \tag{16.78}$$

where ρ is the cluster density, r is cluster radius, R_o is particle radius (particles forming cluster and thus, the characteristic length scale), D is the fractal dimension, and E is the Euclidean embedding dimension (i.e., $D < E$). For a fractal cluster, the density decreases with distance from the point of origin.

Fractal Diffusion Fronts

If we apply methylene blue dye to the top of a sandy soil and add water very slowly in small amounts, a diffusion front begins to appear. Upon close examination, the front appears like the edge of a coastline due to adsorption of the dye to the sand particles. This is another example of percolation. The structure of the diffusion front is fractal and is related to the hull of percolation clusters (Sapoval, Rosso, and Gouyet 1985), but the diffusion process is not fractal. The surface of the diffusion front is called the hull and is a finite fraction of the adsorption sites involved with a fractal dimension. As the particles of dye move from the source, diffusion D is described by the Einstein relation given in equation 16.68 (the particle moves to a neighboring site a distance a, and time step τ). This means that the displacement of a diffusing particle in the x (perpendicular) and y (parallel) directions away from the source are independent of each other.

The position of the hull is given by $p(x_h) = p_c$. Also, if the diffusion has a width of $2L$ instead of L (L is a length over which data is measured), the number of occupied sites belonging to the hull is also doubled, which makes a view of the hull one-dimensional. The hull is a self-similar fractal up to length scales that equal the hull's width. The fractal dimension of the hull can be written as

$$D_h = \frac{1 + v}{v} = 1.75 \tag{16.79}$$

where $v = 4/3$ (controls the divergence of the correlation length ξ at p_c). This was proposed by Sapoval, Rosso, and Gouyet (1985) and proved to be correct by Saleur and Duplantier (1987). This is because the ratio $M_h(L/\ell)\sigma_h/L\ell$ approaches 0.441 as ℓ approaches infinity, where M_h is the number of sites that belong to the hull, L is the length, and ℓ is the diffusion distance defined as the root mean-square displacement of the diffusing particle from its starting point.

At any time step, the particles at the diffusing front occupy sites similar to the percolation process. The probability $p(x)$ of the site being occupied is dependent on distance from the source, and is mathematically described by

$$p(x) = 1 - \frac{2}{\sqrt{\pi}} \int_0^{x/\ell} du \exp(-u^2) \tag{16.80}$$

where u is an element of the set. At the source, $p(x) = 1$ however, it rapidly diminishes for $x > 1$. The internal structure of the diffusing front is also fractal. This fractal structure extends over distances proportional to the diffusion width $\ell = 4Dt^{0.5}$. As we might suspect, this distance diverges with time, which allows the fractal structure to extend over macroscopic distances, even in instances where diffusion is on the molecular scale.

Viscous Fingering in Soil

The principles of viscous flow in soil are exactly similar to the principles which control flow in viscous-flow analogs such as the Hele–Shaw or parallel-plate analogs. Both are well-known devices used for two-dimensional ground-water investigations. The viscous-flow analog is based on the similarity of differential equations that govern flow of a viscous fluid in the narrow space between two parallel plates and those that govern saturated flow in soil. This methodology has been used in investigations of seepage through earthen dams, artificial recharge, drainage, oil production in reservoirs, and in other investigations. The Navier–Stokes equation for a viscous incompressible fluid is given as

$$\frac{DV_x}{Dt} = f_x - \frac{1}{\rho}\frac{\partial p}{\partial x} - v\,\nabla^2 V_x \tag{16.81}$$

where DV_x/Dt represents the hydrodynamic derivative, V_x is the velocity component in the x direction, v is the kinematic viscosity, ρ is the fluid density, f_x is the component of external force per unit mass acting on the liquid, and p is the pressure. Because the fluid flows through a narrow vertical space of width b, $V_y = 0$. For fluids that have very high viscosities or very slow (creeping) motions, viscous forces are much greater than inertial forces. Thus, assuming this type of flow takes place between the parallel plates, the left-hand side of equation 16.81 (the inertial term) can be neglected. Consequently, the only active-body force is gravity with potential gz. Thus, $f_x = -\partial(gz)/\partial x = 0$; force in the y direction is the same, we simply substitute the x component with the y component—however, $f_z = -\partial(gz)/\partial z = -g$. Since b is narrow but also because the fluid adheres to the parallel plates, velocity gradients in the y direction are considerably larger than those in the x or z directions. As a result, we can neglect $\partial V_x/\partial x$, $\partial^2 V_x/\partial x^2$, $\partial V_z/\partial x$, $\partial^2 V_z/\partial x^2$, when compared with $\partial V_x/\partial y$, $\partial^2 V_x/\partial y^2$, $\partial V_z/\partial y$, $\partial^2 V_z/\partial y^2$. Thus, equation 16.81 becomes

$$\frac{\partial(p + \rho gz)}{\partial x} = \mu\,\frac{\partial^2 V_x}{\partial y^2} \tag{16.82}$$

The parameter $(p + \partial gz)$ remains constant in the y direction and thus is equal to zero. Rearranging equation 16.82 for velocity we obtain

$$V_x = -\frac{k_x}{\eta}\nabla(p + \rho gz) \tag{16.83}$$

where k is the permeability of the medium and η is the dynamic viscosity of the fluid. The term k/η is referred to as the mobility—thus, the velocity $V = -M(p + \partial gz)$. For a horizontal position of the parallel plates, $k = b^2/12$ (only for the Hele–Shaw cell; for soil, k is the actual measured permeability). In terms of the Laplace equation, equation 16.83 can be written as

$$\nabla \cdot V = -M\nabla^2(p + \rho gz) = 0 \tag{16.84}$$

where M is κ/η. This equation is characteristic of many diffusion problems that represent potential flows. Normally, these types of analogs are isotropic and in order to find a solution to the equations, the proper boundary conditions have to be specified.

Upon analyzing the displacement of epoxy by air, Feder (1989) finds that there is a relation between the number of air monomers (N) present in fingers within soil, and the distance of the monomer (r_i) from the point of injection. The relation he develops involves the radius of gyration (R_g) and total number of monomers containing air (N_o), such that $R_g = (N_o^{-1}\Sigma_i r_i^2)^{0.5}$. The results of Feder's investigation reveal that there is indeed a number–radius relation, with the resulting data from different fluid types and different times all falling on a simple curve. The number–radius relation is expressed as

$$N(r) = N_o\left(\frac{r}{Rg}\right)^D f\left(\frac{r}{Rg}\right) \tag{16.85}$$

where f is a crossover function that is assumed constant at $x < 1$, and approaches x^{-D} for $x > 1$. As a result, $N(r) \to N_o$ for $r \gg Rg$. Using this method, the results are best fitted with a fractal dimension D, which represents the fingering of the fluid within the media.

There are differences between the results obtained with a Hele–Shaw analog and those observed by viscous fingering. Since the Hele–Shaw has only a small width b (length scale), the fluid flow is controlled by this microscopic length. However, for a soil with a pore diameter equal to b, fluid flow is controlled in all directions by the microscopic length scale. Also, the boundary conditions chosen for each method have a significant influence on the results obtained. For the Hele–Shaw cell, the length plane is set by capillary forces, whereas in a soil the length scale is set by pore diameter. Thus, we can easily ascertain that for the Hele–Shaw analog flow, control is simply a matter of pressure distribution. For a soil, fluid flow or displacement is not simply a pressure difference, but the pressure relative to the capillary pressure at the pore neck. This is basically the same relation as that of the "inkbottle" effect (discussed earlier in the text), where it is more difficult for a fluid to enter a narrow neck versus a wide neck. Because soils are heterogeneous, pore necks vary in both size and shape throughout the system, which introduces a degree of randomness; without this randomness, a fractal structure cannot be produced. In both the Hele–Shaw analog and soil, flow of high viscosity fluids is controlled by the Laplace equation, thus fingering in soil has the components of the Laplace equation as it relates to Darcy's law and pore geometry. The combination of the two components causes the generation of a fractal structure that can be analyzed by the method described by Feder (1989).

Earlier we discussed how the DLA process is one of random walking, where each particle eventually comes to rest and causes the formation of a fractal structure. We assume that the particles diffuse at a constant rate as described by the Einstein relation (equation 16.68). Using this concept, the random walk can be described by the basic diffusion equation

$$\frac{\partial C(r, t)}{\partial t} = D\, \nabla^2 C(r, t) \tag{16.86}$$

where $C(r, t)$ is the concentration of random walkers. By assuming steady state, $\partial C/\partial t = 0$ and equation 16.86 reduces to the Laplace equation. The velocity, as described by Witten and Sander (1983) is given by

$$V_\perp = -Dn \cdot \nabla C\big|_s \tag{16.87}$$

where V is perpendicular to the surface $\perp$, with surface normal n, and the equation is evaluated at the surface, $\big|_s$.

It is well known that if the displacing fluid has a lower viscosity than the resident fluid, wetting-front instability ensues and fingering will occur. For this case, the front moves with velocity $M\nabla(p + \rho gz)$. When the displacing fluid has a higher viscosity, the moving front is stable with a fractal dimension of 1. Also, the capillary number of a fluid greatly influences its ability to form fractal structures. The capillary number (Ca) is defined as

$$Ca = \frac{V\eta}{\sigma} \tag{16.88}$$

where V is the velocity, η the dynamic viscosity, and σ is the interfacial tension. The Ca is a measure of the ratio of capillary to viscous forces. From equation 16.88, it can be seen that Ca is increased by velocity (within a practical limit) or the use of fluids with a small interfacial tension. Thus, with high Ca, DLA accurately describes fingering in two-dimensional soil (i.e., $Ca = 0.05$ to about 0.1). For low Ca ($Ca \approx 10^{-4}$), the fractal structures formed are characteristic of those found using invasion percolation. This means, essentially, that the capillary forces completely dominate the viscous forces due to low velocity, and perhaps high viscosity. In invasion percolation (where Ca drastically decreases), pressure drops in the displacing, resident fluids are neglected, and a simple pressure difference between the two fluids is calculated. This is accomplished by subtracting the pressure of the resident fluid (p_r) from the pressure of the displacing fluid (p_d) as follows

$$(p_d - p_r) = \frac{2\sigma \cos \phi}{r} \tag{16.89}$$

where σ is the interfacial tension between the two fluids, ϕ is the liquid-contact angle between the pore wall and the interface, and r is the pore radius. We note the similarity between equation 16.89 and the height of capillary rise. Perhaps the most prominent example of invasion percolation is the displacement of oil by water—two immiscible fluids.

SUMMARY

We have determined that fractals can provide meaningful answers to some of the questions now confronting water sciences, because they extend the principles and concepts used in geostatistics and scaling. This short discussion was meant to spark the interest of the reader. There are many possibilities to be explored for the use of fractals in the environmental sciences. However, the use of fractal mathematics in the earth sciences is still in its infancy.

ANSWERS TO QUESTIONS

16.1. A normal probability distribution is the result of additive effects of numerous small, random, independent sources of variability. Thus, the properties for which we expect normal distribution are bulk density, water content at saturation, water content at -100 kPa, and the particle-size analysis.

16.2. A log-normal distribution is the result of multiplicative effects due to spatial variability, thus the physical properties we expect to be log-normally distributed are scaling coefficient; saturated and unsaturated conductivity; diffusion coefficient; electrical conductivity; and pore-water velocity. However, in specific cases, the given property may not conform to the distribution given here. The answer, as to what makes a log-normal distribution more appropriate than a normal distribution for certain properties, is left to the student; please consult any standard statistics text.

16.3. **(a)** Sampling error is a difference between a sample's average (resulting from the sampling process), and the population average (due to the heterogeneity of the property being studied). Measurement error is the difference between repeated results of a measuring process applied to a constant uniform object or property. **(b)** Statistical true value is the limiting mean of n measurements (that can be inaccurate due to bias) as n becomes large. Scientific true value is the actual value of an object of measurement by definition of the object. **(c)** scientific bias is a characteristic difference between the statistical true value and the scientific true value due to inadequate specification of the measurement process. Measurement bias is a characteristic tendency of results being either too high or too low, due to deficiencies in sampling apparatus or sampling materials. Sampling bias is the tendency of results to be in error due to inadequate sampling procedures or deficiencies in the sampling process.

16.4. This could be done when the lower-precision method is less biased than the higher-precision method, and also when the lower-precision method is more efficient than the higher-precision method.

ADDITIONAL QUESTIONS

16.5. Prove or show that the ruler dimension equals the similarity dimension in a triadic von Koch curve.

16.6. What are the three basic characteristics of fractals?

16.7. How is the precision of an estimate of a derived value related to the precision of its component measurements?

16.8. What information is carried by the standard error? The coefficient of variation? When would the use of each be appropriate?

16.9. What information is displayed by a correlogram? A variogram?

16.10. Scaling is normally done on a computer; consequently, the method the computer uses to obtain a reduced *SS* is often misunderstood. Two intact soil cores are extracted and the following water retention data is obtained.

Core Sampling Data

Core	h (cm)	S
1	10	0.95
1	20	0.90
2	10	0.94
2	20	0.93

Using the data from the table: **(a)** estimate the reference curve $h_m(s)$ by multiple-linear regression from *SAS* or similar statistical analysis; and **(b)** find coefficients a of the equation that best fit the data. This is accomplished by bringing to one side (set = 0)—that is, residual sums of squares—such that

$$\log (h_r) - a_0 - a_1 S \ldots = 0$$

which is the ideal *SS*. (*Hint:* see section 16.2 and note equations 16.50–16.52.)

APPENDIX 1

Site Characterization and Monitoring Devices

INTRODUCTION

This appendix will acquaint the reader with the site-selection process, including the primary criteria used for selecting a site for waste storage or scientific investigation. We will also discuss the most common monitoring devices used in unsaturated-zone studies, their principles of operation, and the advantages and disadvantages of each.

SITE CHARACTERIZATION

The purpose of site characterization is to determine the biological, chemical, and physical properties at a site that directly affect the movement of contaminants from or within it. However, before site characterization can take place, a site must be selected. The site-selection process can involve a large number of criteria: the availability of land; climatological factors that may bias the outcome of potential accident scenarios; proximity to transportation and population centers; proximity to sensitive natural resources such as aquifers, prevailing winds; and so on. Ideally, the site should be capable of being characterized, analyzed, monitored, and modeled. Because of this, it would be prudent to select several candidate sites based on technical criteria, depending upon the intended use of the site. For example, the criteria necessary for installing a safe landfill would not be as stringent as that for installing a low-level radioactive waste site. In many instances, local, state, or federal agencies already will have established required site guidelines.

The site-selection process usually begins by establishing technical criteria with respect to depth to ground water, acceptable rainfall limits, slope, elevation, and so on. After these have been set, a large-scale reconnaissance of various geographic areas is performed. From these areas, candidate sites are chosen and screened against minimum technical requirements. Site selection can be facilitated by obtaining soil maps and information on the physical properties of soil series from the National Resources Conservation Service (previously the U.S. Soil Conservation Service), and ground water table information from the United States Geological Survey. Once the candidate sites have been screened and a final site selected, it is ready for characterization.

The goal of characterization is three-fold: **(1)** to identify potential pathways for the transport of contaminants from containment areas to sensitive receptors—drinking-water supplies, air, and so on; **(2)** to demonstrate that the site can be characterized, monitored, and modeled, which involves field and laboratory analysis; and **(3)** to confirm that performance objectives (which comprise data analysis and modeling) can be met. Field characterization can be accomplished by in-situ testing, or by collecting disturbed and undisturbed samples for laboratory analysis. In-situ testing is usually preferred, but may not be practical for a large number of parameters. It is also time-intensive, especially for unsaturated-zone properties that occur at greatly reduced rates in comparison to the saturated-zone environment. When designing the characterization effort, the purpose for which the data to be collected will be used must be clear. For example, is the purpose simple data analysis and monitoring, or modeling the site in preparation for a license application to store hazardous waste of some type? Whatever the need, characterization efforts should be designed in cooperation with those who will use the data: regulatory agencies, modelers, and other personnel.

Field characterization can be grouped into the following broad categories: **(1)** chemical properties; **(2)** physical properties; **(3)** flow and transport properties; and **(4)** biological properties. Chemical properties of common interest include pH, chloride and sulfate concentrations, cation exchange capacity, and total salt concentration of the soil solution. Typical physical properties include bulk density, particle density, porosity, and particle-size analysis. Flow properties of interest are saturated hydraulic conductivity, unsaturated hydraulic conductivity (as a function of water content), and soil-moisture characteristic curves that relate water content to matric potential; the transport property of most interest is the dispersion coefficient, D (see chapter 10). Biological properties can include a host of microbiological properties and the properties discussed for physical and chemical parameters. Details on measurement techniques and analysis for each of these properties can be found in Klute (1986).

For modeling purposes, initial and boundary conditions must also be known. These include the vertical profiles over a depth of interest of water content and potential; the amount of water deposited on-site in the form of precipitation (rain and/or snow); and the amount of water leaving the site by overland flow, drainage, and evapotranspiration. For complete characterization, a representative number of samples for each parameter must be collected from the soil surface to the depth of interest (usually the water table). No definitive method has been presented that will provide absolute numbers or locations of samples for analysis, but there are equations that can be used in approximating the necessary number of samples for a chosen parameter (for example, equation 16.21 in this text). As with most scientific research and data collection needs, economics will play a major role in determining the number of samples that can be analyzed.

Frequently, environmental regulations require monitoring of the unsaturated zone for many sites. Ideally, one would wish to minimize cost yet assure reliability, which requires that monitoring devices be installed in the best possible locations. Additionally, the sample volume for most site-selection (or other investigative) purposes is small in comparison to the spacing interval from which the sample is taken. Because of this, there is a significant probability that any anomalies or "hot spots" will not be detected, and such areas are where one would most wish to sample. These are areas where physical, chemical, biological, and flow characteristics can be very different from those that exist in the site as a whole; and spots where a contaminant may be more prone to leak. In order to detect such areas, specific questions must be asked about the site at the initial planning and installation stages: **(1)** what grid spacing is needed to hit a hot spot with a specified confidence; **(2)** what is the probability that a hot spot exists when none are found by sampling on a grid; and **(3)** for a given grid spacing, what is the probability of hitting a hot spot of a specified size? Other questions may also arise, but these are the most important. For further details, the reader is referred to Warrick et al. (1996).

MONITORING DEVICES

This section discusses factors that influence the choice of monitoring devices, and the common types of equipment used for monitoring various soil parameters in unsaturated-zone studies, as well as various measurement techniques. The basic suitability, advantages, and disadvantages of each piece of equipment will be briefly outlined.

Factors Influencing Choice of Devices

The basic factors that influence the choice of monitoring devices include: **(1)** goals of the monitoring program; **(2)** site environment and conditions; **(3)** repeatability of measurements; **(4)** measurement-device resolution and operational range; **(5)** equipment and device durability; **(6)** device installation and replacement; **(7)** remote data acquisition versus manual data collection; and **(8)** device maintenance frequency.

Goals of the monitoring program may include process, compliance or remediation monitoring. *Process monitoring* is commonly used to determine physical processes for establishing background and baseline conditions. This may require high-intensity monitoring near the soil surface to understand periodic changes in data because of seasonal fluctuations. During process monitoring, goals are commonly achieved by strong reliance on data acquisition systems capable of sustained, high-frequency sampling. *Compliance monitoring* is generally performed to confirm waste isolation during operational and post-closure monitoring programs. Specific goals are normally to test the behavior of various parameters and compare them to baseline studies. This will ascertain whether contaminant migration or other significant events are occurring. Frequent data collection may be decreased to daily, weekly, or quarterly intervals depending on the individual parameter. As intervals become less frequent, automated systems may be replaced by manual data collections without accuracy loss. *Remediation monitoring* tracks the success of any cleanup activities that have occurred, and can be short- or long-term depending on the site. This type of monitoring usually combines procedures listed in process and compliance monitoring.

Site and environmental conditions include the soil type and structure; the presence of rocks; layering; depth to water table; macropores; and any obstruction that would affect instrument installation. Seasonally flooded or arid conditions will influence the choice of instrumentation, as well as the depth to the water table. Proximity to a city or town, as well as the availability of AC power should be considered part of the site conditions. This affects the economics, and the choice of manual or automated data sampling and acquisition. The presence of AC power increases the flexibility of instruments that can be used on-site.

Any monitoring study, regardless of duration, must have the ability to obtain repeatable data measurements. These might include electrical conductivity, soil dielectric constant, soil-water energy status, pH, and other parameters. The devices used must be able to obtain reliable, repeatable results.

Measurement-device resolution and operational range must be sufficient to detect changes in soil conditions. The accuracy and precision of each measuring device must be determined, and matched to monitoring goals. The operational range of each device can be used to determine its suitability in the overall monitoring strategy; for example, if the soil is very dry, tensiometers would not be as suitable for measuring soil-water energy status as thermocouple psychrometers. Devices with very narrow measurement ranges should not be relied upon in a long-term monitoring program, unless site conditions are well-understood.

Device life-span, and use in long-term systems, must be considered; that is, whether or not it will extend into post-operational or long-term care periods. If extension into these periods is not possible, the device will only be suitable for establishing baseline conditions.

Devices must be compatible with the specific phase of monitoring, operational or post-operational. Devices or locations where failure could affect the overall integrity of the monitoring program should be avoided.

Because the method of device installation and replacement can affect the measurements of water movement and other parameters, it is important to assess the advantages and disadvantages of horizontal versus vertical installation for each device. The ability to remove, repair, and replace equipment as necessary is an important factor in determining initial installation geometry. For example, at a site where accessibility is easy and long-term, a network of nested devices can be employed by installing them in a caisson of specific diameter and length (see figure A1.1). Depending on site accessibility, installing extra data-access ports adjacent to the site can provide flexibility in the event of instrument failure, the need for replacement, or difficult access after the site is closed.

Remote data acquisition can save valuable time and resources compared to manual data collection. The economics of the labor costs of manual collection versus the higher maintenance costs of telephone modems, solar panels, and computer manipulation must be assessed. Quite often, the maintenance and recalibration of data acquisition systems is complicated by the need to remove a storage module or data-logging device from the field, and return it to the laboratory. However, for various devices capable of automation, remote data acquisition is a very favorable option—although manual data collection allows technical personnel to check on-site equipment and make necessary adjustments. The decision to remotely access a site will depend on the type of data being collected, the frequency that it is needed, the site accessibility, the ability of the remotely accessible device to be incorporated into the overall monitoring scheme, and other factors.

Device maintenance frequency is an important aspect of site monitoring, as high maintenance requirements can affect the long-term viability of the monitoring system and goals, as well as accuracy and personnel costs. As a result, devices that are actively maintained

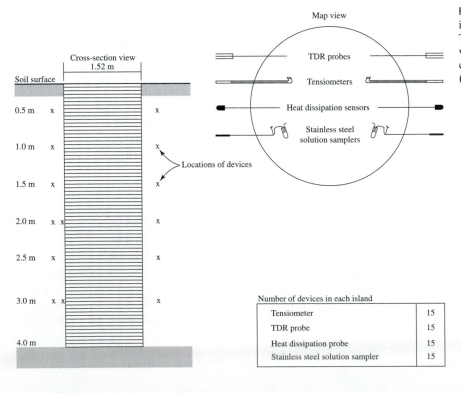

Figure A1.1 Diagram of monitoring island with proposed instrumentation. The caisson shown in the cross-section view is constructed of corrugated culvert pipe. Data from Young et al. (1996)

Number of devices in each island	
Tensiometer	15
TDR probe	15
Heat dissipation probe	15
Stainless steel solution sampler	15

(such as tensiometers), and devices that require passive maintenance (such as pressure transducers), are not as useful for long-term monitoring. The overall durability and maintenance requirements for each device must be considered for the monitoring system and goals to be successful.

Water Content Measuring Devices

Neutron probe   The neutron probe works on the principle of neutron thermalization. It determines water content by releasing high-energy ("fast") neutrons from a radioactive source, such as americium-beryllium. The high-energy neutrons collide with hydrogen atoms in the soil, and form what are known as thermal neutrons. The multiple collisions that take place form a thermal cloud of neutrons whose size is constant, but whose density is dependent on soil-water content. Higher water content leads to increased thermalization and, thus, denser thermal clouds. A "slow" neutron detector, installed adjacent to the source emitter, measures the cloud. The measurement is displayed in the form of a "count ratio," with a higher count denoting a higher water content.

Suitability Criteria: **(1)** possible to merge with other devices; **(2)** durable (greater than 30 years experience); **(3)** horizontal and vertical installation possible; **(4)** after initial tube installation further monitoring scheme disturbance is not necessary.

Advantages and Disadvantages: **(1)** widely accepted with highly reproducible results; **(2)** contains a source of radioactivity and cannot be automated; **(3)** expensive (approximately $8500); and **(4)** requires calibration. Best method is to vertically install neutron tube adjacent to site, take measurements at specific intervals, then destructively obtain three samples directly around the tube in pyramid fashion at each measurement interval. Once this has been accomplished, the samples can be taken back to the laboratory, placed in an oven, and dried at 105 °C for twenty-four hours. A regression equation can then be obtained from the gravimetric water content of the samples versus the count ratio from the neutron probe for each interval (instructors: see final exam in solution manual).

Time domain reflectometry (TDR)   TDR operates on the principle of microwave-pulse travel through a parallel transmission line (the probe). This technology was adapted from the electric power industry, where cable testers are used to determine the location of a break in a power line. The speed of the microwave pulse depends on the dielectric constant K of the medium that surrounds and is in contact with the probe. Because of the significant difference between the dielectric constant of water and those of other constituents in soils, the speed of the pulse down the probe is highly dependent on soil-water content. When the microwave pulse reaches the end of the probe, the remaining energy is reflected back through the line. The apparent dielectric constant Ka of the soil can then be determined by

$$Ka = \left(\frac{tc}{L}\right)^2$$

where L is the length of the probe or "wave guides" (cm), t is the transit time (nanoseconds), and c is the speed of light (cm/nanosecond). The transit time is defined as the time required for the pulse to travel the length of the probe. Depending on the TDR unit, Ka is either calculated manually or, in some units, internally by the use of a zero/reset button and a visual display on the unit panel.

Suitability Criteria: **(1)** possible to merge with other devices; **(2)** durable; **(3)** horizontal and vertical installation possible with some exceptions; and **(4)** after initial tube installation, further monitoring scheme disturbance is not necessary.

Advantages and Disadvantages: **(1)** rapid, reliable, and repeatable; **(2)** minimal soil disturbance; **(3)** probe installation at any orientation; **(4)** no calibration necessary for rough estimates of water content; **(5)** may be automated; **(6)** no radioactive source needed; **(7)** calibration necessary for accurate water content values; **(8)** cable length limited to approximately 50 m; **(9)** expensive ($8500 for cable tester alone); **(10)** has not been in use for more than fifteen years; and **(11)** automation difficult to maintain in field conditions.

Electromagnetic induction (EMI)   An EMI device does not measure water content directly, but measures soil electrical conductivity (mS m^{-1}). Readings are taken manually, or a data logger is incorporated for multiple readings and storage. Once data is collected, it is commonly entered into a spreadsheet program or imported into a database. Water content is then estimated through changes in electrical conductivity. In actuality, the device indicates the position of the wetting front. The greatest benefit of EMI is to quickly show the wetting front to depths of 1–2 m, and to detect anomalies in the soil profile, such as incongruities due to rock or heavy metals, and so on. Therefore, EMI can be an invaluable tool in the final site-selection and characterization process.

An EMI device contains a transmitter coil that induces circular eddy current loops in the soil. The magnitude associated with a loop is directly proportional to the electrical conductivity of the soil around the area of the loop. Each loop generates a secondary electromagnetic field (emf) that is also proportional to the current flowing in the loop. The emf induced by each loop is intercepted by a receiver near the transmitter, which amplifies and forms the emf into an output voltage that is linearly related to soil electrical conductivity. An example of an EMI plot is shown in figure A1.2.

Suitability Criteria: **(1)** possible to merge with other devices; **(2)** durable (several decades experience); and **(3)** no installation required (i.e., nondestructive).

Advantages and Disadvantages: **(1)** rapid, reliable, and repeatable; **(2)** no disturbance; **(3)** only works from soil surface; **(4)** calibration necessary (device must be zeroed at least once for each measurement day); **(5)** cannot be automated; **(6)** no radioactive source needed; **(7)** does not give water content directly; **(8)** expensive (approximately $6500).

Electrical resistivity borehole tomography (ERBT)   ERBT is not a new concept, but is an extention of surface resistivity. However, technological advances in electronics have provided a valuable tool for long-term monitoring. An electrical current is passed from

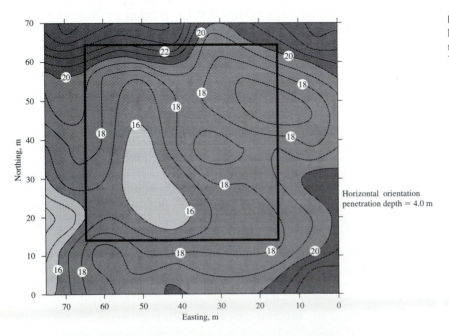

Figure A1.2   Contour map showing horizontal field response of EM-38 manufactured by Geonics Ltd. Data from Young et al. (1996)

inside a pvc source tube, usually through a series of copper plates, at specific depth intervals, to a detecting electrode some distance away. The detectors are usually also copper plates attached to a detection device placed within a similar tube. The detector measures the electrical resistivity of the soil, which is the inverse of electrical conductivity. Measurement is a function of both ionic and water content. Resolution of data is proportional to detection-well spacing; the lower the soil resistivity, the closer the wells must be. A commonly used depth is 1.5 times the distance between wells. For example, wells 10 m apart should be about 15 m deep. The data obtained is sorted via a modeling program and yields a three-dimensional plot of the wetting front. This procedure is valuable for long-term monitoring strategies, but is expensive. It is currently in use by the United States Nuclear Regulatory Commission on its Maricopa, Arizona field-studies evaluation site.

Suitability Criteria: **(1)** possible to merge with other devices; **(2)** durable (however, there is limited experience with the new vertical installation and measurement procedure); **(3)** flexible; and **(4)** service not required.

Advantages and Disadvantages: **(1)** measurements are large scale; **(2)** valuable for long-term studies such as landfills and low-level radioactive waste disposal sites; **(3)** detection wells should be installed during initial site installation; **(4)** measurements are not intrusive; **(5)** maintenance not required on soil probes; **(6)** cross-hole measurements must have sophisticated equipment and data analysis; **(7)** expensive (approximately $10,000); **(8)** requires on-site AC power; and **(9)** commercial units are not readily available.

Matric Potential Measuring Devices

Tensiometers   Tensiometers measure the energy status of water in the soil matrix (i.e., matrix potential). A tensiometer consists of a liquid-filled, unglazed porous ceramic cup connected to a pressure measuring device (such as a vacuum gauge or transducer). Once the ceramic cup is embedded into soil, soil solution can flow into or out of the cup via small pores in the ceramic. The flow will continue until the pressure potential of the liquid inside the cup equals the pressure potential of the soil water around the cup. If the column is completely filled with liquid, the matric potential will be zero or near zero, and no solution will flow into or out of the cup. As the soil dries, the solution will flow out of the cup and the top of the column will recede, creating an air pocket near the pressure measuring device and, thus, a vacuum. The vacuum created will be measured as a pressure or negative suction by the vacuum gauge or transducer. Examples of tensiometers used by the authors are illustrated in figure A1.3.

Suitability Criteria: **(1)** possible to automate and merge with other devices; **(2)** durable, but requires maintenance; **(3)** less flexible installation; **(4)** "Sisson" type tensiometer can be

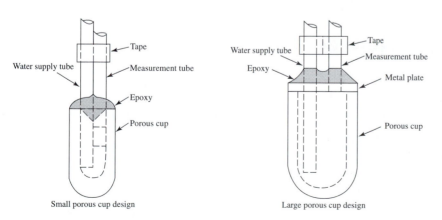

Figure A1.3   Tensiometer design. Small porous cup diameter (outside) is approximately 1 cm. Data from Stannard (1992)

Small porous cup design   Large porous cup design easily serviced and overcomes the depth limitation (less than 2–5 m) of most tensiometers. The Sisson tensiometer is referred to as the "advanced tensiometer," and was developed by personnel at the Idaho National Engineering Laboratory (INEEL). The Sisson design is shown in figure A1.4.

Advantages and Disadvantages: **(1)** works only in the moist range (0 to −85 kPa; 100 kPa = 1 bar); **(2)** nominal cost (approximately $37.00 each, regardless of type); and **(3)** frequent maintenance required.

Heat dissipation sensor (HDS) An HDS measures thermal diffusivity by applying a heat pulse to a heater located within the ceramic cup, then monitoring the temperature in the center of the cup before and after heating. The measurement system is generally a diode bridge circuit that measures electromotive force generated by the change in diode temperature in the sensing element as the heat pulse is applied. The higher the water content, the greater the thermal conductivity and diffusivity of the soils and hence, the lower the measured electromotive force. As the soil drains, causing the ceramic to desorb, thermal conductivity and diffusivity decrease, causing an increase in temperature in the reference matrix material. An HDS is illustrated in figure A1.5.

Suitability Criteria: **(1)** possible to automate and merge into a monitoring system; **(2)** durable; **(3)** flexible installation; and **(4)** not easy to remove or service.

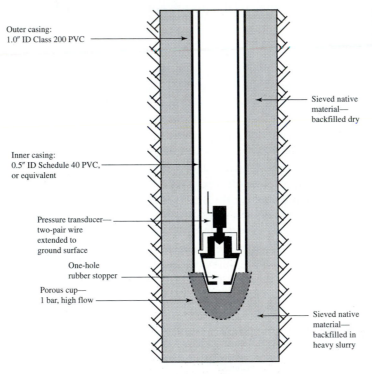

Outer casing:
1.0″ ID Class 200 PVC

Sieved native material—backfilled dry

Inner casing:
0.5″ ID Schedule 40 PVC, or equivalent

Pressure transducer—two-pair wire extended to ground surface

One-hole rubber stopper

Porous cup—1 bar, high flow

Sieved native material—backfilled in heavy slurry

Figure A1.4 Expanded view of "advanced tensiometer" system. Data from Young et al. (1996)

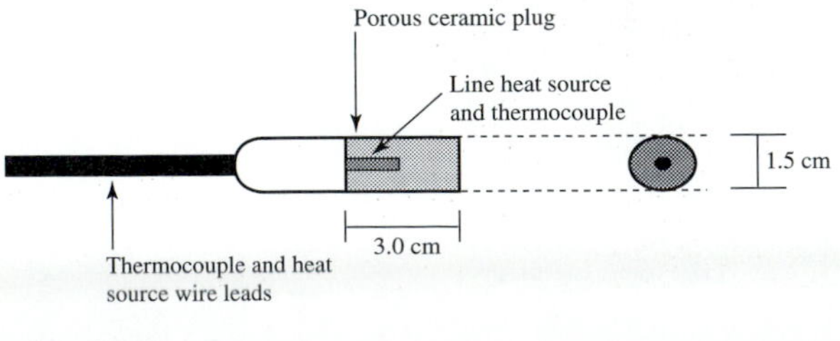

Porous ceramic plug

Line heat source and thermocouple

1.5 cm

3.0 cm

Thermocouple and heat source wire leads

Figure A1.5 Heat dissipation sensor

Advantages and Disadvantages: **(1)** point measurement; **(2)** relatively new, less than five years experience; and **(3)** wide range (10–1500 kPa)—however, some scientists report poorer results in wetter soils (0–0.5 kPa).

Thermocouple psychrometers   These devices are discussed in chapter 9.

Suitability Criteria: **(1)** possible to automate and merge into a monitoring system; **(2)** variable durability with varying soil conditions; **(3)** flexible installation; and **(4)** not easy to remove or service.

Advantages and Disadvantages: **(1)** complex measurement; **(2)** low durability; **(3)** point measurement; and **(4)** only operable in dry soils, from 50–3000 kPa.

Soil Solution Sampling Devices

Solution samplers   A solution sampler obtains a soil water sample through a porous wall. The common name for this device is "suction lysimeter." The solution sampler has two tubes that enter the device; one is an air pressure or vacuum tube, the other a fluid return tube. Once installed, a solution sample is easily obtained. A vacuum equivalent to soil pressure is applied via the vacuum tube, and both tubes are clamped or pinched shut. Once the pressure is equilibrated within the sampler, the clamps on the tubes are removed, a sample bottle is attached to the fluid return tube, and a pressure exerted on the vacuum tube to push the sample from the ceramic cup into the sample bottle. These devices come in both single- and dual-chamber designs, and are made from either ceramic or stainless steel. A dual-chamber, stainless-steel solution sampler is shown in figure A1.6.

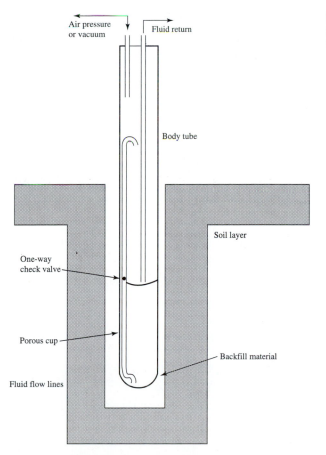

Figure A1.6   Diagram of dual-chamber solution sampler. Data from Wierenga et al. (1993)

Suitability Criteria: **(1)** possible to merge into a monitoring system; **(2)** durable; **(3)** limited installation flexibility; and **(4)** not easy to remove and service.

Advantages and Disadvantages: **(1)** not readily automated; **(2)** small capture area; **(3)** limited to shallow depths, usually less than 10 m; and **(4)** limited to soil water contents characteristic of 0–50 kPa pressure.

Sampling Wells   Sampling wells for saturated zone work may be necessary. However, many other texts present detailed information on this subject. Therefore, the reader is referred to Freeze and Cherry (1979).

Suitability Criteria: **(1)** not easily automated, but possible to merge into a monitoring system; **(2)** durable; **(3)** flexible installation; and **(4)** not easy to remove and service.

Advantages and Disadvantages: **(1)** point measurement; and **(2)** very repeatable.

Pressure Measurement Devices: Differential Pressure Transducers

There are two primary devices used for measuring soil pressure, vacuum gauges and differential pressure transducers. Commonly, each device is attached atop a liquid-filled column, which is in turn attached to a porous cup. Because vacuum gauges are relatively older technology, they shall not be discussed.

Differential pressure transducers are usually four-active-element piezoresistive bridge devices. When installed, and a pressure is applied, a differential output voltage proportional to that pressure is produced. This output voltage is commonly measured with a data logger. A differential transducer connection to the tensiometers illustrated in figure A1.3 is shown in figure A1.7.

Suitability Criteria: **(1)** possible to automate and merge into a monitoring system; **(2)** fairly durable; **(3)** flexible installation; and **(4)** fairly easy to remove and service.

Advantages and Disadvantages: **(1)** point measurement; and **(2)** transducers are delicate and require use of data-logging equipment.

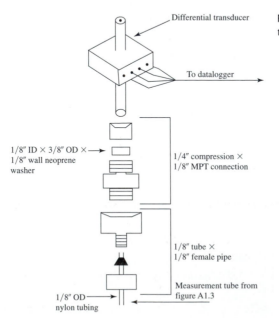

Differential transducer

To datalogger

1/8″ ID × 3/8″ OD × 1/8″ wall neoprene washer

1/4″ compression × 1/8″ MPT connection

1/8″ tube × 1/8″ female pipe

1/8″ OD nylon tubing

Measurement tube from figure A1.3

Figure A1.7   Differential transducer attachment to tensiometer shown in figure A1.3

APPENDIX 2

Mathematics Review

INTRODUCTION

Over the years, we have noticed that an increasing number of students taking our unsaturated zone hydrology courses have come from a broadening variety of scientific disciplines. Many of these students attend our classes to expand their skills in environmental science. Some have been out of school for several years, or are retraining in another discipline. A number of these students have a minimal mathematics background. Consequently, we have experienced a need to review basic math skills, so students can adequately understand the material presented in the text. For those readers who need a refresher in mathematics, we hope that this appendix will serve that purpose.

ALGEBRA

The basic rules for the arithmetical manipulation of a number system are known as algebra. They consist of two operations, addition ($+$) and multiplication ($\cdot$), and eleven laws relating those operations to the number system. For any a, b, or c within a number system, the following properties hold.

1. $a + b = b + a$ *Commutative Law of Addition*
2. $a \cdot b = b \cdot a$ *Commutative Law of Multiplication*
3. $a + (b + c) = (a + b) + c$ *Associative Law of Addition*
4. $a \cdot (b \cdot c) = (a \cdot b) \cdot c$ *Associative Law of Multiplication*
5. $a \cdot (b + c) = (a \cdot b) + (a \cdot c)$ *Distributive Law of Multiplication over Addition*
6. There exists a number a such that $a + b = b$ (and $a = 0$). *Existence of 0, the additive identity*
7. There exists a number a such that $a \cdot b = b \cdot a = b$ (and $a = 1$). *Existence of 1, the multiplicative identity*
8. For any number a, there exists a number b, for a and $b \neq 0$, such that $a \cdot b = 1$ (and $a = 1/b$). *Existence of reciprocals*
9. For any number a, there exists a number b such that $a + b = 0$ (and $b = -a$). *Existence of negative numbers*
10. If a is in the number system, and b is in the number system, then $a + b$ is in the number system. *Closure under addition*
11. If a is in the number system, and b is in the number system, then $a \cdot b$ is in the number system. *Closure under multiplication*

These laws enable one to structure a problem in such a way that it can be solved quickly and consistently. For example, $4 \div 7 \div 3$ can have two different answers depending upon the order in which it is solved, since division is not associative. However, if the problem is stated as $4 \cdot (1/7) \cdot (1/3)$, the answer remains the same no matter what order one chooses to solve it. Likewise, $4 - 7 - 3$ can have two different answers depending upon the order in which it is solved, whereas $4 + (-7) + (-3)$ is consistent. It is these laws of algebra that dictate how equations are set up and solved in the most advanced forms of mathematics.

The smallest algebra possible consists of two numbers (0,1) arranged in a matrix to meet the eleven stated requirements. However, that algebra is of little or no consequence to the applied sciences. The next-smallest number set that meets those requirements is the set of rational numbers. Unfortunately, that set does not contain such important numbers as π or e, or even $\sqrt{2}$; but those numbers *are* contained in the real number system, which is the number system that will be used as a basis for this review. The complex number system, which consists of the real and imaginary numbers (imaginary numbers are those that must be described in terms of i, or $\sqrt{-1}$), also forms an algebra. Although the complex numbers are of considerable importance in fields such as electrical engineering, the real number system is the one that is of primary importance in most applications.

SOME BASIC DEFINITIONS

Since mathematics is the language of numbers, it might help to review some of the basic definitions.

variable   A variable, which is usually represented by a letter of the Roman alphabet, can take on more than one value; or it may represent a single value that is yet to be determined. In the equation $3x + 2 = y$, x is the *independent* variable, and y is the *dependent* variable.

operator   A symbol that is used to denote a mathematical manipulation (operation). In essence, operators are mathematical shorthand: $+, -, \times, \div, \sqrt{}, \int, \nabla$, and Σ are all examples of operators. Likewise, the use of a superscript is a common operator. When one sees x^7, one knows that it is shorthand for the operation $x \cdot x \cdot x \cdot x \cdot x \cdot x \cdot x$.

function   A combination of operators, numbers, and/or variables that lead to a single value. In many cases, the exact mathematical expression may not be given, because it varies considerably or simply is not known. $f(x)$ indicates that the function f involves operations on the single variable x. $G(x, y, z)$ indicates that the function G involves operations on three variables. Neither of these examples gives an indication of the operations to be performed on the variables, only the number of variables involved. However, each function will have only one value for each set of variables. For example, if $G(1, 3, 2)$ is a function, it takes only one value for that number sequence $(1, 3, 2)$. It may have the same value for some other set of numbers, but for the sequence $(1, 3, 2)$ it must always take the same value. It cannot be 7 on Wednesdays and 12 on Fridays; once a value is assigned to a sequence of variables in a function it must remain constant for that sequence of variables. That strictness of definition is not always recognized in the physical sciences, but is dealt with by using a little common sense. As an example, consider Kepler's third law of planetary motion. The law can be stated mathematically as $T^2 = (4\pi^2 a^3 / GM)$; where T is the time for the period of the planetary orbit, a is the mean orbital radius, M is the mass of the planet, and G is the universal gravitational constant. By algebraic manipulation, the equation can be expressed in such a way as to show any of the variables as the dependent variable:

$$T = \left(\frac{4\pi^2 a^3}{GM}\right)^{1/2}; \quad a = \left(\frac{T^2 GM}{4\pi^2}\right)^{1/3}; \quad M = \left(\frac{4\pi^2 a^3}{GT^2}\right) \tag{A2.1}$$

Of the three equations given, two are functions and the other is not! T can have two possible values for any set of values for the independent variables a and M: the negative and the positive square roots of the equation. However, if the problem is set up with the understanding that T can only take positive values, the equation can become a function. $T(M, a)$ is a function for $T > 0$.

monomial	A monomial in x is of the form cx^n, where n is a nonnegative integer and c is a constant; for example, $-3x^5, 4x, 7x^{12}$.
polynomial	A series of monomials added together is a polynomial; for example, $7x^{12} - 3x^5 + 4x$.
linear polynomial	A polynomial is linear if it has the form $ax + b$, where $a \neq 0$.
quadratic polynomial	A polynomial is quadratic if it has the form $ax^2 + bx + c$, where $a \neq 0$. The roots of a quadratic polynomial can be found by using the equation $[-b \pm (b^2 - 4ac)^{1/2}]/2a$.
cubic polynomial	A polynomial is cubic if it has the form $ax^3 + bx^2 + cx + d$, where $a \neq 0$.

Equations are often plotted using the Cartesian coordinate system. When a linear polynomial of the form $y = mx + b$ is plotted with Cartesian coordinates, m is the slope (tangent) of the line, and b is the y-intercept (the value of y where the line crosses the y-axis). Likewise, for any two points $(x_1, y_1), (x_2, y_2)$, the slope of the line between them is $m = (y_2 - y_1)/(x_2 - x_1) = \Delta y/\Delta x$. Changes in b only move the line up or down the y-axis.

Although this definition of a tangent applies to straight lines, and the tangent of a curve can be found through differential calculus, there are often circumstances where the tangent to a curve can be approximated by a line. For that reason, it helps to understand calculus; one can approximate the tangent of a line at just a glance. It also helps one to recognize errors in calculation. Figure A2.1 gives an idea of the ranges of values for the tangents of lines.

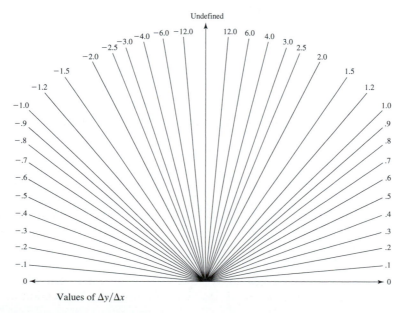

Figure A2.1   Tangent lines of the Cartesian coordinate system

TRIGONOMETRIC RELATIONS

The following discussion utilizes the diagram in figure A2.2. The radius of the circle, r, is 100 mm; the area of the circle is πr^2; the circumference of the circle is $2\pi r$. Angles can be measured in either degrees or radians, and there are $360°$, or 2π radians, in a complete circle. Let α_r = angle measured in radians, and α_o = angle measured in degrees. The relation can be described by: $\alpha°/360° = \alpha_r/2\pi$. This relation allows one to convert degrees to radians, or radians to degrees, for any angle. In figure A2.2, $\alpha°/360° \cdot 2\pi = \alpha_r$, and $\alpha_r \cdot r = z$. The area of the wedge is $\frac{1}{2}\alpha_r r^2 = \frac{1}{2}z \cdot r$. It can readily be seen that angle measurements in radians are important in measuring areas, and areas of portions of circles.

Trigonometric Functions

Considering the triangle formed by the radius r, height y, and base of length x, the trigonometric functions are as follows.

$\sin \alpha = y/r$

$\cos \alpha = x/r$

$\tan \alpha = \sin \alpha/\cos \alpha = y/x$, undefined for vertical line, $\alpha = 90°$, $\pi/2$ rads

$\csc \alpha = 1/\sin \alpha = r/y$, undefined for horizontal line

$\sec \alpha = 1/\cos \alpha = r/x$, undefined for vertical line

$\cot \alpha = 1/\tan \alpha = x/y$, undefined for horizontal line

The area of the triangle is $\frac{1}{2}xy$.

Pythagorean theorem: $x^2 + y^2 = r^2$.

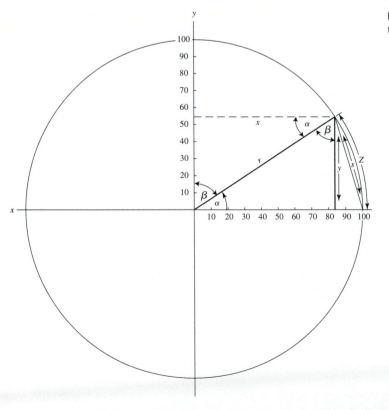

Figure A2.2   Relation of trigonometric shapes

The length of the chord (secant segment) labeled s can be found by using the law of cosines: $s^2 = 2r^2 - 2r^2 \cos \alpha \rightarrow s^2 = 2r^2(1 - \cos \alpha) \rightarrow s = r(2(1 - \cos \alpha))^{1/2}$ for $s \geq 0$.
The area of the triangle formed by the chord and the two radii is $\frac{1}{2} yr$.
The area within the arc and chord is $\frac{1}{2} zr - \frac{1}{2} yr = \frac{1}{2} r(z - y)$.
Important Note: If any value for an angle is not specifically given in degrees, it is assumed to be in radians. For example, sin 30 is assumed to be sin 30 rads, not sin 30°!

The following laws and identities are basic, and are not restricted to figure A2.2.

Law of Cosines

For any triangle with sides of length $a, b,$ and c, if θ represents the angle opposite side c, then $c^2 = a^2 + b^2 - 2ab \cos \theta$.

Law of Sines

For any triangle with sides of lengths $a, b,$ and c, if θ represents the angle opposite c, α represents the angle opposite a, and β represents the angle opposite b, then $a/\sin \alpha = b/\sin \beta = c/\sin \theta$.

For any α, β, θ (except as noted), the following hold.
$\sin^2 \alpha + \cos^2 \alpha = 1$
$\sin(-\theta) = -\sin \theta$
$\cos(-\theta) = \cos \theta$
$\sin(\alpha - \beta) = \sin \alpha \cos \beta - \cos \alpha \sin \beta$
$\cos(\alpha - \beta) = \cos \alpha \cos \beta + \sin \alpha \sin \beta$
$\sin(\alpha + \beta) = \sin \alpha \cos \beta + \cos \alpha \sin \beta$
$\cos(\alpha + \beta) = \cos \alpha \cos \beta - \sin \alpha \sin \beta$

By substituting 90° ($\pi/2$ radians) for α in the preceding four equations, we obtain the following.
$\sin(90° - \beta) = \cos \beta$
$\cos(90° - \beta) = \sin \beta$
$\sin(90° + \beta) = \cos \beta$
$\cos(90° + \beta) = -\sin \beta$

If $\alpha = \beta$ in those same four equations, then the following is true.
$\sin(2\alpha) = \sin \alpha \cos \alpha + \cos \alpha \sin \alpha = 2\sin \alpha \cos \alpha$
$\cos(2\alpha) = \cos^2 \alpha - \sin^2 \alpha$
Dividing $\sin^2 \theta + \cos^2 \theta = 1$ by $\sin^2 \theta \rightarrow 1 + \cot^2 \theta \csc^2 \theta$
Dividing $\sin^2 \theta + \cos^2 \theta = 1$ by $\cos^2 \theta \rightarrow \tan^2 \theta + 1 = \sec^2 \theta$

All of the useful trigonometric identities can be derived by similar substitutions, but these are the most important.

GEOMETRIC RELATIONS

The *area of a triangle* is $\frac{1}{2} bh$, where b is the length of the base and h is the triangle's height. Please note that h is the same length as the triangle's leg only in the case of a right triangle.
A *right triangle* is one in which one of the angles is equal to 90°.

The total inside angles of a triangle sum up to 180°.

An *isosceles triangle* is one in which two of the sides are equal. Note that by the law of sines, that means that two angles are equal also.

An *equilateral triangle* is one in which all three sides are equal. Once again, by the law of sines all three angles must also be equal, and therefore each is 60°.

The *total area of a polygon*, as well as the measurements of its inside angles, can be determined by constructing a series of triangles within the polygon and using the above relations.

A *parallelogram* is a four-sided polygon with its opposing sides parallel and equal in length. It can be formed by two identical triangles, therefore its area is $2(\frac{1}{2}bh) = bh$. Its inside angles total $2 \cdot 180° = 360°$. Squares and rectangles are special cases of parallelograms.

An *ellipse* has the algebraic form $x^2/a^2 + y^2/b^2 = 1$; where a is the length from the center of the ellipse to its furthest point (also known as the "semi-major axis"), and b is the length from the center of the ellipse to the closest point (also known as the "semi-minor axis"). (This definition assumes that the widest portion of the ellipse coincides with the x axis, which is conventional.) It can also be seen that when $a^2 = b^2$, the equation of an ellipse reduces to $x^2 + y^2 = a^2$, which is the equation of a *circle;* showing that the circle is a special case of an ellipse. The geometric definition of an ellipse is that it is a set of points such that the sum of the distances of each point from two fixed points is a nonnegative constant. The two fixed points are known as the foci. It is that definition that leads to the traditional means of constructing an ellipse by using two thumbtacks, a length of string and a pencil. One focus lies on each semi-major axis, and each is the same distance from the center of the ellipse. If the distance of each focus from the center is given as c, then the axes and foci are related by the following equation: $b^2 + c^2 = a^2$. Once again, in a circle, c reduces to zero since the foci coincide at the center. A line that is tangent to an ellipse at any point will form the same angle to a line drawn to either of the two foci. The area of an ellipse is πab.

Volume of an elliptic cylinder = πabh (where h is the height)

Volume of a cone = $\frac{1}{3}\pi r^2 h$.

Surface area of a cone = πrl (where l is the distance along the edge of the cone from its apex to its base).

Volume of a sphere = $\frac{4}{3}\pi r^3$.

Surface area of a sphere = $4\pi r^2$.

POWERS AND ROOTS

One area of mathematics that is actually algebra, but is difficult enough to cause problems even for those who may have taken calculus courses, is the concept of *powers* and *roots*. The following rules are basic to an understanding of the subject.

$x^0 = 1$

$x^1 = x$

$x^2 = x \cdot x$

$x^n = x_1 \cdot x_2 \cdot \cdots \cdot x_n$

Let $\sqrt[n]{a} = x$, then $x^n = a = x_1 \cdot x_2 \cdot \cdots \cdot x_n$, and $x = a^{1/n}$

$x^{-n} = 1/x^n$

Let $r = x^m$ $s = x^n$, then $r \cdot s = x^{m+n}$

$(x^m)^n = x^{m \cdot n}$

For any $y > 0, x > 0$, there is some number n such that $x^n = y$. This fact leads to the concept of logarithms.

LOGARITHMS

For all numbers greater than zero, logarithms provide a simplified way to deal with powers and roots. Logarithms are expressed in terms of a base. In the example given in the last sentence of the previous paragraph, n would be the logarithm of y in base x. The mathematical expression would be $\log_x y = n$. The two most common bases for logarithms are 10 and e. If a number is expressed as $\log y$, it is understood that it is meant as $\log_{10} y$. If a number is expressed as $\ln y$, it is understood that it is meant as $\log_e y$. Therefore, $\log y = n$ means that $10^n = y$, $\ln y = n$ means that $e^n = y$. Some of the rules that apply to logarithms are as follows.

$$\log_x x = 1$$
$$\log_x x^a = a$$
$$\log xy = \log x + \log y$$
$$\log x/y = \log x - \log y$$

The relation between ln and log ($\log_{10}$) is as follows

$$\ln 10 = 2.30258 \ldots$$
$$e^{2.30258} = 10$$
$$e = 10^{0.434294\ldots} \text{ (since } 1/2.30258 \ldots = 0.434294 \ldots \text{)}$$
$$\log e = 0.434294 \ldots$$

For any $x > 0$, $\ln x = 2.30258 \log x$.

This relation is extremely important in the physical sciences. When a differential equation is solved, the answer is often expressed in terms of ln, but experimental data is more conveniently plotted on $\log_{10}$ graphs.

CALCULUS

Differential and Integral Calculus

There are two essential areas of calculus. One is known as *differential calculus,* and is concerned with finding the tangent to a curve at a particular point. The other is known as *integral calculus* and is concerned with finding the area beneath a curve between two points. Although this is a simplification in two dimensions, if one remembers to look at calculus in this simple way it can help to make some problems very easy to solve. That is especially true in the area of applied calculus.

As stated above, differential calculus is concerned with finding the tangent to a curve at a point. The slope of a tangent line at a particular point on a curve is the instantaneous rate of change. The slope of that line is known as the derivative of the equation. The mathematical representation for that instantaneous rate of change, or derivative, is dy/dx. Remember that $\Delta y/\Delta x$ is the representation for the slope of a straight line. There are times when a curve does not change dramatically over a distance; in such an instance dy/dx may actually approach $\Delta y/\Delta x$, making it easier to approximate dy/dx without knowing the actual formula of the curve. That is very important to a scientist, since the actual formula for an experimentally-derived curve may be solvable only through numerical analysis of data. There can be many instances where a scientist may want approximate values before spending long hours developing a computer program to analyze data. A scientist may also desire to assess the data presented by someone else without spending huge amounts of time.

The integral for a curve between the points a and b, which is the area within (beneath) the curve between those points, is represented by $\int_a^b$. In the Cartesian coordinate system, it would be represented by $\int_a^b$ (equation of curve) dx. The representation that shows the two points between which the curve is being evaluated is called the definite integral. The indefinite integral $\int$ (equation) dx, is also called the antiderivative. That is because it undoes the effect of the derivative. The integral of a derivative does not always yield the original equation completely, but it is close enough for the purposes of calculus. It is not the intent of this discussion to go into an explanation of integrals or derivatives, so if a more rigorous explanation or review is needed there are literally scores of calculus books available in any library. The important concept here is that a definite integral represents the area beneath two points on a curve, and the antiderivative "undoes" the effect of the derivative.

If one knows the exact equation of a function, the finding of a derivative at a particular point is a mathematical exercise in differential calculus. However, the purpose of performing experiments is to find unknown values. An experiment may obtain a whole series of values that can be linked together in a curve sketched by the researcher, but unless the curve is very simple it may not be easy to identify its underlying mathematical equation (function). What is a researcher to do in order to find the equation, or its derivative at a certain point? A French engineer named Fourier established that a curve (or even a series of segments of curves) of any shape can be matched against segments of trigonometric functions (sine or cosine), as long as the curve is a function. But Fourier analysis is not a simple exercise to be undertaken with a notepad, pencil, and hand calculator. If all a scientist needed was a value at a particular point, it could be read from the curve and no further calculations would be needed. Yet, quite often the scientist needs to find the derivative at a point, or the derivatives at several points. When a rough approximation is adequate, dy/dx can be obtained by using the chart given in figure A2.1 and finding the value of $\Delta y/\Delta x$ that most closely matches the slope of the curve in the immediate vicinity of the point.

The value of the definite integral can be approximated by plotting the curve of experimental data on quadrille paper and determining the number of squares contained under the curve between the two points. In calculus texts one might find such methods of approximation listed under the category of "trapezoidal approximation."

Consider the task of designing a mathematical function that will describe a trip from the Colorado capitol building in Denver to the summit of Mount Evans (figure A2.3). The capitol will serve as a point of origin, East–West will be the x-axis and geographic North–South the y-axis. It will be impossible to describe the trip as a function in terms of either x or y alone, because the road loops around with various switchbacks as one travels in the mountains; as a consequence East–West or North–South lines can be drawn in such a way that they can pass over the road several times. One may be tempted to make the assumption that the trip can be described in terms of x and y, or $f(x, y)$, as the exact mathematical equation is unknown. But consider what happens when one encounters a cloverleaf intersection. The vehicle turns steeply to the right and within a few seconds one is at precisely the same coordinates as a few seconds before—but not in the same place. The vehicle is approximately eighteen feet above where it was. The contemplated function $f(x, y)$ gives two values at one point; therefore, it cannot be a function after all. In order to meet that difficulty, the function can be changed to $f(x, y, z)$, with z the elevation above, or below, that at the capitol. One still does not know the exact mathematical equation for the journey, but a specific set of values (x_1, y_1, z_1) will give just one location, so there is a function $f(x, y, z)$. For any set of values, one can find the exact position on a topographic map. The topographic map is a two-dimensional representation of that three-dimensional function. The mathematical equation will be tremendously involved and very difficult to determine. Now consider the following philosophical question: Do you really need to know the exact mathematical expression in order to understand the function, or do you actually understand the function better without knowing the exact mathematical expression?

Next consider what is involved in the trip back down from the summit of Mt. Evans. Suppose that one lane has been closed for construction so that the trip back down exactly retraces the route up for at least part of the way. Given a set of coordinates (x_1, y_1, z_1) that

Figure A2.3 Mount Evans, 45 km west of Denver, Colorado (not to scale). Draw imaginary x- and y-axes through highway 5 in proximity of the summit of Mount Evans. If the y-axis is drawn in the appropriate location, one will observe that for each x, there can be multiple locations along y, and vice versa. Map scale is 1:100,000. Source: United States Geological Survey. Map name: Denver West—Colorado

falls somewhere on that section of road; what does $f(x_1, y_1, z_1)$ give for an answer? Is the vehicle on its way to the summit or on its way back down at that point?

There might be a temptation to suggest that time is another variable. But then what happens if the occupants of the vehicle stop to admire the view, or are held up in the construction zone? Or what if someone left a pair of binoculars behind, and the vehicle turns around and heads back up again? If someone told you that it was precisely three hours, twenty-seven minutes and fourteen seconds since the vehicle left the capitol, could you give its exact location? It is obvious that time alone is not what is needed. Reconsider for a moment the question that must be resolved. One only needs to determine whether the vehicle is moving in the direction toward the summit or away from it. Movement implies change in position with time. This is the same as finding a derivative in calculus. In a one-dimensional problem along the x-axis, this would be solved by finding the derivative dx/dy or dx/df; but this is a three-dimensional problem. You might find a short distance where the road is completely level and heads precisely North–South or East–West, but at most times it will not. Consider for a moment what happens when there is vertical change. Unless the driver propels the vehicle over the edge of a cliff there will always be a considerable rate of change along at least one of the other major axes. The variables are interlinked in the underlying function. Change cannot occur in z without change also occurring in x or y (in this case). Even in the x–y plane there are many sharp curves and changes in direction, so their relation is much more involved than that of a straight line, $mx + b = y$. Even if the vehicle were to move at a constant velocity relative to the road surface, its rate of change with respect to any of the three axes would change constantly. If one considered the rates of change along each of the three axes at a particular point, while keeping each of the other variables a constant, one would have found the partial derivatives:

$$\frac{\partial f}{\partial x}, \quad \frac{\partial f}{\partial y}, \quad \frac{\partial f}{\partial z} \tag{A2.2}$$

The tangent to the curve of the road at any point is the set of partial derivatives at that point. In the example given above, if time enters into the format, the relation between the regular derivatives and the partial derivatives would be as follows.

$$\frac{df}{dt} = \frac{\partial f}{\partial x}\frac{dx}{dt} + \frac{\partial f}{\partial y}\frac{dy}{dt} + \frac{\partial f}{\partial z}\frac{dz}{dt} \tag{A2.3}$$

The instantaneous rate of change, df/dt, is the velocity of the vehicle relative to the road surface, and dx/dt, dy/dt, and dz/dt are the regular derivatives relative to the respective axes. The velocity has components that could be measured going off in any direction; the function given above shows where the velocity is greatest—along the path that the vehicle is following. If the vehicle were to follow a straight, level path along an East–West line, then there would cease to be y and z components and there would be no partial derivatives, and df/dt would be equal to dx/dt. This lengthy discussion has led to the subject of vector calculus.

Vector Calculus

Consider for a moment a pollutant moving through the soil or air. It is not possible to know the exact mathematical equation that governs the pollutant's movement, but it is possible to measure its movement at any one point (i.e., as through Δx in figure A2.4). There are an infinite number of points at which the pollutant's movement can be measured. At each one of these points the partial derivatives (the direction of movement measured as components of the three axes) may be different. Taken together, these points are known as a gradient field. The set of partial derivatives at any given point is known as the gradient. The mathematical symbol for the gradient is grad $f(x, y, z)$, or more commonly as $\nabla f(x, y, z)$. It is vital to remember that it is possible to know by experimental measurement the value of $\nabla f(x, y, z)$ at a point without knowing anything at all about the underlying function $f(x, y, z)$.

Nabla (in older usage; del, or del operator in newer usage), symbolized by ∇, is the most important operator in applied vector calculus. $\nabla f(x, y, z) = $ grad $f(x, y, z)$ is the gradient, or the set of partial derivatives at a point. $\nabla \cdot f(x, y, z) = $ div $f(x, y, z)$ is the divergence. ($\nabla \cdot$ is read as "nabla dot"). $\nabla \times f(x, y, z) = $ curl $f(x, y, z)$ is the curl. ($\nabla \times$ is read as "nabla cross"). An extremely simplified explanation of this notation is: $\nabla f(x, y, z)$, which is the tangent to a gradient field at a particular point (the direction of the maximum rate of change broken into its three components). $\nabla \cdot f(x, y, z)$ is the normal to the gradient field (the rate of change 90°

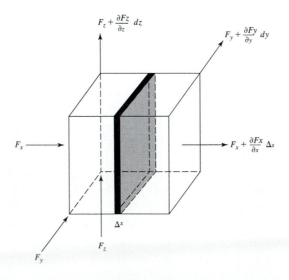

Figure A2.4   Flux across an interval, Δx, within a three-dimensional box

to the gradient field, or the expansion of a field at a particular point). Once again, it would be broken into the three components. If the field is confined within boundaries, then $\nabla \cdot f(x, y, z) \equiv 0$. $\nabla \times f(x, y, z)$ can be considered to be the measure of rotational motion about a point in the gradient field. (Another important, but complicated, use of ∇ is $\nabla^2 f(x, y, z)$, the LaPlacian of a function. The LaPlacian combines the important mathematics of differential equations and vector calculus.)

The actual calculation of these vector quantities using the underlying mathematics would be very difficult. It is very fortunate that the quantities can be measured without even knowing the underlying equation. This very short discussion of vector calculus should enable you to understand what would otherwise be very intimidating mathematical formulas that you will come across. A short example follows. Take a look at it, and see if you can interpret it by yourself.

$$\frac{\partial \theta_v}{\partial t} = \nabla \cdot (D_{\theta v} \nabla \theta_i) + \nabla \cdot (D_{Tv} \nabla T) + E \tag{A2.4}$$

This equation simply says that the partial derivative with respect to the variable time (you will note that it is not concerned with any of the other partials) in the function θ_v is the sum of the other three entries. The two $\nabla \cdot$ components are the expansion of the formula contained within the brackets. Although we do not have the benefit of knowing what each variable stands for, the way that they are being used, as well as their being upper case letters, would seem to indicate that each letter represents some equation that is unknown to the person who created the formula. Therefore, each of these components represents an experimentally derived number that the individual believes to be a function, as indicated by the use of capital letters. ∇T would represent the temperature gradient, a number that should be easily obtained. Likewise, $\nabla \theta$ represents some measurable gradient. Even without knowing what each of the variables means, the overall equation is going to be easily solved. Once the nature of the function is understood, an experiment can be designed to derive an answer, and you will not have to be an expert mathematician in order to solve it!

Differential Equations

One of the most common experimentally obtained results is known as the ordinary differential equation. A differential equation occurs when a measured rate of change is equal to a function of its variables: $dy/dx = f(x, y)$ is an example of how this might be represented in mathematical notation. Differential equations can get very complicated, so this discussion will center on the easiest and (fortunately) most common type, the form $dy/dx = ky$, where k is some constant. By separating the variables one obtains $dy/y = kdx$. Taking the antiderivatives of each side, $\int dy/y = \int kdx \rightarrow \ln |y| = kx + c$. This can also be expressed as $y = e^{kx+c}$, or $y = ce^{kx}$. If one knows the initial condition, one can solve for c.

An equation of the form $y = e^x$ is a hyperbolic function. The slope of its curve changes very dramatically when plotted using Cartesian coordinates. That makes it very difficult, if not impossible, to solve the equation by plotting it on a standard graph. Fortunately, that problem can be overcome by plotting the data on semilogarithmic graph paper. A function of the type under discussion will yield a straight line when plotted on semilogarithmic paper. The semilogpaper is ruled using base 10 cycles (each cycle corresponds to 10 times the value of the previous cycle), but the differential equation is expressed in powers of e. The difference in expressions can be resolved by remembering that very important relation between log and ln: $\ln x = 2.3 \log x$. If that relation is not etched into your memory already, it is a good idea to have it written down in a convenient spot. Whenever you are plotting data on logarithmic or semilogarithmic graph paper in order to solve a differential equation expressed in terms of ln, you will need to substitute $2.3 \log x$ for each place that $\ln x$ appears. If you remember that relation, it becomes a very simple matter to solve such problems graphically. Consider the differential equation $y\Delta t = \ln x_2 - \ln x_1$. This can be simplified to $y\Delta t = \ln x_2/x_1$, or $y = (2.3/\Delta t) \log x_2/x_1$. If $x_2 = 10x_1$, then the equation simplifies to $y = 2.3/\Delta t$, and you have only to read the difference in t across that same cycle in order to solve the problem.

There are also partial differential equations. LaPlace's equation, $\nabla^2 f(x, y, z) = 0$, is an example of a partial differential equation. It is of some importance in solving physics problems, so a few of its forms are given here:

$$\frac{\partial^2 f}{\partial x^2} + \frac{\partial^2 f}{\partial y^2} + \frac{\partial^2 f}{\partial z^2} = 0 \tag{A2.5}$$

$$\frac{\partial}{\partial x}\left(\frac{\partial f}{\partial x}\right) + \frac{\partial}{\partial y}\left(\frac{\partial f}{\partial y}\right) + \frac{\partial}{\partial z}\left(\frac{\partial f}{\partial z}\right) = 0 \tag{A2.6}$$

$$\text{div grad } f = 0 \tag{A2.7}$$

$$\nabla \cdot \nabla = 0 \tag{A2.8}$$

The subject of partial differential equations is very involved, and well beyond the scope of this short mathematics review. The most important thing to remember is that partial differential equations are, as a rule, more complicated than ordinary differential equations, but they still have solutions.

A simple example follows. Find a solution for the LaPlace equation

$$\frac{\partial^2 f}{\partial x^2} + \frac{\partial^2 f}{\partial y^2} + \frac{\partial^2 f}{\partial z^2} = 0 \tag{A2.9}$$

There are an infinite number of solutions for the equation as is stated in this simple form. A simple solution that should be intuitively evident is $x^2 + y^2 - 2z^2 = f(x,y,z)$. To verify, we find

$$\frac{\partial f}{\partial x} = 2x, \qquad \frac{\partial f}{\partial y} = 2y, \qquad \frac{\partial f}{\partial z} = -4z \tag{A2.10}$$

Then,

$$\frac{\partial}{\partial x}\left(\frac{\partial f}{\partial x}\right) = 2, \qquad \frac{\partial}{\partial y}\left(\frac{\partial f}{\partial y}\right) = 2, \qquad \frac{\partial}{\partial z}\left(\frac{\partial f}{\partial z}\right) = -4 \tag{A2.11}$$

so that

$$\frac{\partial^2 f}{\partial x^2} + \frac{\partial^2 f}{\partial y^2} + \frac{\partial^2 f}{\partial z^2} = 2 + 2 - 4 = 0 \tag{A2.12}$$

establishing that our intuition was correct, and the equation is a solution to the LaPlace equation.

It should be evident that an infinite number of functions are solutions to the LaPlace equation. In fact, they form a subset of all possible functions. Keep in mind that LaPlace equations are not all as easily recognized as the one given above.

Vectors

When vectors are written out, they are conventionally given in terms of i, j, and k. These are the unit vectors in the three axis directions, and using i, j, and k avoids any confusion with the variables x, y, and z. If a vector can be expressed as $xi + yj + zk = v$, then $|v| = (x^2 + y^2 + z^2)^{1/2}$ is the magnitude of the vector. The dot product of two vectors u and v is $u \cdot v = |u||v| \cos \theta$, where θ is the angle between the two vectors. The dot product is therefore a scalar, or a nonvector number. The cross product of two vectors is another vector. If $u \times v = w$, then $|w| = |u||v| \sin \theta$, where θ is the angle between u and v. Suppose that $u = ai + bj + ck$ and $v = di + ej + fk$. Then, $u \cdot v = ad + be + cf$; $u \times v$ is more complicated.

It can be expressed as the minors of the determinant:

$$\begin{vmatrix} i & j & k \\ a & b & c \\ d & e & f \end{vmatrix}$$

(A2.13)

which is $(bf - ec)i - (af - dc)j + (ae - db)k$.

Once again, this is intended as a very brief review of the subject. The review problems at the end of this appendix should help prepare you for the course, and help you remember prior mathematics course work. We will begin with a (fanciful) example of some theoretical mathematics, then some easy questions, followed by more difficult problems that come from actual experiments. If you run into difficulty, consult any mathematics book at your disposal.

THE TASTE FUNCTION

Suppose that you have decided to make blueberry pancakes for breakfast, and while doing so it occurs to you that there may be some "taste function" that could exactly describe how you want your pancakes prepared. Is there such a function? Think for a moment: If you could control all of the variables that you might think of, would your blueberry pancakes taste exactly the same each time that you made them? The answer should be yes, or at least that is the presumption made by cookbooks. If you used exactly the same ingredients, mixed together in exactly the same way in the same ratio, cooked in the same pan, over the same heat source for the same amount of time—if you controlled all of the variables—then you would certainly expect the same results each time. It is reasonable to make the beginning assumption that there is a taste function. We can test that assumption by designing a recipe that we will use each time. However, if the same recipe gives us two different results—if the pancakes taste wonderful one time and horrible the next—then it is not a function. (Unless, of course, you have a physiological problem that affected your taste buds, but that is just a problem with the subjective nature of measuring our taste function.)

Let us assume that there is such a taste function. We are all scientists; how would we describe the function in a mathematical sense? First, we would assign some representations for the variables that we believe control the taste of the blueberry pancakes. We could call the blueberries b, the pancake mix p, the milk m, and the eggs e (this is a hypothetical mix, by the way). There will be a specific amount of each ingredient that will be used in the mix, and then the ingredients will be added together. So far this is simple algebra, but since we have not yet determined the amounts, the mix will be abbreviated as $f(b, m, p, e)$. Next, we will want to stir the mix. To simplify things a bit, assume that we are going to slowly pour the ingredients into a bowl while stirring, and that as we stir the mix, it is pouring out of a hole in the bottom of the bowl into a second bowl. The movement of the mix through the first bowl is a gradient field; you could theoretically measure its rate of movement with respect to any of the three axes. The rate of movement toward the second bowl at any one time at any one point would be the gradient, or $\nabla f(x, y, z)$. The mix would be rotated by our stirring it, so that there is a rotational component to the movement of the mix through the bowl. The amount of rotation about any point would be represented by $\nabla \times f(x, y, z)$. Would there be a $\nabla \cdot f$ component? Not really; the mix is constrained by the boundary of the bowl—it is prevented from spreading outward by the sides of the bowl. Suppose that we get to actually heating the pancake. The heat moves upward through the pancake, but it also moves outward as it's moving up. $\nabla \cdot f(x, y, z, T, t)$ is the theoretical expression for that vital component of the taste function.

You can see that we used a mathematical shorthand to describe what is actually a very simple act. If you were to string the components together to form a taste function, it would look very intimidating to a person unfamiliar to the procedure, or unfamiliar with the nature of your variables—and this is not even a well-thought-out model! Do not let the strings of mathematical symbols intimidate you. Think of the equations as abstract shorthand rather than as an actual equation that you have to solve. These equations from vector calculus are just another scientist's way of expressing what he believes to be happening. Feel free to question the logic that was used by the scientist.

MATHEMATICS REVIEW PROBLEMS

Answers to these problems are given in the instructors solution manual

A2.1. How would you go about establishing that the following formula is, or is not, a function?

$$Q_c = \sum_{\alpha=1}^{\infty} - \frac{\overline{v_w}}{4\pi} \left[\int_0^{D^{(\alpha)}} \frac{D^{(\alpha)}}{\varepsilon} \, dD^{(\alpha)} - \frac{(D^{(\alpha)})^2}{2} - T \frac{\partial}{\partial T} \int_0^{D^{(\alpha)}} \frac{D^{(\alpha)}}{\varepsilon} \, dD^{(\alpha)} \right] \frac{s \times 18}{(3.286 \text{ Å})^2 \, N_A} \quad (A2.14)$$

where $D^{(\alpha)}$ is the electric displacement at phase α; ε is the dielectric constant of water; T is temperature; s is surface area per gram of sample (m^2g^{-1}); N_A is Avogadro's number (6.023×10^{23}); and v_w is the partial volume of water.

A2.2. You discover a scientific paper in which the author states that the ability of a soil to conduct water is dependent upon three factors: **(1)** The price of pretzels in Poughkeepsie; **(2)** the number of hairs on the bearded lady's chin; and **(3)** the number of goats on Old McDonald's Farm. Do you consider it likely that this really is a function? Why or why not? How would you go about proving or disproving it? (Although the topic may sound laughable, be as specific as possible in answering the last part of this question.)

A2.3. In order for an equation to be differentiable over an interval, it must be both a function, and continuous over the interval. (The difference between $\int$ and Σ is due to the difference between continuous and discrete phenomena. The amount of space contained within a jar would be determined by using $\int$, because one cannot "count" the space even though it is a very small number. The number of stars contained within the universe would be expressed by using Σ, because even though the number may be infinite, it is still countable—if you lived long enough and could count that high.) Most of the models that are used in this course are calculus-based, with plenty of dx, $\int, \partial, \nabla$, and their friends. If you could slice a cross-section through the area of flow, would the flow of water through a boulder field be continuous? What about a layer of cobbles? Of gravel? Of sand? Obviously, there are an infinite number of points where the water flow could not be measured under each of these circumstances. Yet water flow is one of those factors that is described by using calculus. The question here is, are all of these scientists daft or is there some reason to use calculus to describe a discontinuous function? (This is not a mathematical question, but a logical one. Approach it as would a philosopher.)

A2.4. One of the most important measurable quantities encountered during this course is the hydraulic conductivity of soils—their ability to allow water to be transmitted through them. Henry Darcy designed an apparatus to measure K, the hydraulic conductivity, by filling an enclosed bed with sand and measuring how fast the water flowed through the pipe (chapter 7). What important mathematical operation was negated by this experimental design? Do you think that this might have some effect upon the experimental results?

A2.5. Darcy's experiment required that the soil be saturated in order to measure hydraulic conductivity. The measured result, called the saturated hydraulic conductivity, is represented by K_s. One method of measuring K_s is known as the "falling head method," (you will be spared the details of the method for now, but you can practice the mathematics that are involved). Suppose that $K_s = -a \, \Delta z \, dH/H \, dtA$, where a, Δz, and A are constants. Let H_0 be the value of H at t_0, and H_1 the value of H at t_1. Show that $K_s = 2.3a \, \Delta z/t_1 A$ when $H_0 = 10H_1$ and $t_0 = 0$. This is an ordinary differential equation. If you find this problem difficult you may want to review your mathematics a bit more. This type of differential equation is normally explained in the first or second semester of calculus.

A2.6. Let $\Delta z = 12$ cm, $a = 0.2$ cm^2, $A = 50$ cm^2, $H_0 = 30$ cm. You perform an experiment and obtain the following results.

t (in minutes)	1	5	10	15	20	30	40
H (cm)	28.8	27.3	25.0	23.1	21.1	17.6	15.0

Solve this by graphing the results and using the appropriate formula from problem A2.5. Express K_s in cm sec^{-1}.

A2.7. Another important property encountered in this course is diffusivity (see chapter 8). We will use the laboratory method of Bruce and Klute for obtaining diffusivity values to practice practical mathematics. The Bruce and Klute method uses the equation

$$D(\theta') = -\frac{1}{2t}\left(\frac{dx}{d\theta}\right)\int_{\theta_i}^{\theta_x} x\, d\theta \qquad (A2.15)$$

where t is time in seconds for the entire experiment to be performed; θ is the water content of the porous medium at a sample location (it is a percentage, and therefore unitless); and x is the distance (cm) of the sample from the water source. An experiment obtains the following results.

x	0	1	2	3	5	10	15	20	25	30	32	33	35
θ	.44	.43	.44	.43	.42	.40	.38	.36	.35	.30	.24	.18	.02

θ_i, the water content of the initial "dry" soil, is 0.002; t is forty-five minutes. What is the diffusivity ($D(\theta')$) when the soil has a water content of 40 percent? When it has a water content of 33 percent? 25 percent? (θ_x is shorthand for "the value of θ at x.") This problem is easy to solve if you remember the definitions of $\int$ and the derivative. Final hint: solve it graphically. Feel free to use the values of dy/dx, and a light table, if you need to.

A2.8. As an unsaturated-zone scientist with good mathematics skills, you have been asked to review the paper of a prominent scientist who made a slug test of a well in an unconfined aquifer. Being conscientious, you actually work through the mathematical formulas given, rather than simply peruse them and accept each at face value. Upon your initial reading, you detect that the author chose the wrong formula to integrate, of the two formulas given. You will need to simplify these formulas, and express both so that K is the dependent variable, and y is the independent variable. Underline in an introduction to your answer the basis for your selection of the correct formula that should be used, i.e., either A2.16 or A2.17.

The equilibrium equation for a confined aquifer is

$$T = \frac{Q}{2\pi[h_2 - h_1]}\ln\left(\frac{r_2}{r_1}\right) \qquad (A2.16)$$

and the equilibrium equation for an unconfined aquifer is

$$K = \frac{Q}{\pi[h_2^2 - h_1^2]}\ln\left(\frac{r_2}{r_1}\right) \qquad (A2.17)$$

where $T = Kb$; $h_2 - h_1 = y$ (and $h_2 = b$).

A2.9. (a) Integrate both equations given in problem A2.8; you will likely want to consult a table of integrals in order to do so. (b) What is the difference in the results given by the two equations (i.e., K will be understated by a substantial percentage—what is this percentage?) (c) When will the maximum difference in K between the two formulas occur?

APPENDIX 3

Tables

COMPLEMENTARY ERROR FUNCTION (ERFC)

$$\text{erf}(z) = \frac{2}{\sqrt{\pi}} \int_0^z \exp^{-\beta^2} d\beta$$

$$\text{erf}(-z) = -\text{erf } z$$

$$\text{erfc}(z) = 1 - \text{erf}(z)$$

z	$\text{erf}(z)$	$\text{erfc}(z)$	z	$\text{erf}(z)$	$\text{erfc}(z)$
0	0	1.0	1.1	0.880205	0.119795
0.05	0.056372	0.943628	1.2	0.910314	0.089686
0.1	0.112463	0.887537	1.3	0.934008	0.065992
0.15	0.167996	0.832004	1.4	0.952285	0.047715
0.2	0.222703	0.777297	1.5	0.966105	0.033895
0.25	0.276326	0.723674	1.6	0.976348	0.023652
0.3	0.328627	0.671373	1.7	0.983790	0.016210
0.35	0.379382	0.620618	1.8	0.989091	0.010909
0.4	0.428392	0.571608	1.9	0.992790	0.007210
0.45	0.475482	0.524518	2.0	0.995322	0.004678
0.5	0.520500	0.479500	2.1	0.997021	0.002979
0.55	0.563323	0.436677	2.2	0.998137	0.001863
0.6	0.603856	0.396144	2.3	0.998857	0.001143
0.65	0.642029	0.357971	2.4	0.999311	0.000689
0.7	0.677801	0.322199	2.5	0.999593	0.000407
0.75	0.711156	0.288844	2.6	0.999764	0.000236
0.8	0.742101	0.257899	2.7	0.999866	0.000134
0.85	0.770668	0.229332	2.8	0.999925	0.000075
0.9	0.796908	0.203092	2.9	0.999959	0.0000041
0.95	0.820891	0.179109	3.0	0.999978	0.000022
1.0	0.842701	0.157299			

SERIES EXPANSIONS

Exponential Integral

Let the argument of the exponential integral be x, as in $E_i(x)$. If x is less than or equal to 1.0, then

$$E_i(x) = -\ln x - 0.57721566 + 0.99999193*x - 0.24991055*x^2 + 0.5519968*x^3$$
$$- 0.00976004*x^4 + 0.00107857*x^5$$

If x is greater than 1.0, then

$$E_i(x) = \frac{[\exp(-x)/x]*[0.250621 + x*(2.334733 + x)]}{[1.681534 + x*(3.3306571 + x)]}$$

Error Function

Let the argument of the error function be x, as in $\mathrm{erf}(x)$. If x is greater than 3.0, then

$$\mathrm{erf}(x) = 1.0$$

If x is less than 0.0 (negative), then let $x = -x$; or if x is positive but less than or equal to 0.9, then let $x = x$. Also let $y = x^2$ for x negative or positive. Then,

$$\mathrm{erf}(x) = 1.12838*x*((((((-7.57576E - 4*y + 0.00462963)*y$$
$$- 0.0238095)*y + 0.01)*y) - 0.333333)*y + 1.0)$$

and, if x was negative, then

$$\mathrm{erf}(x) = -\mathrm{erf}(x)$$

If x is greater than 0.9 but less than or equal to 3.0, then $y = x^2$ and

$$T = \frac{1.0}{1.0 + 0.47047*x}$$

Then,

$$\mathrm{erf}(x) = 1.0 - T*((0.7478556*T - 0.0958798)*T + 0.3480242)*\exp(-y)$$

CONVERSION FACTORS

The student may find these conversion units useful for problems encountered in the text, and elsewhere in unsaturated zone hydrology. Most of the values in the following conversion tables have been rounded to the second decimal. For greater accuracy, consult a recent issue of the *CRC Handbook of Chemistry and Physics*. To convert a value from the units in the "From" column to those in the "To" column, multiply the "From" column by the numerical value at the intersection.

Example: 78 gal min^{-1} to m^3 hr^{-1}: 78 gal min^{-1}*0.227 = 17.707 m^3 hr^{-1}; or, in reverse: 17.707 m^3 hr^{-1}*4.404 = 78 gal min^{-1}.

Flow Rate Conversion						
From → To	L sec^{-1}	m^3 min^{-1}	m^3 hr^{-1}	gal min^{-1}	gal hr^{-1}	ft^3 sec^{-1}
L sec^{-1}	1	0.06	3.60	15.85	9.51×10^2	3.53×10^{-2}
m^3 min^{-1}	16.67	1	60	2.64×10^2	1.59×10^4	0.589
m^3 hr^{-1}	0.28	1.67×10^{-2}	1	4.404	2.64×10^2	9.81×10^{-3}
gal min^{-1}	6.31×10^{-2}	3.79×10^{-3}	**0.227**	1	60	2.23×10^{-3}
gal hr^{-1}	1.05×10^{-3}	6.31×10^{-5}	3.79×10^{-3}	1.67×10^{-2}	1	3.71×10^{-5}
ft^3 sec^{-1}	28.32	1.70	1.02×10^2	4.49×10^2	2.69×10^4	1

Concentration Conversion

From → To	mg L^{-1}	g L^{-1}	kg m^{-3}	lb in.$^{-3}$	lb ft^{-3}	lb gal^{-1}	grain gal^{-1}
mg L^{-1}	1	1×10^{-3}	1×10^{-3}	3.61×10^{-8}	6.24×10^{-5}	8.35×10^{-6}	5.84×10^{-2}
g L^{-1}	1×10^{3}	1	1	3.61×10^{-5}	6.24×10^{-2}	8.35×10^{-3}	58.42
kg m^{-3}	1×10^{3}	1	1	3.61×10^{-5}	6.24×10^{-2}	8.35×10^{-3}	58.42
lb in.$^{-3}$	2.77×10^{7}	2.77×10^{4}	2.77×10^{4}	1	1.73×10^{3}	2.31×10^{2}	1.62×10^{6}
lb ft^{-3}	1.60×10^{4}	16.02	16.02	5.79×10^{-4}	1	0.134	9.35×10^{2}
lb gal^{-1}	1.20×10^{5}	1.20×10^{2}	1.20×10^{2}	4.33×10^{-3}	7.48	1	7.0×10^{3}
grain gal^{-1}	17.12	1.71×10^{-2}	1.71×10^{-2}	6.19×10^{-7}	1.07×10^{-3}	1.43×10^{-4}	1

Length Conversion

From → To	micron	mm	cm	m	km	in.	ft	mile
micron	1	1×10^{-3}	1×10^{-4}	1×10^{-6}	1×10^{-9}	3.94×10^{-5}	3.28×10^{-6}	6.21×10^{-10}
mm	1×10^{3}	1	1×10^{-1}	1×10^{-3}	1×10^{-6}	3.94×10^{-2}	3.28×10^{-3}	6.21×10^{-7}
cm	1×10^{4}	10	1	1×10^{-2}	1×10^{-5}	0.394	3.28×10^{-2}	6.21×10^{-6}
m	1×10^{6}	1×10^{3}	1×10^{2}	1	1×10^{-3}	39.37	3.28	6.21×10^{-4}
km	1×10^{9}	1×10^{6}	1×10^{5}	1×10^{3}	1	3.94×10^{4}	3.28×10^{3}	0.621
in.	2.54×10^{4}	25.40	2.54	2.54×10^{-2}	2.54×10^{-5}	1	8.33×10^{-2}	1.58×10^{-5}
ft	3.05×10^{5}	3.05×10^{2}	30.48	0.305	3.05×10^{-4}	12	1	1.89×10^{-4}
mile	1.61×10^{9}	1.61×10^{6}	1.61×10^{5}	1.61×10^{3}	1.61	6.34×10^{4}	5280	1

Area Conversion

From → To	mm^2	cm^2	m^2	km^2	ha	in.^{2}	ft^2	acre	mile^2
mm^2	1	1×10^{-2}	1×10^{-6}	1×10^{-12}	1×10^{-10}	1.55×10^{-3}	1.08×10^{-5}	2.47×10^{-10}	3.86×10^{-13}
cm^2	1×10^{2}	1	1×10^{-4}	1×10^{-10}	1×10^{-8}	0.155	1.08×10^{-3}	2.47×10^{-8}	3.86×10^{-11}
m^2	1×10^{6}	1×10^{4}	1	1×10^{-6}	1×10^{-4}	1.55×10^{3}	10.76	2.47×10^{-4}	3.86×10^{-7}
km^2	1×10^{12}	1×10^{10}	1×10^{6}	1	1×10^{2}	1.55×10^{9}	1.08×10^{7}	2.47×10^{2}	0.386
ha	1×10^{10}	1×10^{8}	1×10^{4}	1×10^{-2}	1	1.55×10^{7}	1.08×10^{5}	2.47	3.86×10^{-3}
in.^{2}	6.45×10^{2}	6.45	6.45×10^{-4}	6.45×10^{-10}	6.45×10^{-8}	1	6.94×10^{-3}	1.59×10^{-7}	2.49×10^{-10}
ft^2	9.29×10^{4}	9.29×10^{2}	9.29×10^{-2}	9.29×10^{-8}	9.29×10^{-6}	144	1	2.30×10^{-5}	3.59×10^{-8}
acre	4.05×10^{9}	4.05×10^{7}	4.05×10^{3}	4.05×10^{-3}	0.405	6.27×10^{6}	4.36×10^{4}	1	1.56×10^{-3}
mile^2	2.59×10^{12}	2.59×10^{10}	2.59×10^{6}	2.59	2.59×10^{2}	4.01×10^{9}	2.79×10^{7}	640	1

Volume Conversion

From → To	cm3	L	m³	in.³	ft³	pint	qt	gal	ac-ft
cm3	1	1×10^{-3}	1×10^{-6}	6.10×10^{-2}	3.53×10^{-5}	2.11×10^{-3}	1.06×10^{-3}	2.64×10^{-4}	8.11×10^{-10}
L	1×10^3	1	1×10^{-3}	61.02	3.53×10^{-2}	2.11	1.06	0.264	8.11×10^{-7}
m³	1×10^6	1×10^3	1	6.10×10^4	35.31	2.11×10^3	1.06×10^3	2.64×10^2	8.11×10^{-4}
in.³	16.39	1.64×10^{-2}	1.64×10^{-5}	1	5.79×10^{-4}	3.46×10^{-2}	1.73×10^{-2}	4.33×10^{-3}	1.33×10^{-8}
ft³	2.83×10^4	28.32	2.83×10^{-2}	1.73×10^3	1	59.84	29.92	7.48	2.30×10^{-5}
pint	4.73×10^2	0.473	4.73×10^{-4}	28.87	1.67×10^{-2}	1	0.500	0.125	3.84×10^{-7}
qt	9.46×10^2	0.946	9.46×10^{-4}	57.74	3.34×10^{-2}	2	1	0.250	7.67×10^{-7}
gal	3.79×10^3	3.785	3.79×10^{-3}	2.31×10^2	0.134	8	4	1	3.07×10^{-6}
ac-ft	1.23×10^9	1.23×10^6	1.23×10^3	7.52×10^7	4.35×10^4	2.61×10^6	1.30×10^6	3.26×10^5	1

Mass Conversion

From → To	mg	g	kg	metric ton	oz.	lb	short ton	grain
mg	1	1×10^{-3}	1×10^{-6}	1×10^{-9}	3.53×10^{-5}	2.21×10^{-6}	1.10×10^{-9}	1.54×10^{-2}
g	1×10^3	1	1×10^{-3}	1×10^{-6}	3.53×10^{-2}	2.21×10^{-3}	1.10×10^{-6}	15.43
kg	1×10^6	1×10^3	1	1×10^{-3}	35.27	2.21	1.10×10^{-3}	1.54×10^4
metric ton	1×10^9	1×10^6	1×10^3	1	3.53×10^4	2.21×10^3	1.10	1.54×10^7
oz.	2.84×10^4	28.35	2.84×10^{-2}	2.84×10^{-5}	1	6.25×10^{-2}	3.13×10^{-5}	437.5
lb	4.54×10^5	453.6	0.454	4.54×10^{-4}	16	1	5.00×10^{-4}	7.00×10^3
short ton	9.07×10^5	9.07×10^5	9.07×10^2	0.907	3.20×10^4	2000	1	1.40×10^7
grain	64.80	6.48×10^{-2}	6.48×10^{-5}	6.48×10^{-8}	2.29×10^{-3}	1.43×10^{-4}	7.14×10^{-8}	1

Pressure Conversion

From → To	dyne cm⁻²	kg m⁻²	Pa	cm Hg	ft H₂O	bar	atm	psi	lb ft⁻²
dyne cm^{-2}	1	1.01×10^{-2}	0.100	7.50×10^{-5}	3.35×10^{-5}	1×10^{-6}	9.87×10^{-7}	1.45×10^{-5}	2.09×10^{-3}
kg m^{-2}	98.07	1	9.81	7.36×10^{-3}	3.28×10^{-3}	9.81×10^{-5}	9.68×10^{-5}	1.42×10^{-3}	0.205
Pa	10.00	0.102	1	7.50×10^{-4}	3.35×10^{-4}	1×10^{-5}	9.87×10^{-6}	1.45×10^{-5}	2.09×10^{-2}
cm Hg	1.33×10^4	1.36×10^2	1.33×10^3	1	0.446	1.33×10^{-2}	1.32×10^{-2}	0.193	27.85
ft H$_2$O	2.99×10^4	3.05×10^2	2.99×10^3	2.241	1	2.99×10^{-2}	2.95×10^{-2}	0.434	62.42
bar	1×10^6	1.02×10^4	1×10^5	75.01	33.46	1	0.987	14.50	2.09×10^3
atm	1.01×10^6	1.03×10^4	1.01×10^5	76.00	33.90	1.013	1	14.70	2.12×10^3
psi	6.90×10^4	7.03×10^2	6.90×10^3	5.171	2.307	6.90×10^{-2}	6.80×10^{-2}	1	144
lb ft^{-2}	4.79×10^2	4.882	47.88	3.59×10^{-2}	1.60×10^{-2}	4.79×10^{-4}	4.73×10^{-4}	6.94×10^{-3}	1

Physical Constants

Atomic mass unit	u	1.661×10^{-27} kg
Avogadro constant	N_A	6.022×10^{23} mol^{-1}
Boltzmann constant	k	1.381×10^{-23} J K^{-1}
Electron rest mass	m_e	9.100×10^{-31} kg
Elementary charge	e	1.602×10^{-19} C (electric charge per mol of electrons)
Electron charge	$-e/m_e$	-1.7588×10^{11} C kg^{-1} or 4.803×10^{-10} abs esu
Faraday constant	F	9.6485×10^4 C mol^{-1}
		$23{,}060$ $\text{cal mol}^{-1} \text{ eV}^{-1}$
Gas constant	R	8.314 $\text{J K}^{-1} \text{ mol}^{-1}$
		0.08314 $\text{L bar K}^{-1} \text{ mol}^{-1}$
		1.987 $\text{cal K}^{-1} \text{ mol}^{-1}$
		0.08206 $\text{L atm K}^{-1} \text{ mol}^{-1}$
Ice point (absolute zero)		273.15 K
Molar volume (ideal gas, 0 °C, 1 atm)	V_m	22.414×10^3 $\text{cm}^3 \text{ mol}^{-1}$
Natural logarithm of 10	ln 10	2.302585
Permittivity of vacuum	ε_o	8.854×10^{-12} $\text{C}^2 \text{ N}^{-1} \text{ m}^{-2}$
	ε_o	8.854×10^{-12} $\text{C}^2 \text{ J}^{-1} \text{ m}^{-2}$
	$1/4\pi\varepsilon_o$	0.8988×10^{10} $\text{N m}^2 \text{ C}^{-2}$
Planck constant	h	6.626×10^{-34} J s
Proton rest mass	m_p	1.673×10^{-27} kg
R ln 10		19.14 $\text{J mol}^{-1} \text{ K}^{-1}$
$RT_{298.15} \ln \chi$		$5706.6 \log \chi$ J mol^{-1} or $1364.1 \log \chi$ cal mol^{-1}
$RTF^{-1} \ln 10$		59.16 mV at 298.15 K
$RTF^{-1} \ln \chi$		$0.05916 \log \chi$, volt at 298.15 K
Rydberg constant	R_∞	1.097×10^7 m^{-1}
Speed of light in vacuum	c	2.998×10^8 m s^{-1}

Useful Conversion Factors

Energy, work, heat

$$
\begin{aligned}
1 \text{ joule} &= 1 \text{ volt-coulomb} = 1 \text{ newton meter} \\
&= 1 \text{ watt-second} = 2.7778 \times 10^{-7} \text{ kilowatt hours} \\
&= 10^7 \text{ erg} \\
&= 9.9 \times 10^{-3} \text{ liter atmospheres} \\
&= 0.239 \text{ calorie} \\
&= 1.0365 \times 10^{-5} \text{ volt-faraday} \\
&= 6.242 \times 10^{18} \text{ eV} \\
&= 5.035 \times 10^{22} \text{ cm}^{-1} \text{ (wave number)} \\
&= 9.484 \times 10^{-4} \text{ BTU (British thermal unit)} \\
&= 3 \times 10^{-8} \text{ kg coal equivalent}
\end{aligned}
$$

Entropy	1 entropy unit, cal mol^{-1} K^{-1} = 4.184 J mol^{-1} K^{-1}
Power	1 watt = 1 kg m^2 s^{-3}
	= 2.39 × 10^{-4} kcal s^{-1} = 0.860 kcal h^{-1}
Pressure	1 atm = 760 torr = 760 mm Hg
	= 1.013 × 10^5 N m^{-2} = 1.013 × 10^5 Pa (Pascal)
	= 1.013 bar

Statistics on water

Density @ 3.98 °C (peak density) = 999.973 kg m^{-3} = 0.999973 g cm^{-3} = 1.000000 g ml^{-1}
Density @ 0 °C = 999.87 kg m^{-3}
Ice Density @ 0 °C = 916.76 kg m^{-3}

Converting *X* number of ppm$_V$ to density (g m^{-3}) units

At STP, 1 mole of gas occupies 22.4 l (V_o), at P_o = 101.325 kPa, T_o = 273.16 K, where a mole of the particular gas has a mass (m_c) in grams/mole.
From the ideal gas law:

$$P_o V_o = m_c R_c T_o \quad \text{and} \quad \rho = \rho_c R_c T$$

Using both equations, solving for ρ_c in terms of volume, and inserting constants:

$$\rho_c[\text{gm}^{-3}] = \left(\frac{X \text{ parts volume}}{10^6 \text{ parts volume}} \right) \left(\frac{m_c[\text{gm mol}^{-1}]}{22.4 \ [\text{L mol}^{-1}]} \right) \left(\frac{1000[\text{L}]}{[\text{m}^3]} \right) \left(\frac{273.15 \text{ K}}{T[\text{K}]} \right) \left(\frac{P[\text{kPa}]}{101.325 \text{ kPa}} \right)$$

where "L" refers to liters.

Converting water in chemical potential units [J kg^{-1}] to pressure potential units [Pa]

Often it is more convenient to work with water potential in pressure units, particularly if water potential is sensed with pressure transducers. Here, a chemical potential of A in [J kg^{-1} or energy per unit mass] is converted to pressure potential [Pa] units by dividing the energy per mass units by the partial molal volume of water (18 × 10^{-6} m^3 mol^{-1}), valid for dilute solutions.

$$\frac{A[\text{J kg}^{-1}]}{18 \times 10^{-6}[\text{m}^3 \text{ mol}^{-1}]} \times 0.018016 \ [\text{kg mol}^{-1}] = A[\text{kPa}]$$

Converting *X* number of pCi (pico Curies) of a radioactive gas to its concentration in parts per million (ppm$_V$) units

Radioactive materials decay at a rate given by the following simple, ordinary differential equation:

$$\frac{dS}{dt} = -kS$$

where S is concentration, t is time, and k is the disintegration constant. Radioactive half-life is defined as the amount of time for half of the mass of radioactive material to decay. Solving the above equation, we can solve for the disintegration constant using half-life:

$$\frac{S}{S_o} = \frac{1}{2} = \exp(-kt)$$

where S_o is the initial concentration of element S. The disintegration rate per unit time $[\text{d s}^{-1}]$ is measured in Curies (C): $C = 3.7 \times 10^{10}\ \text{d s}^{-1}$ or in pico Curies: $pC = 0.037\ \text{d s}^{-1}$. The conversion from $pC\ L^{-1}$ to ppm follows from rearrangement of the above equations:

$$\frac{\dfrac{dS}{dt}}{k} = S = \left(\frac{\dfrac{X[pC]}{[L]}}{k[s^{-1}]} \right)\left(\frac{0.037[\text{d s}^{-1}]}{pC} \right)$$

which yields the result of disintegrations per liter. Since there is one disintegration per atom, the final conversion to ppm_v is:

$$\left(\frac{S\ [\text{atoms}]}{[L]} \right)\left(\frac{1\ [\text{mole}]}{6.023 \times 10^{23}\ [\text{atoms}]} \right)\left(\frac{22.4\ [L]}{[\text{mole}]} \right)\left(\frac{10^6\ \text{parts}\ [L]}{1\ \text{part}\ [L]} \right) = y\,ppm_v$$

This conversion is at STP (standard temperature and pressure).

A conversion example: Radon has a half-life of 3.82 days $= 3.3 \times 10^5$ s. With $t = 3.3 \times 10^5$ s, $k = 2.10 \times 10^{-6}\ \text{s}^{-1}$. If radon concentration is $100\ pC\ l^{-1}$, one may calculate a 1.74×10^6 disintegrations per liter and a concentration of $6.5 \times 10^{-11}\ ppm_v$.

THE INTERNATIONAL SYSTEM OF UNITS (SI)

This unit system is based on the metric system. It was designed to achieve maximum internal consistency and based upon the following defined units:

Physical quantity	Unit	Symbol
Length	meter	m
Mass	kilogram	kg
Time	second	s
Electric current	ampere	A
Temperature	kelvin	K
Luminous intensity	candela	cd
Amount of material	mole	mol

The main derived units are:

Force	newton	N (kg m^{-2})
Energy, work, heat	joule	J (N m)
Pressure	pascal	Pa (N m^{-2})
Power	watt	W (J s^{-1})
Electric charge	coulomb	C (As)
Electric potential	volt	V (W A^{-1})
Electric capacitance	farad	F (As V^{-1})
Electric resistance	ohm	Ω (V A^{-1})
Frequency	hertz	Hz (s^{-1})
Conductance	siemens	S (A V^{-1})

To convert SI units into Non-SI units multiply by	SI unit	Non-SI units	To convert Non-SI units into SI units, multiply by
		Energy, work, quantity of heat	
9.52×10^{-4}	joule, J	British thermal unit, BTU	1.05×10^3
0.239	joule, J	calorie, cal	4.19
10^7	joule, J	erg	10^{-7}
0.735	joune, J	foot-pound	1.36
2.387×10^{-5}	joule per square meter, J m^{-2}	calorie per square centimeter (langley)	4.19×10^4
10^5	newton, N	dyne	10^{-5}
1.43×10^{-3}	watt per square meter, W m^{-2}	calorie per square centimeter minute irradiance, cal cm^{-2} min^{-1}	698
		Transpiration and photosynthesis	
3.60×10^{-2}	milligram per square meter second, mg m^{-2} s^{-1}	gram per square decimeter hour, g dm^{-2} h^{-1}	27.8
5.56×10^{-3}	milligram(H$_2$O) per square meter second, mg m^{-2} s^{-1}	micromole (H$_2$O) per square centimeter second, μmol cm^{-2} s^{-1}	180
10^{-4}	milligram per square meter second, mg m^{-2} s^{-1}	milligram per square centimeter second, mg m^{-2} s^{-1}	10^4
35.97	milligram per square meter second, mg m^{-2} s^{-1}	milligram per square decimeter hour, mg dm^{-2} h^{-1}	2.78×10^{-2}
		Water measurement and flow rate	
9.73×10^{-3}	cubic meter, m^3	acre-inches, acre-in.	102.8
9.81×10^{-3}	cubic meter per hour, m^3 h^{-1}	cubic feet per second, ft^3 s^{-1}	101.9
4.40	cubic meter per hour, m^3 h^{-1}	U.S. gallons per minute, gal min^{-1}	0.227
2.642×10^2	cubic meter per hour, m^3 h^{-1}	U.S. gallons per hour, gal hr^{-1}	3.785×10^{-3}
8.11	hectare-meters, ha-m	acre-feet, acre-ft	0.123
97.28	hectare-meters, ha-m	acre-inches, acre-in.	1.03×10^{-2}
8.1×10^{-2}	hectare-centimeters, ha-cm	acre-feet, acre-ft	12.33
		Concentrations	
1	centimole per kilogram, cmol kg^{-1} (ion exchange capacity)	milliequivalents per 100 grams, meq 100 g^{-1}	1
0.1	gram per kilogram, g kg^{-1}	percent, %	10
1	milligram per kilogram, mg kg^{-1}	parts per million, ppm	1
		Electrical conductivity, electricity, magnetism	
10	siemen per meter, S m^{-1}	millimho per centimeter, mmho cm^{-1}	0.1
10^4	tesla, T	gauss, G	10^{-4}
		Radioactivity	
2.7×10^{-11}	becquerel, Bq	curie, Ci	3.7×10^{10}
2.7×10^{-2}	becquerel per kilogram, Bq kg^{-1}	picocurie per gram, pCi g^{-1}	37
100	gray, Gy (absorbed dose)	rad, rd	0.01
100	sievert, Sv (equivalent dose)	rem (roentgen equivalent man)	0.01
		Plane Angle	
57.3	radian, rad	degrees (angle), °	1.75×10^{-2}

To convert SI units into Non-SI units multiply by	SI unit	Non-SI units	To convert Non-SI units into SI units, multiply by
		Length	
0.621	kilometer, km (10^3 m)	mile, mi	1.609
1.094	meter, m	yard, yd	0.914
3.28	meter, m	foot, ft	0.304
1.0	micrometer, μm (10^{-6} m)	micron, μ	1.0
3.94×10^{-2}	millimeter, mm (10^{-3} m)	inch, in.	25.4
10	nanometer, nm (10^{-9} m)	angstrom, Å	0.1
		Area	
2.47	hectare, ha	acre	0.405
247	square kilometer, km^2 (10^3 m)^2	acre	4.05×10^{-3}
0.386	square kilometer, km^2 (10^3 m)^2	square mile, mi^2	2.590
2.47×10^{-4}	square meter, m^2	acre	4.05×10^3
10.76	square meter, m^2	square foot, ft^2	9.29×10^{-2}
1.55×10^{-3}	square millimeter, mm^2 (10^{-3} m)^2	square inch, in.^{2}	645
		Volume	
9.73×10^{-3}	cubic meter, m^3	acre-inch, acre in	102.8
35.3	cubic meter, m^3	cubic foot, ft^3	2.83×10^{-2}
6.10×10^4	cubic meter, m^3	cubic inch, in.^{3}	1.64×10^{-5}
2.84×10^{-2}	liter, L (10^{-3} m^3)	bushel, bu	35.24
1.057	liter, L (10^{-3} m^3)	quart (liquid), qt	0.946
3.53×10^{-2}	liter, L (10^{-3} m^3)	cubic foot, ft^3	28.3
0.265	liter, L (10^{-3} m^3)	gallon, gal	3.78
33.78	liter, L (10^{-3} m^3)	ounce (fluid), oz	2.96×10^{-2}
2.11	liter, L (10^{-3} m^3)	pint (fluid), pt	0.473
		Mass	
2.20×10^{-3}	gram, g (10^{-3} kg)	pound, lb	454
3.52×10^{-2}	gram, g (10^{-3} kg)	ounce (avdp), oz	28.4
2.205	kilogram, kg	pound, lb	0.454
0.01	kilogram, kg	quintal (metric), q	100
1.10×10^{-3}	kilogram, kg	ton (2000 lb), ton	907
1.102	megagram, Mg (tonne)	ton (U.S.), ton	0.907
1.102	tonne, t	ton (U.S.), ton	0.907
		Yield and Rate	
0.893	kilogram per hectare, kg ha^{-1}	pound per acre, lb acre^{-1}	1.12
7.77×10^{-2}	kilogram per cubic meter, kg m^{-3}	pound per bushel, lb bu^{-1}	12.87
1.49×10^{-2}	kilogram per hectare, kg ha^{-1}	bushel per acre, 60 lb	67.19
1.59×10^{-2}	kilogram per hectare, kg ha^{-1}	bushel per acre, 56 lb	62.71
1.86×10^{-2}	kilogram per hectare, kg ha^{-1}	bushel per acre, 48 lb	53.75
0.107	liter per hectare, L ha^{-1}	gallon per acre, gal acre^{-1}	9.35
893	tonnes per hectare, t ha^{-1}	pound per acre, lb acre^{-1}	1.12×10^{-3}
893	megagram per hectare, Mg ha^{-1}	pound per acre, lb acre^{-1}	1.12×10^{-3}
0.446	megagram per hectare, Mg ha^{-1}	ton (2000 lb) per acre, ton acre^{-1}	2.24
2.24	meter per second, m s^{-1}	mile per hour, mph	0.447
		Specific Surface Area	
10	square meter per kilogram, m^2 kg^{-1}	square centimeter per ram, cm^2 g^{-1}	0.1
1000	square meter per kilogram, m^2 kg^{-1}	square millimeter per gram, mm^2 g^{-1}	0.001

To convert SI units into Non-SI units multiply by	SI unit	Non-SI units	To convert Non-SI units into SI units, multiply by
		Pressure	
9.90	megapascal, MPa (10^6 Pa)	atmosphere	0.101
10	megapascal, MPa (10^6 Pa)	bar	0.1
1.00	megagram, Mg m^{-3}	gram per cubic centimeter, g cm^{-3}	1.00
2.09×10^{-2}	pascal, Pa	pound per square foot, lb ft^{-2}	47.9
1.45×10^{-4}	pascal, Pa	pound per square inch, lb in.$^{-2}$	6.90×10^3
		Temperature	
1.00 (K-273)	kelvin, K	Celsius, °C	1.00 (°C + 273)
(9/5 °C) + 32	Celsius, °C	Fahrenheit, °F	5/9(°F−32)

SI Units for Use in Unsaturated Zone Hydrology

Quantity	Application	Symbol
Concentration	Gas concentration	g m^{-3}
		mol m^{-3}
	Water content	kg kg^{-1}
		m^3 m^{-3}
Density	Particle density	Mg m^{-3}
	Bulk density	
Flux density	Heat flow	W m^{-2}
	Gas diffusion	g m^{-2} s^{-1}
		mol m^{-2} s^{-1}
	Water flow	kg m^{-2} s^{-1}
		m^3 m^{-2} s^{-1}
Gas diffusivity	Gas diffusion	m^2 s^{-2}
Hydraulic conductivity	Water flow	kg s m^{-3}
		m^3 s^{-1}
		m s^{-1}
Potential energy of soil water	Driving force for flow	J kg^{-1}
		kPa
		m
Specific heat	Heat storage	J kg^{-1} K^{-1}

Units for Water Flow Applied to Darcy's Law

Flux density	Hydraulic conductivity	Potential gradient
Potential in energy per unit mass (kg m^{-2} s^{-1})	kg s m^{-3}	J kg^{-1} m^{-1}
Potential in energy per unit volume (m s^{-1})	m^3 s kg^{-1}	Pa m^{-1}
Potential in energy per unit weight (m s^{-1})	m s^{-1}	m m^{-1}

Source: Data from Campbell and Schilfgaarde (1981)

LIST OF SYMBOLS

Symbol	Description	Units
a	Constant or specific volume or albedo	L^3 or decimal
a	Activity	—
a	Sphere radius	m
A	Available energy	$W\,m^{-2}$
A	Measured Carbon-14 radioactivity	—
A	Area	m^2
A	Specific surface area	$m^2\,kg^{-1}$
A	Hamaker constant (see equation 3.48)	ergs
a_i	Activity of species i	$mol\,L^{-1}$
A_o	Wave amplitude	—
A_o	Radioactivity level at some initial time	Bq
A_s	Interstitial surface area of pores	$m^2\,kg^{-1}$
b	Constant	—
b	Complex number	—
b	Width of sample	m
B	Retardation of the soil being used	$J\,m^{-2}$
c	Specific heat of soil	$J\,kg^{-1}\,K^{-1}$
c	Solute concentration	$mol\,L^{-1}$
^{12}C	Carbon 12	—
^{13}C	Carbon 13	—
^{14}C	Carbon 14	—
c	Integration constant	—
C	Curies or concentration	(3.7×10^{10} Bg) or $mg\,L^{-1}$
C	Ion concentration at specified distance from charged surface	$ions\,cm^{-3}$
C_e	Equilibrium aqueous concentration	$mg\,L^{-1}$
C_{gw}	Contaminant concentration in ground water	$\mu g\,L^{-1}$
C_p	Concentration of pollutant	$\mu g\,L^{-1}$
C_i	Specific heat of soil constituent i	$J\,kg^{-1}\,K^{-1}$
C_i	Concentration of species i	$mol\,L^{-1}$
C_o	Ion concentration in bulk solution	$ions\,cm^{-3}$
C_p	Specific heat of air	$J\,kg^{-1}\,K^{-1}$
C_p^o	Heat capacity	$J\,mol^{-1}$
C_s	Molar salt concentration	$M\,L^{-3}$
C_{soil}	Specific heat of soil	$J\,kg^{-1}\,K^{-1}$
d	Depth of soil layer	m
d	Distance, diameter, thickness, or midpoint	m
D	Fractal dimension	—
D	Mass of water lost through drainage out of soil column	kg
D	Hydraulic diffusivity	$L^2\,T^{-1}$
D_H	Hydrodynamic dispersion coefficient	$L^2\,T^{-1}$
D	Width of crack; $2d$ of spherical particle	m
D	Dielectric constant	unitless
D_a	Diffusivity of air	$m\,s^{-1}$
D_{ij}	Binary gas diffusion coefficient	$m\,s^{-1}$
D_i^k	Knudsen diffusion coefficient	$m\,s^{-1}$
D_h	Thermal diffusivity of soil	$m\,s^{-1}$
D_S	Diffusion coefficient of gas species s	$m\,s^{-1}$
D_T	Thermal diffusivity for thermally driven moisture flux	$m\,s^{-1}$
D_V	Molecular diffusivity of water vapor	$m\,s^{-1}$
D_α	Diffusion coefficient in free air	$m\,s^{-1}$
$D^{(\alpha)}$	Electric displacement at phase α	$C\,m^{-2}$
D_θ	Diffusivity for moisture flux due to a moisture gradient	$m\,s^{-1}$
d_j	Thickness of layer	m
d_t	Total depth of considered soil	cm

(continued)

Symbol	Description	Units
d_w	Equivalent depth of soil water if extracted and ponded over soil surface	cm
e	Unit of electronic charge (cgs or esu system)	$4.803*10^{-10}$ esu
e	Unit of electronic charge (SI system)	$1.6021*10^{-19}$ C
e	Void ratio	unitless
e	Vapor pressure	Pa
erf	Error function	—
erfc	Complimentary error function	—
e_s	Saturation vapor pressure	Pa
e_{sw}	Saturated vapor pressure at wet bulb temp.	Pa
E	Evaporation rate	$m\,s^{-1}$
E	Water vapor flux (mass flux or evaporation rate)	$g\,m^{-2}\,s^{-1}$
E	Potential energy per unit volume	$J\,m^{-3}$
E	Young's modulus	$N\,m^{-2}$
E	Electric field strength	N or $Kg\,m\,s^{-2}$
E_c	Charge-balance error (deviation from electroneutrality)	unitless
E_{cum}	Cumulative rate evaporation	$m\,s^{-1}$
E_h	Redox potential	V
E_i	Potential energy of interaction	$erg\,cm^{-2}$
E_{max}	Maximum rate of evaporation	$m\,s^{-1}$
E_o	Free-water evaporation rate	$m\,s^{-1}$
ET	Evapotranspiration (can also use flux units $g\,m^{-2}\,s^{-1}$)	$m\,s^{-1}$
ET_p	Potential evapotranspiration	$m\,s^{-1}$
ET_{eq}	Equilibrium evapotranspiration	$m\,s^{-1}$
e^o	Saturated water vapor pressure	Pa
E^o	Standard redox potential	V
E_s	Induced streaming potential by a thermal gradient	$V\,m^{-1}$
E_s	Electric resistivity	$S\,m^{-1}$
exp	Exponential function	—
f	Fractional amount	decimal
f	Frictional coefficient	unitless
f	Area of the air–water interface associated with the triangular volume	m^2
f	Force or frictional resistance	dynes or N or $Kg\,m\,s^{-2}$
f_i	Volume fraction of a soil constituent	—
F	Specific flux	$g\,m^{-2}\,s^{-1}$
F	Faraday constant	$96490\,C\,mol^{-1}$
F	Partial molal free energy	$J\,g^{-1}$
F	Free energy	$J\,mol^{-1}$ or $ergs\,mol^{-1}$
F	Linear free energy	$J\,mol^{-1}$
F_b	Breaking force applied at the center of the soil sample beam span	N or $Kg\,m\,s^{-2}$
F_c	Critical flocculation concentration value	$mol\,L^{-1}$
F_d	Resistence or drag force on an individual particle	$g\,cm^2\,sec^{-2}$ (dynes)
F^d	Drag force	N or $Kg\,m\,s^{-2}$
F_g	Force of gravity on a soil particle	$g\,cm^2\,sec^{-2}$
F_i	Molar flux of constituent i	$mol\,m^{-2}\,s^{-1}$
F_i^s	Static equilibrium	—
F^s	Static force	N or $Kg\,m\,s^{-2}$
F_w	Partial molal free energy under influence of forces	$J\,g^{-1}$
g	Gravitational acceleration	$m\,s^{-2}$
g	Gravitational constant	$N\,m^2\,kg^{-2}$
G	Soil heat flux	$W\,m^{-2}$
G	Strain energy release rate	$dynes\,cm^{-2}$ or $J\,m^{-2}$
G_C	Change in stored heat	$W\,m^{-2}$
G^E	Excess Gibbs free energy	$J\,mol^{-1}$
G^l	Gibbs free energy of mixing	$J\,mol^{-1}$

(continued)

Symbol	Description	Units
G^M	Energy of mixing	$J\,mol^{-1}$
G^o	Gibbs free energy	$J\,mol^{-1}$
G_x	Soil heat flux measured by a flux plate	$W\,m^{-2}$
G_{total}	Combined measured soil heat flux	$W\,m^{-2}$
G_{stor}	Change in stored soil heat	$W\,m^{-2}$
h	Height	m
h	Planck's constant	erg s
h	Distance below the water table	m
h	Distance	m
h	Suction	$J\,m^{-3}$
^2H	Deuterium	—
^3H	Tritium	—
H	Hurst exponent	unitless
H	Sensible heat flux	$W\,m^{-2}$
H	Hydraulic head	m
H	Minimum distance between two particle surfaces or spheres	m
h_c	Height of fluid column	m
h_f	Pressure drop	m
H_{im}	Heat of immersion	$J\,m^{-2}$
h_m	Matric head of soil water	m
h_o	Osmotic pressure head of soil water	m
h_p	Pressure head	m
h_v	Velocity head	m
i	Infiltration rate	$m\,s^{-1}$
i	Index	—
i	Area for the solid–water interface	m^2
i/n_{io}	Concentration of ionic species in outer solution	$mol\,m^{-2}$
I	Cumulative infiltration	m
I	Intensity of light	$W\,m^{-2}$
I	Strength of applied electric field	$N\,C^{-1}$
I	Ionic strength	$eq\,L^{-1}$
I	Electric current	$J\,C^{-1}$
Ie'	Force acting on charge e'	N or $Kg\,m\,s^{-2}$
I_f	Final infiltration rate	$m\,s^{-1}$
I_o	Initial infiltration rate ($T = 0$)	$m\,s^{-1}$
I_z	z^{th} ionization potential	—
j	Index	—
J	Mass flux	$Kg\,m^{-2}\,s^{-1}$
k	Thermal gradient ratio	—
k	Boltzmann constant	$1.3805*10^{-23}$ $J\,K^{-1}\,mol^{-1}$
k	Intrinsic permeability	m^2
k	Stokes's constant	—
k_a	Air permeability in soil	$m^2\,s^{-1}$
ΔK	Change in kinetic energy	J
K	Equilibrium composition	—
$K(\theta)$	Unsaturated hydraulic conductivity	$m\,s^{-1}$
K	Proportionality constant	$8.988*10^9\,N\,m^2\,C^{-2}$
K	Constant	$8*10^{-36}\,J^2\,L^{-1}$ or $8*10^{-22}$ $ergs^2\,L^{-1}$
K	Hydraulic Conductivity	$m^2\,s^{-1}$
K_s	Saturated hydraulic conductivity	$m^2\,s^{-1}$
K_1	Thermodynamic equilibrium constant (initial condition)	unitless
K_2	Thermodynamic equilibrium constant (final condition)	unitless
K_a	Acid dissociation constant	unitless
K_{app}	Apparent acid dissociation constant	unitless
K_b	Base dissociation constant	unitless
K_c	Thermodynamic equilibrium constant	unitless

(continued)

Symbol	Description	Units
K_c	Exchange equilibrium constant	unitless
K^G	Gapon constant	unitless
K_h	Eddy diffusivity of heat	$m^2\,s^{-1}$
K_H	Henry's law constant	—
k_i^o	Standard partial molar compressibility	—
K_d	Dissociation constant	$mg\,kg^{-1}$
K_e	Effective hydraulic conductivity	$m\,s^{-1}$
K_j	Hydraulic conductivity of each layer	$m\,s^{-1}$
K_{oc}	Linear adsorption coefficient	$mL\,kg^{-1}$
K_{ow}	Octanol–water partition coefficient	unitless
K_p	Thermodynamic equilibrium constant (partial pressure)	unitless
K_s	Salting parameter	unitless
K_v	Eddy diffusivity of vapor	$m^2\,s^{-1}$
K_w	Autodissociation constant of water	unitless
k'	Pore shape factor	—
l	Tube length	m
l	Thickness of clay particle	m
l	Actual length of pore	m
l_e	Effective length of pore	m
L	Point of measurement within the hydrometer	—
L	Length	m
L_f	Distance from soil surface to wetting front	m
L_m	Latent heat of melting	$J\,kg^{-1}\,K^{-1}$
L_s	Latent heat of sublimation	$J\,kg^{-1}\,K^{-1}$
L_v	Latent heat of vaporization	$J\,kg^{-1}\,K^{-1}$
L_e	Effective length	m
m	Molecular weight	$g\,mol^{-1}$
m	Mass of electron	kg
m	Mass	kg
M	Mass of sample	kg
M	Molecular weight of water (mole basis)	$18\,g\,mol^{-1}$
M	Molar concentration	$mol\,L^{-1}$
M	Molar mass of water	g
m_a	Mass of pycnometer filled with air	kg
m_{an}	Molality of the anion	$mol\,g^{-1}$
m_{cat}	Molality of the cation	$mol\,g^{-1}$
m_d	Molecular weight of dry air	$g\,mol^{-1}$
m_i	Mass or number of moles of the various constituents in system	g or mol
m_v	Molecular weight of water	g
ΔM	Change in mass	kg
M_i	Mole fraction of water in inner solution	$mol\,mol^{-1}$
M_i	Mole fraction of the ith constituent	$mol\,mol^{-1}$
$M_{(i)}$	Mass of gas i	kg
M_n	Number of capillaries of radius R_n	unitless
M_o	Mole fraction in the outer solution	$mol\,mol^{-1}$
m_s	Mass of oven dry porous media sample	kg
M_s	Mass of solid	kg
m_{sw}	Mass of pycnometer filled with soil and water	kg
m_w	Mass of pycnometer filled with water	kg
n	Statistical number of clay layers	unitless
n	Electrolyte concentration in solution	$mol\,L^{-1}$ or ions cm^{-3}
n	Integer value	—
^{15}N	Nitrogen 15	—
N	Number of electrons in outermost shell	unitless
N	Number of Carbon 14 atoms	unitless
n^+	Cation concentration	$mol\,L^{-1}$
n^-	Anion concentration	$mol\,L^{-1}$
N_A	Avogadro's number	$6.023*10^{23}$

(*continued*)

Symbol	Description	Units
n_i	Molarity of solute i	mol L^{-1}
n_i	Concentration solutes in inner solution	mol cm^{-3}
n_o	Concentration of ionic species in outer solution	ions cm^{-3}
N_{pv}	Number of pore volumes	—
^{18}O	Oxygen 18	—
p	Vector	unitless
P	Pressure	Pa or Nm^{-2}
p	Potential energy	J kg^{-1}
^{32}P	Phosphorus 32	—
P	Energy consumed in photosynthesis	W m^{-2}
P	Net pressure	Pa
P	Photosynthesis	W m^{-2}
P	Pressure	J m^{-3}
P	Precipitation rate	m s^{-1}
P	Percent media remaining in suspension at reference point	unitless
P	Gas pressure	Pa or N m^{-2}
$\mathscr{P}$	Peclet number	unitless
p_0	Pressure of chamber 2 after isothermic expansion	Pa or N m^{-2}
P_0	Gas pressure required for monolayer saturation at experimental temperature	Pa or N m^{-2}
p_1	Initial pressure	Pa or N m^{-2}
p_2	Final pressure (chamber 1)	Pa or N m^{-2}
Pa	Atmospheric pressure	Pa or N m^{-2}
P_a	Air pressure	Pa or N m^{-2}
P_d	Outer pressure	Pa or N m^{-2}
P_e	Envelope pressure	Pa or N m^{-2}
P_e	External pressure	Pa or N m^{-2}
P_h	Pressure due to standing head in soil column	Pa or N m^{-2}
P_i	Internal force between particles	N or Kg m s^{-2}
p_m	Pressure equal to the matric potential of soil	Pa or N m^{-2}
p_o	Potential energy at reference state	J kg^{-1}
p_o	Vapor pressure at a free surface of pure water under atmospheric conditions	Pa or N m^{-2}
P^o	Standard-state pressure	Pa or N m^{-2}
P_r	Precipitation rate	m s^{-1}
P_s	Swelling pressure	Pa or N m^{-2}
p_t	Tensiometer pressure	Pa or N m^{-2}
P_w	Pore-water pressure	Pa or N m^{-2}
q	Specific humidity	—
q	Flux density (specific discharge)	m s^{-1}
q	Number of atoms per unit of substance making up the particles	unitless
q	Darcy flux	m s^{-1}
q/q'	Electrical charges	C (coulombs)
q_n	Hydraulic flow rate	m^3 s^{-1}
Q	Rate of volume flow	m^3 s^{-1}
Q	Actual composition	—
Q_c	Electric charge	C
Q_e	Dry-weight concentration	mg kg^{-1}
Q_h	Heat of adsorption	cal mol^{-1}
Q_h	Heat given off due to intermolecular potential	cal g^{-1}
Q_{h1}	Heat of adsorption of first layer of adsorbing gas	cal mol^{-1}
Q_{h2}	Heat of gas liquification	cal mol^{-1}
Q_i	Flux density of gas	m s^{-1}
Q_o	Total outflow	m^3 s^{-1}
Q_{10}	Respiration rate factor	—
r	Reflectivity	unitless
r	Tube radius; particle radius	m

(continued)

Symbol	Description	Units
r	Distance between two atoms	m
R	Retardation coefficient	dimensionless
R	Hydraulic resistance	$Pa\ s\ m^{-1}$
R	Universal gas constant	$8.314\ J\ K^{-1}\ mol^{-1}$
R	Runoff rate	$m^3\ s^{-1}$
R_1, R_2	Radii of curvature of meniscus	m
R_a	Corrected hydrometer reading	—
Re	Reynolds's number	dimensionless
R_e	Hydraulic resistence of soil crust	$m\ s^{-1}$
R_n	Net radiation	$W\ m^{-2}$
R_r	Radius of rotation	m
R_{r2}	Depth to point of measurement	m
R_s	Hydraulic resistence of soil subcrust	$m\ s^{-1}$
R_v	Ideal gas constant of vapor	$J\ Kg^{-1}\ K^{-1}$
r_b	Respiration base rate	$mg\ m^{-3}\ s^{-1}$
r_c	Stomatal resistance or areal resistance	$s\ m^{-1}$
r_e	Effective pore radius	m
r_s	Charge separation distance	m
s	Slope of the saturation vapor curve	—
s	Gas constituent	—
s	Degree of saturation	—
s	Specific surface area of soil	$m^2\ kg^{-1}$
s	Triangular volume of the tube	m^3
^{34}S	Sulfur 34	—
S	Solute adsorbed	$\mu g\ g^{-1}$
S	Slope	—
S	Source or sink term of energy or gas	—
S	R/a; reduces to $2 + H/a$	dimensionless
S	Sorption	$m^3\ g^{-1}$
S	Siemens	$A\ V^{-1}$
S	Entropy	$J\ mol^{-1}\ K^{-1}$
s_b	Specific surface area per unit bulk volume of soil	$m^2\ kg^{-1}$
S_e	Effective saturation	decimal
S_i	Concentration of sorbed chemical (i^{th} species)	$mol\ kg^{-1}$
S_{id}	Saturation index	—
s_k	Geometrical shape factor of pore space	—
s_m	Specific surface area per unit mass	$m^2\ kg^{-1}$
SMOW	Standard mean ocean water	—
s_o	Reference distance	m
S^o	Molar solubility	$mg\ L^{-1}$
S_s	Specific storage	m^3
S_{sat}	Soil sorbed contaminant concentration	$\mu g\ kg^{-1}$
S_t	Stored energy	J
s_v	Specific surface area per unit volume	$m^2\ kg^{-1}$
t	Time	s
T	Temperature	K
T_f	Final temperature	K
T	Half life ($\ln 2/\lambda$)	—
T_{DP}	Dew point temperature	°C
T_m	Melting point	K
T_o	Initial or surface temperature	°C
t_p	Time of ponding	s
T_w	Wet-bulb temperature	°C
TU	Tritium units	—
u	Velocity parallel to mean flow	$m\ s^{-1}$
ΔU	Change in potential energy	J
U	Velocity	$m\ s^{-1}$
U	External electric field	N

(*continued*)

Symbol	Description	Units
v	Velocity normal to mean flow	$m\,s^{-1}$
v	Velocity of soil particle	$m\,s^{-1}$
v	Linear velocity	$m\,s^{-1}$
v	Velocity	$m\,s^{-1}$
V	Volume of water in the meniscus	m^3
V	Volume	m^3
V	Molar volume of water	$18\ cm^3\,mol^{-1}$
v_1, v_2	Constant velocity	$m\,s^{-1}$
V_1	Volume of compression chamber 1	m^3
V_2	Volume of compression chamber 2	m^3
V_a	Volume of air	m^3
V_A	Average attraction force per unit of material	$ergs\,cm^{-2}$
v_e	Effective velocity	$m\,s^{-1}$
V_E	Volume rate of flow	$m^3\,s^{-1}$
V_f	Viscous force	N
V_f	Fluid volume	m^3
V_i	Partial molar volumes of i under actual conditions	m^3
V_i^o	Partial molar volumes of i under standard conditions	m^3
V_l	Liquid volume	m^3
V_m	Volume of adsorbed gas forming monomolecular layer on adsorbant	m^3
v_o	London frequency	s^{-1}
V_r	Potential energy due to repulsive forces	$J\,m^{-3}$
V_R	Repulsive force for energy between two spherical particles of equal dimension	$ergs\,cm^{-1}$
$V_R^{\psi\infty}$	Repulsive energy between particles at constant potential	$J\,m^{-2}$
V_s	Solid volume	m^3
V_t	Total volume	m^3
V_T	Total potential energy	$J\,m^{-2}$
v_w	Partial volume of water	m^3
V_w	Partial molal volume of soil water	m^3
V_w	Volume of water in soil pores	m^3
w	Velocity in the vertical direction	$m\,s^{-1}$
w	Mass wetness	g
W	Work	J
W	Work done by the system	J
W	Weight of oven-dried soil	g
W	Hydration energy	$J\,m^{-2}$
W^σ	Reduced repulsive energy at constant charge density	$J\,m^{-2}$
W^ψ	Reduced repulsive energy of two parallel flat plates of constant potential	$J\,m^{-2}$
W_m	Mass of water sorbed to provide one molecular thickness of coverage	g
w_s	Specific volume of water	m^3
x	Distance	m
X	Concentration of each ion in solution	$ions\,cm^{-3}$
X	Negatively charged surface (assuming an adequate number of exchange sites are present)	unitless
x_d	Equals d; distance in equation 3.70	m
x_i	Mole fraction of ith ions	$ions\,mol^{-1}$
y	Distance from reference	m
Y	Concentration of each electrolyte between plates	$mol\,L^{-1}$
z	Distance	m
z	Depth relative to soil surface	m
z	Valence of ion	unitless
z	Gravitational potential	$J\,m^{-3}$
Z_D	Damping depth	m
Z	Depth of soil column or water-table depth	m

(continued)

Symbol	Description	Units
Z	Concentration of dissociated and absorbed cations	ions cm^{-3}
$Z(p)$	Compressibility factor	unitless
z_{soil}	Depth to soil reference point	m
z_{wt}	Depth to water table	m
Z_x	Increment in distance from standard height of point X	m
∇	Laplacian differential operator	unitless
Å	Angstrom	10^{-10} m
∂q	Increment of heat evolved	W m^{-2}
∂_w	Increment of water added	m^3
α	A constant	unitless
α	Dispersivity	cm
α	Aquifer compressibility	unitless
α	Polarizability	m^3
β	Bowen ratio, soil specific constant, fluid compressibility	dimensionless
γ	Psychrometric constant	dimensionless
γ	Temperature change with depth	°C m^{-1}
γ	Surface tension	ergs cm^{-2} or dynes cm^{-1} or J m^{-2} or N m^{-1}
γ_T	Thermodynamic psychrometric constant	unitless
γ_i	Molar activity coefficient of species i	unitless
γ_{LV}	Surface tension of the fluid	J m^{-2} or ergs cm^{-2}
γ_m	Surface tension or energy of the material	J m^{-2}
γ_{YL}	Interfacial tension	J m^{-2} or N m^{-1}
δ	Plate thickness at distance $2d$, depth	m
δ	Net increase of moles of gas in formation reaction	unitless
$\Delta\psi$	Potential difference	J m^{-3}
ΔG^o	Gibbs free energy of formation	kcal mol^{-1}
ΔG^s	Change in surface free energy	kcal mol^{-1}
ΔH	Quantity of heat	J
ΔH_f	Heat of fusion	kJ mol^{-1}
ΔH^o	Change of enthalpy	kJ mol^{-1}
ε	Ratio of molecular weights of water vapor and dry air	unitless
ε	Emissivity	unitless
ε	Dielectric constant (80 for water)	unitless
ε	Permittivity of the solution	esu^2 erg^{-1} cm^{-1}
ε_o	Permittivity constant of vacuum	8.854*10^{-12} Fm
ε_v	Velocity for a potiental gradient of 1 volt	m s^{-1}
ζ	Zeta potential	V
ζ	Temperature change with depth	K
ζ	Density	kg m^{-3}
η	Dynamic viscosity	Pa s^{-1}
η_e	Electrolyte concentration	ion cm^{-3}
θ	Angle	radians
θ	Volumetric water content	m^3 m^{-3}
θ_a	Volume fraction of air	m^3 m^{-3}
θ_i	Initial volumetric water content	m^3 m^{-3}
θ_m	Maximum volumetric water content	m^3 m^{-3}
θ_o	Initial soil-water content	m^3 m^{-3}
θ_r	Residual saturation	m^3 m^{-3}
θ_s	Saturated water content	m^3 m^{-3}
θ_t	Volumetric water content of transition zone during infiltration	m^3 m^{-3}
θ_x	Volumetric water content at distance x	m^3 m^{-3}
$\cos\theta$	Contact angle	degrees
κ	Debye	m^{-1}
κ^{-1}	Effective thickness	m
κ_c	Thermal conductivity	W m^{-1} K^{-1}

(*continued*)

Symbol	Description	Units
κ_i	Thermal conductivity of a soil constituent	W m^{-1} K^{-1}
$\kappa(t)$	Soil-water suction function	J m^3
λ	Wavelength	nm
λ	Pore size distribution index	—
λ	Decay constant for Carbon 14	$1.2*10^{-4}$ yr^{-1}
λ	Wavelength of London frequency	nm
λ'	Specific or electrical conductivity of water	S m^{-1} or ohms cm^{-1}
μ	Rate constant	s^{-1}
μ	Potential energy of medium under consideration	J kg^{-1}
μ_f	Chemical potential at a distance 1/2 particle diameter from solid surface	J kg^{-1}
μ_i	Chemical potential of the ith component	J kg^{-1}
μ_i	Chemical potential between two plates	J kg^{-1}
μ_o	Chemical potential	J kg^{-1}
μ_v	Chemical potential of water (vapor phase)	J kg^{-1}
μ_w	Chemical potential for the soil-water component	J kg^{-1}
$\mu_w(m_i)$	Chemical potential of soil water	J kg^{-1}
μ_w^o	Chemical potential for pure free water at the same temperature and atmospheric pressure	J kg^{-1}
μ_y	Chemical potential at point y	J kg^{-1}
ν	Kinematic viscosity	m^2 s^{-1}
ν_a	Viscosity of air	m^2 s^{-1}
ν_d	Volume drained (fluid)	m^3
Π	Osmotic pressure of solution	J kg^{-1}
Π	Period of time	T
Π	Osmotic potential of solution	J kg^{-1}
π	3.14159...	unitless
π_i	Osmotic coefficient of solute i	J kg^{-1}
ρ	Reaction constant	—
ρ	Density of solution	kg m^{-3}
ρ_a	Dry air density	kg m^{-3}
ρ_α	Density of phase	kg m^{-3}
ρ_f	Fluid density	kg m^{-3}
ρ_g	Gas phase density	kg m^{-3}
ρ_s	Density of solid (particle) phase	kg m^{-3}
ρ_{vapor}	Density of vapor	kg m^{-3}
ρ_v	Water vapor density	kg m^{-3}
ρ_w	Density of water	kg m^{-3}
ρ_b	Bulk density	kg m^{-3}
σ	Stefan–Boltzmann constant	10^{-8}Wm^{-2}K^{-4}
σ	Charge on colloid surface	esu cm^{-2} or meq cm^{-2}
σ	Adsorption per unit area	mol m^{-2}
σ	Substituent constant	—
σ	Surface energy per unit area of fluid–gas interface	J m^{-2}
σ^+	Charge due to a surplus of cations	C m^{-2}
σ^-	Charge due to a deficiency of cations	V
σ_g	Intergranular pressure	Pa
σ_r	Modulus of rupture	dynes cm^{-2}
σ_s	Soil stress	N
σ_s	Limiting stress	N
σ_T	Total charge	V
τ	Tortuosity	m m^{-1}
τ	Transmissivity	m^2 s^{-1}
τ	Tensile stress normal to the plane of the crack	N
$\tau_{1/2}$	Half-life	s
τ_f	Internal frictional force	N

(continued)

Symbol	Description	Units
ε	Geometric term	m^3
ϕ	Polar coordinate	unitless
ϕ	Phase shift	unitless
ϕ	Porosity	unitless
ϕ_α	Volume fraction of considered phase	unitless
ϕ_a	Volume fraction of air in soil pores	unitless
ϕ_e	Effective porosity	unitless
ϕ_g	Volume fraction of gas in soil pores	unitless
ϕ_i	Gain or loss via various source/sink(s)	$kg\,m^{-3}\,s^{-1}$
ϕ_l	Volume of water in media pores	unitless
ϕ_r	Residual water content	unitless
ϕ_s	Volume of solids in soil pores	unitless
ϕ_t	Total volume of soil	unitless
ϕ_v	Volume of all void spaces	unitless
ψ	Effective suction at wetting front	$J\,m^{-3}$
ψ_b	Brooks–Corey bubbling pressure	$J\,m^{-3}$
ψ_o	Saturated suction (matric suction = 0)	$J\,m^{-3}$
ψ	Water potential	$J\,m^{-3}$
ψ	Electrical potential of the colloid at specified distance	V
ψ	Surface charge	V
ψ_a	Pneumatic (air or vapor) potential	$J\,m^{-3}$
ψ_b	Overburden pressure potential	$J\,m^{-3}$
ψ_c	The matric potential for which $K = 1/2K_s$	$J\,m^{-3}$
ψ_d	Electric potential between layers	V
ψ_g	Gravitational potential	$J\,m^{-3}$
ψ_h	Hydrostatic pressure potential	$J\,m^{-3}$
ψ_i	Partial potential	$J\,m^{-3}$
ψ_m	Matric potential	$J\,m^{-3}$
ψ_{mL}	Matric potential of the loaded sample	$J\,m^{-3}$
ψ_{mu}	Matric potential of the unloaded sample	$J\,m^{-3}$
ψ_o	Osmotic potential	$J\,m^{-3}$
ψ_o	Electric potential at surface	$ergs\,esu^{-1}$
ψ_p	Pressure potential	$J\,m^{-3}$
ψ_r	Potential at pore neck (r in meters)	$J\,m^{-3}$
ψ_R	Potential in middle of pore (R in meters)	$J\,m^{-3}$
ψ_t	Total potential	$J\,m^{-3}$
ψ_w	Value of the matric potential when soil stress and external air pressure are zero	$J\,m^{-3}$
ω	Angular frequency	$radians\,s^{-1}$
ω	Angular velocity	$degrees\,s^{-1}$

Note: Units in the symbols list are given as equivalent or derived SI units. However, these units may not be exactly the same as the units for the specified symbol within the text. It should also be noted that we have used $J\,m^{-3}$ for most of our potentials. This is easier to use unless one is concerned with the specific water density being investigated within the unsaturated zone. See pp. 587–595 for further assistance concerning units.

REFERENCES

SUGGESTED READINGS

Adamson, A. W. 1990. *Physical chemistry of surfaces*. New York: John Wiley & Sons.

Alberty, R. A. 1987. *Physical chemistry*. New York: John Wiley & Sons.

Barnsley, M. F., R. L. Devaney, B.B. Mandelbrot, H. O. Peitgen, D. Saupe, and R. F. Voss. *The science of fractal images*. New York: Springer-Verlag.

Bear, Jacob. 1972. *Dynamics of fluid in porous media*. New York: Dover Publications.

Bear, J. and Y. Bachmat. *Introduction to modeling of transport phenomena in porous media*. London: Kluwer Academic Publishers.

Brady, N. C. 1974. *The nature and properties of soils*. New York: Macmillan.

Cushman, J., ed. 1990. *Dynamics of fluids in hierarchical porous media*. New York: Academic Press.

Feder, Jens. 1988. *Fractals*. New York: Plenum Press.

Fetter, C. W. 1994. *Applied hydrogeology*. 3d ed. Englewood Cliffs, NJ: Prentice Hall.

Freeze, R. A. and J. A. Cherry. 1972. *Groundwater*. Englewood Cliffs, NJ: Prentice Hall.

Grimshaw, R. W. 1971. *The chemistry and physics of clays and allied ceramic materials*. New York: Wiley-Interscience.

Hermance, John, F. 1998. *A mathematical primer on groundwater flow: An introduction to the mathematical and physical concepts of saturated flow in the subsurface*. Englewood Cliffs, NJ: Prentice Hall.

Hillel, Daniel. 1982. *Introduction to soil physics*. London: Academic Press.

Kirkham, D. and W. C. Powers. 1972. *Advanced soil physics*. New York: John Wiley & Sons.

Smith, G. D. 1985. *Numerical solutions of partial differential equations: Finite difference methods*. Oxford: Clarendon Press.

Stumm, Werner. 1992. *Chemistry of the solid-water interface: Processes at the mineral-water and particle-water interface in natural systems*. New York: John Wiley & Sons.

Todd, D. K. (1967). *Ground water hydrology*. New York: John Wiley & Sons.

Abeliuk, R., R. Riquelme, and M. Matthey. 1993. Solute transport model: Irrigation with contaminated wastewater. In *Modelling, Measuring and Prediction*. Water Pollution II, 607–14. Boston: Computational Mechanics.

Abriola, L. M. 1987. Modeling contaminant transport in the subsurface: An interdisciplinary challenge. *Rev. Geophys.* 25:125–34.

Abu-El-Sha'r, W., and L. M. Abriola. 1997. Experimental assessment of gas transport mechanisms in natural porous media: L Parameter evaluation. *Water Resour. Res.* 33(4):505–16.

Adamczyk, Z. and J. Petlicki. 1987. Adsorption and desorption kinetics of molecules and colloidal particles. *J. Colloid Interface Sci.* 118:20–49.

Adams, C. D. and E. M. Thurman. 1991. Formation and transport of deethylatrazine in the soil and vadose zone. *J. Env. Qual.* 20:540–47.

Adams, J. E. and R. J. Hanks. 1964. Evaporation from soil shrinkage cracks. *Soil Sci. Soc. Amer. Proc.* 28:281–84.

Adamson, A. W. 1976. *Physical chemistry of surfaces*. 3d ed. New York: Wiley-Interscience.

———. 1990. *Physical chemistry of surfaces*. 5th ed. New York: John Wiley & Sons.

Addiscott, T. M. and R. J. Wagenet. 1985. Concepts of solute leaching in soils: A review of modeling approaches. *J. Soil Sci.* 36:411–24.

Ahlstrom, S. W. and R. G. Baca. 1974. *Transport model user's manual,* BNWL-1716. Richland, WA: Battelle Pacific Northwest Laboratory.

Akamigbo, F. O. R. and J. B. Dalrymple. 1985. Experimental simulation of the results of clay translocation in the B horizons of soils; the formation of intrapedal cutans. *J. Soil Sci.* 36:401–9.

Alberts, J. J., M. A. Wahlgren, D. M. Nelson, and P. J. Jehn. 1977. Submicron particle size and charge characteristics of Pu in natural waters. *Environ. Sci. Technol.* 11:673–76.

Aldon, E. F. 1972. Reactivating soil ripping treatments for runoff and erosion control in the southwestern U.S. *Ann. Arid Zone* 11:154–60.

Allison, G. B., G. W. Gee and S. W. Tyler. 1994. Vadose-zone techniques for estimating groundwater recharge in arid and semiarid regions. *Soil Sci. Soc. Am. J.* 58:6–14.

Aly, S. M. and J. Letey. 1988. Polymer and water quality effects on flocculation of montmorillonite. *Soil Sci. Soc. Am. J.* 52:1453–58.

ASTM (American Society for Testing and Materials). 1994. Standard test method for infiltration rate of soils in field using double-ring infiltrometers. Section 4 in *Construction 1994 Annual Book of ASTM Standards*. Philadelphia: ASTM.

Anderson, D. M. 1986. Heat of immersion. In *Methods of soil analysis*. *See* Klute 1986.

Anderson, E. A. 1968. Development and testing of snow pack energy balance equations. *Water Resour. Res.* 4(1):19–37.

———. 1976. *A point energy and mass balance model of a snow cover*. National Weather Service Technical Report NWS 19. Silver Spring, MD: National Oceanic and Atmospheric Administration.

Anderson, J. L. and J. Buoma. 1977a. Water movement through pedal soils: I. Saturated flow. *Soil Sci. Soc. Am. J.* 41:413–18.

———. 1977b. Water movement through pedal soils: II. Unsaturated flow. *Soil Sci. Soc. Am. J.* 41:419–23.

Andreae, M. O. and D. S. Schimel, eds. 1989. *Exchange of trace gases between terrestrial ecosystems and the atmosphere*. New York: John Wiley & Sons.

Angell, K. G. 1992. In situ remedial methods: Air sparging. *The National Environmental Journal* (January/February):20–23.

Ankeny, M. D., T. C. Kaspar, and R. Horton. 1988. Design for an automated tension infiltrometer. *Soil Sci. Soc. Am. J.* 52:893–96.

Aomine, S. and K. Egashira. 1970. Heat of immersion of allophanic clays. *Soil Sci. Plant Nutr. Tokyo* 16:204–11.

Armstrong, D. E. and J. G. Konrad. 1974. Nonbiological degradation of pesticides. In *Pesticides in soil and water*, ed. W. D. Guenzi, 123–31. American Society of Agronomy.

Atanasiu, N. 1956. Eilhard Alfred Mitscherlich. *Soil Sci.* 82:99–100.

Avnir, D., D. Farin, and P. Pfeifer. 1984. Molecular fractal surfaces. *Nature* 308: 261–63.

Babcock, K. L. 1963. Theory of the chemical properties of soil colloidal systems at equilibrium. *Hilgardia* 34:417–542.

Backer, D. G. 1965. Factors affecting soil temperature. *Minn. Farm Home Sci.* 22: 11–13.

Baehr, A. L. and C. J. Bruell. 1990. Applications of the Stefan-Maxwell equations to determine the limitations of Fick's law when modeling organic vapor transport in sand columns. *Water Resour. Res.* 26(6):1155–63.

Baier, J. H. and Dennis Moran. 1981. *Status report on aldicarb contamination of groundwater as of September 1981*. Hauppauge, NY: Suffolk County Department of Health Services.

Baker, R. S. and D. Hillel. 1990. Laboratory tests of a theory of fingering during infiltration into layered soils. *Soil Sci. Soc. Am. J.* 54:20–30.

Baldocchi, D. D. 1993. Scaling water vapor and carbon dioxide exchange from leaves to a canopy: Rules and tools. In *Physiological processes: Leaf to globe*, ed. J. Ehleringer and C. B. Field, 77–114. London: Academic Press.

Baldocchi, D. D., D. R. Matt, B. A. Hutchison, and R. E. McMillen. 1984. Solar radiation within an oak–hickory forest: An evaluation of the extinction coefficients for several radiation components during fully leafed and leafless periods. *Agric. & For. Meteorol.* 32:307–22.

Baldocchi, D. D. and T. L. Meyers. 1991. Trace gas exchange above the floor of a deciduous forest. I. Evaporation and CO_2 efflux. *J. Geophys. Res.* 96:7271–85.

Ballard, T. M. 1971. Role of humic carrier substances in DDT movement through forest soil. *Soil Sci. Soc. Am. Proc.* 35:145–47.

Barley, K. P. (1962). The effect of mechanical stress on the growth of roots. *J. Exp. Bot.* 13:95–110.

Barnes, C. J. and G. B. Allison. 1988. Tracing of water movement in the unsaturated zone using stable isotopes of hydrogen and oxygen. *J. Hydrol.* 100:143–76.

Barnhisel, R. I. and P .M. Bertsch. 1989. Chorites and hydroxy-interlayered vermiculite and smectite. In *Minerals in soil environments*, eds. J. B. Dixon and S. B. Weed. Madison, WI: SSSA.

Barnsley, M. F., R. L. Devaney, B. B. Mandelbrot, H. O. Peitgen, D. Saupe, and R. F. Voss. 1988. *The science of fractal images*. New York: Springer-Verlag.

Barouch, E., T. H. Wright, and E. Matijevic. 1987. Kinetics of particle detachment. I. General considerations. *J. Colloid Interface Sci.* 118:473–81.

Barr, S. S., R. H. Miller, and J. T. Logan. 1978. Some factors affecting denitrification in soils irrigated with wastewater. *J. Water Pollution Control Federation* 50:709–17.

Bartoli, F. and R. Philippy. 1987. The colloidal stability of variable-charge mineral suspensions. *Clay Minerals* 22:93–107.

Baver, L. D., W. H. Gardner, and W. R. Gardner. 1972. *Soil physics*. New York: John Wiley & Sons.

Baver, L. D. 1928. The relation of exchangeable cations to the physical properties of soils. *J. Am. Soc. Agron.* 20:921–41.

Bear, J. 1972. *Dynamics of fluids in porous media*. New York: Elsevier.

———. 1979. *Hydraulics of groundwater*. New York: McGraw-Hill.

Beckett, R. and N. P. Le. 1990. The role of organic matter and ionic composition in determining the surface charge of suspended particles in natural waters. *Coll. and Surfaces* 44:35–49.

Behara, N., S. K. Joshi, and D. P. Pati. 1990. Root contribution to total soil metabolism in a tropical forest in Orissa, India. *For. Ecol. Manag.* 36:125–45.

Bengtsson, G., C. Enfield, and R. Lindqvist. 1987. Macromolecules facilitate the transport of trace organics. *Sci. Total Env.* 67:159–64.

Benjamin, J. R. and C. A. Cornell. 1970. *Probability, statistics, and decision for civil engineers*. New York: McGraw-Hill.

Bennett, H. H. 1939. *Soil Conservation*. New York: McGraw-Hill.

Benson, S. W. 1978. *Thermochemical kinetics*. 2d ed. New York: John Wiley & Sons.

Benson, S. W., F. R. Cruickshank, D. M. Golden, G. R. Haugen, H. E. O'Neal, A. S. Rogers, R. Shaw, and P. Walsh. 1969. Additivity rules for the estimation of thermo-chemical properties. *Chem. Rev.* 69:279–324.

Bentley, H. W., F. M. Phillips, and S. M. Davis. 1986. Chlorine-36 in the terrestrial environment. In *Handbook of environmental isotope geochemistry*. Volume 1B: *The terrestrial environment*, ed. P. Fritz and J.-C. Foutes, 427–80. New York: Elsevier.

Berg, R. C., Morse, W. J., and Johnson, T. M. 1987. Hydrogeologic evaluation of the effects of surface application of sewage sludge to agricultural land near Rockton, IL. *Environ. Geo. Notes* 119.

Berg, W. A., J. T. Herron, H. P. Harbert, III, and J. E. Kiel. 1979. *Vegetative stabilization of Union Oil Company process B retorted oil shale, 1975–1978*. Department of Agronomy. Fort Collins, CO: Colorado State University.

Bernhard, R. K. and M. Chasek. 1953. Soil density determination by means of radioactive isotopes. *Nondestruct. Text* 11:17–23.

Bertrand, A .R. 1965. Rate of water intake in the field. In *Methods of soil analysis*, ed. C. A. Black, 197–208. Madison, WI: Soil Science Society of America.

Bertrand, A. R. and K. Sor. 1962. The effects of rainfall intensity on soil structure and migration of colloidal materials in soils. *Soil Sci. Soc. Am. Proc.* 26:297–300.

Beven, K. and P. Germann. 1982. Macropores and water flow in soils. *Water Resour. Res.* 18:1311–25.

Biggar, J. W. and D. R. Nielsen. 1962. Miscible Displacement II. Behavior of tracers. *Soil Sci. Soc. Am. Proc.* 26:125–28.

———. 1964. Chloride-36 diffusion during stable and unstable flow through glass beads. *Soil Sci. Soc. Am. Proc.* 28:591–94.

———. 1967. Miscible displacement and leaching phenomenon. In *Irrigation of agricultural lands*, ed. R. M. Hagan, H. R. Haise, and T. W. Edminster, 254–71. Madison, WI: American Society of Agronomy.

———. 1976. Spatial variability of the leaching characteristics of a field soil. *Water Resour. Res.* 12:78–84.

Bird, R. B., W. E. Stewart, and E. N. Lightfoot. 1960. *Transport phenomena*. New York: McGraw-Hill.

Bjerrum, L., J. Moum, and O. Eide. 1967. Application of electro-osmosis to a foundation problem in a Norwegian quick clay. *Geotechnique* 17:214–35.

Black, T. A., W. R. Gardner, and G. W. Thurtell. 1969. The prediction of evaporation drainage and soil water storage for a bare soil. *Soil Sci. Soc. Am. Proc.* 33:655–60.

Blaney, H. F. and W. D. Criddle. 1950, *Determining water requirements in irrigated areas from climatological and irrigation data*. Technical Paper 96. Washington, DC: Soil Conservation Service, U.S. Department of Agriculture.

———. 1957. *Report on irrigation requirements for West Pakistan*. Denver: Tipton and Kalmbach, Inc.

———. 1962, *Determining consumptive use and irrigation water requirements*. Technical Bulletin 1275. Washington, DC: Soil Conservation Service, U.S. Department of Agriculture.

Blaney, H. F. and E. G. Hanson. 1965. *Consumptive use and water requirements in New Mexico*. Technical Report 32. Santa Fe, NM: New Mexico State Engineer.

Boersma, L. 1965. Field measurement of hydraulic conductivity above the water table. In *Methods of soil analysis. See* Klute 1986.

Bolt, G. H. 1976. Soil physics terminology. *Int. Soc. Soil Sci. Bull.* 49:16–22.

———. 1982. Movement of solutes in soils: Principles of adsorption/exchange chromatography. In *Soil chemistry. B. Physico-chemical models*, ed. G. H. Bolt. Amsterdam: Elsevier.

Boltzmann, L. 1894. Zur integration des diffusiongleichung bei variabeln diffusions coefficienten. *Ann. Phys.* 53:959–64.

Bond, R. W., C. R. Cole, and P. J. Gutknecht. 1984. *Unsaturated groundwater flow model (UNSAT1D) computer code manual, CS-2434-CCM*. Palo Alto, CA: Electric Power Research Institute.

Borchardt, G. 1989. Smectites. In *Minerals in soil environments*, ed. J. B. Dixon and S. B. Weed. Madison, WI: Soil Science Society of America.

Bouma, J. 1979. Field measurement of soil hydraulic properties characterizing water movement through swelling clay soils. *J. Hydrol.* 45:149–58.

Bouma, J., L. W. Dekker, and J. C. F. M. Haans. 1979. Drainability of some Dutch clay soils: A case study of soil survey interpretation. *Geoderma* 22:193-203.

Bouwer, H. 1966. Rapid field measurements of air entry value and hydraulic conductivity of soil as significant parameters in flow system analyses. *Water Resources Research* 2(4):729–38.

Bouwman, A. F., I. Fung, E. Matthews, and J. John. 1993. Global analysis of the potential for N_2O production in natural soils. *Glob. Biogeochem. Cy.* 7(3):557–97.

Bouyoucos, G. J. 1917. Measurement of the inactive, or unfree, moisture in the soil by means of the dilatometer method. *J. Agr. Res.* 8:195–217.

———. 1921. A new classification of soil moisture. *Soil Sci.* 11:33–48.

———. 1927a. The hydrometer as a new and rapid method for determining the colloidal content of soils. *Soil Sci.* 23:319–30.

———. 1927b. The hydrometer as a new method for the mechanical analysis of soils. *Soil Sci.* 23:343–49

———. 1947. A new electrical resistance thermometer for soils. *Soil Sci.* 63:291–98.

Bowden, R. D., K. J. Nadelhoffer, R. D. Boone, J. M. Melillo, J. B. Garrison. 1993. Contributions of above-ground litter, below-ground litter, and root respiration to total respiration in a temperature hardwood forest. *Can. J. For. Res.* 23:1402–407.

Bower, C. A. and Goertzen, J. O. 1959. Surface area of soils and clays by an equilibrium ethylene glycol method. *Soil Sci.* 87:289–92.

Brakensiek, D. L. and W. J. Rawls. 1988. Effects of agricultural rangeland management systems on infiltration. In *Modeling Agricultural, Forest, and Rangeland Hydrology*. St. Joseph, MI: American Society of Agricultural Engineers.

Branson, F. A., G. F. Gifford, K. G. Renard, and R. F. Hadley. 1981. *Rangeland hydrology*, 47–72. Dubuque, IA: Kendall/Hunt.

Bremner, J. M. 1967. Nitrogenous compounds. In *Soil biochemistry*, Vol. 1, ed. A. D. McLaren and G. H. Peterson, 19–66. New York: Marcel Dekker.

Brenner, H. 1962. The diffusion model of longitudinal mixing in beds of finite length. Numerical values. *Chem. Eng. Sci.* 17:229–43.

Bresler, E. 1973a. Simultaneous tansport of solutes and water under transient unsaturated flow conditions. *Water Resour. Res.* 9:975–86.

———. 1973b. Anion exclusion and coupling effects in nonsteady transport through unsaturated soil. 1. Theory. *Soil Sci. Soc. Am. Proc.* 37:663–69.

———. 1977. Trickle-drip irrigation: Principles and application to soil-water management. In *Advances in Agronomy*, Vol. 29, ed. N. C. Brady, 343–93. Madison, WI: American Society of Agronomy.

Bresler, E. and G. Dagan. 1979. Solute dispersion in unsaturated heterogeneous soil at field scale. II. Application. *Soil Sci. Soc. Am. J.* 43:467–72.

Bresler, E., W. D. Kemper, and R. J. Hanks. 1969. Infiltration, redistribution and subsequent evaporation of water from soil as affected by wetting rate and hysteresis. *Soil Sci. Soc. Am. Proc.* 33:832–40.

Bresler, E., B. L. McNeal, and D. L. Carter. 1982. Saline and sodic soils: Principles–dynamics–modeling. New York: Springer-Verlag.

Bresler, E., H. Bielorai, and A. Laufer. 1979. Field test of solution flow models in a heterogeneous irrigated cropped soil. *Water Resour. Res.* 15:645-52.

Briggs, L. J. 1897. *The mechanics of soil moisture*. Bulletin 10, USDA Bureau of Soils, Washington, DC: U.S. Department of Agriculture.

———, 1950. Limiting negative pressure of water. *J. Appl. Phys.* 21:721–22.

Briggs, L. J. and J. W. McLane. 1907. *The moisture equivalent of soils*. Bulletin 45, USDA Bureau of Soils. Washington, DC: U.S. Department of Agriculture.

Briggs, L. J. and H. L. Shantz. 1912. *The wilting coefficient for different plants and its indirect determination*. Bulletin 230, USDA Bur. Plant Ind. Washington, DC: U.S. Department of Agriculture.

Broadbridge, P. and I. White. 1988. Constant rate rainfall infiltration: A versatile nonlinear model 1. Analytic solution. *Water Resour. Res.* 24(1):145–54.

Brooks, R. H. and A. T. Corey. 1964. *Hydraulic properties of porous media*. Hydrology Paper No. 3. Fort Collins, CO: Colorado State University.

———. 1966. Properties of porous media affecting fluid flow. *J. Irrig. Drain. Div., Am. Soc. Civil Eng.* 92:61–88.

Brooks, C. S. and W. R. Purcell. 1952. Surface area measurements on sedimentary rocks. *Trans. A.I.M.E.* 195:289–96.

Brown, D. S. and E. W. Flagg. 1981. Empirical prediction of organic pollutant adsorption in natural sediments. *J. Env. Qual.* 10:382–86.

Brown R.W. and B.P. van Haveren, eds. 1971. *Psychrometry in water relations research*. Proceedings, Symposium on Thermocouple Psychrometers. Logan, UT: Utah Agricultural Experiment Station, Utah State University.

Bruce, K., W. Flach, and H. M. Taylor, eds. *Field soil water regime*. SSSA Special Publications Series No. 5. Madison, WI: Soil Science Society of America.

Bruce, R. R. 1972. Hydraulic conductivity evaluation of the soil profile from soil water retention relations. *Soil Sci. Soc. Am. Proc.* 36:555–61.

Bruce, R. R. and A. Klute. 1956. The measurement of soil moisture diffusivity. *Soil Sci. Soc. Am. Proc.* 20:458–62.

———. 1962. Measurements of soil moisture diffusivity from tension plate outflow data. *Soil Sci. Soc. Am. Proc.* 26:18–21.

Bruce, R. R. and F. O. Whisler. 1973. Infiltration of water into layered field soils. In *Physical aspects of soil water and salts in ecosystems*, ed. A. D. Hadas et al. New York: Springer-Verlag.

Brunauer, S. 1945. *The adsorption of gases and vapors*. Vol. 1, Princeton, NJ: Princeton University Press.

Brunauer, S., P. H. Emmett, and E. Teller. 1938. Adsorption of gases in multimolecular layers. *J. Am. Chem. Soc.* 60:309–19.

Brunini, O. and G.W. Thurtell. 1982. An improved thermocouple hygrometer for insitu measurements of soil water potential. *Soil Sci. Soc. Am. J.* 46:900–4.

Brutsaert, W. and D. Chen. 1995. Desorption and the two stages of drying of natural tallgrass prairie. *Water Resour. Res.* 31(5):1305–313.

Buck, A.L. 1981. New equations for computing vapor pressure and enhancement factor. *J. Appl. Meteor.* 20:1527–32.

Buckingham, E. 1904. *Contributions to our knowledge of the aeration of soils*. USDA Bulletin 25. Washington, DC: U.S. Department of Agriculture.

Buckingham, E. 1907. *Studies on the movement of soil moisture*. Bulletin 38, USDA Bureau of Soils. Washington, DC: U.S. Department of Agriculture

Buddemeier, R. W. and J. P. Hunt. 1988. Transport of colloidal contaminants in ground-water: radionuclide migration at the Nevada test site. *Applied Geochem.* 3:535–48.

Budyko, M. I., ed. 1963. *Guide to the atlas of the heat balance of the earth* (translated from Russian), U.S. Department of Commerce, WB\T-106.

Bumb, A. C. 1987. Unsteady-state flow of methane and water in coalbeds. Ph.D. Diss., University of Wyoming, Laramie.

Bumb, A. C., C. R. McKee, R. B. Evans, and L. A. Eccles. 1988. Design of lysimeter leak detector networks for surface impoundments and landfills. *Ground Water Monitoring Review*, Spring 1988.

Bumb, A. C., C. L. Murphy, and L. G. Everett. 1992. A comparison of three functional forms for representing soil moisture characteristics. *Ground Water* 30(2):177–85.

Buol, S. W., F. D. Hole, and R. J. McCracken. 1973. *Soil genesis and classification*. Ames, IA: Iowa State Press.

Camargo, M. N. and F. H. Beinroth, eds. 1978. Proceedings first international soil classification workshop (Brazil). Embrapa, Rio de Janeiro, Brazil.

Cameron, D. R. and A. Klute. 1977. Convective-dispersive solute transport with a combined equilibrium and kinetic adsorption model. *Water Resour. Res.* 13: 183–88.

Campbell, G. S. and Jan van Schilfgaarde. 1981. Use of SI units in soil physics. *J. Agron. Education.* 10:73–74.

Campbell, R. E. 1968. Production capabilities of some Upper Rio Puerco soils of New Mexico. USDA Forest Service Res. Note RM-108. Fort Collins, CO: Rocky Mountain Forest and Range Exp. St.

Carman, P. C. 1937. Fluid flow through a granular bed. *Trans. Inst. Chem. London.* 15:150–56.

Carsel, R. F., C. N. Smith, L. A. Mulkey, J. D. Dean, and P. P. Jowise. 1984. User's manual for the pesticide root zone model (PRZM): Release 1. USEPA EPA-600/3-84-109. Washington, DC: U.S. Government Printing Office.

Carslaw, H. S. and J. C. Jaeger. 1959. *Conduction of heat in solids*. 2d ed. New York: Oxford University Press.

Carter, D. L., M. M. Mortland, and W. D. Kemper. 1986. Specific surface. In *Methods of Soil Analysis*, ed. C. A. Black. Monograph 9. Madison, WI: Am. Soc. Agron.

Casagrande, L. 1959. *A review of past and current work on electro-osmotic stabilization of soils*. Harvard Soil Mechanics Series, No. 45. Cambridge, MA: Harvard University.

Casimir, H. B. G. and D. Polder. 1948. The influence of retardation on the London-van der Waals forces. *Phys. Rev.* 73:360–72.

Celia, M. A., E. T. Bouloutas, and R. Zarba. 1990. A general mass-conservative numerical solution for the unsaturated flow equation. *Water Resour. Res.* 27(7): 1483–96.

CH₂M-Hill. 1985. Soil Contaminant Evaluation Methodology (SOCEM I). In Guidance on Remedial Actions for Contaminated Soils at CERCLA Sites. *Ground Water* (May-June).

Chalky, J. W., J. Cornfield, and H. Park. 1949. A method of estimating volume-surface ratios. *Science* 110:295.

Champ, D. R. 1990. Observations on the movement of colloidal particles in saturated media at the Chalk River nuclear laboratories. *Hazardous Materials Control* (July/August):45–56.

Chapman, D. L. 1913. A contribution to the theory of electro-capillarity. Sixth Series London, Edinburgh and Dublin. *Philosophical Magazine* 25:475–81.

Chapman, H. D. 1965. Cation exchange capacity. In *Methods of Soil Analysis. See* Klute 1986.

Charbeneau, R. J. 1984. Kinematic models for soil moisture and solute transport, *Water Resour. Res.* 20(6):699–706.

———. 1989. Liquid moisture redistribution: Hydrologic simulation and spatial variability. In *Unsaturated flow in hydrologic modeling, theory and practice*. Proceedings of the NATO Advanced Research Workshop on Unsaturated Flow in Hydrologic Modelling Theory and Practice, Arles, France, 13–17 June 1988, ed. H. J. Morel-Seytoux, 127–60. Kluwer Academic Publishers.

Chepil, W.S. (1962). A compact rotary sieve and the importance of dry sieving in physical soil analysis. *Soil Sci. Soc. Am. Proc.* 26:4–6.

Childs, E. C. 1936. Transport of water through heavy clay soils. *I. J. Agr. Sci.* 26:392–405.

———. 1964. The ultimate moisture profile during infiltration in a uniform soil. *Soil Sci.* 97:173–78.

———. 1967. Soil moisture theory. *Hydrosci.* 4:73–117.

———. 1969. *The physical basis of soil water phenomena*. London: John Wiley & Sons.

Childs, E. C. and M. Bybordi. 1969. The vertical movement of water in stratified porous material, 1. Infiltration. *Water Resour. Res.* 5(2).

Childs, E. C. and N. Collis-George. 1948. Soil geometry and soil-water equilibrium. *Discuss. Faraday Soc.* 3:78–85.

———. 1950a. The permeability of porous materials. *Proc. R. Soc. London* A201: 392–405.

———. 1950b. Movement of moisture in unsaturated soils. *Trans. Int. Congr. Soil Sci.*, Amsterdam I:1–4.

Chiou, C. T. 1989. Partition and adsorption on soil and mobility of organic pollutants and pesticides. In *Toxic Organic Chemicals in Porous Media*, eds. Z. Gerstl, Y. Chen, U. Mingelgrin, and B. Yaron, 163–75. New York: Springer-Verlag.

Chiou, C. T., L. J. Peters, and V. H. Freed. 1979. A physical concept of soil-water equilibria for nonionic organic compounds. *Science* 206:831–32.

———. 1989. A physical concept of soil-water equilibria for nonionic organic compounds. *Science* 206:831–32.

Chiou, C. T., P. E. Porter, and D. W. Schmedding. 1983. Partition equilibria of nonionic organic compounds between soil organic matter and water. *Environ. Sci. Technol.* 17:227–31.

Chiou, C. T., R. L. Malcolm, T. I. Brinton, and D. E. Kile. 1986. Water solubility enhancement of some organic pollutants and pesticides by dissolved humic and fulvic acids. *Environ. Sci. Technol.* 20:502–8.

Cho, C. M. 1971. Convective tranport of ammonium with nitrification in soil. *Can. J. Soil Sci.* 51:339–50.

Christensen, R. C., E. B. Johnson, and G. G. Plantz. 1986. Manual for estimating selected streamflow characteristics of natural-flow streams in the Colorado River Basin in Utah. Geological Survey Water-Resources Investigations Report 85-4297, Prepared in cooperation with the U.S. Bureau of Land Management.

Cienciala, E. and A. Lindroth. 1995. Gas exchange and sap flow measurements of willow trees in short-rotation forest. I. Transpiration and sap flow. *Trees* 9:289–94.

Clapp, R. B. and G. M. Hornberger. 1978. Empirical equations for some soil hydraulic properties. *Water Resour. Res.* 14:601–4.

Cleary, R. W. and D. D. Adrian. 1973. Analytical solution of the convective-dispersive equation for cation adsorption in soils. *Soil Sci. Soc. Amer. Proc.* 37:197–99.

Clements, W. E. and M. H. Wilkening. 1974. Atmospheric pressure effects on ²²²Rn transport across the earth-air interface. *J. Geophys. Res.* 79(33):5025–29.

Clothier, B. E. and T. J. Sauer. 1988. Nitrogen transport during drip fertigation with urea. *Soil Sci. Soc. Amer.* 52:345–49.

Coats, K. H. and B. D. Smith. 1964. Dead end pore volume and dispersion in porous media. *Soc. Pet. Eng. J.* 4:73–84.

Code of Federal Regulations (10 CFR Part 60). 1987. Title 10, "Energy," Part 60, "Disposal of high-level radioactive wastes in geologic repositories." Washington, DC: U.S. Government Printing Office.

Cohen, R. M., J. W. Mercer, and J. Matthews. 1993. *DNAPL site evaluation*. Boca Raton, FL: CRC Press.

Cohen, S. 1988. State and USGS activities in New England. *Ground Water Monitoring Review*, Winter 1988.

Collis-George, N. 1955. Hysteresis in moisture content-suction relationships in soils, *Proc. Nat. Acad. Sci. India* 24A:80–85.

———. 1974. A laboratory study of infiltration-advance. *Soil Sci.* 117:282–87.

Conrad, R. 1989. Control of methane production in terrestrial ecosystems. In *Exchange of trace gases between terrestrial ecosystems and the atmosphere*, eds. M. O. Andreae and D. S. Schimel, 39–58. New York: John Wiley & Sons.

———. 1995. Soil microbial processes and the cycling of atmospheric trace gases. *Phil. Trans. R. Soc. Lond.* A351:219–30.

Cook, P. B., I. D. Jolly, F. W. Leaney, G. R. Walker, G. L. Allan, L. K. Fifield, and G. B. Allison. 1994. Unsaturated zone tritium and chlorine 36 profiles from southern Australia: Their use as tracers of soil water movement. *Water Resour. Res.* 30(6): 1709–19.

Cooney, D. O., B. A. Adesanya, and A. L. Hines. 1983. Effects of particle size distribution on adsorption kinetics in stirred batch systems. *Chem. Eng. Sci.* 38:1535–40.

Corey, A. T. 1977. *Mechanics of heterogeneous fluids in porous media*. Ft. Collins, CO: Water Resources Publications.

———. 1986. Air permeability. In *Methods of Soil Analysis, Part I*, ed. A. Klute, 1121–36. Madison, WI: Amer. Soc. Agron. & Soil Sci. Soc. Amer.

Corey, A. T. and A. Klute. 1985. Application of potential concept to soil water equilibrium and transport. *Reviews of Research. Soil Sci. Soc. Am. J.* 49:3–11.

Corey, J. C., D. R. Nielsen, and Don Kirkham. 1967. Miscible displacement of nitrate through soil columns. *Soil Sci. Soc. Am. Proc.* 31:497–501.

Couto, W., C. Sanzonowicz, and A. De O. Barcellos. 1985. Factors affecting oxidation-reduction process in an Oxisol with a seasonal water table. *Soil Sci. Soc. Am. J.* 49:1245–48.

Coutts, J. R. H., M. F. Kandil, J. L. Nowland and J. Tinsley. 1968a. Use of radioactive Fe for tracing soil particle movement. Part I. Field studies of splash erosion. *J. Soil Sci.* 19:311–24.

———. 1968b. Use of radioactive Fe for tracing soil particle movement. Part II. Laboratory studies of labeling and splash displacement. *J. Soil Sci.* 19:325–41.

Craig, P. M. and E. C. Davis. 1985. *Application of the finite element groundwater model FEWA to the engineered test facilities, Oak Ridge National Laboratory*. Publication No. 2581. Oak Ridge, TN: Environmental Sciences Division, Oak Ridge National Laboratory.

Craig, Jr., G. S. and J. G. Rankl. 1978. *Analysis of runoff from small drainage basins in Wyoming*. Geological Survey Water-Supply Paper 2056. Washington, DC: Federal Highway Administration.

Cramer, H. 1955. *Mathematical methods of statistics*. Princeton, NJ: Princeton University Press.

Crank, J. 1956. *The mathematics of diffusion*. London: Oxford Univ. Press.

———. 1975. *The mathematics of diffusion*. Oxford: Clarendon Press.

Croney, D., J. D. Coleman, and W. P. M. Black. 1958. Movement and distribution of water in soil in relation to highway design and performance. In *Water and its conduction in soils*, ed. H. F. Winterkorn. Special Rep. No. 40. Washington, DC: Highway Res. Board.

Cunningham, R. E. and R. J. J. Williams. 1980. *Diffusion in gases and porous media*. New York: Plenum.

Cushman, J. H. 1982. Nutrient transport inside and outside of the root rhizosphere. Theory. *Soil Sci. Soc. Amer. J.* 46:704–9.

Cushman, J., ed. 1990. *Dynamics of fluids in hierarchical porous media*. New York: Academic Press.

Daamen, C. C., L. P. Simmonds, J. S. Wallace, K. B. Laryea, and M. V. K. Sivakumar. 1993. Use of microlysimeters to measure evaporation from sandy soils. *Agric. & For. Meteor.* 65:159–73.

Dagan, G. and E. Bresler. 1983. Unsaturated flow in spatially variable fields. 1. Derivation of models of infiltration and redistribution. *Water Resour. Res.* 19(2): 413–20.

Dalton, J. 1802. Experimental essays on the constitution of mixed gases; on the force of steam or vapor from water and other liquids in different temperatures, both in a Torricellian vacuum and in air; on evaporation; and on the expansion of gases by heat. *Manchester Lit. Phi. Soc. Mem. Proc.* 5:536–602.

Dana, J. D. 1954. *A textbook of mineralogy*. New York: McGraw-Hill.

Dankwaerts, P. V. 1953. Continuous flow systems, *Chemical Engineering Science* 2:1–13.

Darcy, H. 1856. *Les fontaines publiques de la ville de Dijon*. Paris: Dalmont.

Daumas, M. 1958. The chemistry of principles. In *The beginnings of modern science*, ed. R. Taton, 322–23. London: Thames and Hudson.

Davidson, J. M., J. W. Biggar, and D. R. Nielsen. 1963. Gamma attenuation for measuring bulk density and transient water flow in porous materials. *J. Geophys. Res.* 68:4777–83.

Davidson, J. M., L. R. Stone, D. R. Nielsen, and M. E. LaRue. 1969. Field measurement and use of soil water properties. *Water Resour. Res.* 5:1312–21.

Davies, D. B. and D. Payne. 1988. Management of soil physical properties. In *Russel's Soil Conditions & Plant Growth*, ed. Alan Wild, 412–48. Essex, England: Longman Scientific & Technical.

Davies, J. A. 1979. Estimating the surface radiation balance and its components. In *Modification of the Aerial Environment of Crops*, eds. Barfield and Gerber, 183–210. St. Joseph, MI: Amer. Soc. Agric. Engrs.

Davis, J. A. 1982. Adsorption of natural dissolved organic matter at the oxide/water interface. *Geochim. Cosmochim. Acta* 46:2381–93.

Davis, L. A. and S. P. Neuman. 1983. Documentation and user's guide: UNSAT2— Variably saturated flow model. NUREG/CR-3390. Washington, DC: U.S. Nuclear Regulatory Commission.

Davis, L. A. and G. Segol. 1985. Documentation and user's guide: GS2 and GS3— Variably saturated flow and mass transport models. NUREG/CR-3901. Washington, DC: U.S. Nuclear Regulatory Commission.

Davis, L. S. 1980. Computer analysis of seepage and groundwater response beneath tailing impoundments. Report Grant NSF/RA-800054. Washington, DC: National Science Foundation.

Davis, T. L., J. K. Greig, and M. B. Kirkham, 1988. Wastewater irrigation of vegetable crops. *Biocycle* 29:60–63.

Davy, Sir Humphrey. 1813. *Elements of agricultural chemistry*. London: Longman, Hurst, Rees, Orme, and Brown.

Dawson, H. J., B. F. Hrutfiord, R. J. Zasoski, and F. C. Ugolini. 1981. The molecular weight and origin of yellow organic acids. *Soil Sci.* 132:191–99.

Day, P. R. 1953. Experimental confirmation of hydrometer theory. *Soil Sci.* 75:181–86.

Day, P. R. 1965. Particle fractionation and particle-size analysis. In *Methods of Soil Analysis, Part I*, eds. C. A. Black, F. E. Clark, L. G. Ensmiger, D. D. Evans, and J. L. White, 454–567. Madison, WI: Am. Soc. Agron.

de Laat, P. J. M. 1985. MUST, A simulation model for unsaturated flow. Report series no. 16. Delft, The Netherlands: International Institute for Hydraulic and Environmental Engineering.

De Smedt, F. and P. J. Wierenga. 1979a. Mass transfer in porous media with immobile water. *J. Hydrology* 41:59–67.

———. 1979b. A generalized solution for solute flow in soils with mobile and immobile water. *Water Resour. Res.* 15:1137–41.

———. 1984. Solute transfer through columns of glass beads. *Water Resour. Res.* 20: 225–32.

de Vries, D. A. 1963. Thermal properties of soils. In *Physics of plant environment*, ed. W. R. van Wijk, 210–35. Amsterdam: North-Holland.

———. 1975. Heat transfer in soils. In *Heat and mass transfer in the biosphere*, eds. D. A. de Vries and N. H. Afgan, 5–28. Washington, DC: Scripta Book Co.

DeWiest, R. J., ed. 1969. *Flow through porous media*. New York: Acadmic Press.

Debye, P. and E. Huckel. 1923. Theory for electro-chemical double layer. *Physik. Z.* 24:185.

Delin, G. N. and M. K. Landon. 1993. Effects of focused recharge on the transport of agricultural chemicals at the Princeton, Minnesota Management Systems Evaluation Area, 1991–1992. In *Agricultural research to protect water quality*, vol. I, 210–12. Proc. Conf., 21–24 February, Minneapolis, MN. Ankeny, IA: Soil and Water Conservation Society.

Denmead, O. T. 1991. Sources and sinks of greenhouse gases in the soil plant environment. *Vegetatio* 91:73–86.

Derjaguin, B. V. and N. V. Churaev. 1978. On the question of determining the concept of disjoining pressure and its role in the equilibrium and flow of thin films. *J. Colloid Interface Sci.* 66:389–98.

Derjaguin, B. V. and L. Landau. 1941. Theory of the stability of strongly charged lyophobic sols and the adhesion of strongly charged particles in solutions of electrolytes. *Acta Physicochim. URSS* 14:633–62.

Deshpande, T. L., D. J. Greenland, and J. P. Quirk. 1964. Role of iron oxides in the bonding of soil particles. *Nature* 201:107–8.

Deutel, S. K. 1988. Application of the precipitation-charge neutralization model of coagulation. *Environ. Sci. Technol.* 22:825–32.

Diamond, S. (1970). Pore size distribution in clays. *Clays Clay Mineral.* 18:7–24.

Dick, W. A., W. M. Edwards, and F. Haghiri. 1986. Water movement through soil to which no-tillage cropping practices have been continously applied. In *Proc. Agric. Impacts on Ground Water—A Conference*. Dublin, OH: Omaha Nebraska Natl. Water Well Assoc.

Diment, G. A. and K. K. Watson. 1983. Stability analysis of water movement in unsaturated porous materials, 2. Numerical studies. *Water Resour. Res.* 19:1002–10.

———. 1985. Stability analysis of water movement in unsaturated porous materials, 3. Experimental Studies. *Water Resour. Res.* 21:979–84.

Dingman, S.L. 1994. *Physical hydrology*. New York: Macmillan.

Dingman, S. L., W. F. Weeks, and Y.-C. Yen. 1968. The effects of thermal pollution on river ice conditions. *Water Resour. Res.* 4:347–62.

Doorenbos, J. and W. O. Pruitt. 1977. Guidelines for prediction crop water requirements. FAO Irrigation and Drainage Paper No. 24, 2d ed. Rome, Italy: U.N. Food and Agricultural Organization.

Dortignac, E.J. 1960. The Rio Puerco—past, present, and future. *NM Water Conserv. Proc.* 5:45–51.

Dragun, J. 1988. The soil chemistry of hazardous materials. Silver Spring, MD: The Hazardous Materials Control Research Institute.

Ducker, W. A., T. J. Senden, and R. M. Pashley. 1991. Direct measurement of colloidal forces using an atomic force microscope. *Nature* 353:239–41.

Dugas, W. A. 1990. Sap flow in stems. *Remot. Sens. Rev.* 5(1):225–35.

Dunham, R. J. and P. H. Nye. 1973. I. Soil water content gradients near a plane of onion roots. *J. Appl. Ecol.* 10:585–98.

Dwight, H. B. 1961. *Tables of integrals and other mathematical data*. 4th ed. New York: Macmillan Co.

Eckhardt, D. A. and R. J. Wagenet. 1996. Estimation of the potential for atrazine transport in a silt loam soil. ACS Symposium Series 630. 209th National Meeting of Amer. Chem. Soc. April 2–7, 1995. Washington, DC: Amer. Chem. Soc.

Edwards, W. M., J. J. Shipitalo, L. B. Owens, and L. D. Norton. 1990. Effect of *Lumbricus terrestris L.* burrows on hydrology of continuous no-till corn fields. *Geoderma* 46:73–84.

Ehlers, W. 1975. Observations on earthworm channels and infiltration on tilled and untilled loess soil. *Soil Sci.* 119:242–48.

Eichner, M. J. 1990. Nitrous oxide emission form fertilized soils: summary of available data. *J. Environ. Qual.* 19:272–80.

El-Kadi, A. I. 1983. INFIL: A Fortran IV program to calculate infiltration rate and amount, and water content profile at different times. FOS-20. Indianapolis, IN: International Ground Water Modeling Center, Holcomb Research Institute.

Elliott, H. A. 1986. Land application of municipal sewage sludge. *J. Soil Water Conserv.* 1:5–10.

Elrick, D. E. 1963. Unsaturated flow properties of soils. *Aust. J. Soil Res.* 1(1):1–8.

Emerson, C. J., B. Thomas, Jr., and R.J. Luxmoore. 1984. CADIL: Model documentation for chemical adsorption and degradation in land. ORNL/TM-8972. Oak Ridge, TN: Oak Ridge National Laboratory.

Emerson, E. W. 1959. The structure of soil crumbs. *J. Soil Sci.* 10:235.

Emerson, W. W. 1967. A classification of soil aggregates based on their coherence in water. *Aust. J. Soil Res.* 5:47–57.

Enfield, C. G. and G. Bengtsson. 1988. Macromolecular transport of hydrophobic contaminants in aqueous environments. *Ground Water* 26:64–70.

Enfield, C. G., G. Bengtsson, and R. Lindquist. 1989. Influence of macromolecules on chemical transport. *Environ. Sci. Tech.* 23:1278–86.

Enfield, C. G. and R. S. Kerr. 1990. Chemical transport facilitated by colloidal-sized organic molecules. *Hazardous Materials Control.* July/August:50–51.

Entekhabi, D., I. Rodriguez-Iturbe, and F. Castelli. 1996. Mutual interaction of soil moisture state and atmospheric processes. *J. Hydrol.* 184:3–17.

Environmental Science Service Administration (ESSA). 1983. *Climatic atlas of the United States*. Washington, DC: U.S. Department of Commerce.

Environmental Systems and Technologies, Inc. (ES&T) and International Ground Water Modeling Center (IGWMC). 1992. *Petroleum hydrocarbons and organic chemicals in groundwater: Practical models for site assessment and remediation*. A short course organized by ES&T, Inc., Blacksburg, Virginia and the IGWMC, Golden, Colorado. December 1–4. Orlando, Florida.

Environmental Systems & Technologies, Inc. (ES&T) and Hydrogaia, Inc. 1994. VADSAT: A Monte Carlo model for assessing the effects of land-disposed exploration and production wastes on groundwater quality, user and technical guide, Version 2.0. April. Prepared for the American Petroleum Institute, GW-23 Project.

Epstein, L. and G. R. Safir. 1982. Plant diseases associated with municipal wastewater irrigation. In Land treatment of municipal wastewater: Vegetation selection and management. Ann Arbor, MI: Ann Arbor Science.

Ericksson, E. 1971. Compartmental models and reservoir theory. *Annual Review of Ecological Systems* 2:67–84.

Esrig, M. I. and S. Majtenyi. 1965. *A feasibility study of electrokinetic processes for stabilization of soils for military mobility purposes: Report 2—An analysis of the electro-osmotic phenomenon in soil capillary systems*. Ithaca, NY: Cornell University.

Everett, D. H. 1972. Manual of symbols and terminology for physiochemical quantities and units. *Pure Applied Chem.* 31:577–638.

Falta, R. W., I. Javendel, K. Pruess, and P. A. Witherspoon. 1989. Density-driven flow of gas in the unsaturated zone due to the evaporation of volatile organic compounds. *Water Resour. Res.* 25(10):2159–69.

Farrell, D. A., E. L. Greacen, and C. G. Gurr. 1966. Vapor transfer in soil due to air turbulence. *Soil Sci.* 102:305–13.

Fayer, M. H. and G. W. Gee. 1985. UNSAT-H: An unsaturated soil water flow code for use at the Hanford site: Code documentation, PNL-5585. Richland, WA: Battelle Pacific Northwest Laboratory.

Feder, J. 1989. *Fractals*. New York: Plenum Press.

Federer, C. A. 1968. Radiation and snowmelt on a clearcut watershed. In *Proceedings of the Eastern Snow Conference*, 23–42.

Federer, C. A. and D. Lash. 1978. BROOK: A Hydrologic simulation model for eastern forests. Research Report No. 19. Durham, NH: University of New Hampshire Water Resources Research Center.

Fenn, D. G., K. J. Hanley, and T. V. DeGeare. 1975. Use of the water balance method for predicting leachate generation from solid waste disposal sites. EPA/530/SW-168. October. Washington, DC: U.S. Environmental Protection Agency.

Fiering, M. B. and B. B. Jackson. 1971. *Synthetic streamflows*. Water Resources Monograph 1. Washington, DC: American Geophysical Union.

Finlayson-Pitts, B. J. and J. N. Pitts. 1986. *Atmospheric chemistry: fundamentals and experimental techniques*. New York: John Wiley & Sons.

Firestone, M. K. and E. A. Davidson. 1989. Microbiological basis of NO and N_2O production and consumption in soil. In *Exchange of trace gases between terrestrial ecosystems and the atmosphere*, eds. M. O. Andreae and D. S. Schimel, 7–22. New York: John Wiley & Sons.

Fischer, H. B., E. List, R. C. Y. Koh, J. Imberger, and N. H. Brooks. 1979. *Mixing in inland and coastal waters*. New York: Academic Press.

FitzGerald, D. 1886. Evaporation. *Transactions of the American Society of Civil Engineers* 15:581–646.

Fleagle, R. G. and J. A. Businger. 1963. *An introduction to atmospheric physics*. New York: Academic Press.

Fluhler, H. and W. A. Jury. 1983. *Estimating solute transport using nonlinear, rate dependent, two-site adsorption models: An introduction to use explicit and implicit finite difference schemes*. Fortran Program Documentation, Report 245. Birmensdorf, Switzerland: Swiss Federal Institution of Forest Research.

Fokas, A. S. and Y. C. Yortsos. 1982. On the exactly solvable equation occurring in two-phase flow in porous media. *SIAM* 43:318–32.

Food and Agriculture Organization of the United Nations. 1985. *Fertilizer Yearbook*, vol. 35 FAO, Rome Italy.

Foster, R. J. 1969. *General geology*. Columbus, OH: Charles Merrill Publ.

Francis, R. E. 1986. *Phyto-edaphic communities of the Upper Rio Puerco Watershed, New Mexico*. USDA Forest Service Res. Paper RM-272. Fort Collins, CO: Rocky Mountain Forest and Range Exp. Stn.

Freeze, R. A. 1975. A stochastic conceptual analysis of one-dimensional groundwater flow in non-uniform homogeneous media. *Water Resour. Res.* 11:725–41.

———. and J. A. Cherry. 1979. *Groundwater*. Inc. Englewood Cliffs, NJ: Prentice Hall.

Frere, M. H. C. A. Onstad and H. N. Holtan. 1975. *ACTMO, An agricultural chemical transport model*. Washington, DC: U.S. Department of Agriculture, Agriculture Research Service, *ARS-H-3*.

Fresquez, P. R. Francis, R. E. and Dennis, G. L. 1988. Fungal communities associated with phytoedaphic communities in the semi-arid Southwest. *Arid Soil Res. Rehab.* 2:187–202.

Fried, J. J. and M. A. Combarnous. 1971. Dispersion in porous media. *Adv. Hydrosci.* 9:169–282.

Friedman, H. L. and C. V. Krishman. 1973. Aqueous solutions of simple electrolytes. In *Water*, vol 3. (ed. F. Franks, p. 55.) New York: Plenum Press.

Frissel, M. J. P. Poelstra, and P. Reiniger. 1970. Chromatographic transport through soils. III. A simulation model for the evaluation of the apparent diffusion coefficient in undisturbed soils with tritrated water. *Plant and Soil* 33:161–76.

Fritschen, L. J. and L. W. Gay. 1979. *Environmental instrumentation*. New York: Springer-Verlag.

Fritschen, L. J. and C. H. M. van Bavel. 1962. Energy balance components of evaporating surfaces in arid lands. *J. Geophys. Res.* 67:5179–85.

Gao, Z. T. Sun, and X. Qu, 1991. Studies on the land treatment irrigation system of municipal wastewater in Shengyang. *Water Science and Technology* 24:47–53.

Garcia, R. L., J. M. Norman, and D. K. McDermitt. 1990. Measurements of canopy gas exchange using an open chamber system. *Remot. Sens. Rev.* 5(1):141–62.

Gardner, W. 1919. Capillary moisture-holding capacity. *Soil Sci.* 7:319–24.

———. 1920. The capillary potential and its relation to soil-moisture constants. *Soil Sci.* 10:357–59.

———. 1955. Relation of temperature to moisture tension of soil. *Soil Sci.* 79:257–65.

Gardner, W., O. W. Israelsen, N. E. Edlefsen, and H. Clyde. 1922. The capillary potential function and its relation to irrigation practice. *Phys. Rev.* 20:196.

Gardner, W. and J. A. Widstoe. 1921. The movement of soil moisture. *Soil Sci.* 11:215–32.

———. H. R. 1974. Prediction of water loss from a fallow field soil based on soil water flow theory. *Soil Sci. Soc. Am. Proc.* 38:379–82.

Gardner, W. H. and C. Calissendorff. 1967. Gamma-ray and neutron attenuation in measurement of soil bulk density and water content. In *Symposium on use of*

isotope and radiation techniques in soil physics and irrigation studies. Rome: Food and Agriculture Organization.

Gardner, W. H. 1986. Water content. In *Methods of Soil Analysis*. Part I, Physical and Mineralogical methods, 2d. ed. ed. Arnold Klute pp. 493–541. Monograph 9, Am. Soc. Agron., Madison, Wisconsin.

Gardner, W. R. 1956. Calculation of capillary conductivity from pressure plate outflow data. *Soil Sci. Soc. Amer. Proc.* 20:317–320.

———. 1958. Some steady-state solutions of the unsaturated moisture flow equation with application to evaporation from a water table. *Soil Sci.* 85(4):228–32.

———. 1959. Solutions of the flow equation for the drying of soils and other porous media. *Soil Sci. Soc. Amer. Proc.* 30:425–28.

Gardner, W. R. and D. I. Hillel. 1962. The relation of external evaporative conditions to the drying of soils. *J. Geophys. Res.* 67:4319–325.

———. and D. Kirkham. 1952. Determination of soil moisture by neutron scattering. *Soil Sci.* 73:391–401.

Gardner, W. R., D. Hillel, and Y. Benyamini. 1970a. Post-irrigation movement of soil water. 1. Redistribution. *Water Resour. Res.* 6(3):851–61.

———. 1970b, Post-irrigation movement of soil water. 2. Simultaneous Redistribution and Evaporation, *Water Resour. Res.* 6(4):1148–53.

Garratt, J. R. 1992. *The atmospheric boundary layer*. New York: Cambridge Univ. Press.

Garrels, R. M. and C. L. Christ. 1965. *Solutions, minerals and equilibria*. New York: Harper and Row.

Gates, D. M. 1980. *Biophysical ecology*. New York: Springer-Verlag.

Gay, L. W. 1972. On the construction and use of ceramic wick thermocouple psychrometers. In *Psychrometry in water relations research*, eds. R. W. Brown and B. P. van Haveren, pp. 253–58. Proc. Symp. on Thermocouple Psychrometers. Logan UT: Utah St. Univ., Utah Agric. Expt. Sta.

Geiger, R. 1965. *Climate near the ground*. London: Oxford Univ. Press.

Ghildyal, B. P. and R. P. Tripathi. 1987. *Soil phy*. New York: Halsted Press.

Gianessi, L. P. and C. Puffer. 1991. *Herbicide use in the United States*. Washington, DC: Resources for the Future, Quality of the Environment Division.

Gibbs, R. J. 1983. Effect of natural organic coatings on the coagulation of particles. *Environ. Sci. Technol.* 17:237–40.

Gibbs, R. J., M. D. Matthews, and D. A. Link. 1971. The relationship between sphere size and settling velocity, *J. Sed. Petrol.* 41:7–18.

Giddings, J. C. 1965. *Dynamics of chromatography*. Part 1. Principles and Theory. New York: Marcel Dekker.

Gill, W. R. 1979. Tillage. In *The Encyclopedia of Soil Science* Part 1. (R. W. Fairbridge and C. W., Finkl, Jr.) pp. 566–71. Stroudsburg, PA: Dowden, Hutchinson & Ross.

Gillham, R. W., E. A. Sudicky, J. A. Cherry, and E. O. Frind. 1984. An advection-diffusion concept for solute transport in heterogeneous unconsolidated geological deposits. *Water Resour. Res.* 20:369–78.

Gish, T. and A. Shirmohammadi. 1991. *Preferential flow*. St. Joseph, MI: American Society of Agricultural Engineers.

Glass, R. J., T. S. Steenhusi, and J. Y. Parlange. 1989. Mechanism for finger persistence in homogeneous, unsaturated, porous media: Theory and verification. *Soil Sci.* 148:60–70.

Glauber, J. R. Des Teutschlandts Wohlfart (Erster Theil), das dritte Capittel. De concentratione Vegetabilium, Miraculum Mundi. Amsterdam, 1656.

Goldberg, S. and H. S. Forster. 1990. Flocculation of reference clays and arid-zone soil clays. *Soil Sci. Soc. Am. J.* 54:714–18.

Goltz, M. N. 1986. Three-dimensional analytical modeling of diffusion-limited solute transport. Ph. D. Diss. Stanford University, Stanford, CA.

Goulding, K. W. T., B. W. Hutsch, C. P. Webser, T. W. Willison, and D. S. Powlson. 1995. The effect of agriculture on methane oxidation in soil. *Phil. Trans. Roy. Soc. Lond.* A 351:313–25.

Gouy, M. 1910. Sur la Constitution de La charge electrique a'la surface d'un electrolyte. *Jourdal de Physique*, Series 4, 9:457–68.

Grable, A. R. 1966. Soil aeration and plant growth. *Adv. Agron.* 18:57–106.

Gray, D. M. and T. J. Prowse. 1993. "Chapter 7, Snow and Floating Ice," *In Handbook of Hydrology*, ed. D. R. Maidment. New York: McGraw-Hill, Inc.

Green, W. H. and G. A. Ampt. 1911. Studies on soil physics, Part I. Flow of air and water through soils. *J. Agr. Sci.* 4:1–24.

———. 1912. Studies on soil physics, Part II. Permeability of an ideal soil to air and water. *J. Agr. Sci.* 5:1–26.

Greenkorn, R. A. 1983. *Flow phenomena in porous media*. New York: Marcel Dekker, Inc.

Greenkorn, R. A. and D. P. Kessler. 1969. Dispersion in heterogeneous nonuniform anisotropic porous media. *Ind. Eng. Chem.* 61:14–32.

Greenland, D. J. 1977. Soil drainage by intensive arable cultivation: temporary or permanent? *Phil. Trans. Royal Soc.* London, B281:193–208.

Gregory, J. 1989. Fundamentals of flocculation. *Critical Reviews in Environmental Control* 19:185–230.

Grim, R. E. 1953. *Clay mineralogy*. New York: McGraw-Hill Book Company, Inc.

Grove, J. H., C. S. Fowler, and M. E. Sumner. 1982. Determination of the charge character of selected acid soils. *Soil Sci. Soc. Am. J.* 46:32–38.

Grygier, J. C. and J. R. Stedinger. 1990. *SPIGOT, a synthetic streamflow generation software package, technical description, version 2.5*. School of Civil and Environmental Engineering, Cornell University, Ithaca, New York.

Gschwend, P. M. 1990. Formation and mobilization of colloidal particles in contamination plumes. *Hazardous Materials Control* July/August:49.

Gschwend, P. M. and M. D. Reynolds. 1987. Monodisperse ferrous phosphatecolloids in an anoxic groundwater plume. *J. Contam. Hydrol.* 1:309–27.

Gschwend, P. M. and S. Wu. 1985. On the constancy of sediment-water partition coefficients of hydrophobic organic pollutants. *Environ. Sci. Technol.* 19:90–96.

Gupta, S. P. and R. A. Greenkorn. 1974. Determination of dispersion and nonlinear adsorption parameters for flow in porous media. *Water Resour. Res.* 10:839–46.

Gupta, V. K. and R. N. Bhattacharya. 1986. Solute dispersion in multidimensional periodic saturated porous media. *Water Resour. Res.* 22:156–64.

Guven, O. and F.J. Molz. 1986. Deterministic and stochastic analyses of dispersion in an unbounded stratified porous medium. *Water Resour. Res.* 22:1565–574.

Guven, O., R. W. Falta, F. J. Molz, and J. G. Melville. 1985. Analysis and interpretation of single well tracer tests in stratified aquifers. *Water Resour. Res.* 21:676–84.

Gvirtzman, H. and M. Margaritz. 1986. Investigation of water movement in the unsaturated zone under an irrigated area using environmental tritium. *Water Resour. Res.* 22:635–42.

Hachum, A. Y. and J. F. Alfaro. 1980. Rain infiltration into layered soils: Prediction. *Journal of Irrigation and Drainage Engineering* 106:311–21.

Hagedorn, C., E. L. McCoy and T. M. Rahe. 1981. The potential for groundwater contamination from septic effluents. *J. Environ. Qual.* 10:1–8.

Hagen, G. H. L. 1839. Ueber die bewegung des wassers in engen cylindrischen rohren. *Poggend. Annal.* 46:423–42.

Hayduk, W. and H. Laudie. 1974. Prediction of diffusion coefficients for non-electrolysis in dilute aqueous solutions. *AIChE J.* 20:611–15.

Haith, D. A. 1983. Planning model for land application of sewage sludge. *J. Env. Eng.* 109:66–69.

Haith, D. A., J. E. Reynolds, P. T. Landre, and T. L. Richard. 1992. Sludge loading rates for forest land. *J. Env. Eng.* 118:196–202.

Hance, R. J. 1969. An empirical relationship between chemical structure and the sorption of some herbicides by soils. *J. Agric. Food Chem.* 17:667–68.

Harbeck, G. E., Jr. et al. 1954. *Water-loss investigations, Vol. 1, Lake Hefner studies, technical report.* Washington, DC: U.S. Geological Survey Professional Paper 269.

Harbeck, G. E., Jr., M. A. Kohler, and G. E. Koberg. 1958. *Water-loss investigations, Lake Mead studies,* Washington, DC: U.S. Geological Survey Professional Paper. 298.

Harvey, R. W., L. H. George, R. L. Smith, and D. R. LeBlanc. 1989. Transport of microspheres and indigenous bacteria through a sandy aquifer: Results of natural and forced gradient tracer experiments. *Environ. Sci. Tech.* 23:51–56.

Hausenbuiller, R. L. 1978. *Soil science: Principles and practices.* Dubuque, IA: William C. Brown.

Hayes, M. H. B. and R. S. Swift. 1978. The chemistry of soil organic colloids. In The chemistry of soil constituents. eds. D. J. Greenland and M. H. B. Hayes 179–320, New York: John Wiley and Sons.

———. 1981. Organic colloids and organo-mineral associations. *Bull. Internl. Soc. Soil Sci.* 60:67–74.

Healy, R. W. 1990. *Simulation of solute transport in variably saturated porous media with supplemental information on modifications to the U.S. Geological Survey's computer program VS2D.* U.S. Geological Survey Water-Resources Investigations Report 90-4025. Washington, DC: U.S. Government Printing Office.

Healy, R. W., R. G. Striegl, T. F. Russell, G. L. Hutchinson, and G. P. Livingston. 1996. Numerical evaluation of static-chamber measurements of soil-atmosphere gas exchange: Identification of physical processes. *Soil Sci. Soc. Amer. J.* 60:740–47.

Helfferich, F. 1962. *Ion exchange.* New York: McGraw-Hill.

Helmholtz von, H. 1879. Studien über elektrische grenzschichten. *Annalen der Physik und Chemie, Neue Folge Band VII,* No. 7, Wiedemann 7, p. 337.

Herron, J. T., W. A. Berg and H. P. Harbert III. 1980. Vegetation and lysimeter studies on decarbonized oil shale, technical bulletin 136. Fort Collins, CO, Colorado State University Experiment Station.

Higuchi, T., K. Yoda, and K. Tensho. 1984. Further evidence for gaseous CO_2 transport in relation to root uptake of CO_2 in rice plant. *Soil Sci. Plant Nutr.* 30(2): 125–36.

Hilgard, W. W. 1906 (1921 ed.). Soils, their formation, properties, composition, and relations to climate and plant growth in the humid and arid regions. New York: The Macmillan Co.

Hill, D. E. and J. Y. Parlange. 1972. Wetting front instability in layered soils. *Soil Sci. Soc. Am. J.* 36:697–702.

Hill, T. L. 1986. *An introduction to statistical thermodynamics.* New York: Dover Publications.

Hillel, D. 1971. *Soil and water, physical principles and processes.* New York: Academic Press.

———. 1980. *Applications of soil physics.* New York: Academic Press.

Hillel, D. and R. S. Baker. 1988. A descriptive theory of fingering during infiltration into layered soils. *Soil Sci.* 146:51–56.

Hillel, D. and J. Mottes. 1966. Effect of plate impedance, wetting method and aging on soil moisture retention. *Soil Sci.* 102:135–40.

Hillel, D. and W. R. Gardner. 1970a. Transient infiltration into crust-topped profiles. *Soil Sci.* 109:69–70.

———. 1970b. Measurement of unsaturated conductivity and diffusivity by infiltration through an impeding layer. *Soil Sci.* 109:149–53.

Hillel, Daniel and David E. Elrick, eds. 1990. *Scaling in soil physics: principles and applications.* Soil Sci. Soc. Am. Special Publication Number 25.

Hoffman, M. R., E. C. Yost, S. J. Eisenreich, and W. J. Maier. 1981. Characterization of soluble and colloidal-phase metal complexes in river water by ultrafiltration: A mass-balance approach. *Environ. Sci. Technol.* 15:655–61.

Hohn, M. E. 1988. *Geostatistics and petroleum geology.* New York: Van Nostrand Reinhold.

Holtan, H. N. 1961. *A concept for infiltration estimates in watershed engineering.* U.S. Department of Agriculture Bulletin 41–51. Washington, DC: U.S. Department of Agriculture.

Holtan, H. N., C. B. England, G. P. Lawless and G. A. Schumaker. 1968. *Moisture tension data for selected soils on experimental watersheds.* Publication RS41–144. Washington, DC: U.S. Department of Agriculture.

Holtan, H. N., G. J. Stiltner, W. H. Henson and N.C. Lopez. 1975. *USDAHL-74 revised model of watershed hydrology.* Agricultural Research Service Technical Bulletin 1518. Washington, DC: U.S. Department of Agriculture.

Honig, E. P. and P. M. Mul. 1971. Tables and equations of the double layer repulsion at constant potential and at constant charge. *J. Colloid Interface Sci.* 36:258–72.

Hopmans, J. W. 1987. A comparison of various methods to scale soil hydraulic properties. *J. Hydro.* 93:241–56.

Horton, R. E. 1917. A new evaporation formula developed. *Engineering News-Record* 78(4):196–99.

———. 1939. Analysis of runoff-plot experiments with varying infiltration-capacity. *Transactions American Geophysical Union* 20:693–711.

———. 1940. An approach toward a physical interpretation of infiltration capacity. *Soil Science Society of America Proceedings.* 5:399–417.

Houghton, R. A. 1995. Land-use change and the carbon cycle. *Glob. Chng. Biol.* 1:275–87.

Howard, P. H. 1991. *Handbook of environmental fate and exposure data for organic chemicals,* vol. 2: *Solvents.* Chelsea, MI: Lewis Publishers.

Hunter, R. J. and A. E. Alexander. 1963. Surface properties and flow behavior of kaolinite. Part I. Electrophoretic mobility and stability of kaolinite soils. *J. Colloid Sci.* 18:820–32.

Hurst, H. E. 1951. Long-term storage capacity of reservoirs. *Trans. Am. Soc. Civ. Eng.* 116:770–808.

Huyakorn, P. S., J. B. Kool, and Y.S. Wu. *VAM2D-Variably saturated analysis model in two dimensions: version 5.2 with hysteresis and chained decay transport, documentation and user's guide.* NUREG/CR-5352. Washington, DC: U.S. Nuclear Regulatory Commission.

Huyakorn, P. S. and G. F. Pinder. 1983. *Computational methods in subsurface flow.* New York: Academic Press.

Huyakorn, P. S. and T. D. Wadsworth. 1985. FLAMINCO: *A Three-Dimensional Finite Element Code for Analyzing Water Flow and Solute Transport in Saturated-Unsaturated Porous Media.* Technical Report for the U.S. Department of Agriculture, Northwest Watershed Research Center, Boise, Idaho. Sterling, Virginia: GeoTrans, Inc.

Idso, S. B., J. K. Asae, and R. D. Jackson. 1975. Net radiation— Soil heat flux relations as influenced by soil water content variations. *Bound. Lay. Meteorol.* 9:113–22.

Israelachvili, J. N. and G. E. Adams. 1978. Measurement of forces between two mica surfaces in aqueous electrolyte solutions in the range 0–100nm. *J. Chem. Soc., Faraday Trans. 1,* 74:975–1001.

Israelachvili, J. N. and D. Tabor. 1972. The treatment of van der Waals dispersion forces in the range of 1.5 to 130 nm. *Proc. Royal Soc.* London. A331:19–38.

Istok, J. 1989. Groundwater modeling by the finite element method, AGU water resources monograph 13. Washington, DC: American Geophysical Union.

Jackson, R. D., B. A. Kimball, R. J. Reginato, and F. S. Nakayama. 1973. Diurnal soil-water evaporation: Time-depth flux patterns. *Soil Sci. Soc. Am. Proc.* 37: 505–509.

Jacob, C. E. 1940. On the flow of water in an elastic artesian aquifer. *Trans. Amer. Geophys. Union.* 2:574–86.

Jamison, V. C. 1945. The penetration of irrigation and rain water into sandy soils in central Florida. *Soil Sci. Soc. Am. Proc.* 10:25–29.

Janert, H. 1934. The application of heat of wetting measurements to soil research problems. *J. Agric. Sci.* 24:136–50.

Jardine, P. M., N. L. Weber, and J. F. McCarthy. 1989. Mechanisms of dissolved organic carbon adsorption on soil. *Soil Sci. Soc. Am. J.* 53:1378–85.

Jardine, P. M., N. Weber, and J. F. McCarthy. 1990. Chemical and hydrologic factors controlling the transport of organic colloids. *Hazardous Materials Control.* July/August:44–46.

Jarvis, Nicholas. 1994. *The MACRO model (version 3.1): Technical description and sample simulations.* Department of Soil Science Reports and Dissertations, No. 19. Uppsala: Swedish University of Agricultural Sciences.

Jarvis, P. G., G. B. James, and J. J. Landsberg. 1976. Coniferous forests. In *Vegetation and the Atmosphere,* Vol. 2. (ed. J. L. Montieth), 171–240. New York: Academic Press.

Javendel, I., C. Doughty and C. F. Tsang. 1984. Groundwater transport: Handbook of mathematical models, AGU water resources monograph No. 10. Washington, DC: American Geophysical Union.

Jekel, M. R. 1986. The stabilization of dispersed mineral particles by adsorption of humic substances. *Water Res.* 20:1543–54.

Jenne, E. A. 1968. "Controls on Mn, Fe, Co, Ni, Cu, and Zn concentrations in soils and water: the significant role of hydrous Mn and Fe oxides." In *Trace Inorganics in Water,* ed. R. F. Gould. *Adv. Chem. Ser.* 73:337–387. Washington, DC: ACS.

Jennings, A. A., D. J. Kirkner, and T. L. Theis. 1982. Multicomponent equilibrium chemistry in groundwater quality models. *Water Resour. Res.* 18:1089–1096.

Jennings, G. D. and D. L. Martin. 1990. Solute leaching due to chemigation. In *Visions of the future*. St. Joseph, MI: ASAE Publications, Amer. Soc. Ag. Engineers.

Jenny, H. and G. D. Smith. 1935. Colloid chemical aspects of clay pan formation in soil profiles. *Soil Sci.* 39:377–89.

Jensen, M. E., R. D. Burman, and R. G. Allen. 1990. Evapotranspiration and irrigation water requirements. ASCE Manuals and Reports on Engineering Practice No. 70. New York: *Amer. Soc. Civil Eng.*

Jensen, M. E. and H. R. Haise. 1963. Estimating evapotranspiration from solar radiation. *Journal of the Irrigation and Drainage Division*, ASCE 89:15–41.

Jensen, M. E., J. L. Wright and B. J. Pratt. 1971. Estimating soil moisture depletion from climate, crop, and soil data, transactions. *ASAE* 14:954–59.

Johnson, W. M. and J. E. McClelland. 1960. Classification and description of soil pores. *Soil Sci.* 89:319–21.

Johnson, P. C., M. W. Kemblowski and J. C. Colthart. 1991b. Quantitative analysis for the cleanup of hydrocarbon contaminated soils, by in situ soil venting. *Ground Water* 28(3):413–29.

Johnson, P. C., C. C. Stanley, M. W. Kemblowski, D. D. Byers, and J. C. Colthart. 1990a. A practical approach to design, operation, and monitoring of in situ soil-venting systems. *Groundwater Monitoring Review* (Spring):159–77.

Jones, K. C., J. A. Stratford, K. S. Waterhouse, and N.B. Vogt. 1989. Organic compounds in Welsh soils: polynuclear aromatic hydrocarbons. *Environ. Sci. and Tech.* 23:540–50.

Jongerius, A. 1957. Morphological investigations of soil structure. Bedomkundige studiea, no. 2. Wageningen, Denmark: Meded van der Stickting Bodenkartiering.

Journel, A. G. and C.J. Huijbregts. 1978. *Mining geostatistics.* New York: Academic Press.

Jump, R. and B. R. Sabey. 1983. *Field studies on Union Oil Company Process B decarbonized retorted oil shales—valley site.* Department of Agronomy, Colorado State University, Fort Collins, CO. June.

Jurinak, J. J. 1963. Multilayer adsorption of water by kaolinite. *Soil Sci. Soc. Am. Proc.* 27:269–72.

Jury, W. A. 1982. Simulation of solute transport using a transfer function model. *Water Resour. Res.* 18:363–68.

———. 1986. Volatilization from soil. In *Vadose Zone Modeling of Organic Pollutants*, eds. S. C. Hern and S. M. Melancon, 159–76. Chelsea, MI: Lewis Publishers, Inc.

Jury, W. A., H. Elabd, and M. Resketo. 1986. Field study of napropamide movement through unsaturated soil. *Water Resour. Res.* 22:749–55.

Jury, W. A., W. R. Gardner and W. H. Gardner. 1991. *Soil physics.* 5th ed. New York: John Wiley & Sons.

Jury, W. A., W. R. Gardner, P. G. Saffigna and C. B. Tanner. 1976. Model for predicting simultaneous movement of nitrate and water through a loamy sand. *Soil Sci.* 122:36–43.

Jury, W. A., J. Letey, and T. Collins. 1982. Analysis of chamber methods used for measuring nitrous oxide production in the field. *Soil Sci. Soc. Amer. J.* 46:250–56.

Jury, W. A., David Russo, and Garrison Sposito. 1987. II. Scaling models of water transport. *Hilgardia* 55:33–57.

Jury, W. A. and K. Roth. 1990. *Transfer Functions and Solute Movement through Soil, Theory and Applications.* Basel, Switzerland: Birkhäuser Verlag.

Jury, W. A., L. A. Stolzy and P. Shouse. 1982. A field test of the transfer function model for predicting solute transport. *Water Resour. Res.* 18:369–375.

Kallay, N. E. Barouch and E. Matijevic. 1987. Diffusional detachment of colloidal particles from solid/solution interfaces. *Adv. Colloid Interface Sci.* 27:1–42.

Kalma, J. D. 1970. Some aspects of the water balance of an irrigated orange plantation. Ph.D. thesis, Volcani Institute Agric. Res., Bet Dagan, Israel.

Kan, A. T. and M. B. Tomason. 1990. Groundwater transport of hydrophobic organic compounds in the presence of dissolved organic matter. *Environ. Toxicol. Chem.* 9:253–63.

Karickhoff, S. W. 1981. Semi-empirical estimation of sorption of hydrophobic pollutants on natural sediments and soils. *Chemosphere* 10:833–46.

Karickhoff, S. W., D. S. Brown, and T. A. Scott. 1979. Sorption of hydrophobic pollutants on natural sediments. *Water Res.* 13:241–48.

Kashkuli, H. A. 1981. A numerical linked model for the prediction of the decline of groundwater mounds developed under recharge. Ph.D. thesis, Department of Civil Engineering, Colorado State University, Fort Collins, CO.

Kaspar, T. C., D. G. Wooley, and H. M. Taylor. 1981. Temperature effect on the inclination of lateral roots of soybeans. *Agron. J.* 73:383–85.

Katz, D. L., D. Cornell, R. Kobayashi, F. Poettmann, J. A. Vary, J. R. Elenbaas, and C. F. Weinaug. 1959. Handbook of natural gas engineering. New York: McGraw-Hill.

Kaufman, S. S. and D. A. Haith. 1986. Probabilistic analysis of sludge land application. *J. Env. Eng.* 112:1041–1045.

Kayser, R. 1988. Use of biologically treated wastewater together with excess sludge for irrigation. In *Treatment and use of sewage effluent for irrigation.* London: Butterworths.

Kazo, B. and L. Gruber. 1962. The investigation of microsolifluxion with the aid of tagged isotopes. *Int. Assoc. Sci. Hydrol. Publ. 59 Symp. Bari.* 62–6.

Keen, B. A. 1926. The physicist in agriculture with special reference to soil problems (with introduction by Sir Daniel Hall). *Physics in Industry* 4:27.

Keen, B. A. 1931. *The physical properties of the soil.* London: Longmans, Green and Co.

Keen, B. A. and E. J. Russell. 1921. The factors determining soil temperature. *J. Arg. Sci.* 11:211–39.

Keeney, D. R., 1982. Nitrogen management for maximum efficiency and minimum pollution. In *Nitrogen in agricultural soils,* Agronomy monograph 22, ed. F. J. Stevenson, 605–98. Madison, WI: American Society of Agronomy.

Kellog, C. E. 1938. *Soil and society.* In USDA Yearbook of Agriculture, ed. H. G. Knight. Washington, DC: U.S. Government Printing Office.

Kenaga, E. E. and C. A. I. Goring. 1980. *Relationship between water solubility, soil sorption, octanol-water partitioning, and bioconcentration of chemicals in biota.* American Society for Testing and Materials, Third Aquatic Toxicology Symposium, October 17–18, New Orleans, LA. Philadelphia, PA: ASTM, Special Technical Publication (STP) 707.

Khaleel, R., K. R. Reddy, and M. R. Overcash. 1981. Changes in soil physical properties due to organic waste applications: A review. *J. Environ. Qual.* 10:133–41.

Khaleel, R. and T.-C. Yeh. 1985. A galerkin finite element program for simulating unsaturated flow in porous media. *Ground Water* 23(1):90–96.

Kia, S. F., H. S. Fogler, and M.G. Reed. 1987. Effect of pH on colloidally induced fines migration. *J. Colloid Interface Sci.* 118:158–168.

Kilkelly, K. M. 1981. *Field studies on Union Oil Company Process B and decarbonized retorted oil shale, 1979–1980.* Colorado State University, Department of Agronomy, Disturbed Lands Research. July.

Kimball, B. A. 1983. Canopy gas exchange: Gas exchange with soil, In *Limitations to efficient water use in crop production.* Madison, WI: Amer. Soc. of Agron.

Kimball, B. A. and R. D. Jackson. 1971. Seasonal effects on soil drying after irrigation: Hydrology and water resources of Arizona and the South-West. *Proc. Ariz. Soc., Amer. Water Resour. Assn. and the Hydrol. Sec. Ariz. Acad. Sci.* 1:85–98.

Kimball, B. A. and E. R. Lemon. 1971. Air turbulence effects upon soil gas exchange. *Soil Sci. Soc. Amer. Proc.* 35:16–21.

Kirda, C., D. R. Nielsen, and J.W. Biggar. 1973. Simultaneous transport of chloride and water during infiltration. *Soil Sci. Soc. Amer. Proc.* 37:339–45.

———. 1974. The combined effects of infiltration and redistribution of leaching. *Soil Sci.* 117:323–30.

Kirkby, L. J. 1984. *Radionuclide distributions and migration mechanisms at shallow land burial sites.* NUREG/CR-3607. Washington, DC: U.S. Nuclear Regulatory Commision.

Kirkby, M. J. 1988. Hillslope runoff processes and models. *Journal of Hydrology* 100:315–39.

Kirkham, D. 1961. Soil physics: 1936–61 and a look ahead. *Soil Sci. Soc. Am. Proc.* 25:423–27.

Kirkham, D. 1971. Problems and trends in drainage research, mixed boundary conditions. *Soil Sci.* 113:285–93.

Kirkner, D. J., A. A. Jennings, and T. L. Theis. 1985. Multisolute mass transport with chemical interaction kinetics. *J. Hydrol.* 76:107–117.

Klinkenberg, L. J. 1941. The permeability of porous media to liquids and gases. In *Drilling and Production Practice,* pp. 200–213. New York: Amer. Petrol. Inst.

Klute, A. 1952a. A numerical method for solving the flow equation for water in unsaturated materials. *Soil Sci.* 73:105–116.

———. 1952b. Some theoretical aspects of the flow of water in unsaturated soils. *Soil Sci. Soc. Am. Proc.* 16:144–48.

———. 1972. The determination of the hydraulic conductivity and diffusivity of unsaturated soils. *Soil Sci.* 113:264–76.

———. 1986. *Methods of soil analysis part I: Physical and mineralogical methods,* monograph #9. Madison WI: Soil Science Society of America.

Klute, A. and L. A. Richards. 1962. Effect of temperature on relative vapor pressure of water in soil. Apparatus and preliminary measurements. *Soil Sci.* 93:391–97.

Knudtsen, K. and G. A. O'Connor. 1987. Characterization of iron and zinc in Albuquerque sewage sludge. *J. Environ. Qual.* 16:85–90.

Kohler, M. A., T. J. Nordenson and W. E. Fox. 1955. *Evaporation from pans and lakes.* Research Paper 38. Washington, DC: U.S. Weather Bureau.

Kohnke, H. 1968. *Soil physics.* New York: McGraw-Hill.

Kolpin, D. W. and E. M. Thurman. 1993. *Postflood occurrence of selected agricultural chemicals and volatile organic compounds in near-surface unconsolidated aquifers in the upper Mississippi River Basin, 1993.* U.S. Geological Survey Circular 1120-G. Washington, DC: U.S. Government Printing Office.

Korfiatis, G. P. 1984. Modeling the moisture transport through solid waste landfills. Ph.D. Thesis. Rutgers, The State University of New Jersey, New Brunswick, NJ.

Kostiakov, A. N. 1932. On the dynamics of the coefficient of water-percolation in soils and on the necessity of studying it from a dynamic point of view for purposes of amelioration. *Trans. of the Sixth Comm. of the Int. Soc. of Soil Sci.,* Part A, pp. 17–31.

Kowal, N. E., H. R. Pahren, and E. W. Akin. 1981. Microbiological health effects associated with the use of municipal wastewater for irrigation. In *Municipal Wastewater in Agriculture.* New York: Academic Press.

Kozeny, J. 1927. Uber Kapillare Leitung des Wassers im Boden, Sitzungsber. *Akad. Wiss. Wien* 136:271–306.

Krahn, J., D. G. Fredlund, L. Lam and S. L. Barbour. 1989. PC-SEEP: A finite element program for modeling seepage, Geo-Slope Programming, Ltd., Calgary, Alberta, Canada.

Kramer, P. J. and J. S. Boyer. 1995. *Water relations of plants and soils.* New York: Academic Press.

Kramer, S. N. 1958. *History Begins at Sumer.* New York: Doubleday.
———. 1963. *The Sumerians, their history, culture, and character.* Chicago Univ. Chicago Press.
Kraner, H. W., G. L. Schroeder and R. D. Evans. 1964. Measurements of the effects of atmospheric variables on Radon-222 flux and soil-gas concentrations. In *The Natural Radiation Environment,* eds. J. A. S. Adams and W. M. Lowder, 191–215. Univ. of Chicago Press: William Marsh Rice Univ.
Kumada, K. 1987. *Chemistry of soil organic matter. Developments in soil science 17.* Japan Scientific Societies Press. Tokyo: Elsevier.
Kunkel, J. R. and R. B. Murphy. 1983. Water balance estimates of the reclamation zone of a retorted oil-shale disposal pile, paper no. 83–2504. American Society of Agricultural Engineers, 1983 Winter Meeting, Chicago, Illinois, December 13–16, 1983.
Lai, S. H. and J. J. Jurniak. 1971. Numerical approximation of cation exchange in miscible displacement through soil columns. *Soil Sci. Soc. Amer. Proc.* 35:894–99.
Lai, S. H. and J. J. Jurinak. 1972. Cation adsorption in one-dimensional flow through soils: A numerical solution. *Water Resour. Res.* 8:99–107.
Lal, R., N. R. Fansey, and D. J. Eckert. 1995. Land-use and soil management effects of emissions of radiatively active gases from two soils in Ohio. In *Soil management and the greenhouse effect,* eds. R. Lal, J. Kimble, E. Levine, and B. A. Stewart, 41–59. Boca Raton, FL: Lewis Publishers.
Laliberte, G. E. 1969, A mathematical function for describing capillary pressure-desaturation data. *Bulletin of the International Association of Scientific Hydrology (IASH)* 14:2.
Laliberte, G. E., A. T. Corey, and R. H. Brooks. 1966. Properties of unsaturated porous media, hydrology report no. 17. Colorado State University, Fort Collins, CO.
Landon, J. R., ed. 1984. *Booker tropical soil manual.* London Booker Agric. Int. Ltd.
Lane, W. L. 1991. Synthetic streamflows for global climate change, USGS open file report 91–244. Proceedings of the US-PRC Bilateral Symposium on Droughts and Arid Region Hydrology, September 16–20, Tucson, AZ.
Lane, W. L. and D. K. Frevert. 1990. *Applied Stochastic Techniques.* Personal computer version 5.2, user's manual. Denver, CO: U.S. Bureau of Reclamation.
Lapidus, L. and N. R. Amundson. 1952. Mathematics of adsorption in beds. VI. The effect of longitudinal diffusion in ion exchange and chromatographic columns. *J. Phys. Chem.* 56:984–88.
LaRocque, A. 1957. The admirable discourses of Bernard Palissy (with biographical introduction). Urbana: Univ. Illinois Press.
Lau, L. S. 1981. *Wastewater use for irrigation: A case history in Hawaii.* In *Collected reprints,* volume v: 1978–1981, 302–312. June 1984. Honolulu, HI: Water Resources Research Center.
Lauren, J. G., Wagenet, R. J. Bouma, J. and Wosten, J. H. M. 1988. Variability of saturated hydraulic conductivity in a Glossaquic Hapludalf with macropores. *Soil Sci.* 145:20–28.
Lavy, T. L. 1977. Herbicide transport in soil under center pivot irrigation systems. PB-270 777. Springfield, VA: National Technical Information Service.
Lawes, J. B., H. H. Gilbert, and R. Warington. 1882. On the amount and composition of the rain and drainage water collected at Rothamstead. London: Williams Clowes and Sons, Ltd. Originally published in *J. Royal Agr. Soc. Of England XVII*(1881):241–79, 311–50; *XVIII*(1882):1–71.
Lawrence Livermore National Laboratory (LLNL). 1990. *CERCLA feasibility study for the LLNL Livermore site.* Environmental Protection, Environmental Restoration Series, UCRL-AR-104040, Draft Final, Appendix G. Livermore, CA.
Lawrence Livermore National Laboratory (LLNL). 1995. *Recommendations to improve the cleanup process for California's leaking underground fuel tanks (LUFTs).* UCRL-AR-121762. Livermore, CA.
Lee, J., M. M. Mortland, C. T. Chiou, and S. A. Boyd. 1989. Shape selective adsorption of aromatic molecules from water by tetramethylammonium-smectite. *J. Chem. Soc. Faraday. Trans.* 85:2953–62.
Lee, L. S., P. S. C. Rao, M. L. Brusseau, and R. A. Ogwada. 1988. Nonequilibrium sorption of organic contaminants during flow through columns of aquifer materials. *Environ. Toxicology Chem.* 7:799–803.
Lee, Y. W. and M. Fl. Dahab. 1992. Nitrate risk management under uncertainty. *J. Water Resources, Planning, and Management* 118:151–65.
Leenheer, J. A. and H. A. Stuber. 1981. Migration through soil of organic solutes in an oil-shale process water. *Environ. Sci. Technol.* 15:1467–75.
Leij, F. J., W. B. Russell, and S. M. Lesch. 1997. Closed-form expressions for water retention and conductivity data. *Ground Water* 35(5):848–58.
Lemon, E. R. 1962. Soil aeration and plant relations. I. Theory. *Agron. J.* 54:176–80.
Lemon, E., and C. C. Wiegand. 1962. Soil aeration and plant root relations. II. Root respiration. *Agron. J.* 54:171–75.
Lenhard, R. J. and J. C. Parker. 1987. Measurement and prediction of saturation-pressure relationships in air-organic liquid-water porous media systems. *Journal of Contaminant Hydrology* 1(4):395–406.
Leverett, M. C. 1941. Capillary behavior of porous solids, transactions. *Society of Petroleum Engineers of AIME* 142:152–69.
Leyton, L. 1978. *Some thermodynamic concepts of water movement. Fluid behavior in biological systems.* New York: Oxford Univ. Press.
Li, R-M. K. G. Eggert, and K. Zachmann. 1983. Parallel processor algorithm for solving three-dimensional ground water flow equations. Washington, DC: National Science Foundation.

Lide, D. R. 1992. *CRC handbook of chemistry and physics.* Boca Raton, FL: CRC Press.
Lindstrom, F. T. 1976. Pulse dispersion of trace chemical concentrations in a saturated sorbing porous medium. *Water Resour. Res.* 12:229–38.
Lindstrom, F. T., R. Hague, V. H. Freed, and L. Boersma. 1967. Theory on the movement of some herbicides in soils: linear diffusion and convection of chemicals in soils. *Environ. Sci. Technol.* 1:561–65.
Lindstrom, F. T., D. E. Garfield and L. Boersma. 1988. *CTSPAC: Mathematical model for coupled transport of water, solutes, and heat in the soil-plant-atmosphere continuum, EPA/600/3-88/030,* Corvallis, OR: U.S. Environmental Protection Agency, Environmental Research Laboratory.
Lindzen, R. S. 1990. Some coolness concerning global warming. *Bulletin Amer. Meteorol. Soc.* 71:288–99.
Loman, D. O. and A. W. Warrick. 1978. Linearized moisture flow with loss at the soil surface. *Soil Science Society of America Journal* 42:396–99.
Looney, B. B., M. E. Newman, and A. W. Elzerman. 1990. Colloid facilitated transport in groundwater: Laboratory and field studies. *Hazardous Materials Control* (July/August):47–49.
López-Bakovic, I. L. and J. L. Nieber. 1989. Analytic steady-state solution to one-dimensional unsaturated water flow in layered soils. In *Unsaturated Flow in Hydrologic Modeling, Theory and Practice.* Proceedings of the NATO Advanced Research Workshop on Unsaturated Flow in Hydrologic Modeling Theory and Practice, Arles, France, 13–17 June 1988, ed. H. J. Movel-Seytoux, 471–80. Kluwer Academic Publishers.
Loskot, C. L., J. P. Rousseau, and M. A. Kurzmack. 1994. Evaluation of a 6-wire thermocouple psychrometer for determination of in-situ water potentials, pp. 2084–2091. Proceedings High-Level Radioactive Waste Management, Fifth Intnl. Conf., Las Vegas, NV., May 22–26.
Low, P. F. 1979. Nature and properties of water in montmorillonite. *Soil Sci. Soc. Am. J.* 43:651–58.
Lowham, H. W. 1976. Techniques for estimating flow characteristics of Wyoming streams, Geological Survey Water-Resources Investigations 76–112. Prepared in Cooperation with the Wyoming Highway Department, November, 83 p.
Luxmoore, R. J. 1981. Micro-, meso-, and macroporosity of soil. *Soil Sci. Soc. Am. J.* 45:671–72.
Lyklema, J. 1978. Surface chemistry of colloids in connection with stability. In *The scientific basis of flocculation.* ed. K. J. Ives, 3–36. The Netherlands: Sijthoff and Noordhoff.
Lyman, W. J., F. R, William, and D. H. Rosenblatt. 1990. *Handbook of chemical property estimation methods.* Washington, DC: Amer. Chem. Soc.
MacCracken, M. C. and F. M Luther, eds. 1985. Detecting the climatic effects of increasing carbon dioxide. Report DOE/ER 0235. U.S. Dept. Energy, Wash., DC.
Mackay, D. and W. Shui. 1981. A critical review of Henry's law constants for chemicals of environmental interests. *J. Phys. Chem. Ref. Data* 10(4):1175–99.
Magnusson, T. 1992. Studies of soil atmosphere and related physical site characteristics in mineral forest soils. *J. Soil Sci.* 43:767–90.
Male, D. H. and D. M. Gray. 1981. *Snowcover ablation and runoff.* In: *Handbook of Snow,* eds. D. H. Male and D. M. Gray. New York: Pergamon Press.
Mandelbrot, B. B. 1982. *The fractal geometry of nature.* New York: W. H. Freeman.
Mandelbrot, B. B. 1984. Fractals in physics: squig clusters, diffusions, fractal measures, and the unicity of fractal dimensionality. *J. Stst. Phys.* 34:895–930.
———. 1985. Self-affine fractals and the fracal dimension. *Physica Scripta* 32:257–60.
Marencik, J. 1991. State-by-state summary of cleanup standards. *Soils* (November-December 1991). 14–16, 18–23, 48–51.
Marle, C., P. Simandoux, J. Pacsirsky, and C. Gaulier. 1967. Etude du deplacement de fluides miscibles en milieu poreux stratifiei. *Rev. Inst. Fr. Petrol.* 22:272–94.
Marshall, C. E. 1964. *The physical chemistry and mineralogy of soils.* New York: John Wiley & Sons.
Marshall, T. J. 1958. A relation between permeability and size distribution of pores. *J. Soil Sci.* 9:1–8.
Marshall, T. J. and J. W. Holmes. 1979. *Soil physics.* New York: Cambridge Univ. Press.
Martin, T. L. 1921. *Decomposition of green manure at different stages of growth.* New York (Cornell) Agr. Exp. Sta. Bul. 406.
———. 1925. Effect of straw on accumulation of nitrates and crop growth. *Soil Sci.* 20:159–164.
Martinez, M. J. 1985. FEMTRAN—A finite element computer program for simulating radionuclide transport thorough porous media, SAND84-0747. Albuquerque, NM: Sandia National Laboratory.
Masch, F. D. 1984. *Hydrology, Federal Highway Administration hydrologic engineering circular no. 19.* Washington, DC: U.S. Department of Transportation.
Mason, E. A. and A. P. Malinauskas. 1983. *Gas transport in porous media: The dusty-gas model.* Chem. Eng. Monogr. #17. New York: Elsevier.
Mason, E. A., A. P. Malinauskas, and R.B. Evans, III. 1967. Flow and diffusion of gases in porous media. *J. Chem. Phys.* 46:3199–216.
Massman, J. and D.F. Farrier. 1992. Effects of atmospheric pressures on gas transport in the vadose zone. *Water Resour. Res.* 28(3):777–91.
Mathur, S.P., and R.S. Farnham. 1985. Geochemistry of humic substances in natural and cultivated peatlands. In *Humic substances in soil, sediment, and water: Geochemistry, isolation, and characterization,* eds. G. R. Aiken, D. M. McKnight, and R. L. Wershaw, 53–86. New York: John Wiley & Sons.

McAneny, C. C., P. G. Tucker, J. M. Morgan, C. R. Lee, M. F. Kelley and R. C. Horz. 1985. Covers for uncontrolled hazardous waste sites. Vicksburg, MS: U.S. Army Engineer Waterways Experiment Stations.

McBride, M. B. 1989. Surface chemistry of soil minerals. In *Minerals in soil environments*, eds. J. B. Dixon and S. B. Weed. Madison, WI: Soil Sci. Soc. Am.

McCarthy, J. F. 1990. The mobility of colloidal particles in the subsurface. *Hazardous Materials Control.* July/August:38–43.

McCarthy, J. F. and B. D. Jimenez. 1985. Interactions between polycyclic aromatic hydrocarbons and dissolved humic material: binding and dissociation. *Environ. Sci. Technol.* 19:1072–76.

McCarthy, J. F. and J. M. Zachara. 1989. Subsurface transport of contaminants. *Environ. Sci. Tech.* 23:496–502.

McCaslin, B. D. and O'Connor, G. A. 1982. Potential fertilizer value of gamma-irradiated sewage sludge on calcareous soil. *New Mexico Agric. Exp. Stn. Bull.* 692.

McDowell-Boyer, L. M., J. R. Hunt, and N. Sitar. 1986. Particle transport through porous media. *Water Resour. Res.* 22:1901–21.

McKee, C. R. and A. C. Bumb. 1984. *The importance of unsaturated flow parameters in designing a monitoring system for a hazardous waste site.* Proceedings of the National Conference on Hazardous Wastes and Environmental Emergencies. The Hazardous Materials Control Research Institute. Houston, TX, March 1984, pp. 50–58.

McKee, C. R., A. C. Bumb and T. L. Deshler. 1983. A three-dimensional computer model to aid in selecting monitor locations in the Vadose Zone, Proceedings of the NWWA/U.S. EPA Conference on Characterization and Monitoring of the Vadose (Unsaturated) Zone. National Water Well Association, Worthington, OH, Las Vegas, December 8, pp. 133–61.

McKee, C. R. and A. C. Bumb. 1987. Flow-testing coalbed methane production wells in the presence of water and gas. SPE Formation Evaluation, December, 599–608.

———. 1988. A three-dimensional analytical model to aid in selecting monitoring locations in the Vadose Zone. *Ground Water Monitoring Review,* Spring 1988.

McLeod, A. I. and K. W. Hipel. 1978. Simulation procedures for Box-Jenkins models. *Water Resources Research* 14(5):969–75.

McNeal, B. L. and N. T. Coleman. 1966. Effect of solution composition on soil hydraulic conductivity. *Soil Sci. Soc. Amer. Proc.* 308–17.

McWhorter, D. B. and B. H. Kueper. Undated. *Validation of a DNAPL flow model using an analytical solution.* Unpublished manuscript.

McWhorter, D. B. and D. K. Sunada. 1990. Exact integral solutions for two phase flow. *Water Resour. Res.* 26(3):399–414.

Means, J. C. and R. Wijayaratne. 1982. Role of natural colloids in the transport of hydrophobic pollutants. *Science* 215:968–70.

Means, J. C., S. G. Wood, J. J. Hassett and W. L. Banwart. 1980. Sorption of polynuclear aromatic hydrocarbons by sediments and soils. *Environmental Science and Technology* 14:1524–28.

Meller, E. J. and R. W. Dalzell. 1982. Mine air quality control. In *Mine ventilation and air conditioning*, ed. H. L. Hartman, 37–130. New York: John Wiley & Sons.

Mengel, K. and E. A. Kirkby. 1982. *Principles of plant nutrition.* Worblufen-Bern, Switzerland: International Potash Institute.

Meyer, A. F. 1915. Computing run-off from rainfall and other physical data. *Transactions of the American Society of Civil Engineers* 79:1056–155.

Meyer, J. L. and C. M. Tate. 1983. The effects of watershed disturbance on dissolved organic carbon dynamics of a stream. *Ecology.* 64:33–44.

Miller, C. T. and W. J. Weber. 1986. Sorption of hydrophobic organic pollutants in saturated soil systems. *J. Contam. Hydrol.* 1:243.

Miller, D. E. and W. H. Gardner. 1962. Water infiltration into stratified soil. *Soil Sci. Soc. Amer. Proc.* 26:115–19.

Miller, E. J. and R. W. Dalzell. 1982. Mine air quality control. In *Mine Ventilation and Air Conditioning*, ed. H. L. Hartman 37–130. New York: John Wiley & Sons.

Miller, E. E. and R. D. Miller. 1956. Physical theory for capillary flow phenomena. *Journal Applied Physics.* 27:324–32.

Miller, R. H., S. S. Brar, and J. T. Logan, 1977. *Effect of spray irrigation of municipal wastewater on nitrogen transformations in soil.* National Technical Information Service, Springfield, VA, PB-276 661. Columbus, OH: Ohio Water Resources Center. Project Completion Report No. 495X, 59 p.

Miller, R. H. and J. F. Wilkinson. 1977. Nature of the organic coating on sand grains of nonwettable golf greens. *Soil Sci. Soc. Am. J.* 41:1203–04.

Miller, W. P. and M. K. Baharuddin. 1986. Relationship of soil dispersibility to infiltration and erosion of southeastern soils. *Soil Sci.* 142:235–40.

Miller, W. P. and D. E. Radcliffe. 1992. Soil crusting in the southeastern United States. In *Soil crusting, chemical and physical processes*, eds. M. E. Sumner and B. A. Stewart 233–66. Boca Raton FL: Lewis Publishers.

Millington, R. J. and J. P. Quirk. 1959. Permeability of porous media. *Nature* (London) 183:387–88.

———. 1960. Transport in porous media. Trans. Seventh Int. Congr. *Soil Sci.* I:97–106.

———. 1961. Permeability of porous solids. *Trans. Faraday Society* 57:1200–07.

Milly, P. C. D. 1982. Moisture and heat transport in hysteretic, inhomogeneous porous media: A matric head-based formulation and a numerical model. *Water Resour. Res.* 18:489–98.

———. 1984. A simulation analysis of thermal effects on evaporation from soil. *Water Resour. Res.* 20:1087–1098.

Mitchell, J. K. 1993. Fundamentals of soil behavior, 2d. ed. New York: John Wiley & Sons.

Mitscherlich, E. A. 1901. Untersuchungen uber die physikalischen boden-eigenshaften. *Landw. Jahrb.* 30:360–445.

———. 1905. *Bodenkunds fur land-und-forstwirthe.* Berlin: P. Parey.

———. 1930. *Die bestimmung des dungebedurfnisses des bodens.* Berlin: P. Parey.

Monteith, J. L. 1965. Evaporation and environment. In *The state and movement of water in living organisms.* Nineteeth Symp. Soc. Exp. Biol. New York: Academic Press.

———. 1973. *Principles of environmental physics.* New York: Elsevier.

Monteith, J. L. and P. C. Owen. 1958. A thermocouple method for measuring relative humidity in the range 95–100%. *J. Sci. Instrum.* 35:443–46.

Moore, C. J. 1986. Frequency response corrections for eddy correlation systems. *Bound. Lay. Meteorol.* 37: 17–35.

Moore, R. E. 1941. The relation of soil temperature and soil moisture: pressure potential relation and infiltration rate. *Soil Science Society of America Proceedings* 5:61–64.

Morel-Seytoux, H. J. 1973. Systematic treatment of infiltration with applications. Completion report series no. 50, Environmental Research Center. Fort Collins, CO: Colorado State University.

———. 1979. *Analytical results for predictions of variable rainfall infiltration, Hydrowar program, CEP 79-80HJM37.* Department of Civil Engineering, Colorado State University, Fort Collins, CO.

———. 1987. "Multiphase flows in porous media." In Developments in Hydraulic Engineering—4, ed. P. Novak, 103–43. London: Elsevier Applied Science Publishers, Ltd.

Morel-Seytoux, H. J. and J. Khanji. 1974. Derivation of an Equation of Infiltration. *Water Resour. Res.* vol. 10(4):795–800.

Moreno, J. C. 1981. Agricultural land irrigation with wastewater in the Mezquital Valley. In *Municipal wastewater in agriculture.* New York: Academic Press.

Morrison, R. 1993. Hydrocarbon Transport in Soils. *The National Environmental Journal* September/October:52–56.

Mosier, A. R., W. J. Parton, D. W. Valentine, D. S. Ojima, D. S. Schimel, and O. Heinemeyer. 1997. CH_4 and N_2O fluxes in the Colorado shortgrass steppe. 2. Long-term impact of land use change. *Glob. Biogeochem. Cy.* 11(1): 29–42.

Mott, C. J. B. 1988. Surface chemistry of soil particles. In *Russell's soil conditions and plant growth*, 11th ed., ed. A. Wild. New York: John Wiley & Sons, Inc.

Mualem, Y. 1976. Hysteretical models for prediction of the hydraulic conductivity of unsaturated porous media. *Water Resour. Res.* 12:1248–54.

Mueller, D. K. and D. R. Helsel. 1996. Nutrients in the nation's waters—too much of a good thing? U.S. Geological Survey Circular 1136. Washington, DC: U.S. Govt. Printing Office.

Murphy, J. W., A. C. Bumb and C. R. McKee. 1987. *Vadose model of gasoline Leak,* (unpublished Manuscript). Laramie, WY: In-Situ, Inc.

Musgrave, G. W. 1955. How much of the rain enters the soil?, In *Water yearbook of agriculture*, 151–59. Washington, DC: U.S. Department of Agriculture.

Musgrave, G. W. and H. N. Holtan. 1964. Section 12, Infiltration. In *Handbook of applied hydrology*, V. T. Chow. New York: McGraw-Hill Book Company.

Nachabe, M. H., T. H. Illangasekare, H.J. Morel-Seytoux, and L.R. Ahuja. 1997. Infiltration over heterogeneous watershed: Influence of rain excess. *Journal of Hydrologic Engineering* 2(3):140–43.

Nakane, K., M. Yamamoto, and H. Tsubota. 1983. Estimation of root respiration in a mature forest ecosystem. *Japn. J. Ecol.* 33:397–408.

Nakayama, F. S. 1990. Soil respiration. *Remot. Sens. Rev.* 5(1):311–20.

Nakshabandi, G. A. and H. Kohnke. 1965. Thermal conductivity and diffusivity of soils as related to moisture tension and other physical properties. *Agric. Meteor.* 2:271–79.

Narasimhan, T. N., S. N. Neuman and P. A. Witherspoon. 1978. Finite element method for subsurface hydrology using a mixed explicit-implicit iterative scheme. *Water Resour. Res.* 14(5):863–77.

Nazeroff, W. W., S. R. Lewis, S. M. Doyle, B. A. Moed, and A. E. Nero. 1987. Experiments on pollutant transport from soil into residential basements by pressure-driven airflow. *Env. Sci. Tech.* 21:459–66.

Neal, O. R., L. A. Richards, and M. B. Russell. 1937. Observations on moisture conditions in lysimeters. *Soil Sci. Soc. Am. Proc.* 2:35–44.

Neilsen, G., D. S. Stevenson, J. J. Fitzpatrick, and C. H. Brownlee. 1989. Yield and plant nutrient content of vegetables trickle-irrigated with municipal wastewater. *Hortscience* 24:249–52.

Nelson, D. M. and K. A. Orlandini. 1986. Environmental Research Division progress report 1984–1985. Anl-86-15. Argonne, IL: Argonne Nat'l Laboratory.

Nelson, D. M., W.R. Penrose, J. O. Kaltunen, and P. Mehlhaf. 1985. Effects of dissolved organic carbon on the adsorption properties of plutonium in natural waters. *Environ. Sci. Technol.* 19:127–31.

Neretnieks, I. 1983. A note on fracture flow dispersion mechanisms in the ground. *Water Resour. Res.* 19:364–70.

Neumann, H. H., and G. W. Thurtell. 1972. A peltier cooled thermocouple dewpoint hygrometer for in situ measurement of water potentials. In *Psychrometry in water relations research*, eds. R. W. Brown and B. P. van Haveren, 103–112, Proc. Symp. on Thermocouple Psychrometers, Utah St. Univ., Utah Agric. Expt. Sta., Logan, UT.

Neumann, H. H., G. W. Thurtell, and K. R. Stevenson. 1974. In situ measurements of leaf water potential and resistance to water flow in corn, soybean, and sunflower at several transpiration rates. *Can. J. Plant Sci.* 54:175–84.

Nicolaides, G. and C. Eckert. 1978. Optimal representation of binary liquid mixture nonidealities. *Ind. Eng. Chem. Fundam.* 17:331.

Nielsen, D. R. and J. W. Biggar. 1962. Miscible displacement. III. Theoretical considerations. *Soil Sci. Soc. Amer. Proc.* 26:216–21.

Nielsen, D. R., J. W. Biggar, and J. M. Davidson. 1962. Experimental consideration of diffusion analysis in unsaturated flow problems. *Soil Sci. Soc. Amer. Proc.* 26:107–11.

Nielsen, D. R., D. Kirkham, and E. R. Perrier. 1960. Soil capillary conductivity: Comparison of measured and calculated values. *Soil Sci. Soc. Am. Proc.* 24:157–60.

Nielsen, D. R., D. Kirkham, and W. R. Van Wijk. 1959. Measuring water stored temporarily above field moisture capacity. *Soil Sci. Soc. Am. Proc.* 23:408–512.

Nielsen, D. R., J. W. Biggar, and K. T. Erb. 1973. Spatial variability of field-measured soil-water properties. *Hilgardia* 42:215–59.

Nightingale, H. I. and W. C. Bianchi. 1977. Ground-water turbidity resulting from artificial recharge. *Ground Water* 15:146–52.

Nilson, R. H., E. W. Paterson, K.H. Lie, N.R. Burkhard, and J.R. Hearst. 1991. Atmospheric pumping: A mechanism causing vertical transport of contaminated gases through fractured permeable media. *J. Geophys. Res.* 96(B13):21933–48.

Nkedi-Kizza, P., J. W. Biggar, M. Th. Van Genuchten, P. J. Wierenga, H. M. Selim, J. M. Davidson, and D. R. Nielsen. 1983. Modeling tritium and chloride 36 transport through an aggregated Oxisol. *Water Resour. Res.* 19:691–700.

Nkedi-Kizza, P., P. S. C. Rao, R. E. Jessup, and J. M. Davidson. 1982. Ion exchange and diffusive mass transfer during miscible displacement through an aggregated oxisol. *Soil Sci. Soc. Am. J.* 46:471–78.

Nofziger, D. L. 1985. *Interactive simulation of one-dimensional water movement in soils.* User's Guide, Circular 675, Software in Soil Science, Florida Cooperative Extension Service, University of Florida, Gainesville, Florida.

Nofziger, D. L. and J. R. Williams. 1988. *Interactive simulation of the fate of hazardous chemicals during land treatment of oily wastes.* RITZ User's Guide. Ada, OK: U.S. Environmental Protection Agency, Robert S. Kerr Environmental Research Laboratory, EPA/600/8-88/001, January, 61 p.

Nolan, T. K. R. Srinivasan, and H. S. Fogler. 1989. Dioxin sorption by hydroxy-aluminum-treated clays. *Clays and Clay Minerals* 37:487–92.

Norel, G. 1967. Analyse quantitative par absorption ou fluourescence d'un traceur stable. *Suppl. Bull. Inform. ATEN* 65:8–11.

Norman, J. M. 1979. Modeling the complete crop canopy, In *Modification of the aerial environment of crops,* Barfield and Gerber, eds., 249–280. St. Joseph, MI: Amer. Soc. Agric. Eng.

Norrish, K. and J. A. Rausell-Colom. 1963. Low-angle X-ray diffraction studies of the swelling of montmorillonite and vermiculite. *Clays Clay Miner.* 10:123–49.

Novich, B. E. and T. A. Ring. 1984. Colloid stability of clays using photon correlation spectroscopy. *Clays Clay Miner.* 32(5):400–406.

Nuclear Waste Policy Act (Section 113). 1988. Site characterization plan: Overview—Yucca mountain site, Nevada research and development area, Nevada. U.S. Department of Energy, Office of Civilian Radioactive Waste Management. Washington, DC: U.S. Govt. Printing Office.

Nylan, J. W., B. J. Drennan, W. V. Abeele, M. L. Wheeler, W. D. Purtymun, G. Trujillo, W. J. Herrera, and J. W. Booth. 1985. Distribution of plutonium and americium beneath a 33-yr-old liquid waste disposal site. *J. Environ. Qual.* 14:501–509.

Oades, J. M. 1989. An introduction to organic matter in mineral soils. In *Minerals in soil environments,* eds. J. B. Dixon and S. B. Weed. Madison, WI: Soil Sci. Soc. Am.

O'Brien, G. M. 1993. *Earthquake-induced water-level fluctuations at Yucca Mountain, Nevada, June 1992.* U.S. Geological Survey Open-File Report 93-73. Washington, DC: U.S. Govt. Printing Office.

Oke, T. R. 1987. *Boundary Layer Climates.* London: Methune & Co. Ltd.

Olsen, H. W. 1966. Darcy's law in saturated kaolinite. *Water Resour. Res.* 2:287–95.

O'Melia, C. R. 1980. Aquasols: the behavior of small particles in aquatic systems. *Environ. Sci. Technol.* 14:1052–60.

Oron, G., Y. DeMalach, Z. Hoffman, Y. Keren, and H. Hartman. 1991. Wastewater disposal by sub-surface trickle irrigation. *Water Science & Tech.* 23:2149–158.

Osburn, R. C. and C. E. Burkhead. 1992. Irrigating vegetables with reclaimed wastewater. *Water Environ. & Tech.* 4:38–43.

Otton, J. K., L. C. S. Gunderson, and R. R. Schumann. 1993. *The geology of radon.* Bull. 356–733. Washington, DC: U.S. Govt. Printing Office.

Ozima, M. and F. A. Podosek. 1983. *Noble gas geochemistry.* New York: Cambridge University Press.

Page, A. L. and A. C. Chang, 1981. Trace metals in soils and plants receiving municipal wastewater irrigation. In *Municipal wastewater in agriculture,* 351–72. New York: Academic Press.

Page, J. B. 1948. Advantages of the pressure pycnometer for measuring the pore space in soils. *Soil Sci. Soc. Am. Proc.* 12:81–84.

Palissy, B. 1957. *Admirable Discourses.* Translated by Aurele LaRocque. Urbana: Univ. Illinois Press.

Park, I. S., J. C. Parker, and A. J. Valocchi. 1986. Constraints on the validity of equilibrium and first-order kinetic transport models in structured soils. *Water Resour. Res.* 22:339–43.

Parker, G. W. 1977. Methods for determining selected flow characteristics for streams in Maine. Geological Survey Open-File Report 78–871.

Parker, J. C., R. J. Lenhard, and T. Kuppusamy. 1987. A parametric model for constitutive properties governing multiphase flow in porous media, *Water Resour. Res.* 23(4):618–24.

Parker, J. C. and M. Th. Van Genuchten. 1984a. Determining transport parameters from laboratory and field tracer experiments. *Virg. Agric. Exp. Stat. Bull.* 84–3.

———. 1984b. Flux-averaged concentrations in continuum approaches to solute transport. *Water Resour. Res.* 20:866–72.

Paschis, J. A., J. R. Kunkel, and R. A. Keonig. 1988. *Wellfield installation and investigations, Creston study area,* Eastern Washington, NUREG/CR-5251. Washington, DC: U.S. Nuclear Regulatory Commission, Division of Engineering, Office of Nuclear Regulatory Research, NRC FIN D1163.

Passioura, J. B. 1971. Hydrodynamic dispersion in aggregated media. 1. *Theory. Soil Sci.* 111:339–44.

Passioura, J. B. and D. A. Rose. 1971. Hydrodynamic dispersion in aggregated media. 2. Effects of velocity and aggregated size. *Soil Sci.* 111:345–51.

Peck, A. J., C. D. Johnston and D. R. Williamson. 1981. Analysis of Solute Distributions in Deeply Weathered Soils. *Agricultural Water Management* 4:83–102.

Peck, A. J., R. J. Luxmoore, and Janice L. Stolzy. 1977. Effects of spatial variability of soil hydraulic properties in water budget modeling. *Water Resour. Res.* 13:348–54.

Peitgen, H.O. and D. Saupe. 1988. *The science of fractal images.* New York: Springer-Verlag.

Penman, H. L. 1940. Meteorological and soil factors affecting evaporation from fallow soil. *Quart. J. R. Meteorol. Soc.* 66:401–10.

———. 1948a. Natural evapotranspiration from open water, bare soil, and grass. *Proc. Roy. Soc. London, Ser. A.* 193: 120–45.

———. 1948b. Physics in agriculture. *Sci. Instrum.* 25:425–32.

———. 1951. The role of vegetation in meteorology, soil mechanics and hydrology. *Brit. J. Appl. Phys.* 2:145–51.

Penrose, W. R., W. L. Polzer, E. H. Essington, D. M. Nelson, and K. A. Orlandini. 1990. Mobility of plutonium and americium through a shallow aquifer in a semiarid region. *Environ. Sci. Technol.* 24:228–34.

Perfect, E. and B.D. Kay. 1995. Applications of fractals in soil and tillage research: a review. *Soil Till. Res.* 36:1–20.

Perrier, E. R. and A. C. Gibson. 1982. Hydraulic Simulation of Solid Waste Disposal Sites, Office of Solid Waste and Emergency Response. Washington, DC: U.S. Environmental Protection Agency.

Perroux, K. M. and I. White. 1988. Designs of Disc Permeameters. *Soil Sci. Soc. Am. J.* 52:1205–15.

Peterson, A. and G. Bubenzer. 1986. Intake rate sprinkler infiltrometer. In *Methods of Soil Analysis. See* Klute 1986.

Pfannkuch, H. O. 1963. Contribution a l'etude des deplacements de fluides dans un milieu poreux. *Rev. Inst. Fr. Petr.* 18:215–70.

Philip, J. R. 1957a. The theory of infiltration: 1. The infiltration equation and its solution. *Soil Science* 83:345–57.

———. 1957b. The theory of infiltration: 2. The profile of infinity. *Soil Science* 83:435–48.

———. 1957c. The theory of infiltration: 3. Moisture profiles and relation to experiment. *Soil Science* 84:163–74.

———. 1957d. The theory of infiltration: 4. Sorptivity and algebraic infiltration equations. *Soil Science* 84:257–64.

———. 1968. Diffusion, dead-end-pores and linearized absorption in aggregated media. *Aust. J. Soil Res.* 6:21–30.

———. 1968. The theory of absorption in aggregated media. *Aust. J. Soil Res.* 6:1–19.

———. 1969. Theory of infiltration, In *Advances in Hydroscience,* vol. 5, ed. V. T. Chow, 215–305. New York: Academic Press.

———. 1971. Hydrology of swelling soils. In *Salinity and water use,* eds. T. Talsma and J. R. Philip. London: Macmillan.

———. 1974. Water movement in soil. In: *Heat and mass transfer in the biosphere,* eds. D.A. deVries and N.H Afgan, 29–47. New York: Halsted Press-Wiley.

———. 1975a. Stability analysis of infiltration. *Soil Sci. Soc. Amer. Proc.* 39: 1042–49.

———. 1975b. The growth of disturbances in unstable infiltration flows. *Soil Sci. Soc. Amer. Proc.* 39:1049–53.

Philip, J. R. and D. A. deVries. 1957. Moisture movement in porous materials under temperature gradients. *Trans. Amer. Geophys. Union* 38:222–32.

Phillips, F. M. 1994. Environmental tracers for water movement in desert soils of the American Southwest. *Soil Science Society of America Journal.* 58:15–24.

Phillips, F. M., J. L. Mattick, T. A. Duval, D. Elmore and P. W. Kubik. 1988. Chlorine-36 and tritium from nuclear weapons fallout as tracers for long-term liquid and vapor movement in desert soils. *Water Resour. Res.* 24:1877–91.

Pierotti, G., C. Deal, and E. Derr. 1959. Activity coefficients and molecular structure. *Ind. Eng. Chem.* 51:95.

Pinder, G. F. and W. G. Gray. 1977. Finite elements in surface and subsurface hydrology. New York: Academic Press.

Pochop, L. O., J. Borrell and R. D. Burman. 1984. Elevation-A bias in SCS Blaney Criddle ET estimates. *Transactions of the American Society of Agricultural Engineers* 27:125–28.

Poirria, M. A., B. R. Bordelon and J.L. Laseter. 1972. Adsorption and concentration of dissolved carbon-14 DDT by coloring colloids in surface waters. *Environ. Sci. Technol.* 6:1033–35.

Poiseuille, J. L. 1840–1841. Recherches experimentales sur le mouvement des liquides dans les tubes de tres petits diametres. *Compt. Rend.* 11:961–67, 1041–48; 12:112–15.

Polmann, D. J., E. G. Vomvoris, D. McLaughlin, E. M. Hammick, and L.W. Gelhar. 1988. Application of stochastic methods to the simulation of large-scale unsaturated flow and transport. NUREG/CR-5094. Washington, DC: U.S. Nuclear Regulatory Commission.

Powlson, D. S., K. W. T. Goulding, T. W. Willison, C. P. Webster, and B. W. Hutsch. 1997. The effect of agriculture on methane oxidation in soil. *Nutr. Cy. in Agroecosys.* 49:59–70.

Priestely, C. H. B. and R. J. Taylor. 1972. On the assessment of surface heat flux and evaporation using large-scale parameters. *Month. Weather Rev.* 100:81–92.

Pye, V., R. Patrick, and J. Quarles. 1983. *Ground water quality in the United States.* Philadelphia: University of Pennsylvania Press.

Querner, E. P. 1986. *An integrated surface and ground-water flow model for the design and operation of drainage systems.* Proceedings of the International Conference on Hydraulic Design in Water Resources Engineering: Land Drainage, Southampton, UK, April 16-18, 1986, pp. 101-108, Report No. 15. Wageningen, The Netherlands: Institute for Land and Water Management Research.

Quirk, J. P. 1978. Some physico-chemical aspects of soil structure stability—a review. In *Modification of Soil Structure,* eds. W. W. Emerson, R. D. Bond, and A. R. Dexter. New York: John Wiley & Sons.

Quirk, J. P. and R. K. Schofield. 1955. The effect of electrolyte concentration on soil permeability. *J. Soil Sci.* 6:163–78.

Raats, P. A. C. 1972. Steady infiltration from point sources, cavities, and basins. *Soil Sci. Soc. Am. Proc.* 35:689–94.

———. 1973. Unstable wetting fronts in uniform and nonuniform soils. *Soil Sci. Soc. Am. Proc.* 37:681–85.

———. 1974. Steady flow of water and salt in uniform profile with plant roots. *Soil Sci. Soc. Am. J.* 38:717–22.

———. 1978. Convective transport of solutes by steady flows, I. General theory, *Agricultural Water Management* 1:201–18.

———. 1984. Tracing parcels of water and solutes in unsaturated zones. In *Pollutants in Porous Media,* eds. B. Yaron, G. Datgan, and J. Goldschmidt. New York: Springer-Verlag.

Rahe, T. M., C. Hagedorn, E. L. McCoy and G. F. Kling. 1978. Transport of antibiotic-resistant Escherichia coli through western Oregon hillslope soils under conditions of saturated flow. *J. Environ. Qual.* 7:487–94.

Raloff, J. 1990. The colloid threat. *Science News.* 137:169–70.

Rantz, S. E. 1964. Snowmelt hydrology of a Sierra Nevada stream, Geological Survey Water-Supply Paper 1779-R. Prepared in cooperation with California Department of Water Resources. Washington, DC: U.S. Government Printing Office.

Rao, P. S. C. and J. M. Davidson. 1980. Estimation of pesticide retention and transformation parameters required in nonpoint source pollution models. In *Environmental impact of nonpoint source pollution,* eds. M. R. Overcash and J. M. Davidson. Ann Arbor, MI: Ann Arbor Science Publishers, Inc.

Rao, P. S. C., R. E. Jessup, and T. M. Addiscott. 1981. Experimental and theoretical aspects of solute diffusion in spherical and nonspherical aggregates. *Soil Sci.* 133:342–47.

Rao, P. S. C., R. E. Jessup, D. E. Rolston, J. M. Davidson, and D. P. Kilcrease. 1980a. Experimental and mathematical description of nonadsorbed solute transfer by diffusion in spherical aggregates. *Soil Sci. Soc. Am. J.* 44:684–88.

Rao, P. S. C., D. E. Rolston, R. E. Jessup, and J. M. Davidson. 1980a. Solute transport in aggregated porous media. Theoretical and experimental evaluation. Soil Sci. Soc. Am. J. 44:1139–46.

Rao, P. S. C., D. E. Rolston, R. E. Jessup, and J. M. Davidson. 1980b. Solute transport in aggregated porous media: Theoretical and experimental evaluation. *Soil Sci. Soc. Am. J.* 44:1139–44.

Rasmuson, A. 1982. Transport processes and conversion in isothermal fixed-bed catalytic reactor. *Chem. Eng. Sci.* 37:411–16.

Rasmuson, A. and I. Neretnieks. 1981. Migration of radionuclides in fissured rock: The influence of micropore diffusion and longitudinal dispersion. *J. of Geophys. Res.* 86:3749–58.

Rasmusson, E. M., R. E. Dickinson, J. E. Kutzback, and M. K. Cleaveland. 1993. Chapter 2, Climatology, In *Handbook of Hydrology,* ed. D. R. Maidment. New York: McGraw-Hill, Inc.

Ravina, I. and P. F. Low. 1972. Relation between swelling, water properties and b-dimension in montmorillonite-water systems. *Clays Clay Miner.* 20:109–23.

Rawlins, S. L. and G. S. Campbell. 1986. Water potential: Thermocouple psychrometry. In *Methods of Soil Analysis, Part I.* Physical and Mineralogical Methods, Agronomy Monograph (9)597–618. Madison, WI: Amer. Soc. Agron.

Rawls, W. J., L. R. Ahuja, D. L. Brakensiek and A Shirmonhammadi. 1993. Chapter 5 - Infiltration and soil water movement. In *Handbook of Hydrology,* ed. D. R. Maidment. New York: McGraw-Hill, Inc.

Rawls, W. J. and D. L. Brakensiek. 1983. A Procedure of predict Green Ampt infiltration parameters. Proceedings of the Symposium on Advances in Infiltration, American Society of Agricultural Engineers, St. Joseph, MI., pp. 102–12.

Rawls, W. J., D. L. Brakensiek and R. Savabi. 1989. Infiltration parameters for rangeland soils. *Journal of Range Management* 42(2):139–42.

Rawls, W. J., D. L. Brakensiek, and K. E. Saxton. 1982. Estimation of soil water properties. *Transactions of the American Society of Agricultural Engineers* 25:1316–20, 1328.

Razavi, M. -S., B. J. McCoy, and R. G. Carbonell. 1978. Moment theory of breakthrough curves for fixed-bed adsorbers and reactors. *Chem. Eng. J.* 16:211–22.

Rees, T. F. 1987. A review of light-scattering techniques for the study of colloids in natural waters. *J. Contaminant Hydrol.* 1:425–39.

Reid, R. C., J. M. Prausnitz, and T. K. Sherwood. 1977. The properties of gases and liquids, 3d ed. New York: McGraw-Hill, Inc.

Reisenauer, A. E., K. T. Key, T. N. Narasimhan, and R. W. Nelson. 1982. TRUST: A Computer program for variably saturated flow in multidimensional, deformable media, NUREG/CR-2360. Washington, DC: U.S. Nuclear Regulatory Commission.

Rengasamy, P. 1983. Clay dispersion in relation to the changes in the electrolyte composition of dialysed red-brown earths. *J. Soil Sci.* 34:723–32.

Rengasamy, P., R. S. B. Greene, G. W. Ford, and A. H. Mehanni. 1984. Identification of dispersive behavior in the management of red-brown earths. *Aust. J. Soil Res.* 22:413–22.

Rengasamy, P. and M. E. Sumner. Forthcoming. In *Sodic Soils: Distribution, Processes, Management, and Environmental Consequences,* ed. M. E. Sumner and R. Naidu. New York: Oxford University Press.

Reynolds, O. 1883. An experimental investigations of the circumstances which determine whether the motion of the water shall be direct or sinuous, and of the law of resistance in parallel channels. *Proc. R. Soc. London* 35:84–99.

Richards, L. A. 1931. Capillary conduction of liquids through porous mediums. *Physics* 1:318–33.

———. 1941. A pressure-membrane extraction apparatus for soil solution. *Soil Sci.* 51:377–86.

———. 1942. Soil moisture tensiometer materials and construction. *Soil Sci.* 53:241–48.

———. 1948. Porous plate apparatus for measuring moisture retention and transmission by soil. *Soil Sci.* 66:105–10.

———. 1949. Methods of measuring soil moisture. *Soil Sci.* 68:95–112.

———. 1950. Laws of soil moisture. *Trans. Am. Geophys. Union* 31:750–56.

———. 1955. Water content changes following the wetting of bare soil in the field. *Soil Sci. Fla. Proc.* 15:142–48.

———. 1960. Advances in soil physics. Trans Seventh Int. Congr. *Soil Sci.* 1:67–79.

Richards, L. A. and Milton Fireman. 1943. Pressure plate apparatus for measuring moisture sorption and transmission by soils. *Soil Sci.* 56:395–404.

Richards, L. A. and W. Gardner. 1936. Tensiometers for measuring the capillary tension of soil water. *J. Am Soc. Agron.* 28:352–58.

Richardson, C. W. and D. A. Wright, 1984, WGEN: A model for generating daily weather variables, ARS-8, Agricultural Research Service, U.S. Department of Agriculture, 83 p.

Ripple, C. D., J. Rubin, and T. E. A. van Hylckama. 1972. *Estimating steady-state evaporation rates from bare soils under conditions of high water table.* Water Supply Paper 2019-A. Washington, DC: U.S. Geological Survey.

Risken, H. 1984. The Fokker-Planck Equation. Methods of solution and application. New York: Springer-Verlag.

Robbins, C. W., J. J. Jurinak, and R. J. Wagenet. 1980a. Calculating cation exchange in a salt transport model. *Soil Sci. Soc. Amer. J.* 44:1195–29.

Robbins, C. W., R. J. Wagenet, and J. J. Jurinak. 1980b. A combined salt transport-chemical equilibrium model for calcareous and gypsiferous soils. *Soil Sci. Soc. Amer. J.* 44:1191–94.

Robbins, G. A., S. Wang, and J. D. Stuart. 1993. Using the static headspace method to determine Henry's law constants. *Analy. Chem.* 65(21):3113–18.

Romell, L. G. 1922. Luftvaxlingen i marken s.ekol. faktor. *Medd. Statens Skogsforskningsinst* 19:125–359.

Rose, D. A. and J. B. Passioura. 1971a. The analysis of experiments on hydrodynamic dispersion. *Soil Sci.* 111:252–57.

———. 1971b. Gravity segregation during miscible displacement. *Soil Sci.* 111:258–65.

Rosenberg, N. J., B. L. Blad, and S. B. Verma. 1983. *Microclimate.* New York: John Wiley & Sons.

Ross, D. L. and H. J. Morel-Seytoux. 1982. *User's Manual for SOILMOP: A Fortran IV program for prediction of infiltration and water content profiles under variable rainfall conditions,* Interim Report for FY 1981–82, DER-82-DLR-HJM45, Department of Civil Engineering, Colorado State University, Fort Collins, Colorado.

Rosswall, T. 1989. What regulates production and consumption of trace gases in ecosystems: Biology or physiochemistry? In *Exchange of trace gases between terrestrial ecosystems and the atmosphere,* eds. M. O. Andreae and D. S. Schimel, 73–96. New York: John Wiley & Sons.

Rower, C. 1931. *Evaporation from free water surfaces.* U.S. Department of Agriculture Technical Bulletin 27.

Ruben, S., M. Randall, and J. L. Hyde. 1941. Heavy oxygen (O^{18}) as a tracer in the study of photosynthesis. *J. Amer. Chem. Soc.* 63:837–79.

Rubin, J. 1983. Transport of reacting solutes in porous media: Relation between mathematical nature of problem formulation and chemical nature of reactions. *Water Resour. Res.* 19:1231–52.

Rubin, J. and R. V. James. 1973. Dispersion-affected transport reacting solutes in saturated porous media: Galerkin method applied to equilibrium-controlled exchange in unidirectional steady water flow. *Water Resour. Res.* 9:1332–56.

Russell, E. J. 1912. *Soil conditions and plant growth* (8th ed., 1950, recast and rewritten by E. Walter Russell). London: Longmans, Green.

———. 1957. *The world of the soil.* London: Collins.

Russell, M. B. 1949. Methods of measuring soil structure and aeration. *Soil Sci.* 68, 25–35.

———. 1952. Soil aeration and plant growth. In *Soil physical conditions and plant growth,* ed. B. T. Shaw, 253–301. New York: Academic Press.

Ryan, J. N. and P. M. Gschwend. 1990. Colloid mobilization in two Atlantic coastal plain aquifers: Field studies. *Water Resour. Res.* 26:307–22.

Salas, J. D. and M. Markus. 1992. Modeling and generation of multivariate seasonal hydrologic data (programs CSU003 and CSU004), technical report 3. Fort Collins, CO: Computing Hydrology Laboratory, Engineering Research Center, Colorado State University.

Saleur, H. and B. Duplantier. 1987. Exact determination of the percolation hull exponent in two dimensions. *Phys. Rev. Lett.* 58:2325–28.

Salisbury, F. B. and C. W. Ross. 1978. *Plant physiology.* Belmont, CA: Wadsworth.

Salmon, S. C. and A. A. Hanson. 1964. *The principles and practice of agricultural research.* London: Leonard Hill.

Sapoval, B., M. Rosso, and J. F. Gouyet. 1985. The fractal nature of a diffusing front and the relation to percolation. *J. Phys. Lett.* 46:L149–56.

Sass, J. H., A. H. Lachenbruch, W. W. Dudley, Jr., S. S. Priest, and R. J. Munroe. 1988. *Temperature, thermal conductivity, and heat flow near Yucca Mountain, Nevada: Some tectonic and hydrological implications.* Open-file report 87-649. Washington, DC: U.S. Geological Survey.

Saugier, B., A. Granier, J. Y. Pontailler, E. Dufrene, and D. D. Baldocchi. 1997. Transpiration of a boreal pine forest measured by branch bag, sap flow, and micrometeorological methods. *Tree Physiol.* 17:511–19.

Savage, M. J., A. Cass, and J. M. de Jager. 1983. Statistical assessment of some errors in thermocouple hygrometric water potential measurement. *Agric. Meteor.* 30:83–97.

Savage, M. J. and H. H. Wiebe. 1987. Voltage endpoint determination for thermocouple psychrometers and the effect of cooling time. *Agric. & For. Meteor.* 39:309–17.

Saxen, U. 1892. Electrokinetic potentials. In *Textbook of physical chemistry,* ed. S. Glasstone. New York: Van Nostrand.

Sayles, R. S. and T. R. Thomas. 1978. Surface topography as a nonstationary random process. *Nature* 271:431–34.

Scanlon, B. R. and P. C. D. Milly. 1994. Water and heat fluxes in desert soils: 2. Numerical simulations. *Water Resour. Res.* 30(3):721–33.

Scheidegger, A. E. 1961. General theory of dispersion in porous media. *J. Geophys. Res.* 66:3273–78.

———. 1974. *The physics of flow through porous media.* Toronto: University of Toronto Press.

Schenkel, J. H. and J. A. Kitchener. 1960. A test of the Derjaguin-Verwey-Overbeek theory with a colloidal suspension. *Trans. Faraday Soc.* 62:161–73.

Scheunert, I. and W. Klein. 1985. Predicting the movement of chemicals between environmental compartments (air-water-soil-biota). In *Appraisal of tests to predict the environmental behavior of chemicals,* eds. P. Sheehan, F. Korte, W. Klein and Ph. Bordeau 285–331. New York: John Wiley & Sons Ltd.

Schirado, T., I. Vergara, E. B. Schalscha, and P. F. Pratt. 1986. Evidence for movement of heavy metals in a soil irrigated with untreated wastewater. *J. Env. Quality* 15:9–12.

Schlicting, H. 1968. *Boundary-layer theory.* New York: McGraw-Hill.

Schnitzer, M. 1976. The chemistry of humic substances. In *Environmental biochemistry,* Vol. 1. ed. J. O. Nriagn, 89–107. Ann Arbor, MI: Ann Arbor Sci.

Schofield, R. K. 1935. The pF of the water in soil. *Trans.* 3rd Int. Cong. *Soil Sci.* 2:37–48.

———. 1946. Ionic forces in thick films of liquid between charged surfaces. *Trans. Faraday Soc.* 42B:219–25.

———. 1950. Soil moisture and evaporation. Trans. Fourth Intr. Congress. *Soil Sci.* 2:20–28.

Schroeder, P. R., T. S. Dozier, P. A. Zappi, B. M. McEnroe, J. W. Sjostrom, and R. L. Payton. 1994b. *The Hydrologic evaluation of landfill performance (HELP) model, Engineering documentation for version 3.* EPA\600\R-94\168b, September. Cincinnati, OH: U.S. Environmental Protection Agency.

Schroeder, P. R., C. M. Lloyd, P. A. Zappi, and N. M. Aziz. 1994a. *Hydrologic evaluation of landfill performance (HELP) model: User's guide for version 3.* EPA/600/R-94/168a, Risk Reduction Engineering Laboratory, 84 p., Appendix A. Cincinnati, OH: U.S. Environmental Protection Agency.

Schwartzenback, R. P. and J. Westall. 1981. Transport of nonpolar organic compounds from surface water to groundwater, laboratory sorption studies. *Environmental Science and Technology* 19:61–68.

Schwille, F. 1988. Dense chlorinated solvents in porous and fractured media. Chelsea MI: Lewis Publishers.

Scotter, D. R. 1978. Preferential solute movement through larger soil voids. I. Some computations using simple theory. *Aust. J. Soil Res.* 16:257–67.

Selim, H. M., J. M. Davidson, and R. S. Mansell. 1976a. Evaluation of a two-site adsorption-desorption model for describing solute transport in soil, 444–48. *Proc. Summer Conf.,* Washington DC.

Selim, H. M., R. S. Mansell, and A. Elzeftway. 1976b. Distribution of 2,4-D and water in soil during infiltration and redistribution. *Soil Sci.* 121:176–83.

Selim, H. M., R. Schulin, and H. Fluhler. 1987. Transport and ion exchange in an aggregated soil. *Soil Sci. Soc. Amer. J.* 51:876–84.

Sellers, W. D. 1965. *Physical climatology.* Chicago: University of Chicago Press.

Sennett, Paul, and Olivier, J. P. 1965. Colloidal dispersions and the concept of zeta potential. *Ind. Eng. Chem.* 57:32–50.

Sharma, M. L., G. A. Gander, and C. G. Hunt. 1980. Spatial variability of infiltration in a watershed. *J. Hydrol.* 45:101–22.

Shaw, R. H., K. T. Paw U, X. J. Zhang, W. Gao, G. den Hartog, and H. H. Neumann. 1990. Retrieval of turbulent pressure fluctuations at the ground surface beneath a forest. *Bound. Lay. Meteor.* 50:319–38.

Sheppard, J. C., M. J. Campbell, T. Cheng, and J. A. Kittrick. 1980. Retention of radionuclides by mobile humic compounds and soil particles. *Environ. Sci. Technol.* 14:1349–53.

Short, T. E. 1985. *Movement of contaminants from oily waste during land treatment.* Proceedings of the Conference on Environmental and Public Health Effects of Soils Contaminated with Petroleum Products, Amherst, MA.

Shorter, J. 1972. The separation of polar, steric and resonance effects by the use of linear free energy relationships. Chapter 2 in *Advances in linear free energy relationships,* eds. N. B. Chapman and J. Shorter. New York: Plenum Press.

Shouse, P., J. Scisson, G. de Rooij, J. Jobes, and M. Th. Van Genuchten. 1989. *Application of fixed-gradient methods for estimating soil hydraulic conductivity.* Proceedings of International Workshop on Indirect Methods for Estimating the Hydraulic Properties of Unsaturated Soils, Riverside, California, October 11–13. Ed. M Th. Van Genuchten et al. Riverside, CA: Univ. of California.

Shuttleworth, W. J. 1993. Chapter 4, Evaporation. In *Handbook of hydrology,* ed. D. R. Maidment. New York: McGraw-Hill, Inc.

Shuval, H. I. 1993. Investigation of typhoid fever and cholera transmission by raw wastewater irrigation in Santiago, Chile. *Water Science & Tech.* 27:167–74.

Shuval, H. I., A. Adin, B. Fattal, E. Rawitz, and P. Yekutiel. 1986b. Wastewater irrigation in developing countries: Health effects and technical solutions. World Bank Technical Paper No. 51. Washington, DC: The World Bank.

Shuval, H. I., N. Buttman-Bass, J. Applebaum, and B. Fattal. 1989. Aerosolized enteric bacteria and viruses generated by spray irrigation of wastewater. *Water Science & Tech.* 21:131–35.

Shuval H. I., P. Yekutiel, and B. Fattal. 1985. Epidemiological evidence for helminth and cholera transmission by vegetables irrigated with wastewater: Jerusalem—a case study. *Water Science & Tech.* 17:433–42.

———. 1986a. Epidemiological model of the potential health risk associated with various pathogens in wastewater irrigation. *Water Science & Tech.* 18: 191–98.

Sisson, J. B., A. H. Ferguson and M. Th. van Genuchten. 1980. Simple methods for predicting drainage from field plots. *Soil Sci. Soc. Am. J.* 44:1147–52.

Skaggs, R. W. and R. Khaleel. 1982. Infiltration, In *Hydrologic modeling of small watersheds,* Monograph 5, ed. C. T. Haan, 4–166. St. Joseph, MI: American Society of Agriculture Engineers.

Slater, J. G. and J. G. Kirkwood. 1931. The van der Waals forces in gases. *Phys. Rev.* 37:682–97.

Slatyer, R. O. and I. C. McIlroy. 1961. *Practical microclimatology.* Australia: CSIRO.

Slichter, C. S. 1898. *Theoretical investigation of the motion of ground waters.* Nineteenth Annual Rep., Pt. 2:295–384. Washington, DC: U.S. Geological Survey.

Smiles, D. E. and J. R. Philip. 1978a. Solute transport during absorption of water by soil: Laboratory studies and their practical implications. *Soil Sci. Soc. Amer. J.* 42:537–44.

Smiles, D. E., J. R. Philip, J. H. Knight, and D. E. Elrick. 1978b. Hydrodynamic dispersion during absorption of water by soil. *Soil Sci. Soc. Amer. J.* 42:229–34.

Smith, J. A., P. R. Jaff, C. T. Chiou. 1990. Effect of ten quaternary ammonium cations on tetrachloromethane sorption to clay from water. *Environ. Sci. Technol.* 24:1167–72.

Smith, M. S., A. W. Thomas, R. E. White, and D. Ritonga. 1985. Transport of *Escherichia coli* through intact and disturbed soil columns. *J. Environ. Qual.* 14:87–91.

Smith, R. E. and J. -Y. Parlange. 1978. A parameter-efficient hydrologic infiltration model. *Water Resources Research* 14(3):533-38.

Sophocleus, M. and C. A. Perry, 1985, Experimental studies in natural groundwater recharge dynamics: The analysis of observed recharge events. *Journal of Hydrology,* 81:297-332.

Sopper, W. E. and S. N. Kerr, eds. 1979. Utilization of municipal sewage effluent and sludge on forest and disturbed land. University Park, PA: The Pennsylvania State University Press.

Sopper, W. E. and J. L. Richenderfer. 1978. Effects of spray irrigation of municipal wastewater on the physical properties of the soil. National Technical Information Service, VA, PB-290 732. Institute for Research on Land and Water Resources, Final Tech. Report, 14-31-0001-4205, 188 p.

———. 1979. Effect of municipal wastewater irrigation on the physical properties of the soil. In *Utilization of municipal sewage effluent and sludge on forest and disturbed land,* pp. 179–95. University Park, PA: The Pennsylvania State University Press.

Sorey, M. L. 1971. Measurement of vertical ground-water velocity from temperature profiles in wells. *Water Resour. Res.* 7(4):963–70.

Spanner, D. C. 1951. The Peltier effect and its use in the measurement of suction pressure. *J. Exptl. Botany* 2: 145–68.

Sposito, G. 1984. *The surface chemistry of soils.* New York: Oxford University Press.

———. 1989. *The chemistry of soils.* New York: Oxford University Press.

Sposito, G. and S. W. Mattigod. 1977. On the chemical foundation of the sodium adsorption ratio. *Soil Sci. Soc. Am. J.* 41:323–29.

Srinivasan, K. R. and H. S. Fogler. 1989. Use of modified clays for the removal and disposal of chlorinated dioxins and other priority pollutants from industrial wastewater. *Chemosphere* 18:333–42.

———. 1990a. Use of inorgano-organo-clays in the removal of priority of pollutants from industrial wastewaters: structural aspects. *Clays Clay Min.* 38:277–86.

———. 1990b. Use of inorgano-organo-clays in the removal of priority of pollutants from industrial wastewaters: Adsorption of Benzo(a)pyrene and chlorophenol. *Clays and Clay Minerals* 38:287–93.

Stannard, D. I. 1992. Tensiometers-theory, construction, and use. *Geotech. Testing J.* 15:48–58.

———. 1993. Comparison of Penman-Monteith, Shuttleworth-Wallace, and modified Priestly-Taylor evapotranspiration models for wildland vegetation in semiarid rangeland. *Wat. Resour. Res.* 29(5):1379–92.

Stannard, D. I., J. H. Blanford, W. Kustas, S. Amer, T. Schmugge, and M. Weltz. 1994. Interpretation of surface flux measurements in heterogeneous terrain during the Monsoon '90 experiment. *Water Resour. Res.* 30(5):1227–39.

Starr, R. C., R. W. Gillham, and E. A. Sudicky. 1985. Experimental investigation of solute transport in stratified porous media. 2. The reactive case. *Water Resour. Res.* 21:1043–50.

State of California. 1986. Site mitigation decision tree manual: California department of health services—toxic substance control division.

Steenhuis, T. S., J. Y. Parlange, and M. S. Andreini. 1990. A numerical model for preferential solute movement in structured soils. *Geoderma* 46:193–208.

Stephens, D. B. and R. Knowlton. 1986. Soil water movement and recharge through sand at a semi-arid site in New Mexico. *Water Resources Research* 22:881–89.

Stern, O. 1924. Zur theorie der elektrolytischen doppel-schicht. Zeitschrift fur elektrochemie der ange-wandte physikalische chemie. *Leipziq-Berlin.* 30:508–16.

Stevens, D. K., W. J. Grenney, and Z. Yan. 1991. A model for the evaluation of hazardous substances in the soil. Version 3.0. Department of Civil and Environmental Engineering, Utah State University, Logan, Utah.

Stevenson, F. J. 1976. Organic matter reactions involving pesticides in soil. *Am. Chem. Soc. Symp. Ser.* 29:180–207.

———. 1982a. Humus chemistry; genesis, composition, reactions. New York: Wiley-Interscience.

———. 1982b. Nitrogen in agricultural soils. Agronomy monograph 22. *Soil Sci. Soc. Am.* Madison, WI.

———. 1985. Geochemistry of soil humic substances. In *Humic substances in soil, sediment, and water: Geochemistry, isolation, and characterization,* eds. G. R. Aiken, D. M. McKnight, and R. L. Wershaw, 13–52. New York: John Wiley & Sons.

Stokes, Sir G. G. 1898. Stream-line motion of a viscous film. II. Mathematical proof of the identity of the stream lines obtained by means of a viscous film with those of a perfect fluid moving in two dimensions. Report of the 68th meeting of the British Assoc. Adv. Sci., Bristol. John Murray, London, pp. 143–44.

Streeter, V. L. 1962. *Fluid mechanics.* New York: McGraw-Hill.

Stumm, W. and J. J. Morgan. 1981. Aquatic chemistry: An introduction emphasizing chemical equilibria in natural waters. New York: John Wiley & Sons.

———. 1996. *Aquatic chemistry: An introduction emphasizing chemical equilibria in natural waters,* 3d ed. New York: John Wiley & Sons.

Su, C. and R. H. Brooks. 1975. Soil hydraulic properties from infiltration tests. Watershed Management Proceedings, Irrigation and Drainage Division, August 11-13, pp. 516–42, ASCE, Logan, Utah.

Sudicky, E. A., R. W. Gillham, and E. O. Frind. 1985. Experimental investigation of solute transport in stratified porous media. 1. The nonreactive case. *Water Resour. Res.* 21:1035–41.

Summers, K., S. Gherini, and C. Chen. 1980. Methodology to evaluate the potential for groundwater contamination from geothermal fluid releases, EPA/600/7-80/117. Cincinnati, OH: U.S. Environmental Protection Agency.

Sumner, M. E. 1992. The electrical double layer and clay dispersion. In *Soil crusting: chemical and physical processes,* eds. M. E. Sumner and B. A. Stewart, 1–32. Advances in soil science. Boca Raton, FL: CRC Press.

Sumner, M. E. and Stewart, B. A. 1992. *Soil crusting: chemical and physical processes.* Boca Raton, FL: Lewis Publishers.

Tabor, D. and R. H. S. Winterton. 1968. Surface forces: direct measurement of normal and retarded van der Waals forces. *Nature* 219:1120–121.

Talsma, T. 1977. Measurement of the overburden component of total potential in swelling field soils. *Aust. J. Soil Res.* 15:95–102.

Tang, D. H., E. O. Frind, and E. A. Sudicky. 1981. Contaminant transport in fractured porous media: An analytical solution for a single fracture. *Water Resour. Res.* 17:555–64.

Terzaghi, K. 1943. *Theoretical soil mechanics.* New York: John Wiley & Sons.

Tessens, E. 1984. Clay migration in upland soils in Malaysia. *J. Soil Sci.* 35:615–24.

Tetens, O. 1930. Uber einige meteorologische begriffe. *Z. Geophys.* 6:297–309.

Thom, A. S. and H. R. Oliver. 1977. On Penman's equation for estimating regional evapotranspiration. *Quart. J. Roy. Meteor. Soc.* 103:345–57.

Thomas, G. W. and R. E. Phillips. 1979. Consequences of water movement in macropores. *J. Environ. Qual.* 8:149–52.

Thorburn, P. J., T. J. Hatton, and G. R. Walker. 1993. Combining measurements of transpiration and stable isotopes of water to determine groundwater discharge from forests. *J. Hydrol.* 150:563–87.

Thornthwaite, C. W. and J. R. Mather. 1955. The water balance. Centerton, NJ, Drexel Institute of Technology. Laboratory of Climatology. *Publications in Climatology,* v. 8, no. 1.

———. 1957. Instructions and tables for computing potential evapotranspiration and the water balance. Centerton, NJ, Drexel Institute of Technology. Laboratory of Climatology. *Publications in Climatology,* v. 10, no. 3.

Thorstenson, D. C. and D. W. Pollock. 1989a. Gas transport in unsaturated porous media: The adequacy of Fick's law. *Rev. Geophys.* 27(1):61–78.

———. 1989b. Gas transport in unsaturated porous media: Multicomponent systems and the adequacy of Fick's law. *Water Resour. Res.* 25(3):477–507.

Thurman, E. M. 1985a. *Organic geochemistry of natural waters.* Boston: Nijhoff.

———. 1985b. Humic substances in groundwater. In *Humic substances in soil, sediment, and water: Geochemistry, isolation, and characterization,* eds. G. R. Aiken, D. M. McKnight, and R. L. Wershaw, 87–104. New York: John Wiley & Sons.

Tillotson, Patricia M. and Donald R. Nielsen. 1984. Scale factors in soil science. *Soil Sci. Soc. Am. J.* 48:953–59.

Timpson, M. E., J. L. Richardson, L. P. Keller and G. J. McCarthy. 1986. Evaporite mineralogy associated with saline seeps in southwestern North Dakota. *Soil Sci. Soc. Am. J.* 50:490–493.

Tindall, J. A., K. Hemmen, and J. F. Dowd. 1992. An improved method for field extraction and laboratory analysis of large, intact soil cores. *J. Environ. Qual.* 21:259–63.

Tindall, J. A., K. J. Lull, and N. G. Gaggiani. 1994. Effects of land disposal of municipal sewage sludge on fate of nitrates in soil, streambed sediment and water quality. *J. Hydrology* 163:147–85.

Tindall, J. A., R. L. Petrusak, and P. B. McMahon. 1995. Nitrate transport and transformation processes in unsaturated porous media. *J. Hydrology* 169:51–94.

Tindall, J. A. and W. K. Vencill. 1993. Preferential transport of atrazine in well structured soils. In *Agricultural research to protect water quality* (vol. I), 154–56. Proc. Conf., 21–24 February, Minneapolis, MN. Ankeny, IA: Soil and Water Conservation Society.

———. 1995. Transport of atrazine, 2,4-D, and dicamba through preferential flowpaths in an unsaturated claypan soil near Centralia, Missouri. *J. Hydrology* 166:37–59.

Tipping, E. and D. C. Higgins. 1982. The effect of adsorbed humic substances on the colloid stability of haematite particles. *Colloid Surf.* 5:85–92.

Tisdale, S. L. and W. L. Nelson. 1975. *Soil fertility and fertilizers.* New York: Macmillan.

Todd, D. K. 1967. *Ground water hydrology,* 6th ed. New York: John Wiley & Sons.

Topp, G. C. 1969. Soil water hysteresis measured in a sandy loam and compared with the hysteresis domain model. *Soil Sci. Soc. Am. Proc.* 33:645–51.

Toride, N., F. J. Leij, and M. Th. Van Genuchten. 1995. The CXTFIT code for estimating transport parameters from laboratory or field tracer experiments. Version 2.0. Research report no. 137. U.S. Salinity Laboratory, Agricultural Research Service. Riverside, CA: U.S. Department of Agriculture.

Toth, S. T. and R. B. Alderfer. 1960. A procedure for tagging water stable aggregates with Co-60. *Soil Sci.* 89:36–37.

Traina, S. J., D. A. Spontak, and T. J. Logan. 1989. Effects of cations of complexation of napthalene by water-soluble organic carbon. *J. Environ. Qual.* 18:221–27.

Travis, B. 1984. TRACER3D: A model of flow and transport in porous/fractured media, Los Alamos National Laboratory, Report LA-9667-MS. Los Alamos, New Mexico.

Tripathi, R. P. and B. P. Ghildyal. 1975. Validity of soil water evaporation versus square root of time relation. *J. Indian Soc. Soil Sci.* 23:158–62.

Tull, J. 1731. *The horse hoeing husbandry.* London: Jethro Tull.

Turcotte, D. L. 1986. Fractals and fragmentation. *J. Geophys. Res.* 91(b2):1921–26.

Tyler, S. W. and G. R. Walker. 1994. Root Zone Effects on Tracer Migration in Arid Zones. *Soil Sci. Soc. Am. J.* 58(January-February):25–31.

U.S. Army Corps of Engineers. 1956. Snow Hydrology, Summary Report of the Snow Investigations, North Pacific Division. Portland, OR, June 30.

U.S. Army. 1987. Single Pathway Pollutant Limit Values and Preliminary Pollutant Limit Values (SPPPLV and PPLV). In *Inventory of cleaning criteria and methods to select criteria.* Unpublished Document by G. M. Richardson, Industrial Programs Branch, Environment Canada.

U.S. Bureau of Reclamation. 1984. SBIR Phase I Final Report, Modeling Physics and Chemistry of Contaminant Transport in Three-Dimensional Unsaturated Ground-Water Flow, Final Report, Contract 4-CR-93-00010, (NTIS access # PB85-160683). Washington, DC: U.S. Department of the Interior.

U.S. Department of Agriculture. 1984. User's Guide for the CREAMS Model: Washington Computer Center Version, USDA-SCS Engineering Division Technical Release 72. Washington, DC: Soil Conservation Service.

———. 1977. The 1978 Fertilizer Situation Publication FS-8. Washington DC: Econ. Res. Serv.

———. 1960. USDA yearbook of agriculture: Soils and men. Washington, DC: U.S. Government Printing Office.

U.S. Environmental Protection Agency (EPA). 1989. Determining Soil Response Action Levels Based on Potential Contaminant Migration to Ground Water: A compendium of Examples, EPA/540/2-89/057. Washington, DC: Office of Emergency and Remedial Response.

———. 1991a. Superfund Innovative Technology Evaluation (SITE) Program: EPA-540/8-91/005, Spring update.

———. 1991b. Soil Vapor Extraction Technology Reference Handbook: Office of Research and Development, EPA-540/2-91/003, February.

———. 1991c. Innovative Treatment Technologies: Overview and Guide to Information Sources: EPA-540/9-91/002, October.

————. 1989a. Determining Soil Response Action Levels Based on Potential Contaminant Migration to Ground Water: A Compendium of Examples: Office of Emergency and Remedial Response, Washington, DC, EPA-540/2-89/057, October, 144 p.

————. 1989b. Guidance for conducting remedial investigations and feasibility studies: EPA-540/G-89/004.

————. 1981. Process design manual for land treatment of municipal wastewater, Center for Environmental Research Information, EPA 625/1-81-013, October.

————. 1978. Sludge treatment and disposal. Vol. 2. Environmental Research Information Center, No. 625/4-78-012, Cincinnati, OH.

————. 1986. Quality criteria for water 1986. Washington, DC: USEPA Report 440/5-86-001.

U.S. Salinity Laboratory Staff. 1954. *Diagnosis and improvement of saline and alkali soils.* Handbook 60, ed. L. A. Richards. Riverside, CA: U. S. Salinity Lab.

U.S. Soil Conservation Service (SCS). 1980. National engineering handbook, Section 4, Hydrology. Washington, DC: U.S. Department of Agriculture.

————. 1964. National engineering handbook, Section 15, Irrigation, Chapter I, Soil-Plant-Water Relationships. Washington, DC: U.S. Department of Agriculture.

————. 1986. Urban hydrology for small watersheds, Technical release 55 (210-VI-TR-55, 2d Ed., June 1986), 6 Chapters, Appendices A through F.

————. 1970. Irrigation water requirements, Technical release No. 21, Engineering Division, September, 88 p.

————. 1971. National engineering handbook, Chapter 16, Drainage of agricultural land. Washington, DC: U.S. Department of Agriculture.

————. 1985. National engineering handbook—Section 4—Hydrology. Washington, DC: U.S. Department of Agriculture.

Valdares, J. M., Gal, M., Mingelgrin, U. and Page, A. L. 1983. Some heavy metals in soils treated with sewage sludge, their effects on yield, and their uptake by plants. *J. Environ. Qual.* 12:49–57.

Valocchi, A. J., R. L. Street, and P. V. Roberts. 1981. Transport of ion-exchanging solutes in groundwater: chromatographic theory and field simulation. *Water Resour. Res.* 17:1517–27.

Vandenberg, A. 1985. A physical model of vertical infiltration, drain discharge and surface runoff, Ottawa, Ontario, Canada: National Hydrology Research Institute, Inland Waters Directorate.

Van Bavel, C. H. M. 1967. Changes in canopy resistance to water loss from alfalfa induced by soil water depletion. *Agric. Meteor.* 4:165–76.

————. 1967. Changes in canopy resistance to water loss from alfalfa induced by soil water depletion. *Agric. Meteor.* 4:165–76.

Van der Heijde, P. K. M. 1994. Identification and compilation of unsaturated/vadose zone models, EPA/600/R-94/028, R. S. Kerr Environmental Research Laboratory, Office of Research and Development. U.S. Environmental Protection Agency, Ada, Oklahoma.

Van der Pol, R. M., P. J. Wierenga, and D. R. Nielsen. 1977. Solute movement in a field soil. *Soil Sci. Soc. Am. J.* 41:10–13.

Van Doren, D. M., Jr. and R. R. Allmaras. 1978. Effect of residue management practices on the soil physical environment, microclimate, and plant growth. In *Crop Residue Management Systems,* ed. W. R. Oschwald. ASA Spec. Pub. No. 31. Amer. Soc. of Agron., Madison, WI.

Van Genuchten, M. Th. 1978. Numerical solutions of the one-dimensional saturated/unsaturated flow equation. Report 78-WR-9, Water Resources Program, Princeton, New Jersey: Department of Civil Engineering, Princeton University.

————. 1980. A closed-form equation for predicting the hydraulic conductivity of unsaturated soils, *Soil Sci. Soc. Am. J.* 44:892–98.

————. 1981. Non-equilibrium transport parameters from miscible displacement experiments. Research Report No.119 Riverside, CA: U.S. Salinity Laboratory.

————. 1985. A general approach for modeling solute transport in structured soils. In *Proc. Hydrogeology of rocks of Low Hydraulic Conductivity. Mem. IAH* 17:513–15.

Van Genuchten, M. Th. and W. J. Alves. 1982. Analytical solutions of the one-dimensional convective-dispersive solute transport equation. *USDA Tech. Bulletin.* Washington, DC: No. 1661.

Van Genuchten, M. Th. and R. W. Cleary. 1982. Movement of solutes in soil: Computer-simulated and laboratory results. In *Soil chemistry.* B. Physico-chemical models, ed. G. H. Bolt. Amsterdam: Elsevier.

Van Genuchten, M. Th. and W. A. Jury. 1987. Progress in unsaturated flow and transport modeling. *IUGG Rev.* 25:135–40.

Van Genuchten, M. Th., F. J. Leij and S. R. Yates. 1991. The RETC code for quantifying the hydraulic functions of unsaturated soils, U.S. Environmental Protection Agency, EPA/600/2-91-065, December, Appendices A and B. Washington, DC: Office of Research and Development.

Van Genuchten, M. Th., D. H. Tang, and R. Guennelon. 1984. Some exact solutions for solute transport through soils containing large cylindrical macropores. *Water Resour. Res.* 20:335–46.

Van Genuchten, M. Th. and P. J. Wierenga. 1976. Mass transfer studies in sorbing porous media. I. Analytical solutions. *Soil Sci Soc. Amer. J.* 40:473–80.

————. 1977. Mass transfer studies in sorbing porous media. II. Experimental evaluation with tritium (3H20). *Soil. Sci. Soc. Amer. J.* 41:272–78.

————. 1986. Solute dispersion coefficients and retardation factors. In Methods of Soil Analysis. I. Physical and Mineralogical Methods. Madison, WI: Soil Sci. Soc. of Amer.

Van Genuchten, M. Th., P. J. Wierenga, and G. A. O'Connor. 1977. Mass transfer studies in sorbing porous media. III. Experimental evaluation with 2,4,5-T. *Soil Sci. Soc. Amer. J.* 41:278–84.

Van Olphen, H. 1963. *An introduction to clay colloid chemistry.* New York: Interscience.

————. 1969. Thermodynamics of interlayer adsorption of water in clays: II. *Proc. Int. Clay Cong. Tokyo* 1:649–57; 2:161.

————. 1977. *An introduction to clay colloid chemistry,* 2d ed. New York: John Wiley & Sons.

Van Zyl, D. J. A., I. P. G. Hutchinson and J. E. Kiel eds. 1988. *Introduction to Evaluation, Design and Operation of Precious Metal Heap Leaching Projects.* Littleton, CO: Society of Mining Engineers, Inc.

Vehrencamp, J. E. 1953. Experimental investigation of heat transfer at an air-earth interface. *Trans. Amer. Geophys. Union* 34:22–30.

Verma, S. B. 1990. Micrometeorological methods for measuring surface fluxes of mass and energy. In *Instrumentation for studying vegetation canopies for remote sensing in optical and thermal infrared regions,* eds. N. S. Goel and J. M. Norman, *Remote Sensing Reviews* 5(1):99–115.

Verwey, E. J. W. and J. T. G. Overbeek. 1948. *Theory of the stability of lyophobic colloids.* New York: Elsevier.

Viehmeyer, F. J. and F. A. Brooks. 1954, Measurements of cumulative evaporation from bare soil. *Transactions of the American Geophysical Union.* 35:601–607.

Vinten, A. J. A. and P. H. Nye. 1985. Transport and deposition of dilute colloid suspensions in soils. *J. Soil Sci.* 36:531–41.

Voet, A. 1936. Ionic radii and heat of hydration. *Trans. Faraday Soc.* 32:1301–04.

Von Reichenbach, H. and G. von der Bussche. 1963. Investigation on strontium sorption in the soils of Schleswig Holstein. *Z. PfiErnahr* 101:24–33.

Voss, C. I. 1984. SUTRA: A finite element simulation model for saturated-unsaturated fluid density-dependent ground water flow with energy transport or chemically reactive single species solute transport. Water-resources investigations report 84-4369. Reston, VA: U.S. Geological Survey.

Wackernagel, H. 1995. *Multivariate geostatistics.* New York: Springer-Verlag.

Wagenet, R. J. and B. K. Rao. 1983. Description of nitrogen movement in the presence of spatially-variable soil hydraulic properties. *Agricultural Water Management* 6:227–42.

Waggoner, P. E., P. M. Miller, and H. C. DeRoo. 1960. Plastic mulching-principles and benefits. Bulletin No. 634. New Haven, CT: Conn. Agric. Expt. Sta.

Walker, C. D. and J. -P. Brunel. 1990. Examining evapotranspiration in a semi-arid region using stable isotopes of hydrogen and oxygen. *J. Hydrol.* 118:55–75.

Walker, G. R., I. D. Jolly, M. H. Stadter, F. W. Leaney, R. F. Davie, L. K. Fifield, T. R. Ophel, and J. R. Bird. 1992. Evaluation of the use of chlorine-36 in recharge studies. In *Isotope Techniques in Water Resources Development 1991.* Proceedings of an International Symposium on Isotope Techniques in Water Resources Development, IAEA, Vienna, pp. 19–29.

Wallis, P. M. 1979. Sources, transportation, and utilization of dissolved organic matter in groundwater and streams. Inland water directorate scientific series #100. Ottawa, Canada: Environment Canada.

Wang, K. 1992. Estimation of ground surface temperatures from borehole temperature data. *J. Geophys. Res.* 97:2095–99.

Ward, J. C. 1964. Turbulent flow in porous media. *Proc. Amer. Soc. Civil Eng. No. Hy5* (90):1–12.

Warrick, A. W. 1974. Solution to the one-dimensional linear moisture flow equation with water extraction. *Soil Sci. Soc. Am. Proc.* 38:573–76.

Warrick, A. W., J. W. Biggar, and D. R. Nielsen. 1971. Simultaneous solute and water transfer for an unsaturated soil. *Water Resour. Res.* 7:1216–25.

Warrick, A. W., D. O. Lomen, and A. L. Islas. 1990. An Analytical Solution to Richards' Equation for a Draining Soil Profile. *Water Resources Research* 26(2):253–58.

————. 1991. An analytical solution to Richards' Equation for time-varying infiltration. *Water Resources Research* 27(5):763–66.

Warrick, A. W., G. J. Mullen, and D. R. Nielsen. 1977. Scaling field-measured soil hydraulic properties using a similar media concept. *Water Res. Res.* 13:355–62.

Warrick, A. W., M. H. Young, S. A. Musil, P. J. Wierenga, and L. L. Hofmann. 1996. Probability of intersecting hot spots with alternative subsurface sampling patterns.

Watson, K. K. and M. J. Jones. 1982. Hydrodynamic dispersion during absorption in a fine sand. 2. The constant flux case. *Water Res. Research* 18:1435–1443.

Watson, K. W. and R. J. Luxmoore. 1986. Estimating macroporosity in a forest watershed by use of a tension infiltrometer. *Soil Sci. Am. J.* 50:578–87.

Way, S. C., A. C. Bumb, R. A. Koenig, C. R. McKee and J. M. Reverand. 1985. Hydrologic characterization of coal seams for methane recovery, activities 5 and 7 progress report: Review of single-phase hydrologic testing in coalbeds and development of unsaturated-flow well test procedures, Topical Report (June 1983–December 1984). Report prepared by In-Situ, Inc. for the Gas Research Institute, Report No. GRI-85/0046, Appendices A through D, May. Chicago, IL: Gas Research Institute.

Webb, E. K., G. I. Pearman, and R. Leuning. 1980. Correction of flux measurement for density effects due to heat and water vapor transfer. *Quart. J. Roy. Meteor. Soc.* 106:85–100.

Weeks, E. P. 1978. Field determination of vertical permeability to air in the unsaturated zone. U. S. Geol. Surv. Prof. Pap. 1051.

————. 1987. Effect of topography on gas flow in unsaturated fractured rock: Concepts and observations. In *Flow and transport through unsaturated fractured rock,* eds. D. D. Evans and T. J. Nicholson, 165–70. Geophys. Monogr. 42, Amer. Geophys. Un., Washington, DC.

———. 1993. Does the wind blow through Yucca Mountain? U.S. Nuclear Reg. Comm. Rpt. NUREG/CP-0049, pp. 43–53. Washington, DC: U.S. Nuclear Reg. Comm.

Weil, R. R., R. A. Weismiller, and R. S. Turner. 1990. Nitrate contamination of groundwater under irrigated coastal plain soils. *J. Environ. Qual.* 19:441–48.

Wells, P. R. 1968. *Linear free energy relations.* New York: Academic Press.

Wesseling, J. G., P. Kabat, B. H. van den Broek, and R. A. Feddes. 1989. *SWACROP: Simulating the dynamics of the unsaturated zone and water limited crop production.* Wagenigen, The Netherlands: Winand Staring Centre, Department of Agrohydrology.

West, L. T., S. C. Chiang, and L.D. Norton. 1992. The morphology of surface crusts. In *Soil crusting: chemical and physical processes,* eds. Summer and Stewart. Boca Raton, FL: Lewis Publishers.

White, R. E. 1985. The influence of macropores on the transport of dissolved and suspended matter through soil. In *Advances in soil science,* Vol. 3, ed. B. A. Stewart, 95–120. New York: Springer-Verlag.

Whitford, W., Aldon, E. F., Freckman, D. W., Steinberger, Y. and Parker, L. W. 1989. The effects of organic amendments on soil biota on a degraded rangeland. *J. Range Management* 42:56–60.

Whittig, L. D. and W. R. Allardice. 1986. X-ray diffraction techniques for mineral identification and mineralogical composition. In *Methods of Soil Analysis,* Part I, Chap. 49, pp. 331–359. ASA, Madison, Wisconsin.

Wickland, K. and R. Striegl. 1997. *Measurements of soil carbon dioxide and methane concentrations and fluxes, and soil properties at four ages of Jack Pine forest in the southern study area of the Boreal Ecosystem study, Saskatchewan, Canada, 1993–5.* U.S. Geological Survey Open File Reprt. 97–49, Denver, CO.

Widstoe, J. A. 1911. *Dry farming.* New York: The Macmillan Co.

———. 1914. *The principles of irrigation practice.* New York: Macmillan.

Widstoe, J. A. and W. W. McLaughlin. 1902. Irrigation experiments in 1901 (on the college farm). Bulletin 80, Utah Agr. Exp. Sta., Logan, UT.

———. 1902. *The movement of water in irrigated soils.* Utah Agr. Exp. Sta. Bulletin. 115, p. 224.

Wierenga, P. J., M. H. Young, G. W. Gee, R. G. Hills, C. T. Kincaid, T. J. Nicholson, and R.E. Cady. 1993. *Soil characterization methods for unsaturated low-level waste sites.* NUREG/CR-5988, PNL-8480. Washington, DC: U.S. Nuclear Reg. Comm.

Wiersum, L. K. (1957). The relationship of the size and structural rigidity of pores to their penetration by roots. *Pl. Soil* 9:75–85.

Wild, A. 1988. *Russell's soil conditions and plant growth.* 11th ed. Essex, England: Longman Scientific and Technical. Co-published in the U.S.; New York: John Wiley & Sons, Inc.

Wild, S. R., S. P. McGrath, and J. C. Jones. 1990. The polynuclear aromatic hydrocarbon (PAH) content of archived sewage sludge. *Chemosphere* 20:703–16.

Wilemski, G. 1982. Weak repulsive interaction between dissimilar electrical double layers. *J. Colloid Interface Sci.* 88:111–16.

Wilkening, M. H. 1974. Radon 222 from the island of Hawaii: Deep soils are more important than lava fields or volcanoes. *Science* 183:413.

Wilkening, M. H., W. E. Clements, and D. Stanley. 1974. Radon flux in widely separated regions. In *Natural Radiation Environment II.* Technical Information Division. Oak Ridge, TN: U.S. Atomic Energy Commission.

Wilkinson, G. E. and A. Klute. 1962. The temperature effect on the equilibrium energy status of water held by porous media. *Soil Sci. Soc. Am. Proc.* 17:326–329.

Williams, J. K. and R. A. Dawe. 1986. Fractals—An overview of potential applications to transport in porous media. *Transport Porous Media* 1:201–10.

Williams, J. R., T. E. Short, C. L. Eddington and C. G. Enfield. 1988. Contaminant transport and fate in unsaturated porous media in the presence of both mobile and immobile organic material. Ada, OK: U.S. Environmental Protection Agency, Robert S. Kerr Environmental Research Laboratory.

Williams, R. J., K. Boersma, and A. L. van Ryswyk. 1978. Equilibrium and actual evapotranspiration from a very dry vegetative surface. *J. Appl. Meteor.* 17:1827–32.

Williamson, S. J. 1973. Fundamentals of air pollution. Reading, MA: Addison-Wesley.

Winterkorn, H. F. and L. D. Baver. 1934. Sorption of liquids by soil colloids: I. Liquid intake and swelling by soil colloidal materials. *Soil Sci.* 38:291–98.

Witten, T. A. and L. M. Sander. 1983. Diffusion-limited aggregation. *Phys. Rev. B* 27:5686–97.

Wium-Andersen, S. 1971. Photosythetic uptake of free CO_2 by the roots of Lobelia dortmanna. *Physiol. Plant.* 25:245–48.

Wofsy, S. C., M. L. Goulden, J. W. Munger, S. -M. Fan, P. S. Bakwin, B. C. Daube, S. L. Bassow, and F. A. Bazzaz. 1993. Net exchange of CO_2 in a mid-latitude forest. *Science* 260:1314–1317.

Wollney, M. E. 1878. Untersuchungen uber den einfluss der pflanzendecke und der beschttung auf die physikalischen eigenschaften des bodens. *Forsch. Gebiete Agr. Phys.* 6:197–256.

Wollum, A. G. and D. K. Cassel. 1978. Transport of microorganisms in sand columns. *Soil Sci. Soc. Am. J.* 42:72–76.

Woolridge, D. D. 1965. Tracing soil particle movement with Fe-59. *Soil Sci. Soc. Am. Proc.* 29:469–72.

Wymore, I. F. 1979. Draft report, *Estimated water balance for the proposed Union Oil Company shale disposal pile.* Report Prepared for Union Oil Company of California, February.

Yao, K. M., M. T. Habibian and C. R. O'Melia. 1971. Water and waste water filtration: concepts and applications. *Environ. Sci. Technol.* 5:1105–12.

Yeh, G. T. and R. H. Strand. 1982. FECWATER: User's manual of a finite-element code for simulating water flow through saturated-unsaturated porous media. ORNL/TM 7316, Oak Ridge National Laboratory, Oak Ridge, Tennessee.

Yeh, G. T. and D. S. Ward. 1980. FEMWATER: A finite-element model of water flow through saturated-unsaturated porous media. ORNL-5567, Oak Ridge National Laboratory, Oak Ridge, Tennessee.

Young, M. H., P. J. Wierenga, A. W. Warrick, L. L. Hofmann, S. A. Musil, B. R. Scanlon, and T. J. Nicholson. 1996. *Field testing plan for unsaturated zone monitoring and field studies.* NUREG/CR-6462. Washington, DC: Nuclear Reg. Comm.

Youngs, E. G. 1964. An infiltration method measuring the hydraulic conductivity of unsaturated porous materials. *Soil Science* 97:307–11.

Zimmerman, U., D. Ehhalt, and K. O. Munnich. 1967. Soil-water movement and evapotranspiration: Changes in the isotropic composition of water. Proceedings of the Symposium on Isotope Hydrology, Vienna, November 14–18, 1966, pp. 567–85, IAEA, Vienna, Austria.

INDEX

A

A horizon, 14
AB horizon, 14
AC horizon, 14
acid dissociation, 137–139
activity coefficient, 131–134, 137
adhesive forces, 98
adsorption coefficient, 127
adsorption isotherms, 290
advective dispersion equation (ADE), 280
aerobic respiration, 146
aggregate stability, 85
aggregated soils, 306–309
aggregate formation, 81–84
aggregation, 79–81, 82
air conductivity, 221
air diffusivity, 221
air entry value, 418
air-filled porosity, 32–33
air sparging, 515–517
air temperature, 380, 381
albedo, 212
allophanes, 12, 13, 56
allowable concentration model, 504
algebra, 571
alumina octahedron, 12
AMC, 384, 386
amorphous clays, 10
amorphous material, 8
analytical models, 407, 429
analytical solution, 411, 413, 416
    three-dimensional, 434
anion exclusion, 289
anisotropic, 172
antecedent soil moisture conditions, 384
areal measurements, 372–373
    detention flow technique, 372
    time-condensation technique, 372
    block techniques, 372
attractive forces, 98
Avogadro's number, 41

B

b dimension, 69
B horizon, 14
balloon pressure, 87
Bateman's equation, 304
Bernoulli's equation, 153–155
BET equation, 35
Bingham liquid, 171
biochemical degradation, 126
bioconcentration factor, 291
biodegradation, 127
biological treatment, 518
bioremediation, 127
Blaney–Criddle, 392
Boltzmann constant, 41, 44, 48
    distribution, 50, 431, 432
    equation, 48, 458
    transformation, 195–197, 288

boundary conditions, 295–299, 303
Bowen-ratio-energy-balance (BREB), 241
box dimension, 541–543
Boyle's law, 245
breakthrough curve, 276–288
Brooks–Corey, 422
    variables, 356
Brownian motion, 21, 72, 549
    fractal, 551
Buck's equation, 206
bubbling pressure, 418
bulk density, 30, 85

C

C horizon, 14
calculus, 577–582
cap design, 512
capillarity, 98–101
capillary fringe, 483
capillary conductivity, 188
capillary number, 558
capillary potential, 102–103
capping, 511–513
cation exchange capacity (CEC), 33, 63–64
Cauchy condition, 295
cementing agents, 87
center-of-mass technique, 461
centrifugation, 22
cgs unit system, 41
chamber systems, 268
Charles' law, 245
chemical crust, 87
chemical potential, 56, 113, 115
chemical saturation, 145
chloride, 463
Chlorine-36, 463
Clapp and Hornberger constant, 500
Clausius–Clapeyron equation, 204
clays, 7–12
cleavage plane, 88
clod, 17, 79
colloid
    migration, 318
    transport, 322–325
colloidal facilitated transport, 318–325
complimentary error function (erfc), 586
compressibility factor, 112
compression chamber, 29
concretion, 18, 79
condensation, 202–204
conduction, 201
consolidation, 53
constant charge, 56–58
constant surface potential, 56–58
contact angle, 97, 98, 99, 120
contaminant profile model, 506
continuity equation, 175, 185, 417, 421
coulombic interaction, 64
Coulomb's law, 41
coupled processes, 128–129

coupled transport, 258–260
critical zeta potential, 70
crusted soils, 365–367
crystal lattice, 8
cubic polynomial, 573
cumulative infiltration, 354, 360

D

Dalton's law, 204, 246, 390
Darcy's law, 107, 165, 171, 184, 234, 479
    limits of, 170
    validity, 185
Debye–Hückel, 47, 132
dead end pores, 31
decision-tree process, 507
deflocculation, 19, 72
degree of saturation, 31
denitrification, 147
depositional crust, 85
deterministic flow models, 409–429
deuterium, 464
devil's staircase, 548
dielectric constant, 41
differential equations, 581–582
differential heat of wetting, 41
differential pressure transducer, 570
diffusion, 142–143
    coefficient, 143, 307
    equation, 178–179
    ordinary, 255
    nonequimolar, 255
diffusion-limited aggregation (DLA), 554–556
diffusivity, molecular, 258
Dirichlet condition, 295
discontinuity, 187
dispersion, 19, 39, 72–73, 305
    longitudinal, 276
dispersion coefficient, 286–288
    calculating, 285
dispersivity, 288
displacing front, calculating, 285
dissolved species, 140–142
DLVO theory, 39, 54–63, 319
Donan equilibrium, 68
Dorn effect, 51
double-layer potential, 62
double-layer theory, 47, 61
drag forces, 275
drainage cycle, 117
Dupuit assumption, 482
Dupuit–Forchheimer theory, 478

E

eddy covariance, 242
    deployment, 243
Einstein equation, 550, 558
    Brownian motion, 323
electric dipole, 96
electrical potential, 52

electrical resistivity borehole tomography, 566
electromagnetic energy spectrum, 211
electromagnetic induction, 566
electroneutrality, 136–137
electroosmosis, 51–53
electroosmotic
    hydraulic conductivity, 52
    dewatering, 52, 53
    velocity, 50
electrophoresis, 50, 53, 54
electrostatic attractive forces, 58
elutriation, 22
emission spectra, 214
emissivity, 212
energy balance, 215–221, 401, 407
    components, 217
    earth, 213
energy budget, 216, 217
    measuring terms, 240
energy exchange, 210–212
energy transfer, 201–202
entrapped air, 84, 120, 169
equilibrium chromatography, 292–294
equilibrium constant, 63
equipotential lines, 484
equivalent curve numbers, 387
equivalent depth of water, 31
error function, 281, 587
Eulerian method, 294
evaporation, 202–204, 389–399
    requirements, 303
    from shallow water table, 233
evapotranspiration, 238, 391
    model estimates, 238
exchange sites, types, 301
exchangeable sodium percentage (ESP), 65, 85, 86
exit boundary condition, 297
exponential distribution, 534
extended Debye-Hückel, 132
extraction function, root, 466

F

falling rate stage, 232
FAO–Penman, 395
feldspar, 15
Fermi distribution, 431, 432
Fickian diffusion, 253–254
Fick's first law, 279
field radiation, 401
field water balance, 379–401
fingering, 312, 313
first-order decay coefficient, 303
    factor, 445
flocculation, 39, 60, 70–72, 81
    stages of, 39, 40, 72
Florida Everglades, case study, 491
flow, fluid, 478
    between parallel ditches, 480–481
    nets, 485–488
    toward well, 479–480
    in unconfined aquifer, 478–479
    with uniform recharge, 481–482
fluidity, 166
flux density, 166, 167
foliage density, 219
Fourier's law, 223

fractal(s), 534
    Brownian motion, 551
    construction, 543–545
    diffusion fronts, 556
    dimension, 539, 543, 545
    global dimension, 551
    grid dimension, 546
    point dimension, 546–547
    self-affine, 548–554
    self-similar, 545–547
fractionation, 19
fractured media, 309–310
fragment, 17, 79
Freundlich equation, 291, 497, 498
frictional coefficient, 143
Frumkin, type equation, 141
fulvic acids, 73–76
function, definition, 572
furrow techniques, 374

G

Gapon equation, 64
gas expansion, 29
gas flow, 251
geometric mean, 172
geometric relations, 575
geostatistics, 529
Gibbs, free energy, 55, 135, 136
    function, 116
Gouy–Chapman theory, 63
gravity segregation, 286
Green–Ampt infiltration model, 353, 356, 361
    approach, 353
    cumulative infiltration, 357
    equation, 355
greenhouse gasses, 200, 201, 214
Griffith theory (crack propagation), 89
Guoy–Chapman theory, 47, 49
Guoy distribution, 49

H

Hagan–Pouiselle's law, 278
Hamaker constant, 59, 60
Hammett correlation, 138
Hausdorff–Besicovitch dimension, 542
heat capacity, 130
heat dissipation sensor, 568
heat flux, 241
    latent, 241
    sensible, 241
heat of immersion, 42, 43, 45
heat of wetting, 42, 43
heat transport, 224
    sinusoidal solution to, 224
Heaviside unit function, 304
height of capillary rise, 100
Helmont, Jan Baptiste van, 3
Helmholtz theory, 49, 52
Henry's law, constant, 441, 445, 446
    coefficients, 261
Hofmeister series, 64
Holtan's equation, 358
Hooghoudt's equation, 482, 489
horizontal diffusivity, equation, 195
horizontal infiltration rate, 360
Horton equation, 356–358

hull, 556
humic acid, 73–76, 83, 126
humins, 74
humus, 13, 73–79
Hurst, exponent, 548
    empirical law, 548–549
hydration, 39, 65–70
hydraulic conductivity, 52, 166–170, 416, 477
hydraulic diffusivity, 192–195
hydraulic potential, 111, 275
hydraulic pressure, 477
hydrodynamic dispersion, 276, 277, 278
    coefficient, 229, 289
hydrodynamic flux, 279
hydrogen bonding, 11, 83
hydrolysis, 139–140
hydrometer, 22
hydrophilic, 75
hydrophobic, 75
hydroxy-interlayered minerals (HIM), 320
hydroxyl groups, 63, 141
hyperbolic distribution, 554
hysteresis, 33, 117–121

I

ideal gas law, 245
illite, 11
immiscible displacement, 275
infiltration, 346
    cumulative, 368
    point measurements, 373
    rate, 348, 364
    theories, 352–362
infinite dilution activity coefficient, 133–134
initial conditions, 295–299, 303
ink bottle effect, 118
inlet boundary condition, 297
in-situ vitrification, 521
interactive repulsive forces, 55
interception, forest, 379
interfacial tension, 98
intergranular pressure, 111–113
internal drainage, 370
International System of Units (SI), 592
intra-aggregate, 307
ion complexes, 140–142
ion exchange, 63, 305
    types of, 293
irrotational flow, 153
isomorphous substitution, 46, 47
isotropy, 172

J

Jacobian, 527
Jensen–Haise, 392, 395, 396

K

kaolinite, 11, 44, 56
Khanji model, 361
kinematic profile, 420, 421–426
    moisture, 423
Kirchhoff's law, 212
    integral transformation, 409
Klinkenberg effect, 256
Knudsen-diffusion coefficient, 256

Kostiakov equation, 356, 358
Kozeny's equation, 191
Kriging, 530–531

L

Lagrangian method, 294
laminar flow, 21, 159, 160
landfills, 331
Langmuir equation, 35, 75
    isotherms, 290
Laplacian differential operator, 42
Laplace's equation, 175–178, 536
law of cosines, 575
law of sines, 575
layered media, 310–312
layered soils, 363–365, 413
linear polynomial, 573
liquid, properties of, 152, 350
logarithms, 577
log-normal distribution, 533
    PDF, 533
London frequency, 60
lyotropic series, 65
lysimeter, 334

M

macropores, 312, 313, 362
    flow, 313
macroporosity factor, 363
mass action, 129–131
mathematics review, 571–585
measurement error, 532
mechanical agitation, 19
mechanical analysis, 16
mechanical dispersion, 278
Menger sponge, 545
mercury injection, 29
mesopores, 313
methane profiles, 265
mica, 15
microbial mediation, 147–148
microbial sealing, 169
micropores, 313
mineralogical composition, 7, 8
miscible displacement, 275
modulus of rupture, 87
moisture characteristic curve, 117, 456
molar solubility, 144
molecular diffusion, 279–280
    coefficient, 280
monitoring, 511
    devices, 561–570
monochromatic radiation, 212
monomial, 573
montmorillonite, 11, 33
Morel–Seytoux model, 361
multiphase transport, 260–262

N

Navier–Stokes equation, 557
net radiation, 241
Neumann condition, 295
neutron probe, 565
Newtonian liquid, 171
Newton's law, 274
nitrous oxide, 477
no action, 510

no-flow boundary, 444
non-Darcy flow, 185
nonequilibrium conditions, 301–305
nonreactive solutes, 288
nonuniform distribution, 315
number of samples, calculation, 528
numerical solutions, 299–301
    equilibrium exchange, 299
    models, 408, 427, 448, 459

O

O horizon, 14
octanol–water
    activity coefficients, 132
    partition coefficients, 132, 502–503
on-site thermal treatment, 520–521
operator, mathematical, 572
osmotic pressure, 67, 107
osmotic swelling, 68
overburden pressure, 110, 112
oxidation, reactions, 145
oxygen hole, 11
oxygen-18, 464

P

pan-evaporation method, 397
particle-size analysis, 16, 19–24
particle-size distribution, 19
particle density, 16, 24–26
passive bioremediation, 510
Peclet number, 288, 298, 299, 441, 442
ped, 17, 79
Penman model, 239
Penman–Monteith, 239
perimeter area dimension, 547
permittivity, 40, 50
pH, 147
Philip model, 360–361
photosynthesis, 235
physical constants
    for unsaturated zone, 590
    for selected liquids, 454
physical properties, 7
pipette method, 24
piston flow, 277
Poiseuille's law, 158, 159, 190, 536
Poisson distribution, 534
Poisson's equation, 42, 50
polynomial, 573
pore compression coefficient, 112
potential(s)
    gravitational, 102, 104, 105, 113
    hydraulic pressure, 110
    hydrostatic pressure, 110
    intermolecular, 43
    matric, 106, 109–110, 188
    osmotic, 105, 106
    pressure, 103, 113
    vapor, 107–108
power-law distribution, 531–534
power-law relation, 410
power spectral density, 553
powers, in math, 576
precipitation, 380
predictive equations for K_{oc}, 499

preferential flow, 312
    paths, 312–318
    modeling, 315–318
pressure equivalent, 103
pressure field, 250
Priestly–Taylor, 239
probability density function, 470, 471
profile-moisture distribution, 347–352
psychrometers, 206, 208
psychrometric chart, 205
    equation, 206
pycnometer, 25, 29

Q

quicksand, 111
quadratic polynomial, 573

R

R horizon, 14
radiation, 201
    longwave, 229
    solar variation, 229
    terrestrial, 214
radiation balance, 212–215, 401–402
    solar, 213
radionuclides, 305
    distribution, 332–338
    properties, 305
radius of curvature, 99, 100
radon gas, 201
    transport, 266
rainfall simulators, 373
random variables, 531
random walks, 549–550
Raoult's law, 114
recharge rate, 477
redistribution, 369–370
    processes, 417
    water, 369
redox reaction, 145–147
relative error, 469
relative humidity, 204
remediation monitoring, 563
remote data aquisition, 564
repulsive force, 39, 57, 59
rescaled range, 552
resistivity, 167, 191
retardation coefficient, 291, 442
retardation constant (R), 59
Reynolds number, 161, 170, 248
Richards' equation, 292, 418, 429
    derivation of, 339
ring/cylinder infiltrometers, 373
RITZ model, 509
root extraction function, 466, 467
    uniform, 468
root zone, 1
roughness/length, 552
ruler dimension, 547
runoff, 367–369, 383–389
    curve number, 385
    volume, 368

S

sampling wells, 570
saturated flow, 173–175

saturated vapor pressure, 204
scaling, 534–540
scanning curve, 118
Schultz–Hardy rule, 71, 73
SCS curve number, 384
sedimentary crusts, 86
sedimentation, 20
self-similarity, 540
semivariogram, 530
semivolatile organic compounds, 126
separates, 15
sesquioxides, 12
Sierpinski carpet, 544
silica tetrahedra, 8, 12
silicate clays, 9
similarity dimension, 541
similitude analysis, 536
site characterization, 561–562
skin seal, 86
slaking, 83, 84
slipping plane, 50, 70
Smith–Parlange model, 361
snowmelt, 387–389
sodium adsorption ratio (SAR), 64, 65
soil(s)
    air composition, 262
    air measurement, 207
    characteristics, 349, 384
    classes, 18–19
    cracking, 88–90
    crusting, 84–88
    excavation, 518–519
    formation, 13
    forming factors, 8
    flushing, 517–518
    frequency distribution, 524
    gas transport, 245–258
    heat flux, 223, 240
    heat transfer, 221–226
    horizon, 13, 14
    humidity, measurement, 208
    hydrologic grouping, 359
    measuring gas flux, 267
    moisture characteristic curve, 117, 188,
        232, 400
    moisture diffusivity, 3
    moisture distribution, 238
    moisture evaporation, 230–234
    permeability, 166, 169, 173, 246
    –plant–atmosphere continuum, 235
    porosity, 26–29, 246
    pressure, 95
    profile, 13–14
    remediation, 494
    shrinking, 120
    stages of drying, 230, 231
    structure, 17
    swelling, 65, 66, 120
    temperature, 218
        diurnal variation, 228
    temperature distribution, 226–230
    temperature profiles, 227
    texture, 15–18, 85
    textural classification, 388

textural triangle, 18
thermal conductivity, 222
    calculation of, 223
thermal properties, 202
    venting, 513–515
    water flux, 258
solar radiation, 380
solubility, 143–145
solute movement, 273, 274
solution samplers, 569
sorption curve, 117
sorption of contaminants, 76–79
sorptivity, 354, 360
sorption reactions, 289–292
spatial moments, 311
spatial variability, 524
specific surface area, 33–35
SPPPLV model, 506
static forces, 275
steady-state solution, 436
Stephan–Boltzmann constant, 213
Stern layer, 44
Stern's double-layer theory, 47, 49
stochastic models, 470–474
Stoke's law, 20, 21, 50
stomatal function, 237
streaming potential, 54
streamlines, 484
structural crusts, 86
structure of water, 95–97
submergence potential, 109
Summers model, 504
    modified, 505
surface charge, 46
surface tension, 98, 99
swelling pressure, 66, 68

T

taste function, 583–584
tensiometer, 567
    pressure potential, 111
tension infiltrometers, 374
tetrahedral lattice, layers, 11
Tetson's equation, 206
thermocouple, 207
three-dimensional solution, 439
    deterministic flow model, 448
time domain reflectometry, 565
Torricelli's law, 155–158
tortuosity, 167
    diagram of, 252
total petroleum hydrocarbons
    (TPH), 494
total potential, 103, 104, 106
total potential energy, 60–62
tracer flux, 465
tracers, 460–470
transfer function model, 462, 464, 470–474
transition zone, 348
transmission zone, 348
transpiration, 236, 389–399
    measuring, 244
transport parameters, 309

triadic Von Koch curves, 540
trigonometric functions, 574
tritium, 463
turbulent flow, 159
two-dimensional solution, 439
two-phase air–NAPL system, 452

U

uniform distribution, 315
unsaturated flow equations, 3, 189–192
unsaturated hydraulic conductivity,
    185–189

V

vacuum extraction system, 513
vadose zone, definition, 1
van der Waals–London forces, 9, 46, 55, 58,
    59, 60, 78
Van Laar equation, 133
vapor
    pressure, 108
    diffusion, 105
variable charge, 9
variogram, 553
vector, 582
vector calculus, 580–581
velocity gradient, 160, 161
vermiculite, 11
viscous fingering, 554, 556–559
viscous flow, 247, 249, 250, 251
Voet equation, 44
void ratio, 26, 28
volatile organic compounds
    (VOC), 126, 513

W

Washburn–Bunting porosimeter, 29
Water
    balance, 380
    budget, 239
    content, 30–31
        profiles, 351, 352
    film, 95
    imbibition, 65
    immobile, 305
    mobile, 305
    retention, 33
    table, 483
water-soluble organic carbon
    (WSOC), 78, 321
wetland definition, 477
wetting zone, 347, 348

Y

Young's equation, 98
    modulus, 89
Yucca Mountain, 330

Z

zeta potential, 50, 70, 71
zone of aeration, 1